Advances in Experimental Medicine and Biology

Volume 1373

Advances in Experimental Medicine and Biology provides a platform for scientific contributions in the main disciplines of the biomedicine and the life sciences. This series publishes thematic volumes on contemporary research in the areas of microbiology, immunology, neurosciences, biochemistry, biomedical engineering, genetics, physiology, and cancer research. Covering emerging topics and techniques in basic and clinical science, it brings together clinicians and researchers from various fields.

Advances in Experimental Medicine and Biology has been publishing exceptional works in the field for over 40 years, and is indexed in SCOPUS, Medline (PubMed), Journal Citation Reports/Science Edition, Science Citation Index Expanded (SciSearch, Web of Science), EMBASE, BIOSIS, Reaxys, EMBiology, the Chemical Abstracts Service (CAS), and Pathway Studio.

2020 Impact Factor: 2.622

Julien Santi-Rocca
Editor

Periodontitis

Advances in Experimental Research

Springer

Editor
Julien Santi-Rocca
Science and Healthcare for Oral Welfare
Toulouse, France

ISSN 0065-2598 ISSN 2214-8019 (electronic)
Advances in Experimental Medicine and Biology
ISBN 978-3-030-96880-9 ISBN 978-3-030-96881-6 (eBook)
https://doi.org/10.1007/978-3-030-96881-6

This Springer imprint is published by the registered company Springer Nature Switzerland AG
The registered company address is: Gewerbestrasse 11, 6330 Cham, Switzerland

Preface

Research in periodontology is a blooming field: half the articles about periodontitis referenced in Pubmed were published after 2009 (search: "periodont*[title/abstract]"). This intensification of research came together with its diversification and with the specialization of its growing branches. As periodontitis is an inflammatory and infectious disease, fundamental research pays a special attention to microbiology and physiopathology, as reflected in this book. Furthermore, the crosstalk between periodontitis and other diseases through their inflammatory and infectious components is being intensively studied, highlighting the interest for integrative approaches for the prevention and the treatment of all these conditions.

In evidence-based medicine, recommendations integrate results from research and provide the practitioners safe and efficient tools for defined clinical situations. However, when guidelines cannot be applied or fail, solutions must be found, and the health practitioner should choose the best option according to the state of the art. Clinicians must thus be literate in immunology, microbiology, pathophysiology, and clinical research. As health practitioners are not experts in all these fields, they should get intelligible data from fundamental researchers to understand the guidelines and to be able to propose alternatives when they do not fit their practice or when the treatment is not adapted to the patient. It is important that researchers also get feedback from clinicians so their observations, in particular about therapeutic failure, can be scientifically analyzed to lead to the evolution towards more adapted guidelines including alternative options.

The aim of this book is to provide insights into the recent advances in experimental research to make the bridge between current treatment guidelines and future directions, promoting both experimental and clinical research, as well as public engagement with science. Thus, we have invited specialists in experimental research on periodontitis to review the recent literature in their field of expertise and to provide chapters presenting solid scientific evidence to reach the widest audience through dedicated sections: "Highlights" for researchers, "Considerations for Practice" for clinicians, and "Patient Summary" for patients and non-specialists.

Chapters were peer-reviewed in a double-blinded process: manuscripts were anonymized before submission to reviewers, and the identities of the authors and reviewers were kept secret until final acceptance, when reviewers could choose to be cited in the chapters.

We thank the authors and the reviewers for their work, which provides a strong frame to develop new research and to make decisions for periodontitis therapeutic and prophylactic management.

Toulouse, France Julien Santi-Rocca

List of Reviewers

Chapter 2

Thomas Van Dyke
Clinical and Translational Research, Forsyth Institute, Cambridge, MA, USA
Oral Medicine, Infection, and Immunity, Harvard School of Dental Medicine, Boston, MA, USA, tvandyke@forsyth.org
Nailê Damé-Teixeira
Department of Dentistry, School of Health Sciences, University of Brasilia, Brasília-DF, Brazil; nailedame@hotmail.com

Chapter 3

Reviewers asked to remain anonymous

Chapter 4

Reviewers asked to remain anonymous

Chapter 5

Mark Bonner
International Institute of Periodontology Victoriaville, QC, Canada; mark.bonnerdmd@hotmail.com

Chapter 6

Cristiane Yumi Koga-Ito
Institute of Science and Technology, São Paulo State University/UNESP, São José dos Campos, SP, Brazil. cristiane.koga-ito@unesp.br.

Chapter 7

Paulo Henrique Braz Silva
Department of Stomatology, School of Dentistry, University of São Paulo, São Paulo, Brazil
Laboratory of Virology (LIM-52), Institute of Tropical Medicine of São Paulo, School of Medicine, University of São Paulo, São Paulo; pbraz@usp.br

Chapter 8

Torsten Sterzenbach
Clinic of Operative and Pediatric Dentistry, Medical Faculty Carl Gustav Carus, Technische Universität Dresden, Fetscherstraße 74, 01307, Dresden, Germany; torsten.sterzenbach@uniklinikum-dresden.de

Chapter 9

Massimo Costalonga
Division of Periodontology, Department of Developmental and Surgical Sciences, School of Dentistry, University of Minnesota, Minneapolis, Minnesota, USA; costa002@umn.edu

Chapter 10

Reviewers asked to remain anonymous

Chapter 11

Reviewers asked to remain anonymous

Chapter 12

José Lopez-López
Department of Odontoestomatology, Faculty of Medicine and Health Sciences (Dentistry), Bellvitge Campus, University of Barcelona, Barcelona, Spain.
Oral Health and Masticatory System Group, Institut D'Investigació Biomédica de Bellvitge (IDIBELL, Bellvitge Institute of Biomedical Research), L'Hospitalet de Llobregrat, Barcelona, Spain.
Odontology Hospital University of Barcelona (HOUB), Barcelona, Spain; 18575jll@gmail.com.

Chapter 13

Reviewers asked to remain anonymous

Chapter 14

Reviewers asked to remain anonymous

Chapter 15

Reviewers asked to remain anonymous

Chapter 16

Marie-Eve Bezzina
Private practice, Nice, France; marieeve.bezzina@gmail.com

Chapter 17

Reviewers asked to remain anonymous

Chapter 18

Reviewers asked to remain anonymous

Chapter 19

Rok Gašperšič
Department of Oral Medicine and Periodontology, Faculty of Medicine, University of Ljubljana, Ljubljana, Slovenia; rok.gaspersic@mf.uni-lj.si

Chapter 20

Pierre-Yves Collard-Dutilleul
LBN, Université de Montpellier, Montpellier, France; pierre-yves.collart-dutilleul@umontpellier.fr
Jeffrey Gimble
Obatala Sciences Inc., New Orleans, LA, USA.

Department of Structural and Cell Biology, Tulane University School of Medicine, New Orleans, LA, USA
Department of Medicine, Tulane University School of Medicine, New Orleans, LA, USA.
Department of Surgery, Tulane University School of Medicine, New Orleans, LA, USA; jgimble@tulane.edu

Contents

Part II Periodontitis Pathophysiology

Part III Periodontitis-Associated Conditions

Part IV Therapeutic Management of Periodontitis

1 Advances in Experimental Research About Periodontitis: Lessons from the Past, Ideas for the Future

Julien Santi-Rocca

Abstract

Periodontitis is an inflammation-driven disease triggered by microbes that affects the tissues surrounding the teeth and eventually destroys the alveolar bone, leading to tooth loss. This is probably the most precise and consensual definition of periodontitis that can be given to date. How to deal with a disease whose aetiology, pathophysiology, and therapeutic strategies are still subject to debate? In the book "Periodontitis: Advances in Experimental Research", 20 chapters aim at clarifying chosen questions left unresolved from the twentieth century, with methodologies from the twenty-first century. Here we expose the concerns authors, reviewers, and editors have raised during the writing of this book. The aim of this collaborative work is to present to a large audience the current state of experimental periodontal research, and to help decision makers think out of the box while defining future research.

Keywords

Periodontal diseases · Periodontitis · Gingivitis · Microbiota

J. Santi-Rocca (✉)
Science and Healthcare for Oral Welfare, Toulouse, France
e-mail: jsr@periodontitis.show

1.1 Past Achievements in Periodontal Research

1.1.1 New Concepts to Understand Periodontitis Pathophysiology

1.1.1.1 Periodontitis: A Microbial Disease

From early times after the identification of possible periodontopathogens, researchers have tried to validate Koch's postulates to prove the etiological link of these microbes with periodontitis. From the late nineteenth century, bacteria and protozoa were suspected to take part in the pathophysiology of periodontitis (reviewed in (Meyer 1917)). While the first postulate – identification of the pathogen exclusively and in all the disease cases – was easy to test, the second one – obtention of a pure culture of the microbial agent – was not completed. Historically, 6 organisms from the *Cytota* superdomain have been particularly considered *bona fide* candidates for the causation of periodontitis: 2 parasites and 4 bacteria.

1.1.1.2 Bacterial Candidates: Isolation and Axenization

The first formally identified possible periodontopathogen is the protist *Entamoeba gingivalis* (Gros 1849), which is still not cultured axenically. The second is also a protist, *Trichomonas tenax*, first described in the eighteenth century (Müller 1773), then unified as a species in the

J. Santi-Rocca (ed.), *Periodontitis*, Advances in Experimental Medicine and Biology 1373,
https://doi.org/10.1007/978-3-030-96881-6_1

twentieth century (Dobell 1939), and whose culture has been axenized (Diamond 1962). The bacterium *Spirochaete denticola* (Flügge et al. 1886) was renamed *Treponema denticola* by Brumpt in 1925 and was later isolated in pure cultures (Cheng and Chan 1983). A complex of bacteria was cultured and identified as the species *Bacterium melaninogenicum* (Oliver and Wherry 1921), whose subspecies *Bacterium melaninogenicum asaccharolyticus* was redefined as *Bacteroides asaccharolyticus* (Holdeman and Moore 1970); from this polyphyletic taxon was extracted the culture-amenable *Bacteroides gingivalis* (Coykendall et al. 1980), later renamed *Porphyromonas gingivalis* (Shah and Collins 1988). The identification of *Bacteroides forsythus*, then named *Tannerella forsythia*, was concomitant with its isolation (Tanner et al. 1985). And last, *Aggregatibacter actinomycetemcomitans*, which was already characterized in other pathologies (Holm 1954), was identified under its former name *Actinobacillus actinomycetem comitans* in pure cultures (Baehni et al. 1979) (see Chap. 3 by Hbibi and collaborators). More periodontopathogen candidates are often identified, while other microbial actors whose role might be milder or more complex in the periodontal environment remain unsuspected due to the experimental setups that are used.

1.1.1.3 Periodontitis: A Biofilm-Based Disease

Indeed, the periodontal environments are never monophyletic, and models using microbial community and biofilms are closer to reality. Researchers try to recreate *in vitro*, or to study *ex vivo*, the interactions within the biofilms (see Chap. 8 by Meneguzzo Prado and colleagues). Biofilms are a kind of symbiotic multispecies community, whose thickness (Suarez et al. 2019) and three-dimensional organization (Ramsing et al. 1993) impact their interaction with the environment, and thus the stream of nutrients and information, as well as the penetrance of molecules from the exterior, like antibiotics or oxygen. This seems to be the case for the "hedgehog" structures, which may result from the positioning of an outer shell of aerophilic or aerotolerant bacteria, and centripetal dwelling of anaerobic bacteria (Mark Welch et al. 2016). By their *de facto* collaborative metabolism, these low-motility, spatially organized trophic networks generate by-products, that may follow polarized streams according to the abundance of bacterial species. The impact of the human host is also to be better considered, and analyses of metabolomics, a flourishing field (Baima et al. 2021a; Baima et al. 2021b), will allow to integrate and interpret metatranscriptomics data (Nguyen et al. 2020).

1.1.1.4 Periodontitis: A Microbiota Disease

The organization in biofilms has a strong impact on host response: inflammation is exacerbated, even by species that were linked to health in genetics-based screening (Redanz et al. 2021). This calls once again for caution about the over-interpretation and about the extrapolation of data obtained from descriptive explorations. This result also questions the relevance of progressing to animal research for the study of periodontitis pathophysiology while our understanding of the aetiology is still partial, and that the inoculation with planktonic microorganisms will not mimic the naturally occurring infection. By refining the scientific questions and using appropriate animal models (Oz and Puleo 2011), some conclusions can be drawn within the frame of the experimental conditions. These explorations, with their obvious conceptual limitations, have helped understanding the potential of some bacteria, like was the case for *A. actinomycetemcomitans*(Irving et al. 1975). However, 2 years after the latter study, one of its authors commented (Socransky 1977):

> *It is difficult to interpret the animal pathogenicity testing data. One feels that destruction induced by an organism reveals the potential of the organism to lead to loss of supporting structures. However, the gnotobiotic animal is a somewhat artificial situation, in that the animal is exposed to a pure culture of a human isolate without the possible ameliorating effects of other microorganisms.*

Indeed, the saturation of a system in which the importance of the holobiont is neglected can be compared to the colonization of the intestine or the mouth by *Candida albicans* after an antibiotherapy: while the yeast is ubiquitous, the liberation of a niche promotes its growth and induces phenotypical changes. Considering these experimental situations as models for pathophysiology study will be circumvented thanks to a deeper knowledge of the disease.

1.1.1.5 Periodontitis: An Inflammatory Disease

Inflammation has often been overlooked as a mere consequence of microbial, in particular bacterial, colonization of the periodontal sulcus. However, in a recent model, inflammation was proposed to fuel the remodeling of the periodontal microbiota, allowing the overgrowth of pathogens and the reaching of a dysbiotic state (Van Dyke et al. 2020). These bacterial profiles are linked with lipid mediators, underlining the interest of complementary approaches targeting both inflammation and microbiota to treat periodontitis (Lee et al. 2021).

1.1.2 New Technologies to Reach New Fields of Research

1.1.2.1 New Technical Possibilities Generate New Scientific Questions

Periodontology research, as all the biomedical fields, has been impacted by the emergence of systems biology, rendered possible by the exponentially improving *in silico* solutions and by the development of next-generation biotechnologies. In particular, *de novo* bulk identifications and unsupervised modelling have changed and magnified inductive research: a high number of parameters can be screened at the same time, which are later analyzed by deep learning. When carefully prepared and reinterpreted in the context of the state of the art, the data generated by these tools can answer complex scientific questions and be a source of information that opens unexpected fields (Finak and Gottardo 2016). In periodontology research, the advances using metagenomics and metataxonomics are far beyond the hypotheses that could be formulated before the results were obtained (see Chap. 2 by Bing and collaborators).

1.1.2.2 New Methods Generate New Readouts

Historically, microbiology in periodontitis relied on the study of the presence or absence of microbial entities in periodontal pockets. For instance, the association of species in the founding article of Socransky and colleagues was assessed by the positive or negative detection by DNA-DNA hybridization of the presence of bacterial species (Socransky et al. 1998). More recently, diagnosis by end-point PCR also yielded binary results and the prevalence of periodontopathogen candidates in health and disease has been frequently used to evaluate their pathogenicity. The quantitative PCR methods introduced continuous data about target abundance, which questioned the sensitivity of previous analyses and the relevance of a species' prevalence without determining its abundance.

However, determining thresholds to convert a continuous variable to a binary categorical variable (positive/negative) has been a real struggle to allow the comparison of metagenomics/metataxonomics data with former methods. Indeed, the relevance of the detection of a single or few copies of targets in a large pool of sequences is dubious. Machine learning and statistical models have been useful to identify confounders in RNA sequencing data (Schlieben et al. 2021).

1.1.2.3 New Analyses Generate New Concepts

Developing new methods does not aim at interpreting them with old readouts. In the field of periodontology, the transition from individual binary readouts to association studies, and finally to concomitant relative quantification of species by next-generation sequencing has changed the way the microbiota was considered. The fine quantification of the abundance of a whole set of micro-organisms, or of gene expression, has opened new ways for genuine "omics" approaches and speeded up the obtention of results with multivariate outcomes (Yost et al. 2015; Meuric et al. 2017; Yost et al. 2017). The available data allowed to switch from a simple question with a binary answer to the modelling of interaction networks that can be compared to results obtained in ecological studies, which already provide tools like the diversity Shannon index (Shannon 1948).

1.1.3 New Tools for Therapeutic Management

1.1.3.1 Biofilm Control

The ecology of the biofilm is a main target for therapeutic management of periodontitis. In the case of treating biofilms with chemical antimicrobial agents, the choice of their spectra and delivery sites (periodontal, oral, systemic) is of central importance (see Chap. 16 by Vinel and collaborators). Furthermore, penetration of chemical antimicrobials within biofilms is reduced and may face a collaborative community with detoxifying functions, as is the case for oxygen, as previously discussed. Biofilms can be fought by mechanical disruption, which is efficient and leads to partial restauration of health-related bacterial communities (Johnston et al. 2021). Mechanical strategies can be aided by the local release of molecules, in particular antibiotics like doxycycline (Nastri et al. 2019; Ranch et al. 2021). Innovative materials allow the control of delivery length and dose (Zupancic et al. 2018). Finally, biofilm control can be assisted by photonic adjunctive therapies (see Chap. 18 by Giannelli and Bani).

In order to guide recolonization of the periodontal sulcus by health-related species, supplementation by probiotics has been proposed (see Chap. 19 by Furlaneto and collaborators). Furthermore, probiotics could be used for prevention and treatment since they can disrupt the balance in dysbiotic biofilms (Myneni et al. 2020; Zupancic et al. 2018). Beside probiotic strategies involving one or few species, microbiota transplants are being studied and have produced satisfying results in the dog model (Beikler et al. 2021).

It is important to remember that the community does not agree on the aetiology and the pathophysiology of periodontitis. This hinders the consensual choice of microbial targets for treatment. Moreover, the tools to specifically treat one microbial species are scarce, if not inexistent. Phage therapies can solve this specificity issue. Lytic bacteriophages are already used for some therapies; their suitability for periodontitis is still questioned due to the lack of *bona fide* bacterial targets (Szafranski et al. 2017). However, the rapid personalization of phage library for a broad-spectrum precision medicine approach is a real hope for cases refractory to conventional treatments. Lysogenic phages can also be of interest to hijack the protein-producing machinery of some bacteria and make them secrete antimicrobial compounds or peptides with vaccine purposes (Fathima and Archer 2021).

1.1.3.2 Modulation of Host Responses

Some vaccine strategies have already been tested in animal models, using the main periodontopathogen candidates, in an individual way (reviewed in (Myneni et al. 2020)). However, their efficiency is limited in controlling the polymicrobial biofilm formation and thus their relevance for periodontitis prevention or cure is questioned.

Beside the development of a protective host response, the control of a deleterious inflammation can be an efficient strategy. Instead of diminishing the magnitude of inflammation, which is reversible when the treatment is discontinued, new molecules aim at providing “end signals” to the inflammatory response, leading to its resolution (reviewed in (Van Dyke 2020)). Resolving inflammation provokes a modification of the periodontal microbiota in animal models (Lee et al. 2016). These underlines that the interplay between the host and the microbial biofilms is not only one-way and that considering only one side of the coin may not facilitate therapeutic strategies.

1.1.3.3 Tissue Engineering

As damage caused by periodontitis is defined as “irreversible”, regeneration was often considered a post-therapy option. Beside the scaffolding role for filling materials in reconstruction, biomaterials can also serve for the regeneration of functional tissues, as well as for treatment strategies (Mariani et al. 2019). For instance, implants can bear at their surface bioactive molecules and then modulate the interaction with the host, improving their biocompatibility (Guo et al. 2022). Bioactive molecules can also be incorporated into filling materials, helping to heal and to control the biofilm (Emanuel et al. 2020; Qiu et al. 2021). Tissue engineering can also be based on cell therapy, like transfer of stem cells (see Chap. 20 by Dubuc and colleagues). Finally, the reorientation of host responses by the transfer of autologous re-programmed or selected cells is a promising field, as already successfully experimented in animal models (Miao et al. 2020).

1.2 Considerations for the Future

1.2.1 Variability: More Than Just a Statistical Matter

1.2.1.1 Genetic Diversity in Patients

The genetic diversity of humans is obviously a source of variability in clinical experiments. For instance, the systematic analyses and high content methods used for genomics allowed the identification of genetic profiles linked with the disease, showing the population is not homogeneous and some individuals respond in a different way to the disease (see Chap. 11 by Schaefer). Rare events can have a strong link with the disease: thinking in terms of central tendency and making categories may be part of an analytical approach but should not preclude the use of multivariate statistics, the resort to machine learning, and the critical review of possibly meaningful outliers.

Interestingly, the diversity at the genetic level has been associated to a diversity at the microbiological and clinical level (Offenbacher et al. 2016).

1.2.1.2 Diversity in Clinical Manifestations and in Micro-niches

The American Academy of Periodontology and the European Federation of Periodontology recognizes three forms of periodontal diseases: necrotizing periodontitis, periodontitis as a direct manifestation of systemic diseases, and periodontitis *sensu stricto* (Tonetti et al. 1998). This diversity is also linked to a diversity in the methods used for the diagnosis. For instance, the use of different cytokines as oral biomarkers in the diagnosis of periodontitis can produce diverging results (see Chap. 15 by Blanco-Pintos and collaborators).

In periodontal microenvironments, bacteria are organized in three-dimensional structures, and several types of these structures exist (Mark Welch et al. 2016). According to the proportion of each individual structure that are sampled, the relative proportion of bacteria will greatly differ. The detected composition thus reflects the sum of different microenvironments rather than an average one by itself. Future analyses will consider samples are not homogeneous and will determine their components and their relative abundance.

1.2.1.3 Diversity in Research

The diversity in the concepts and methods used for the study of periodontitis can greatly affect the understanding of the disease. Though they are not always investigated in some studies, some non-bacterial components of the microbiota are linked to periodontitis: archaea (see Chap. 4 by de Cena and collaborators), parasites (see Chap. 5 by Martin-Garcia and collaborators), fungi (see Chap. 6 by Karkowska-Kuleta and collaborators), and viruses (see Chap. 7 by Jakovljevic and collaborators). This highlights that some fields of research can be neglected; the resulting vision of periodontitis pathophysiology cannot thus be holistic. In addition, variability in the aetiology of periodontitis *sensu stricto* is not reflected in the recommendations for clinical practice, probably because the present literature does not provide sufficient evidence to propose a differential therapeutic management (Sanz et al. 2020).

1.2.1.4 Only One Periodontitis?

In most of the studies about periodontitis, there is one definition of the disease, which is exclusive and excluding, meaning that all cases that are not included are considered healthy. Gingivitis is often claimed to form a continuum between health and periodontitis (Abusleme et al. 2021). Gingivitis can also be seen as an independent condition with a more diverse microbiota which does not produce irreversible damage (Kharitonova et al. 2021). This example highlights the fact that, according to the classification we apply, the clinical and microbial features may differ. Indeed, in some studies, *Trichomonas tenax* prevalence in periodontitis does not seem to differ from this in the healthy, while splitting of the periodontitis group according to severity allows to evidence differences: the parasite is more often detected in the most severe cases (Bisson et al. 2018). The two resulting subgroups characterized by their positivity for *T. tenax* are *de facto* more homogeneous for the parameters defining the severity of the disease and may be less variant for some parameters.

These considerations call for caution about the heritage from early research in periodontitis, in which only a few clinical pictures allowed an easy classification into a few categories relying on basic clinical features. At the beginning of the twentieth century, periodontitis was still called "*pyorrhoea alveolaris*" (or "Riggs' disease"), which means "purulent discharge of the tooth socket" and reminds us that the frequency of this type of presentations of the disease may have been different back then, as may have been their aetiologies, as recently proposed (Shiba et al. 2021).

1.2.1.5 Perspectives for Clinical Management

It is obvious but should be reaffirmed: a complex of diseases affecting half the humankind will not be summed up by one simplistic model. The effort of AAP-EFP to provide a classification tool gives an outstanding base for the future (Tonetti et al. 2018). In addition, there is a need to define the core characteristics of all the different types of periodontal diseases. The differences in symptomatology of periodontal diseases are perhaps subtle and arduous to diagnose, but they exist and should be considered. Finally, the model of continuous progression towards more severe forms of periodontal diseases is questioned, and "inactive" pockets should not be considered as definitely cured, but rather possibly paused or less severe sites.

1.2.2 Emerging and Expanding Fields

1.2.2.1 Peri-Implantitis and Peri-Implant Mucositis

Peri-implantitis and peri-implant mucositis can be considered mirror pathologies for periodontitis and gingivitis, respectively, and thus are a great opportunity for comparative studies. However, due to the presence of the implants, which are foreign bodies, some obvious peculiarities of these models can be of interest for fundamental research. Indeed, implants interact during a long time with the patients and their composition can impact their surroundings and the pathophysiology of these diseases. For instance, the microbial community structure differs between periodontitis and peri-implantitis (Komatsu et al. 2020). Furthermore, the composition or coating of implants can modulate the secretion of cytokines (Soskolne et al. 2002) and the activation of the complement system (Pham et al. 2020). Interestingly, Langerhans cell abundance in the implant vicinity is reduced, while their differentiation from bone marrow precursors is impaired *in vitro* by Titanium ions (Heyman et al. 2018). These myeloid dendritic cells have been seen as sempiternal resident sentinels orchestrating the immune response (reviewed in (Magan-Fernandez et al. 2020)), which calls for cautiousness about the definition of biocompatibility and about the central place of immunocompatibility in this concept.

1.2.2.2 The Immune System: More Than a Passive Sentinel

The mouth epithelium has long been considered impermeable when healthy (Squier and Hopps

1976). Later studies highlighted permeability to proteins at the periodontal sulcus, more precisely at the level of the junctional epithelium (Romanowski et al. 1988). Immune surveillance at the periodontal sulcus involves "para-inflammatory" features (Fine et al. 2016), in particular in neutrophils, whose neutrophil extracellular traps (NET) can play a pivotal role in inflammation resolution of aggravation (reviewed in (Vitkov et al. 2020)). Their recruitment and activation state has been recently hypothesized to be linked to stromal response to microbes (Williams et al. 2021), which sheds a new light on the immune role of "non-immune" cells.

Other cells of interest are the antigen-presenting cells, bridging innate and adaptive immunity. During periodontitis pathophysiology, Langerhans cells are less abundant while the number of plasmacytoid dendritic cells increase (Sharawi et al. 2021), reminding the effect of Titanium ions (Heyman et al. 2018).

Monocyte/macrophage homeostasis and polarization are important and intertwined with lymphocyte profiles and responses (see Chap. 10 by Cavalla and Hernández). Interestingly, the role of B cells is not limited to antibody production: they exert regulatory functions on the immune system, impacting the outcome of the disease (see Chap. 9 by Demoersman and Pers).

While a pro-inflammatory response can lead to damage due to oxidative stress on mucosal cells, breaching the epithelium and promoting invasion by pathogens, it can also directly impact bone dynamics, displacing it towards an osteoclastic activity (AlQranei and Chellaiah 2020). In vitro experiments suggest that autophagy inhibition leads to osteoclastogenesis via an increased secretion of IL-6 (Mayr et al. 2021). This underlines that immune responses leading to periodontal damage cannot be restrained to direct toxicity by immune effector cells.

1.2.2.3 Molecular Biology Beyond Genomics and Taxonomics

Identification of microbial species and of susceptibility factors in the human host is fundamental for the definition of the periodontal ecology in health and disease. The exploration of the human genome and of its association with periodontitis is still going on (see Chap. 11 by Schaefer). In addition, epigenetics studies have identified transcriptional regulations that can cause chronic damage. Indeed, long-term bone damage can involve epigenetic events (Ferreira et al. 2021), which could last even after the initial trigger has been suppressed. These responses are evidenced by transcriptomics, which can reach unprecedented level of precision using single-cell RNA sequencing (Qian et al. 2021). However, gene expression at the phenotypic level can also respond to post-transcriptional regulation, for instance by non-coding RNAs (Assis et al. 2021). Altogether, these mechanisms call for caution about the analysis by omics of the impact of microbial triggers on host response.

Taxonomics and metataxonomics are now routinely used along with metagenomics to determine the species composition in the dental plaque (see Chap. 2 by Bing and collaborators). Metagenomes represent the integrality of genomic information from a sample, while pangenomes group the totality of genomic information at a given taxonomic level. Metapangenomics, integrating both focuses, permit to map genes responding to environmental pressure and selection within a taxon, but also to evidence genetic interaction networks (Utter et al. 2020). In the oral sphere, this approach highlighted biogeographical restrictions for populations of some bacterial genera, suggesting that habitat adaptation can lead to the emergence of cryptic subpopulations and thus questioning the concept of species, as discussed by the authors (Utter et al. 2020).

1.2.2.4 New Models

Beside the three core members of the red complex (*P. gingivalis*, *T. denticola*, *T. forsythia*), other organisms gain importance, as well as their synergetic contributions for each other, as previously discussed. Current evolutions of periodontitis research have implied biofilms and focused on the interactions within them, as well as with host cells during the early steps of disease development (Redanz et al. 2021). The organization of these biofilms is probably shaped by oxygen pressure (Mark Welch et al. 2016). Other factors are obviously involved and thus some species or the clusters they form are site-specialists (Mark Welch et al. 2019). As the plaque mature along time, as the environment change, their composition and their genetic expression should be modulated. This produces new niches for possible superinfections by opportunistic colonizers or by contextual pathobionts that are already present but cannot express their pathogenic traits in an unpropitious environment. This does not exclude previous models for dysbiosis (Hajishengallis and Lamont 2012), but rather calls for the consideration of low abundance microorganisms from all the kingdoms, superkingdoms, and even non-cellular organisms like viruses. Interestingly, it has been shown that periodontopathogens could interact with host targets at a distance, using extracellular vesicles (Ha et al. 2020; Nara et al. 2021).

In parallel to the refinement of microbial models, new host-mimicking models have been developed. First, cultures leading to differentiated structures, the organoids, are efficiently produced *in vitro* (Basu et al. 2019; Wang et al. 2021; Banavar et al. 2021; Chu et al. 2021). Second, ex vivo models have been used for a long time, like cultured immune cells from patients or biopsies/explants. However, recent advances have been made in *in vitro* tissue engineering including cell lines or *ex vivo* cell cultures (Dommisch et al. 2021), allowing the integration of macrophages and making this model immunoresponsive (Ollington et al. 2021). And last, recently developed animal models open new perspectives for the study of periodontitis, like humanized mouse models (Rojas et al. 2021).

1.2.3 Considering the Patient Beyond Their Mouth

1.2.3.1 Considering the Patient as a Whole

Genetic predisposition of individuals and exposure to environmental threats, like infection with microbes, are only a part of the risk factors that may impact the outcome of the disease. Risk factors – elements correlated to the disease – are not compulsory causative elements. Several non-oral diseases are risk factors for periodontitis (for diabetes, see Chap. 12 by Salhi and Reners; for Alzheimer disease, see Chap. 13 by Harding and collaborators, for cardiovascular diseases, see Chap. 14 by Hansen and Holmstrup). In the case of diabetes, the influence is bidirectional, and the underlying mechanisms are not fully understood. For Alzheimer disease, evidence is piling up about the direct role of *P. gingivalis* in brain damage. The link between rheumatoid arthritis and periodontitis is hypothesized to be direct through the triggering of autoimmunity by bacterium-induced citrullination, and not circumscribed to the oral site of infection due to circulation of autoantibodies (reviewed in (Rooney et al. 2020) and (Gonzalez-Febles and Sanz 2021). Finally, inflammatory bowel disease and periodontitis seem to be linked by inflammation (de Mello-Neto et al. 2021). Altogether, these hypotheses and conclusions remind that periodontal health should be part of prevention strategies for several conditions.

Though aging has been considered for a long time a risk factor for periodontitis, the impact of periodontitis on biological aging is a recent concept (Baima et al. 2021c). Besides, some forms of periodontitis in children and adolescents have been hypothesized to have different causes as compared to the generally described adult periodontitis (Kinane et al. 2001). Periodontal health during pregnancy is also an issue (Jakovljevic et al. 2021), and the oral microbial communities change, as in supragingival plaque in women (Lin et al. 2018) and in subgingival plaques in animals (Walkenhorst et al. 2020).

It is important to consider, during treatment, that underlying parameters may modify the pathophysiology of the disease, or its impact on the patients' health, and thus the relevance of therapeutic and prophylactic strategies. These underlying parameters are of central interest to understand the possible link between periodontitis and cancers beyond the oral sphere (Radaic et al. 2000).

1.2.3.2 Considering the Patient's Environment

The mouth is an open system, impacted by almost constant inward and outward flows towards the exterior environment and the respiratory tract. It is also serves for the ingestion of solids and liquids. Microorganisms can enter the mouth while breathing and eating, but also when microbe-bearing surfaces enter in contact with the mouth, like when touching it or during intimate contacts. Surprisingly, the composition of the periodontal microbiota is quite stable (reviewed in (Joseph and Curtis 2021)) and may be due to the resilience of the polymicrobial communities (reviewed in (Wade 2000)). This questions the events leading to disruption of the healthy equilibrium leading to the transition to a new balanced state, the "dysbiotic" periodontal environment.

In experimental animal models, horizontal and vertical microbiota transmission triggers the development of periodontitis, suggesting "that an oral dysbiotic microbial community structure is stable to transfer and can act in a similar manner to a conventional transmissible infectious disease agent with concomitant effects on pathology", as commented elsewhere (Payne et al. 2019). The intra-group transmission of microbiota in human settings has been difficult to assess due to the complexity of human interactions, and due to the intertwined influence of shared genetical and environmental factors. However, recent data suggest that the weight of the environment should not be underestimated when considering the genetics, not excluding its participation in shaping a highly individualized microbiota (Mukherjee et al. 2021). These considerations raise the issue of the status of periodontitis, classified by the World Health Organization as a "non-communicable disease".

In addition to inter-human interactions, animals, in particular pets, can be a source of periodontal microbiota. For instance, cats naturally develop periodontitis and bear bacteria that could take part in periodontitis pathogenesis in humans (Harris et al. 2015). Transmission of these possible periodontopathogens has been suspected from dog to owners (Oh et al. 2015). The role of this pet-inherited microbiota in dysbiosis and its persistence in the human host are still to be investigated.

1.2.3.3 Considering the Patient's Mind

The mechanisms leading to bacterial contamination from the environment and from other individuals are studied, unlike the transmission of components of the oral microbiota belonging to other kingdoms of life. However, prevention guidelines are to be set, including individual interventions by health professionals and public health strategies (Cota et al. 2021). This comprises promoting health literacy in patients, which is correlated with protection or lower grades of periodontitis (Wehmeyer et al. 2014; Holtzman et al. 2017; Batista et al. 2017; Baskaradoss 2018; Timkova et al. 2020). Oral hygiene instructions by the health professionals are beneficial for the course of the therapy (see Chap. 17 by Salhi and collaborators).

Periodontitis has an impact on the psychology of the patient. First, stress predisposes and exacerbates periodontitis (Spector et al. 2020). Second, periodontitis negatively impacts the quality of life, also in its psychological component (Al-Harthi et al. 2013). Then, depression and periodontitis interact both ways, aggravating one another (Dumitrescu 2016).

To manage the psychological aspects of periodontitis management, efforts must be made in fundamental and clinical research. Reliable scientific results are needed to guide the practitioners beyond their own experience. Indeed, while the technical practice should be based on evidence, the lack of guidelines for patient motivation, education, and psychological support is noteworthy.

Future health intervention policies should aim at empowering the patients, so they can be an active part of their preventive and curative strategies against periodontitis.

1.3 Concluding Remarks

Experimental research in periodontitis is expending and diversifying, with increasing connections with other basic research topics. In this book, the variety of treated subjects reflects that current experimental research in periodontitis explores diverse fields covering clinical and fundamental research. This enriching complexity was also reflected when inviting authors and reviewers: a particular attention was paid to balance the clinical and fundamental focus of the people who worked on each chapter, allowing a fair and constructive reviewing process.

Furthermore, these chapters were double-blinded peer-reviewed to avoid conflicts of interest: the authors and the reviewers did not know their respective identities during the reviewing process (Copsey et al. 2021; Smith 2021). The reviewers then had the choice to reveal or not their identity.

Finally, authors were asked to provide specialized sections targeting specific readerships: scientificallyoriented "highlights", "considerations for practice", and a "patient summary". The "highlights" section permits to sum up the main messages coming from experimental research contained in the chapter. In the "considerations for practice" section, the authors present the actual or hypothetic impact of the results presented in the chapter for the clinical management of periodontitis. The objective of this section is not to provide guidelines to drive evidence-based medicine, but rather to provide practitioners with scientific take-home messages, to assist decision makers in considering original perspectives, and to seed new ideas and hypotheses for research. Finally, the "patient summary", should help the patients understand the content of the chapter. It should also give the clinicians an idea about how a specialist would speak to a broad audience about this subject, which can be valuable to improve patient literacy for a better therapeutic management of periodontitis, and more generally to stimulate "public engagement with science" in this field (Stilgoe et al. 2014; Verran et al. 2020).

References

Abusleme L, Hoare A, Hong B-Y, Diaz PI (2021) Microbial signatures of health, gingivitis, and periodontitis. Periodontology 86(1):57–78. https://doi.org/10.1111/prd.12362

Al-Harthi LS, Cullinan MP, Leichter JW, Thomson WM (2013) The impact of periodontitis on oral health-related quality of life: a review of the evidence from observational studies. Aust Dent J 58(3):274–277. https://doi.org/10.1111/adj.12076. quiz 384

AlQranei MS, Chellaiah MA (2020) Osteoclastogenesis in periodontal diseases: possible mediators and mechanisms. J Oral Biosci 62(2):123–130. https://doi.org/10.1016/j.job.2020.02.002

Assis RIF, Feltran GDS, Silva MES, Palma I, Rovai ES, Miranda TB, Ferreira MR, Zambuzzi WF, Birbrair A, Andia DC, da Silva RA (2021) Non-coding RNAs repressive role in post-transcriptional processing of RUNX2 during the acquisition of the osteogenic phenotype of periodontal ligament mesenchymal stem cells. Dev Biol 470:37–48. https://doi.org/10.1016/j.ydbio.2020.10.012

Baehni P, Tsai CC, McArthur WP, Hammond BF, Taichman NS (1979) Interaction of inflammatory cells and oral microorganisms. VIII. Detection of leukotoxic activity of a plaque-derived gram-negative microorganism. Infect Immun 24(1):233–243. https://doi.org/10.1128/iai.24.1.233-243.1979

Baima G, Corana M, Iaderosa G, Romano F, Citterio F, Meoni G, Tenori L, Aimetti M (2021a) Metabolomics of gingival crevicular fluid to identify biomarkers for periodontitis: a systematic review with meta-analysis. J Periodontal Res 56(4):633–645. https://doi.org/10.1111/jre.12872

Baima G, Iaderosa G, Citterio F, Grossi S, Romano F, Berta GN, Buduneli N, Aimetti M (2021b) Salivary metabolomics for the diagnosis of periodontal diseases: a systematic review with methodological quality assessment. Metabolomics 17(1):1. https://doi.org/10.1007/s11306-020-01754-3

Baima G, Romandini M, Citterio F, Romano F, Aimetti M (2021c) Periodontitis and accelerated biological aging: a Geroscience approach. J Dent Res 220345211037977. https://doi.org/10.1177/00220345211037977

Banavar SR, Rawal SY, Pulikkotil SJ, Daood U, Paterson IC, Davamani FA, Kajiya M, Kurihara H, Khoo SP, Tan EL (2021) 3D clumps/extracellular matrix complexes of periodontal ligament stem cells ameliorate the attenuating effects of LPS on proliferation and osteogenic potential. J Pers Med 11(6). https://doi.org/10.3390/jpm11060528

Baskaradoss JK (2018) Relationship between oral health literacy and oral health status. BMC Oral Health 18(1):172. https://doi.org/10.1186/s12903-018-0640-1
Basu A, Rothermund K, Ahmed MN, Syed-Picard FN (2019) Self-assembly of an organized cementum-periodontal ligament-like complex using scaffold-free tissue engineering. Front Physiol 10:422. https://doi.org/10.3389/fphys.2019.00422
Batista MJ, Lawrence HP, Sousa M (2017) Oral health literacy and oral health outcomes in an adult population in Brazil. BMC Public Health 18(1):60. https://doi.org/10.1186/s12889-017-4443-0
Beikler T, Bunte K, Chan Y, Weiher B, Selbach S, Peters U, Klocke A, Watt RM, Flemmig TF (2021) Oral microbiota transplant in dogs with naturally occurring periodontitis. J Dent Res 100(7):764–770. https://doi.org/10.1177/0022034521995423
Bisson C, Lec PH, Blique M, Thilly N, Machouart M (2018) Presence of trichomonads in subgingival biofilm of patients with periodontitis: preliminary results. Parasitol Res 117(12):3767–3774. https://doi.org/10.1007/s00436-018-6077-2
Cheng S-L, Chan ECS (1983) The routine isolation, growth, and maintenance of the intermediate-size anaerobic oral spirochetes from periodontal pockets. J Periodontal Res 18(4):362–368. https://doi.org/10.1111/j.1600-0765.1983.tb00371.x
Chu J, Pieles O, Pfeifer CG, Alt V, Morsczeck C, Docheva D (2021) Dental follicle cell differentiation towards periodontal ligament-like tissue in a self-assembly three-dimensional organoid model. Eur Cell Mater 41:20–33. https://doi.org/10.22203/eCM.v042a02
Copsey M, Allen J, Cooper AI (2021) Double-anonymised peer review: a new option for authors at chemical science. Chem Sci 12(25):8586–8588. https://doi.org/10.1039/d1sc90122b
Cota LOM, Villar CC, Vettore MV, Campos JR, Amaral G, Cortelli JR, Cortelli SC (2021) Periodontal diseases: is it possible to prevent them? A populational and individual approach. Braz Oral Res 35(Supp 2):e098. https://doi.org/10.1590/1807-3107bor-2021.vol35.0098
Coykendall A, Kaczmarek F, Slots J (1980) Genetic heterogeneity in Bacteroides asaccharolyticus (Holdeman and Moore 1970) Finegold and Barnes 1977 (approved lists, 1980) and proposal of Bacteroides gingivalis sp. nov. and Bacteroides macacae (Slots and Genco) comb. nov. Int J Syst Evol Microbiol 30:559–564
de Mello-Neto JM, Nunes JGR, Tadakamadla SK, da Silva Figueredo CM (2021) Immunological traits of patients with coexistent inflammatory bowel disease and periodontal disease: a systematic review. Int J Environ Res Public Health 18(17):doi:10.3390/ijerph18178958
Diamond LS (1962) Axenic cultivation of *Trichomonas tenax*, the oral flagellate of man. I Establishment of cultures. J Protozool 9:442–444. https://doi.org/10.1111/j.1550-7408.1962.tb02650.x
Dobell C (1939) The common flagellate of the human mouth, *Trichomonas tenax* (O.F.M.): its discovery and its nomenclature. Parasitology 31(1):138–146. https://doi.org/10.1017/S0031182000012671
Dommisch H, Stolte KN, Jager J, Vogel K, Müller R, Hedtrich S, Unbehauen M, Haag R, Danker K (2021). Characterization of an ester-based core-multishell (CMS) nanocarrier for the topical application at the oral mucosa. Clin oral Investig 25(10):5795–5805. https://doi.org/10.1007/s00784-021-03884-x
Dumitrescu AL (2016) Depression and inflammatory periodontal disease considerations-an interdisciplinary approach. Front Psychol 7:347. https://doi.org/10.3389/fpsyg.2016.00347
Emanuel N, Machtei EE, Reichart M, Shapira L (2020) D-PLEX500: a local biodegradable prolonged release doxycycline-formulated bone graft for the treatment for peri-implantitis. A randomized controlled clinical study. Quintessence Int 51(7):546–553. https://doi.org/10.3290/j.qi.a44629
Fathima B, Archer AC (2021) Bacteriophage therapy: recent developments and applications of a renaissant weapon. Res Microbiol 172(6):103863. https://doi.org/10.1016/j.resmic.2021.103863
Ferreira RS, Assis RIF, Feltran GDS, do Rosario Palma IC, Francoso BG, Zambuzzi WF, Andia DC, da Silva RA (2021) Genome-wide DNA (hydroxy) methylation reveals the individual epigenetic landscape importance on osteogenic phenotype acquisition in periodontal ligament cells. J Periodontol. https://doi.org/10.1002/JPER.21-0218
Finak G, Gottardo R (2016) Promises and pitfalls of high-throughput biological assays. Methods Mol Biol 1415:225–243. https://doi.org/10.1007/978-1-4939-3572-7_12
Fine N, Hassanpour S, Borenstein A, Sima C, Oveisi M, Scholey J, Cherney D, Glogauer M (2016) Distinct oral neutrophil subsets define health and periodontal disease states. J Dent Res 95(8):931–938. https://doi.org/10.1177/0022034516645564
Flügge C, Fox W, Royal College of Physicians of L (1886) Die Mikroorganismen: mit besonderer Berücksichtigung der Ätiologie der Infektionskrankheiten, vol 1. Zweite völlig umgearbeitete Auflage der "Fermente und Mikroparasiten" .. edn. Verlag von F. C. W. Vogel, Leipzig
Gonzalez-Febles J, Sanz M (2021) Periodontitis and rheumatoid arthritis: what have we learned about their connection and their treatment? Periodontology 87(1):181–203. https://doi.org/10.1111/prd.12385
Gros G (1849) Fragments d'helminthologie et de physiologie microscopique. Bulletin de la Société Impériale des Naturalistes de Moscou 22:549–573
Guo X, Bai J, Ge G, Wang Z, Wang Q, Zheng K, Tao H, Zhang L, Zhang H, Wang D, Zhang X, Li H, Pan G, Geng D (2022) Bioinspired peptide adhesion on Ti implants alleviates wear particle-induced inflammation and improves interfacial osteogenesis. J Colloid Interface Sci 605:410–424. https://doi.org/10.1016/j.jcis.2021.07.079
Ha JY, Choi SY, Lee JH, Hong SH, Lee HJ (2020) Delivery of periodontopathogenic extracellular vesicles to brain monocytes and microglial IL-6 promotion by

RNA cargo. Front Mol Biosci 7:596366. https://doi.org/10.3389/fmolb.2020.596366
Hajishengallis G, Lamont RJ (2012) Beyond the red complex and into more complexity: the polymicrobial synergy and dysbiosis (PSD) model of periodontal disease etiology. Mol Oral Microbiol 27(6):409–419. https://doi.org/10.1111/j.2041-1014.2012.00663.x
Harris S, Croft J, O'Flynn C, Deusch O, Colyer A, Allsopp J, Milella L, Davis IJ (2015) A pyrosequencing investigation of differences in the feline subgingival microbiota in health, gingivitis and mild periodontitis. PloS One 10(11):e0136986. https://doi.org/10.1371/journal.pone.0136986
Heyman O, Koren N, Mizraji G, Capucha T, Wald S, Nassar M, Tabib Y, Shapira L, Hovav AH, Wilensky A (2018) Impaired differentiation of Langerhans cells in the murine Oral epithelium adjacent to titanium dental implants. Front Immunol 9:1712. https://doi.org/10.3389/fimmu.2018.01712
Holm P (1954) The influence of carbon dioxide on the growth of Actinobacillus actinomycetemcomitans (bacterium actinomycetem comitans Klinger 1912). Acta Pathol Microbiol Scand 34(3):235–248. https://doi.org/10.1111/j.1699-0463.1954.tb00821.x
Holtzman JS, Atchison KA, Macek MD, Markovic D (2017) Oral health literacy and measures of periodontal disease. J Periodontol 88(1):78–88. https://doi.org/10.1902/jop.2016.160203
Irving JT, Newman MG, Socransky SS, Heely JD (1975) Histological changes in experimental periodontal disease in rats mono-infected with a gram-negative organism. Arch Oral Biol 20(3):219–220. https://doi.org/10.1016/0003-9969(75)90013-8
Jakovljevic A, Sljivancanin Jakovljevic T, Duncan HF, Nagendrababu V, Jacimovic J, Aminoshariae A, Milasin J, Dummer PMH (2021) The association between apical periodontitis and adverse pregnancy outcomes: a systematic review. Int Endod J 54(9):1527–1537. https://doi.org/10.1111/iej.13538
Johnston W, Rosier BT, Artacho A, Paterson M, Piela K, Delaney C, Brown JL, Ramage G, Mira A, Culshaw S (2021) Mechanical biofilm disruption causes microbial and immunological shifts in periodontitis patients. Sci Rep 11(1):9796. https://doi.org/10.1038/s41598-021-89002-z
Joseph S, Curtis MA (2021) Microbial transitions from health to disease. Periodontology 86(1):201–209. https://doi.org/10.1111/prd.12377
Kharitonova M, Vankov P, Abdrakhmanov A, Mamaeva E, Yakovleva G, Ilinskaya O (2021) The composition of microbial communities in inflammatory periodontal diseases in young adults Tatars. AIMS Microbiol 7(1):59–74. https://doi.org/10.3934/microbiol.2021005
Kinane DF, Podmore M, Murray MC, Hodge PJ, Ebersole J (2001) Etiopathogenesis of periodontitis in children and adolescents. Periodontology 26:54–91. https://doi.org/10.1034/j.1600-0757.2001.2260104.x
Komatsu K, Shiba T, Takeuchi Y, Watanabe T, Koyanagi T, Nemoto T, Shimogishi M, Shibasaki M, Katagiri S, Kasugai S, Iwata T (2020) Discriminating microbial community structure between Peri-Implantitis and periodontitis with integrated metagenomic, Metatranscriptomic, and network analysis. Front Cell Infect Microbiol 10:596490. https://doi.org/10.3389/fcimb.2020.596490
Lee CT, Teles R, Kantarci A, Chen T, McCafferty J, Starr JR, Brito LC, Paster BJ, Van Dyke TE (2016) Resolvin E1 reverses experimental periodontitis and dysbiosis. J Immunol 197(7):2796–2806. https://doi.org/10.4049/jimmunol.1600859
Lee CT, Li R, Zhu L, Tribble GD, Zheng WJ, Ferguson B, Maddipati KR, Angelov N, Van Dyke TE (2021) Subgingival microbiome and specialized pro-resolving lipid mediator pathway profiles are correlated in periodontal inflammation. Front Immunol 12:691216. https://doi.org/10.3389/fimmu.2021.691216
Lin W, Jiang W, Hu X, Gao L, Ai D, Pan H, Niu C, Yuan K, Zhou X, Xu C, Huang Z (2018) Ecological shifts of supragingival microbiota in association with pregnancy. Front Cell Infect Microbiol 8:24. https://doi.org/10.3389/fcimb.2018.00024
Magan-Fernandez A, Rasheed Al-Bakri SM, O'Valle F, Benavides-Reyes C, Abadia-Molina F, Mesa F (2020) Neutrophil extracellular traps in periodontitis. Cell 9(6):doi:10.3390/cells9061494
Mariani E, Lisignoli G, Borzi RM, Pulsatelli L (2019) Biomaterials: foreign bodies or tuners for the immune response? Int J Mol Sci 20(3):10.3390/ijms20030636
Mark Welch JL, Rossetti BJ, Rieken CW, Dewhirst FE, Borisy GG (2016) Biogeography of a human oral microbiome at the micron scale. Proc Natl Acad Sci U S A 113(6):E791–E800. https://doi.org/10.1073/pnas.1522149113
Mark Welch JL, Dewhirst FE, Borisy GG (2019) Biogeography of the oral microbiome: the site-specialist hypothesis. Annu Rev Microbiol 73:335–358. https://doi.org/10.1146/annurev-micro-090817-062503
Mayr A, Marciniak J, Eggers B, Blawat K, Wildenhof J, Bastos Craveiro R, Wolf M, Deschner J, Jager A, Beisel-Memmert S (2021) Autophagy induces expression of IL-6 in human periodontal ligament fibroblasts under mechanical load and overload and effects osteoclastogenesis in vitro. Front Physiol 12:716441. https://doi.org/10.3389/fphys.2021.716441
Meuric V, Le Gall-David S, Boyer E, Acuna-Amador L, Martin B, Fong SB, Barloy-Hubler F, Bonnaure-Mallet M (2017) Signature of microbial dysbiosis in periodontitis. Appl Environ Microbiol 83(14). https://doi.org/10.1128/AEM.00462-17
Meyer KF (1917) The present status of dental bacteriology. J Natl Dent Assoc 4(9):966–996. https://doi.org/10.14219/jada.archive.1917.0233

Miao Y, He L, Qi X, Lin X (2020) Injecting immunosuppressive M2 macrophages alleviates the symptoms of periodontitis in mice. Front Mol Biosci 7:603817. https://doi.org/10.3389/fmolb.2020.603817

Mukherjee C, Moyer CO, Steinkamp HM, Hashmi SB, Beall CJ, Guo X, Ni A, Leys EJ, Griffen AL (2021) Acquisition of oral microbiota is driven by environment, not host genetics. Microbiome 9(1):54. https://doi.org/10.1186/s40168-020-00986-8

Müller OF (1773) Vermium terrestrium et fluviatilium, seu animalium infusoriorum, helminthicorum et testaceorum, non marinorum, succincta historia. Heineck and Faber, Copenhagen and Leipzig

Myneni SR, Brocavich K, Howard HW (2020) Biological strategies for the prevention of periodontal disease: probiotics and vaccines. Periodontology 84(1):161–175. https://doi.org/10.1111/prd.12343

Nara PL, Sindelar D, Penn MS, Potempa J, Griffin WST (2021) Porphyromonas gingivalis outer membrane vesicles as the major driver of and explanation for neuropathogenesis, the cholinergic hypothesis, iron dyshomeostasis, and salivary lactoferrin in Alzheimer's disease. J Alzheimers Dis 82(4):1417–1450. https://doi.org/10.3233/JAD-210448

Nastri L, De Rosa A, De Gregorio V, Grassia V, Donnarumma G (2019) A new controlled-release material containing metronidazole and doxycycline for the treatment of periodontal and Peri-implant diseases: formulation and in vitro testing. Int J Dent 2019:9374607. https://doi.org/10.1155/2019/9374607

Nguyen T, Sedghi L, Ganther S, Malone E, Kamarajan P, Kapila YL (2020) Host-microbe interactions: profiles in the transcriptome, the proteome, and the metabolome. Periodontology 82(1):115–128. https://doi.org/10.1111/prd.12316

Offenbacher S, Divaris K, Barros SP, Moss KL, Marchesan JT, Morelli T, Zhang S, Kim S, Sun L, Beck JD, Laudes M, Munz M, Schaefer AS, North KE (2016) Genome-wide association study of biologically informed periodontal complex traits offers novel insights into the genetic basis of periodontal disease. Hum Mol Genet 25(10):2113–2129. https://doi.org/10.1093/hmg/ddw069

Oh C, Lee K, Cheong Y, Lee SW, Park SY, Song CS, Choi IS, Lee JB (2015) Comparison of the oral microbiomes of canines and their owners using next-generation sequencing. PLoS One 10(7):e0131468. https://doi.org/10.1371/journal.pone.0131468

Oliver WW, Wherry WB (1921) Notes on some bacterial parasites of the human mucous membranes. J Infect Dis 28(4):341–344. https://doi.org/10.1093/infdis/28.4.341

Ollington B, Colley HE, Murdoch C (2021) Immunoresponsive tissue-engineered Oral mucosal equivalents containing macrophages. Tissue Eng Part C Methods 27(8):462–471. https://doi.org/10.1089/ten.TEC.2021.0124

Oz HS, Puleo DA (2011) Animal models for periodontal disease. J Biomed Biotechnol 2011:754857. https://doi.org/10.1155/2011/754857

Payne MA, Hashim A, Alsam A, Joseph S, Aduse-Opoku J, Wade WG, Curtis MA (2019) Horizontal and vertical transfer of oral microbial dysbiosis and periodontal disease. J Dent Res 98(13):1503–1510. https://doi.org/10.1177/0022034519877150

Pham MH, Haugen HJ, Reseland JE (2020) Fluoride modification of titanium surfaces enhance complement activation. Materials (Basel) 13(3):10.3390/ma13030684

Qian SJ, Huang QR, Chen RY, Mo JJ, Zhou LY, Zhao Y, Li B, Lai HC (2021) Single-cell RNA sequencing identifies new inflammation-promoting cell subsets in Asian patients with chronic periodontitis. Front Immunol 12:711337. https://doi.org/10.3389/fimmu.2021.711337

Qiu G, Wu H, Huang M, Ma T, Schneider A, Oates TW, Weir MD, Xu HHK, Zhao L (2021) Novel calcium phosphate cement with biofilm-inhibition and platelet lysate delivery to enhance osteogenesis of encapsulated human periodontal ligament stem cells. Mater Sci Eng C Mater Biol Appl 128:112306. https://doi.org/10.1016/j.msec.2021.112306

Radaic A, Ganther S, Kamarajan P, Grandis J, Yom SS (2000) Kapila YL (2021) paradigm shift in the pathogenesis and treatment of oral cancer and other cancers focused on the oralome and antimicrobial-based therapeutics. Periodontology 87(1):76–93. https://doi.org/10.1111/prd.12388

Ramsing NB, Kühl M, Jørgensen BB (1993) Distribution of sulfate-reducing bacteria, O_2, and H_2S in photosynthetic biofilms determined by oligonucleotide probes and microelectrodes. Appl Environ Microbiol 59(11):3840–3849. https://doi.org/10.1128/aem.59.11.3840-3849.1993

Ranch KM, Maulvi FA, Koli AR, Desai DT, Parikh RK, Shah DO (2021) Tailored doxycycline hyclate loaded in situ gel for the treatment of periodontitis: optimization, in vitro characterization, and antimicrobial studies. AAPS PharmSciTech 22(3):77. https://doi.org/10.1208/s12249-021-01950-x

Redanz U, Redanz S, Treerat P, Prakasam S, Lin LJ, Merritt J, Kreth J (2021) Differential response of oral mucosal and gingival cells to corynebacterium durum, streptococcus sanguinis, and porphyromonas gingivalis multispecies biofilms. Front Cell Infect Microbiol 11:686479. https://doi.org/10.3389/fcimb.2021.686479

Rojas C, Garcia MP, Polanco AF, Gonzalez-Osuna L, Sierra-Cristancho A, Melgar-Rodriguez S, Cafferata EA, Vernal R (2021) Humanized mouse models for the study of periodontitis: an opportunity to elucidate unresolved aspects of its immunopathogenesis and analyze new immunotherapeutic strategies. Front Immunol 12:663328. https://doi.org/10.3389/fimmu.2021.663328

Romanowski AW, Squier CA, Lesch CA (1988) Permeability of rodent junctional epithelium to exogenous protein. J Periodontal Res 23(2):81–86. https://doi.org/10.1111/j.1600-0765.1988.tb01338.x

Rooney CM, Mankia K, Emery P (2020) The role of the microbiome in driving RA-related autoimmunity. Front Cell Dev Biol 8:538130. https://doi.org/10.3389/fcell.2020.538130

Sanz M, Herrera D, Kebschull M, Chapple I, Jepsen S, Beglundh T, Sculean A, Tonetti MS, Participants EFPW, Methodological C (2020) Treatment of stage I-III periodontitis-the EFP S3 level clinical practice guideline. J Clin Periodontol 47(Suppl 22):4–60. https://doi.org/10.1111/jcpe.13290

Schlieben LD, Prokisch H, Yepez VA (2021) How machine learning and statistical models advance molecular diagnostics of rare disorders via analysis of RNA sequencing data. Front Mol Biosci 8:647277. https://doi.org/10.3389/fmolb.2021.647277

Shah H, Collins M (1988) Proposal for reclassification of Bacteroides asaccharolyticus, Bacteroides gingivalis, and Bacteroides endodontalis in a new genus, Porphyromonas. Int J Syst Evol Microbiol 38:128–131

Shannon CE (1948) A mathematical theory of communication. Bell Syst Tech J 27(3):379–423. https://doi.org/10.1002/j.1538-7305.1948.tb01338.x

Sharawi H, Heyman O, Mizraji G, Horev Y, Laviv A, Shapira L, Yona S, Hovav AH, Wilensky A (2021) The prevalence of gingival dendritic cell subsets in periodontal patients. J Dent Res 100(12):1330–1336. https://doi.org/10.1177/00220345211004864

Shiba T, Komatsu K, Sudo T, Sawafuji R, Saso A, Ueda S, Watanabe T, Nemoto T, Kano C, Nagai T, Ohsugi Y, Katagiri S, Takeuchi Y, Kobayashi H, Iwata T (2021) Comparison of periodontal bacteria of edo and modern periods using novel diagnostic approach for periodontitis with micro-CT. Front Cell Infect Microbiol 11:723821. https://doi.org/10.3389/fcimb.2021.723821

Smith EM (2021) Reimagining the peer-review system for translational health science journals. Clin Transl Sci 14(4):1210–1221. https://doi.org/10.1111/cts.13050

Socransky SS (1977) Microbiology of periodontal disease – present status and future considerations. J Periodontol 48(9):497–504. https://doi.org/10.1902/jop.1977.48.9.497

Socransky SS, Haffajee AD, Cugini MA, Smith C, Kent RL Jr (1998) Microbial complexes in subgingival plaque. J Clin Periodontol 25(2):134–144

Soskolne WA, Cohen S, Sennerby L, Wennerberg A, Shapira L (2002) The effect of titanium surface roughness on the adhesion of monocytes and their secretion of TNF-alpha and PGE2. Clin Oral Implants Res 13(1):86–93. https://doi.org/10.1034/j.1600-0501.2002.130111.x

Spector AM, Postolache TT, Akram F, Scott AJ, Wadhawan A, Reynolds MA (2020) Psychological stress: a predisposing and exacerbating factor in periodontitis. Curr Oral Health Rep 7(3):208–215. https://doi.org/10.1007/s40496-020-00282-2

Squier CA, Hopps RM (1976) A study of the permeability barrier in epidermis and oral epithelium using horseradish peroxidase as a tracer in vitro. Br J Dermatol 95(2):123–129. https://doi.org/10.1111/j.1365-2133.1976.tb00814.x

Stilgoe J, Lock SJ, Wilsdon J (2014) Why should we promote public engagement with science? Public Underst Sci 23(1):4–15. https://doi.org/10.1177/0963662513518154

Suarez C, Piculell M, Modin O, Langenheder S, Persson F, Hermansson M (2019) Thickness determines microbial community structure and function in nitrifying biofilms via deterministic assembly. Sci Rep 9(1):5110. https://doi.org/10.1038/s41598-019-41542-1

Szafranski SP, Winkel A, Stiesch M (2017) The use of bacteriophages to biocontrol oral biofilms. J Biotechnol 250:29–44. https://doi.org/10.1016/j.jbiotec.2017.01.002

Tanner AC, Strzempko MN, Belsky CA, McKinley GA (1985) API ZYM and API an-ident reactions of fastidious oral gram-negative species. J Clin Microbiol 22(3):333–335. https://doi.org/10.1128/jcm.22.3.333-335.1985

Timkova S, Klamarova T, Kovalova E, Novak B, Kolarcik P, Madarasova Geckova A (2020) Health literacy associations with periodontal disease among Slovak adults. Int J Environ Res Public Health 17(6):doi:10.3390/ijerph17062152

Tonetti MS, Imboden MA, Lang NP (1998) Neutrophil migration into the gingival sulcus is associated with transepithelial gradients of interleukin-8 and ICAM-1. J Periodontol 69(10):1139–1147. https://doi.org/10.1902/jop.1998.69.10.1139

Tonetti MS, Greenwell H, Kornman KS (2018) Staging and grading of periodontitis: framework and proposal of a new classification and case definition. J Periodontol 89(Suppl 1):S159–S172. https://doi.org/10.1002/JPER.18-0006

Utter DR, Borisy GG, Eren AM, Cavanaugh CM, Mark Welch JL (2020) Metapangenomics of the oral microbiome provides insights into habitat adaptation and cultivar diversity. Genome Biol 21(1):293. https://doi.org/10.1186/s13059-020-02200-2

Van Dyke TE (2020) Shifting the paradigm from inhibitors of inflammation to resolvers of inflammation in periodontitis. J Periodontol 91(Suppl 1):S19–S25. https://doi.org/10.1002/JPER.20-0088

Van Dyke TE, Bartold PM, Reynolds EC (2020) The nexus between periodontal inflammation and Dysbiosis. Front Immunol 11:511. https://doi.org/10.3389/fimmu.2020.00511

Verran J, Jackson S, Scimone A, Kelly P, Redfern J (2020) Biofilm control strategies: engaging with the public. Antibiotics (Basel) 9(8):doi:10.3390/antibiotics9080465

Vitkov L, Minnich B, Knopf J, Schauer C, Hannig M, Herrmann M (2020) NETs are double-edged swords with the potential to aggravate or resolve periodontal inflammation. Cell 9(12):doi:10.3390/cells9122614

Wade WG (2000) (2021) resilience of the oral microbiome. Periodontology 86(1):113–122. https://doi.org/10.1111/prd.12365

Walkenhorst MS, Reyes L, Perez G, Progulske-Fox A, Brown MB, Phillips PL (2020) A uniquely altered Oral microbiome composition was observed in pregnant rats with Porphyromonas gingivalis induced peri-

odontal disease. Front Cell Infect Microbiol 10:92. https://doi.org/10.3389/fcimb.2020.00092
Wang A, Weldrick PJ, Madden LA, Paunov VN (2021) Biofilm-infected human Clusteroid three-dimensional Coculture platform to replace animal models in testing antimicrobial nanotechnologies. ACS Appl Mater Interfaces 13(19):22182–22194. https://doi.org/10.1021/acsami.1c02679
Wehmeyer MM, Corwin CL, Guthmiller JM, Lee JY (2014) The impact of oral health literacy on periodontal health status. J Public Health Dent 74(1):80–87. https://doi.org/10.1111/j.1752-7325.2012.00375.x
Williams DW, Greenwell-Wild T, Brenchley L, Dutzan N, Overmiller A, Sawaya AP, Webb S, Martin D, Genomics NN, Computational Biology C, Hajishengallis G, Divaris K, Morasso M, Haniffa M, Moutsopoulos NM (2021) Human oral mucosa cell atlas reveals a stromal-neutrophil axis regulating tissue immunity. Cell 184 (15):4090-4104:e4015. https://doi.org/10.1016/j.cell.2021.05.013
Yost S, Duran-Pinedo AE, Teles R, Krishnan K, Frias-Lopez J (2015) Functional signatures of oral dysbiosis during periodontitis progression revealed by microbial metatranscriptome analysis. Genome Med 7(1):27. https://doi.org/10.1186/s13073-015-0153-3
Yost S, Duran-Pinedo AE, Krishnan K, Frias-Lopez J (2017) Potassium is a key signal in host-microbiome dysbiosis in periodontitis. PLoS Pathog 13(6):e1006457. https://doi.org/10.1371/journal.ppat.1006457
Zupancic S, Rijavec T, Lapanje A, Petelin M, Kristl J, Kocbek P (2018) Nanofibers with incorporated autochthonous bacteria as potential probiotics for local treatment of periodontal disease. Biomacromolecules 19(11):4299–4306. https://doi.org/10.1021/acs.biomac.8b01181

Part I

Periodontal Microbiota

Microbiota in Periodontitis: Advances in the Omic Era

2

Shao Bing Fong, Emile Boyer, Martine Bonnaure-Mallet, and Vincent Meuric

Abstract

The complexity of the oral microbiome continues to astound researchers even with the advancement of multi-disciplinary strategies being used to study these microorganisms in relation to the human body. There is extensive literature available that explains how oral bacterial communities exist within the biofilm and maintains a balance with the host immune system, but when this balance is tipped disease can occur. The purpose of this review is to highlight the subgingival microbial compositions during health and periodontal disease using next generation sequencing techniques, as well as determining the types of functional activities that partake during these states. The subgingival microbiota is a fluid structure that can adapt accordingly to the environment and the identification of signature biomarkers may aid in the assessment of risk and disease severity in an individual to complement clinical diagnosis in the future.

Keywords

Periodontitis · Metagenomics · Metataxonomics · NGS

Abbreviations

BOP	bleeding on probing
GO	Gene ontology
JNK	c-Jung N-terminal kinase
MAPK	mitogen-activated protein kinase
NGS	next generation sequencing
PCT	periodontal complex traits

Highlights

- The importance in experimental design can influence the interpretation of subgingival microbiota composition.
- NGS analyses of microbiota reveal specific bacteria are associated with health or disease.
- The use of NGS helps in the identification of novel bacteria and biological systems important in disease development.

S. B. Fong (✉)
University of Rennes 1, Rennes, France
e-mail: shao-bing.fong@univ-rennes1.fr

E. Boyer · M. Bonnaure-Mallet · V. Meuric
University of Rennes 1, Rennes, France

INSERM, INRAE, Univ Rennes 1, CHU de Rennes, Nutrition Metabolisms and Cancer, Rennes, France
e-mail: emile.boyer@gmail.com; martine.bonnaure@univ-rennes1.fr; vincent.meuric@univ-rennes1.fr

J. Santi-Rocca (ed.), *Periodontitis*, Advances in Experimental Medicine and Biology 1373, https://doi.org/10.1007/978-3-030-96881-6_2

Considerations for Practice

- Next generation sequencing techniques are rapid and can be used complementary in clinical measures to identify individuals with a higher disease risk.
- Identification of key bacteria species allows profiling samples as healthy or diseased.
- The use of adjunct antibiotics in standard periodontal interventions may not necessarily be beneficial in reducing disease-associated bacteria.

Patient Summary

Periodontal disease is a common oral infection that affects 20–50% of the global population, caused by an imbalance between the local microorganisms and the immune system. Genetics, habitual and environmental factors can influence the susceptibility of an individual to infection, but some others may not develop disease. Next generation sequencing technologies have enabled clinicians and scientists to understand the mechanisms behind oral disease development and the interplay of bacterial and immune responses. By identifying key biological factors, the future of these technologies may help in assessing the risk of disease and facilitate treatment plans.

2.1 Introduction

Chronic periodontitis is a destructive inflammatory condition with intermittent periods of remission and relapse. It is associated with the activation of the host inflammatory response to bacterial challenge at the periodontal pocket, resulting in a progressive loss of the alveolar bone supporting the teeth (Hancock and Newell 2001; Holt and Ebersole 2005) (Fig. 2.1). The severity of periodontitis may be affected by environmental, physical and genetic risk factors, predisposing an individual to disease susceptibility (Albandar 2002; Eke et al. 2012). A new classification of periodontitis has been adopted where forms of the disease previously categorized as "chronic" or "aggressive" (Armitage 1999) are now classified collectively as "periodontitis", with a further characterization based on a staging and grading system (Papapanou et al. 2018). This system first determines the severity of disease upon presentation (e.g.: clinical attachment loss, tooth loss and probing depth) and the disease management, followed by supplementary information which takes into consideration aspects such as the biological features of the disease, patient history, and multiple risk factors. The gold standard for evaluating periodontal disease is through probing of the periodontal pocket. The presence of bleeding on probing (BOP) has been shown to correlate with an increased inflammatory infiltrate in the adjacent gingival tissues (Greenstein et al. 1981) and deemed at risk for disease progression (Tanner et al. 2007). However, pocket depth measurements or current microbiological tests are unable to differentiate gingival sulci that are in recovery or are at risk of periodontitis. It is therefore imperative to develop a more specific microbiological measure to facilitate in periodontal diagnosis and treatment when required.

The pathogenesis of periodontitis has been revisited multiple times since the nineteenth century (Fig. 2.2). Decades of research have been carried out to determine the interplay of oral bacteria and inflammation during periodontitis development. Early studies using experimental animals have shown that specific bacteria are the cause of disease development (Fitzgerald et al. 1960), and thereafter demonstrated in humans (Löe et al. 1965). As the presence of dental biofilm is directly associated with disease development and plaque removal re-establishes health, the "Non-specific Plaque Hypothesis" was introduced by Miller W. D. in 1890 (Theilade 1986). However, many instances have described that particular species existing within the microbial biofilm are the leading cause, which results in a "Specific Plaque Hypothesis" where an alteration in the microbiota shifts from healthy state to dis-

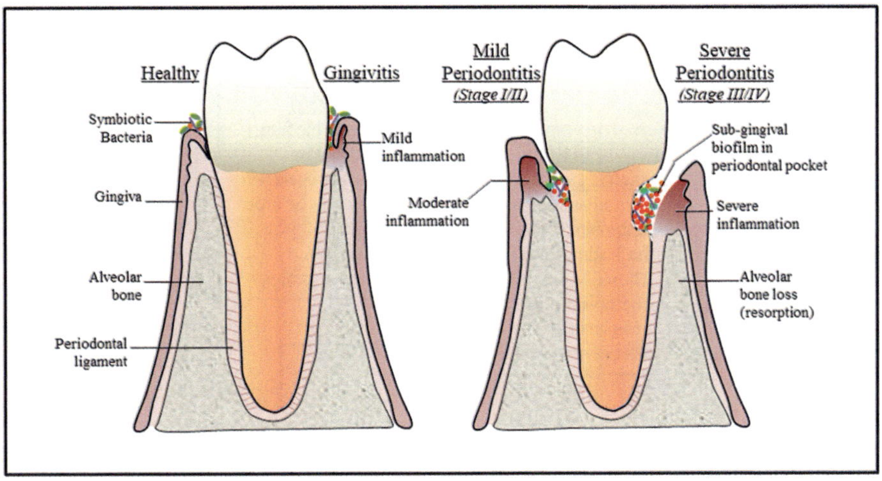

Fig. 2.1 Development of gingivitis and the progression to periodontitis. During healthy conditions, the host immune system and commensal bacteria maintain a symbiotic state. Mild inflammation can occur in the gingival tissue, resulting to gingivitis which can be reversible. If gingivitis persists, the inflammation can advance into a moderate state (mild-moderate periodontitis), initiating tissue and bone destruction. Chronic inflammation occurs when subgingival bacteria in the periodontal pocket contribute to extensive bone resorption and tissue loss, resulting in periodontitis

ease progression (Loesche 1976). The downside of this hypothesis was the difficulty in removing these specific pathogens as many of these bacteria either exist as commensals and are opportunistic pathogens (pathobionts), or are uncultivatable. Theilade (1986) then proposed an updated version of the non-specific hypothesis, stating that, while the collective biofilm plays a role in disease, one should take into consideration of specific bacterial species that can be more virulent than others and thus creating a pathogenic ecology (Theilade 1986).

Based on these early hypotheses, Marsh (1994) proposed the "Ecological Plaque Hypothesis" where an imbalance of the microflora under ecological stress results in an enrichment of certain pathogenic microorganisms, leading to disease progression (Marsh 1994). By this stage, it is clear that bacteria are a prerequisite for etio-pathogenesis, and the presence of specific species certainly encourages the progression. However, there have been some patients or periodontal sites reported to experience gingivitis without proceeding into periodontitis, while others suffer from chronic inflammatory periodontal disease. Indeed, periodontitis is a multi-factorial disease and its establishment is dependent on bacterial growth and metabolism, its environment (e.g. nutrients, pH, redox potential, etc.) and the host response (Marsh 2003). In the last decade, as a deeper understanding in individual oral bacteria is formulated, the "Keystone-Pathogen Hypothesis" was introduced to help explain the role of distinct periodontal pathogens. While seemingly in small proportions within the biofilm periodontal pocket under normal circumstances, they are able to dysregulate the host immune response and facilitate disease progression (Hajishengallis 2014; Hajishengallis et al. 2012). This occurs when such low abundance pathogenic bacteria proliferate to higher numbers that are sufficient to trigger inflammation (Byrne et al. 2009), thereby tipping the symbiotic environment to dysbiosis (Hajishengallis and Lambris 2011). A recent review brought together evidence to demonstrate the relationship between microbiology and inflammation, and concluded that it is the inflammatory continuum in bid to contain the polymicrobial infection which fuels the disease process, while the microbiological aspect plays a role only in the later stages when inflammation is dysregulated, forming a favourable environment

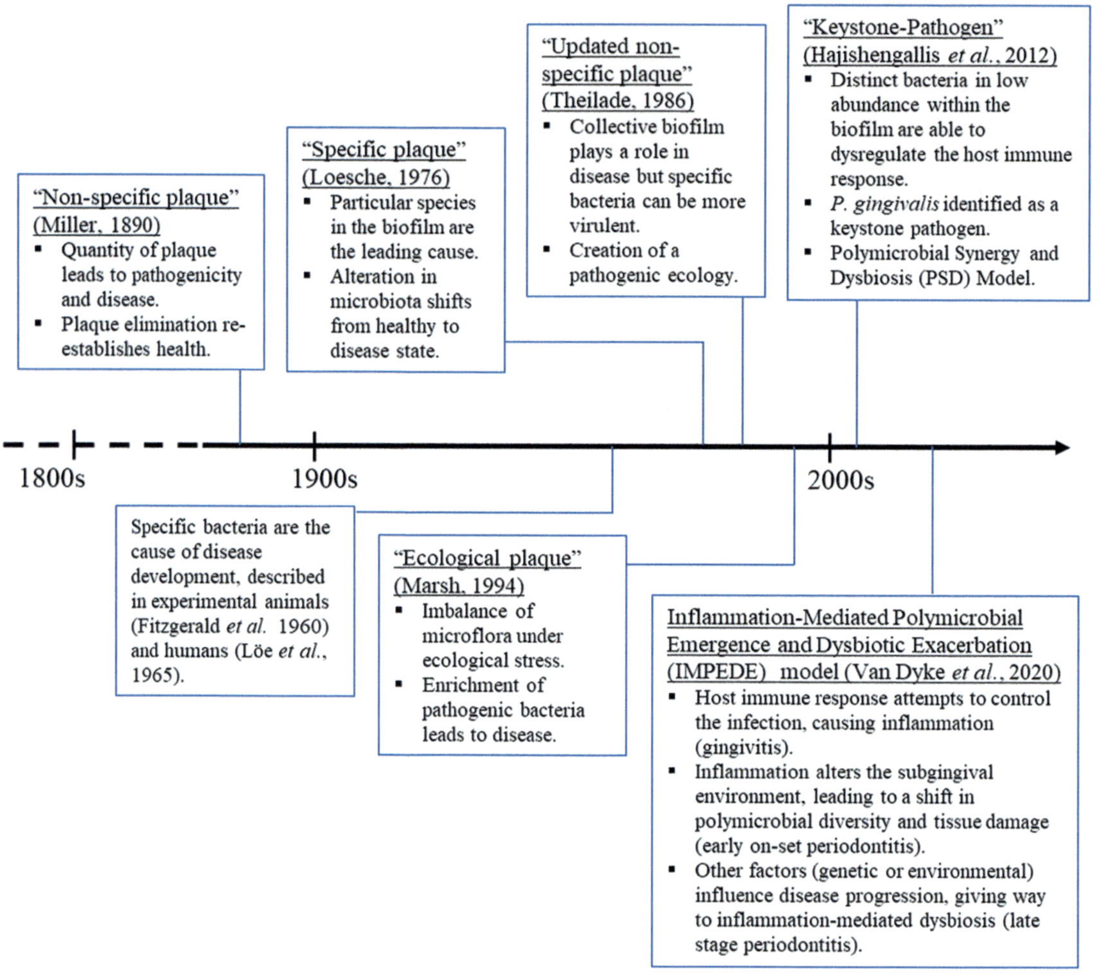

Fig. 2.2 **The evolution of hypotheses in periodontal disease and proposed models**. The timeline shows how various hypotheses and models were developed based on observations made upon the relationship of bacteria within the oral biofilm, the environment and the host immune response. (Fitzgerald et al. 1960; Hajishengallis et al. 2012; Löe et al. 1965; Loesche 1976; Marsh 1994; Theilade 1986; Van Dyke et al. 2020)

for periodontal pathogens to thrive (Van Dyke et al. 2020).

It is well-established that the microbiota associated with periodontitis differs from that in healthy conditions (Socransky and Haffajee 2005). The "complex theory" was developed by Dr. Sigmund Socransky et al. (1998) which classified periodontal bacteria into coloured groups based on their association with disease severity (Socransky et al. 1998) (Fig. 2.3). Early colonizers such as *Capnocytophaga, Streptococcus spp.* and *Veillonella* are categorized under green, yellow and purple complexes, respectively. These bacteria form the base of colonization on the sulcus for further bacterial association within the biofilm. Following, bacteria of the orange complex (e.g. *Fusobacterium nucleatum* and *Prevotella spp.*) are known as "bridging species", as they form a link between the early colonizers and the later colonizers (i.e.: the red complex). Members of the red complex, namely, *Porphyromonas gingivalis*, *Treponema denticola* and *Tannerella forsythia*, have been found highly associated with periodontitis (Socransky et al. 1998). However, there is inadequate information on the total diversity of the subgingival microbi-

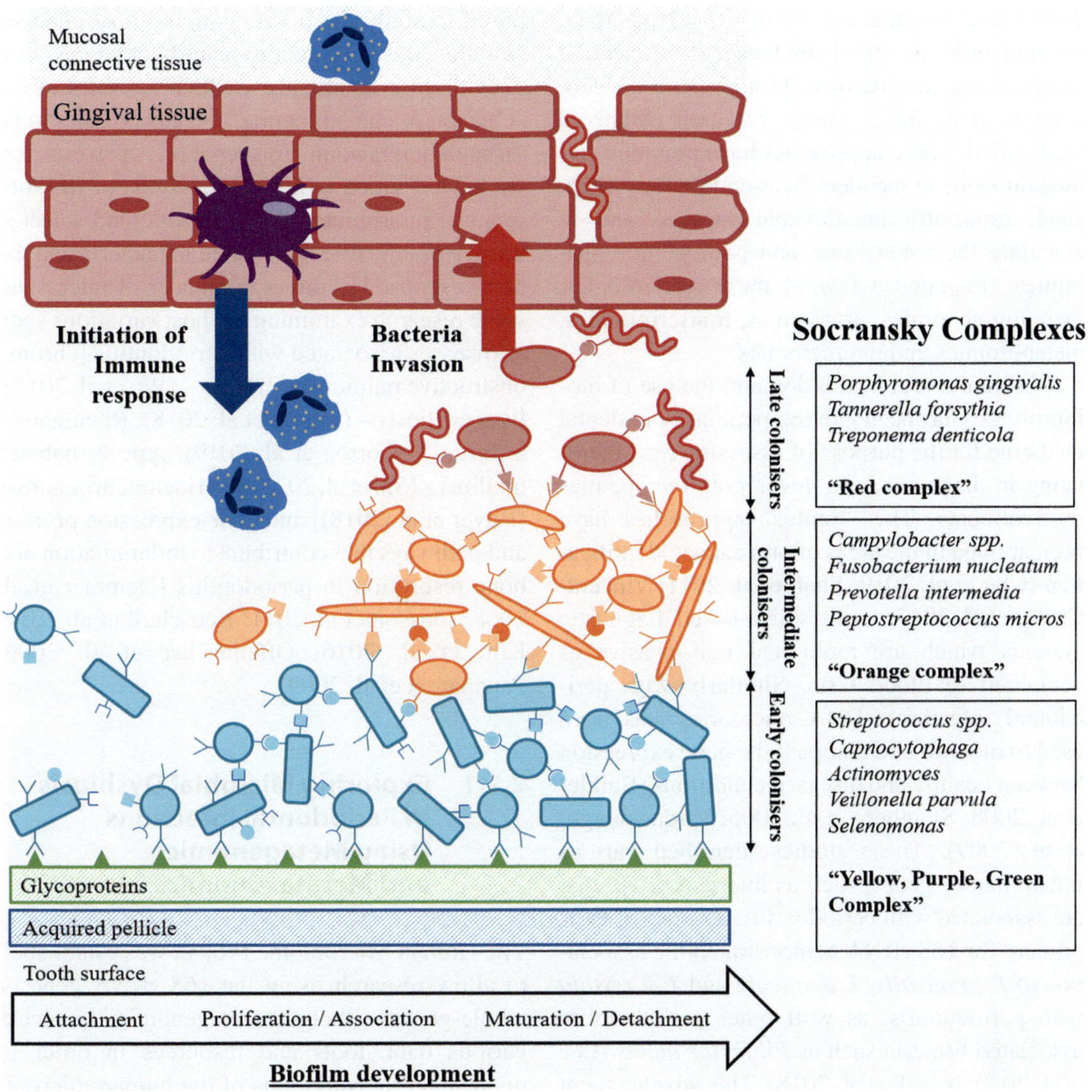

Fig. 2.3 **The development of an oral biofilm.** Planktonic bacteria attach to an acquired pellicle of various glycolipids and glycoproteins on the tooth surface. Early colonizing bacteria including *Streptococcus oralis* and *Streptococcus sanguinis* express adhesins and receptors to allow bacterial co-aggregation and the subsequent arrival of intermediate colonizers such as *Prevotella denticola* and *Actinomyces israelii*, to the expanding bacteria micro-colonies. During this stage, *Fusobacterium nucleatum* adheres within the biofilm to provide a bridge for the attachment of late colonizers including *Tannerella forsythia*, *Treponema denticola*, *Porphyromonas gingivalis*. Bacterial multiplication occurs within the biofilm until it matures and bacterial detachment causes the bacteria and its contents to be released for tissue invasion or re-colonization

ota using culture-based studies, partly due to variable growth requirements and thus many taxa was unsupported in culture. Although molecular approaches such as DNA hybridisation and PCR-based assays enabled the study of individual species (Colombo et al. 2009; Olson et al. 2011; Socransky et al. 2004), understanding how bacteria function in a community is still lacking. Owing to the revolution of high-throughput sequencing technologies, >700 oral bacterial species have now been identified (Dewhirst et al. 2010), including many uncultivable species with unknown roles in its environment. Since the end of the last century, there has been an expansion in digital and computer technology that leads to an advancement in the knowledge of molecular mechanisms in diseases through various subdisciplines of the "-omics" research (Manzoni et al. 2016). These approaches have provided significant value in the identification of microorganisms, especially uncultivable species, and to elucidate their metabolic and pathogenic capabilities. To name a few of these technologies, there are genomics, proteomics, transcriptomics, metabolomics and interactomics.

Within the oral health domain, the use of bioinformatics has been gaining popularity in dental medicine for the purpose of assessing risk, facilitating in diagnoses, and developing therapeutics or treatments. The "-omics" approaches have been applied in the areas of oral cancer identification (Chu et al. 2019; Shah et al. 2011; Vincent-Chong et al. 2017), using saliva-based diagnostic systems which are rapid and non-invasive as compared to blood tests. Similarly with periodontal infections, DNA microarray has been used to monitor and compare the gene expression between healthy and diseased conditions (Beikler et al. 2008; Steinberg et al. 2006; Vardar-Sengul et al. 2009). These studies identified various inflammatory genes, such as interleukin-1β, that are associated with periodontitis. The use of PCR primers for 16S rRNA demonstrated the association of *P. gingivalis*, *T. denticola* and *T. forsythia* with periodontitis, as well other periodontitis-associated bacteria such as *Filifactor alocis* (Lee et al. 2020; Nayak et al. 2018). This advancement of culture-independent 16S rRNA gene amplification and high-throughput DNA sequencing (e.g. 454 pyrosequencing), also termed as "next generation sequencing" (NGS), facilitated the collection of thousands of sequences per sample (Keijser et al. 2008; Zaura et al. 2009). In turn, researchers are able to comprehensively analyse bacterial composition (even at the species level) and behaviour within its community and the environment (Abusleme et al. 2013; Dewhirst et al. 2010; Griffen et al. 2012; Kumar et al. 2005).

The pathogenesis of periodontal diseases is a two-way street that involves the role of microorganisms and the response of the host. It was proposed to study the microbiota and host changes simultaneously in order to identify at which point does disease commences, which is essentially a "chicken or egg" dilemma, a nexus proving to be difficult at this point. To keep this review concise, we have decided to limit discussions of the subgingival microbiota at the genomic and selected transcriptomic findings related to bacteria and the host response. To direct the reader, if interested, some research examining the host variations such as diseases associated with periodontitis [chronic obstructive pulmonary disease – (Wu et al. 2017), liver cirrhosis – (Jensen et al. 2018), Rheumatoid arthritis – (Corrêa et al. 2019), type 2 diabetes mellitus – (Shi et al. 2020) and Haemochromatosis (Boyer et al. 2018)] and gene expression profiles and pathways that contribute to inflammation and bone resorption in periodontitis (Demmer et al. 2008; Jönsson et al. 2011; Kebschull et al. 2014; Kim et al. 2016; Offenbacher et al. 2009; Papapanou et al. 2009).

2.1.1 Exploring Microbial Dysbiosis in Periodontal Infections Using Metagenomics and Metataxonomics

The Human Microbiome Project was established to allow research using the 16S rRNA gene or whole-genome shotgun metagenomics to access various data, tools and resources in order to understand the variations of the human microbiome during health and disease (Peterson et al. 2009). A drawback in comparing different studies is the use of different methodologies (i.e.: clinical examination and diagnosis of the health status, sample collection protocols, DNA extraction protocols, and analysis of the hypervariable regions of the 16S rRNA gene). Typical ecological studies include measurements of bio-diversity (i.e.: species richness, evenness, alpha- and beta-diversities), as in the oral cavity where different periodontal states are compared. The consensus is varied across published research; for example, while some studies reported higher richness and increased evenness associated with periodontitis (Abusleme et al. 2013; Griffen et al. 2012;

Papapanou et al. 2020), no significant difference in the diversity metrics was observed between periodontitis and healthy in some others (Kirst et al. 2015). Furthermore, with periodontal interventions, the richness and abundance of pathogenic species may decrease, while simultaneously those of the health-associated species may also increase (Zhang et al. 2020). Thus, it can be difficult to measure how the alpha-diversity of a microbial community changes at times. Additional differences can be due to technical variations such as sampling choices (paper points, curettes), different 16S rRNA regions (V1–V3, V3–V4), sequencing chemistry and read lengths. In addition, the origins of samples (Europe, North America, South America, or Asia), host genetics and other environmental factors can play a part in these variations (Table 2.1). The choice of the 16S rRNA variable gene region may play a role in influencing the resultant ecological metrics and that sequencing the full 16S rRNA gene would provide the best taxonomic resolution (Birtel et al. 2015; Johnson et al. 2019), however constraints in resources may not allow it. Indeed, the selection of the regions used in analyses is dependent mostly upon published or in-house protocols rather than the nature of the sample. Furthermore, it has been reported that genomic shot-gun sequencing may generate 16S rRNAs more randomly and thus, resulting in differences in the microbial community structure (Wade 2011).

The routine assessment of clinically-active disease involves pocket depth measurements and BOP. While many early clinical studies have reported a positive association of certain species with BOP (Armitage et al. 1982; Socransky et al. 1991; Socransky and Haffajee 2005), not all sites with periodontal destruction (pockets >5 mm) have increased inflammation (Abusleme et al. 2013). No significant difference between periodontitis sites with or without BOP have been observed, but a significantly higher alpha-diversity (species richness and evenness) was observed in sites with periodontitis than in healthy sites (Abusleme et al. 2013). Even within diseased sites, subtle differences have been reported. Kirst et al. measured a significantly

Table 2.1 List of sample properties and experimental elements that can contribute to differences in metataxonomic results

Variable element	Examples	References
Location in the oral cavity	Tongue, tooth surface, buccal mucosa, tonsils, soft/hard palate, lip vestibule, saliva, subgingival crevice	Caselli et al. (2020) and Socransky and Haffajee (2005)
Sampling techniques	Paper points, curettes	Beyer et al. (2017)
Origin of samples	Europe, North America, South America, Asia	Socransky and Haffajee (2005)
Probing depths	Shallow sites (≤3–4) vs deep sites (≥6 mm)	Kirst et al. (2015), Pérez-Chaparro et al. (2018), and Socransky and Haffajee (2005)
Primers for 16S rRNA regions	V1–V3, V3–V4…	Johnson et al. (2019)
Other health complications	Oral related : dental caries, periodontal diseases, endodontic lesions, odontogenic infections, dry socket, halitosis, etc	Genco and Sanz (2020) and Socransky and Haffajee (2005)
	Systemic diseases: pancreatic cancer, heart disease, diabetes, etc	Genco and Sanz (2020)
NGS platforms	454 pyrosequencing	Mardis (2008)
	Sequencing by Oligo Ligation Detection (SOLiD; Applied Biosystems)	Mardis (2008)
	Ion Torrent (Applied Biosystems)	Kageyama et al. (2019)
	Illumina (HiSeq, NextSeq, MiSeq)	Lazarevic et al. (2009) and Mardis (2008)
	Single molecule real-time (SMRT) technology (PacBio)	Sadowsky et al. (2017)

(continued)

Table 2.1 (continued)

Variable element	Examples	References
	MinION technology (Oxford Nanopore)	Mikheyev and Tin (2014)
Read lengths	Varies, depending on the sequencing technology	Benn et al. (2018)
Bioinformatics tools for analysis (most frequently used for microbiota analyses)	Quantitative Insights Into Microbial Ecology (QIIME)	Caporaso et al. (2010)
	Human Oral Microbe Identification Microarray (HOMIM) and Human Oral Microbe Identification using Next Generation Sequencing (HOMINGS)	Mougeot et al. (2016)
	MOTHUR	Schloss (2020) and Schloss et al. (2009)

higher diversity and species richness in 6 mm pocket depths as compared to depths >7–8 mm (Kirst et al. 2015). It was thought that the difference in microbial diversity according to pocket depths may suggest a transitional microbiota state that occurs during invasion resulting in a highly diverse community of healthy- and disease-associated microbes. Over time, the microbial community develops into a predominantly disease-associated microbiota which appears to be more homogenous than in healthy sites (Kirst et al. 2015). Table 2.1 provides a few examples of elements that needs considerations when conducting a 16S rRNA analysis. Notably, there is a long list of factors in experimental design and analyses that can influence the interpretation of the biology. For this reason, a number of recent reviews suggest standardizations on how one may conduct an analysis on microbiomes (Knight et al. 2018; Wade and Prosdocimi 2020).

The earliest experimental gingivitis model developed by Löe and colleagues in the 1960s described that the bacterial flora accumulating in plaque during gingivitis commence from gram-positive cocci, to filamentous forms and rods, and finally transitioned to vibrios and spirochetes with a prominent presence of gram-negative cocci (Löe et al. 1965). Further on, the microbial ecology and oral biofilm development have been scrutinized, leading to the definition of the Socranksy microbial complexes (Socransky et al. 1998) (Fig. 2.3). This progressive microbial transition from periodontal health to gingivitis has not changed present-day, but 16S rRNA sequencing techniques can rapidly provide more details than phenotypic information on the microbiome composition in tissue samples. Periodontal healthy sites were found to harbour aerobic and facultative anaerobic gram-positive cocci and rods, such as from the genera *Actinomyces*, *Actinobacteria*, *Firmicutes Haemophilus*, *Rothia* and *Streptococcus* (Chen et al. 2018; Griffen et al. 2012; Kirst et al. 2015; Kistler et al. 2013). While sites associated with gingivitis and BOP showed a higher abundance of gram-negative bacteria from the genera *Campylobacter*, *Fusobacterium*, *Leptotrichia*, *Lautropia*, *Selenomonas*, *Porphyromonas*, and *Tannerella* (Kistler et al. 2013). Similarly in periodontitis, the phyla Bacteroidetes (a majority in deeper pockets), Spirochetes and Synergistetes, as well as genera *Treponema*, *Prevotella*, *Fusobacterium* and *Porphyromonas* were predominant (Abusleme et al. 2013; Chen et al. 2018; Griffen et al. 2012; Kirst et al. 2015). Certain discrepancies were also reported between studies, such as the absence of TM7 in the study by Tsai et al. (2018), where this difference was suggested to be due to geographic variability, the sampling depth or technical variations in the laboratory (Tsai et al. 2018). In addition, it is known that *A. actinomycetemcomitans* is highly associated with localized aggressive periodontitis, which is found infrequently in older adults and thus should be expected for the absence in samples from an older demographic (Genco et al. 2019).

As expected, the red complex (*P. gingivalis*, *T. denticola* and *T. forsythia*), *F. nucleatum*, *P. inter-*

media, have been reported in a wide number of studies analysing the subgingival bacterial profiles, to be associated with periodontitis samples, as well as the Gram-positive bacterium *F. alocis*, which was found highly abundant and strongly associated with disease. Table 2.2 is a non-exhaustive list of bacterial species most abundantly found with a relative abundance of at least 1%, reported in healthy, gingivitis or periodontitis. In addition, many of the disease-associated species are found not only in deep sites, but also in shallow pockets which suggests an intermedi-

Table 2.2 List of bacterial species found most abundant (relative abundance of at least 1%) in periodontal states, reported by the respective studies – (Abusleme et al. 2013; Chen et al. 2018; Deng et al. 2017, 2018; Griffen et al. 2012; Hong et al. 2015; Kistler et al. 2013; Offenbacher et al. 2016; Szafrański et al. 2015)

Species	Cited in healthy	Cited in Gingivitis	Cited in Periodontitis
Acinetobacter junii	Griffen (2012)		Griffen (2012)
Actinomyces naelundii	Griffen (2012), Abusleme (2013), Kistler (2013), and Deng et al. (2018)		
Actinomyces sp. Oral taxon 169/170/171	Abusleme (2013), Hong (2015), and Kirst (2015)		
Aggregatibacter actinomycetemcomitans			Offenbacher (2016) and Chen (2018)
Campylobacter gracilis	Chen (2018)	Kistler (2013)	Kistler (2013)
Campylobacter rectus		Deng et al. (2017)	Szafrański (2015) and Offenbacher (2016)
Campylobacter showae	Chen (2018)	Kistler (2013)	
Capnocytophaga leadbetteri	Chen (2018)	Kistler (2013)	
Capnocytophaga sputigena	Hong (2015)	Kistler (2013)	
Corynebacterim durum	Abusleme (2013) and Hong (2015)		
Corynebacterium matruchotii	Abusleme (2013), Kistler (2013), and Hong (2015)	Kistler (2013)	Kistler (2013) and Hong (2015)
Desulfobulbus sp. Oral taxon 041			Abusleme (2013) and Kirst (2015)
Eubacterium brachy			Griffen (2012) and Abusleme (2013)
Filifactor alocis			Griffen (2012), Abusleme (2013), Kirst (2015), Szafrański (2015), and Chen (2018)
Fusobacterium nucleatum ss. *animalis*	Abusleme (2013), Hong (2015), and Chen (2018)	Kistler (2013)	Abusleme (2013), Kistler (2013), Hong (2015), and Chen (2018)
Fusobacterium nucleatum ss. *nucleatum*	Abusleme (2013) and Deng et al. (2017)	Deng et al. (2017)	Abusleme (2013), Offenbacher (2016), and Chen (2018)
Fusobacterium nucleatum ss. *polymorphum*	Kistler (2013)	Kistler (2013)	Kistler (2013), Hong (2015), and Chen (2018)
Fusobacterium nucleatum ss. *vincentii*	Abusleme (2013), Kistler (2013), Hong (2015), and Chen (2018)	Kistler (2013)	Abusleme (2013), Kistler (2013), Hong (2015), Kirst (2015), and Chen (2018)
Gemella haemolysans	Kirst (2015) and Szafrański (2015)		
Gemella morbillorum	Griffen (2012) and Chen (2018)	Kistler (2013)	Kistler (2013)

(continued)

Table 2.2 (continued)

Species	Cited in healthy	Cited in Gingivitis	Cited in Periodontitis
Granulicatella adiacens	Griffen (2012), Kirst (2015), and Szafrański (2015)		Kistler (2013)
Lautropia mirabilis	Abusleme (2013), Kistler (2013), and Hong (2015)	Kistler (2013)	Abusleme (2013)
		Abusleme (2013)	
Mogibacterium timidum			Abusleme (2013) and Kirst (2015)
Neisseria elongata	Hong (2015)		Abusleme (2013)
Neisseria flavescens	Chen (2018)		
Neisseria subflava			Abusleme (2013)

ate asymptomatic stage during pathogenesis and this would be an avenue for pre-disease detection (Griffen et al. 2012). *T. denticola* was found in high abundance in subjects with severe periodontitis, but bacteria normally associated with disease such as *T. forsythia*, *P. gingivalis* and *A. actinomycetemcomitans* were observed to be at low abundance in severe periodontitis. Although an overlap of bacterial taxa have been noted across phenotypes (healthy, mild periodontitis, moderate periodontitis and severe periodontitis) and pocket depths, certain bacteria were noticeably more abundant in severe disease states and deeper pockets than in healthy samples and shallower sites, namely, *P. gingivalis*, *F. alocis*, *Fretibacterium fastidiosum* and TM7 (Papapanou et al. 2020). The reverse was true with species such as *Veillonella parvula*, *Rothia dentocariosa* and *Lautropia mirabilis*. As such, these bacteria should be regarded as constituents of the resident microbiota (core) and dysbiosis occurs when there is a progressive enrichment of certain resident bacterial species rather than an exogenous pathogen that invades the system through a niche (Papapanou et al. 2020).

The keystone hypothesis (Fig. 2.2) describes that low abundance microbial pathogens can exist discreetly during healthy conditions, and when the environment becomes favourable, they can initiate or contribute to dysbiosis and disease development (Hajishengallis et al. 2012). A number of studies have provided evidence of this hypothesis. One example is *Treponema* which normally exists at ~1% in health, but can dramatically increase to a mean relative abundance of ~20% in periodontitis. In contrast, *Rothia* and *Actinomyces* showed the reverse, measuring ~20% in health and dropping to ~1% in periodontitis (Abusleme et al. 2013). Furthermore, the authors observed that, while the relative abundance of *Actinomyces* decreased from health to periodontitis, the total load was not altered. This suggests that during disease, the biomass of health-associated species such as *Actinomyces* is not lost but it is the increase of biomass of other (disease-associated) community members, resulting in a depleted relative abundance of *Actinomyces* over the total biomass of the microbiota (Abusleme et al. 2013). This supports the same observation made in the Griffen study (2012) where health-associated species were not lost but remain as a smaller fraction of the total community (Griffen et al. 2012). On the same note, the relative abundance of *F. nucleatum* tend not differ between health and disease states, which may confirm its functional role in providing structural support for the microbial community (Abusleme et al. 2013).

Metagenomic analyses on the oral microbiota has been extended to elderly subjects (Genco et al. 2019; LaMonte et al. 2019; Papapanou et al. 2020), and other lifestyle changes such as smoking habit (Bizzarro et al. 2013; Mason et al. 2015; Yu et al. 2017) and periodontal treatments (Schwarzberg et al. 2014; Shi et al. 2015), that may play a role in shaping the subgingival microbiome community. Few studies have described the effects of standard periodontal treatment with antibiotics on the subgingival microbiota in periodontitis patients (Han et al. 2017; Jünemann et al. 2012; Laksmana et al. 2012; Schwarzberg et al. 2014), but these are small-scale case studies and interpersonal variability in the microbiome between patients suggests a need to determine more consistent taxa before and after therapy as predictive indicators of disease progression or remission. Bizzarro et al. (2016) investigated the impact of periodontal treatment with and without antibiotics in the subgingival microbiota composition at baseline, 3-month, 6-month, and 12-month post-treatment study. In general, periodontal treatments yielded a decline in disease-associated taxa (e.g., *Porphyromonas*, *Filifactor*, *Synergistaceae*, *Treponema* and *Tannerella*) and an increase in health-associated bacteria (e.g., *Neisseria*, *Rothia*, *Veillonella* and *Capnocytophaga*). The efficiency of adjunctive antibiotics was only observed at 3-months post-treatment, but not at the further time-points. Patients presented with a more diverse microbiota at baseline resulted in a better clinical outcome, while patients with higher levels of *Pseudomonadaceae* (non-typical oral taxa such as *Pseudomonas aeruginosa*) at baseline were found to have poorer responses (Bizzarro et al. 2016). A prospective cohort study on subjects that underwent third molar surgery with or without antibiotic treatment post-extraction revealed that, although antibiotic treatment reduced the relative abundance of genera *Actinobacteria*, *Bacteriodetes* and *Fusobacterium* up to 6 months, it did not create a significant impact in the overall microbial profile long-term (Menon et al. 2019). In a recent study, individuals with periodontitis who had periodontal treatment and maintained a healthy periodontal state, were found to harbour a different subgingival microbial composition of higher diversity and contained more disease-associated bacteria genera including *Actinomyces*, *Leptotrichia* and *Fusobacterium*, and presented a higher abundance of host destructive functions, as compared to healthy subjects (Lu et al. 2020). Taken together, these studies suggest that the use of adjunct antibiotics in periodontal treatment may not be advantageous, especially when a dysbiotic microbiome may still be present post-treatment. Furthermore, long-term antibiotic use can also lead to a risk for developing antibiotic resistance, inducing the dysbiosis of the oral and gut microbiota (Sweeney et al. 2004; Zaura et al. 2015).

2.1.2 Microbial Classification for Prediction in Periodontal Dysbiosis

Clustering analyses to associate bacterial species into different periodontal states have facilitated in studying relationships within the bacterial community. Duran-Pinedo and colleagues identified two distinct clusters (each made up of a number of bacterial network modules) associated with healthy individuals and one major community associated with periodontitis (Duran-Pinedo et al. 2011). Interestingly, there was no overlap of bacterial modules between the two healthy communities, suggesting that bacterial species are able to interact with different partners in various environmental conditions. Healthy cluster networks were less centralized and more stable, whereas in disease clusters, highly centralized nodes can be removed and cause failure in the network during environmental challenges. This is reminiscent of the 'keystone pathogen' idea whereby removing a disease-associating species (or encouraging the growth of a health-associating species) may alter the structure of the community (Duran-Pinedo et al. 2011).

A notable study by Offenbacher et al. created distinct models based on periodontal complex traits (PCT) using genome-wide associations in order to characterize the subgingival microbiome in clinical periodontal disease (Offenbacher et al. 2016). This study is based on the idea that the manifestation of periodontitis is complex, and genetic variations can lead to various clinical states as well as the influence from the interactions between the immune system and microflora. As such, the interplay of various factors may result in different levels of dysbiosis and altered biological signature in each individual. For instance, individuals that had high correlation and positive loading of *P. gingivalis*, *P. intermedia*, *P. nigrescens*, *T. forsythia*, *Treponema denticola*, *F. nucleatum*, *C. rectus*, and *A. actino-*

mycetemcomitans were classified as "Socransky trait" (PCT1); samples that had a high positive of GCF-IL-β and *A. actinomycetemcomitans* were labelled "Aa trait" (PCT3); and those with the highest loading of *P. gingivalis* were classified "Pg trait" (PCT5) (Offenbacher et al. 2016). Using a larger cohort of samples and an independent sample from a German study revealed that there was insufficient validation for clinically defined chronic periodontitis and aggressive periodontitis. The authors suggested that there may be other means of pathogenic mechanisms, or the relevance of the identified gene loci may be associated with the disease in a specific microbial/immunological situation, and thus the possibility of different subgroups of clinical periodontitis (Offenbacher et al. 2016). While this study is unable to clearly define clinical periodontitis and therefore no diagnostic utility in this context, the loci determined through genome-wide associations may facilitate the identification of new pathogenic pathways and thus the identification of candidate genes for genetic susceptibility in periodontitis.

The current prediction models to assess periodontitis incidence and progression have been identified lack validation and limited for clinical applications (Du et al. 2018). These models are either risk assessments or based on clinical predictors. Alternatively, a couple of clinical studies analysing the subgingival microbiome formulated prediction models based on the microbial classification associated with the periodontal state. Shi et al. (2015) classified the clinical states based on the subgingival microbiome profile of 21 independent samples (13 diseased, 1 resolved and 7 healthy) and built co-occurrence networks, which were then tested to an accuracy of 81% in assigning the samples to their respective clinical states (Shi et al. 2015). In another study, eight bacteria genera (*Aggregatibacter, Gracilibacteria, Megasphaera, Mycoplasma, Agrobacterium, Veillonellaceae, Capnocytophaga, Fusobacterium*) were used as biomarkers to generate a classification model and tested to be 83% accurate (Lu et al. 2020). This latter study was based on well-maintained individuals as there may be a recurrence of disease in patients with a history of periodontitis, which is highly relevant considering the high prevalence of periodontitis in the world. While the authors recognised the small sample size as a limitation, the findings supported the need to pay consideration on residual pathogenicity due to the presence of a more dysbiotic community compared to healthy subjects (Lu et al. 2020).

By taking subgingival samples from patients with chronic periodontitis and healthy subjects, together with data made available by other published studies, Meuric et al. pooled all samples and subjected them to a microbiota analysis using the same pipeline to identify microbiota specific to health and disease (Meuric et al. 2017). This large number of samples (>600 subgingival) treated under the same bioinformatics analysis would potentially overcome individual and technical variations normally faced by small sample sizes and the use of different methods. Data was found clustered into five groups according to the clinical status or the sampling site. Healthy subgingival samples were primarily distributed into two main clusters and the majority of periodontitis samples (90%) clustered together. Interestingly, 10% of the periodontitis samples were found in the healthy clusters, and 16% of the healthy subgingival samples were found in the periodontitis cluster. The subgingival microbial richness was observed to be for the majority of the sample lower in periodontitis. However, the microbial diversity of one of the healthy clusters was greater than that the other healthy cluster and the periodontitis (CP) cluster (Meuric et al. 2017).

When analysing the microbial communities, healthy subgingival samples were dominated by 8 major genera, namely, *Fusobacterium, Actinomyces, Streptococcus, Neisseria, Capnocytophaga, Prevotella, Corynebacterium* and *Rothia*. There were also a number of minor genera worth mentioning: *Leptotrichia, Veillonella, Porphyromonas, Granulicatella, Kingella* and *Gemella*. In periodontitis samples, the major genera were less abundant, namely *Treponema, Porphyromonas, Prevotella* and *Fusobacterium*, followed by *Streptococcus, Eubacterium, Tannerella* and *Campylobacter* (Fig. 2.4). Using the relative abundance of genera highly prevalent (>95%) in periodontitis samples

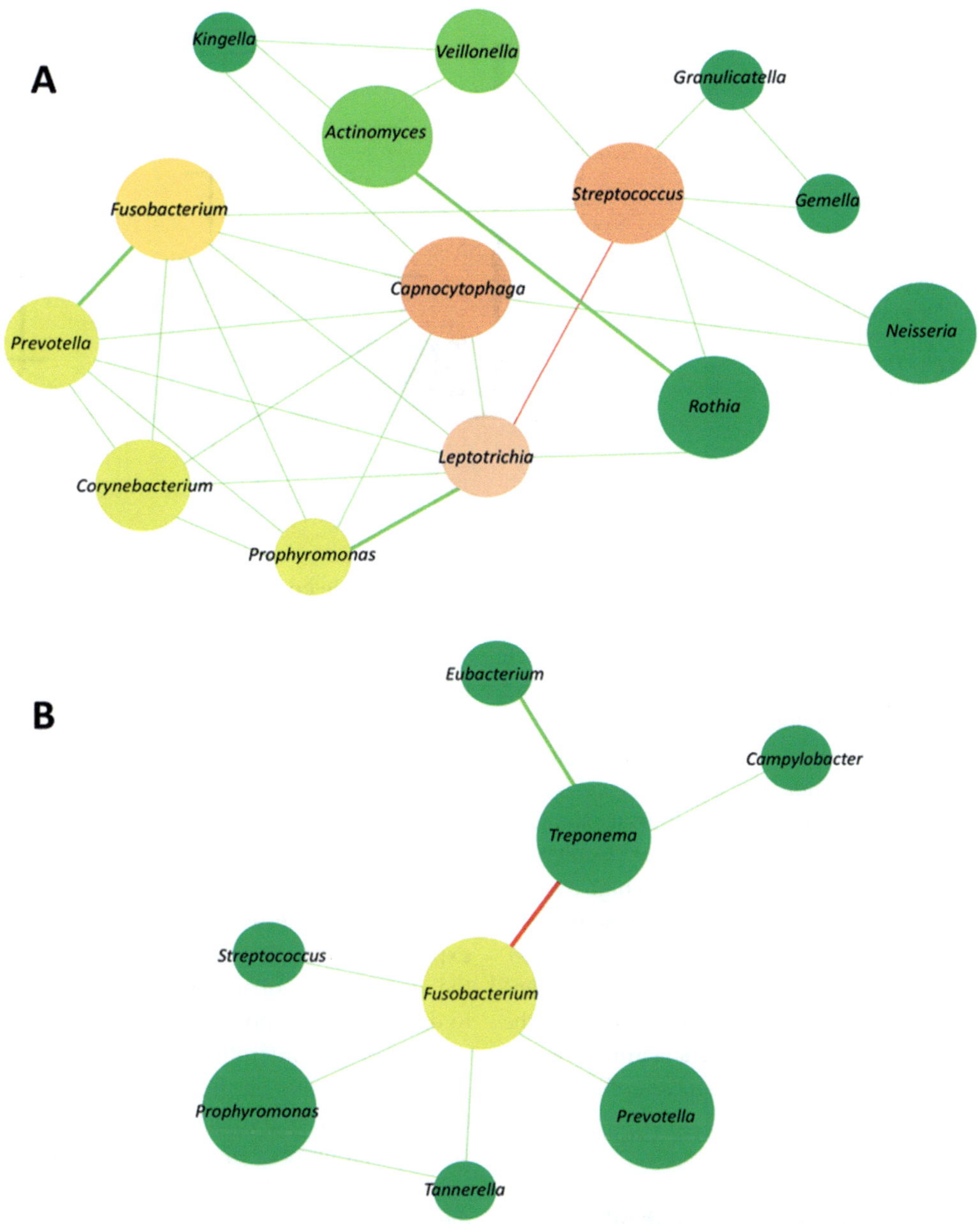

Fig. 2.4 Patterns of subgingival microbial communities. (**a**) Patterns of genera present in at least 95% of all healthy subgingival samples. (**b**) Patterns of genera present in at least 95% of all chronic periodontitis samples. Edges represent 1 (thin line) or 2 or 3 (thick line) significant correlations between genera (green – negative; red – negative). Node colors represent the number of partners, ranging from 1 (green) to 7 (dark orange). Node sizes represent the abundance of each taxon. (Figure was reproduced from Meuric et al. 2017)

(*Eubacterium*, *Campylobacter*, *Treponema* and *Tannerella*) versus genera highly prevalent (>95%) in healthy samples (*Veillonella*, *Neisseria*, *Rothia*, *Corynebacterium* and *Actinomyces*), a dysbiosis formula was generated (Fig. 2.5). Results revealed a clear distinction between healthy samples from diseased samples (Meuric et al. 2017). A different dataset containing samples from Bizzarro et al. (2016), which were not included in the initial analysis, were used to validate and to simplify the dysbio-

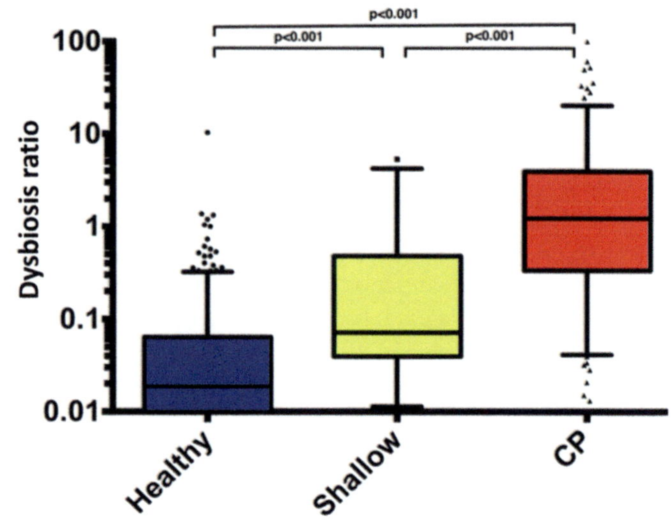

Fig. 2.5 **Subgingival dysbiosis ratios of *Eubacterium*, *Campylobacter*, *Treponema*, and *Tannerella* to *Veillonella*, *Neisseria*, *Rothia*, *Corynebacterium*, and *Actinomyces*.** Comparison between healthy, shallow and chronic periodontitis (CP) samples from all clusters. (Figure was reproduced from Meuric et al. 2017)

sis ratio (*Porphyromonas*, *Treponema* and *Tannerella* versus *Rothia* and *Corynebacterium*). A correlation was found with pocket depth, which confirmed the link between microbial dysbiosis and the depth of the periodontal pocket. The advantage of using the dysbiosis ratio during analysis can provide further information. For instance, a calculated low dysbiosis ratio from the subgingival sample at deep sites could indicate that the patient is microbiologically on the mend, or a calculated high dysbiosis ratio at shallow sites may suggest a potential risk of periodontitis (Fig. 2.5) (Meuric et al. 2017). Thus, by identifying the periodontal pathogens responsible for the imbalance, also termed as the "dysbiotic signature", it may be a starting point as useful prediction tool in the diagnoses of periodontal disease.

2.1.3 Understanding Microbial Dysbiosis and Host Responses Through Metatranscriptomics

While metagenomics/metataxonomic studies allow the identification of community members (taxonomic profile) of the microbiome, metatranscriptomics can enable the achievement of a functional profile to understand how the microbial community respond to changes in the environment over time. Thus, it is possible to identify the active functional profile of the microbiome under specific conditions, often dependent on the status of the host. As many studies reported a substantial overlap of microbial communities between active and non-progressing lesions, the periodontal status of sites (healthy, diseased or at risk of periodontitis) could not be determined solely by the observed differences in the subgingival microbial composition measured. With the ease of high-throughput sequencing, additional transcriptomic data becomes easily attainable to provide a genome-wide mapping of altered gene expression when health transits to disease. While the identification of signature pathogens that lead to disease is of importance, the key to the pathogenesis of polymicrobial diseases such as periodontitis depends on the molecular mechanisms (host and microbial) that contribute to the dysbiosis. A small number of studies have analysed the transcriptome of the oral microbiome to define the microbial expression in periodontal infections, but the question of why in certain cases disease progresses while others remain stable remains unresolved.

It has been shown that the composition and structure of the microbiota can differ between individuals, and even at different sites within the same oral cavity (Ge et al. 2013; Simón-Soro et al. 2013). Similar to identifying key pathogenic bacteria, the examination of crucial metabolic or functional activities that render a healthy subject susceptible to disease can be carried out.

The characterization of alterations within the ecosystem may reveal how commensal bacteria can switch to virulence. A metabolism model by gene expression of the microbiota during health and disease revealed that ~18% of the metabolic enzymes were differentially expressed during disease (Jorth et al. 2014). Lysine fermentation to butyrate, histidine catabolism, nucleotide biosynthesis and pyruvate formation were the pathways found to be upregulated during disease. Notably, it was identified that *F. nucleatum* was the only bacterium in diseased samples that degraded lysine to butyrate, while a number of different bacteria, varying in different patients, were found to express genes associated with histidine degradation and pyruvate fermentation during disease (Jorth et al. 2014). This activity of butyrate production by *F. nucleatum* may be correlated with a reduction of oxygen levels in deeper periodontal pockets (Diaz et al. 2000). As expected, key species such as *T. forsythia* and *P. gingivalis* express genes encoding for virulence factors such as proteases (e.g. collagenase, gingipains) which can also vary at different expression levels among patients. Despite the high inter-patient variability in microbial composition, conserved changes in the metabolism of the microbial community were identified, suggesting that multiple organisms are able to carry out specific metabolic roles that characterise a microbiota to be health- or disease-associated (Jorth et al. 2014).

Gene ontology (GO) terms associated with pathogenesis, oxidative stress response, cobalamin biosynthesis, proteolysis and ferrous iron transport were found actively expressed in periodontally active sites (Yost et al. 2015). Furthermore, metabolic activities such as citrate transport, iron transport, potassium transport, amino acid transport, isoprenoid biosynthesis and flagellar motility were found to be significantly highly expressed in progressing sites, microbial functional activities that have been reported in the initial virulent stages leading to the progression of disease (Nowicki et al. 2018; Yost et al. 2015). Interestingly, there was an over-representation of GO terms related to potassium ion transport (Yost et al. 2015). This importance of potassium in the periodontal pocket during dysbiosis was further explored (Yost et al. 2017). *In vitro* experiments involving the incubation of dental plaque with potassium showed that activities associated with disease including iron ion transport, flagellum assembly and proteolysis. Other virulence factors were upregulated, such as hemolysis, were significantly expressed by *P. nigrescens* and *S. mitis*. Furthermore, using a three-dimensional gingival multi-layered tissue model, the presence of potassium ion induced the expression of pro-inflammatory cytokines, IL-6 and TNF-α, as well as decreased the expression levels of human beta defensin-3, an antibacterial component known to maintain periodontal homeostasis (Yost et al. 2017).

In progressing sites with clinical breakdown, a high level of expression of putative virulence factors by both red and orange complex members was observed, however, at baseline, only *P. gingivalis*, *S. constellatus* and *P. intermedia* were actively expressing these virulence factors (Yost et al. 2015). The authors also observed that members of the red complex had an increased response to beta-lactamase, while *Porphyromonas* and *Prevotella* (particularly, *P. gingivalis*, *P. nigrescens* and *P. intermedia*) upregulated genes for conjugative transposons for horizontal transfer of antibiotic resistance. However, it is still unclear at this point the role of these antibiotic-related gene activities as the patients were not under antibiotic therapy at the time of the study. More interestingly, bacterial normally associated with periodontal health, such as *V. parvula*, *S. mitis*, *Staphylococcus intermedius* and *V. parvula* and have been observed to partake in the upregulation of putative virulence factors in active sites. With particular mention of *Pseudomonas fluorescens*, a species not normally associated with periodontitis but have been found as one of the top producers of virulence factors, both in baseline and active sites. When relating clinical parameters with gene expression profiles, increasing probing depth and increasing clinical attachment level were highly associated with proteolytic activity and potassium ion transport. The results demonstrated by Yost and colleagues suggest that the whole microbiome community, including previously identified health-associated bacteria, are responsible for the increase in virulence that can contribute to the progression of disease (Yost et al. 2015).

Szafrański et al. 2015 compared the gene expression profiles of samples from healthy and chronic periodontitis subjects and identified three functional biomarker genes that clearly differentiates between health and disease (Szafrański et al. 2015). These highly expressed biomarkers genes encodes: the gene for haem binding protein HmuY in *P. gingivalis*, the gene for flagella filament core protein FlaB3 in *T. denticola*, and the gene for a repeat protein of unknown function in *F. alocis*. To validate these biomarkers, the authors applied them to a separate raw dataset previously published by Jorth et al. (2014) and showed that all but one sample was correctly classified, which was identified as an outlier. Nevertheless, this validation determined that the difference in methods relating to sample collection, processing and analysis did not influence the result and the three identified biomarkers can be used as "true signatures" for classifying health and disease samples (Szafrański et al. 2015). In a further study by the same group, transcriptional profiles of the three most prominent microbial members during periodontitis, *P. gingvalis*, *T. denticola* and *F. nucleautum*, were studied and compared between *in vivo* and laboratory culture states (Deng et al. 2018). *P. gingivalis* from *in vivo* samples actively expressed genes associated with protein metabolism, translation, cell adhesion and pathogenesis, most particularly the *hmu* hemin/heme uptake gene locus, as well as other nutrient uptake transporters. In contrast, *P. gingivalis* grown in culture expressed more genes related to DNA methylation, thiamine biosynthesis and cell wall organisation. *In vivo T. denticola* upregulates genes associated with iron transporters but downregulates genes coding for hemolysin suggesting that during disease *T. denticola* do not actively lyse erythrocytes but scavenges iron from the environment readily made available by other bacteria such as *F. nucleatum*. Both *in vivo* and culture *T. denticola* were found to actively express genes for flagella filament proteins, the Msp protein and dentilisin. The most strongly expressed gene by *in vivo F. nucleatum* was found to be a hemin receptor as well as the Fad adhesin, which was also shown to be expressed in culture as well (Deng et al. 2018). The metabolic activities of each organism complement other species in the community which consequently encourages the growth and virulence during disease (Fig. 2.6). This metabolic cooperativity has been well-studied between *P. gingivalis* and *T. denticola*; free glycine is made available to the latter through hydrolysis of glycine-containing peptides by *P. gingivalis*, providing an energy source for *T. denticola* (Kin et al. 2020; Tan et al. 2014). By unravelling the metabolic activities of key players of pathogenesis, it is possible to map out the relationships within the micro-diversity of the periodontal pocket. Thus, by interfering with certain genes that supports their survival and co-dependence within the microbial community may help to shift the dysbiosis towards eubiosis.

A small handful of studies have examined the host gene expression profiles and pathways that contribute to inflammation and bone resorption in periodontitis. Genes found highly expressed during disease were noted to be immunoglobulin-related genes, chemokines, matrix metalloproteinases, cell adhesion molecules and tumour necrosis factor signalling (Demmer et al. 2008; Papapanou et al. 2009). Other metabolic pathways [e.g. p38, mitogen-activated protein kinase (MAPK), and c-Jun N-terminal kinase (JNK)] that lead to proliferation, inflammation and apoptosis were also found significantly upregulated in diseased tissues (Demmer et al. 2008; Papapanou et al. 2009). GO analysis of biological processes was conducted with regards to the subgingival colonisation of selected subgingival species. Antigen processing and presentation was the highest differentially regulated activity associated with *A. actinomycetemcomitans*, *P. gingivalis*, *T. denticola* and *T. forsyth*ia, followed by regulation of cell differentiation with the members of the red complex (Papapanou et al. 2009). A study by Kim et al. (2016) showed that genes associated with defence and immunity proteins, receptor signalling, proteases and signalling molecules were highly enriched during periodontitis, while genes of cytoskeletal structural proteins were more expressed during healthy conditions

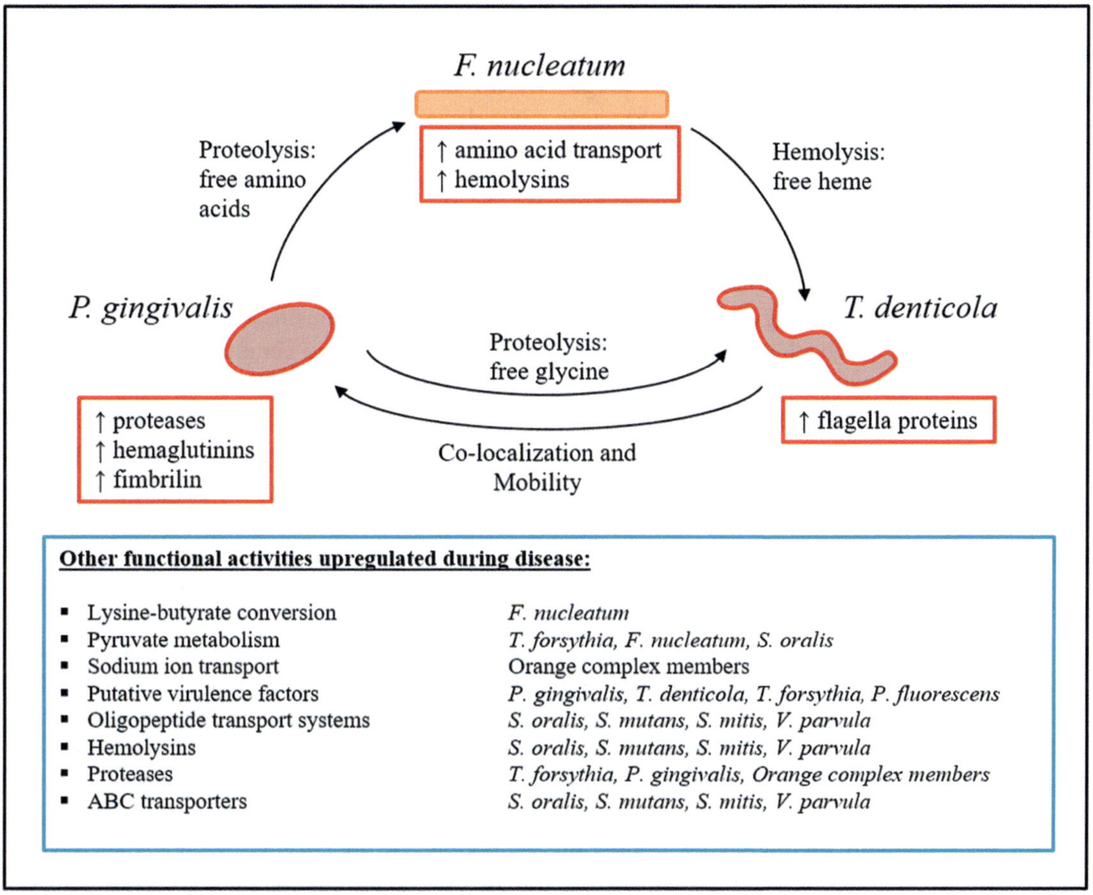

Fig. 2.6 A schematic diagram showing the synergistic functional activities between *F. nucleatum*, *T. denticola* and *P. gingivalis*, and some examples of metabolic activities up-regulated by oral bacteria during disease. During disease conditions, *F. nucleatum* increases hemolysin expression which helps to release heme from the environment. Free heme is taken up by *T. denticola*, which upregulates flagella proteins facilitating mobility. *T. denticola* and *P. gingivalis* are known to associate together through adherence proteins such as fimbrilin and thus allowing *P. gingivalis* to be transported within the biofilm. *P. gingivalis* upregulates protease and hemagglutinin genes to cause proteolysis and the release of amino acids or glycine, which are then available for uptake by *F. nucleatum* and *T. denticola*, respectively. Other bacteria within the oral biofilm have been shown to upregulate various metabolic pathways and putative virulence genes that play a role in bacterial survival and disease progression. (Figure was adapted using data from Deng et al. 2018; Jorth et al. 2014; Szafrański et al. 2015; Tan et al. 2014; Yost et al. 2015)

(Kim et al. 2016). Furthermore, genes associated with the IL-17 cytokine pathway were also found upregulated in diseased gingival tissues (Kebschull et al. 2014; Kim et al. 2016). IL-17 is one of the most important cytokines involved in the early innate immune response, produced by Th17 cells (a subset of CD4 T cells, NKT cells and γδ T cells) to activate neutrophils and the release of IL-1β and TNF-α.

Two separate studies analysed the sequential gene expression changes in gingival tissue samples of periodontal-healthy individuals, at various time points from induction of experimental gingivitis and resolution to periodontal health (Jönsson et al. 2011; Offenbacher et al. 2009). There was a consensus of upregulated genes associated with cell adhesion, leukocyte trans-endothelial migration, chemokines and innate immunity activation. As gingivitis progressed to resolution, genes previously observed to be upregulated during infection were downregulated, notably chemokines and factors for host immune cell activation (Jönsson et al. 2011; Offenbacher et al. 2009). Interestingly, both studies also found that pathways and genes associated with neural activation, tight junctions and adherens junctions were differentially regulated between health and gingivitis (Jönsson et al. 2011; Offenbacher et al. 2009). Furthermore, Kim et al. (2016) described an ion channel-forming member of the claudin family, *CLDN10*, which is a constituent of tight junction (Krug et al. 2014) to be highly overexpressed during periodontitis, while genes for keratin and late cornified envelope were actively downregulated (Kim et al. 2016). Taken together, these findings suggest that epithelial barrier activity may be an important biologic process involved during the pathogenesis of oral infections.

Despite being the lesser common tool as compared to metagenomics/metataxonomics, the area of metatranscriptomics has flourished in the last decade with an increasing number of metatranscriptomic datasets being deposited in the National Center for Biotechnology Information (NCBI) Sequence Read Archive (SRA) database annually; and less than 10% of these datasets comprise of human/animal samples, where a large proportion of human samples are related to the gut and oral microbiome (Human Microbiome Project Consortium 2012; Shakya et al. 2019). However, there are still obstacles that allow a successful metatranscriptomic analysis, including the lack of adequate reference metadata as well as the need for longer reads to provide higher resolution of transcripts. Nevertheless, there does exist a large collection of bioinformatics tools and workflows to enable the analyses of metatranscriptomics data and continual developments should be underway to support the next generation of metatranscriptomics (Shakya et al. 2019).

2.2 Conclusion and Perspectives

NGS technologies have been around for the last two decades, and tremendous progress have been made using the "omics" tools to comprehend the root of disease, assess risks and target specific key pathways. One drawback of these technologies is the inability to provide details at the subspecies level, as well as a difficulty in studying non-bacterial organisms (i.e.: fungi, viruses) of the oral microbiome simultaneously. The use of oligo-typing have enabled the exploration in diversity within defined microbial species between health and disease samples, such as between *P. gingivalis* and *P. intermedia* (Genco et al. 2019). Other methods such as whole genome sequencing could lend a hand in this domain, as shown in a recent study that comprehensively analysed the oral microbiome during healthy conditions (Caselli et al. 2020). The usefulness of these ever-evolving high throughput systems has allowed rapid analyses and identification of bacterial species newly-associated with periodontitis, including *Desulfobulbus*, *F. alocis* and TM7 (now known as *Saccharibacteria*). The importance of identifying low-abundance pathogens has been highlighted by Hajishengallis (2011), whereby low-abundance species can play roles in the initiation and progression of disease, which are largely unknown at the moment. Even within well-known phyla such as Bacteroidetes, previously-uncultured *Tannerella* sp. BU063 which is a relative of *T. forsythia*, have been identified by single-cell genomic sequencing, to exist predominantly in healthy oral communities and lacks disease-associated virulence genes (Beall et al. 2014). Single-cell genomics have enabled the study of uncultured microorganisms from a wide range of environments including the oral cavity, which have enabled scientists to culture these obscure microorganisms under controlled environments in order to understand their role in the community (Balachandran et al. 2020; Cross et al. 2019). This technology has been used to identify other uncultured bacteria such as *Deltaproteobacteri*a (Campbell et al. 2013a),

Saccharibacteria (TM7) (Marcy et al. 2007), and *Absconditabacteria* (SR1) (Campbell et al. 2013b).

New types of sequencers such as the MinION (Oxford Nanopore Technologies) have been developed as a portable, cost-effective sequencer that can handle long-reads to generate whole genome sequences. Despite certain limitations like sequence fidelity, the MinION sequencer has already been used in various applications (Kerkhof et al. 2017). Computational tools in the NGS analysis workflow have also been upgraded gradually to overcome computational errors and bias, as well as becoming more powerful in deciphering the multiple layers of data sets generated (Pereira et al. 2020). As newer approaches in NGS become more affordable and readily accessible, massive amount of data generated will need to be well-integrated and standardized to fully comprehend the significance of each aspect of the study. This is especially important as the time required for data analysis commonly forms a “bottleneck” in NGS studies (Vincent et al. 2017). Certainly, this would also pave the way for better designed multi-omics longitudinal clinical studies and improve clinical interventions. Conducting multi-omics studies is not a new idea as shown by studies of the gut (Quinn et al. 2016) and the periodontal pocket (Califf et al. 2017). Despite the small sample sizes, these studies create the foundation for careful development of larger projects. Further opportunities include exploring the roles of genes, proteins, and metabolites within the ecosystem not only with bacteria, but also archaea, protozoa, fungi, and viruses.

The fields of other omic technologies such as metaproteomics, metabolomics and metalipidomics have not been discussed in this chapter. The use of these technologies in dental studies produces a vast amount of data that can characterize the oral microbiome more accurately as they identify the reaction, communication and signaling by-products within the oral ecosystem. However, the interpretation of data in these areas is far more challenging and thus, they are less widespread than metagenomics and metatranscriptomics. A recent review provided a detailed collection of clinical studies using metaproteomics/metabolomics profiling (Bostanci et al. 2021).

Considering the new 2018 classification of periodontal diseases, while the extent of periodontitis generally remains unchanged, the inclusion of information such as general health status, other exposures including smoking, diabetes and response to treatments allows the grading of disease progression risk. A 2014 study clustering patients with chronic periodontitis or aggressive periodontitis based on the clinical severity and subgingival colonization compared the gene expression profiles of gingival tissues (Kebschull et al. 2014). One cluster presented more severe periodontitis and was shown to express pathways related to chemotaxis, proliferation, differentiation, immune activation, and cytotoxicity, whereas another cluster expressed pathways associated with cell proliferation, growth and maintenance. Although these results did not follow the previous classification of chronic and aggressive periodontitis, the distinctive gene expression signatures related to the severity of the diseased gingival tissues suggests a possible basis for re-classification of periodontitis (Kebschull et al. 2014). As such, it would be interesting to re-analyze microbiome data previously published with regards to the new 2018 classification, which may result in a clearer association between clinical observation and biological presentation. Undeniably, periodontitis is a complex disease, and a key challenge will be to identify reliable biomarker signatures that can classify an individual at risk for disease. This would likely be formulated with microbial and host elements as these aspects are highly interlinked. Although current prediction models presented in this review are still insufficient for use in the clinical setting, there is potential in developing these ideas into promising solutions in the near future.

References

Abusleme L, Dupuy AK, Dutzan N, Silva N, Burleson JA, Strausbaugh LD, Gamonal J, Diaz PI (2013) The subgingival microbiome in health and periodontitis and its relationship with community biomass and inflammation. ISME J 7:1016–1025. https://doi.org/10.1038/ismej.2012.174

Albandar JM (2002) Global risk factors and risk indicators for periodontal diseases. Periodontol 2000(29):177–206. https://doi.org/10.1034/j.1600-0757.2002.290109.x

Armitage GC (1999) Development of a classification system for periodontal diseases and conditions. Ann Periodontol 4:1–6. https://doi.org/10.1902/annals.1999.4.1.1

Armitage GC, Dickinson WR, Jenderseck RS, Levine SM, Chambers DW (1982) Relationship between the percentage of subgingival spirochetes and the severity of periodontal disease. J Periodontol 53:550–556. https://doi.org/10.1902/jop.1982.53.9.550

Balachandran M, Cross KL, Podar M (2020) Single-cell genomics and the oral microbiome. J Dent Res 99:613–620. https://doi.org/10.1177/0022034520907380

Beall CJ, Campbell AG, Dayeh DM, Griffen AL, Podar M, Leys EJ (2014) Single cell genomics of uncultured, Health-Associated Tannerella BU063 (Oral Taxon 286) and comparison to the closely related pathogen Tannerella forsythia. PLOS ONE 9:e89398. https://doi.org/10.1371/journal.pone.0089398

Beikler T, Peters U, Prior K, Eisenacher M, Flemmig TF (2008) Gene expression in periodontal tissues following treatment. BMC Med Genomics 1:30. https://doi.org/10.1186/1755-8794-1-30

Benn AML, Heng NCK, Broadbent JM, Thomson WM (2018) Studying the human oral microbiome: challenges and the evolution of solutions. Aust Dent J 63:14–24. https://doi.org/10.1111/adj.12565

Beyer K, Brandt BW, Buijs MJ, Brun JG, Crielaard W, Bolstad AI, Zaura E (2017) Comparison of paper-point and curette sampling of subgingival microbiome composition as analyzed by 16S rRNA gene amplicon sequencing. J Oral Microbiol 9. https://doi.org/10.1080/20002297.2017.1325260

Birtel J, Walser J-C, Pichon S, Bürgmann H, Matthews B (2015) Estimating bacterial diversity for ecological studies: methods, metrics, and assumptions. PLos One 10:e0125356. https://doi.org/10.1371/journal.pone.0125356

Bizzarro S, Loos BG, Laine ML, Crielaard W, Zaura E (2013) Subgingival microbiome in smokers and non-smokers in periodontitis: an exploratory study using traditional targeted techniques and a next-generation sequencing. J Clin Periodontol 40:483–492. https://doi.org/10.1111/jcpe.12087

Bizzarro S, Laine ML, Buijs MJ, Brandt BW, Crielaard W, Loos BG, Zaura E (2016) Microbial profiles at baseline and not the use of antibiotics determine the clinical outcome of the treatment of chronic periodontitis. Sci Rep 6:20205. https://doi.org/10.1038/srep20205

Bostanci N, Grant M, Bao K, Silbereisen A, Hetrodt F, Manoil D, Belibasakis GN (2021) Metaproteome and metabolome of oral microbial communities. Periodontol 2000(85):46–81. https://doi.org/10.1111/prd.12351

Boyer E, Le Gall-David S, Martin B, Fong SB, Loréal O, Deugnier Y, Bonnaure-Mallet M, Meuric V (2018) Increased transferrin saturation is associated with subgingival microbiota dysbiosis and severe periodontitis in genetic haemochromatosis. Sci Rep 8:15532. https://doi.org/10.1038/s41598-018-33813-0

Byrne SJ, Dashper SG, Darby IB, Adams GG, Hoffmann B, Reynolds EC (2009) Progression of chronic periodontitis can be predicted by the levels of Porphyromonas gingivalis and Treponema denticola in subgingival plaque. Oral Microbiol Immunol 24:469–477. https://doi.org/10.1111/j.1399-302X.2009.00544.x

Califf KJ, Schwarzberg-Lipson K, Garg N, Gibbons SM, Caporaso JG, Slots J, Cohen C, Dorrestein PC, Kelley ST (2017) Multi-omics analysis of periodontal pocket microbial communities pre- and posttreatment. mSystems 2:e00016. https://doi.org/10.1128/mSystems.00016-17

Campbell AG, Campbell JH, Schwientek P, Woyke T, Sczyrba A, Allman S, Beall CJ, Griffen A, Leys E, Podar M (2013a) Multiple single-cell genomes provide insight into functions of uncultured Deltaproteobacteria in the Human Oral Cavity. PLos One 8:e59361. https://doi.org/10.1371/journal.pone.0059361

Campbell JH, O'Donoghue P, Campbell AG, Schwientek P, Sczyrba A, Woyke T, Soll D, Podar M (2013b) UGA is an additional glycine codon in uncultured SR1 bacteria from the human microbiota. Proc Natl Acad Sci 110:5540–5545. https://doi.org/10.1073/pnas.1303090110

Caporaso JG, Kuczynski J, Stombaugh J, Bittinger K, Bushman FD, Costello EK, Fierer N, Peña AG, Goodrich JK, Gordon JI, Huttley GA, Kelley ST, Knights D, Koenig JE, Ley RE, Lozupone CA, McDonald D, Muegge BD, Pirrung M, Reeder J, Sevinsky JR, Turnbaugh PJ, Walters WA, Widmann J, Yatsunenko T, Zaneveld J, Knight R (2010) QIIME allows analysis of high-throughput community sequencing data. Nat Method 7:335–336. https://doi.org/10.1038/nmeth.f.303

Caselli E, Fabbri C, D'Accolti M, Soffritti I, Bassi C, Mazzacane S, Franchi M (2020) Defining the oral microbiome by whole-genome sequencing and resistome analysis: the complexity of the healthy picture. BMC Microbiol 20:120. https://doi.org/10.1186/s12866-020-01801-y

Chen W-P, Chang S-H, Tang C-Y, Liou M-L, Tsai S-JJ, Lin Y-L (2018) Composition analysis and feature selection of the oral microbiota associated with Periodontal disease [WWW Document]. BioMed Res Int. https://doi.org/10.1155/2018/3130607

Chu H-W, Chang K-P, Hsu C-W, Chang IY-F, Liu H-P, Chen Y-T, Wu C-C (2019) Identification of salivary biomarkers for oral cancer detection with untargeted and targeted quantitative proteomics approaches. Mol. Cell. Proteomics MCP 18:1796–1806. https://doi.org/10.1074/mcp.RA119.001530

Colombo APV, Boches SK, Cotton SL, Goodson JM, Kent R, Haffajee AD, Socransky SS, Hasturk H, Dyke TEV, Dewhirst F, Paster BJ (2009) Comparisons of subgingival microbial profiles of refractory Periodontitis, severe Periodontitis, and Periodontal health using the human oral microbe identification microarray. J Periodontol 80:1421–1432. https://doi.org/10.1902/jop.2009.090185

Corrêa JD, Fernandes GR, Calderaro DC, Mendonça SMS, Silva JM, Albiero ML, Cunha FQ, Xiao E, Ferreira GA, Teixeira AL, Mukherjee C, Leys EJ, Silva TA, Graves DT (2019) Oral microbial dysbiosis linked to worsened periodontal condition in rheumatoid arthritis patients. Sci Rep 9:8379. https://doi.org/10.1038/s41598-019-44674-6

Cross KL, Campbell JH, Balachandran M, Campbell AG, Cooper SJ, Griffen A, Heaton M, Joshi S, Klingeman D, Leys E, Yang Z, Parks JM, Podar M (2019) Targeted isolation and cultivation of uncultivated bacteria by reverse genomics. Nat Biotechnol 37:1314–1321. https://doi.org/10.1038/s41587-019-0260-6

Demmer RT, Behle JH, Wolf DL, Handfield M, Kebschull M, Celenti R, Pavlidis P, Papapanou PN (2008) Transcriptomes in healthy and diseased gingival tissues. J Periodontol 79:2112–2124. https://doi.org/10.1902/jop.2008.080139

Deng K, Ouyang XY, Chu Y, Zhang Q (2017) Subgingival microbiome of gingivitis in Chinese undergraduates. Chin J Dent Res 20(3):145–152. https://doi.org/10.3290/j.cjdr.a38769

Deng Z-L, Sztajer H, Jarek M, Bhuju S, Wagner-Döbler I (2018) Worlds apart – transcriptome profiles of key oral microbes in the periodontal pocket compared to single laboratory culture reflect synergistic interactions. Front Microbiol 9:124. https://doi.org/10.3389/fmicb.2018.00124

Dewhirst FE, Chen T, Izard J, Paster BJ, Tanner ACR, Yu W-H, Lakshmanan A, Wade WG (2010) The human oral microbiome. J Bacteriol 192:5002–5017. https://doi.org/10.1128/JB.00542-10

Diaz PI, Zilm PS, Rogers AH (2000) The response to oxidative stress of Fusobacterium nucleatum grown in continuous culture. FEMS Microbiol Lett 187:31–34. https://doi.org/10.1111/j.1574-6968.2000.tb09132.x

Du M, Bo T, Kapellas K, Peres MA (2018) Prediction models for the incidence and progression of periodontitis: a systematic review. J Clin Periodontol 45:1408–1420. https://doi.org/10.1111/jcpe.13037

Duran-Pinedo AE, Paster B, Teles R, Frias-Lopez J (2011) Correlation network analysis applied to complex Biofilm communities. PLos One 6:e28438. https://doi.org/10.1371/journal.pone.0028438

Eke PI, Dye BA, Wei L, Thornton-Evans GO, Genco RJ (2012) Prevalence of Periodontitis in Adults in the United States: 2009 and 2010. J Dent Res 91:914–920. https://doi.org/10.1177/0022034512457373

Fitzgerald RJ, Jordan HV, Stanley HR, Poole WL, Bowler A (1960) Experimental caries and Gingival pathologic changes in the Gnotobiotic Rat. J Dent Res 39:923–935. https://doi.org/10.1177/00220345600390052701

Ge X, Rodriguez R, Trinh M, Gunsolley J, Xu P (2013) Oral microbiome of deep and shallow dental pockets in chronic periodontitis. PloS One 8:e65520. https://doi.org/10.1371/journal.pone.0065520

Genco RJ, Sanz M (2020) Clinical and public health implications of periodontal and systemic diseases: an overview. Periodontol 2000(83):7–13. https://doi.org/10.1111/prd.12344

Genco RJ, LaMonte MJ, McSkimming DI, Buck MJ, Li L, Hovey KM, Andrews CA, Sun Y, Tsompana M, Zheng W, Banack HR, Murugaiyan V, Wactawski-Wende J (2019) The subgingival microbiome relationship to Periodontal disease in older women. J Dent Res. https://doi.org/10.1177/0022034519860449

Greenstein G, Caton J, Polson AM (1981) Histologic characteristics associated with bleeding after probing and visual signs of inflammation. J Periodontol 52:420–425. https://doi.org/10.1902/jop.1981.52.8.420

Griffen AL, Beall CJ, Campbell JH, Firestone ND, Kumar PS, Yang ZK, Podar M, Leys EJ (2012) Distinct and complex bacterial profiles in human periodontitis and health revealed by 16S pyrosequencing. ISME J 6:1176–1185. https://doi.org/10.1038/ismej.2011.191

Hajishengallis G (2014) Immunomicrobial pathogenesis of periodontitis: keystones, pathobionts, and host response. Trends Immunol 35:3–11. https://doi.org/10.1016/j.it.2013.09.001

Hajishengallis G, Lambris JD (2011) Microbial manipulation of receptor crosstalk in innate immunity. Nat Rev Immunol 11:187–200. https://doi.org/10.1038/nri2918

Hajishengallis G, Liang S, Payne MA, Hashim A, Jotwani R, Eskan MA, McIntosh ML, Alsam A, Kirkwood KL, Lambris JD, Darveau RP, Curtis MA (2011) Low-abundance biofilm species orchestrates inflammatory periodontal disease through the commensal microbiota and complement. Cell Host Microbe 10:497–506. https://doi.org/10.1016/j.chom.2011.10.006

Hajishengallis G, Darveau RP, Curtis MA (2012) The keystone-pathogen hypothesis. Nat Rev Microbiol 10:717–725. https://doi.org/10.1038/nrmicro2873

Han J, Wang P, Ge S (2017) The microbial community shifts of subgingival plaque in patients with generalized aggressive periodontitis following non-surgical periodontal therapy: a pilot study. Oncotarget 8:10609–10619. https://doi.org/10.18632/oncotarget.12532

Hancock EB, Newell DH (2001) Preventive strategies and supportive treatment. Periodontol 2000(25):59–76. https://doi.org/10.1034/j.1600-0757.2001.22250105.x

Holt SC, Ebersole JL (2005) Porphyromonas gingivalis, Treponema denticola, and Tannerella forsythia: the "red complex", a prototype polybacterial pathogenic consortium in periodontitis. Periodontol 2000(38):72–122. https://doi.org/10.1111/j.1600-0757.2005.00113.x

Hong B-Y, Araujo MVF, Strausbaugh LD, Terzi E, Ioannidou E, Diaz PI (2015) Microbiome profiles in Periodontitis in relation to host and disease characteristics. PLOS ONE 10:e0127077. https://doi.org/10.1371/journal.pone.0127077

Human Microbiome Project Consortium (2012) A framework for human microbiome research. Nature 486:215–221. https://doi.org/10.1038/nature11209

Jensen A, Ladegaard Grønkjær L, Holmstrup P, Vilstrup H, Kilian M (2018) Unique subgingival microbiota associated with periodontitis in cirrhosis patients. Sci Rep 8:10718. https://doi.org/10.1038/s41598-018-28905-w

Johnson JS, Spakowicz DJ, Hong B-Y, Petersen LM, Demkowicz P, Chen L, Leopold SR, Hanson BM, Agresta HO, Gerstein M, Sodergren E, Weinstock GM (2019) Evaluation of 16S rRNA gene sequencing for species and strain-level microbiome analysis. Nat Commun 10:5029. https://doi.org/10.1038/s41467-019-13036-1

Jönsson D, Ramberg P, Demmer RT, Kebschull M, Dahlén G, Papapanou PN (2011) Gingival tissue transcriptomes in experimental gingivitis. J Clin Periodontol 38:599–611. https://doi.org/10.1111/j.1600-051X.2011.01719.x

Jorth P, Turner KH, Gumus P, Nizam N, Buduneli N, Whiteley M (2014) Metatranscriptomics of the human oral microbiome during health and disease. mBio 5. https://doi.org/10.1128/mBio.01012-14

Jünemann S, Prior K, Szczepanowski R, Harks I, Ehmke B, Goesmann A, Stoye J, Harmsen D (2012) Bacterial community shift in treated periodontitis patients revealed by ion torrent 16S rRNA gene amplicon sequencing. PloS One 7:e41606. https://doi.org/10.1371/journal.pone.0041606

Kageyama S, Takeshita T, Takeuchi K, Asakawa M, Matsumi R, Furuta M, Shibata Y, Nagai K, Ikebe M, Morita M, Masuda M, Toh Y, Kiyohara Y, Ninomiya T, Yamashita Y (2019) Characteristics of the salivary Microbiota in patients with various digestive tract cancers. Front Microbiol 10:1780. https://doi.org/10.3389/fmicb.2019.01780

Kebschull M, Demmer RT, Grün B, Guarnieri P, Pavlidis P, Papapanou PN (2014) Gingival tissue transcriptomes identify distinct periodontitis phenotypes. J Dent Res 93:459–468. https://doi.org/10.1177/0022034514527288

Keijser BJF, Zaura E, Huse SM, van der Vossen JMBM, Schuren FHJ, Montijn RC, ten Cate JM, Crielaard W (2008) Pyrosequencing analysis of the oral microflora of healthy adults. J Dent Res 87:1016–1020. https://doi.org/10.1177/154405910808701104

Kerkhof LJ, Dillon KP, Häggblom MM, McGuinness LR (2017) Profiling bacterial communities by MinION sequencing of ribosomal operons. Microbiome 5:116. https://doi.org/10.1186/s40168-017-0336-9

Kim Y-G, Kim M, Kang JH, Kim HJ, Park J-W, Lee J-M, Suh J-Y, Kim J-Y, Lee J-H, Lee Y (2016) Transcriptome sequencing of gingival biopsies from chronic periodontitis patients reveals novel gene expression and splicing patterns. Hum Genomics 10:28. https://doi.org/10.1186/s40246-016-0084-0

Kin LX, Butler CA, Slakeski N, Hoffmann B, Dashper SG, Reynolds EC (2020) Metabolic cooperativity between Porphyromonas gingivalis and Treponema denticola. J Oral Microbiol 12:1808750. https://doi.org/10.1080/20002297.2020.1808750

Kirst ME, Li EC, Alfant B, Chi Y-Y, Walker C, Magnusson I, Wang GP (2015) Dysbiosis and alterations in predicted functions of the subgingival microbiome in chronic Periodontitis. Appl. Environ. Microbiol. 81:783–793. https://doi.org/10.1128/AEM.02712-14

Kistler JO, Booth V, Bradshaw DJ, Wade WG (2013) Bacterial community development in experimental gingivitis. PloS One 8:e71227. https://doi.org/10.1371/journal.pone.0071227

Knight R, Vrbanac A, Taylor BC, Aksenov A, Callewaert C, Debelius J, Gonzalez A, Kosciolek T, McCall L-I, McDonald D, Melnik AV, Morton JT, Navas J, Quinn RA, Sanders JG, Swafford AD, Thompson LR, Tripathi A, Xu ZZ, Zaneveld JR, Zhu Q, Caporaso JG, Dorrestein PC (2018) Best practices for analysing microbiomes. Nat Rev Microbiol 16:410–422. https://doi.org/10.1038/s41579-018-0029-9

Krug SM, Schulzke JD, Fromm M (2014) Tight junction, selective permeability, and related diseases. Semin Cell Dev Biol 36:166–176. https://doi.org/10.1016/j.semcdb.2014.09.002

Kumar PS, Griffen AL, Moeschberger ML, Leys EJ (2005) Identification of candidate periodontal pathogens and beneficial species by Quantitative 16S clonal analysis. J Clin Microbiol 43:3944–3955. https://doi.org/10.1128/JCM.43.8.3944-3955.2005

Laksmana T, Kittichotirat W, Huang Y, Chen W, Jorgensen M, Bumgarner R, Chen C (2012) Metagenomic analysis of subgingival microbiota following non-surgical periodontal therapy: a pilot study. Open Dent J 6:255–261. https://doi.org/10.2174/1874210601206010255

LaMonte MJ, Genco RJ, Buck MJ, McSkimming DI, Li L, Hovey KM, Andrews CA, Zheng W, Sun Y, Millen AE, Tsompana M, Banack HR, Wactawski-Wende J (2019) Composition and diversity of the subgingival microbiome and its relationship with age in postmenopausal women: an epidemiologic investigation. BMC Oral Health 19:246. https://doi.org/10.1186/s12903-019-0906-2

Lazarevic V, Whiteson K, Huse S, Hernandez D, Farinelli L, Osterås M, Schrenzel J, François P (2009) Metagenomic study of the oral microbiota by Illumina high-throughput sequencing. J Microbiol Method 79:266–271. https://doi.org/10.1016/j.mimet.2009.09.012

Lee JS, Spooner R, Chowdhury N, Pandey V, Wellslager B, Atanasova KR, Evans Z, Yilmaz Ö (2020) In situ intraepithelial localizations of opportunistic pathogens, porphyromonas gingivalis and Filifactor alocis, in Human Gingiva. Curr Res Microb Sci 1:7–17. https://doi.org/10.1016/j.crmicr.2020.05.001

Löe H, Theilade E, Jensen SB (1965) Experimental Gingivitis in Man. J Periodontol 36:177–187. https://doi.org/10.1902/jop.1965.36.3.177
Loesche WJ (1976) Chemotherapy of dental plaque infections. Oral Sci Rev 9:65–107
Lu H, He L, Xu J, Song W, Feng X, Zhao Y, Meng H (2020) Well-maintained patients with a history of periodontitis still harbor a more disbiotic microbiome than health. J Periodontol 91:1584–1594. https://doi.org/10.1002/JPER.19-0498
Manzoni C, Kia DA, Vandrovcova J, Hardy J, Wood NW, Lewis PA, Ferrari R (2016) Genome, transcriptome and proteome: the rise of omics data and their integration in biomedical sciences. Brief Bioinform 19:286–302. https://doi.org/10.1093/bib/bbw114
Marcy Y, Ouverney C, Bik EM, Lösekann T, Ivanova N, Martin HG, Szeto E, Platt D, Hugenholtz P, Relman DA, Quake SR (2007) Dissecting biological "dark matter" with single-cell genetic analysis of rare and uncultivated TM7 microbes from the human mouth. Proc Natl Acad Sci 104:11889–11894. https://doi.org/10.1073/pnas.0704662104
Mardis ER (2008) The impact of next-generation sequencing technology on genetics. Trends Genet 24:133–141. https://doi.org/10.1016/j.tig.2007.12.007
Marsh PD (1994) Microbial ecology of dental plaque and its significance in health and disease. Adv Dent Res 8:263–271. https://doi.org/10.1177/08959374940080022001
Marsh PD (2003) Are dental diseases examples of ecological catastrophes? Microbiol Read Engl 149:279–294. https://doi.org/10.1099/mic.0.26082-0
Mason MR, Preshaw PM, Nagaraja HN, Dabdoub SM, Rahman A, Kumar PS (2015) The subgingival microbiome of clinically healthy current and never smokers. ISME J 9:268–272. https://doi.org/10.1038/ismej.2014.114
Menon RK, Gomez A, Brandt BW, Leung YY, Gopinath D, Watt RM, Crielaard W, Nelson KE, Botelho MG (2019) Long-term impact of oral surgery with or without amoxicillin on the oral microbiome-A prospective cohort study. Sci Rep 9:18761. https://doi.org/10.1038/s41598-019-55056-3
Meuric V, Le Gall-David S, Boyer E, Acuña-Amador L, Martin B, Fong SB, Barloy-Hubler F, Bonnaure-Mallet M (2017) Signature of microbial dysbiosis in Periodontitis. Appl Environ Microbiol 83:e00462–e00417. https://doi.org/10.1128/AEM.00462-17
Mikheyev AS, Tin MMY (2014) A first look at the Oxford Nanopore MinION sequencer. Mol Ecol Resour 14:1097–1102. https://doi.org/10.1111/1755-0998.12324
Mougeot J-LC, Stevens CB, Cotton SL, Morton DS, Krishnan K, Brennan MT, Lockhart PB, Paster BJ, Bahrani Mougeot FK (2016) Concordance of HOMIM and HOMINGS technologies in the microbiome analysis of clinical samples. J Oral Microbiol 8:1–9. https://doi.org/10.3402/jom.v8.30379
Nayak A, Bhat K, Shivanaikar S, Pushpa P, Kugaji M, Kumbar V (2018) Detection of red complex organisms in chronic periodontitis by multiplex polymerase chain reaction. J Adv Clin Res Insights 5:139–144. https://doi.org/10.15713/ins.jcri.231
Nowicki EM, Shroff R, Singleton JA, Renaud DE, Wallace D, Drury J, Zirnheld J, Colleti B, Ellington AD, Lamont RJ, Scott DA, Whiteley M (2018) Microbiota and metatranscriptome changes accompanying the onset of Gingivitis. mBio 9:e00575–e00518. https://doi.org/10.1128/mBio.00575-18
Offenbacher S, Barros SP, Paquette DW, Winston JL, Biesbrock AR, Thomason RG, Gibb RD, Fulmer AW, Tiesman JP, Juhlin KD, Wang SL, Reichling TD, Chen K-S, Ho B (2009) Gingival transcriptome patterns during induction and resolution of experimental gingivitis in humans. J Periodontol 80:1963–1982. https://doi.org/10.1902/jop.2009.080645
Offenbacher S, Divaris K, Barros SP, Moss KL, Marchesan JT, Morelli T, Zhang S, Kim S, Sun L, Beck JD, Laudes M, Munz M, Schaefer AS, North KE (2016) Genome-wide association study of biologically informed periodontal complex traits offers novel insights into the genetic basis of periodontal disease. Hum Mol Genet 25:2113–2129. https://doi.org/10.1093/hmg/ddw069
Olson JC, Cuff CF, Lukomski S, Lukomska E, Canizales Y, Wu B, Crout RJ, Thomas JG, McNeil DW, Weyant RJ, Marazita ML, Paster BJ, Elliott T (2011) Use of 16S ribosomal RNA gene analyses to characterize the bacterial signature associated with poor oral health in West Virginia. BMC Oral Health 11:7. https://doi.org/10.1186/1472-6831-11-7
Papapanou PN, Behle JH, Kebschull M, Celenti R, Wolf DL, Handfield M, Pavlidis P, Demmer RT (2009) Subgingival bacterial colonization profiles correlate with gingival tissue gene expression. BMC Microbiol 9:221. https://doi.org/10.1186/1471-2180-9-221
Papapanou PN, Sanz M, Buduneli N, Dietrich T, Feres M, Fine DH, Flemmig TF, Garcia R, Giannobile WV, Graziani F, Greenwell H, Herrera D, Kao RT, Kebschull M, Kinane DF, Kirkwood KL, Kocher T, Kornman KS, Kumar PS, Loos BG, Machtei E, Meng H, Mombelli A, Needleman I, Offenbacher S, Seymour GJ, Teles R, Tonetti MS (2018) Periodontitis: consensus report of workgroup 2 of the 2017 world workshop on the classification of periodontal and peri-implant diseases and conditions. J Periodontol 89(Suppl 1):S173–S182. https://doi.org/10.1002/JPER.17-0721
Papapanou PN, Park H, Cheng B, Kokaras A, Paster B, Burkett S, Watson CW-M, Annavajhala MK, Uhlemann A-C, Noble JM (2020) Subgingival microbiome and clinical periodontal status in an elderly cohort: the WHICAP ancillary study of oral health. J Periodontol 91:S56–S67. https://doi.org/10.1002/JPER.20-0194
Pereira R, Oliveira J, Sousa M (2020) Bioinformatics and computational tools for next-generation sequencing analysis in clinical genetics. J Clin Med 9:132. https://doi.org/10.3390/jcm9010132
Pérez-Chaparro PJ, McCulloch JA, Mamizuka EM, Moraes A d CL, Faveri M, Figueiredo LC, Duarte PM, Feres M (2018) Do different probing depths

exhibit striking differences in microbial profiles? J Clin Periodontol 45:26–37. https://doi.org/10.1111/jcpe.12811

Peterson J, Garges S, Giovanni M, McInnes P, Wang L, Schloss JA, Bonazzi V, McEwen JE, Wetterstrand KA, Deal C, Baker CC, Di Francesco V, Howcroft TK, Karp RW, Lunsford RD, Wellington CR, Belachew T, Wright M, Giblin C, David H, Mills M, Salomon R, Mullins C, Akolkar B, Begg L, Davis C, Grandison L, Humble M, Khalsa J, Little AR, Peavy H, Pontzer C, Portnoy M, Sayre MH, Starke-Reed P, Zakhari S, Read J, Watson B, Guyer M (2009) The NIH human microbiome project. Genome Res 19:2317–2323. https://doi.org/10.1101/gr.096651.109

Quinn RA, Navas-Molina JA, Hyde ER, Song SJ, Vázquez-Baeza Y, Humphrey G, Gaffney J, Minich JJ, Melnik AV, Herschend J, DeReus J, Durant A, Dutton RJ, Khosroheidari M, Green C, da Silva R, Dorrestein PC, Knight R (2016) From sample to multi-omics conclusions in under 48 hours. mSystems 1:e00038. https://doi.org/10.1128/mSystems.00038-16

Sadowsky MJ, Staley C, Heiner C, Hall R, Kelly CR, Brandt L, Khoruts A (2017) Analysis of gut microbiota – an ever changing landscape. Gut Microbes 8:268–275. https://doi.org/10.1080/19490976.2016.1277313

Schloss PD (2020) Reintroducing mothur: 10 years later. Appl Environ Microbiol 86:e02343. https://doi.org/10.1128/AEM.02343-19

Schloss PD, Westcott SL, Ryabin T, Hall JR, Hartmann M, Hollister EB, Lesniewski RA, Oakley BB, Parks DH, Robinson CJ, Sahl JW, Stres B, Thallinger GG, Horn DJV, Weber CF (2009) Introducing mothur: open-source, platform-independent, community-supported software for describing and comparing microbial communities. Appl Environ Microbiol 75:7537–7541. https://doi.org/10.1128/AEM.01541-09

Schwarzberg K, Le R, Bharti B, Lindsay S, Casaburi G, Salvatore F, Saber MH, Alonaizan F, Slots J, Gottlieb RA, Caporaso JG, Kelley ST (2014) The personal human oral microbiome obscures the effects of treatment on periodontal disease. PloS One 9:e86708. https://doi.org/10.1371/journal.pone.0086708

Shah FD, Begum R, Vajaria BN, Patel KR, Patel JB, Shukla SN, Patel PS (2011) A review on salivary genomics and proteomics biomarkers in oral cancer. Indian J Clin Biochem IJCB 26:326–334. https://doi.org/10.1007/s12291-011-0149-8

Shakya M, Lo C-C, Chain PSG (2019) Advances and challenges in metatranscriptomic analysis. Front Genet 10:904. https://doi.org/10.3389/fgene.2019.00904

Shi B, Chang M, Martin J, Mitreva M, Lux R, Klokkevold P, Sodergren E, Weinstock GM, Haake SK, Li H (2015) Dynamic changes in the subgingival microbiome and their potential for diagnosis and prognosis of periodontitis. mBio 6:e01926–e01914. https://doi.org/10.1128/mBio.01926-14

Shi B, Lux R, Klokkevold P, Chang M, Barnard E, Haake S, Li H (2020) The subgingival microbiome associated with periodontitis in type 2 diabetes mellitus. ISME J 14:519–530. https://doi.org/10.1038/s41396-019-0544-3

Simón-Soro A, Tomás I, Cabrera-Rubio R, Catalan MD, Nyvad B, Mira A (2013) Microbial geography of the oral cavity. J Dent Res 92:616–621. https://doi.org/10.1177/0022034513488119

Socransky SS, Haffajee AD (2005) Periodontal microbial ecology. Periodontol 2000(38):135–187. https://doi.org/10.1111/j.1600-0757.2005.00107.x

Socransky SS, Haffajee AD, Smith C, Dibart S (1991) Relation of counts of microbial species to clinical status at the sampled site. J Clin Periodontol 18:766–775. https://doi.org/10.1111/j.1600-051x.1991.tb00070.x

Socransky SS, Haffajee AD, Cugini MA, Smith C, Kent RL (1998) Microbial complexes in subgingival plaque. J Clin Periodontol 25:134–144. https://doi.org/10.1111/j.1600-051x.1998.tb02419.x

Socransky SS, Haffajee AD, Smith C, Martin L, Haffajee JA, Uzel NG, Goodson JM (2004) Use of checkerboard DNA–DNA hybridization to study complex microbial ecosystems. Oral Microbiol Immunol 19:352–362. https://doi.org/10.1111/j.1399-302x.2004.00168.x

Steinberg T, Dannewitz B, Tomakidi P, Hoheisel JD, Müssig E, Kohl A, Nees M (2006) Analysis of interleukin-1beta-modulated mRNA gene transcription in human gingival keratinocytes by epithelia-specific cDNA microarrays. J Periodontal Res 41:426–446. https://doi.org/10.1111/j.1600-0765.2006.00884.x

Sweeney LC, Dave J, Chambers PA, Heritage J (2004) Antibiotic resistance in general dental practice–a cause for concern? J. Antimicrob. Chemother. 53:567–576. https://doi.org/10.1093/jac/dkh137

Szafrański SP, Deng Z-L, Tomasch J, Jarek M, Bhuju S, Meisinger C, Kühnisch J, Sztajer H, Wagner-Döbler I (2015) Functional biomarkers for chronic periodontitis and insights into the roles of Prevotella nigrescens and Fusobacterium nucleatum; a metatranscriptome analysis. NPJ Biofilms Microbiomes 1:15017. https://doi.org/10.1038/npjbiofilms.2015.17

Tan KH, Seers CA, Dashper SG, Mitchell HL, Pyke JS, Meuric V, Slakeski N, Cleal SM, Chambers JL, McConville MJ, Reynolds EC (2014) Porphyromonas gingivalis and Treponema denticola exhibit metabolic symbioses. PLoS Pathog 10:e1003955. https://doi.org/10.1371/journal.ppat.1003955

Tanner ACR, Kent R, Kanasi E, Lu SC, Paster BJ, Sonis ST, Murray LA, Van Dyke TE (2007) Clinical characteristics and microbiota of progressing slight chronic periodontitis in adults. J Clin Periodontol 34:917–930. https://doi.org/10.1111/j.1600-051X.2007.01126.x

Theilade E (1986) The non-specific theory in microbial etiology of inflammatory periodontal diseases. J Clin Periodontol 13:905–911. https://doi.org/10.1111/j.1600-051X.1986.tb01425.x

Tsai C-Y, Tang CY, Tan T-S, Chen K-H, Liao K-H, Liou M-L (2018) Subgingival microbiota in individuals with severe chronic periodontitis. J Microbiol Immunol Infect Wei Mian Yu Gan Ran Za Zhi 51:226–234. https://doi.org/10.1016/j.jmii.2016.04.007

Van Dyke TE, Bartold PM, Reynolds EC (2020) The nexus between periodontal inflammation and dysbiosis. Front Immunol 11:511. https://doi.org/10.3389/fimmu.2020.00511
Vardar-Sengul S, Arora S, Baylas H, Mercola D (2009) Expression profile of human gingival fibroblasts induced by interleukin-1beta reveals central role of nuclear factor-kappa B in stabilizing human gingival fibroblasts during inflammation. J Periodontol 80:833–849. https://doi.org/10.1902/jop.2009.080483
Vincent AT, Derome N, Boyle B, Culley AI, Charette SJ (2017) Next-generation sequencing (NGS) in the microbiological world: how to make the most of your money. J. Microbiol. Methods, What's next in microbiology methods? Emerg Method 138:60–71. https://doi.org/10.1016/j.mimet.2016.02.016
Vincent-Chong VK, Salahshourifar I, Woo KM, Anwar A, Razali R, Gudimella R, Rahman ZAA, Ismail SM, Kallarakkal TG, Ramanathan A, Wan Mustafa WM, Abraham MT, Tay KK, Zain RB (2017) Genome wide profiling in oral squamous cell carcinoma identifies a four genetic marker signature of prognostic significance. PloS One 12:e0174865. https://doi.org/10.1371/journal.pone.0174865
Wade WG (2011) Has the use of molecular methods for the characterization of the human oral microbiome changed our understanding of the role of bacteria in the pathogenesis of periodontal disease? J Clin Periodontol 38(Suppl 11):7–16. https://doi.org/10.1111/j.1600-051X.2010.01679.x
Wade WG, Prosdocimi EM (2020) Profiling of oral bacterial communities. J Dent Res 99:621–629. https://doi.org/10.1177/0022034520914594
Wu X, Chen J, Xu M, Zhu D, Wang X, Chen Y, Wu J, Cui C, Zhang W, Yu L (2017) 16S rDNA analysis of periodontal plaque in chronic obstructive pulmonary disease and periodontitis patients. J Oral Microbiol 9:1324725. https://doi.org/10.1080/20002297.2017.1324725
Yost S, Duran-Pinedo AE, Teles R, Krishnan K, Frias-Lopez J (2015) Functional signatures of oral dysbiosis during periodontitis progression revealed by microbial metatranscriptome analysis. Genome Med. 7:27. https://doi.org/10.1186/s13073-015-0153-3
Yost S, Duran-Pinedo AE, Krishnan K, Frias-Lopez J (2017) Potassium is a key signal in host-microbiome dysbiosis in periodontitis. PLoS Pathog 13:e1006457. https://doi.org/10.1371/journal.ppat.1006457
Yu G, Phillips S, Gail MH, Goedert JJ, Humphrys MS, Ravel J, Ren Y, Caporaso NE (2017) The effect of cigarette smoking on the oral and nasal microbiota. Microbiome 5:3. https://doi.org/10.1186/s40168-016-0226-6
Zaura E, Keijser BJF, Huse SM, Crielaard W (2009) Defining the healthy "core microbiome" of oral microbial communities. BMC Microbiol 9:259. https://doi.org/10.1186/1471-2180-9-259
Zaura E, Brandt BW, de Mattos MJT, Buijs MJ, Caspers MPM, Rashid M-U, Weintraub A, Nord CE, Savell A, Hu Y, Coates AR, Hubank M, Spratt DA, Wilson M, Keijser BJF, Crielaard W (2015) Same exposure but two radically different Responses to antibiotics: resilience of the salivary microbiome versus long-term microbial shifts in Feces. mBio 6. https://doi.org/10.1128/mBio.01693-15
Zhang Y, Qi Y, Lo ECM, McGrath C, Mei ML, Dai R (2020) Using next-generation sequencing to detect oral microbiome change following periodontal interventions: a systematic review. Oral Dis 27:1073–1089. https://doi.org/10.1111/odi.13405

Aggregatibacter actinomycetemcomitans: From Basic to Advanced Research

3

Abdelhadi Hbibi, Amal Bouziane, Badiaa Lyoussi, Mimoun Zouhdi, and Driss Benazza

Abstract

Aggregatibacter actinomycetemcomitans is a major periodontal pathogen that was identified firstly in actinomycotic lesions and later in advanced forms of periodontal diseases as well as in oral cavity of healthy subjects. The particular pathogenicity of this specie makes it a target for extensive studies both at fundamental and practical scales. The current advances in experimental and clinical research related to this bacterium focus the light on epidemiologic features, virulence, and invasiveness aspects as well as on identification challenges, bacterial susceptibility, and anti-virulence strategies. The present chapter provide to scientists and periodontal researchers a comprehensive overview on the main advances made in this field with a special focus on epidemiologic dissemination, microbial diagnosis, virulence factors and clinical implementations of such progress.

Keywords

Aggregatibacter actinomycetemcomitans · Periodontitis · Microbial identification · PCR · Microbial diagnosis · Anti-microbial · Anti-virulence

Abbreviations

A.	*Aggregatibacter*
AI-2	autoinducer-2
bp	base pair
cdt	cytolethal distending factor
DNA	deoxyribonucleic acid
ELISA	enzyme-linked immunosorbent assay

A. Hbibi (✉)
Research laboratory in oral biology and biotechnology, Faculty of dental medicine, Mohammed V University in Rabat, Rabat, Morocco

Centre des Soins Dentaires, Hôpital Provincial Moulay Abdellah, Salé, Morocco
e-mail: abdelhadi.hbibi@um5s.net.ma; abdlhadi_h@hotmail.com

A. Bouziane
Department of Periodontology, Faculty of Dental Medicine, Biostatistical Clinical and Epidemiological Research Laboratory, Faculty of Medicine and Pharmacy, Mohammed V University in Rabat, Rabat, Morocco

B. Lyoussi
Laboratory of Physiology, Pharmacology and Environmental Health, Faculty of Science, University Sidi Mohammed Ben Abdellah, Fez, Morocco

M. Zouhdi
Department of Microbiology, Faculty of Medicine and Pharmacy, Mohammed V University in Rabat, Rabat, Morocco

D. Benazza
Department of Periodontology, Faculty of Dental Medicine, Mohammed V University in Rabat, Rabat, Morocco

J. Santi-Rocca (ed.), *Periodontitis*, Advances in Experimental Medicine and Biology 1373,
https://doi.org/10.1007/978-3-030-96881-6_3

Ig	immunoglobulins
LAgP	localized aggressive periodontitis
LFA-1	lymphocyte function-associated antigen-1
LJP	localized juvenile periodontitis
LPS	lipopolysaccharides
LtxA	Leukotoxin
MALDI-TOF	matrix-assisted laser desorption/ionization time-of-flight mass spectrometry
MIC	minimum inhibitory concentration
MMPs	matrix metallo-proteinases
OMV	outer membrane vesicles
PACs	cranberry proanthocyanidins
PBS	phosphate-buffered saline
PCR	polymerase chain reaction
RANKL	Receptor activator of NF-kappaB Ligand
RNA	ribonucleic acid
rPCR	rapid PCR;
RS	Ringer's solution
RTF	reduced transport fluid
TNF-α	tumor necrosis factor α
TSBV	tryptic soy-serum-bacitracin-vancomycin

Highlights

- *Aggregatibacter actinomycetemcomitans* is one of the key periopathogens that are involved in the pathogenesis of destructive forms of periodontal diseases.
- Numerous studies have focused on drawing the dissemination cartography of this pathogen and defining the geographic distribution of the most prevalent serotypes and genotypes.
- Advances in molecular biology and biotechnology resulted in more accurate and less time-consuming tools to identify this germ facilitating therefore a credible microbial diagnosis.
- Current therapeutic solutions are based on the use of synthetic and natural antimicrobial agents, while promising research are targeting the utilization of anti-virulence strategies.

Considerations for Practice

- For clinicians Periodontal patient might harbour non-virulent or virulent serotypes of A. actinomycetemcomitans, with possible horizontal and/or vertical transmission.
- Microbial diagnosis might be useful if used as a part of personalized approach taking into consideration the local epidemiologic features and the clinical parameters.
- Despite the wide range of antimicrobial and anti-virulence possibilities, the individual plaque control and mechanical debridement should be a constant part of any therapeutic strategy.

Patient Summary

Periodontal patients are required to ask their periodontists on the interest of providing microbial test to detect the presence of *A. actinomycetemcomitans*. In case of its detection, positive subjects should not attribute the severity of their form of periodontal disease to the only infection by this germ. Diseased individuals should understand the multifactorial aspect of this pathology, and the important role that play bacterial plaque and plaque related factors in the onset and progression of periodontal diseases. However, those patients need to know that such bacterial screening might enhance therapeutic procedures, guide antimicrobial solutions and reduce bacterial resistance risks.

3.1 Introduction

Research on periodontal microbiology exhibited, in the last decades, a large interest from researchers trying to understand the role of microbial factor in the pathogenesis and treatment of periodontal disease. The concept and the impact of bacterial biofilm in these processes were the main conclusions of these investigations. In this biofilm, Socransky stipulated the presence of many subgingival bacterial agglomerations grouped in five « major complexes » based on nutritional and ecologic interactions (Socransky et al. 1998). Among these bacterial communities there is one specie called afterwards *Aggregatibacter actinomycetemcomitans* which benefited from intensive research due to its particular virulence and its presumed role in aggressive forms of periodontal diseases (Haubek et al. 1996; Tsai et al. 2018). Longitudinal studies have shown that presence of *A. actinomycetemcomitans* in subgingival plaque of periodontally healthy adolescents is significantly associated with a future attachment loss (Van der Velden et al. 2006; Fine et al. 2007; Haubek et al. 2008; Höglund Åberg et al. 2014). This association was further strengthen if the bacteria were from the highly leukotoxic genotype of *A. actinomycetemcomitans* (Haubek et al. 2008; Höglund Åberg et al. 2014). These aspects have made this germ an attractive target for clinical and fundamental studies that have focused on taxonomy, identification, dissemination, virulence factors, antimicrobial susceptibilities and recently on anti-virulence strategies. In the current chapter, we synthetize the latest findings regarding these fields with special focus on clinical implementations of such advances. The aim of this work is to provide to researchers a comprehensive analysis of this specie and discuss some critical concerns related to its screening and clinical impacts on periodontal health and disease.

3.2 From Sero-Epidemiology to Microbial Susceptibility: Advances and Challenges

3.2.1 Taxonomy, Dissemination, and Organization

3.2.1.1 Taxonomy

A. actinomycetemcomitans is a unique periodontal pathogen which was described for the first time by Klinger in 1912 (Klinger 1912) as a coccobacillus under "Bacterium actinomycetemcomitans" designation due to what was considered as a close phenotypic likeness to *Actinobacillus lignieresii*. The specie has been reclassified later as *Actinobacillus actinomycetemcomitans* (Topley 1929), then affiliated to *Haemophilus* genus as *Haemophilus actinomycetemcomitans* in 1985 (Potts et al. 1986). *Actinobacillus actinomycetemcomitans* and *Haemophilus actinomycetemcomitans* were considered as the same specie for decades. They were described as a gram negative facultative anaerobe and capnophilic coccobacillus (Newman et al. 1976). Both *Haemophilus* and *Actinobacillus actinomycetemcomitans* were replaced by *Aggregatibacter actinomycetemcomitans* since 2006 by Norskov-Lauritsen (Nørskov-Lauritsen and Kilian 2006). This author highlighted the close molecular taxonomy between the three *Aggregatibacter* species which compose *Aggregatibacter* genus: *A. actinomycetemcomitans*, *A. aphrophilus*, and *A. segnis*. The "*Aggregatibacter*" term was created to indicates "a rod-shaped bacterium that aggregates" (Figs. 3.1 and 3.2).

This genus has long been considered to be indigenous to man until the isolation of *A. actinomycetemcomitans* from old world nonhuman primates and rhesus macaque (Nørskov-Lauritsen and Kilian 2006; Karched et al. 2012a, b). In the human clinical context, this bacteria is more likely to be isolated from periodontal and actino-

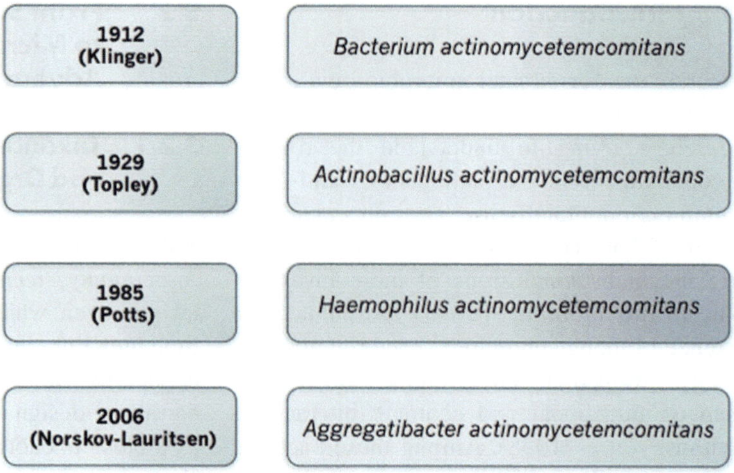

Fig. 3.1 Taxonomy history of *Aggregatibacter actinomycetemcomitans*

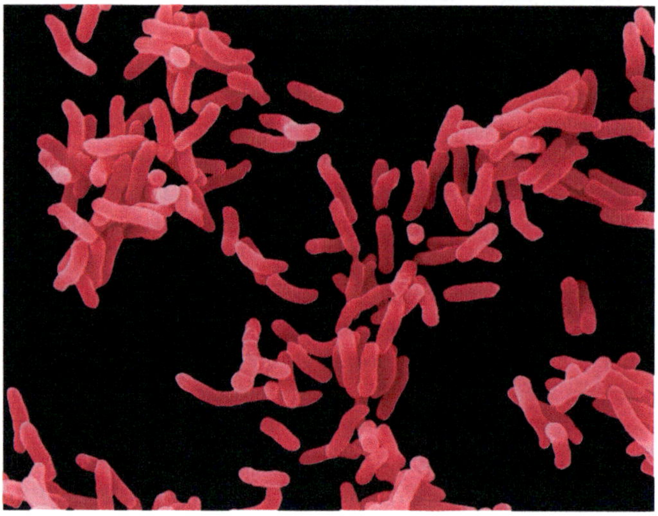

Fig. 3.2 Coloured scanning electron micrograph of Aggregatibacter actinomycetemcomitans (magnification × 2280). Science Photo Library (rights managed)

mycotic lesions (Socransky et al. 1998; Zijlstra et al. 1992), as from respiratory and urinary infections (Storms et al. 2017; Komaru et al. 2018), brain abscess (Moazzam et al. 2015) and endocarditis (Sharara et al. 2016). This pathogen is usually associated with aggressive forms of periodontal disease (Montenegro et al. 2020).

3.2.1.2 Serotypes and Dissemination

Based on differences in antigenic properties of the different strains of *A. actinomycetemcomitans*, molecular biology techniques have defined 7 serotypes of this bacteria (a, b, c, d, e, f and g), as well as non-serotypeable ones (Perry et al. 1996a, b; Kaplan et al. 2001; Tsuzukibashi et al. 2014). Subjects who are positives for *A. actinomycetemcomitans* could express healthy or pathologic periodontal condition including chronic and aggressive forms. In fact, in a limited serotyped cohort (63 subjects) in India, 12 positive patients, for *A. actinomycetemcomitans*, had a healthy periodontium (19%) while 51 subjects who were positives (80.9%) had a chronic periodontitis (Joshi et al. 2017). In Brazil, up to 57.8% of healthy included participants harboured *A. actinomycetemcomitans* (Arenas Rodrigues et al. 2018). Based on serotypes status, in the same country, chronic periodontitis was associated with serotype c while healthy periodontium was associated with serotype b (Teixeira et al.

2006). Controversially, this serotype was most associated with "aggressive periodontitis" in other populations (Mínguez et al. 2016; Yang et al. 2004).

The large diversity of serotypes among studied populations, and their association with different periodontal conditions, may be affected by differences in ethnic and geographic origins (Brígido et al. 2014). This could explain likewise the heterogeneity of different co-colonisations of mixed serotypes which were highlighted in the literature. Positive patients for A. actinomycetemcomitans might be colonized predominantly by one serotype, while less prevalent subjects could be affected by two to four serotypes of this bacterium at the same time (Akrivopoulou et al. 2017; Roman-Torres et al. 2010). Globally, the most predominant serotypes are serotypes a, b and c which is most prevalent worldwide(Brígido et al. 2014). Otherwise, a limited cohort revealed a combined colonization by any serotype of *A. actinomycetemcomitans* and herpesvirus (cytomegalovirus and Epstein-Barr virus) (Joshi et al. 2017).

3.2.1.3 Genotyping and Dissemination

Among the pre-cited serotypes, serotype b has attracted more attention from researchers due to its specific genotypes' variety (Claesson et al. 2017). In this context, the JP2 clone was more studied because of its particular virulence and was more likely to be associated with advanced periodontal lesions (Haubek et al. 1996; Tsai et al. 2018). This virulence is linked, among others, to the excessive liberation of an exotoxin (leukotoxin), and therefore designated as a highly leukotoxic strain, following mutations in the leucotoxin promoter structure (Hritz et al. 1996). These mutations resulted in many highly leucotoxic strains among serotype b including JP2 strain (following 530-base pair (bp) deletion in the ltxCABD promoter region), strain with a 640-bp deletion, other strains with full length promoter such as strain Y4, and two recently discovered strains, one with a 172-bp duplication, and another one with 230-bp deletion (Claesson et al. 2020).

JP2 genotype has been suggested to appear as a specific strain of North-African zone since 2400 years ago (Haubek et al. 2007), then spread to West Africa and over seas. Its current geographic dissemination seems to be widely enlarged and was therefore detected in many other countries with distinct tropism to individuals of African descent (Haubek et al. 2007; de Araujo Neris et al. 2015; Burgess et al. 2017). However, some limited cases of JP2 positive hosts among Caucasian populations have been reported even without any history of close contact with Africans or known positive subjects, while no cases have been reported among Asian populations (Stähli et al. 2020; Claesson et al. 2011, 2017) (Table 3.1).

3.2.1.4 *A. actinomycetemcomitans*: Commensal or Pathogen?

In the light of the taxonomic features above, many research discussed the ecologic status of *A. actinomycetemcomitans*. In fact, the high detected prevalence of this bacteria among Chinese population, regardless of its serotyping, suggested that this specie could be a part of the ordinary oral microbiota of this ethnicity (Haubek et al. 1996). Otherwise, this genus was firstly suggested to be opportunistic pathogen in the aggressive forms of periodontitis worldwide, and exogenous pathogen (JP2 clone) in diseased hosts originated from African continent (Haubek et al. 1996, 2007). Later reports have described distinct transmission models of *A. actinomycetemcomitans* for various serotypes and not specifically for JP2 clone (Doğan et al. 2008; Al Yahfoufi 2017). In fact, Van Winkelhoff highlighted variable frequencies of horizontal transmission in spouses between 14% and 60% while vertical transmission was described between 30% and 60% (Van Winkelhoff and Boutaga 2005). This transmission from mother or caregiver to child focused the light on the impact of *A. actinomycetemcomitans* as early colonizer in predentate children. Such early presence in oral mucosa, especially buccal epithelial cells, suggests that those cells represent a possible reservoir for the further bacterial colonization of oral tissues (Fine et al. 2010; Lamell et al. 2000; Tanner et al. 2002).

Other studies in eastern countries were in the line with such finding pointing that members of a

Table 3.1 Global distribution of *A. actinomycetemcomitans* serotypes by continent

Continent	Investigated countries	Total sample size*	Global found serotypes	Global found genotypes	Studies
Africa	Ghana	798	b	JP2, non JP2, cdt+	Åberg et al. (2012), Höglund Åberg et al. (2013) and Dahlén et al. (2014)
	Morocco	1643	b, a, e	JP2, non JP2,cdt+,cdt–	Ennibi et al. (2012), Haubek et al. (2008), Haubek et al. (2001), Jensen et al. (2016) and Mínguez et al. (2016)
	Sudan	66	b	JP2, non JP2	Elamin et al. (2011) and Elabdeen et al. (2015)
America	Brazil	535	a, b, c, e	NI	Teixeira et al. (2006) and Roman-Torres et al. (2010)
	United States	1381	a, b, c, d, e, f	Y4, JP2	Celenligil and Ebersole (1998), Chen et al. (2010) and Fine et al. (2007)
Australia	NI	NI	NI	NI	NI
Asia	China	179	a, b, c, e, f,	Non JP2, cdt+, cdt–	Leung et al. (2005), Mombelli et al. (1999) and Tan et al. (2001)
	India	143	a, b, c, d, e	Non JP2	Joshi et al. (2017) and Suprith et al. (2018)
	Indonesia	94	a, b, c, d, e	NI	van der Reijden et al. (2008)
	Japan	384	a, b, c, d, e	NI	Yoshida et al. (2003) and Thiha et al. (2007)
	Thailand	453	a, b, c, f	Non JP2, cdt+, cdt–	Bandhaya et al. (2012)
Europe	Germany	99	a, b, c, d, e, f	Non JP2, JP2,cdt+,cdt–	Jentsch et al. (2012)
	Greece	228	a, b, c, e	Non JP2	Sakellari et al. (2011)
	Spain	701	a, b, c, d	Non JP2	Mínguez et al. (2014)
	Sweden	26	a, b, c, e	Non JP2, cdt+, cdt–	Aberg et al. (2009) and Sjödin et al. (1995)
	Turkey	142	a, b, c, f	Y4, JP2	Celenligil and Ebersole (1998) and Doğan et al. (2016)
	United Kingdom	50	a, b, c, e	NI	Akrivopoulou et al. (2017)

*Total sample size describes the total number of subjects, included in the corresponding studies, who benefited from bacterial and serotyping sampling; *cdt* cytolethal distending factor, *NI* not identified

same family might harbour and share one to many clonal types of *A. actinomycetemcomitans* (Doğan et al. 2008; Al Yahfoufi 2017). Such transmission could be explained either by oral hygiene and personal habits or by the common genetic and immuno-inflammatory factors in the same family.

3.2.1.5 Oral Habitats, Organization and Interaction

A. actinomycetemcomitans could be organized in planktonic state in gingival crevicular fluid and saliva, and as biofilm colonizer in subgingival and supragingival biofilm including on gingivae, tongue, cheek, tonsil, and hard palate (Ennibi et al. 2019; Asikainen and Chen 1999; Müller et al. 2001; Slots et al. 1980). The highest level of presence of this micro-organism was found in periodontal pockets and more interestingly in subgingival plaque of first molars and maxillary third molars (Van der Weijden et al. 1994; Müller et al. 2001). This presence within dental plaque engages this specie in the possibility of interacting with other microorganisms of the oral microflora. In fact, the biofilm formation of *Candida albicans* could be inhibited by a molecule named autoinducer-2 (AI-2) synthetized by *A. actinomycetemcomitans* (Bachtiar et al. 2014). In contrast, the growth of this bacterium was documented to be inhibited by many commensal oral *lactobacil-*

lus and *streptococcus* strains among healthy individuals (van Essche et al. 2013; Herrero et al. 2016; Teughels et al. 2007).

3.2.2 Isolation and Identification

3.2.2.1 Samples Collection, Transport and Storage

The main principle of collecting specimens in bacteriology is to target this process in the concerned infected site and avoid assembling unnecessary species surrounding this infectious aera. This technical challenge is also present during the collecting procedure of *A. actinomycetemcomitans* with some variations related to natural habitats of this specie. In fact, this bacterium, as described above, could be present in many oral sites including narrow sites (periodontal pockets) and larges ones (e.g. saliva, tongue). For the first category, the use of absorbent endodontic paper is mandatory to collect *A. actinomycetemcomitans*, while for the other one the swabs are usually used. Such processes need to be made before any use of oral antimicrobial agents (antibiotics and antiseptics) by the patients during 2–3 last months. This deadline is deduced from reported periods of bacterial re-colonization after periodontal therapy, which is estimated between 9 and 11 weeks (Greenstein 1992; Allen et al. 2008). This period is necessary to sustain a detectable level of targeted bacteria for the further in vitro culture. Sampling process is recommended to be performed using saliva isolation and before any clinical measurements, especially probing procedure, to enhance accuracy and detectability in the further culture (Nguyen-Hieu 2013).

The transport media is another issue to be highlighted since this *Aggregatibacter* species is commonly fastidious and needs to maintain its viability until laboratory handling. In this issue, the best procedure is to process to the culture of clinical samples without any delay, especially if the clinical institution and laboratory are too close. However, this closeness is not always available, and samples usually need some hours, or even days to reach laboratories. In this case, many transport medias were suggested such as phosphate-buffered saline (PBS) and Ringer's solution (RS) (Piccolomini et al. 1998), AaTM media (Piccolomini et al. 1998), Reduced Transport Fluid (RTF) (Bollen 1999), and VMGA III (Dahlén et al. 1993). Although few studies have focused on comparison of the effectiveness of those media on *A. actinomycetemcomitans* viability, it seems that AaTM, VMGA III, and RTF are the best media for better viability of this pathogen (Piccolomini et al. 1998) (Dahlén et al. 1993; Syed and Loesche 1972). Storage conditions were even less studied. However, such survival was reported to be enhanced at 8–12 °C, during a maximum 24 h (Piccolomini et al. 1998). The duration of exposition of extracted samples to air, before culture, needs also to be taken into consideration since this factor was reported to limit the recovery chances of this micro-organism in the laboratory (Paolantonio et al. 1996). Serotypes a and c were shown to be more sensitive to air exposition then serotype b (Piccolomini et al. 1998).

3.2.2.2 Isolation

The culture and isolation of *A. actinomycetemcomitans* require enriched selective media using vancomycin or bacitracin, and needs to be incubated in a 5–10% CO^2 atmosphere (James H. Jorgensen et al. 2015). In this field, the most popular reported media is Dentaid-1, which was described for the first time by Alsina et al. in 2001 (Alsina et al. 2001). This media was presented as a « low-cost, non-inhibitory formula » for primary isolation of *A. actinomycetemcomitans* without using blood or serum. The presence of vancomycin among its main components gives it the potential to inhibit the growth of other germs within the same culture. Hence, Dentaid-1 media was reported to be more sensitive then previous medias, especially tryptic soy-serum-bacitracin-vancomycin (TSBV) in detecting *A. actinomycetemcomitans* among subgingival samples (Alsina et al. 2001; Rurenga et al. 2013).

Macroscopically, on agar support, the colonies are small (1–2 mm at 48 h), opaque, slightly grayish to yellowish, and characterized in the first isolation by their rough and adherent surface with slimly irregular edges (sessile adherent form). This primary aspect is usually transformed to a smooth and non-adherent morphology with regular borders (planktonic form), by the loss of fimbriae, after multiple subcultures (Fine et al. 1999). This morphologic evolution during repeated passages in laboratory might modify some virulence factors of this bacteria as would be described furtherly in this chapter.

By microscopic examination, following gram stain, this bacterium shows a coccoid to rod-shaped form (coccobacillus) mostly arranged in single and pairs and rarely expressing filamentous forms, and organized in small and large colonies.

3.2.2.3 Identification

By its isolation, the identification step is mandatory to confirm within the isolated germs correspond to *A. actinomycetemcomitans*. In this issue, biochemical and/or molecular reactions are the mean tools to reach this objective. In this context, conventional enzymatic used tests are:

- Catalase and oxidase activities;
- Carbohydrate fermentation: glucose, sucrose, lactose, xylose, maltose; and mannitol;
- Indole, ornithine decarboxylase and urease production;
- Dependence on X and V factors.

By using the identification process above, confusions occur frequently with *Haemophilus* species. In this situation, the four main tests which could differentiate between *A. actinomycetemcomitans sp.* and *Haemophilus* spp. are the reactions for production of ornithine decarboxylase, indole and urease, and the dependence on X factor. In fact, *A. actinomycetemcomitans* is usually negative for those reactions while *Haemophilus* species are at least positive for one of the mentioned tests (James H. Jorgensen et al. 2015).

Besides the culture technique, other methods were suggested to identify periodontal pathogens such as DNA probes, PCR (Polymerase Chain Reaction) methods and immunofluorescence assay. In this context, French et al. were the firsts to propose DNA probes to identify *A. actinomycetemcomitans* (French et al. 1986). This specific and sensitive technique involved DNA-DNA hybridization based on the genomic library of this bacterium with relatively low detectable level (10^3–10^4 cells). This detectability was afterwards questioned by other authors due to the cross-reactivity of the whole genomic probe of *A. actinomycetemcomitans* with haemophilus species (van Steenbergen et al. 1996; Strzempko et al. 1987). Therefore more accurate assays were developed using PCR methods based particularly on genes like 16Sr DNA and leukotoxin production genes (Kraig et al. 1990; Furcht et al. 1996). However, the conventional PCRs were judged to be a time consuming methods (two to many hours) and this factor was the raison to develop innovative rapid PCR (rPCR) techniques, which were reported to detect *A. actinomycetemcomitans* within 35 minutes (Furcht et al. 1996). The mono-identification aspect of those molecular techniques has been overcome later using the Multiplex PCR which demonstrated promising results in the simultaneous detection of *A. actinomycetemcomitans* and other periodontal pathogens (Coffey et al. 2016; Marin et al. 2018; Lochman et al. 2019).

Immunoassays using antibodies were also suggested to detect *A. actinomycetemcomitans*. In fact, immunofluorescence methods were tested to identify this pathogen, and their sensitivity was correlated with the initial microbial load of this bacterium in periodontal pockets and gingival tissue (Christersson et al. 1987). Indirect immunofluorescence revealed to be a rapid and highly sensitive technique compared to cultures, but relatively expansive (Bonta et al. 1985; Listgarten et al. 1995). Enzyme-linked immunosorbent assay (ELISA) test using antibody titers (for IgG and IgA) in saliva and serum was also reported to be a successful method for the screening of positive patients for *A. actinomycetemcomitans* with more focus on burden of antibody response (Gadekar et al. 2018).

Mass spectrometry tools, especially, MALDI-TOF (matrix-assisted laser desorption/ionization time-of-flight mass spectrometry) system have also shown a high performance to identify *A.*

actinomycetemcomitans among clinical isolates with a high confidence score (Couturier et al. 2011; Gürcan et al. 2016).

Besides the reactions above, many commercial biochemical systems were suggested to improve rapid presumptive identification of *A. actinomycetemcomitans* with differences in sensitivity. In this issue, API ZYM (bioMérieux Vitex) and RapID NH (Remel) systems were one of the firsts to be proposed and studied. In fact, RapID NH, failed to identify 85% of clinical isolates of this bacterium (Doğan et al. 1999). The false reactions to proline p-nitroanilide were the main raison for this poor detection ability. Interestingly, API ZYM kit differentiated this pathogen from *H. aphrophilus and H. paraphrophilus,* with lack of data concerning the global sensitivity to *A. actinomycetemcomitans* compared to PCR (Doğan et al. 1999). Afterwards, more accurate kits were advocated such as, VITEK 2 (bioMerieux). The first study evaluating VITEK 2 showed a correct identification of *A. actinomycetemcomitans* among more then 95% of clinical isolates (Rennie et al. 2008). Nevertheless, misidentifications have been highlighted when comparing VITEK 2 to MALDI-TOF system, which showed better performance for *A. actinomycetemcomitans* screening (Hussain et al. 2020; Gürcan et al. 2016).

In this identification section we can conclude that among all the described detection methods, PCRs followed by mass spectrometry methods and immunoassays are the most accurate and sensitive ones to identify *A. actinomycetemcomitans.* Cultures, even selective ones, as well as commercial enzymatic systems will rather serve as means of presumptive identification, which need to be controlled by molecular methods or mass spectrometry techniques (Fig. 3.3).

3.2.3 Virulence

Independently of the fact that bacterial virulence is obviously variable among *A. actinomycetemcomitans* genotypes, cumulative evidence links this pathogenicity to numerous virulence factors such as exotoxins, endotoxins, and membrane vesicles. Such factors might induce the onset and progression of periodontal disease.

3.2.3.1 Exotoxins

A. actinomycetemcomitans is the only oral bacterium which produces two exotoxins: leukotoxin (LtxA) and Cytolethal Distending Toxin (CDT). LtxA Can lead to the death of macrophages, degranulation of neutrophils with liberation of matrix metallo-proteinases (MMPs) and lysosomes, and the activation of cytokines secretion by the host (Åberg et al. 2015; Belibasakis et al. 2019; Claesson et al. 2002; Kelk et al. 2005, 2008). Genetically, the production of this toxin is linked to the presence of correspondent operon which contains four encoding genes designated as *ltxC, ltxA, ltxB* and *ltxD* (Haubek 2010). The regulation of expression of those genes is under the control of a specific promoter region which the deletion of the portion 530 base pair (bp) is responsible of excessive secretion of this exotoxin among JP2 strains (Haubek 2010). This deletion was suggested to be more likely associated with the presence of cagE gene among JP2 strains (Johansson et al. 2019). Three other genetic modifications were also reported to be associated with high production of LtxA: 640 bp deletion among the ltx promoter, 230-bp deletion, and an acquisition of 886 bp resulting in four highly leukotoxic genotypes (He et al. 1999; Claesson et al. 2015; Claesson et al. 2020) Interestingly, limited clinical isolates, recognized as a subgroup of serotype b (non-JP2 genotype), demonstrated high leukotoxicity despite the intact aspect of the promoter region (Höglund Åberg et al. 2014). This reveals that the virulence model of LtxA is more complicated and seems to be related to the large diversity of genotypes (Jensen et al. 2020). In addition, the growth conditions such as anaerobic environment and the availability of adequate substrates might influence the production of such toxin (Hritz et al. 1996; Kolodrubetz et al. 2010).

3.2.3.2 Cytolethal Distending Toxin

Apart from LtxA, CDTs are the second toxins family which could be secreted by *A. actinomycetemcomitans.* Those exotoxin proteins are

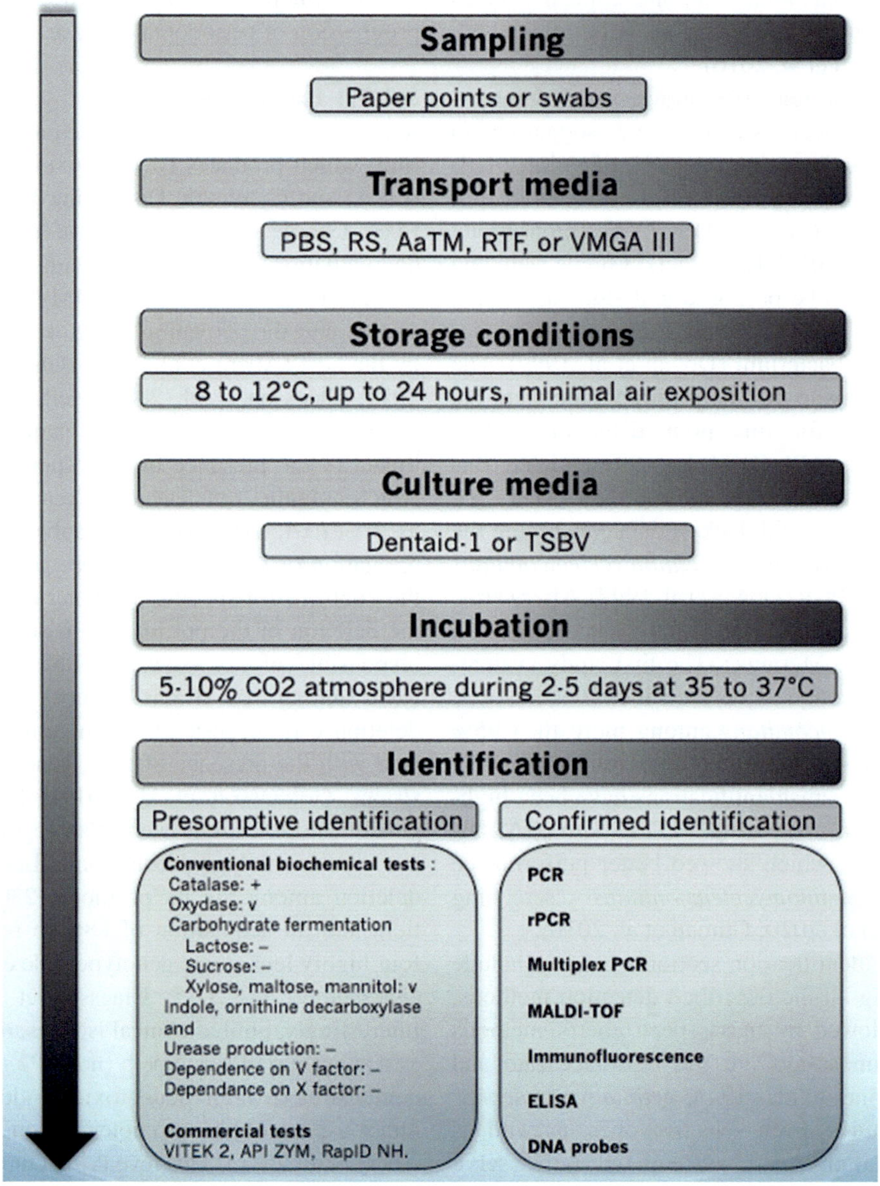

Fig. 3.3 From sampling to identification of *A. actinomycetemcomitans*: global strategy and key implements; **+:** positive reaction; **−**: negative reaction

commonly produced by various Gram-negative bacteria and are known to induce DNA damages of host cells leading to cells cycle arrest then cells death (Lara-Tejero and Galán 2000). Some reports have even highlighted a possible implication of CDT in cancer pathogenesis (Faïs et al. 2016). The genotoxic activity of CDT in periodontal tissues is expanded to gingival fibroblasts, ligament, and immune cells (Chen et al. 2017). In fact, CDT enhances RANKL expression among human gingival fibroblasts promoting therefore osteoclasts activity and consequent bone resorption (Belibasakis et al. 2005). This multisubunit protein is composed of three subunits designed as CdtA, CdtB, and CdtC. Among all oral bacterial species, *A. actinomycetemcomi-*

tans is the only one to produce such toxin by many of its strains. The prevalence of the gene encoding for this exotoxin is variable within populations and its association with the progression and severity of periodontal disease is increasingly evident (Höglund Åberg et al. 2013; Faïs et al. 2016).

3.2.3.3 Endotoxins

Similarly to other Gram-negative bacteria, lipopolysaccharides (LPS) are the main endotoxins related to *A. actinomycetemcomitans*. LPS activity targets immune, gingival and epithelial cells as well as osteoblasts and dendritic cells (Suga et al. 2013) (Naqvi et al. 2014; Sosroseno et al. 2009; Díaz-Zúñiga et al. 2014). Such toxin, may induce a high production of inflammatory mediators like IL-6, IL-8, IL-12 and TNF-α, and engender alveolar bone destruction (Díaz-Zúñiga et al. 2014; Gualtero et al. 2019; Ridwan et al. 2018). Interestingly, the extent of activating the release of those cytokines seems to be related to *A. actinomycetemcomitans* serotypes. In fact, LPS originated from serotype b resulted in a higher production of IL-12 and TNF-α by dendritic cells (Díaz-Zúñiga et al. 2014).

3.2.3.4 Cytokine-Binding Molecules and Outer Membrane Vesicles

Cytokines are ones of many inflammatory mediators released in the context of host immune response. During this reaction, *A. actinomycetemcomitans* is reported to be up to bind many cytokines and internalize them (Ahlstrand et al. 2017). Many outer membrane proteins are implicated in this process including interleukin-1β-binding protein and secretin HofQ (Paino et al. 2012; Ahlstrand et al. 2018). Bacterial membrane harbours also, like all Gram negative, outer membrane vesicles (OMVs), which participate in *A. actinomycetemcomitans* virulence. Firstly, those vesicles serve as a delivery tool by exporting toxins to host tissues and excreting bacterial RNAs that infect host cells and promote TNF-α production in macrophages. Immune invasion might be facilitated by such vesicles, which internalize into host cells and altering their functions (Han et al. 2019; O'Donoghue and Krachler 2016). Serum resistance, a characteristic that allows bacteria to cause bacteremia, is another trait that outer membrane vesicles promote in *A. actinomycetemcomitans* invasiveness. In fact, it has been shown that OMVs can initiate complement activation and reinforce *A. actinomycetemcomitans* serum resistance. Moreover, outer membrane proteins such as OmpA1 and OmpA2 were shown to be critical for the survival of this specie in human serum (Lindholm et al. 2020).

The virulence ability of *A. actinomycetemcomitans* strains might be subject of modification during laboratory handling (Fine et al. 2020). In fact, the usual rough aspect of clinical isolates after primary isolated cultures is rapidly lost after repeated subcultures reducing therefore the adherence to saliva-coated hydroxyapatite. In addition, the transformed strains – with smooth trait – demonstrated a reduced fibroblast proteinase and endotoxin activity then fresh isolates (Fine et al. 1999). The restricted data in this issue limits the consideration for any practical implementation but could represent a possible bias for the assessment of microbial behaviour toward antimicrobials, and more experiments are needed to assess the impact of such trait transformation on trials findings in the regard of this bacterium.

3.2.3.5 Impact of *A. actinomycetemcomitans* Infection on the Onset and Progression of Periodontal Disease

The role of *A. actinomycetemcomitans* in the onset of periodontal disease has been discussed more then three decades ago especially for the highly leucotoxic strains. In fact, JP2 strains were reported to be harboured by too young individuals (mean age 12.7 years) expressing what was classified as localized juvenile periodontitis (LJP) and later as localized aggressive periodontitis (LAgP) (Zambon et al. 1996). *A. actinomycetemcomitans* was also pointed to be involved in the pathogenesis of the generalized form of "prepubertal periodontitis" (López 1992). In addition, many authors reported continuous attachment loss in periodontal sites which harboured this pathogen and linked therefore the progression of

LJP to *A. actinomycetemcomitans* infection (Kim et al. 1992). Recent findings indicate that periodontium of incisors and first molars expresses rapid onset and progression in the context of LAgP among adolescents and young adults from African descent while rapid recovery and lack of disease progression have been reported from treated and long-term maintained cases of LAgP (Fine et al. 2019; Miller et al. 2017). Otherwise, *A. actinomycetemcomitans* infection might interact with other risk factors of disease progression such as age, ethnic origin, calculus, general conditions and existing deep pockets resulting thereby in additional tissue damages (Van Dyke and Sheilesh 2005). The co-infection by *A. actinomycetemcomitans, Porphyromonas gingivalis and Prevotella intermedia* and the presence of bleeding on probing were also evoked in retrospective studies since decades as predictors of initiation and progression of periodontal disease (Slots et al. 1986; Rodenburg et al. 1990). Thus, progression mechanisms are rather complex and there are no firm conclusions for any unique or decisive role of *A. actinomycetemcomitans* in disease activity despite the strong evidence of the role of this microorganism as important part among a large bacterial consortium linked to destructive forms of periodontal diseases (Miller et al. 2017). Therefore, the cumulative evidence links the disease progression to a multifactorial process involving the interaction of periodontal microflora and host factors. Interestingly, the silent initiation and rapid progression of aggressive forms of periodontitis involve an emphasis on prematurely diagnostic and close follow-up to reverse disease activity.

3.2.4 Issues in Antimicrobial Susceptibility

In the light of the multifactorial virulence of *A. actinomycetemcomitans* as described above, and its ability to invade periodontal tissues independently of providing conventional mechanical periodontal treatment, many consensus reports focused on the interest of adjuncting antimicrobials to the mechanical therapy in periodontal diseases involving this pathogen (Sanz and Teughels 2008; Lang et al. 2015; Pretzl et al. 2019). In fact, several molecules and formulation have been tested as local or systemic application or combination of both of them to target this pathogen. Systemic formulations used firstly, tetracycline as adjunct to nonsurgical periodontal therapy (Genco 1981), while most topical applications were based on minocycline, tetracycline or doxycycline (Baker et al. 1985). The further resistance emergence of *A. actinomycetemcomitans* to tetracycline, metronidazole and amoxicillin in many cohorts – due among others to the lack of antibiotic stewardships – focused the interest of the use of other formulations such as, amoxicillin + metronidazole, amoxicillin/clavulanic acid and more recently cephalosporins and fluoroquinolones especially ciprofloxacin, moxifloxacin, cefotaxime and ceftriaxone which induced, in some trials, 100% growth inhibition (Akrivopoulou et al. 2017; Madinier et al. 1999; Dabija-Wolter et al. 2018; Bhat et al. 2019). Thereby, it should be stated that there is no current evidence which supports a consensual antimicrobial strategy toward this specie. This is particularly due to the fact that antimicrobial susceptibility profile of this pathogen is as dependant to geographic localization and microbial diversity as other bacterial species around the world (Brook et al. 2013). Current Efforts are focusing on enlarging antimicrobial solutions by testing many common antibiotics usually used for other infectious diseases (Dabija-Wolter et al. 2018; Bhat et al. 2019). Furthermore, trials on antimicrobial benefits of natural products such as essential oils and bee products reveal interestingly promising results in this context (Akkaoui et al. 2020; Hbibi et al. 2020). In other respects, there is an increasing need to identify genes and factors involved in the spread of resistant strains of *A. actinomycetemcomitans* in each population. In this field, a unique report highlighted that antibiotics might paradoxically induce this bacterium to form a biofilm and disclosed genetic fitness and resistance determinants using a laboratory and clinical strains in American context (Narayanan et al. 2017). This limited finding should not be extrapolated to all *A. actinomycetemcomitans* strains and more similar works are required to define resistance factors and aspects in the regard of antimicrobial use. In a Swiss sample, remarkable stability of antibiosusceptibility of *A. actinomycetemcomitans* was noticed with constancy among MIC values of azithromycin, amoxicillin, amoxicillin/clavulanic acid, ertapenem, moxifloxacin, doxycycline, clindamycin, and metronidazole over 37 years (1980–2017) (Kulik et al. 2019). In the same report, many *Porphyromonas gingivalis* strains have developed resistances following mutation among certain genes (ermF and gyrA). Such resistance interested mainly moxifloxacin. This particular study confirms the possible occurrence of changes of bacterial behaviour toward antibiotics even if they belong to quite recent drug generations like moxifloxacin (third generation of quinolones). Overall, such stability of microbial susceptibility of *A. actinomycetemcomitans* and other oral pathogens in Switzerland was related to the limited and cautious use of antimicrobials by local care providers. This restrictive prescription approach should be a model to apply in daily periodontal practice.

3.3 Advances in Therapy and Prevention

3.3.1 Microbiological Diagnostic: Is It Useful for Patients?

The interest of microbial diagnosis and follow-up for periodontal patients has been widely discussed among the literature. In fact, the screening of periopathogens has been proposed for decades as a diagnostic tool adjunctively to clinical and radiographic examination (Socransky et al. 1999; D'Ercole et al. 2008). However, the multimicrobial aspect of microflora associated with periodontal diseases make the interest of such tool questionable since no specific specie could be a synonymous of any clinical form of the disease (Belstrøm et al. 2017). Moreover, fastidious bacteria like *A. actinomycetemcomitans* are hardly cultivable, and their detection in culture could not be a reliable indicator of disease activity since this aspect is linked to more complex factors, and even healthy subjects might harbor this germ. Otherwise, the ultimate objective to cultivate this specie is to explore its antibiotic sensitivity in order to guide therapeutic choices and enhance clinical outcomes, while this goal is challenged by the fact that bacterial response to antimicrobials could be different depending on whether this germ is organized or not in biofilm (Socransky and Haffajee 2002).

It is worthy of note that the earlier classification of periodontal diseases of 1999 highlighted the interest of microbial diagnosis as a secondary feature – after clinical and radiographic investigation – to point a possible elevated proportion of *A. actinomycetemcomitans* as a sign of « aggressive periodontitis » (Armitage 2004). Whereas the last classification of periodontal and periimplant conditions of 2017 has delinked the presence of this bacteria with any form of periodontitis based on recent microbiological findings (Papapanou et al. 2018). However, it is obvious that the absence of *A. actinomycetemcomitans* in periodontal sites complementary to the absence of attachment loss and bleeding of probing would be an indicator of periodontal health. Likewise, a recurrent bleeding on probing and continuous attachment loss in specific sites after conventional therapy or during periodontal maintenance could reasonably justify microbial screening for *A. actinomycetemcomitans* as for other major periodontopathogens with keeping in mind that such unresponsiveness to conventional therapy might be attributed to the virulence of other microbial components of biofilm and host related factors.

3.3.2 Therapeutic Targeting and Prevention Strategies

In the light of the pre-cited considerations, an individual approach might be more accurate to take advantage of microbial screening. Firstly, such implement needs to be performed after comprehensive analysis of periodontal parameters, host factors and epidemiologic context. The presence of attachment loss and severe periodontal lesions in patients from African descent might suggest an infection by JP2 clone. Similarly, advanced forms of periodontitis in Chinese context could presume infection by many periopathogens other then *A. actinomycetemcomitans* since this specie was found to be commensal in this population (Goh and Ong 2019).

The inclusion of microbial diagnosis in daily periodontal practice requires the selection of practical chair-side microbial tools and reduced time-consuming implements in the laboratory. Preliminary microscopic assessment of dental plaque in office would be interesting but does not allow antimicrobial susceptibility testing. DNA probes could be performed easily by clinician and might offer higher sensitivity and lower threshold detection then culture (Al Yahfoufi et al. 2015) but have as disadvantage a cross-reactivity with *haemophilus spp* (van Steenbergen et al. 1996). Immuno-assays are also highly sensitive in *A. actinomycetemcomitans* screening, but the need of monoclonal antibodies limits their effectiveness (D'Ercole et al. 2008). PCR, rPCR and Multiplex PCR testing are highly effective alternatives but need specific training and well-developed technical platform. Such technics might be evenly useful in detecting resistance genes and could be beneficial in the monitoring of antibiosusceptibility during maintenance therapy. In fact, the chronic and recurrent aspect of periodontal infection prompts clinicians to monitor the presence of pathogens like *A. actinomycetemcomitans* among intermittent bleeding periodontal pockets during supportive periodontal therapy in order to orientate the antimicrobial strategy and target appropriate antibiotic prescription. This microbial screening and monitoring is justified by the fact that the presence or absence of some pathogens has been considered as a marker of the onset, progression and/or stability of the infection (Suchett-Kaye et al. 2001). Actually, consistent evidence supports the role of JP2 clone in the early onset and severity of aggressive forms of periodontitis (Haubek et al. 2001; Haubek et al. 2002; Jensen et al. 2016). The significant association between this clone and attachment loss in adolescents and adults of African decent evokes the possible consideration of JP2 clone as a risk factor or risk indicator of "aggressive periodontitis". Such consideration involves the interest of including the screening of this specie in the preventive approach among at-risk populations and subjects. In this context, microbiological follow-up of *A. actinomycetemcomitans* status among patients has been raised as a part of predictive markers to be used in the preventive strategy to control general conditions such as rheumatoid arthritis (Cheng et al. 2017).

Anti-virulence strategies are currently under investigation to limit and reduce tissue damages induced by virulence factors of *A. actinomycetemcomitans*. Those innovative paths include anti-LtxA approach and preventing internalization of LtxA in immune cells (Krueger and Brown 2020; Webb et al. 2016). Such strategies are based on the interaction between LtxA, cholesterol – as lipid of cell membrane – and lymphocyte function-associated antigen-1 (LFA-1) (Krueger and Brown 2020). In fact, a membrane-permeable nuclear stain called DRAQ5™ – originally developed as anti-cancer molecule – has been reported to inhibit LtxA internalization but not association among Jurkat T cell line by interfering with cholesterol membrane decreasing therefore LtxA binding to cells (Webb et al. 2016). The main challenge with DRAQ5™ is its reported cardiotoxicity which limited its use as anti-cancer agent. However, its application as therapeutic molecule in periodontology remains promising since the dosage in periodontal use would be broadly lower then in cancerology. Otherwise, CD11a-based peptides have shown unneglectable potential to block LFA-1 receptor disrupting therefore the LtxA binding process (Krueger and Brown 2020).

The inhibition of LtxA association to host cells was suggested to be equally possible by using some polyphenols derived from fruits and vegetables, such as catechins, which were found to inhibit LtxA binding by altering its structure that leads to the inability of this protein to bind to membrane's cholesterol (Chang et al. 2019, 2020). Extracts of *Psidium guajava* have shown similar action to neutralize LtxA activity regarding host cells as well as inactivating the pro-inflammatory response of human immune cells (Kwamin et al. 2012).

The latest reported anti-Ltxa strategy is focusing on disruption of Ltxa gene expression among highly leucotoxic *A. actinomycetemcomitans* strains. This field interests the use of cranberry proanthocyanidins (PACs) to attenuate the expression of *ltxB* and *ltxC* genes as well as the P2X7 receptor and NALP3 related genes which are involved in cell death (Ben Lagha et al. 2019; Kelk et al. 2011). Interestingly, the significant ability of LtxA to induce cell death has inspired researchers to develop an anti-cancer LtxA-based agent (Leukothera®) to overcome hematological malignancies since this toxin might initiate the degenerative cycle of cancerous cells (Vega et al. 2019).

3.4 Conclusions

In the view of the above, it can be stated that *A. actinomycetemcomitans* plays substantial role, in conjunction with other pathogens, in the onset, progression and severity of periodontal disease among several populations. Its screening in diseased patients – in addition to clinical and radiographic examination in the pre-cited specific situations – would facilitate an accurate diagnosis and bacteriology-driven periodontal treatment in the respect of the ecologic and epidemiologic contexts and avoid random prescription of antimicrobials. However, many critical issues require specific attention. Firstly, the culture and isolation techniques of this fastidious germ need additional development in order to improve its detection using simplified and efficient chair-side implements. PCR tests – as the most current accurate tools – are not fully adapted for at-in office use and their threshold detection level is still non-consensual since these techniques might detect even dead and non clinically significant bacterium number which would be amplified following PCR procedure leading therefore to a possible overestimation of the clinical context. Thirdly, conclusions related to microbiological tests are linked to the microbial status of sampled sites and their extrapolation to the global periodontal context remains questionable. Current research axes are focusing on the selection of suitable antimicrobials and on anti-Ltxa approaches. Better-designed investigations are needed to enlarge anti-virulence possibilities since Ltxa is one among many virulence factors of *A. actinomycetemcomitans*. The main challenge to all these promising solutions is the eradication of this pathogen with evasion to any form of re-colonization.

References

Åberg CH, Kelk P, Johansson A (2015) Aggregatibacter actinomycetemcomitans: virulence of its leukotoxin and association with aggressive periodontitis. Virulence 6(3):188–195. https://doi.org/10.4161/21505594.2014.982428

Åberg CH, Kwamin F, Claesson R, Johansson A, Haubek D (2012) Presence of JP2 and Non-JP2 Genotypes of Aggregatibacter actinomycetemcomitans and attachment loss in adolescents in Ghana. J Periodontol 83(12):1520–1528. https://doi.org/10.1902/jop.2012.110699

Aberg CH, Sjödin B, Lakio L, Pussinen PJ, Johansson A, Claesson R (2009) Presence of Aggregatibacter actinomycetemcomitans in young individuals: a 16-year clinical and microbiological follow-up study. J Clin Periodontol 36(10):815–822. https://doi.org/10.1111/j.1600-051X.2009.01457.x

Ahlstrand T, Torittu A, Elovaara H, Välimaa H, Pöllänen MT, Kasvandik S, Högbom M, Ihalin R (2018) Interactions between the Aggregatibacter actinomycetemcomitans secretin HofQ and host cytokines indicate a link between natural competence and interleukin-8 uptake. Virulence 9(1):1205–1223. https://doi.org/10.1080/21505594.2018.1499378

Ahlstrand T, Tuominen H, Beklen A, Torittu A, Oscarsson J, Sormunen R, Pöllänen MT, Permi P, Ihalin R (2017) A novel intrinsically disordered outer membrane lipoprotein of Aggregatibacter actinomycetemcomitans binds various cytokines and plays a role in biofilm response to interleukin-1β and interleukin-8. Virulence 8(2):115–134. https://doi.org/10.1080/21505594.2016.1216294

Akkaoui S, Johansson A, Yagoubi M, Haubek D, El Hamidi A, Rida S, Claesson R, Ennibi O (2020) Chemical composition, antimicrobial activity, in vitro cytotoxicity and leukotoxin neutralization of essential oil from Origanum vulgare against Aggregatibacter actinomycetemcomitans. Pathogens (Basel, Switzerland) 9(3). https://doi.org/10.3390/pathogens9030192

Akrivopoulou C, Green IM, Donos N, Nair SP, Ready D (2017) Aggregatibacter actinomycetemcomitans serotype prevalence and antibiotic resistance in a UK population with periodontitis. J Glob Antimicrob Resist 10:54–58. https://doi.org/10.1016/j.jgar.2017.03.011

Al Yahfoufi Z (2017) Prevalence of periodontal destruction and putative periodontal pathogens in the same

lebanese family. J Contemp Dent Pract 18(10):970–976. https://doi.org/10.5005/jp-journals-10024-2158
Al Yahfoufi Z, Hadchiti W, Berberi A (2015) Deoxyribonucleic acid probes analyses for the detection of periodontal pathogens. J Contemp Dent Pract 16(9):727–732. https://doi.org/10.5005/jp-journals-10024-1748
Allen E, Ziada H, Irwin C, Mullally B, Byrne PJ (2008) Periodontics: 10. Maintenance in periodontal therapy. Dental update 35(3):150–152–154–156. https://doi.org/10.12968/denu.2008.35.3.150
Alsina M, Olle E, Frias J (2001) Improved, low-cost selective culture medium for Actinobacillus actinomycetemcomitans. J Clin Microbiol 39(2):509–513. https://doi.org/10.1128/jcm.39.2.509-513.2001
Arenas Rodrigues VA, de Avila ED, Nakano V, Avila-Campos MJ (2018) Qualitative, quantitative and genotypic evaluation of Aggregatibacter actinomycetemcomitans and Fusobacterium nucleatum isolated from individuals with different periodontal clinical conditions. Anaerobe 52:50–58. https://doi.org/10.1016/j.anaerobe.2018.05.015
Armitage GC (2004) Periodontal diagnoses and classification of periodontal diseases. Periodontology 2000(34):9–21. https://doi.org/10.1046/j.0906-6713.2002.003421.x
Asikainen S, Chen C (1999) Oral ecology and person-to-person transmission of Actinobacillus actinomycetemcomitans and Porphyromonas gingivalis. Periodontology 2000 20:65–81. https://doi.org/10.1111/j.1600-0757.1999.tb00158.x
Bachtiar EW, Bachtiar BM, Jarosz LM, Amir LR, Sunarto H, Ganin H, Meijler MM, Krom BP (2014) AI-2 of Aggregatibacter actinomycetemcomitans inhibits Candida albicans biofilm formation. Front Cell Infect Microbiol 4:94. https://doi.org/10.3389/fcimb.2014.00094
Baker PJ, Evans RT, Slots J, Genco RJ (1985) Susceptibility of human oral anaerobic bacteria to antibiotics suitable for topical use. J Clin Periodontol 12(3):201–208. https://doi.org/10.1111/j.1600-051x.1985.tb00917.x
Bandhaya P, Saraithong P, Likittanasombat K, Hengprasith B, Torrungruang K (2012) Aggregatibacter actinomycetemcomitans serotypes, the JP2 clone and cytolethal distending toxin genes in a Thai population. J Clin Periodontol 39(6):519–525. https://doi.org/10.1111/j.1600-051X.2012.01871.x
Belibasakis GN, Johansson A, Wang Y, Chen C, Kalfas S, Lerner UH (2005) The cytolethal distending toxin induces receptor activator of NF-kappaB ligand expression in human gingival fibroblasts and periodontal ligament cells. Infect Immun 73(1):342–351. https://doi.org/10.1128/iai.73.1.342-351.2005
Belibasakis GN, Maula T, Bao K, Lindholm M, Bostanci N, Oscarsson J, Ihalin R, Johansson A (2019) Virulence and pathogenicity properties of Aggregatibacter actinomycetemcomitans. Pathogens (Basel, Switzerland) 8(4). https://doi.org/10.3390/pathogens8040222
Belstrøm D, Sembler-Møller ML, Grande MA, Kirkby N, Cotton SL, Paster BJ, Holmstrup P (2017) Microbial profile comparisons of saliva, pooled and site-specific subgingival samples in periodontitis patients. PLoS One 12(8):e0182992. https://doi.org/10.1371/journal.pone.0182992
Ben Lagha A, Howell A, Grenier D (2019) Cranberry proanthocyanidins neutralize the effects of Aggregatibacter actinomycetemcomitans leukotoxin. Toxins 11(11). https://doi.org/10.3390/toxins11110662
Bhat KG, Khot P, Patil S, Pattar G, Majukar S (2019) Antimicrobial susceptibility pattern of oral isolates of Aggregatibacter actinomycetemcomitans. J Oral Maxillofac Pathol JOMFP 23(2):231–235. https://doi.org/10.4103/jomfp.JOMFP_123_19
Bollen C (1999) The sampling of plaque specimens in oral microbiology. Revue belge de medecine dentaire 54(3):190–194
Bonta Y, Zambon JJ, Genco RJ, Neiders ME (1985) Rapid identification of periodontal pathogens in subgingival plaque: comparison of indirect immunofluorescence microscopy with bacterial culture for detection of Actinobacillus actinomycetemcomitans. J Dent Res 64(5):793–798. https://doi.org/10.1177/00220345850640050201
Brígido JA, da Silveira VR, Rego RO, Nogueira NA (2014) Serotypes of Aggregatibacter actinomycetemcomitans in relation to periodontal status and geographic origin of individuals-a review of the literature. Medicina oral, patologia oral y cirugia bucal 19(2):e184–e191. https://doi.org/10.4317/medoral.19304
Brook I, Wexler HM, Goldstein EJ (2013) Antianaerobic antimicrobials: spectrum and susceptibility testing. Clin Microbiol Rev 26(3):526–546. https://doi.org/10.1128/cmr.00086-12
Burgess D, Huang H, Harrison P, Aukhil I, Shaddox L (2017) Aggregatibacter actinomycetemcomitans in African Americans with localized aggressive periodontitis. JDR Clin Transl Res 2(3):249–257. https://doi.org/10.1177/2380084417695543
Celenligil H, Ebersole JL (1998) Analysis of serum antibody responses to periodontopathogens in early-onset periodontitis patients from different geographical locations. J Clin Periodontol 25(12):994–1002. https://doi.org/10.1111/j.1600-051x.1998.tb02404.x
Chang EH, Giaquinto P, Huang J, Balashova NV, Brown AC (2020) Epigallocatechin gallate inhibits leukotoxin release by Aggregatibacter actinomycetemcomitans by promoting association with the bacterial membrane. Mol Oral Microbiol 35(1):29–39. https://doi.org/10.1111/omi.12275
Chang EH, Huang J, Lin Z, Brown AC (2019) Catechin-mediated restructuring of a bacterial toxin inhibits activity. Biochim Biophys Acta Gen Subj 1863(1):191–198. https://doi.org/10.1016/j.bbagen.2018.10.011
Chen C, Wang T, Chen W (2010) Occurrence of Aggregatibacter actinomycetemcomitans serotypes in subgingival plaque from United States subjects.

Mol Oral Microbiol 25(3):207–214. https://doi.org/10.1111/j.2041-1014.2010.00567.x

Chen HP, Li L, Chen X, Yang MF, Ye Y, Wang XQ, Xu Y (2017) The mechanism of Jurkat cells apoptosis induced by Aggregatibacter actinomycetemcomitans cytolethal distending toxin. Apoptosis 22(6):841–851. https://doi.org/10.1007/s10495-017-1357-3

Cheng Z, Meade J, Mankia K, Emery P, Devine DA (2017) Periodontal disease and periodontal bacteria as triggers for rheumatoid arthritis. Best Pract Res Clin Rheumatol 31(1):19–30. https://doi.org/10.1016/j.berh.2017.08.001

Christersson LA, Wikesjö UM, Albini B, Zambon JJ, Genco RJ (1987) Tissue localization of Actinobacillus actinomycetemcomitans in human periodontitis. II Correlation between immunofluorescence and culture techniques. J Periodontol 58(8):540–545. https://doi.org/10.1902/jop.1987.58.8.540

Claesson R, Chiang HM, Lindholm M, Höglund Åberg C, Haubek D, Johansson A, Oscarsson J (2020) Characterization of Aggregatibacter actinomycetemcomitans Serotype b Strains with Five Different, Including Two Novel, Leukotoxin Promoter Structures. Vaccine 8(3). https://doi.org/10.3390/vaccines8030398

Claesson R, Gudmundson J, Åberg CH, Haubek D, Johansson A (2015) Detection of a 640-bp deletion in the Aggregatibacter actinomycetemcomitans leukotoxin promoter region in isolates from an adolescent of Ethiopian origin. J Oral Microbiol 7:26974. https://doi.org/10.3402/jom.v7.26974

Claesson R, Höglund-Åberg C, Haubek D, Johansson A (2017) Age-related prevalence and characteristics of Aggregatibacter actinomycetemcomitans in periodontitis patients living in Sweden. J Oral Microbiol 9(1):1334504. https://doi.org/10.1080/20002297.2017.1334504

Claesson R, Johansson A, Belibasakis G, Hänström L, Kalfas S (2002) Release and activation of matrix metalloproteinase 8 from human neutrophils triggered by the leukotoxin of Actinobacillus actinomycetemcomitans. J Periodontal Res 37(5):353–359. https://doi.org/10.1034/j.1600-0765.2002.00365.x

Claesson R, Lagervall M, Höglund-Aberg C, Johansson A, Haubek D (2011) Detection of the highly leucotoxic JP2 clone of Aggregatibacter actinomycetemcomitans in members of a Caucasian family living in Sweden. J Clin Periodontol 38(2):115–121. https://doi.org/10.1111/j.1600-051X.2010.01643.x

Coffey J, Choudhry M, Shlossman M, Makin IRS, Singh VK (2016) Multiplex real-time PCR detection and relative quantification of periodontal pathogens. Clin Exp Dent Res 2(3):185–192. https://doi.org/10.1002/cre2.37

Couturier MR, Mehinovic E, Croft AC, Fisher MA (2011) Identification of HACEK clinical isolates by matrix-assisted laser desorption ionization-time of flight mass spectrometry. J Clin Microbiol 49(3):1104–1106. https://doi.org/10.1128/jcm.01777-10

D'Ercole S, Catamo G, Piccolomini R (2008) Diagnosis in periodontology: a further aid through microbiological tests. Crit Rev Microbiol 34(1):33–41. https://doi.org/10.1080/10408410701693317

Dabija-Wolter G, Al-Zubaydi SS, Mohammed MMA, Bakken V, Bolstad AI (2018) The effect of metronidazole plus amoxicillin or metronidazole plus penicillin V on periodontal pathogens in an in vitro biofilm model. Clin Exp Dent Res 4(1):6–12. https://doi.org/10.1002/cre2.96

Dahlén G, Claesson R, Aberg CH, Haubek D, Johansson A, Kwamin F (2014) Subgingival bacteria in Ghanaian adolescents with or without progression of attachment loss. J Oral Microbiol 6. https://doi.org/10.3402/jom.v6.23977

Dahlén G, Pipattanagovit P, Rosling B, Möller AJ (1993) A comparison of two transport media for saliva and subgingival samples. Oral Microbiol Immunol 8(6):375–382. https://doi.org/10.1111/j.1399-302x.1993.tb00614.x

de Araujo NM, Cortelli SC, Aquino DR, de Miranda TB, Costa Fde O, Cortelli JR (2015) A. actinomycetemcomitans profile and red complex bacterial species of an Afro-Brazilian community: a comparative study. Arch Oral Biol 60(5):753–759. https://doi.org/10.1016/j.archoralbio.2015.01.013

Díaz-Zúñiga J, Yáñez JP, Alvarez C, Melgar-Rodríguez S, Hernández M, Sanz M, Vernal R (2014) Serotype-dependent response of human dendritic cells stimulated with Aggregatibacter actinomycetemcomitans. J Clin Periodontol 41(3):242–251. https://doi.org/10.1111/jcpe.12205

Doğan B, Asikainen S, Jousimies-Somer H (1999) Evaluation of two commercial kits and arbitrarily primed PCR for identification and differentiation of Actinobacillus actinomycetemcomitans, Haemophilus aphrophilus, and Haemophilus paraphrophilus. J Clin Microbiol 37(3):742–747. https://doi.org/10.1128/jcm.37.3.742-747.1999

Doğan B, Chen J, Çiftlikli SY, Huang J, Kadir T, Alnıak AK, Chen C (2016) Occurrence and serotype distribution of Aggregatibacter actinomycetemcomitans in subjects without periodontitis in Turkey. Arch Oral Biol 61:125–129. https://doi.org/10.1016/j.archoralbio.2015.10.023

Doğan B, Kipalev AS, Okte E, Sultan N, Asikainen SE (2008) Consistent intrafamilial transmission of Actinobacillus actinomycetemcomitans despite clonal diversity. J Periodontol 79(2):307–315. https://doi.org/10.1902/jop.2008.070270

Elabdeen HR, Mustafa M, Hasturk H, Klepac-Ceraj V, Ali RW, Paster BJ, Van Dyke T, Bolstad AI (2015) Subgingival microbial profiles of Sudanese patients with aggressive periodontitis. J Periodontal Res 50(5):674–682. https://doi.org/10.1111/jre.12250

Elamin A, Albandar JM, Poulsen K, Ali RW, Bakken V (2011) Prevalence of Aggregatibacter actinomycetemcomitans in Sudanese patients with aggressive periodontitis: a case-control study.

J Periodontal Res 46(3):285–291. https://doi.org/10.1111/j.1600-0765.2010.01337.x
Ennibi OK, Benrachadi L, Bouziane A, Haubek D, Poulsen K (2012) The highly leukotoxic JP2 clone of Aggregatibacter actinomycetemcomitans in localized and generalized forms of aggressive periodontitis. Acta Odontol Scand 70(4):318–322. https://doi.org/10.3109/00016357.2011.642002
Ennibi OK, Claesson R, Akkaoui S, Reddahi S, Kwamin F, Haubek D, Johansson A (2019) High salivary levels of JP2 genotype of Aggregatibacter actinomycetemcomitans is associated with clinical attachment loss in Moroccan adolescents. Clin Exp Dent Res 5(1):44–51. https://doi.org/10.1002/cre2.156
Faïs T, Delmas J, Serres A, Bonnet R, Dalmasso G (2016) Impact of CDT toxin on human diseases. Toxins 8(7). https://doi.org/10.3390/toxins8070220
Fine DH, Armitage GC, Genco RJ, Griffen AL, Diehl SR (2019) Unique etiologic, demographic, and pathologic characteristics of localized aggressive periodontitis support classification as a distinct subcategory of periodontitis. J Am Dent Assoc (1939) 150(11):922–931. https://doi.org/10.1016/j.adaj.2019.07.024
Fine DH, Furgang D, Schreiner HC, Goncharoff P, Charlesworth J, Ghazwan G, Fitzgerald-Bocarsly P, Figurski DH (1999) Phenotypic variation in Actinobacillus actinomycetemcomitans during laboratory growth: implications for virulence. Microbiology 145(Pt 6):1335–1347. https://doi.org/10.1099/13500872-145-6-1335
Fine DH, Markowitz K, Furgang D, Fairlie K, Ferrandiz J, Nasri C, McKiernan M, Gunsolley J (2007) Aggregatibacter actinomycetemcomitans and its relationship to initiation of localized aggressive periodontitis: longitudinal cohort study of initially healthy adolescents. J Clin Microbiol 45(12):3859–3869. https://doi.org/10.1128/jcm.00653-07
Fine DH, Markowitz K, Furgang D, Velliyagounder K (2010) Aggregatibacter actinomycetemcomitans as an early colonizer of oral tissues: epithelium as a reservoir? J Clin Microbiol 48(12):4464–4473. https://doi.org/10.1128/jcm.00964-10
Fine DH, Schreiner H, Velusamy SK (2020) Aggregatibacter, a low abundance pathobiont that influences biogeography, microbial dysbiosis, and host defense capabilities in periodontitis: the history of a bug, and localization of disease. Pathogens (Basel, Switzerland) 9(3). https://doi.org/10.3390/pathogens9030179
French CK, Savitt ED, Simon SL, Eklund SM, Chen MC, Klotz LC, Vaccaro KK (1986) DNA probe detection of periodontal pathogens. Oral Microbiol Immunol 1(1):58–62. https://doi.org/10.1111/j.1399-302x.1986.tb00320.x
Furcht C, Eschrich K, Merte K (1996) Detection of Eikenella corrodens and Actinobacillus actinomycetemcomitans by use of the polymerase chain reaction (PCR) in vitro and in subgingival plaque. J Clin Periodontol 23(10):891–897. https://doi.org/10.1111/j.1600-051x.1996.tb00508.x
Gadekar NB, Hosmani JV, Bhat KG, Kotrashetti VS, Nayak RS, Babji DV, Pattanshetty SM, Joshi VM, Bansode RA (2018) Detection of antibodies against Aggregatibacter actinomycetemcomitans in serum and saliva through ELISA in periodontally healthy individuals and individuals with chronic periodontitis. Microb Pathog 125:438–442. https://doi.org/10.1016/j.micpath.2018.10.007
Genco RJ (1981) Antibiotics in the treatment of human periodontal diseases. J Periodontol 52(9):545–558. https://doi.org/10.1902/jop.1981.52.9.545
Goh EXJ, Ong MMA (2019) Anatomical, microbiological, and genetic considerations in treatment of Chinese periodontal patients. J Investig Clin Dent 10(1):e12381. https://doi.org/10.1111/jicd.12381
Greenstein G (1992) Periodontal response to mechanical non-surgical therapy: a review. J Periodontol 63(2):118–130. https://doi.org/10.1902/jop.1992.63.2.118
Gualtero DF, Lafaurie GI, Fontanilla MR (2019) Differential responses of endothelial cells on three-dimensional scaffolds to lipopolysaccharides from periodontopathogens. Mol Oral Microbiol 34(5):183–193. https://doi.org/10.1111/omi.12263
Gürcan Ş, Ünlü S, Kuloğlu F, Karadenizli A, Kuşkucu MA (2016) Microbiological approach to a possible infective endocarditis case caused by Aggregatibacter actinomycetemcomitans. Mikrobiyoloji Bulteni 50(2):315–321. https://doi.org/10.5578/mb.10803
Han EC, Choi SY, Lee Y, Park JW, Hong SH, Lee HJ (2019) Extracellular RNAs in periodontopathogenic outer membrane vesicles promote TNF-α production in human macrophages and cross the blood-brain barrier in mice. FASEB J 33(12):13412–13422. https://doi.org/10.1096/fj.201901575R
Haubek D (2010) The highly leukotoxic JP2 clone of Aggregatibacter actinomycetemcomitans: evolutionary aspects, epidemiology and etiological role in aggressive periodontitis. APMIS Suppl 130:1–53. https://doi.org/10.1111/j.1600-0463.2010.02665.x
Haubek D, Ennibi OK, Abdellaoui L, Benzarti N, Poulsen S (2002) Attachment loss in Moroccan early onset periodontitis patients and infection with the JP2-type of Actinobacillus actinomycetemcomitans. J Clin Periodontol 29(7):657–660. https://doi.org/10.1034/j.1600-051x.2002.290711.x
Haubek D, Ennibi OK, Poulsen K, Poulsen S, Benzarti N, Kilian M (2001) Early-onset periodontitis in Morocco is associated with the highly leukotoxic clone of Actinobacillus actinomycetemcomitans. J Dent Res 80(6):1580–1583. https://doi.org/10.1177/00220345010800062001
Haubek D, Ennibi OK, Poulsen K, Vaeth M, Poulsen S, Kilian M (2008) Risk of aggressive periodontitis in adolescent carriers of the JP2 clone of Aggregatibacter (Actinobacillus) actinomycetemcomitans in Morocco: a prospective longitudinal cohort study. Lancet (London, England) 371(9608):237–242. https://doi.org/10.1016/s0140-6736(08)60135-x

Haubek D, Poulsen K, Kilian M (2007) Microevolution and patterns of dissemination of the JP2 clone of Aggregatibacter (Actinobacillus) actinomycetemcomitans. Infect Immun 75(6):3080–3088. https://doi.org/10.1128/iai.01734-06

Haubek D, Poulsen K, Westergaard J, Dahlèn G, Kilian M (1996) Highly toxic clone of Actinobacillus actinomycetemcomitans in geographically widespread cases of juvenile periodontitis in adolescents of African origin. J Clin Microbiol 34(6):1576–1578. https://doi.org/10.1128/jcm.34.6.1576-1578.1996

Hbibi A, Sikkou K, Khedid K, El Hamzaoui S, Bouziane A, Benazza D (2020) Antimicrobial activity of honey in periodontal disease: a systematic review. J Antimicrob Chemother 75(4):807–826. https://doi.org/10.1093/jac/dkz527

He T, Nishihara T, Demuth DR, Ishikawa I (1999) A novel insertion sequence increases the expression of leukotoxicity in Actinobacillus actinomycetemcomitans clinical isolates. J Periodontol 70(11):1261–1268. https://doi.org/10.1902/jop.1999.70.11.1261

Herrero ER, Slomka V, Bernaerts K, Boon N, Hernandez-Sanabria E, Passoni BB, Quirynen M, Teughels W (2016) Antimicrobial effects of commensal oral species are regulated by environmental factors. J Dent 47:23–33. https://doi.org/10.1016/j.jdent.2016.02.007

Höglund Åberg C, Antonoglou G, Haubek D, Kwamin F, Claesson R, Johansson A (2013) Cytolethal distending toxin in isolates of Aggregatibacter actinomycetemcomitans from Ghanaian adolescents and association with serotype and disease progression. PLoS One 8(6):e65781. https://doi.org/10.1371/journal.pone.0065781

Höglund Åberg C, Haubek D, Kwamin F, Johansson A, Claesson R (2014) Leukotoxic activity of Aggregatibacter actinomycetemcomitans and periodontal attachment loss. PLoS One 9(8):e104095. https://doi.org/10.1371/journal.pone.0104095

Hritz M, Fisher E, Demuth DR (1996) Differential regulation of the leukotoxin operon in highly leukotoxic and minimally leukotoxic strains of Actinobacillus actinomycetemcomitans. Infect Immun 64(7):2724–2729. https://doi.org/10.1128/iai.64.7.2724-2729.1996

Hussain M, Yang A, Yassin M, Arbulu R, Phan T (2020) MALDI-TOF vs. VITEK 2 for identification of Aggregatibacter actinomycetemcomitans chest wall abscess. IDCases 20:e00749. https://doi.org/10.1016/j.idcr.2020.e00749

James HJ, Pfaller MA, Carroll KC, Funke G, Landry ML, Richter SS, Warnock DW (2015) Manual of clinical microbiology. ASM Press, Washington, DC, pp 652–666

Jensen AB, Ennibi OK, Ismaili Z, Poulsen K, Haubek D (2016) The JP2 genotype of Aggregatibacter actinomycetemcomitans and marginal periodontitis in the mixed dentition. J Clin Periodontol 43(1):19–25. https://doi.org/10.1111/jcpe.12486

Jensen AB, Isidor F, Lund M, Væth M, Johansson A, Lauritsen NN, Haubek D (2020) Prevalence of Aggregatibacter actinomycetemcomitans and periodontal findings among 14 to 15-year old danish adolescents: a descriptive cross-sectional study. Pathogens (Basel, Switzerland) 9(12). https://doi.org/10.3390/pathogens9121054

Jentsch H, Cachovan G, Guentsch A, Eickholz P, Pfister W, Eick S (2012) Characterization of Aggregatibacter actinomycetemcomitans strains in periodontitis patients in Germany. Clin Oral Investig 16(6):1589–1597. https://doi.org/10.1007/s00784-012-0672-x

Johansson A, Claesson R, Höglund Åberg C, Haubek D, Lindholm M, Jasim S, Oscarsson J (2019) Genetic profiling of Aggregatibacter actinomycetemcomitans serotype B isolated from periodontitis patients living in Sweden. Pathogens (Basel, Switzerland) 8(3). https://doi.org/10.3390/pathogens8030153

Joshi VM, Bhat KG, Kugaji MS, Shirahatti R (2017) Characterization and serotype distribution of Aggregatibacter actinomycetemcomitans: relationship of serotypes to herpesvirus and periodontal status in Indian subjects. Microb Pathog 110:189–195. https://doi.org/10.1016/j.micpath.2017.06.041

Kaplan JB, Perry MB, MacLean LL, Furgang D, Wilson ME, Fine DH (2001) Structural and genetic analyses of O polysaccharide from Actinobacillus actinomycetemcomitans serotype f. Infect Immun 69(9):5375–5384. https://doi.org/10.1128/iai.69.9.5375-5384.2001

Karched M, Furgang D, Planet PJ, DeSalle R, Fine DH (2012a) Genome sequence of Aggregatibacter actinomycetemcomitans RHAA1, isolated from a rhesus macaque, an Old World primate. J Bacteriol 194(5):1275–1276. https://doi.org/10.1128/jb.06710-11

Karched M, Furgang D, Sawalha N, Fine DH (2012b) Rapid identification of oral isolates of Aggregatibacter actinomycetemcomitans obtained from humans and primates by an ultrafast super convection based polymerase chain reaction. J Microbiol Methods 89(1):71–75. https://doi.org/10.1016/j.mimet.2012.01.016

Kelk P, Abd H, Claesson R, Sandström G, Sjöstedt A, Johansson A (2011) Cellular and molecular response of human macrophages exposed to Aggregatibacter actinomycetemcomitans leukotoxin. Cell Death Dis 2(3):e126. https://doi.org/10.1038/cddis.2011.6

Kelk P, Claesson R, Chen C, Sjöstedt A, Johansson A (2008) IL-1beta secretion induced by Aggregatibacter (Actinobacillus) actinomycetemcomitans is mainly caused by the leukotoxin. Int J Med Microbiol IJMM 298(5–6):529–541. https://doi.org/10.1016/j.ijmm.2007.06.005

Kelk P, Claesson R, Hänström L, Lerner UH, Kalfas S, Johansson A (2005) Abundant secretion of bioactive interleukin-1beta by human macrophages induced by Actinobacillus actinomycetemcomitans leukotoxin. Infect Immun 73(1):453–458. https://doi.org/10.1128/iai.73.1.453-458.2005

Kim KJ, Kim DK, Chung CP, Son S (1992) Longitudinal monitoring for disease progression of localized juvenile periodontitis. J Periodontol 63(10):806–811. https://doi.org/10.1902/jop.1992.63.10.806

Klinger R (1912) Untersuchungen u 'ber menschliche Aktinomykose. Zentralbl Bakteriol 62:191–200

Kolodrubetz D, Phillips L, Burgum A (2010) Repression of aerobic leukotoxin transcription by integration host factor in Aggregatibacter actinomycetemcomitans. Res Microbiol 161(7):541–548. https://doi.org/10.1016/j.resmic.2010.05.001

Komaru Y, Ishioka K, Oda T, Ohtake T, Kobayashi S (2018) Nephritis-associated plasmin receptor (NAPlr) positive glomerulonephritis caused by Aggregatibacter actinomycetemcomitans bacteremia: a case report. Clin Nephrol 90(2):155–160. https://doi.org/10.5414/cn109173

Kraig E, Dailey T, Kolodrubetz D (1990) Nucleotide sequence of the leukotoxin gene from Actinobacillus actinomycetemcomitans: homology to the alpha-hemolysin/leukotoxin gene family. Infect Immun 58(4):920–929. https://doi.org/10.1128/iai.58.4.920-929.1990

Krueger E, Brown AC (2020) Aggregatibacter actinomycetemcomitans leukotoxin: from mechanism to targeted anti-toxin therapeutics. Mol Oral Microbiol 35(3):85–105. https://doi.org/10.1111/omi.12284

Kulik EM, Thurnheer T, Karygianni L, Walter C, Sculean A, Eick S (2019) Antibiotic susceptibility patterns of Aggregatibacter actinomycetemcomitans and Porphyromonas gingivalis strains from different decades. Antibiotics (Basel, Switzerland) 8(4). https://doi.org/10.3390/antibiotics8040253

Kwamin F, Gref R, Haubek D, Johansson A (2012) Interactions of extracts from selected chewing stick sources with Aggregatibacter actinomycetemcomitans. BMC Res Notes 5:203. https://doi.org/10.1186/1756-0500-5-203

Lamell CW, Griffen AL, McClellan DL, Leys EJ (2000) Acquisition and colonization stability of Actinobacillus actinomycetemcomitans and Porphyromonas gingivalis in children. J Clin Microbiol 38(3):1196–1199. https://doi.org/10.1128/jcm.38.3.1196-1199.2000

Lang N, Feres M, Corbet E, Ding Y, Emingil G, Faveri M, Humagain M, Izumi Y, Kamil W, Kemal Y, Mahanonda R, Rajpal J, Sakellari D, Tan WC, Yamazaki K (2015) Group B. Consensus paper. Non-surgical periodontal therapy: mechanical debridement, antimicrobial agents and other modalities. J Int Acad Periodontol 17(1 Suppl):34–36

Lara-Tejero M, Galán JE (2000) A bacterial toxin that controls cell cycle progression as a deoxyribonuclease I-like protein. Science (New York, NY) 290(5490):354–357. https://doi.org/10.1126/science.290.5490.354

Leung WK, Ngai VK, Yau JY, Cheung BP, Tsang PW, Corbet EF (2005) Characterization of Actinobacillus actinomycetemcomitans isolated from young Chinese aggressive periodontitis patients. J Periodontal Res 40(3):258–268. https://doi.org/10.1111/j.1600-0765.2005.00805.x

Lindholm M, Metsäniitty M, Granström E, Oscarsson J (2020) Outer membrane vesicle-mediated serum protection in Aggregatibacter actinomycetemcomitans. J Oral Microbiol 12(1):1747857. https://doi.org/10.1080/20002297.2020.1747857

Listgarten MA, Wong MY, Lai CH (1995) Detection of Actinobacillus actinomycetemcomitans, Porphyromonas gingivalis, and Bacteroides forsythus in an A. actinomycetemcomitans-positive patient population. J Periodontol 66(2):158–164. https://doi.org/10.1902/jop.1995.66.2.158

Lochman J, Zapletalova M, Poskerova H, Izakovicova Holla L, Borilova Linhartova P (2019) Rapid multiplex real-time PCR method for the detection and quantification of selected cariogenic and periodontal bacteria. Diagnostics (Basel, Switzerland) 10(1). https://doi.org/10.3390/diagnostics10010008

López NJ (1992) Clinical, laboratory, and immunological studies of a family with a high prevalence of generalized prepubertal and juvenile periodontitis. J Periodontol 63(5):457–468. https://doi.org/10.1902/jop.1992.63.5.457

Madinier IM, Fosse TB, Hitzig C, Charbit Y, Hannoun LR (1999) Resistance profile survey of 50 periodontal strains of Actinobacillus actinomyectomcomitans. J Periodontol 70(8):888–892. https://doi.org/10.1902/jop.1999.70.8.888

Marin MJ, Ambrosio N, Herrera D, Sanz M, Figuero E (2018) Validation of a multiplex qPCR assay for the identification and quantification of Aggregatibacter actinomycetemcomitans and Porphyromonas gingivalis: in vitro and subgingival plaque samples. Arch Oral Biol 88:47–53. https://doi.org/10.1016/j.archoralbio.2018.01.012

Miller KA, Branco-de-Almeida LS, Wolf S, Hovencamp N, Treloar T, Harrison P, Aukhil I, Gong Y, Shaddox LM (2017) Long-term clinical response to treatment and maintenance of localized aggressive periodontitis: a cohort study. J Clin Periodontol 44(2):158–168. https://doi.org/10.1111/jcpe.12640

Mínguez M, Ennibi OK, Pousa X, Lakhdar L, Abdellaoui L, Sánchez M, Sanz M, Herrera D (2016) Characterization of A. actinomycetemcomitans strains in subgingival samples from periodontitis subjects in Morocco. Clin Oral Investig 20(7):1809–1818. https://doi.org/10.1007/s00784-015-1653-7

Mínguez M, Pousa X, Herrera D, Blasi A, Sánchez MC, León R, Sanz M (2014) Characterization and serotype distribution of Aggregatibacter actinomycetemcomitans isolated from a population of periodontitis patients in Spain. Arch Oral Biol 59(12):1359–1367. https://doi.org/10.1016/j.archoralbio.2014.07.021

Moazzam AA, Rajagopal SM, Sedghizadeh PP, Zada G, Habibian M (2015) Intracranial bacterial infections of oral origin. J Clin Neurosci 22(5):800–806. https://doi.org/10.1016/j.jocn.2014.11.015

Mombelli A, Gmür R, Lang NP, Corbert E, Frey J (1999) Actinobacillus actinomycetemcomitans in Chinese adults. Serotype distribution and analysis of the leukotoxin gene promoter locus. J Clin Periodontol 26(8):505–510. https://doi.org/10.1034/j.1600-051x.1999.260803.x

Montenegro SCL, Retamal-Valdes B, Bueno-Silva B, Duarte PM, Faveri M, Figueiredo LC, Feres M (2020) Do patients with aggressive and chronic periodontitis exhibit specific differences in the subgingival microbial composition? A Syst Rev J Periodontol. https://doi.org/10.1002/jper.19-0586

Müller HP, Heinecke A, Fuhrmann A, Eger T, Zöller L (2001) Intraoral distribution of Actinobacillus actinomycetemcomitans in young adults with minimal periodontal disease. J Periodontal Res 36(2):114–123. https://doi.org/10.1034/j.1600-0765.2001.360208.x

Naqvi AR, Fordham JB, Khan A, Nares S (2014) MicroRNAs responsive to Aggregatibacter actinomycetemcomitans and Porphyromonas gingivalis LPS modulate expression of genes regulating innate immunity in human macrophages. Innate Immun 20(5):540–551. https://doi.org/10.1177/1753425913501914

Narayanan AM, Ramsey MM, Stacy A, Whiteley M (2017) Defining genetic fitness determinants and creating genomic resources for an oral pathogen. Appl Environ Microbiol 83(14). https://doi.org/10.1128/aem.00797-17

Newman MG, Socransky SS, Savitt ED, Propas DA, Crawford A (1976) Studies of the microbiology of periodontosis. J Periodontol 47(7):373–379. https://doi.org/10.1902/jop.1976.47.7.373

Nguyen-Hieu T (2013) Microbial sampling process can change results of microbiological analysis in periodontitis diagnosis. A minireview J Investig Clin Dent 4(3):151–159. https://doi.org/10.1111/jicd.12010

Nørskov-Lauritsen N, Kilian M (2006) Reclassification of Actinobacillus actinomycetemcomitans, Haemophilus aphrophilus, Haemophilus paraphrophilus and Haemophilus segnis as Aggregatibacter actinomycetemcomitans gen. nov., comb. nov., Aggregatibacter aphrophilus comb. nov. and Aggregatibacter segnis comb. nov., and emended description of Aggregatibacter aphrophilus to include V factor-dependent and V factor-independent isolates. Int J Syst Evol Microbiol 56(Pt 9):2135–2146. https://doi.org/10.1099/ijs.0.64207-0

O'Donoghue EJ, Krachler AM (2016) Mechanisms of outer membrane vesicle entry into host cells. Cell Microbiol 18(11):1508–1517. https://doi.org/10.1111/cmi.12655

Paino A, Lohermaa E, Sormunen R, Tuominen H, Korhonen J, Pöllänen MT, Ihalin R (2012) Interleukin-1β is internalised by viable Aggregatibacter actinomycetemcomitans biofilm and locates to the outer edges of nucleoids. Cytokine 60(2):565–574. https://doi.org/10.1016/j.cyto.2012.07.024

Paolantonio M, di Girolamo G, Pedrazzoli V, di Murro C, Picciani C, Catamo G, Cattabriga M, Piccolomini R (1996) Occurrence of Actinobacillus actinomycetemcomitans in patients wearing orthodontic appliances. A cross-sectional study. J Clin Periodontol 23(2):112–118. https://doi.org/10.1111/j.1600-051x.1996.tb00543.x

Papapanou PN, Sanz M, Buduneli N, Dietrich T, Feres M, Fine DH, Flemmig TF, Garcia R, Giannobile WV, Graziani F, Greenwell H, Herrera D, Kao RT, Kebschull M, Kinane DF, Kirkwood KL, Kocher T, Kornman KS, Kumar PS, Loos BG, Machtei E, Meng H, Mombelli A, Needleman I, Offenbacher S, Seymour GJ, Teles R, Tonetti MS (2018) Periodontitis: consensus report of workgroup 2 of the 2017 World Workshop on the classification of periodontal and peri-implant diseases and conditions. J Periodontol 89(Suppl 1):S173–s182. https://doi.org/10.1002/jper.17-0721

Perry MB, MacLean LL, Gmür R, Wilson ME (1996a) Characterization of the O-polysaccharide structure of lipopolysaccharide from Actinobacillus actinomycetemcomitans serotype b. Infect Immun 64(4):1215–1219. https://doi.org/10.1128/iai.64.4.1215-1219.1996

Perry MB, MacLean LM, Brisson JR, Wilson ME (1996b) Structures of the antigenic O-polysaccharides of lipopolysaccharides produced by Actinobacillus actinomycetemcomitans serotypes a, c, d and e. Eur J Biochem 242(3):682–688. https://doi.org/10.1111/j.1432-1033.1996.0682r.x

Piccolomini R, Catamo G, Di Bonaventura G, Picciani C, Paolantonio M (1998) Laboratory and clinical comparison of preservation media and transport conditions for survival of Actinobacillus actinomycetemcomitans. J Med Microbiol 47(8):743–748. https://doi.org/10.1099/00222615-47-8-743

Potts TV, Mitra T, O'Keefe T, Zambon JJ, Genco RJ (1986) Relationships among isolates of oral haemophili as determined by DNA-DNA hybridization. Arch Microbiol 145(2):136–141. https://doi.org/10.1007/bf00446770

Pretzl B, Sälzer S, Ehmke B, Schlagenhauf U, Dannewitz B, Dommisch H, Eickholz P, Jockel-Schneider Y (2019) Administration of systemic antibiotics during non-surgical periodontal therapy-a consensus report. Clin Oral Investig 23(7):3073–3085. https://doi.org/10.1007/s00784-018-2727-0

Rennie RP, Brosnikoff C, Shokoples S, Reller LB, Mirrett S, Janda W, Ristow K, Krilcich A (2008) Multicenter evaluation of the new Vitek 2 Neisseria-Haemophilus identification card. J Clin Microbiol 46(8):2681–2685. https://doi.org/10.1128/jcm.00449-08

Ridwan RD, Sidarningsih KT, Salim S (2018) Effect of lipopolysaccharide derived from surabaya isolates of Actinobacillus actinomycetemcomitans on alveolar bone destruction. Veterinary world 11(2):161–166. https://doi.org/10.14202/vetworld.2018.161-166

Rodenburg JP, van Winkelhoff AJ, Winkel EG, Goené RJ, Abbas F, de Graff J (1990) Occurrence of Bacteroides gingivalis, Bacteroides intermedius and Actinobacillus actinomycetemcomitans in severe periodontitis in relation to age and treatment history. J Clin Periodontol 17(6):392–399. https://doi.org/10.1111/j.1600-051x.1990.tb00036.x

Roman-Torres CV, Aquino DR, Cortelli SC, Franco GC, Dos Santos JG, Corraini P, Holzhausen M, Diniz

MG, Gomez RS, Cortelli JR (2010) Prevalence and distribution of serotype-specific genotypes of Aggregatibacter actinomycetemcomitans in chronic periodontitis Brazilian subjects. Arch Oral Biol 55(3):242–248. https://doi.org/10.1016/j.archoralbio.2010.01.008

Rurenga P, Raangs E, Singadji Z, Wekema-Mulder G, Veloo AC, van Winkelhoff AJ (2013) Evaluation of three selective media for isolation of Aggregatibacter actinomycetemcomitans. J Periodontal Res 48(5):549–552. https://doi.org/10.1111/jre.12037

Sakellari D, Katsikari A, Slini T, Ioannidis I, Konstantinidis A, Arsenakis M (2011) Prevalence and distribution of Aggregatibacter actinomycetemcomitans serotypes and the JP2 clone in a Greek population. J Clin Periodontol 38(2):108–114. https://doi.org/10.1111/j.1600-051X.2010.01649.x

Sanz M, Teughels W (2008) Innovations in non-surgical periodontal therapy: Consensus Report of the Sixth European Workshop on Periodontology. J Clin Periodontol 35(8 Suppl):3–7. https://doi.org/10.1111/j.1600-051X.2008.01256.x

Sharara SL, Tayyar R, Kanafani ZA, Kanj SS (2016) HACEK endocarditis: a review. Expert Rev Anti-Infect Ther 14(6):539–545. https://doi.org/10.1080/14787210.2016.1184085

Sjödin B, Arnrup K, Matsson L, Wranne L, Carlsson J, Hänström L (1995) Periodontal and systemic findings in children with marginal bone loss in the primary dentition. J Clin Periodontol 22(3):214–224. https://doi.org/10.1111/j.1600-051x.1995.tb00137.x

Slots J, Bragd L, Wikström M, Dahlén G (1986) The occurrence of Actinobacillus actinomycetemcomitans, Bacteroides gingivalis and Bacteroides intermedius in destructive periodontal disease in adults. J Clin Periodontol 13(6):570–577. https://doi.org/10.1111/j.1600-051x.1986.tb00849.x

Slots J, Reynolds HS, Genco RJ (1980) Actinobacillus actinomycetemcomitans in human periodontal disease: a cross-sectional microbiological investigation. Infect Immun 29(3):1013–1020

Socransky SS, Haffajee AD (2002) Dental biofilms: difficult therapeutic targets. Periodontology 2000 28:12–55. https://doi.org/10.1034/j.1600-0757.2002.280102.x

Socransky SS, Haffajee AD, Cugini MA, Smith C, Kent RL Jr (1998) Microbial complexes in subgingival plaque. J Clin Periodontol 25(2):134–144. https://doi.org/10.1111/j.1600-051x.1998.tb02419.x

Socransky SS, Haffajee AD, Ximenez-Fyvie LA, Feres M, Mager D (1999) Ecological considerations in the treatment of Actinobacillus actinomycetemcomitans and Porphyromonas gingivalis periodontal infections. Periodontology 2000 20:341–362. https://doi.org/10.1111/j.1600-0757.1999.tb00165.x

Sosroseno W, Bird PS, Seymour GJ (2009) Nitric oxide production by a human osteoblast cell line stimulated with Aggregatibacter actinomycetemcomitans lipopolysaccharide. Oral Microbiol Immunol 24(1):50–55. https://doi.org/10.1111/j.1399-302X.2008.00475.x

Stähli A, Sculean A, Eick S (2020) JP2 genotype of Aggregatibacter actinomycetemcomitans in Caucasian patients: a presentation of two cases. Pathogens (Basel, Switzerland) 9(3). https://doi.org/10.3390/pathogens9030178

Storms I, van den Brand M, Schneeberger P, van't Hullenaar N (2017) Aggregatibacter actinomycetemcomitans pneumonia with chest and abdominal wall involvement. BMJ Case Rep 2017. https://doi.org/10.1136/bcr-2016-217377

Strzempko MN, Simon SL, French CK, Lippke JA, Raia FF, Savitt ED, Vaccaro KK (1987) A cross-reactivity study of whole genomic DNA probes for Haemophilus actinomycetemcomitans, Bacteroides intermedius, and Bacteroides gingivalis. J Dent Res 66(10):1543–1546. https://doi.org/10.1177/00220345870660100601

Suchett-Kaye G, Morrier JJ, Barsotti O (2001) Clinical usefulness of microbiological diagnostic tools in the management of periodontal disease. Res Microbiol 152(7):631–639. https://doi.org/10.1016/s0923-2508(01)01242-6

Suga T, Mitani A, Mogi M, Kikuchi T, Fujimura T, Takeda H, Hishikawa T, Yamamoto G, Hayashi J, Ishihara Y, Noguchi T (2013) Aggregatibacter actinomycetemcomitans lipopolysaccharide stimulated epithelial cells produce interleukin-15 that regulates T cell activation. Arch Oral Biol 58(10):1541–1548. https://doi.org/10.1016/j.archoralbio.2013.06.020

Suprith SS, Setty S, Bhat K, Thakur S (2018) Serotypes of Aggregatibacter actinomycetemcomitans in relation to periodontal status and assessment of leukotoxin in periodontal disease: a clinico-microbiological study. Journal of Indian Society of Periodontology 22(3):201–208. https://doi.org/10.4103/jisp.jisp_36_18

Syed SA, Loesche WJ (1972) Survival of human dental plaque flora in various transport media. Appl Microbiol 24(4):638–644

Tan KS, Woo CH, Ong G, Song KP (2001) Prevalence of Actinobacillus actinomycetemcomitans in an ethnic adult Chinese population. J Clin Periodontol 28(9):886–890. https://doi.org/10.1034/j.1600-051x.2001.028009886.x

Tanner AC, Milgrom PM, Kent R Jr, Mokeem SA, Page RC, Liao SI, Riedy CA, Bruss JB (2002) Similarity of the oral microbiota of pre-school children with that of their caregivers in a population-based study. Oral Microbiol Immunol 17(6):379–387. https://doi.org/10.1034/j.1399-302x.2002.170608.x

Teixeira RE, Mendes EN, Roque de Carvalho MA, Nicoli JR, Farias Lde M, Magalhães PP (2006) Actinobacillus actinomycetemcomitans serotype-specific genotypes and periodontal status in Brazilian subjects. Can J Microbiol 52(3):182–188. https://doi.org/10.1139/w05-121

Teughels W, Kinder Haake S, Sliepen I, Pauwels M, Van Eldere J, Cassiman JJ, Quirynen M (2007) Bacteria interfere with A. actinomycetemcomitans colonization. J Dent Res 86(7):611–617. https://doi.org/10.1177/154405910708600706

Thiha K, Takeuchi Y, Umeda M, Huang Y, Ohnishi M, Ishikawa I (2007) Identification of periodontopathic bacteria in gingival tissue of Japanese periodontitis patients. Oral Microbiol Immunol 22(3):201–207. https://doi.org/10.1111/j.1399-302X.2007.00354.x

Topley WWC, Wilson GS (1929) The principles of bacteriology and Immunity. Pp. 1–587

Tsai CC, Ho YP, Chou YS, Ho KY, Wu YM, Lin YC (2018) Aggregatibacter (Actinobacillus) actimycetemcomitans leukotoxin and human periodontitis – a historic review with emphasis on JP2. Kaohsiung J Med Sci 34(4):186–193. https://doi.org/10.1016/j.kjms.2018.01.014

Tsuzukibashi O, Saito M, Kobayashi T, Umezawa K, Nagahama F, Hiroi T, Hirasawa M, Takada K (2014) A gene cluster for the synthesis of serotype g-specific polysaccharide antigen in Aggregatibacter actinomycetemcomitans. Arch Microbiol 196(4):261–265. https://doi.org/10.1007/s00203-014-0965-3

van der Reijden WA, Bosch-Tijhof CJ, van der Velden U, van Winkelhoff AJ (2008) Java project on periodontal diseases: serotype distribution of Aggregatibacter actinomycetemcomitans and serotype dynamics over an 8-year period. J Clin Periodontol 35(6):487–492. https://doi.org/10.1111/j.1600-051X.2008.01218.x

Van der Velden U, Abbas F, Armand S, Loos BG, Timmerman MF, Van der Weijden GA, Van Winkelhoff AJ, Winkel EG (2006) Java project on periodontal diseases. The natural development of periodontitis: risk factors, risk predictors and risk determinants. J Clin Periodontol 33(8):540–548. https://doi.org/10.1111/j.1600-051X.2006.00953.x

Van der Weijden GA, Timmerman MF, Reijerse E, Wolffe GN, Van Winkelhoff AJ, Van der Velden U (1994) The prevalence of a. actinomycetemcomitans, P. gingivalis and P. intermedia in selected subjects with periodontitis. J Clin Periodontol 21(9):583–588. https://doi.org/10.1111/j.1600-051x.1994.tb00747.x

Van Dyke TE, Sheilesh D (2005) Risk factors for periodontitis. J Int Acad Periodontol 7(1):3–7

van Essche M, Loozen G, Godts C, Boon N, Pauwels M, Quirynen M, Teughels W (2013) Bacterial antagonism against periodontopathogens. J Periodontol 84(6):801–811. https://doi.org/10.1902/jop.2012.120261

van Steenbergen TJ, Timmerman MF, Mikx FH, de Quincey G, van der Weijden GA, van der Velden U, de Graaff J (1996) Discrepancy between culture and DNA probe analysis for the detection of periodontal bacteria. J Clin Periodontol 23(10):955–959. https://doi.org/10.1111/j.1600-051x.1996.tb00518.x

Van Winkelhoff AJ, Boutaga K (2005) Transmission of periodontal bacteria and models of infection. J Clin Periodontol 32(Suppl 6):16–27. https://doi.org/10.1111/j.1600-051X.2005.00805.x

Vega BA, Belinka BA Jr, Kachlany SC (2019) Aggregatibacter actinomycetemcomitans leukotoxin (LtxA; Leukothera(®)): mechanisms of action and therapeutic applications. Toxins 11(9). https://doi.org/10.3390/toxins11090489

Webb JN, Koufos E, Brown AC (2016) Inhibition of bacterial toxin activity by the nuclear stain, DRAQ5™. J Membr Biol 249(4):503–511. https://doi.org/10.1007/s00232-016-9892-3

Yang HW, Asikainen S, Doğan B, Suda R, Lai CH (2004) Relationship of Actinobacillus actinomycetemcomitans serotype b to aggressive periodontitis: frequency in pure cultured isolates. J Periodontol 75(4):592–599. https://doi.org/10.1902/jop.2004.75.4.592

Yoshida Y, Suzuki N, Nakano Y, Shibuya K, Ogawa Y, Koga T (2003) Distribution of Actinobacillus actinomycetemcomitans serotypes and Porphyromonas gingivalis in Japanese adults. Oral Microbiol Immunol 18(3):135–139. https://doi.org/10.1034/j.1399-302x.2003.00034.x

Zambon JJ, Haraszthy VI, Hariharan G, Lally ET, Demuth DR (1996) The microbiology of early-onset periodontitis: Association of highly toxic Actinobacillus actinomycetemcomitans strains with localized juvenile periodontitis. J Periodontol 67(Suppl 3S):282–290. https://doi.org/10.1902/jop.1996.67.3s.282

Zijlstra EE, Swart GR, Godfroy FJ, Degener JE (1992) Pericarditis, pneumonia and brain abscess due to a combined Actinomyces—Actinobacillus actinomycetemcomitans infection. J Infect 25(1):83–87. https://doi.org/10.1016/0163-4453(92)93633-2

Meta-analyses on the Periodontal Archaeome

4

Jéssica Alves de Cena, Yuri Silvestre-Barbosa, Aline Belmok, Cristine Miron Stefani, Cynthia Maria Kyaw, and Nailê Damé-Teixeira

Abstract

Recently, we have published a scoping review on the oral archaeome, showing that these microorganisms inhabit various oral niches, including periodontal sites. In order to reinforce the importance of the *Archaea* domain and alert the scientific community about the importance of inter-domain relationships in oral dysbiosis, we have performed meta-analyses evaluating the prevalence of archaea in periodontal diseases (PROSPERO protocol: CRD42020213109). A systematic search in the literature was conducted in several databases and in grey literature, retrieving 30 reports on periodontal archaeome, published from 1980 to 2020. The methodological quality of included studies and the certainty of evidence were evaluated by using validated tools. Most studies focused on the detection of methanogens, revealing that the diversity of the periodontal archaeome is currently underestimated. Two meta-analyses concluded that individuals with periodontitis are prone to have archaeal-positive subgingival biofilms when compared to periodontally healthy individuals (OR 6.68, 95% CI 4.74–9.41 for 16S rRNA gene analysis and OR 9.42, 95% CI 2.54–34.91 for *mcrA* gene analysis). Despite the archaeal enrichment in sites with periodontitis, less than half of the individuals with periodontitis tested positive for archaeal DNA (general estimative of 46%; 95% CI 36–56%). Conventional treatment for periodontitis reduced the archaeal population, but systemic antibiotics used as adjunctive therapy did not increase its effectiveness. Hence, it could conceivably be hypothesised that archaea are secondary colonizers of areas with dysbiosis, probably flourishing in the inflammatory environment. Due to their lower prevalence, archaeal cells are probably underestimated by the current detection protocols. It may also be speculated that archaea do not have a single central role in the infection, with bacterial cells directly involved in that role. New studies are necessary, with different methodological approaches, to explore the underestimated diversity of the oral archaeome.

Keywords

Oral microbiology · *Archaea* domain · Periodontal diseases

J. A. de Cena · Y. Silvestre-Barbosa · C. M. Stefani
N. Damé-Teixeira (✉)
Department of Dentistry, School of Health Sciences, University of Brasilia, Brasilia, Brazil
e-mail: jessica.a.cena@gmail.com; yurisilver5@gmail.com; cmstefani@unb.br; nailedame@unb.br

A. Belmok · C. M. Kyaw
Department of Cell Biology, Institute of Biology, University of Brasilia, Brasilia, Brazil
e-mail: abelmokadias@gmail.com; cmkyaw@gmail.com

J. Santi-Rocca (ed.), *Periodontitis*, Advances in Experimental Medicine and Biology 1373,
https://doi.org/10.1007/978-3-030-96881-6_4

Highlights

- Most studies of the periodontal archaeome focused on the detection of methanogens, resulting in an underestimation of the periodontal archaeome diversity
- Meta-analysis showed a positive association between periodontal disease and the presence of archaea
- 46% of the individuals with periodontitis tested positive for archaeal DNA, although this estimative should be underestimated and highly impacted by the methodological limitations for identifying archaea

Considerations for Practice

- Archaea are prevalent in periodontitis and their role in periodontitis development and progression is still lacking
- Adjunctive therapy with systemic antibiotics may not reduce archaeal load after periodontal therapy
- More studies employing methodological approaches directed to a broader archaeal diversity detection are necessary to confirm the role of archaeal cells on the periodontal archaeome

Patient Summary

The presence of archaea in human sites was recognized in the beginning of the 20th century. At that time, it was believed that methanogens colonized only the human gastrointestinal tract, but later it was demonstrated that these archaea are also found on the oral cavity. Later, other archaeal phyla were also detected on humans. In this review we gathered all available data from the literature, and performed meta-analyses. It could be hypothesized that archaea may act as secondary pathogens in areas in dysbiosis and be favoured in the inflammatory environment. Their higher detection in different pathological conditions, when compared to healthy sites suggests a putative inter-domain interaction between different archaeal and bacterial species. More studies, employing methodological approaches directed to a broader archaeal diversity detection are necessary to confirm the role of archaeal cells on the periodontal archaeome.

4.1 Introduction

Over the past few years, a putative contribution of *Archaea* in periodontal disorders has been suggested, even though these microorganisms were shown to represent a small fraction of the total prokaryotic load in oral samples (Horz and Conrads 2011). The proposal of the *Archaea* domain is relatively recent and was based on the unique phylogenetic, structural, physiological, and molecular features of these prokaryotes, which clearly distinguish them from bacteria (Woese and Fox 1977).

Although initially associated to inhospitable habitats, since the first recognized *Archaea* members mostly thrived in extremely hot, salty, or acidic conditions (Woese and Fox 1977; Magrum et al. 1978), the increased use of culture-independent methods, particularly the detection of markers genes such as the 16S rDNA directly from environmental and clinical samples, revealed the ubiquity of these organisms (DeLong 1998). The presence of *Archaea* in the human microbiome has been increasingly reported, with members already detected in gastrointestinal and respiratory tracts, skin, and oral cavity samples (Matarazzo et al. 2012; Koskinen et al. 2017; Moissl-Eichinger et al. 2017; Wagner Mackenzie et al. 2020).

Archaea were first identified in the human oral cavity in 1987, through the enrichment of methanogens from subgingival plaque samples (Brusa et al. 1987). Since then, the knowledge about the oral archaeome has been significantly expanded. A scoping review recently published by our group revealed that *Archaea* members have been described on various oral niches, including saliva, different portions of the tongue, periodontal and endodontic sites (Belmok et al. 2020). Furthermore, the putative involvement of archaea in oral diseases, especially periodontitis, is being increasingly pointed out, with studies suggesting that, although these organisms are part of the indigenous oral microbiota, their abundance rises along with the increase of periodontitis severity (Li et al. 2014; Grine et al. 2018; Gohler et al. 2018; Belmok et al. 2020).

The contribution of archaea to periodontal dysbiosis may be related to their ability to provide conditions for the growth of pathogenic bacteria in subgingival sites (Lepp et al. 2004). Methanogens are the most representative archaeal group in periodontal samples, where they may play a central role in the removal of hydrogen excess from different electron donors, such as methanol, methylamine, acetate, ethanol or formate. Methanogens metabolize the H_2 generated during the fermentation of carbohydrates in methane gas, raising the local pH and rendering the microenvironment favourable to the anaerobic subgingival bacterial population, including opportunistic pathogens. This kind of event represents a classical syntrophic behaviour, with cross-feeding among prokaryotic species within the anaerobic microniches in a mature biofilm (Horz and Conrads 2010). Thus, it can be hypothesized that the increase in methanogenic activity leads to a reduction in the availability/concentration of acidic products, acting as "acid sinks", and consequently help the homeostasis maintenance for non-aciduric organisms growth, also allowing the enrichment of the proteolytic organisms (Matarazzo et al. 2011).

Due to this environmental change mediated by inter-domain interactions, some methanogenic species have been considered putative keystone pathogens of periodontal diseases (Vianna et al. 2009; Matarazzo et al. 2012). However, even though the role of methanogens in oral health is becoming increasingly clear, recent molecular findings suggest that the diversity of archaea inhabiting human oral sites may be currently underestimated. DNA sequences affiliated to the non-methanogenic class *Thermoplasmata* have been detected in patients with different stages of periodontitis (Li et al. 2009; Horz et al. 2012; Li et al. 2014). Moreover, metatranscriptomic analyses performed in periodontal pocket samples from individuals with chronic periodontitis and periodontally healthy subjects revealed that only five out of ten archaeal reads were classified in methanogenic groups, with transcripts associated to non-methanogenic *Euryarchaeota*, and ammonia-oxidizing *Thaumarchaeota* taxa (Deng et al. 2017). Recently, 16S rDNA sequences of ammonia-oxidizing thaumarchaeal groups have also been identified in both supragingival and carious biofilms (Dame-Teixeira et al. 2020).

Understanding the context of archaea within oral biofilms and how their presence could modulate oral conditions is imperative. Although oral archaeome is an increasingly discussed research topic, to date no consensus regarding their role in oral diseases has been reached (Belmok et al. 2020). To shed some light on the periodontal archaeome comprehension, we performed meta-analyses on the prevalence of these organisms in periodontitis sites comparing it with healthy sites, as well as after periodontitis treatments. Some hypotheses on the role of archaeal species that comprise the periodontal archaeome were also discussed.

4.2 Results

4.2.1 Narrative Synthesis and the Quality Assessment of Individual Studies

The inclusion of other databases revealed three new studies, resulting in a total of 30 studies that were used in the meta-analyses of the periodontal archaeome (28 cross-sectional; 2 randomized clinical trials). Those studies comprised 1312 individuals with periodontitis, and 441 periodontally healthy individuals sampled for archaeal detection.

Table 4.1 shows the results of the quality assessment of individual studies, type and method of sample collection, and DNA extraction method

Table 4.1 The quality of individual study, sampling, and DNA extraction method of the periodontal archaeome studies (*NA* not available)

Author, year	Quality assessment of individual studies	Type of sample (sample collection method)	DNA extraction method
Aleksandrowicz, 2020	+	Subgingival biofilm samples (curettes)	Genomic Mini kit (A&A Biotechnology)
Ashok, 2013	+	Subgingival biofilm samples: periodontitis (curette); healthy controls = gingival crevices.	Chemical lysis with Tween 20, Nonidet and proteinase K protocol.
Belay, 1988	-	Subgingival plaque (curette).	**NA**
Bringuier, 2013	+	Subgingival plaque from 3- to 12-mm periodontal pockets.	FastPrep-24 incubation with acid-washed beads, followed by extraction with Qiagen EZ1 DNA tissue kit (Qiagen, France).
Brusa, 1987	+	Pool of subgingival plaque from gingival crevice (curettes).	**NA**
Brusa, 1993	-	Dental biofilm pools from pockets 3 to 7 mm deep, one in each quadrant (curettes).	
Dabdoub, 2016	+	Healthy subjects = pools from 15 mesial sites on teeth (paper-points). Periodontitis = pools of subgingival plaque from four nonadjacent proximal sites were collected using 15 paper points.	Qiagen DNA MiniAmp kit (Qiagen, Valencia, CA, USA)
Deng, 2017	-	Several sites (two paper points per site).	According to Szafranski (2015).
Fermiano, 2011	++	Periodontitis = subgingival biofilm from 9 interproximal sites (curette); Healthy = 9 sites (curette).	Manual extraction.
Ferrari, 1994	-	Subgingival samples (curettes), according to Socransky et al.	**NA**
Göhler, 2014	++	Tongue biofilm from the middle third of the tongue dorsum with a sterile spatula.	MagNA Pure LC platform.
Göhler, 2018	++	Tongue biofilm from the middle third of the tongue dorsum with a sterile spatula; Subgingival plaque from the mesiobuccal pocket of the most distally located, clinically examined, upper tooth in the periodontally examined quadrants (paper points were inserted until the pocket base for 10 seconds).	
Horz, 2012	-	Pool of subgingival plaque from the four deepest periodontal pockets (paper points).	QIAamp DNA Mini kit ("tissue protocol", Qiagen, Germany).
Horz, 2015	-	According to Horz (2012).	
Huynh, 2015a	+	**NA**	FastPrep-24 incubation with acid-washed beads, followed by extraction with Qiagen EZ1 DNA tissue kit (Qiagen, France).
Huynh, 2015b	+	Subgingival dental plaque = all periodontal pockets (curette).	**NA**
Huynh, 2017	++	Subgingival dental plaque = all periodontal pockets (curette).	
Kulik, 2001	+	Pool of plaque samples.	Phenol-chlorophorm.
Lepp, 2004	+	Subgingival plaque samples from 6–12 periodontal pockets (curettes).	
Li, 2009	+	Subgingival samples from periodontal pockets with a probing depth >4 mm (curettes).	
Li, 2014	+	Subgingival plaque sample from the bottom of the deepest pocket (curette).	Genomic DNA Extraction Kit (Tiangen, Beijing, China).
Lira, 2013	++; **	Subgingival plaque (curettes).	
Matarazzo, 2011	+	Subgingival samples (curette), after removing supragingival biofilm.	Qiamp DNA minikit (Qiagen, Germany).
Ramiro, 2018	++; **	Subgingival plaque (curettes) collected after supragingival plaque removal -- baseline and at 6 months post-SRP from 6 non-contiguous interproximal sites per subject.	Masterpure RNA & DNA purification kit (Epicentre).
Robichaux, 2003a	-	Subgingival plaque from interproximal sites with pocket depths of 1-4 mm, after a removal of all supragingival plaque and isolation with cotton wool (curette).	**NA**
Robichaux, 2003b	-	Tissue samples (curette).	**NA**
Sogodogo, 2019	+	Subgingival plaque samples (curettes). Swabs from the abscesses.	Tissue DNA Kit.
Vianna, 2008	++	Pools of subgingival plaque samples from the four deepest periodontal pockets (paper points). Healthy subjects = pools of plaque from vestibular sulcus of first molars from all quadrants (paper points).	Qiamp DNA Mini kit (with modifications: addition of zirconia-silica beads and FastPrep before proteinase K).
Vianna, 2009	+	Pools of subgingival plaque from the four deepest periodontal pockets (paper points).	Qiamp DNA minikit (Qiagen, Germany).
Yamabe, 2008	+	Plaque samples from the periodontal pockets (paper points).	InstaGene Matrix (Bio-Rad).

(-) low quality of individual study; (+) moderate quality; (++): high quality; (**) RCT studies analyzed with RoB 2 of the selected reports. Low methodological approaches were detected in seven studies. However, eight were classified with a high quality and fifteen with a moderate quality.

In most studies, samples of subgingival biofilms were collected using sterile curettes, while in others the crevicular fluid was collected using endodontic paper points, corresponding to pooled or specific sites samples. Unquestionably, this approach may not represent the actual microbial content, as this sampling technique collects mainly the fluids surrounding the biofilm, while scraping the subgingival biofilm with a curette displaces the sticky biofilm from both, the periodontal pocket and the tooth. Since several microbial species, mostly the anaerobic ones, grow inside the biofilm, the paper point sampling method will retrieve preferentially the organisms present in the biofilm surface, leaving behind important members of the subgingival microbiome.

Various DNA extraction kits and protocols were employed in the different studies. The use of commercial kits is considered a faster and easier alternative when high quality nucleic acids are desired, especially when samples containing a low number of organisms are processed (Ghavami et al. 2015). It is worth mentioning that this step may hamper the analyses of archaeal DNA (Pausan et al. 2019; Mahnert et al. 2018), since the current DNA extraction protocols have been established and optimized for bacterial cells, and do not consider the peculiarities of the archaeal cell wall (Bang and Schmitz 2015). Therefore, a more realistic view of the diversity and quantity of archaea associated with the human microbiome would greatly benefit from improvements and standardizations of archaeal nucleic acids extraction methods.

In this review, only one study (Aleksandrowicz et al. 2020) adopted the current classification established by the American Academy of Periodontology and the European Federation of Periodontology in 2018, which proposes a multidimensional periodontitis staging and grading system (Papapanou et al. 2018). It is important to point out that most of the included studies were performed from 1999 to 2018, and adopted the classification proposed by Armitage (Armitage 1999) that assessed, at the time, two different conditions: chronic (Ashok et al. 2013; Bringuier et al. 2013; Dabdoub et al. 2016; Deng et al. 2017; Fermiano et al. 2011; Horz and Conrads 2010; Horz et al. 2012; Horz et al. 2015; Huynh et al. 2015a, 2017; Lepp et al. 2004; Li et al. 2009; Li et al. 2014; Ramiro et al. 2018; Sogodogo et al. 2019; Vianna et al. 2008, 2009; Yamabe et al. 2008) and aggressive periodontitis (Fermiano et al. 2011; Lira et al. 2013; Matarazzo et al. 2011; Yamabe et al. 2008). New insights about these conditions have proposed to be part of the same disease group. This is justified, among other factors, by the common end result presented from a pathophysiologic point of view and the similar metabolic end-products stemming from microbial complex (Fine et al. 2018). On this basis, this review coupled those results in the meta-analyses regardless the periodontitis classification (shown below). Besides, another study followed the classification proposed by the World Workshop in Clinical Periodontics in 1989 (AAP 1989) and classified the participants as juvenile, adult, rapidly progressing and refractory periodontitis (Kulik et al. 2001), conditions which were not differentiated in this study as well. Lastly, a glossary from American Dental Association (AAP 1977) was referred as the criteria for diagnosis in three studies (Belay et al. 1988; Robichaux et al. 2003a, b) (Table 4.2).

The first studies describing the detection of methanogenic archaea in periodontal sites employed culture-based methods. Afterwards, culture-independent studies based on PCR employed several pairs of primers directed to the archaeal 16S rRNA genes, as well as primers targeting other specific housekeeping genes, such as *mcrA* (Methyl-Coenzyme M reductase, found in methanogens) and *cnp60* (heat shock chaperone 60) from *Methanobrevibacter oralis* have become more frequent. Recurrently, more than one target gene was used to assess the archaeal diversity in the same sample. Also, other approaches, such as fluorescence in situ hybridization (FISH), real-time quantitative PCR (RT-qPCR), and less frequently, metagenomics and metatranscriptomics, were used to estimate the proportion of archaeal-

Table 4.2 Periodontal diseases classification across the selected studies (n = 30)

Author, year	Periodontitis classification
Aleksandrowicz et al. (2020)	New Classification of the American Academy of Periodontology.
Huynh et al. (2015a), Lira et al. (2013) and Ramiro et al. (2018)	American Academy of Periodontology.
Ashok et al. (2013)	Healthy sites: sulcus (PD ≤ 3 mm), shallow pocket (4 mm ≤ PD ≤ 5 mm) and deep pocket (PD ≥ 6 mm).
Belay et al. (1988), Robichaux et al. (2003a) and Robichaux et al. (2003b)	According to the criteria of the American Dental Association.
Bringuier et al. (2013)	Chandra (2007), using BOP, PD, plaque index, and radiographs.
Brusa (1987)	According to Socransky et al. (1963).
Dabdoub et al. (2016)	Attachment loss ≥5, PD ≥5, gingival index > 1 in 30% or more sites.
Deng et al. (2017)	According to Szafranski et al. (2015).
Fermiano et al. (2011)	PCrG Group: have at least 20 teeth, excluding third molars, 30% of the sites with PS and NCI ≥ 5mm and age ≥ 30 yearsPAgG Group: have at least 20 teeth, excluding third molars, have at least 6 teeth with at least 1 interproximal site presenting non-contiguous PD and NCI ≥ 5mm, located in the region of incisors and molars and another 6 teeth with the same clinical characteristics located in other groups of teeth and age <30 yearsSP Group: have at least 20 teeth, excluding third molars; not presenting sites with OS and / or NCI ≥ 4mm; no clinical signs of generalized gingivitis. Do not present more than 10% of the sites with BOP and age ≥ 18 years.
Göhler et al. (2014)	Mean clinical attachment levels and mean probing depths ranged within the highest quartile calculated separately within sex and 5-year-age categories.
Göhler et al. (2018)	PD and clinical attachment levels = four sites per tooth (disto/mid/ mesiobuccal and midlingual) according to the half-mouth method, alternating on the left or right side, excluding third molars.
Horz et al. (2012)	Different stages of severity of chronic periodontitis.
Horz et al. (2015)	PD and clinical attachment level recorded at six sites per tooth.
Huynh et al. (2015b)	According to Chandra (2007).
Huynh et al. (2017)	Generalized dental calculus, generalized BOP and pockets with a depth of 7 mm in tooth 38, 6 mm in teeth 16 and 27 and 5 mm in teeth 16, 15, 13, 12, 25, 26, 38, 37, 44 and 47. Radiography showed bone loss along the apex of 16 and up to the third center of 15 and 13–27.
Kulik et al. (2001)	Followed the classification proposed by the World Workshop in Clinical Periodontics in 1989 (American Academy of Periodontology, 1989) and classified the participants as juvenile, adult, rapidly progressing and refractory periodontitis.
Lepp et al. (2004)	Clinical attachment loss to the nearest millimeter at the mesiobuccal, buccal, distobuccal, mesiolingual, lingual, and distolingual sites around each tooth.
Li et al. (2009)	Based on radiographic evidence, with an absence of periapical periodontitis and other oral soft tissue diseases.
Li et al. (2014)	Presence of BOP, suppuration and visible plaques, PD, and clinical attachment loss.
Matarazzo et al. (2011)	According to Faveri et al. (2011) = presence to disc-shaped bone defects of >3 mm, a PD of >5 mm and an inflamed mucosa showing BOP and / or suppuration.
Sogodogo et al. (2019)	According to Armitage et al. (1999).
Vianna et al. (2008), Vianna et al. (2009) and Yamabe et al. (2008)	Healthy sites = PD <3 mm and no BOP; Periodontitis = PD ≥4 mm.

(continued)

Table 4.2 (continued)

Author, year	Periodontitis classification
Brusa et al. (1993) and Ferrari et al. (1994)	**NA**

PD probing depth, *NCI* clinical insertion level, *BOP* Bleeding on probing, *NA* not available

positive subgingival biofilms from individuals with and without periodontitis (Table 4.3).

4.2.2 The Taxonomy of the *Archaea* domain in Periodontal Sites

As described before, several approaches were applied to detect archaeal cells or DNA in periodontal niches. However, most investigations are focused on protocols which were directed preferably to the detection of methanogenic archaea. Consequently, the most frequently reported archaeal taxa in human periodontal samples are *Methanobrevibacter* species, particularly *M. oralis* (Yamabe et al. 2008; Vianna et al. 2009; Matarazzo et al. 2011). However, sequences affiliated to *Thermoplasmata*, a class of *Euryarchaeota*, were also found when species-specific primers were used (Li et al. 2009; Horz et al. 2012). Interestingly, previously described organisms belonging to this class are usually acidophiles, growing optimally at pH below 2 and lacking a cell wall (Ahn et al. 2012). In this case, it is more reasonable to assume that this organism could thrive on supragingival biofilms, where an acidic environment is habitual. Thus, physiological characteristics associated with *Thermoplamata* phylotypes encountered in subgingival niches must be further explored to confirm this taxonomy.

4.2.3 The Prevalence of *Archaea* in Periodontal Sites and Its Association with Periodontitis

Although the differences in both sampling and methodologies employed for the detection of archaea led to divergent data, it was possible to meta-analyse the results from these studies, and, not surprisingly, all revealed higher proportions of individuals with periodontitis positive for archaea than periodontally healthy ones. The data of approximately 1000 individuals with different degrees of periodontitis from 24 studies reporting the number of periodontally-diseased individuals with subgingival biofilms positive for archaea were grouped, regardless the method to detect archaea, reaching a general estimative of 46% of individuals positive for archaea (95%CI 36%–56%; I^2 = 94%) (Fig. 4.1). Sensitivity analysis was performed, removing the smaller samples studies (Ashok et al. 2013; Bringuier et al. 2013; Fermiano et al. 2011; Lepp et al. 2004; Matarazzo et al. 2011; Yamabe et al. 2008), but the same tendency persisted (pooled prevalence 39%, 95% CI 0.28–0.52, I^2 = 94%). Data from the baseline of the longitudinal studies were used.

In healthy sites, some divergences have been noticed. While some studies have not detected archaeal DNA in subgingival biofilm samples from periodontally healthy subjects (Li et al. 2009; Li et al. 2014; Yamabe et al. 2008; Horz et al. 2015), others described archaea as a common member of oral samples from individuals without periodontitis (Faveri et al. 2011; Ashok et al. 2013; Grine et al. 2018; Brusa et al. 1993). Prevalence of archaea in periodontally health individuals was 13% (95%CI 0.02–0.28, I^2 = 92%). When periodontally healthy and diseased individuals were compared, a significant positive association between periodontitis and the detection of archaea in subgingival biofilms was found. For that, two meta-analyses were performed including studies based on culture-independent methods (Fig. 4.2a = 16S rRNA gene – 10 studies; 2B = *cnp*60 gene – 2 studies). The first meta-analysis was sub-grouped according to the pair of primers used to amplify the 16S rRNA gene. Individuals with periodontitis were 6.68 fold (95% CI = 4.74–9.41) and 9.42 fold (95% CI = 2.54–34.91) more likely to test positive for archaea than the periodontally healthy individuals, as shown in Figs. 4.2a, b, respectively. The results were consistent in terms of magnitude and direction, despite the high v in

Table 4.3 Sample size, methods of archaea detection, and results of the selected studies (n = 30)

Author, year	N periodontal diseases group / type of periodontal disease	N control group	Methods of archaea detection	Proportion of individuals with archaea periodontal diseases group / or Relative Abundance (RA)	Proportion of individuals with archaea control group / or Relative Abundance (RA)
Aleksandrowicz et al. (2020)	61 = periodontitis	NA	Sequencing of the 16S rRNA gene.	Mild periodontitis = 53%; Moderate/advanced periodontitis = 64%.	None.
Ashok et al. (2013)	34 = chronic periodontitis	16	PCR for 16S rRNA gene.	Chronic periodontitis = 29.41%.	12.50% (2/12).
Belay et al. (1988)	36 = some degree of periodontal disease	**NA**	Enrichment cultures in anaerobic conditions; methane detection by gas-chromatographic analysis, transmission electron microscopy and antigenic fingerprinting.	Methanogenic = 9 patients; Moderate = 26%; advanced-periodontitis = 76%; Early periodontitis (7%).	7%.
Bringuier et al. (2013)	22 = periodontitis	10	RT-qPCR for *cnp60* gene with *M. oralis*-specific and probe.	*M. oralis* = 54% individuals, *M. oralis*-negative = 100% individuals with a clinical score of <20; and positive = 63% patients with a clinical score of >20.	*M. oralis* = 30%.
Brusa et al. (1987)	10 = periodontitis	**NA**	Enrichment cultures in anaerobic conditions; methane detection by gas-chromatographic analysis; fluorescence microscopy.	Methane production and observation of methanogens in 3 of the 10 cultures.	**NA**
Brusa et al. (1993)	20 = periodontitis	**NA**	Enrichment cultures in anaerobic conditions; methane detection by gas-chromatographic analysis; fluorescence microscopy; methanogenic archaea counts.	45% individuals.	**NA**
Dabdoub et al. (2016)	25 = generalized moderate to severe chronic periodontitis	25	Metagenome (Illumina MiSeq).	RA ≤ 0.003	**NA**

Deng et al. (2017)	4 = chronic periodontitis	10	Metatranscriptome.	RA 0.11% = periodontitis; transcripts of archaea = 0.22% of total putative mRNA reads in the metatranscriptome. The 10 most abundant archaea accounted for more than 72% of archaea reads on average. *Methanosarcina vacuolata* was the most abundant archaeal species in periodontitis for 21.5% of total archaeal sequences.	RA = 0.27%; *M. vacuolata* was the most abundant archaeal species = 62.5% of total archaeal sequences.
Fermiano et al. (2011)	15 = generalized chronic periodontitis (GCrP); 15 = generalized aggressive periodontitis (GAgP)	15	PCR for 16S rRNA gene with euryarchaeal primers.	SP Group: 12 (80%).	PCrG Group: 11 (73.3%); PAgG Group: 9 (60%).
Ferrari et al. (1994)	**NA**	10	Enrichment cultures in anaerobic conditions; methane detection by gas-chromatographic analysis; fluorescence microscopy; physiological tests and DNA dot plot hybridization.	**NA**	Methane production = 40%.
Göhler et al. (2014)	88 = periodontitis	88	RT-qPCR for 16S rRNA.	23 (26.1%).	17 (19.3%).
Göhler et al. 2018	237 samples	**NA**	RT-qPCR for 16S rRNA.	26 (65.0%) / RA: 2.63%.	**NA**
Horz et al. (2012)	30 = different stages of severity of chronic periodontitis	**NA**	PCR for *mcrA* gene with two primer sets for all methanogens; three Thermoplamatales-clade; PCR for 16S rRNA gene with three Thermoplasmatales-clade; RT-qPCR for *mcrA* gene.	16S rRNA and MM_Mx3Fw_/MM_Mx3_Rv = 3/30 samples; RT-qPCR: RA 0.5% of Thermoplamatales; 2/3 samples with *M. oris*.	**NA**

(continued)

Table 4.3 (continued)

Author, year	N periodontal diseases group / type of periodontal disease	N control group	Methods of archaea detection	Proportion of individuals with archaea periodontal diseases group / or Relative Abundance (RA)	Proportion of individuals with archaea control group / or Relative Abundance (RA)
Horz et al. (2015)	125 = various stages of chronic periodontitis	25	PCR pre-amplification for 16S rRNA gene with universal archaeal primers; RT-qPCR for *M. oralis*; sequencing.	56 (45%) positive for methanogens; Methanogens RA = 5.1%.	None.
Huynh et al. (2015a)	17 = chronic periodontitis	**NA**	Enrichment cultures in anaerobic conditions; methane detection by gas-chromatographic analysis; detection of *M. oralis* in methane positive cultures by RT-qPCR for *cnp60* gene with *M. oralis*; multispacer sequence typing – PCR for intergenic spacers of *M. oralis*.	MST4 was the most frequent genotype, which was found in 4/17 isolates in two individuals with generalised severe periodontitis;17 clinical isolates exhibited different genotypes from that of MST1 of the reference *M. oralis* DSM 7256; MST9 was found in 1/17 isolates and was obtained from a patient with generalized severe periodontitis; two individuals harbored three different *M. oralis* genotypes.	**NA**
Huynh et al. (2015b)	65 = periodontitis	15	Enrichment cultures in anaerobic conditions; methane detection by gas-chromatographic analysis; detection of *M. oralis* in methane positive cultures by RT-qPCR for *cnp60* gene with *M. oralis*; multispacer sequence typing – PCR for intergenic spacers of *M. oralis*.	36/65 (55.38%) positive for methane; Positive PCR amplification in 31/36 (86.11%) methane-producing;Cultures= *M. smithii* (n = 2 subjects); *Methanobrevibacter* sp. strain N13 (n = 3 subjects with severe periodontitis.	Control group methane detection = 1/15 (6.67%).Cultures = *M. oralis* (n = 1 subject).

Huynh et al. (2017)	1 = chronic, severe generalized periodontitis	**NA**	Same as Huynh et al. 2015b.	An 18-month culture yielded a co-culture between a new archaea species (propesed to be called "Methanobrevibacter massiliense"), identified by 16S rRNA and mcrA gene sequencng, and Pyramidobacter piscolens, identified by 16S rRNA gene sequencing	**NA**
Kulik et al. (2001)	48 (37 with periodontitis; 8 with rapidly progressing periodontitis; 1 with localized juvenil periodontitis; 1 with refractory periodontitis; 1 with epilepsy and periodontitis)	**NA**	PCR for 16S rRNA gene with euryarchaeal primers; Hybridization of PCR products in agarose gels with probe ARCH 915; cloning, Sanger sequencing.	Positive amplification and probe hybridization = 37 (77%).	**NA**
Lepp et al. (2004)	50 = periodontitis	8	PCR for 16S rRNA gene and RT-PCR for 16S rRNA gene; Cloning, Sanger sequencing; FISH (cloned amplified 16S rDNA).	Archaeal SSU rDNA: 36% = periodontitis patients (76.6% sites); RA in relation to total prokaryotic SSU rDNA > severe and moderate periodontitis; < RA post vs. before treatment; FISH = [18.5% (≥6 mm); 7.2% (4–5 mm); 0.4% (2–3 mm)].	None.
Li et al. (2009)	41 = chronic periodontitis	15	PCR for 16S rRNA gene with 3 different primers; Cloning, Sanger sequencing.	Primer set 1 = 31 (73.2%); Primer set 2 = (70.7%).	None.
Li et al. (2014)	49 = chronic periodontitis	45	PCR for 16S rRNA gene with universal archaeal primers; Nested PCR for 16S rRNA (including Thermoplasmata specific primers); RT-PCR for 16S rRNA.	Universal archaeal primers: 34 samples; 49 subjects. Prevalence of archaea phylotypes = 69.4%.**Nested PCR**: Thermoplasmata in 9 samples (prevalence = 18.4%);**RT-PCR**: 87.8%; **RA** = 0.01% to 7.53%, with a median value of 0.459% and an average value of 0.91%; Thermoplasmata ranged from 0.83% to 32.22%, with a median value of 3.05% and an average value of 8.21%.	**Nested PCR**: No Thermoplasmata DNA. **RT-PCR**: 3 samples; Thermoplasmata RA = 0.20%, 0.11%, and 0.07%.

(continued)

Table 4.3 (continued)

Author, year	N periodontal diseases group / type of periodontal disease	N control group	Methods of archaea detection	Proportion of individuals with archaea periodontal diseases group / or Relative Abundance (RA)	Proportion of individuals with archaea control group / or Relative Abundance (RA)
Lira et al. (2013)	15 = Test group SRP + systemic MTZ at the dosage of 400 mg and AMX at the dosage of 500 mg.	15 = Control group SRP + Placebo	PCR for 16S rRNA gene with euryarchaeal primers.	TEST GROUP: Baseline: 9; 6mo: 3; Proportion of sites colonized by archaea: Baseline: 53.3 ± 22.5; 6 mo.: 19.3 ± 10.5.	CONTROL GROUP: Baseline: 9; 6mo: 4; Proportion of sites colonized by archaea baseline: 59.2 ± 18.4; 6 mo.: 22.6 ± 12.4
Matarazzo et al. (2011)	30 = generalized aggressive periodontitis (4 samples each)	30	PCR for 16S rRNA gene with euryarchaeal primers; RT-qPCR for 16S rRNA; cloning, sequencing.	GAgP = 27 (68% of the sites); **qPCR**: 11.2 _ 104 _ 6.6 _ 104 copies of the 16S rRNA gene;**Sequencing:** *M. oralis* = 82% of the clones identified in the samples from the GAgP group;*M. curvum*/congolense phylotype, = 7.2% clones;*Methanosarcina mazeii* = 10.8% clones;RA in relation to the total prokaryotes = 0.08%.	26 (58.3% of the sites); **qPCR:** Lower levels (0.6 _ 104 _ 0.2 _ 104) copies of the 16S rRNA gene;**Sequencing:***M. oralis* = 70.1%;M. curvum/ congolense = 17.9%;Methanosarcina mazeii = 12%;RA in relation to the total prokaryotes = 0.02%.
Ramiro et al. (2018)	TEST 1 = 20 patients (SRP + MTZ); TEST 2 = 19 patients (SPP + TZ + AMX)	40 = SRP (control)	PCR for 16S rRNA gene with primers 931f and m1100r.	TEST GROUPS: Baseline: 25 (adding the two test groups)6 mo: 10 (adding the two test groups); Proportion of sites colonized by archaea: TEST 1: Baseline: 36.2 ± 31.7; 6 mo.: 11.2 ± 16.1; TEST 2: Baseline: 39.9 ± 28.3; 6 mo.: 12.3 ± 13.4	CONTROL GROUP: Baseline: 146 mo: 7; Proportion of sites colonized by archaea: Baseline: 42.5 ± 31.7; 6 mo.: 28.3 ± 32.0
Robichaux et al. (2003a)	8	**NA**	Methane detection by gas chromatographic analysis; enrichment cultures of sulfate-reducing bacteria and methanogenic.	6 subjects, type III periodontal disease = > methanogens.	**NA**

Robichaux et al. (2003b)	17 = varying degree of periodontitis	**NA**	Enrichment cultures in anaerobic conditions; isolation in pure culture and characterization of a methanogenic by microscopic techniques and physiological test.	Detection of microbial growth after five transfers in only one culture from type IV periodontal patient's samples; detection of methanogenic archaea in pure culture.	**NA**
Sogodogo et al. (2019)	2 = abcesses; 29 = periodontitis	NA	16S rRNA gene PCR-sequencing-based detection of methanogens; fluorescent in situ hybridization detection of methanogens.	Methanogens in 26% samples (*M. oralis*, *M. smithii*, *M. massiliense*; 1 co-infection with *M. oralis, M. massiliense*)	**NA**
Vianna et al. (2008)	102 samples = different stages of severity of chronic periodontitis	65	RT-qPCR for mcrA gene; direct sequencing of qPCR amplicons.	Methanogens = 43.1% periodontitis; Absolute abundance of methanogens = 0.26% [0.5% (≥6 mm) and 0.1% (<6 mm)].	None.
Vianna et al. (2009)	44 = periodontitis	**NA**	T-RFLP analysis of methanogens (*mcrA* gene); 16S rRNA PCR; cloning and Sanger sequencing.	*M. oralis* = the sole methanogenic organism in 39 samples;*mcrA* gene = 3 samples;Coexistence of *M. oralis* and *mcrA* gene = 2 samples.	**NA**
Yamabe et al. (2008)	49 = periodontitis (17 aggressive periodontitis)	17	16S rRNA PCR; cloning, Sanger sequencing; Western immunoblotting – detection of humoral immune response to the archaeal components. *M. oralis* and *M. smithii* used as antigens.	11 patients (22.4%);20.6% in plaque samples from pocket depth > or = 6 mm;29.4% = aggressive periodontitis.	None.

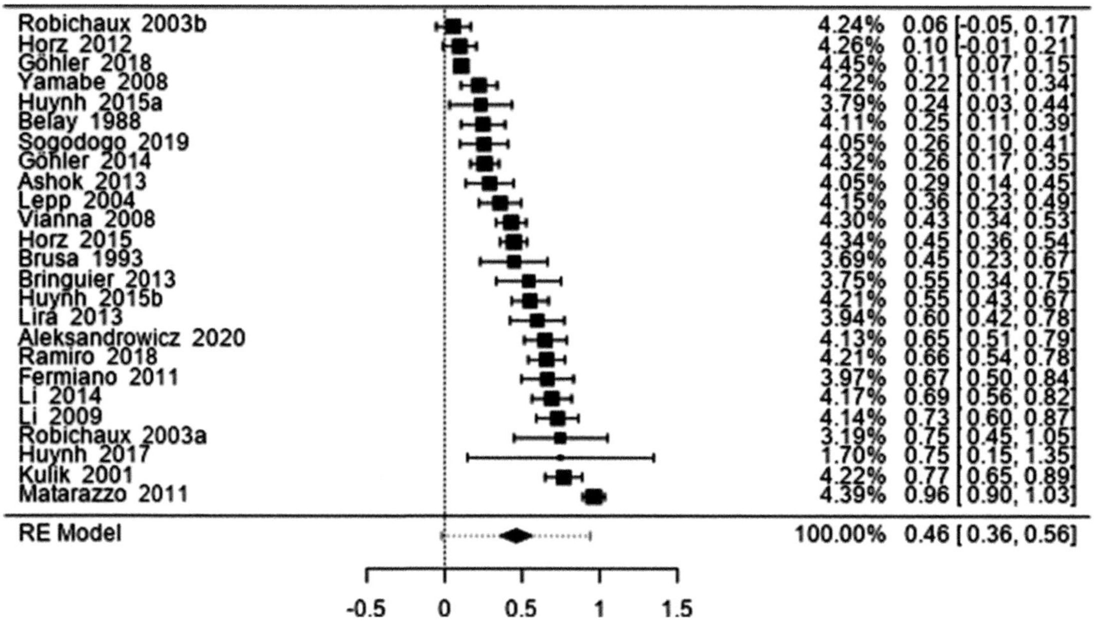

Fig. 4.1 Prevalence of the *Archaea* domain detection in individuals with periodontitis across 24 studies on periodontal archaeome

both cases (Fig. 4.2c). Sensitivity analysis, removing one study at a time (one-by-one), revealed that no study could significantly modify these results. Studies using culture-based methods were not meta-analysed due to the absence of a control group.

4.2.4 Periodontal Treatment

Two studies evaluated samples from individuals positive for archaea, before and after different treatments for periodontal diseases (Lira et al. 2013; Ramiro et al. 2018). Two meta-analyses were performed, comparing the scaling and root planing (SRP) (conventional non-surgical periodontal treatment) alone or in combination with systemic antibiotics (amoxicillin and metronidazole association) (Fig. 4.3a, b). The articles showed a low heterogeneity ($I^2 = 0\%$), however, just as in the primary studies, the meta-analyses found no significant differences between treatments, suggesting that the addition of systemic antibiotics as adjunct to conventional periodontal treatment does not significantly contribute to reduce archaeal cells. Since archaeal cell walls do not possess peptidoglycan, antibiotics directed against the process of cell wall synthesis are ineffective against *Archaea* (Kandler and König 1998). However, it has been shown that methanogens can be sensitive to antibiotics that interfere with nucleic acids synthesis (Dridi et al. 2011) and, perhaps this could explain the small tendency of reduction in archaea numbers in subjects using metronidazole.

4.2.5 Certainty of Evidence

Although the same trend was observed in all included studies, the certainty of evidence from outcomes assessed by the GRADE system was low for the prevalence of archaea in patients with and without periodontal disease, since most of the included studies were of observational design and were evaluated as moderate methodological quality. Also, a low certainty of evidence was graded to the reduction of archaea

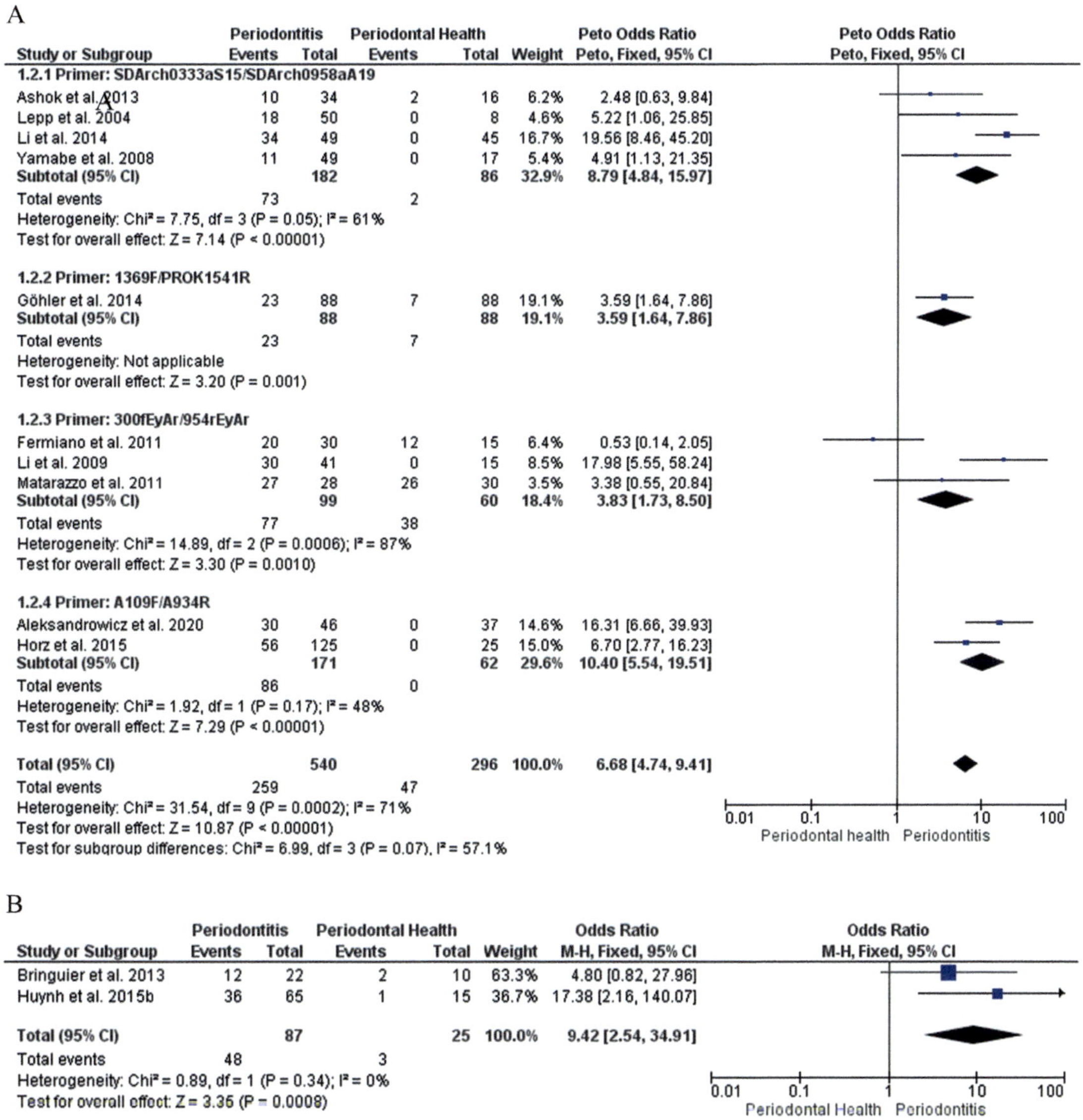

Fig. 4.2 Forest plot of the prevalence of archaea-positive periodontitis *vs.* periodontally healthy individuals. (**a**): 16S rRNA gene with different pairs of primers; (**b**) PCR of the *cnp*60 gene. (**c**): Funnel plot representing publication bias for studies of the meta-analysis (no evidence of publication bias)

prevalence when the treatments SRP and SRP associated with systemic antibiotics were compared (Table 4.4), mainly due to imprecision (low number of treated individuals in included studies). In this sense, we believe that more studies, employing more stringent protocols are necessary to increase the level of evidence in this field.

4.3 Discussion

There are still many unanswered questions about the roles of archaea in periodontitis, even after nearly 30 years of the first archaea detection on periodontal sites. In this study we compiled the available data on periodontal archaeome in health

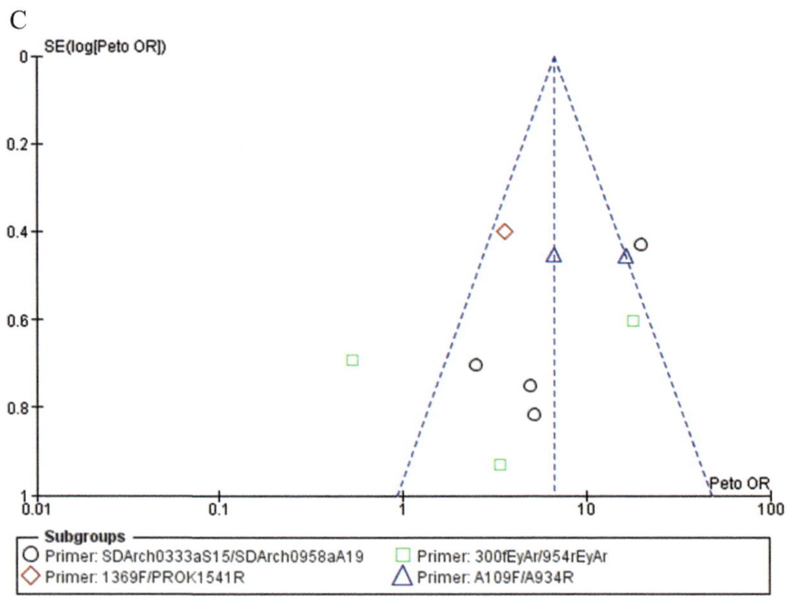

Fig. 4.2 (continued)

A

Study or Subgroup	SRP + MTZ/AMX Events	SRP + MTZ/AMX Total	SRP Events	SRP Total	Weight	Risk Ratio IV, Random, 95% CI
Lira et al. 2013	3	9	4	9	35.0%	0.75 [0.23, 2.44]
Ramiro et al. 2018	5	13	7	14	65.0%	0.77 [0.32, 1.83]
Total (95% CI)		**22**		**23**	**100.0%**	**0.76 [0.38, 1.53]**
Total events	8		11			

Heterogeneity: Tau² = 0.00; Chi² = 0.00, df = 1 (P = 0.97); I² = 0%
Test for overall effect: Z = 0.76 (P = 0.45)

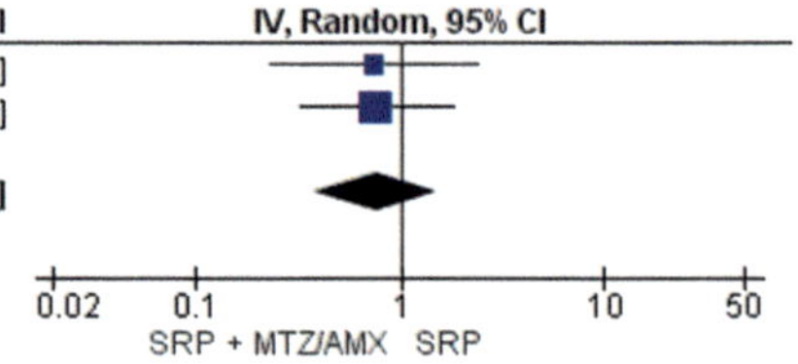

B

Study or Subgroup	SRP + MTZ/AMX Mean	SRP + MTZ/AMX SD	SRP + MTZ/AMX Total	SRP Mean	SRP SD	SRP Total	Weight	Mean Difference IV, Random, 95% CI
Lira et al. 2013	-34	8.275	9	-36.6	7.396	9	50.5%	2.60 [-4.65, 9.85]
Ramiro et al. 2018	-27.6	8.68	13	-14.2	12.038	14	49.5%	-13.40 [-21.28, -5.52]
Total (95% CI)			**22**			**23**	**100.0%**	**-5.32 [-21.00, 10.36]**

Heterogeneity: Tau² = 113.08; Chi² = 8.58, df = 1 (P = 0.003); I² = 88%
Test for overall effect: Z = 0.67 (P = 0.51)

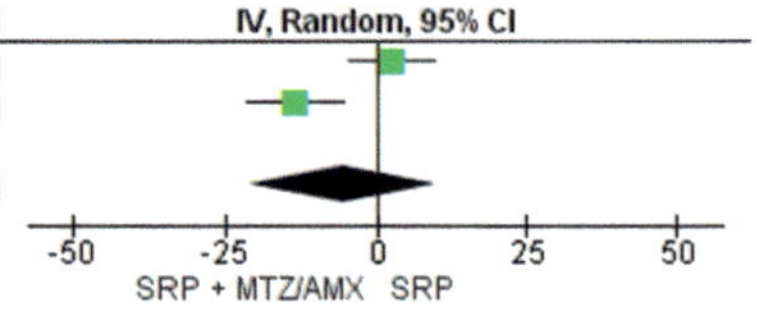

Fig. 4.3 Forest plot of the prevalence (**a**) and relative abundance average (**b**) of *Archaea* domain in treatments for periodontal disease with scaling and root planing (SRP) + metronidazole and amoxicillin (MTZ/AMX) vs. SRP alone

and disease conditions, as well as after different treatment protocols, and the ultimate conclusion is clear: the urgent need for more research on this topic. The detection of archaea in periodontal sites has frequently been related to disease, leading to the proposition by Horz and Conrads (Horz and Conrads 2011) that the presence of archaea in the oral cavity would be a predictive factor for this pathological condition. The meta-analyses performed in this study confirmed that *Archaea* is enriched in periodontitis, and 46% of individuals with periodontitis tested positive for archaea in subgingival biofilms.

Other aspect worth mentioning is the probable underestimation of archaeal diversity in oral niches since most of the studies focused on methanogenic organisms. The methods used to detect archaeal DNA on oral samples are also questionable, espe-

Table 4.4 Certainty of the evidence evaluated by the GRADE system

Certainty assessment							№ of patients		Effect		Certainty	Importance
№ of studies	Study design	Risk of bias	Inconsistency	Indirectness	Imprecision	Other considerations	Intervention	Control	Relative (95% CI)	Absolute (95% CI)		
Archaea prevalence periodontitis versus healthy individuals												
12	Observational	Serious [b]	Not serious	Not serious	Not serious	Strong association	298/612 (48.7%)	50/321 (15.6%)	OR 6.93 (5.02 to 9.57)	**405 more per 1.000** (from 325 more to 483 more)	⊕⊕◯◯ LOW	IMPORTANT
Archaea prevalence after periodontitis treatment (AMX + MTZ + SRP versus SRP) (follow up: Median 3 months)												
2	RCT	Not serious	Not serious	Not serious	Very serious [a]	None	8/22 (36.4%)	11/23 (47.8%)	RR 0.76 (0.38 to 1.53)	**115 fewer per 1.000** (from 297 fewer to 253 more)	⊕⊕◯◯ LOW	IMPORTANT

CI Confidence interval, *RR* Risk ratio, *OR* Odds ratio; Explanations: a. Optimal Information Size (OIS) not achieved, IC reaching significant risk (1.53); b. Included studies with moderate methodological quality. AMX + MTZ + SRP= scaling and root planing + metronidazole and amoxicillin; *SRP* scaling and root planing cially in those articles that adopt conventional PCR assays to estimate the archaeal diversity. Primer bias is a recognized problem with PCR-based approaches, rending the reliability of the species identification and level of diversity questionable. Furthermore, the detection of archaea is strongly sensitive to methodological issues such as DNA extraction protocols, primer choice and sequence data processing pipelines (Koskinen et al. 2017) that can lead to underestimated values, or result in erroneous classification as artefacts or contamination (Lazarevic et al. 2016). These issues may contribute to the relatively low prevalence of archaea-positive individuals (46%).

Another hypothesis to that would explain the global archaeal prevalence is the fact that archaea are not officially recognized as 'keystone' pathogens, with their enrichment more associated to inflammatory sites, in accordance to the ecological hypothesis proposed by Marsh (Marsh 1994). For instance, *P. gingivalis* widely recognized as a keystone pathogen in periodontitis has also been found in low prevalence, identified in only 23 out of 73 periodontitis samples in combination with the other 'red complex' bacteria in a study using NGS approach (Dabdoub et al. 2016), undoubtedly less sensitive to bias than the techniques used to detect archaea thus far. The diversity and complexity of periodontal biofilms represent an unique component of pathogenicity, in addition to bacterial partners that colonize that surfaces. Specific associations among species in dental biofilms are dependent of the biofilm complexity. As an example, cells of the genera *Prevotella* are among those showing the most interspecies associations, suggesting a central role in establishing and maintaining biofilm complexity. A significant positive correlation between methanogens abundance and the amount of *P. intermedia* suggested that, if the metabolic activity of methanogens promotes the bacterial species from the red and orange complex growth in periodontitis, this role is probably mediated through direct or indirect interactions with *P. intermedia* (Horz et al. 2015). As oral biofilms become more complex and mature, they are joined or replaced by other species (Socransky and Haffajee 2002), and different colonizing organisms will produce biofilms with varying pathogenic potential. In addition, the low global archaeal prevalence is expected, since these organisms comprehend a small fraction of the human microbiome, as described recently by Koskinen (Koskinen et al. 2017).

4.3.1 *Archaea* Detection in Periodontitis and Healthy Subgingival Sites

Despite the expressive differences observed between periodontally healthy and diseased individuals, these were not considered statistically significant in some primary studies. Such findings were previously linked to ethnic and dietary aspects of the subjects (Brusa et al. 1993). However, in studies employing small sample sizes of healthy groups (see Table 4.3), caution must be taken to interpret this hypothesis, as the absence of archaea can be related to the poor discriminating power of the study. On the other hand, sensitivity analyses that excluded small sample studies (less than 25 periodontally healthy individuals) returned even more marked results (OR: 8.52, CI = 95% 5.64–12.87). The meta-analyses performed in the present study reinforce the conclusions of most studies on periodontal archaeome and represent a breakthrough in questions regarding the ecological potential, function, and structural interactions with the host and other microorganisms.

To better contextualize our analyses, we present a brief timeline of the periodontal archaeome discoveries. In the past few years, archaea were described as common inhabitants of the oral cavity, and these findings add another variable to be taken into account in studies of the etiopathogenesis of oral diseases (Grine et al. 2018). As described above, the first reports describing the periodontal archaeome were based on the cultivation of methanogens from subgingival plaque samples (Brusa et al. 1987, 1993; Ferrari et al. 1994). Only after 14 years, the first article describing the detection of archaea in patients with periodontitis using molecular techniques was published (Kulik et al. 2001). The genus

Methanobrevibacter was the first methanogen identified and characterized in the oral cavity (Belay et al. 1988) and until now is the genus most related to oral dysbiosis (Maeda et al. 2013; Dabdoub et al. 2016). Although the predominance of methanogens in oral sites has been shown, these results should be interpreted with caution, since other archaeal phyla have been described in the last years (Deng et al. 2017; Koskinen et al. 2017).

4.3.2 The Methanogens in Subgingival Biofilms

Currently, methanogens are the only archaeal group which putative role on oral diseases is relatively known and debated. These organisms thrive in anaerobic microniches, which are frequently found on biofilms, or formed as a result of oxygen consumption by facultative anaerobic microorganisms (Dridi et al. 2011), probably playing important roles in the carbon cycle (Thauer and Shima 2008). The co-occurrence of methanogenic archaea and sulfate-reducing bacteria (SRB), as well as other microorganisms in periodontal samples (Vianna et al. 2008), and a positive correlation between archaea and *Porphyromonas gingivalis* and *Tannerella forsythia* levels in subgingival biofilms from subjects with chronic periodontitis (Matarazzo et al. 2012) suggest a syntrophic metabolism in these oral biofilms. Theoretically, this role of methanogens can be replaced by SRB and reductive acetogens in subgingival biofilms negative for archaea, since both groups are able to grow in H_2 with H_2S and acetate as final metabolic products. In this context, the antibiotic support therapy could be valuable, since it may reduce the bacterial population, but not the archaeal one. Conventional treatment for periodontitis with SRP would reduce the whole biofilm and, consequently, the archaeal content.

The syntrophic interactions of methanogens and fermentative bacteria are beneficial to both, since they sustain their growth. Depending on the bacterial species involved in those interactions, methanogenic archaea can be viewed as secondary pathogens of the human microbiota (Horz and Conrads 2010). When interacting with some bacterial species, methanogens collaboratively degrade organic substances, such as acetate, propionate, and butyrate, producing methane under anaerobic conditions (Horz and Conrads 2011). In addition to the co-occurrence of archaea with bacterial periodontal pathogens, their physical interaction with other organisms may result in relevant outcomes. A curious interaction between *Methanothermobacter thermautotrophicus*, a non-oral methanogenic archaeon and the bacterium *Pelotomaculum thermopropionicum*, results in the stimulation of methanogenesis. When the bacterial flagellar tip touches the surface of the archaeal cell, the propionate produced by the bacterium is converted to methane, restoring the environmental pH, which is positive to both organisms (Shimoyama et al. 2009). It is tempting to speculate that this kind of interaction can also happen in other complex microenvironments. The discovery of such kind of detoxifying mechanism in subgingival biofilms would not be surprising since the removal of the H^+ produced by many bacterial species would also restore the microenvironment pH, favouring the colonization of bacteria linked to periodontitis.

The well-known oral methanogens, *Methanobrevibacter* and *Methanosphaera* species can also produce diverse glycosyltransferases, enzymes that play a major role in maintaining the integrity of oral biofilms by producing the biofilm's extracellular matrix (Samuel et al. 2007). Furthermore, experimental evidence has shown that methanoarchaeal species commonly associated with human mucosa are able to form biofilms on different surfaces (Bang and Schmitz 2015), which could indicate archaeal participation in subgingival biofilm structures, conferring beneficial properties in this microenvironment as well as another manner of network with bacterial cells (Borrel et al. 2020).

The occurrence of interactions between archaea and humans is still under debate, as well as if archaea have their own virulence factors, since studies associating these organisms with specific pathogenesis are scarce. Probably, the better explored cases linked them to periodontal diseases (Aminov 2013). Subsequently, we will discuss some hypotheses for the enrichment of archaea other than methanogens in periodontitis.

4.3.3 Putative Roles of Other Archaeal Groups in Periodontitis

Although previous studies have concluded that archaea constitute only a minor component of the oral microbiome and that their diversity is restricted to methanogens (Wade 2013), other species can also compose this microbiota, as recently determined by molecular surveys and metagenomic analyses (Deng et al. 2017; Horz et al. 2012; Li et al. 2009, 2014).

Some archaeal groups which characteristics would allow them to be part of the periodontal microbiota are discussed below. However, it is important to highlight that the information about these groups are still extremely limited and further investigations are needed to test our hypotheses in the context of the periodontal archaeome.

4.3.3.1 Thermoplasmatales

These heterotrophic and thermoacidophilic organisms are phylogenetically grouped with methanogenic and halophilic archaea in the Euryarchaeota phylum. They were first reported in subgingival samples in 2009 (Li et al. 2009) and further detected in other studies investigating archaeal presence in similar samples (Horz et al. 2012; Li et al. 2014), as well as in ancient calculus specimens (Huynh et al. 2015a). Since these non-methanogenic organisms were identified by PCR studies using primers directed to the *mcrA* genes, it was speculated the existence of a new sister group of Thermoplasmatales present in the oral cavity and intestine, probably able to produce methane (Horz and Conrads 2011).

4.3.3.2 Ammonia-Oxidizing Archaea

Besides euryarchaeotes, other archaeal phyla have been shown to play crucial roles in environmental nutrients cycling. Members of Thaumarchaeota have been found to greatly contribute to nitrogen cycling, performing oxidation of ammonia to nitrite, and subsequently to nitrate in numerous marine and terrestrial environments (Cavicchioli et al. 2007). Ammonia oxidizing thaumarchaeotes of subgroup I.1b were found to be the most abundant archaea in human skin, with active physiological status and potential for ammonia oxidation further suggested by FISH visualization and *amoA* genes detection (Probst et al. 2013). Although the clinical relevance of Thaumarchaeota in the human skin remains unclear, it has been speculated that they might impact it by lowering the pH and removing certain nitrogen compounds (Moissl-Eichinger et al. 2017; Probst et al. 2013). Interestingly, DNA sequences affiliated to I.1b thaumarchaeotes have recently been identified in supragingival and carious biofilms, suggesting that ammonia-oxidizing archaea may inhabit oral niches (Dame-Teixeira et al. 2020).We can speculate that the nitrate produced by these could be used as the final electron acceptor by oral anaerobic bacteria during their respiration.

4.3.3.3 Halophilic Archaea

Halophilic archaea have already been identified in the human digestive tract (Oxley et al. 2010), but their impact on the host is still unclear. Since haloarchaea have mostly been detected by DNA sequencing approaches (Oxley et al. 2010), has been hypothesized that these archaea transitorily pass the human gut as a result of the ingestion of salty foods, which can be heavily colonized by these organisms, (Oxley et al. 2010). However, recent comprehensive studies based on genomics, culturomics and FISH analyses of fecal samples suggest that halophilic archaea may be residents of the human gastrointestinal tract and not only transient passengers(Seck et al. 2019; Kim et al. 2020). Considering the close association of the gastrointestinal tract and the oral cavity, the ability of these archaeal groups to colonize oral niches has been previously hypothesized (Horz and Conrads 2011) and further reinforced by the recent detection of haloarchaea sequences in a metatranscriptomic analysis of periodontitis samples (Deng et al. 2017). Despite salinity requirements, haloarchaea present very diverse physiological and metabolic features, thriving in aerobic and anaerobic conditions and being able to catabolize numerous compounds (Falb et al. 2008). Furthermore, the formation of biofilms by different haloarchaeal species has already been described, evidencing that these archaea are able to strongly attach surfaces and produce varied biofilm structures (Fröls et al. 2012).

4.4 Conclusion and Perspectives

It could be hypothesised that archaea may act as secondary pathogens in areas in dysbiosis and be favoured in the inflammatory environment. Their higher detection in different pathological conditions, when compared to healthy sites suggest a putative inter-domain interaction between different archaeal and bacterial species. This kind of speculation is reinforced since until now no virulence factor has been identified in archaeal cells. Although some halophiles could secrete some bacteriocins, no true virulence factors for human cells have been detected.

The archaeal diversity on the oral cavity must be better characterized, since recent works have described new genera and classes, previously not described, such as halophiles and thermoplasmas.

There is an urgent need to better characterize the physiology of archaeal species that are members of the human microbiome, since only after this kind of studies, their actual role in the oral cavity may be clarified. An interesting perspective would be the development of a worldwide network of researchers interested in pioneering the oral archaeome. This kind of project could provide important data concerning the diversity of oral archaeas in subjects of different ages, diets, and ethnicities. The studies could also provide a worldwide picture of the different oral pathologies in which archaea could be isolated.

4.5 Methods

4.5.1 Studies Eligibility and Search Strategy

Meta-analyses were performed to answer two questions: (1) what is the prevalence of archaea-positive subjects at subgingival periodontal sites in health and disease? (2) Do periodontal treatments reduce the prevalence of archaea in periodontal sites?

Details on the search strategy and studies selection were described elsewhere (Belmok et al. 2020). Briefly, a systematic search was conducted in five databases (MEDLINE via PubMed, Cochrane Library, Scopus, LILACS, and Livivo), as well as the grey literature (Google Scholar, OpenGrey) and hand search of included studies references lists. General controlled vocabulary (MeSH Terms) and keywords (Archaea AND Cavity oral AND Periodontitis) were chosen and the searches had no language, year or publication type restriction. Here, the search strategy was updated to gather any new study published in 2020. New databases (Embase, Proquest) were included at the updated search.

Eligibility criteria were also described by Belmok et al. (2020). Only observational/clinical studies where the target population consisted of humans of any age who were donors of periodontal samples (biofilms, crevicular fluid) from oral cavity were included. Studies could evaluate archaea using methanogenic cultures, methods of PCR amplification and sequencing, DNA-DNA hybridization, next-generation sequencing, etc. Either animal, *in vitro*, or not original studies were excluded. Two reviewers independently screened the eligibility in two steps (1: reading titles and abstract; and 2: reading the full text). A subsample of identified studies on periodontal sites was used to these meta-analyses. The full screening of the articles was revised in the updated search.

4.5.2 Quality Assessment of Individual Studies

Two reviewers independently assessed the methodological quality of individual studies, using validated tools according to the type of study: the revised Cochrane tool for randomised controlled trials (RoB2) (Sterne et al. 2019) and the Joanna Briggs Institute (JBI) Critical Appraisal Checklist for Analytical Cross-Sectional Studies (Moola et al. 2017).

For JBI, due to the design of included articles, besides all eight questions are considered important, four of them were considered critical domains to this systematic review (criteria 1–4). Criteria related to the outcome were considered non-critical (criteria numbers 5–8). At least one "no" and one "unclear" or two "unclear" in critical domains, or two "unclear" and one or more "no" in non-critical domains represented low methodological quality. Decision on critical and non-critical domains and classification system was discussed with research team before the application of the instrument, as described at JBI Reviewer's Manual (Moola et al. 2017).

The risk of bias of included randomized clinical trials was assessed through RoB2 tool independently by two calibrated authors, that scored each item as "yes", "probably yes", "probably no", "no" and "no information". A third reviewer solved disagreements, and overall risk of bias was calculated using the RoB2 tool algorithm.

4.5.3 Data Extraction and Qualitative Analysis

Data from the subsample of studies including subgingival samples was orderly in an Excel file to analyse the periodontal archaeome. The following information was collected: author, year; country; sample size; characteristics of periodontal sites (index used for periodontal diseases diagnosis); type of sample collection (paper point, manual instrument); DNA extraction method; methods of archaea detection; sets of primers used; the proportion of individuals with archaea detection in each group; and the relative abundance of archaea in each group.

A combination of quantitative and qualitative approaches was used in the data analysis. A narrative synthesis of the findings regarding the molecular techniques and the clinical sampling and periodontal diagnosis was performed, as well as a qualitative synthesis was performed.

4.5.4 Meta-analyses and Certainty of Evidence

Continuous variables were compared through the DerSimonian & Laird random-effects meta-analytic model and presented as mean differences and 95% confidence interval. Dichotomous variables (prevalence of archaea comparison between periodontal health and periodontitis) were compared through Peto Odds Ratio, with fixed effect model and 95% confidence interval. Peto Odds Ratio was chosen due to prevalence of archaea in healthy individuals was rare and some studies returned zero event (Deeks et al. 2019). Archaea global prevalence was meta-analysed through Restricted Maximum Likelihood model for Raw Proportions with 95% CI (jamovi software version 1.6 retrieved from https://www.jamovi.org/). Heterogeneity between studies was estimated by Cochran's Q test and the inconsistency by I^2 statistic. Study characteristics considered as potential sources of heterogeneity were analysed through sensitivity analysis (removing on study at a time – one-by-one approach, and removing small sample studies) and subgroup analysis regarding the type of primers used to archaea detection (meta-analysis comparing periodontally diseased and healthy individuals). Publication bias was evaluated through funnel plot analysis. The certainty of evidence was evaluated by using GRADE (Grading of Recommendations, Assessment, Development and Evaluation) approach, through the analysis of risk of bias, inconsistency, imprecision, indirectness and publication bias (Schünemann 2019).

Acknowledgements The authors thank the University of Brasilia. Jéssica Alves de Cena acknowledge a scholarship from CNPq.

References

Ahn J-H, Song J, Kim B-Y, Kim M-S, Joa J-H, Weon H-Y (2012) Characterization of the bacterial and archaeal communities in rice field soils subjected to long-term fertilization practices. J Microbiol 50(5):754–765

Aleksandrowicz P, Brzezińska-Błaszczyk E, Dudko A, Agier J (2020) Archaea occurrence in the subgingival biofilm in patients with Peri-implantitis and periodontitis. Int J Periodont Rest Dent 40(5)

American Academy of Periodontology (1977) Current procedural terminology for periodontics, 4th edn. American Academy of Periodontology (AAP), Chicago, p 20

American Academy of Periodontology (1989) Proceedings of the World Workshop in Clinical Periodontics. The American Academy of Periodontology (AAP), Chicago, pp I/23-I/24

Aminov RI (2013) Role of archaea in human disease. Front Cell Infect Microbiol 3:42

Armitage GC (1999) Development of a classification system for periodontal diseases and conditions. Ann Periodontol 4(1):1–6

Ashok N, Warad S, Singh VP, Chaudhari H, Narayanan A, Rodrigues J (2013) Prevalence of archaea in chronic periodontitis patients in an Indian population. Indian J Dent Res 24(3):289

Bang C, Schmitz RA (2015) Archaea associated with human surfaces: not to be underestimated. FEMS Microbiol Rev 39(5):631–648

Belay N, Johnson R, Rajagopal BS, Conway de Macario E, Daniels L (1988) Methanogenic bacteria from human dental plaque. Appl Environ Microbiol 54(2):600–603

Belmok A, de Cena J, Kyaw C, Damé-Teixeira N (2020) The oral archaeome: a scoping review. J Dent Res 99(6):630–643

Borrel G, Brugere J-F, Gribaldo S, Schmitz RA, Moissl-Eichinger C (2020) The host-associated archaeome. Nat Rev Microbiol 18(11):622–636

Bringuier A, Khelaifia S, Richet H, Aboudharam G, Drancourt M (2013) Real-time PCR quantification of Methanobrevibacter oralis in periodontitis. J Clin Microbiol 51(3):993–994

Brusa T, Conca R, Ferrara A, Ferrari A, Pecchioni A (1987) The presence of methanobacteria in human subgingival plaque. J Clin Periodontol 14(8):470–471

Brusa T, Canzi E, Allievi L, Del Puppo E, Ferrari A (1993) Methanogens in the human intestinal tract and oral cavity. Curr Microbiol 27(5):261–265

Cavicchioli R, Demaere MZ, Thomas T (2007) Metagenomic studies reveal the critical and wide-ranging ecological importance of uncultivated Archaea: the role of ammonia oxidizers. BioEssays 29(1):11–14. https://doi.org/10.1002/bies.20519

Chandra RV (2007) Evaluation of a novel periodontal risk assessment model in patients presenting for dental care. Oral Health Prev Dent 5:39–48

Dabdoub SM, Ganesan SM, Kumar PS (2016) Comparative metagenomics reveals taxonomically idiosyncratic yet functionally congruent communities in periodontitis. Sci Rep 6:38993

Dame-Teixeira N, de Cena JA, Côrtes DA, Belmok A, dos Anjos Borges LG, Marconatto L et al (2020) Presence of Archaea in dental caries biofilms. Arch Oral Biol 110:104606

Deeks JJ, Higgins JP, Altman DG, Group CSM (2019) Analysing data and undertaking meta-analyses. In: Cochrane handbook for systematic reviews of interventions, pp 241–284

DeLong EF (1998) Everything in moderation: archaea as 'non-extremophiles'. Curr Opin Genet Dev 8(6):649–654

Deng Z-L, Szafrański SP, Jarek M, Bhuju S, Wagner-Döbler I (2017) Dysbiosis in chronic periodontitis: key microbial players and interactions with the human host. Sci Rep 7(1):1–13

Dridi B, Fardeau ML, Ollivier B, Raoult D, Drancourt M (2011) The antimicrobial resistance pattern of cultured human methanogens reflects the unique phylogenetic position of archaea. J Antimicrob Chemother 66(9):2038–2044. https://doi.org/10.1093/jac/dkr251

Falb M, Müller K, Königsmaier L, Oberwinkler T, Horn P, von Gronau S et al (2008) Metabolism of halophilic archaea. Extremophiles 12(2):177–196

Faveri M, Goncalves LF, Feres M, Figueiredo LC, Gouveia LA, Shibli JA et al (2011) Prevalence and microbiological diversity of Archaea in peri-implantitis subjects by 16S ribosomal RNA clonal analysis. J Periodontal Res 46(3):338–344. https://doi.org/10.1111/j.1600-0765.2011.01347.x

Fermiano D, Lira EAGD, Sampaio FR, Figueiredo LCD, Feres M, Faveri M (2011) Prevalência de archaea em indivíduos com doença periodontal e saúde periodontal. Periodontia 86–93

Ferrari A, Brusa T, Rutili A, Canzi E, Biavati B (1994) Isolation and characterization of Methanobrevibacter oralis sp. nov. Curr Microbiol 29(1):7–12

Fine DH, Patil AG, Loos BG (2018) Classification and diagnosis of aggressive periodontitis. J Clin Periodontol 45:S95–S111

Fröls S, Dyall-Smith M, Pfeifer F (2012) Biofilm formation by haloarchaea. Environ Microbiol 14(12):3159–3174

Ghavami, S. B., Sephay, A. A., Aghdaei, H. A., Sattari, T. N., & Zali, M. R. 7 (2015) 'Comparison study on effect of different methods on DNA extraction of Methanobrevibacter smithii' Biological Forum' . Research Trend, p. 549 2

Gohler A, Hetzer A, Holtfreter B, Geisel MH, Schmidt CO, Steinmetz I et al (2014) Quantitative molecular detection of putative periodontal pathogens in clinically healthy and periodontally diseased subjects. PLoS One 9(7):e99244

Gohler A, Samietz S, Schmidt CO, Kocher T, Steinmetz I, Holtfreter B (2018) Comparison of Oral microbe

quantities from tongue samples and subgingival pockets. Int J Dent 2018:2048390. https://doi.org/10.1155/2018/2048390
Grine G, Terrer E, Boualam MA, Aboudharam G, Chaudet H, Ruimy R et al (2018) Tobacco-smoking-related prevalence of methanogens in the oral fluid microbiota. Sci Rep 8(1):9197. https://doi.org/10.1038/s41598-018-27372-7
Horz HP, Conrads G (2010) The discussion goes on: what is the role of Euryarchaeota in humans? Archaea 2010:967271. https://doi.org/10.1155/2010/967271
Horz HP, Conrads G (2011) Methanogenic Archaea and oral infections – ways to unravel the black box. J Oral Microbiol 3. https://doi.org/10.3402/jom.v3i0.5940
Horz HP, Seyfarth I, Conrads G (2012) McrA and 16S rRNA gene analysis suggests a novel lineage of Archaea phylogenetically affiliated with Thermoplasmatales in human subgingival plaque. Anaerobe 18(3):373–377. https://doi.org/10.1016/j.anaerobe.2012.04.006
Horz HP, Robertz N, Vianna ME, Henne K, Conrads G (2015) Relationship between methanogenic archaea and subgingival microbial complexes in human periodontitis. Anaerobe 35(Pt A):10–12. https://doi.org/10.1016/j.anaerobe.2015.02.008
Huynh HT, Nkamga VD, Drancourt M, Aboudharam G (2015a) Genetic variants of dental plaque Methanobrevibacter oralis. Eur J Clin Microbiol Infect Dis 34(6):1097–1101. https://doi.org/10.1007/s10096-015-2325-x
Huynh HT, Pignoly M, Nkamga VD, Drancourt M, Aboudharam G (2015b) The repertoire of archaea cultivated from severe periodontitis. PLoS One 10(4):e0121565
Huynh HT, Pignoly M, Drancourt M, Aboudharam G (2017) A new methanogen "Methanobrevibacter massiliense" isolated in a case of severe periodontitis. BMC Res Notes 10(1):657
Kandler O, König H (1998) Cell wall polymers in Archaea (Archaebacteria). Cell Mol Life Sci CMLS 54(4):305–308
Kim JY, Whon TW, Lim MY, Kim YB, Kim N, Kwon M-S, et al (2020) The human gut archaeome: identification of diverse haloarchaea in Korean subjects
Koskinen K, Pausan MR, Perras AK, Beck M, Bang C, Mora M et al (2017) First insights into the diverse human Archaeome: specific detection of archaea in the gastrointestinal tract, lung, and nose and on skin. MBio 8(6). https://doi.org/10.1128/mBio.00824-17
Kulik EM, Sandmeier H, Hinni K, Meyer J (2001) Identification of archaeal rDNA from subgingival dental plaque by PCR amplification and sequence analysis. FEMS Microbiol Lett 196(2):129–133. https://doi.org/10.1111/j.1574-6968.2001.tb10553.x
Lazarevic V, Gaïa N, Girard M, Schrenzel J (2016) Decontamination of 16S rRNA gene amplicon sequence datasets based on bacterial load assessment by qPCR. BMC Microbiol 16(1):73
Lepp PW, Brinig MM, Ouverney CC, Palm K, Armitage GC, Relman DA (2004) Methanogenic Archaea and human periodontal disease. Proc Natl Acad Sci U S A 101(16):6176–6181. https://doi.org/10.1073/pnas.0308766101
Li CL, Liu DL, Jiang YT, Zhou YB, Zhang MZ, Jiang W et al (2009) Prevalence and molecular diversity of Archaea in subgingival pockets of periodontitis patients. Oral Microbiol Immunol 24(4):343–346. https://doi.org/10.1111/j.1399-302X.2009.00514.x
Li CL, Jiang YT, Liu DL, Qian J, Liang JP, Shu R (2014) Prevalence and quantification of the uncommon Archaea phylotype Thermoplasmata in chronic periodontitis. Arch Oral Biol 59(8):822–828. https://doi.org/10.1016/j.archoralbio.2014.05.011
Lira E, Ramiro F, Chiarelli F, Dias R, Feres M, Figueiredo L et al (2013) Reduction in prevalence of Archaea after periodontal therapy in subjects with generalized aggressive periodontitis. Aust Dent J 58(4):442–447
Maeda H, Hirai K, Mineshiba J, Yamamoto T, Kokeguchi S, Takashiba S (2013) Medical microbiological approach to Archaea in oral infectious diseases. Jpn Dent Sci Rev 49(2):72–78
Magrum LJ, Luehrsen KR, Woese CR (1978) Are extreme halophiles actually "bacteria"? J Mol Evol 11(1):1–8
Mahnert A, Blohs M, Pausan M-R, Moissl-Eichinger C (2018) The human archaeome: methodological pitfalls and knowledge gaps. Emer Top Life Sci 2(4):469–482
Marsh PD (1994) Microbial ecology of dental plaque and its significance in health and disease. Adv Dent Res 8(2):263–271
Matarazzo F, Ribeiro AC, Feres M, Faveri M, Mayer MP (2011) Diversity and quantitative analysis of Archaea in aggressive periodontitis and periodontally healthy subjects. J Clin Periodontol 38(7):621–627. https://doi.org/10.1111/j.1600-051X.2011.01734.x
Matarazzo F, Ribeiro AC, Faveri M, Taddei C, Martinez MB, Mayer MP (2012) The domain Archaea in human mucosal surfaces. Clin Microbiol Infect 18(9):834–840. https://doi.org/10.1111/j.1469-0691.2012.03958.x
Moissl-Eichinger C, Probst AJ, Birarda G, Auerbach A, Koskinen K, Wolf P et al (2017) Human age and skin physiology shape diversity and abundance of Archaea on skin. Sci Rep 7(1):1–10
Moola S, Munn Z, Tufanaru C, Aromataris E, Sears K, Sfetcu R et al (2017) Checklist for analytical cross sectional studies. Joanna Briggs Institute Reviewer's Manual [Internet]. The Joanna Briggs Institute
Oxley AP, Lanfranconi MP, Würdemann D, Ott S, Schreiber S, McGenity TJ et al (2010) Halophilic archaea in the human intestinal mucosa. Environ Microbiol 12(9):2398–2410
Papapanou PN, Sanz M, Buduneli N, Dietrich T, Feres M, Fine DH et al (2018) Periodontitis: consensus report of workgroup 2 of the 2017 world workshop on the classification of periodontal and peri-implant diseases and conditions. J Periodontol 89:S173–S182
Pausan MR, Csorba C, Singer G, Till H, Schöpf V, Santigli E et al (2019) Exploring the archaeome: detection of archaeal signatures in the human body. Front Microbiol 10:2796

Probst AJ, Auerbach AK, Moissl-Eichinger C (2013) Archaea on human skin. PloS One 8(6):e65388

Ramiro F, de Lira E, Soares G, Retamal-Valdes B, Feres M, Figueiredo L et al (2018) Effects of different periodontal treatments in changing the prevalence and levels of Archaea present in the subgingival biofilm of subjects with periodontitis: a secondary analysis from a randomized controlled clinical trial. Int J Dent Hyg 16(4):569–575

Robichaux M, Howell M, Boopathy R (2003a) Methanogenic activity in human periodontal pocket. Curr Microbiol 46(1):0053–0058

Robichaux M, Howell M, Boopathy R (2003b) Growth and activities of sulfate-reducing and methanogenic bacteria in human oral cavity. Curr Microbiol 47(1):12–16. https://doi.org/10.1007/s00284-002-3930-3

Samuel BS, Hansen EE, Manchester JK, Coutinho PM, Henrissat B, Fulton R et al (2007) Genomic and metabolic adaptations of Methanobrevibacter smithii to the human gut. Proc Natl Acad Sci 104(25):10643–10648

Schünemann H (2019) Chapter 14: Completing 'Summary of findings' tables and grading the certainty of the evidence. In: Higgins JP, Thomas J, Chandler J, Cumpston M, Li T, Page MJ et al (eds) Cochrane Handbook for Systematic Reviews of Interventions version 6.0 (updated July 2019). Available from www.training.cochrane.org/handbook

Seck E, Senghor B, Merhej V, Bachar D, Cadoret F, Robert C et al (2019) Salt in stools is associated with obesity, gut halophilic microbiota and Akkermansia muciniphila depletion in humans. Int J Obes 43(4):862–871

Shimoyama T, Kato S, Ishii S i, Watanabe K (2009) Flagellum mediates symbiosis. Science 323(5921):1574–1574

Socransky SS, Haffajee AD (2002) Dental biofilms: difficult therapeutic targets. Periodontol 2000(28):12–55. https://doi.org/10.1034/j.1600-0757.2002.280102.x

Sogodogo E, Doumbo O, Aboudharam G, Kouriba B, Diawara O, Koita H et al (2019) First characterization of methanogens in oral cavity in Malian patients with oral cavity pathologies. BMC Oral Health 19(1):232

Sterne JA, Savović J, Page MJ, Elbers RG, Blencowe NS, Boutron I et al (2019) RoB 2: a revised tool for assessing risk of bias in randomised trials. BMJ 366

Thauer RK, Shima S (2008) Methane as fuel for anaerobic microorganisms. Ann N Y Acad Sci 1125(1):158–170

Vianna ME, Holtgraewe S, Seyfarth I, Conrads G, Horz HP (2008) Quantitative analysis of three hydrogenotrophic microbial groups, methanogenic archaea, sulfate-reducing bacteria, and acetogenic bacteria, within plaque biofilms associated with human periodontal disease. J Bacteriol 190(10):3779–3785. https://doi.org/10.1128/jb.01861-07

Vianna ME, Conrads G, Gomes BP, Horz HP (2009) T-RFLP-based mcrA gene analysis of methanogenic archaea in association with oral infections and evidence of a novel Methanobrevibacter phylotype. Oral Microbiol Immunol 24(5):417–422. https://doi.org/10.1111/j.1399-302X.2009.00539.x

Wade WG (2013) The oral microbiome in health and disease. Pharmacol Res 69(1):137–143. https://doi.org/10.1016/j.phrs.2012.11.006

Wagner Mackenzie B, West AG, Waite DW, Lux CA, Douglas RG, Taylor MW et al (2020) A novel description of the human sinus archaeome during health and chronic rhinosinusitis. Front Cell Infect Microbiol 10:398

Woese CR, Fox GE (1977) Phylogenetic structure of the prokaryotic domain: the primary kingdoms. Proc Natl Acad Sci 74(11):5088–5090

Yamabe K, Maeda H, Kokeguchi S, Tanimoto I, Sonoi N, Asakawa S et al (2008) Distribution of Archaea in Japanese patients with periodontitis and humoral immune response to the components. FEMS Microbiol Lett 287(1):69–75

Parasites in Periodontal Health and Disease: A Systematic Review and Meta-analysis

5

David Felipe Martin-Garcia, Malik Sallam, Gabriela Garcia, and Julien Santi-Rocca

Abstract

Background: Periodontitis is an inflammatory disease triggered by the infection of the periodontal sulcus by microbes. Together with the abundant eubacterial microbiota, at least two parasites have often been identified: the amoeba *Entamoeba gingivalis* and the flagellate *Trichomonas tenax*. The role of these protists in the pathophysiology of periodontal disease remains to be deciphered. A high diversity in their measured prevalence, mainly due to methodological concerns, prevents further analysis of the aetiological link between these parasites and periodontitis.

Methods: To determine *E. gingivalis* and *T. tenax* prevalence in periodontal pockets as compared to healthy sulci, we have conducted a systematic review, searching 3 remote databases (Pubmed, LILACS, and Google Scholar), restricting to papers in which the diagnostic of the parasite was made using molecular methods. A total of 5 studies for E. gingivalis and 2 studies for *T. tenax* were included for the meta-analysis.

Results: In the periodontal pockets, the prevalence of parasites is 76.9% (95%-CI: 71.5–81.7%) for *E. gingivalis* and 38.6% (95%-CI: 27.2–50.0%) for *T. tenax*. Both parasites are more abundant in periodontal pockets as compared to healthy sulci, with a risk ratio of 3.96 (95%-CI: 1.57–9.98) for *E. gingivalis* and 21.82 (95%-CI: 6.71–70.96) for *T. tenax*. The two subtypes of *E. gingivalis* exhibited the same risk ratio: 3.30 (95%-CI: 1.27–8.55) for ST1 and 3.30 (95%-CI: 0.42–26.03) for ST2, but ST1 was more prevalent (70.6%, 95%-CI: 65.0–76.2%) than ST2 (43.9%, 95%-CI: 35.5–52.4%) in periodontal pockets.

Conclusion: Altogether, the data show that parasites are more prevalent in the diseased than in the healthy. However, the differences in prev-

D. F. Martin-Garcia
Universidad Pontificia Comillas, Madrid, Spain
e-mail: dfmartin@icade.comillas.edu

M. Sallam
Department of Pathology, Microbiology and Forensic Medicine, School of Medicine, The University of Jordan, Amman, Jordan
e-mail: malik.sallam@ju.edu.jo

G. Garcia
Department of Clinical Laboratories and Forensic Medicine, Jordan University Hospital, Amman, Jordan

Department of Translational Medicine, Faculty of Medicine, Lund University, Malmö, Sweden

Laboratorio de Microbiología Molecular, Departamento de Microbiología y Parasitología, Facultad de Medicina, Universidad Nacional Autónoma de México (UNAM), Ciudad de México, Mexico
e-mail: garciap@unam.mx

J. Santi-Rocca (✉)
Science and Healthcare for Oral Welfare, Toulouse, France
e-mail: jsr@periodontitis.show

J. Santi-Rocca (ed.), *Periodontitis*, Advances in Experimental Medicine and Biology 1373,
https://doi.org/10.1007/978-3-030-96881-6_5

alence between species and subtypes call for more studies to be able to conclude about their individual contributions in the pathophysiology of periodontal diseases. The heterogeneity in prevalence estimation should be investigated further, in particular to make out biological from methodological heterogeneity.

Keywords

Entamoeba gingivalis · *Trichomonas tenax* · Periodontal diseases · Periodontitis · Meta-analysis

Highlights

- The prevalence of *E. gingivalis* in periodontal pockets is 76.9%
- The prevalence of *T. tenax* in periodontal pockets is 38.6%
- *E. gingivalis* is 3.96 more frequent in periodontitis than in healthy sites
- *T. tenax* is 21.82 more frequent in periodontitis than in healthy sites

Considerations for Practice

- Parasites are frequently detected in gingival sulci
- Parasite detection is correlated with periodontal diseases

Patient Summary

The parasites *Entamoeba gingivalis* and *Trichomonas tenax* are unicellular eukaryotic organisms that are frequently detected in the gingival sulci. Researchers have previously shown they own virulence factors, which casts doubts about their possible commensal interactions with the human host. Their more frequent detection in patients with periodontitis suggests they may be involved in the development of this disease. They may become future targets for diagnosis and therapeutic strategies.

5.1 Introduction

Periodontal diseases are characterized by the inflammation of the tissues surrounding the tooth (gingivitis) and the later destruction of the alveolar bone (periodontitis), eventually leading to tooth loss. They are ubiquitous and affect more than half the adults worldwide (Petersen and Ogawa 2005). Periodontal diseases exhibit different stages and grades, in other words, extents and progression rates (Tonetti et al. 2018). Severe forms of the disease affect approximately a fifth of adults between 35 and 44 years old, this proportion increasing with age (Petersen and Ogawa 2005). Age cannot account by its own for all the grades and stages of periodontal diseases: among the factors, the genetic of the host and the microbial traits of the infection are to consider (Offenbacher et al. 2016).

Inflammation in periodontal diseases is triggered by microbes and, along history, different models have tried to expose why a single causative agent could not explain the aetiology of this disease deemed unique (Hajishengallis and Lamont 2012; Hajishengallis 2014; Hajishengallis et al. 2012). Indeed, while some species of interest have been identified, like *Porphyromonas gingivalis*, *Treponema denticola*, and *Tannerella forsythia* from the red complex (Socransky et al. 1998) and *Aggregatibacter actinomycetemcomitans* (Haubek et al. 1996), their participation cannot be considered at the individual level but rather within a given microbial environment in interaction with the host. The three-dimensional arrangements of these biofilms can support a community-like cooperative association (Esteves et al. 2021). This higher grade of organization may sustain synergy and a gain of new functionalities, in particular at the metabolic level through metabolite exchange after specialization in a given step of a metabolic way (Yost et al. 2015; Deng et al. 2018).

Alongside the intensively studied bacteria, other organisms have been identified in the periodontal environments: parasites, fungi, archaea, and viruses. Among the parasites, two species have been described as frequent colonizers of the gingival sulcus: the amoeba *Entamoeba gingiva-*

lis and the flagellate *Trichomonas tenax*. However, their role in the pathophysiology of periodontal diseases is still to be elucidated.

In 1849, *Entamoeba gingivalis* was the first amoeba identified in samples from humans (Gros 1849). The close resemblance with *Entamoeba histolytica*, then recently identified as a pathogen causing dysentery (Lösch 1875), raised suspicion about its link to periodontitis in a late nineteenth century marked by the emergence of Infectiology (Kartulis 1893). The virulence of *E. gingivalis* was and is still questioned, though the scientific community calls for caution when classifying it as a commensal or a saprophyte (Bonner et al. 2018). Indeed, some traits of the amoeba from the mouth are very similar to some virulence mechanisms of *E. histolytica*: phagocytosis of human cells (Prowazek 1904), tissue destruction, and inflammation triggering (Bao et al. 2020; Bao et al. 2021). Microscopic diagnosis has revealed, with a great variability between users, that *E. gingivalis* was ubiquitous around the world and very frequent in periodontitis (reviewed in (Badri et al. 2021)). Several species were suspected to colonize the periodontal sulcus, but were soon unified under the name "*Endameba gingivalis*" (Smith and Barrett 1915). Interestingly, a second subtype of *E. gingivalis*, ST2, has been identified by molecular methods and seems genetically distant from *E. gingivalis* ST1, which leaves the question open about its identity as a new species (Garcia et al. 2018b). Furthermore, data from one publication suggest that these two taxonomic entities are differentially present between gingivitis and periodontitis patients (Garcia et al. 2018a).

Virulence can also vary between variants (strains) of *T. tenax* (Benabdelkader et al. 2019). The parasite is shown or hypothesized to possess numerous virulence factors, which was recently reviewed (Bisson et al. 2019). Furthermore, the presence of the flagellate has been correlated to the disease itself (Marty et al. 2017) and to the severity of periodontal disease, more precisely tooth mobility, accumulation of plaque, and clinical attachment loss (Bisson et al. 2018). It is noteworthy that the prevalence of parasite greatly varies according to the definition of the disease and according to the method used (Santi-Rocca 2020). The most accurate methods to identify organisms in a diagnosis setup using periodontal material involve genetic identification, like PCR, to avoid misidentification and user inference, as already discussed (Bonner et al. 2014).

A better comprehension of the pathophysiology of periodontal diseases requires a deeper understanding of the role of the whole microbiota, including the parasites. And to that end, more homogeneity and consensus in their detection is an important methodological pre-condition for future research. We propose here to conduct a systematic review and a meta-analysis of the prevalence of both parasites and their identifiable subtypes by molecular methods.

5.2 Materials and Methods

This systematic review and meta-analysis was performed in accordance with the PRISMA guidelines (Moher et al. 2009). The protocol received the PROSPERO registration number CRD42021224592.

5.2.1 Systematic Review

Three databases were used: Pubmed (https://pubmed.ncbi.nlm.nih.gov/), LILACS (https://lilacs.bvsalud.org/), and Google Scholar (https://scholar.google.com/). The searches were performed on June 30th 2021.

5.2.1.1 Pubmed Search Strategies and Retrieval

For *E. gingivalis*: (Stomatognathic Diseases [mh:exp] OR periodont*[title/abstract] OR gingiv*[title/abstract]) AND (entamoeba[title/abstract] OR endamoeba[title/abstract])

For *T. tenax*: (Stomatognathic Diseases [mh:exp] OR periodont*[title/abstract] OR gingiv*[title/abstract]) AND (trichomonas[title/abstract])

Results were exported in .csv format and then opened in Microsoft Excel for further treatment.

5.2.1.2 LILACS Search Strategies

For *E. gingivalis*: (entamoeba*) AND ((mh:"stomatognathic diseases") OR (mh:"periodontitis") OR (mh:"periodontal disease") OR (mh:"gingivitis") OR buccal OR oral OR mouth OR gingivalis)

For *T. tenax*: (trichomon*) AND ((mh:"stomatognathic diseases") OR (mh:"periodontitis") OR (mh:"periodontal disease") OR (mh:"gingivitis") OR buccal OR oral OR mouth OR tenax)

Results were exported in .csv format and then opened in Microsoft Excel for further treatment.

5.2.1.3 Google Scholar

For *E. gingivalis*: "Entamoeba gingivalis"

For *T. tenax*: "Trichomonas tenax"

Results were retrieved using the Publish or Perish software (v7.33) (Harzing and Adams 2017) and then opened in Microsoft Excel for further treatment.

5.2.1.4 Selection of Publications

After organizing the columns so the databases from the three sources were compatible and before merging them, the publications were tagged in a column with the source database name. Other 2 columns were added: 1 for the classification index attributed by the researcher and 1 for their comments.

The classification index were the following:

- DUP = duplicated entry
- LAN = publication in a language not understood by the researcher
- +1 = peer-reviewed original research article about the concerned parasite, infecting humans, with obvious diagnosis data from title or abstract. Correspond to the "screening" step.
- −1 = publication not entering the "+1" category and reviews, theses.
- +2 = "+1" publication with numerical epidemiological data obtained using molecular methods, as seen in the full text. Corresponds to the "eligibility" step.
- −2 = "+1" publication not entering the "+2" category
- +3 = "+2" publication concerning adults, with a defined group with periodontitis and a group with healthy controls. The diagnosis must have been done on one site. "Correspond to the "inclusion" step.
- −3 = "+2" publication not entering the "+3" category. The reasons for the exclusion at this step are discussed in the review.

The previous steps were independently performed by GG, MS, and JSR. Results were then compared, and discrepancies were solved by discussing them.

5.2.2 Meta-analysis

We analyzed separately and successively: the set of studies selected for *E. gingivalis* (any strain), for *E. gingivalis* ST1, for *E. gingivalis* ST2; and finally, for *T. tenax*. We used the risk ratio (RR) – and its associated 95%-confidence interval – as a summary statistic for each of the studies. RR refers here to the ratio of the risk of being infected by a given parasite type among patients with periodontitis versus the risk of being infected among a control group of healthy patients. We performed meta-analyses to estimate the summary effect of periodontitis on the prevalence of the parasite, the estimate being calculated as a weighted average of the effects estimated in the individual studies.

We systematically assessed the heterogeneity of the studies following a triple approach. First, we applied a traditional Chi^2 to assess whether observed differences in results across studies were compatible with chance alone: a low p-value provides evidence of heterogeneity. We also measured the inconsistency I^2, which provides the percentage of the variability in effect estimates that is due to heterogeneity rather than sampling error (chance). And finally, we measured the extent of this variation, referred to as τ^2, or Tau^2, which is the estimated variance of underlying effects across studies.

We relied of the inverse-variance method approach to compute the summary effect as a weighted average of the estimated effects from their respective set of eligible studies. According to this approach, the weight attributed to each study is the inverse of the variance of the effect estimate. Considering the heterogeneous set of studies, we implemented systematically both a fixed-effect method and a random-effects method of meta-analysis, in order to compare and discuss them. A fixed-effect meta-analysis assumes that all effect estimates are estimating the same underlying "intervention effect" (here, the effect of periodontal disease on parasite presence), however, the result of the meta-analysis can be interpreted without making such an assumption (Rice et al. 2018). In such a model the weight of each individual estimate is the inverse of the standard error of that estimate. On the other hand, a random-effects model assumes that the different studies are estimating different, yet related and normally distributed, effects (Higgins et al. 2009), in that case, still following the inverse-variance method, the variance now includes the between-studies variance, tau-squared, in addition to the within-studies variance given by the individual standard errors. This translates into attributing more weight to smaller studies than the previous method would.

The meta-analyses and the forest plots of the data were performed with R 3.5.2, resorting to the "Metafor 3.0-2" and "Meta 4.18-2" packages.

5.3 Results

5.3.1 *Entamoeba gingivalis*

A total of 1207 studies were identified: 1120 in Google Scholar, 53 in Pubmed, 34 in Lilacs. After applying the filters presented in the Methods section and in Fig. 5.1, 14 studies were identified as peer-reviewed articles including epidemiological data generated by molecular methods.

It is noteworthy that the founder study by Kikuta and colleagues was excluded because the definition of the disease included gingivitis (Kikuta et al. 1996). Other four studies were excluded because there was no healthy control group (Cembranelli et al. 2013; Hussian 2017; Huang et al. 2020; Stensvold et al. 2021). In a study, the patients were not classified into groups corresponding to their periodontal status (Sharifi et al. 2020). Was also rejected a study about children (Zahraa Abd Alhammza Abbass 2020). The sole study using metagenomics was excluded because the samples were chosen according to previous microbial outcomes in former studies (Deng et al. 2017). Finally, another study was rejected because of technical incongruencies: while the EGO-1 and EGO-2 primers from the Kikuta study were used, the amplicon was not from the expected size: 250 bp instead of the expected 1.4 kb (Rahdar et al. 2019).

The characteristics of the included studies are presented in Table 5.1. The 5 studies were published between 2011 and 2020 by different research teams. It is noteworthy that the definition of periodontitis greatly varies between the studies. The only comparable parameter is the periodontal pocket depth (PD) that has a threshold value ranging from 3 to 6 mm between the studies, corresponding to stage I and stages III-IV in the current AAP-EFP classification, respectively. For instance, in the Bonner study, patients could be at any stage of the disease, while patients from the Garcia study could only be at stages III-IV (Bonner et al. 2014; Garcia et al. 2018a). Variability was also observed for exclusion criteria, as for the way patients were enrolled (see comments in Table 5.1) and for the molecular diagnosis itself, from the sampling from patient to the primers that were used. The risks of biases were evaluated (Table 5.2) and was relatively low either at the study or at the domain level.

The prevalence for *E. gingivalis* in periodontal pockets is 76.9% (95%-CI: 71.5–81.7%), and 29.6% in controls (95%-CI: 24.4–35.3%) pooling all the individuals from the different studies without applying weights. In the same way, ST1 is highly prevalent in periodontal pockets 70.6% (95%-CI: 65.0–76.2%) and much less in healthy sites 28.7% (95%-CI: 23.4–34.1%). In contrast, ST2 was moderately prevalent in periodontal pockets 43.9% (95%-CI: 35.5–52.4%), and close to the prevalence observed for ST1 in healthy controls 23.0% (95%-CI: 15.9–30.1%).

Using meta-analysis methods, the overall relative risk is 2.44 according to a fixed-model approach. But, with regards to the heterogeneity between the studies (I^2 = 92%, $p < 0.01$), conducted in different countries and conditions, we consider wiser to conclude with the results of the

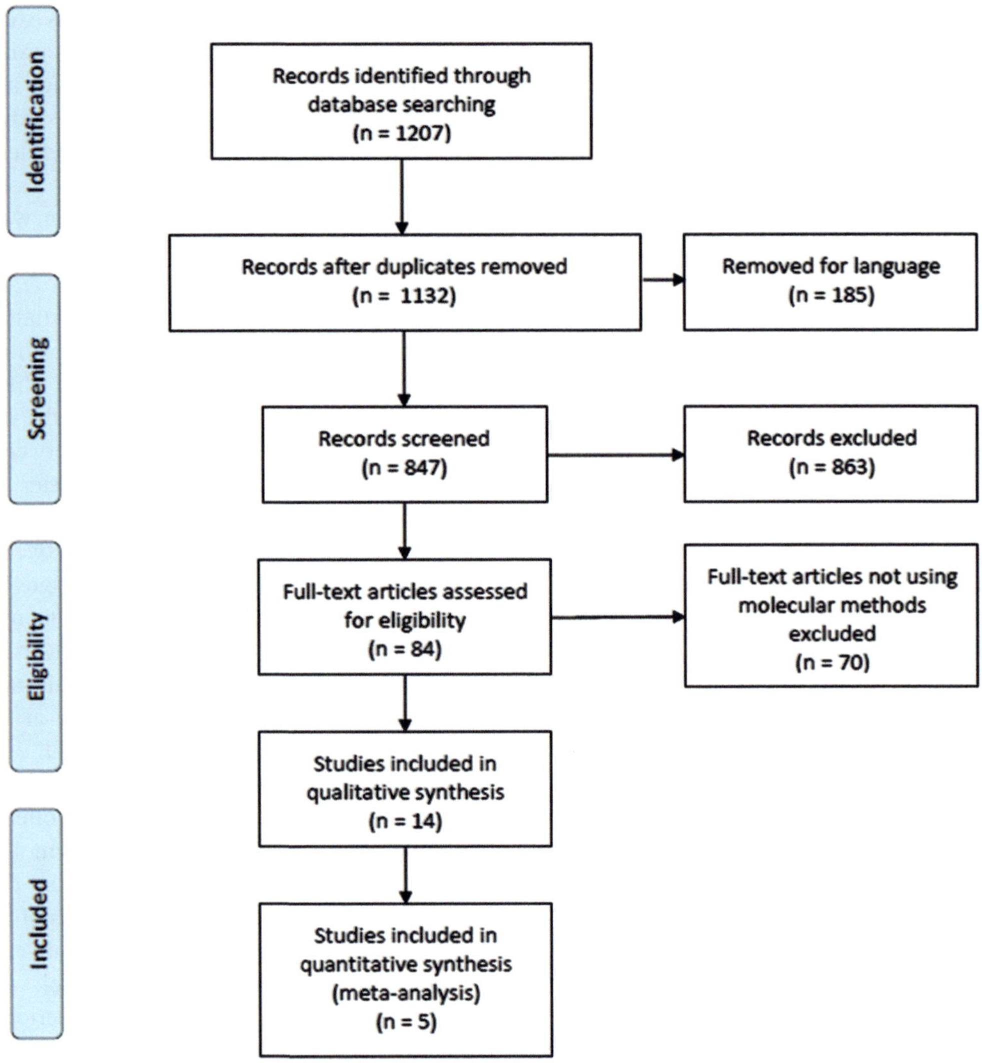

Fig. 5.1 Prisma flow diagram for the selection of studies about *E. gingivalis*

random effects model claiming that the relative risks in different contexts are log-normaly distributed around 3.96 (95%-CI: 1.57–9.98), meaning that the prevalence is predicted to be around 3.96 times as high in the periodontal pockets as in healthy sulci (Fig. 5.2 Panel A). Interestingly, two studies (Trim and Dubar) exhibit elevated relative risks and very low or null prevalence in the control group, accounting for most of between-studies heterogeneity: the biological (or merely methodological) reasons for such null prevalences in health would be an important issue to solve in further research.

Interestingly, PCR primers allowed to indistinctly detect both ST1 and ST2 in the Trim's study, as evidenced in another one (Zaffino et al. 2019). In the Garcia and Dubar studies, ST1 and ST2 were detected separately. Exclusive detection of ST1 by the Bonner's PCR primers was determined by the same methodological study and were re-used in the Bao's study. Though low numbers of studies were then used, we chose to distinguish the two subtypes.

For ST1, the overall relative risk using a random effects model is distributed around 3.30 (95%-CI: 1.27–8.55) (Fig. 5.2 Panel B). The

Table 5.1 Characteristics of the 2 studies included for *E. gingivalis* meta-analysis, and presentation of their results

Study		First author	Trim	Bonner	Garcia	Dubar	Bao
		Year	2011	2014	2018	2019	2020
Characteristics of the studies	Recruitment	Case	>4 mm	2 or more positive: oedema, bleeding, PD ≥ 3 mm, GR ≥ 1 mm, mobility ≥ 2	PD ≥6 mm	Stage II-IV, grade A-C, 2 pockets ≥ 5mm	"periodontitis as defined by the 2018 AAP-EFP classification" (no indications of stage or grade)
		Control	<3 mm	Not positive	Not positive, without treatment	Not positive	No inflammation
		Exclusion	Antibiotics in the last 6 weeks	Antibiotics in the last few months, chronic disease, pregnancy	Not stated	Medication modifying microbiota, antibiotics, root planing in the last 6 months, pregnancy	Not stated
		Other patient characteristics	Not stated	Recorded but not presented	Recorded but not presented	Not stated	Recorded in the periodontitis group
	PCR	Comments	Controls were taken either in healthy sites in the mouths of 7 cases, or in 5 healthy individuals	Multi-site		Different smoking habits between groups	Controls were staff volunteers who self-reported as healthy before clinical check
		Sampling	Paper point	Curettage	Collection brush	Paper point	Curettage/swab
		Blinded treatment of samples (ascertainment)	Coded	Coded	Transferred to another service	Coded	Coded
		Purification	QIAamp DNA mini	Phenol-chloroform extraction and alcohol precipitation	QIAamp DNA mini	QIAamp DNA mini	Phenol-chloroform extraction and alcohol precipitation
		Method	cPCR	PCR	Nested PCR	qPCR	PCR
		Conclusion (espèce/Subtype)	ST1 + ST2 (Zaffino et al. 2019)	ST1 (Zaffino et al. 2019)	ST1 + ST2	ST1 + ST2	ST1 (Zaffino et al. 2019)
		Normalisation	no	n/a	n/a	no	n/a
		Control of (i) inhibitors and (ii) degradation/purifcation	(ii) amplification of eukaryotic and prokaryotic DNA	(i) spiking with limiting dilutions of Eg DNA, (ii) amplification of human DNA	No	No	(ii) human actin gene amplification

(continued)

Table 5.1 (continued)

ST1+ST2	case	Age	Not stated	Recorded but not presented	49.8	51	61
		Sex ratio	Not stated	Recorded but not presented	0.5	0.76	0.89
		Positive	18	58	75	26	39
		Total	26	72	102	30	51
		Prevalence	69.2%	80.6%	73.5%	86.7%	76.5%
	control	Age	Recorded but not presented	Recorded but not presented	24.3	55	42
		Sex ratio	Not stated	Recorded but not presented	0.36	0.76	0.84
		Positive	0	11	57	1	16
		Total	12	33	105	30	107
		Prevalence	0.0%	33.3%	54.3%	3.3%	15.0%
ST1	case	Positive		58	59	24	39
		Total		72	102	30	51
		Prevalence		80.6%	57.8%	80.0%	76.5%
	control	Positive		11	51	1	16
		Total		33	105	30	107
		Prevalence		33.3%	48.6%	3.3%	15.0%
ST2	case	Positive			51	7	
		Total			102	30	
		Prevalence			50.0%	23.3%	
	control	Positive			31	0	
		Total			105	30	
		Prevalence			29.5%	0.0%	

Table 5.2 Modified Newcastle-Ottawa quality assessment Scale, presenting the 5 studies included for *E. gingivalis* meta-analysis

Study	First author Year	Trim 2011	Bonner 2014	Garcia 2018	Dubar 2019	Bao 2020	Score (/domain)
Selection	Adequacy of case definition	+	+	+	+	possible bias	4/5
	Representativeness of cases	+	+	+	+	+	5/5
	Selection of controls	possible bias	+	+	+	possible bias	3/5
	Definition of controls	+	+	+	+	+	5/5
Comparability	Comparability of cases and controls	possible bias	possible bias	+ age bias	++	+ age bias	4/10
Exposure	Ascertainment of exposure	+	+	+	+	+	5/5
	Same method for cases and controls	+	+	+	+	+	5/5
	Score (/study)	5/8	6/8	7/8	8/8	5/8	

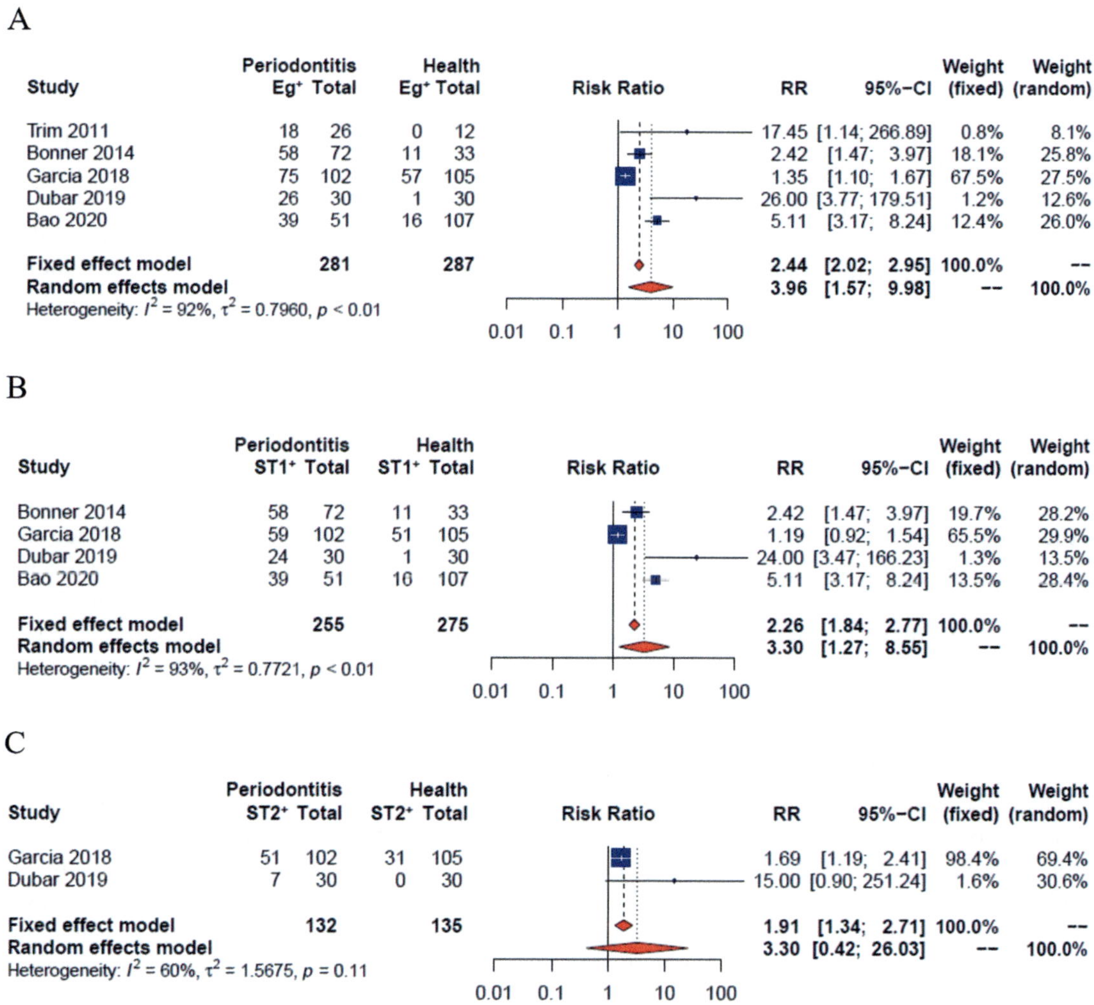

Fig. 5.2 Results of the meta-analysis, presented as forest plots, for *E. gingivalis*. Panel A: results without differentiating ST1 and ST2. Panel B: results for ST1. Panel C: results for ST2

results are heterogeneous between the 4 studies ($I^2 = 93\%$, $p < 0.01$). Yet, the Dubar study is an outlier with an elevated relative risk (with a very wide 95% confidence interval) and a very low prevalence in the control group. One could be tempted to minimize the representativity (and weight) of this outlier and bet on the fixed-effect model outcome, concluding on a lower universal risk ratio for ST1 of 2.26 (95%-CI: 1.84–2.77).

For ST2, only two studies are included (Garcia and Dubar); the overall relative risk using a random effects model is 3.30 (95%-CI: 0.42–26.03) (Fig. 5.2 Panel B). The Dubar study has again an elevated relative risk (with a very wide 95% confidence interval) and a null prevalence in the control group. In that case, meta-analysis is not so meaningful and instructive beyond single studies until we gain further insight in the understanding of the respective biological and methodological context of the two studies.

Altogether, these results show that *E. gingivalis*, may it be the whole complex, ST1 or ST2, has a consistently higher prevalence in periodontal pockets than in healthy sulci. Despite heterogeneity, all single studies account for a risk ratio significantly superior to 1.

5.3.2 *Trichomonas tenax*

A total of 1261 studies were identified: 1187 in Google Scholar, 45 in Pubmed, 29 in Lilacs. After applying the filters presented in the Methods section and in Fig. 5.3, 7 studies were identified as peer-reviewed articles including epidemiological data generated by molecular methods.

Three studies were excluded because the patients were children (Mehr et al. 2015; Abbass et al. 2020; Hamad 2021). Two other studies were excluded because the patients were not classified into groups corresponding to their periodontal status (Dybicz et al. 2018; Jaffer et al. 2020) ; and another one did not include healthy controls (Bracamonte-Wolf et al. 2019). Finally, a study was rejected because of a strong methodological specificity: several sites were sampled and pooled (Benabdelkader et al. 2019). This operation changes the outcome of the diagnosis: the prevalence for various sites – or even of the whole mouth – is measured, which consistently differs with the prevalence for a particular site, taking into consideration that parasites are not homogeneously scattered in a mouth.

The characteristics of the two included studies are presented in Table 5.3. The definition of the inclusion and exclusion parameters was not clearly stated in the Athari paper, resulting in a consistent risk of bias for the study (Table 5.4).

The prevalence for *T. tenax* in periodontal pockets is 38.6% (95%-CI: 27.2–50.0%), and 1.6% in controls (95%-CI: 0.0–3.3%) pooling all the individuals from the different studies without applying weights.

Using meta-analysis methods, the overall relative risk using a fixed effects model is 21.82 (95%-CI: 6.71–70.96), meaning that the prevalence is predicted to be 21.82 times as high in the periodontal pockets as in healthy sulci (Fig. 5.4). The results are homogeneous between the studies ($I^2 = 0\%$, $p = 0.96$). The high convergence should be moderated by the limited size of the sample of studies.

5.4 Discussion

In this meta-analysis, we have shown congruent results from analyses allowing to conclude that *E. gingivalis* and *T. tenax* are more abundantly detected in the periodontal environment during periodontitis than in healthy setups. It is noteworthy that there was no inconsistency in the direction of the effects since all the included studies have evidenced a higher prevalence in the sick.

However, a certain heterogeneity between the studies has been detected and reported in the Tables. In particular, the definition of the disease, which is diverse, should be unified using the world consensus published by AAP-EFP as a starting point (Tonetti et al. 2018) even if the authors can provide supplementary parameters to

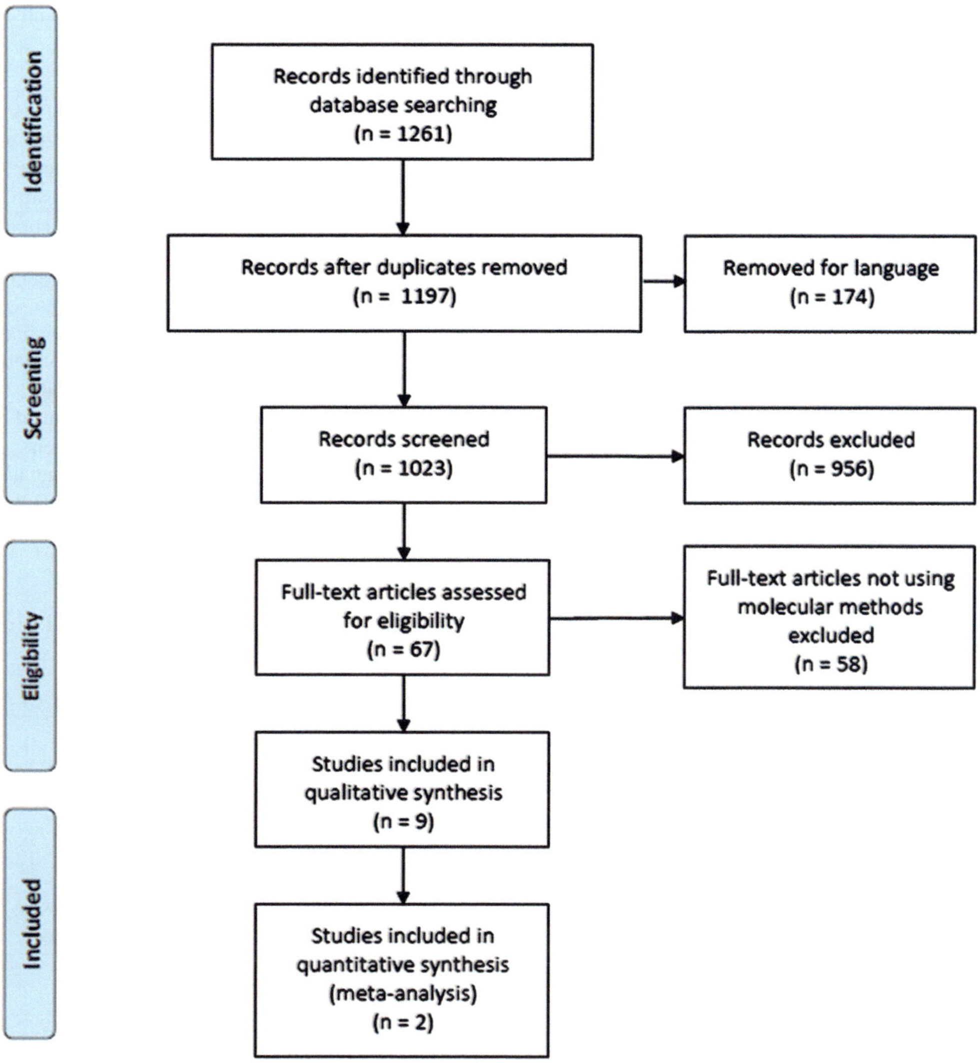

Fig. 5.3 Prisma flow diagram for the selection of studies about *T. tenax*

refine this classification afterwards. This is particularly obvious in the case of *T. tenax* because the parasite detection is correlated to disease severity (Bisson et al. 2018). Furthermore, some *T. tenax* strains are linked to the severity of the disease (Benabdelkader et al. 2019) and *E. gingivalis* ST1 and ST2 may have different prevalence patterns among periodontal diseases (see (Garcia et al. 2018a) and discussion below), meaning that the understanding of the interactions between parasites and their host cannot be summed up by an analysis of 3 interacting partners in two health states. Considering that the full periodontal microbiota could be a combination of several traits impacted by human genetics (Offenbacher et al. 2016), the Sisyphean task to build intelligible models of periodontitis pathophysiology integrating all the kingdoms of the living may be in its infancy.

The Garcia and Dubar studies allowed us to split the results obtained for *E. gingivalis* ST1 and ST2. Interestingly, even if the RR are equal

Table 5.3 Characteristics of the 2 studies included for *T. tenax* meta-analysis, and presentation of their results

Study		First author / Year	Athari 2007	Dubar 2019
Characteristics of the studies	Recruitment	Case	"periodontitis"	Stage II-IV, grade A-C, 2 pockets ≥ 5mm
		Control	"healthy"	Not positive
		Exclusion	Not stated	Medication modifying microbiota, antibiotics, root planing in the last 6 months, pregnancy
		Other patient characteristics	Not stated	Not stated
		Comments		Different smoking habits between groups
	PCR	Sampling	Curettage	Paper point
		Blinded treatment of samples (ascertainment)	Transferred to another service	Coded
		Purification	Phenol-chloroform extraction and alcohol precipitation	QIAamp DNA mini
		Method	PCR	qPCR
		Normalisation	n/a	no
		Control of (i) inhibitors and (ii) degradation/purifcation	no	no
Results	Case	Age	est. 20–60	51
		Sex ratio	est. 0.88	0.76
		Positive	16	11
		Total	40	30
		Prevalence	40.0%	36.7%
	Control	Age	est. 20–60	55
		Sex ratio	est. 0.88	0.76
		Positive	3	0
		Total	160	30
		Prevalence	1.9%	0.0%

Table 5.4 Modified Newcastle-Ottawa quality assessment scale, presenting the 2 studies included for *T. tenax* meta-analysis

Study	First author / Year	Athari 2007	Dubar 2019	Score (/domain)
Selection	Adequacy of case definition	Possible bias	+	1/2
	Representativeness of cases	Possible bias	+	2/2
	Selection of controls	+	+	2/2
	Definition of controls	Possible bias	+	1/2
Comparability	Comparability of cases and controls	Possible bias	++	2/4
Exposure	Ascertainment of exposure	+	+	2/2
	Same method for cases and controls	+	+	2/2
	Score (/study)	3/8	8/8	

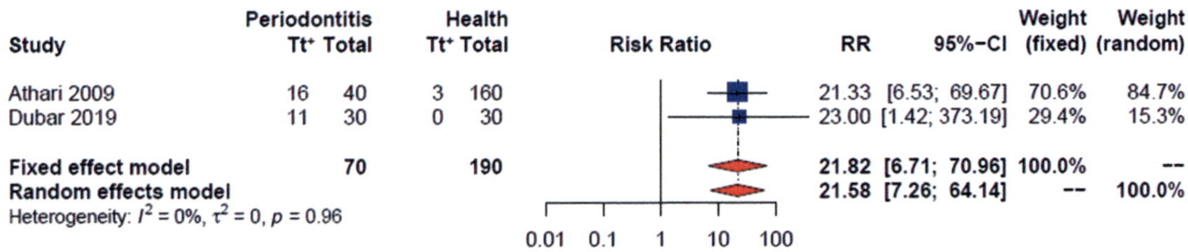

Fig. 5.4 Results of the meta-analysis, presented as a forest plot, for *T. tenax*

for the two subtypes, the prevalence is not. Indeed, for ST1, the prevalence in the sick was 80.6% (Bonner), 57.8% (Garcia), 80.0% (Dubar), and 76.5% (Bao), while it was 50.0% (Garcia) and 23.3% (Dubar) for ST2 (see Table 5.1 for details). The higher prevalence of ST1 or ST1+ST2 in the healthy of the Garcia study is noteworthy (48.6% and 54.3%, respectively) and may reveal either methodological biases or specificities in the studied population. In this study, gingivitis patients with orthodontics were included and revealed a higher prevalence of ST2 than in the controls (73.8% and 29.5%) and a lower prevalence of ST1 (47.5% in gingivitis patients and 48.6% in the healthy). The difference in infection patterns between these two subtypes suggests that there may thrive on different environments or take part in their ecological equilibrium.

Some differences were also observed in previous studies for *T. tenax*, as reviewed by Bisson and colleagues (Bisson et al. 2019), the prevalence of the parasite ranging from 7.1% by microscopy (Ferrara et al. 1986) to 28.2% by PCR (Benabdelkader et al. 2019) in the healthy, from 0% by microscopy (Ferrara et al. 1986) to 32.3% by microscopy (Feki et al. 1981) in gingivitis sites, and from 32.3% by microscopy (Ferrara et al. 1986) to 42.5% by PCR (Benabdelkader et al. 2019) in periodontal patients. Though different detection techniques were used, putting these data together allows to extract a trend, which must be moderated by the fact that the prevalence of *T. tenax* is expected to be variable in the periodontitis group according to the severity of the disease (Bisson et al. 2018). In the case of *E. gingivalis* detection, a high variability in the prevalence in health groups calls for caution about the definition of health, which should not be simply defined as the cases excluded from the disease group but should respond to specific inclusion criteria. This would allow to differentiate between repetitive seeding with harmful parasites, taking into account incubation and early events of invasion, and commensal behavior, as already discussed elsewhere (Bonner et al. 2014; Bonner et al. 2018). Further studies should include finer classification and more clinical parameters to better appreciate the ecology of the environments hosting and infected by both parasites, and a great effort is needed to define the environments devoid of them.

In a parallel manner, when looser inclusion criteria are applied and various diagnosis techniques are included, the prevalence in the sick appears lower (37%, 95%-CI: 29–46) (Badri et al. 2021). Unfortunately, the methodology did not preclude the inclusion of studies without control groups: a relative risk is thus impossible to calculate. It is noteworthy that two studies here included were excluded from their analysis: the Bonner and Garcia studies, while a preprint of another one was included, in which Dr. Santi-Rocca is also a co-author (Yaseen et al. 2020), highlighting the importance to precisely define inclusion and exclusion criteria, and to discuss all the eligible studies in the systematic review.

This incongruence between meta-analyses pinpoints the importance of a precise definition of the inclusion criteria in the studies and in the meta-analyses themselves. First, at the level of the studies, this allows the selection of a sharp subcategory within the periodontal diseases, which will be more reproducible. In the present meta-analysis, the Deng study was excluded because patients were chosen according to their bacterial profiles (Szafranski et al. 2015). This exclusion was motivated by the impossibility to relate to other studies but may be a valid way to make more homogeneous subgroups in order to define a precise type of periodontitis. Indeed, though conclusions with such low numbers of patients should be regarded with caution, it is noteworthy that 100% of periodontal cases were positive for *E. gingivalis*, with very relatively high abundance of E. gingivalis transcripts. This may be relevant at the pathophysiological level.

Making sub-categories seems to be a double-edged sword since it will not be representative of all the situations. But after a century of experiments and models pre-supposing that the disease was driven by a single ecological environment with a unique immune response from the host, the scientific and medical community agrees on the fact that the disease evolves in several steps and might be of several types (Offenbacher et al. 2016). The scientific community must wisely and cautiously dissect all these steps and should not yield to the temptation of oversimplification, which is, in the case of this complex disease, deleterious for the comprehension of its pathophysiology.

In the present meta-analysis, the choice between a wider inclusion and a loss of precision was not made, resulting in a very low number of included papers. This is the main weakness of this study, which highlights the need for the scientific community to gather and unify the methods and define the contours of the future research in this field. A consensus about a common direction is needed.

New data about the pathogenicity of *E. gingivalis* suggest that the amoeba may not be saprophytic as formerly described (Bao et al. 2020; Bao et al. 2021) but invite us to reassess its role in periodontitis pathophysiology (Bonner et al. 2018). The pathogenicity of *T. tenax* is also questioned (Marty et al. 2017; Bisson et al. 2019).

It is too early to determine the role of the parasites in the pathophysiology of periodontitis, but their presence is undoubted and unequivocal. The amoeba *E. gingivalis* and the flagellate *T. tenax* can be formally defined as risk factors for periodontitis. Future studies will determine which part they take in the pathophysiology of the disease and if they are *bona fide* prognosis markers and possible targets to control the disease.

5.5 Author Contribution

Study design: JSR
Data acquisition: DD; MS, JSR
Analysis and interpretation of data: GG, DM, MS, JSR
Drafting of the manuscript: JSR
Reviewing of the manuscript: GG, DM, MS, JSR
Study supervision: JSR

References

Abbass ZAA, Hanoon SA, Kadhim TA (2020) The prevalence of Entamoeba gingivalis and Trichomonas tenax in children treated with orthodontic appliances in AL Muthanna Province, Iraq. Indian Journal of Public Health 11(02):1983

Badri M, Olfatifar M, Abdoli A, Houshmand E, Zarabadipour M, Abadi PA, Johkool MG, Ghorbani A, Eslahi AV (2021) Current global status and the epidemiology of *Entamoeba gingivalis* in humans: a systematic review and meta-analysis. Acta Parasitol 66:1102–1113. https://doi.org/10.1007/s11686-021-00423-2

Bao X, Wiehe R, Dommisch H, Schaefer AS (2020) *Entamoeba gingivalis* causes oral inflammation and tissue destruction. J Dent Res 99(5):561–567. https://doi.org/10.1177/0022034520901738

Bao X, Weiner J 3rd, Meckes O, Dommisch H, Schaefer AS (2021) *Entamoeba gingivalis* exerts severe pathogenic effects on the Oral Mucosa. J Dent Res 100(7):771–776. https://doi.org/10.1177/00220345211004498

Benabdelkader S, Andreani J, Gillet A, Terrer E, Pignoly M, Chaudet H, Aboudharam G, La Scola B (2019) Specific clones of *Trichomonas tenax* are associated with periodontitis. PLoS One 14(3):e0213338. https://doi.org/10.1371/journal.pone.0213338

Bisson C, Lec PH, Blique M, Thilly N, Machouart M (2018) Presence of trichomonads in subgingival biofilm of patients with periodontitis: preliminary results. Parasitol Res 117(12):3767–3774. https://doi.org/10.1007/s00436-018-6077-2

Bisson C, Dridi SM, Machouart M (2019) Assessment of the role of *Trichomonas tenax* in the etiopathogenesis of human periodontitis: a systematic review. PLoS One 14(12):e0226266. https://doi.org/10.1371/journal.pone.0226266

Bonner M, Amard V, Bar-Pinatel C, Charpentier F, Chatard JM, Desmuyck Y, Ihler S, Rochet JP, Roux de La Tribouille V, Saladin L, Verdy M, Girones N, Fresno M, Santi-Rocca J (2014) Detection of the amoeba *Entamoeba gingivalis* in periodontal pockets. Parasite 21:30. https://doi.org/10.1051/parasite/2014029

Bonner M, Fresno M, Girones N, Guillen N, Santi-Rocca J (2018) Reassessing the role of *Entamoeba gingivalis* in periodontitis. Front Cell Infect Microbiol 8:379. https://doi.org/10.3389/fcimb.2018.00379

Bracamonte-Wolf C, Orrego PR, Munoz C, Herrera D, Bravo J, Gonzalez J, Varela H, Catalan A, Araya JE (2019) Observational cross-sectional study of *Trichomonas tenax* in patients with periodontal disease attending a Chilean university dental clinic. BMC Oral Health 19(1):207. https://doi.org/10.1186/s12903-019-0885-3

Cembranelli SB, Souto FO, Ferreira-Paim K, Richinho TT, Nunes PL, Nascentes GA, Ferreira TB, Correia D, Lages-Silva E (2013) First evidence of genetic intraspecific variability and occurrence of *Entamoeba gingivalis* in HIV(+)/AIDS. PLoS One 8(12):e82864. https://doi.org/10.1371/journal.pone.0082864

Deng ZL, Szafranski SP, Jarek M, Bhuju S, Wagner-Dobler I (2017) Dysbiosis in chronic periodontitis: key microbial players and interactions with the human host. Sci Rep 7(1):3703. https://doi.org/10.1038/s41598-017-03804-8

Deng ZL, Sztajer H, Jarek M, Bhuju S, Wagner-Dobler I (2018) Worlds apart – transcriptome profiles of key oral microbes in the periodontal pocket compared to single laboratory culture reflect synergistic interactions. Front Microbiol 9:124. https://doi.org/10.3389/fmicb.2018.00124

Dybicz M, Perkowski K, Sedzikowska A, Baltaza W, Chomicz L (2018) Studies on prevalence of infection with *Trichomonas tenax* identified by molecular techniques – in respect to oral health of patients with various systemic disease requiring immunosuppressive therapy. Ann Parasitol 64(3):193–197

Esteves GM, Pereira JA, Azevedo NF, Azevedo AS, Mendes L (2021) Friends with benefits: an inside look of periodontal microbes' interactions using fluorescence in situ hybridization—scoping review. Microorganisms 9(7):1504

Feki A, Molet B, Haag R, Kremer M (1981) Protozoa of the human oral cavity (epidemiological correlations and pathogenic possibilities). J de biologie buccale 9(2):155–161

Ferrara A, Conca R, Grassi L Jr, de Carneri I (1986) Possible pathogenic role of *Trichomonas tenax* in chronic periodontitis. Ann Ist Super Sanita 22(1):253–255

Garcia G, Ramos F, Maldonado J, Fernandez A, Yanez J, Hernandez L, Gaytan P (2018a) Prevalence of two *Entamoeba gingivalis* ST1 and ST2-kamaktli subtypes in the human oral cavity under various conditions. Parasitol Res 117(9):2941–2948. https://doi.org/10.1007/s00436-018-5990-8

Garcia G, Ramos F, Martinez-Hernandez F, Hernandez L, Yanez J, Gaytan P (2018b) A new subtype of *Entamoeba gingivalis*: "*E. gingivalis* ST2, kamaktli variant". Parasitol Res 117(4):1277–1284. https://doi.org/10.1007/s00436-018-5798-6

Gros G (1849) Fragments d'helminthologie et de physiologie microscopique. Bulletin de la Société Impériale des Naturalistes de Moscou 22:549–573

Hajishengallis G (2014) Immunomicrobial pathogenesis of periodontitis: keystones, pathobionts, and host response. Trends Immunol 35(1):3–11. https://doi.org/10.1016/j.it.2013.09.001

Hajishengallis G, Lamont RJ (2012) Beyond the red complex and into more complexity: the polymicrobial synergy and dysbiosis (PSD) model of periodontal disease etiology. Mol Oral Microbiol 27(6):409–419. https://doi.org/10.1111/j.2041-1014.2012.00663.x

Hajishengallis G, Darveau RP, Curtis MA (2012) The keystone-pathogen hypothesis. Nat Rev Microbiol 10(10):717–725. https://doi.org/10.1038/nrmicro2873

Hamad HK (2021) Prevalence of oral trichomoniasis in Down syndrome children with gingivitis in Baghdad. Heart Dis 11(7):64

Harzing A-W, Adams D (2017) Publish or perish: realising Google Scholar's potential to democratise citation analysis. In: Google Scholar day: changing current evaluation paradigms (Cybermetrics Lab) Madrid, Spain. Seminar.

Haubek D, Poulsen K, Westergaard J, Dahlèn G, Kilian M (1996) Highly toxic clone of *Actinobacillus actinomycetemcomitans* in geographically widespread cases of juvenile periodontitis in adolescents of African origin. J Clin Microbiol 34(6):1576–1578. https://doi.org/10.1128/jcm.34.6.1576-1578.1996

Higgins JPT, Thompson SG, Spiegelhalter DJ (2009) A re-evaluation of random-effects meta-analysis. J R

Stat Soc Ser A (Stat Soc) 172(1):137–159. https://doi.org/10.1111/j.1467-985X.2008.00552.x
Huang JM, Ting CC, Chen YC, Yuan K, Lin WC (2020) The first study to detect co-infection of *Entamoeba gingivalis* and periodontitis-associated bacteria in dental patients in Taiwan. J Microbiol Immunol Infect 54:745–747. https://doi.org/10.1016/j.jmii.2020.08.001
Hussian RSH (2017) Molecular detection of *Entamoeba gingivalis* using polymerase chain reaction. Pak J Biotechnol 14(3):351–354
Jaffer NT, Al-Noori AS, Salih AM, Zuhdi SS (2020) Molecular detection of oral *Trichomonas tenax* among individuals attending dental care units using PCR in Duhok city – Kurdistan region. J Duhok Univ 22(2):195–202. https://doi.org/10.26682/sjuod.2019.22.2.22
Kartulis S (1893) Ueber pathogene Protozoën bei dem Menschen. Zeitschrift für Hygiene und Infektionskrankheiten 13
Kikuta N, Yamamoto A, Goto N (1996) Detection and identification of *Entamoeba gingivalis* by specific amplification of rRNA gene. Can J Microbiol 42(12):1248–1251. https://doi.org/10.1139/m96-161
Lösch F (1875) Massenhafte Entwicklung von Amöben im Dickdarm. Virchow Arch 65:196–211
Marty M, Lemaitre M, Kemoun P, Morrier JJ, Monsarrat P (2017) *Trichomonas tenax* and periodontal diseases: a concise review. Parasitology 144(11):1417–1425. https://doi.org/10.1017/S0031182017000701
Mehr AK, Zarandi A, Anush K (2015) Prevalence of oral *Trichomonas tenax* in periodontal lesions of Down syndrome in Tabriz, Iran. J Clin Diagn Res 9(7):Zc88–Zc90. https://doi.org/10.7860/jcdr/2015/14725.6238
Moher D, Liberati A, Tetzlaff J, Altman DG (2009) PRISMA Group. Preferred reporting items for systematic reviews and meta-analyses: the PRISMA statement. PLoS Med 6(7):e1000097. https://doi.org/10.1371/journal.pmed.1000097. Epub 2009 Jul 21. PMID: 19621072; PMCID: PMC2707599
Offenbacher S, Divaris K, Barros SP, Moss KL, Marchesan JT, Morelli T, Zhang S, Kim S, Sun L, Beck JD, Laudes M, Munz M, Schaefer AS, North KE (2016) Genome-wide association study of biologically informed periodontal complex traits offers novel insights into the genetic basis of periodontal disease. Human Mol Genet 25(10):2113–2129. https://doi.org/10.1093/hmg/ddw069
Petersen PE, Ogawa H (2005) Strengthening the prevention of periodontal disease: the WHO approach. J Periodontol 76(12):2187–2193. https://doi.org/10.1902/jop.2005.76.12.2187
Prowazek SV (1904) *Entamoeba buccalis* n. sp. Arb ad kaiserl Gesundheitsamte 21:42–44
Rahdar M, Abolfazli-Karizi S, Pedram H (2019) The comparison of *Entamoeba gingivalis* presence in healthy and Periodontitis patients by using direct examination and PCR methods. Jundishapur J Health Sci 11(1):e86066. https://doi.org/10.5812/jjhs.86066
Rice K, Higgins JPT, Lumley T (2018) A re-evaluation of fixed effect(s) meta-analysis. J R Stat Soc Ser A (Stat Soc) 181(1):205–227. https://doi.org/10.1111/rssa.12275
Santi-Rocca J (2020) The Protozoome of the Periodontal Sulcus: from health to disease. In: First Eukaryome congress at Institut Pasteur, Institut Pasteur, Paris, October 16–18, 2019 2020. Eukaryome Impact on Human Intestine Homeostasis and Mucosal Immunology. Springer, pp 113–131. https://doi.org/10.1007/978-3-030-44826-4_9
Sharifi M, Jahanimoghadam F, Babaei Z, Mohammadi MA, Sharifi F, Hatami N, Danesh M, Poureslami P, Poureslami H (2020) Prevalence and associated-factors for *Entamoeba gingivalis* in adolescents in Southeastern Iran by culture and PCR, 2017. Iran J Publ Health 49(2):351–359
Smith AJ, Barrett MT (1915) The parasite of oral Endamebiasis, *Endameba gingivalis* (Gros). J Parasitol 1(4):159–174. https://doi.org/10.2307/3270804
Socransky SS, Haffajee AD, Cugini MA, Smith C, Kent RL Jr (1998) Microbial complexes in subgingival plaque. J Clin Periodontol 25(2):134–144
Stensvold CR, Nielsen M, Baraka V, Lood R, Fuursted K, Nielsen HV (2021) *Entamoeba gingivalis*: epidemiology, genetic diversity and association with oral microbiota signatures in North Eastern Tanzania. J Oral Microbiol 13(1):1924598. https://doi.org/10.1080/20002297.2021.1924598
Szafranski SP, Wos-Oxley ML, Vilchez-Vargas R, Jauregui R, Plumeier I, Klawonn F, Tomasch J, Meisinger C, Kuhnisch J, Sztajer H, Pieper DH, Wagner-Dobler I (2015) High-resolution taxonomic profiling of the subgingival microbiome for biomarker discovery and periodontitis diagnosis. Appl Environ Microbiol 81(3):1047–1058. https://doi.org/10.1128/AEM.03534-14
Tonetti MS, Greenwell H, Kornman KS (2018) Staging and grading of periodontitis: Framework and proposal of a new classification and case definition. J Periodontol 89(Suppl 1):S159–S172. https://doi.org/10.1002/JPER.18-0006
Yaseen A, Mahafzah A, Dababseh D, Tayem D, Hamdan AA, Al-Fraihat E, Hassona Y, Şahin GÖ, Santi-Rocca J, Sallam M (2020) Oral colonization by *Entamoeba gingivalis* and *Trichomonas* tenax: PCR-based study in health, gingivitis, and periodontitis. medRxiv:2020.2007.2016.20155861. https://doi.org/10.1101/2020.07.16.20155861

Yost S, Duran-Pinedo AE, Teles R, Krishnan K, Frias-Lopez J (2015) Functional signatures of oral dysbiosis during periodontitis progression revealed by microbial metatranscriptome analysis. Genome Med 7(1):27. https://doi.org/10.1186/s13073-015-0153-3

Zaffino M, Dubar M, Debourgogne A, Bisson C, Machouart M (2019) Development of a new TaqMan PCR assay for the detection of both *Entamoeba gingivalis* genotypes. Diagn Microbiol Infect Dis 95(4):114886. https://doi.org/10.1016/j.diagmicrobio.2019.114886. Epub 2019 Aug 11. PMID: 31522801

Zahraa Abd Alhammza Abbass SAHTAK (2020) The prevalence of *Entamoeba gingivalis* and *Trichomonas tenax* in children treated with Orthodontic appliances in AL Muthanna Province, Iraq. Indian J Publ Health Res Dev 11(2):1977–1982. https://doi.org/10.37506/ijphrd.v11i2.3301

Fungi—A Component of the Oral Microbiome Involved in Periodontal Diseases

6

Justyna Karkowska-Kuleta, Dorota Satala, Magdalena Smolarz, Marcin Zawrotniak, and Maria Rapala-Kozik

Abstract

The human oral cavity is a diverse ecological niche favorable for colonization by hundreds of different species of microorganisms. They include not only bacteria but also numerous species of fungi, many of which are able to cause opportunistic infections when the host's immunity is impaired, predominantly by systemic and chronic diseases like diabetes, pulmonary diseases, renal disorders, or acquired immunodeficiency syndrome. Within the dental biofilm and subgingival sites, fungi of the genus *Candida* are often found, also in individuals affected with periodontitis. Moreover, fungal species of other genera, including *Malassezia*, *Aspergillus*, *Penicillium*, and *Rhodotorula* were identified in the oral cavity as well. The wide range of various virulence factors and mechanisms displayed by fungal pathogens allows them effectively invading host tissues during periodontal infections. These pathogenicity-related mechanisms include firstly the fungal ability to adhere successfully to the host tissues closely related to the formation of hyphae, the increase in the surface hydrophobicity, and the surface display of a wide variety of adhesins. Further mechanisms include biofilm formation and secretion of an armory of hydrolytic enzymes and toxins enabling the attack on host cells, modulation of the local inflammatory state, and evading the host immune system. In the pathogenesis of periodontitis, the significant role of fungal co-existence with key bacterial periodontopathogens has been demonstrated, and such interactions were primarily confirmed for *Candida albicans* and *Porphyromonas gingivalis*, where the presence of fungi ensured the survival of strictly anaerobic bacteria under unfavorable aerobic conditions. However, several other mechanisms, including those related to the production of quorum sensing molecules, might also be indicated as particularly important for synergistic or antagonistic interactions with a variety of bacterial species within mixed biofilms. These interactions constitute an extraordinary challenge for applying effective methods of combating biofilm-related infections in the periodontium without the risk of the development of drug resistance, the recur-

Authors Justyna Karkowska-Kuleta and Dorota Satala have equally contributed to this chapter.

J. Karkowska-Kuleta · D. Satala · M. Smolarz
M. Zawrotniak · M. Rapala-Kozik (✉)
Department of Comparative Biochemistry and Bioanalytics, Faculty of Biochemistry, Biophysics and Biotechnology, Jagiellonian University in Krakow, Kraków, Poland
e-mail: justyna.karkowska@uj.edu.pl; dorota.satala@uj.edu.pl; magdalena.smolarz@doctoral.uj.edu.pl; marcin.zawrotniak@uj.edu.pl; maria.rapala-kozik@uj.edu.pl

J. Santi-Rocca (ed.), *Periodontitis*, Advances in Experimental Medicine and Biology 1373,
https://doi.org/10.1007/978-3-030-96881-6_6

rence of disease symptoms, and the progress of life-threating systemic complications.

Keywords

Fungi · *Candida* · *Aspergillus* · Virulence factors · Biofilm · Fungal-bacterial interaction · Periodontium

Abbreviations

Als	Agglutinin-like sequence protein family
ECM	Extracellular matrix
EPM	Extracellular polysaccharide matrix
GAG	Galactosaminogalactan
G-CSF	Granulocyte colony-stimulating factor
GM-CSF	Granulocyte-macrophage colony-stimulating factor
Hwp1	Hyphal wall protein 1
NET	Neutrophil extracellular trap
NO	Nitric oxide
ROS	Reactive oxygen species
Sap	Secreted aspartic proteinase
TLR	Toll-like receptor
TNF-α	Tumor necrosis factor α
QSM	Quorum-sensing molecules

Highlights

- Several fungal species from genera *Candida*, *Rhodotorula*, *Penicillium*, *Aspergillus*, and *Malassezia* belong to the human oral microbiome and some of them may be important in the pathogenesis of periodontitis
- Conditions associated with decreased immunity may lead to the development of periodontal disease associated with the presence of fungi in subgingival sites
- Several fungal virulence factors, including adhesins, toxins, and hydrolytic enzymes, may contribute to the development of periodontitis
- The coexistence of fungi with key bacterial periodontopathogens within the periodontium may affect the pathogenesis of periodontitis

Considerations for Practice

- The documented relationship between periodontal diseases and a healthy lifestyle confirms that proper oral hygiene and smoking cessation may increase the chances of long-term oral health
- Controlling and systematically removing the biofilm from supragingival sites is an effective method of maintaining healthy gums
- In patients with chronic diseases like diabetes mellitus, chronic pulmonary diseases, or chronic kidney diseases, the prevention and treatment of periodontitis should be planned individually

Patient Summary

The most common periodontal diseases are caused by various species of bacteria that colonize the oral cavity along with a broad spectrum of fungal pathogens. Bearing in mind the noticeable role of fungi in the pathogenesis of periodontal diseases, the effective treatment should consider the increased resistance of mixed subgingival dental biofilm to antibiotics and antimycotics. The use of appropriately selected therapy increases the chances of a complete cure for periodontal diseases and reduces future recurrence risk.

6.1 Fungal Species as a Part of the Microbiome of the Human Oral Cavity

The human oral cavity has one of the most diverse microbiomes, comprising about six hundred species of bacteria and about one hundred fungal species (Ghannoum et al. 2010; Dewhirst et al. 2010; Baumgardner 2019). Its humid, warm, and nutrient-rich environment is divided into several niches, including the tongue, cheeks, palate, tonsils, gingival pockets, teeth, and saliva, which enable the coexistence of numerous microorganisms within the complex and protecting structure which is a biofilm. A well-known example is a dental biofilm in which various bacterial species may collectively coexist and form communities both above the gingival line (supragingival sites) and below the gingival line (subgingival sites) (Könönen et al. 2019). Collaborating, they can modulate the host's defense response, thus ensuring successful colonization associated often with the development of two of the most common oral cavity disorders—caries and periodontitis (Marsh 2006; Murakami et al. 2018; Valm 2019).

Contrary to bacteria, fungal microbiome profile has been so far poorly characterized in particular with regard to species belonging to genera other than *Candida*. Furthermore, their presence in the oral cavity is often limited to diseases associated with mucosal infections (Hellstein and Marek 2019). The first study characterizing the fungal microbiome in healthy people's oral cavity was presented a decade ago and identified 101 species of fungi from oral wash samples using molecular biology methods (Ghannoum et al. 2010). The most frequently isolated genera were *Candida*, *Cladosporium*, *Aureobasidium*, *Saccharomycetales*, *Aspergillus*, *Fusarium*, and *Cryptococcus*. The later studies additionally indicated *Penicillium*, *Schizophyllum*, *Rhodotorula*, and *Gibberella* as fungi commonly found in the oral cavity of healthy people (Monteiro-da-Silva et al. 2014; Peters et al. 2017).

Despite the widespread use of oral hygiene products, the periodontal disease remains one of the most common diseases affecting adults and children (Kinane et al. 2017). Supragingival microbes are responsible for the mild course of gingivitis, and root caries disease. Subgingival species accelerate the destruction of the tissues that support the teeth and might cause severe periodontitis (Shi et al. 2018a; Könönen et al. 2019). The main species in the supragingival sites are *Candida*, *Malassezia*, *Cryptococcus*, *Saccharomyces*, *Trichoderma*, and *Cladosporium*, but their abundance fluctuates when caries occurs (Fechney et al. 2019; Baraniya et al. 2020). Despite the diverse microbiome of individual niches in the oral cavity, the only species isolated from the subgingival sites of healthy individuals were *Candida albicans* and *C. dubliniensis* (Reynaud et al. 2001; Urzúa et al. 2008; Canabarro et al. 2013; Babitha et al. 2018), which suggests that periodontal pockets are not a favorable site for fungal microorganisms (Urzúa et al. 2008; Jewtuchowicz et al. 2008). However, in patients with chronic periodontitis, *C. dubliniensis*, *C. parapsilosis*, *C. tropicalis*, and *C. glabrata* were abundantly isolated (Babitha et al. 2018; De-la-Torre et al. 2018), but there was no correlation between the degree of colonization and the depth of periodontal pockets (Babitha et al. 2018). Moreover, species such as *C. parapsilosis*, *C. dubliniensis*, *C. tropicalis*, and *Rhodotorula* spp. have been identified in the subgingival biofilm of patients with severe chronic periodontitis but always as associated with *C. albicans* (Canabarro et al. 2013). The diversity of fungal communities in the oral cavity is summarized in Table 6.1.

6.2 Factors Predisposing to Fungal Infections in the Periodontium

Among several factors predisposing to the development of periodontitis associated with an oral fungal infection should be considered specific genetic conditions, primarily increasing the susceptibility of particular individuals to the progression of this inflammatory-related infectious disease. However, other risk factors are important, including acquired systemic diseases that weaken the human organism's ability to defend

Table 6.1 The most common species of fungi isolated from selected niches within the oral cavity

	Dental biofilm	Subgingival sites	Root canals
Candida			
C. albicans	+ [a]	+, # [b-k]	+ [l, m]
C. glabrata		# [b, f-i, k, n, o]	+ [p, r]
C. tropicalis		# [b, e, h, i, k, n, o]	
C. parapsilosis		# [b, e, k, n]	
C. dubliniensis	+ [a, s]	+, # [b, c, e, t]	
C. krusei	# [u]	# [k, o]	
other *Candida* spp.	# [u]	# [k]	+ [p, r]
Malassezia	+ [a]		
Saccharomyces	+ [a]		
Aspergillus	+ [a]	# [j]	# [w, x]
Penicillium	+ [a, s]		# [w]
Rhodotorula	+ [a, s]	# [e]	+ [m]

+ fungi identified in healthy people's oral cavity, # fungi identified in patients with periodontal disease, [a](O'Connell et al. 2020); [b](Babitha et al. 2018); [c](Urzúa et al. 2008); [d](Reynaud et al. 2001); [e](Canabarro et al. 2013); [f](Melton et al. 2010); [g](Hammad et al. 2013); [h](Sardi et al. 2011b); [i](Matic Petrovic et al. 2019); [j](Kamma et al. 1999); [k](Brusca et al., 2010); [l](Baumgartner et al. 2000); [m](Egan et al. 2002); [n](De-la-Torre et al. 2018); [o](Santhana Krishnan et al. 2020); [p](Waltimo et al. 1997); [r](Waltimo et al. 2003); [s](Fechney et al. 2019); [t](Jewtuchowicz et al. 2008); [u](Brusca et al. 2010); [w](Gomes et al. 2010); [x](Gomes et al. 2015)

against oral opportunistic pathogens. The specific environmental risk factors should not be underestimated either (Fig. 6.1) (Page and Kornman 1997; Johansson and Dahlén 2018).

One of the main medical conditions that might contribute to the onset or exacerbation of pre-existing symptoms of fungal-related periodontal disorders is the infection with the human immunodeficiency virus (HIV) and the subsequent development of acquired immunodeficiency syndrome (AIDS) (Aas et al. 2007). The analysis of samples taken from HIV-positive patients with periodontal lesions showed intense colonization of the oral cavity by yeasts *C. albicans* and *S. cerevisiae* (Aas et al. 2007).

The development of periodontal disease may also be influenced by cancer diseases leading to general immunosuppression. Qualitative and quantitative changes in the oral microbiome have been reported in patients with hematological neoplasms, where, in addition to *C. albicans*, have been found other *Candida* species—*C. glabrata*, *C. tropicalis*, *C. krusei*, *C. parapsilosis* or *C. kefyr* (Schelenz et al. 2011; Aslani et al. 2018).

The incidence of prolonged inflammation related to periodontitis has been frequently reported as associated with chronic systemic diseases, and the observed relationship was often suggested to be bidirectional, where the occurrence and severity of one condition correlated with an increased risk of the other disorder (Wahid et al. 2012; Casanova et al. 2014; Hou et al. 2017). Although more extensive research on this issue is still required due to the high heterogeneity of the previously reported studies, the interdependence of periodontitis and diabetes mellitus, chronic pulmonary diseases, or chronic kidney diseases has been suggested so far (Taylor et al. 2004; Shi et al. 2018b; Zhao et al. 2018; Kapellas et al. 2019; Norhammar et al. 2019). Under such concomitant conditions weakening the immune system, *Candida* yeasts were often identified within the group of microorganisms isolated from the subgingival dental biofilm in individuals suffering from periodontitis (Javed et al. 2009; Bastos et al. 2011; Sardi et al. 2012a; Matić Petrović et al. 2015). In the case of patients with well-controlled and poorly controlled diabetes mellitus, both the most common *Candida* species—*C. albicans*—and other species of the genus, including *C. glabrata*, were isolated from periodontal tissues affected by inflammation (Melton et al. 2010; Hammad et al. 2013). It was also previously reported by Sardi et al. (2011b) that patients with co-morbid chronic periodontitis and diabetes showed greater colonization of the periodontal pockets and furcation sites by various species of the genus *Candida*, with the predominance of *C. dubliniensis* and *C. albicans*, in comparison to non-diabetics. Also, in these studies, two other *Candida* species—*C. tropicalis* and *C. glabrata*—were identified in subgingival biofilm exclusively in diabetic patients (Sardi et al. 2011b). Other studies also indicated that the prevalence of *Candida* species in subgingival sites was higher in diabetics with poor glycoregulation, and the most frequent species isolated

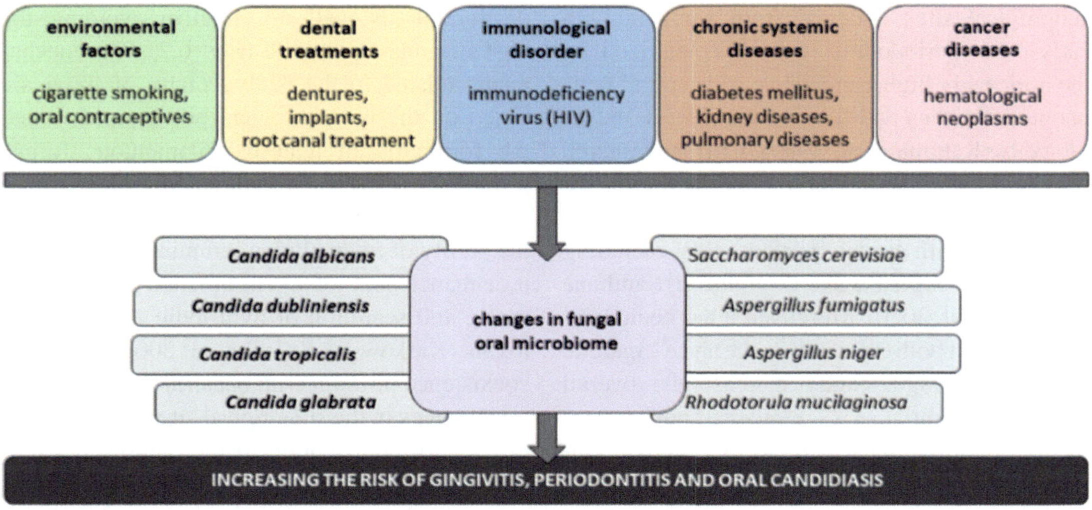

Fig. 6.1 Factors predisposing to oral fungal infections Environmental factors, dental treatments, chronic and immunological diseases contribute to the differentiation of the oral microbiome profile. Qualitative and quantitative changes in the fungi present in the oral cavity are associated with an increased risk of periodontal diseases. The species of fungi whose relationship with the development of periodontal diseases is discussed in the text below are in bold

from subgingival areas in such patients was *C. albicans* followed by *C. glabrata* and *C. tropicalis* (Matic Petrovic et al. 2019). Furthermore, it was also shown that in the case of patients with chronic kidney disease, periodontitis was characterized by greater severity and increased frequency of *C. albicans* and red-complex bacteria, compared with a control group of patients (Bastos et al. 2011).

Other factors that increase periodontal disease risk are dental procedures, such as endodontic treatment, which expose recessed dental structures. The most commonly isolated species from infected root canals were *C. albicans*, *R. mucilaginosa*, *C. glabrata*, *C. guilliermondii* and *C. inconspicua*; however significant correlation with the presence of these yeasts in saliva was found (Waltimo et al. 1997, 2003; Baumgartner et al. 2000; Egan et al. 2002). *C. albicans*, as a dentinophilic species with an invasive affinity for the dentin smear layer and dental tubules, can form a complex biofilm in root canals. The possibility to dentin colonization provides yeasts access to nutrients and protects them from the action of antifungal drugs, thus these species becoming one of the causes of persistent cases of apical periodontitis (Siqueira and Sen 2004; Ghogre 2014). In addition to the *C. albicans* species, in root canals with pulp necrosis and periapical lesions, filamentous fungi of the genus *Aspergillus*, *Penicillium*, and *Fusarium* were also identified (Gomes et al. 2010, 2015); however, they have not yet been identified as dentinophilic species (Siqueira et al. 2002). Besides, the implant sites have been shown to provide an attractive microenvironment that fosters uncontrolled microbial biofilms (Øilo and Bakken 2015); however, the fungal profile depends on the implant's clinical condition. *C. dubliniensis* and *Cladosporium cladosporioides* predominate in healthy implantation sites. In contrast, in samples taken from peri-implantation sites, dominate *C. albicans*, *C. boidinii*, *Penicillium* spp., *R. laryngis*, *Paecilomyces* spp., *Saccharomyces* and *Cl. cladosporioides*. The coexistence of fungi with *Parvimonas micra* and *Tannerella forsythia* was found in healthy and diseased peri-implant sites (Schwarz et al. 2015).

One of the most important risk factors for periodontitis associated with environmental conditions and the chosen lifestyle is cigarette smoking (Tomar and Asma 2000). The samples of subgingival sites obtained from smokers with early-onset periodontitis were characterized by the noticeably higher incidence of bacterial periodontopathogens and fungi *A. fumigatus* and *C. albicans* (Kamma et al. 1999). Current research confirms the presence of other species of the *Candida* genus in the subgingival sites and saliva of smokers with chronic periodontitis, including *C. krusei*, *C. tropicalis* and *C. glabrata* (Santhana Krishnan et al. 2020). Moreover, it has been demonstrated that both cigarette smoke and e-cigarette vapor might significantly increase the overall virulence potential of *C. albicans* and enhance the adhesion of fungal cells to the gingival epithelial cells and fibroblasts, thereby increasing the pathogenicity of fungi during infections localized in the oral cavity and periodontium (Alanazi et al. 2014, 2019).

Other environmental factors may also contribute to the increased colonization of subgingival sites by *Candida* fungi, including changes in the hormonal balance caused by oral contraceptives. As reported by Brusca et al. (Brusca et al. 2010), the oral administration of ethinyl estradiol, gestodene, and drospirenone, combined with cigarette smoking, caused a statistically significant increase in the incidence of severe periodontitis, and under such conditions, different species of the genus *Candida*, including *C. parapsilosis*, *C. glabrata*, *C. tropicalis*, *C. krusei*, *C. albicans*, and *C. guilliermondii* were isolated from the gingival pockets. In most cases, yeast isolates were identified together with bacterial periodontal pathogens, such as *Actinobacillus actinomycetemcomitans*, *Prevotella intermedia* and *Porphyromonas gingivalis* (Brusca et al. 2010).

6.3 Virulence Factors Involved in the Fungal Invasion of Periodontal Tissue

The participation of fungi in the development of periodontal disease might be related to the exploitation of a number of virulence factors by fungal pathogens that allow them to efficacious adhesion to host tissues followed by their invasion and destruction as well as effective avoiding the host's immune response (Fig. 6.2). The mechanisms related to the pathogenicity of fungi are based on the morphological polymorphism and the formation of adhesive filamentous forms, production of protective cellular envelopes, exposition of surface adhesins, biofilm formation and the ability of intercellular communication within it, enhancement of environmental stress resistance, and secretion of hydrolytic enzymes and toxins (Karkowska-Kuleta et al. 2009). Also, the coexistence of fungi with bacterial key periodontopathogens in the subgingival sites is considered as a factor contributing to the pathogenicity of fungi identified in the periodontal lesions (O'Donnell et al. 2015; Delaney et al. 2018).

6.3.1 Adhesion and Biofilm Formation

The adhesion of pathogen cells to the host tissues or the surface of artificial materials located in the oral cavity is one of the first and crucial steps in initiating colonization and further invasion by a microorganism (Nikawa et al. 2006). In fungi, there are several mechanisms associated with the enhancement in cell adhesiveness and surface hydrophobicity, including the presentation of a wide range of different adhesive proteins and N-linked glycans on their cell surface (Glee et al. 1995; Masuoka and Hazen 2004; El-Kirat-Chatel et al. 2015; Hoyer and Cota 2016). These properties are also associated with the microbial ability to form biofilms as structures more resistant to adverse environmental conditions and the host's immune system activity (Kucharíková et al. 2015; Silva-Dias et al. 2015).

The most frequently described yeast adhesins, that are typically multidomain proteins equipped with N- and O-linked polysaccharide chains and a glycosylphosphatidylinositol (GPI) anchor, are agglutinins and flocculins of *S. cerevisiae*, lectin-like Epa (epithelial adhesin) protein family of 17 members, six adhesins Awp (adhesin-like wall proteins) and seven Pwp proteins (PA14-containing wall protein) of *C. glabrata* and *C.*

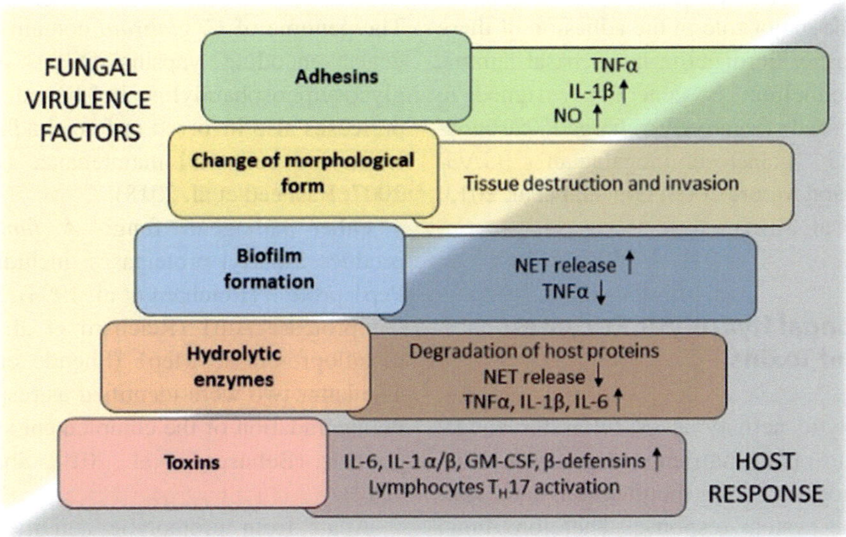

Fig. 6.2 Host responses induced by the fungal virulence factors
The mechanisms related to the pathogenicity of fungi include morphological polymorphism, production of surface adhesins, biofilm formation, and secretion of hydrolytic enzymes and toxins. These mechanisms influence the invasion of the host's tissues, their destruction, and modulation of the host's immune system

albicans adhesins from Als (agglutinin-like sequence) protein family consisting of eight members (Als1–7 and Als9), (Hoyer 2001; Verstrepen and Klis 2006; Dranginis et al. 2007; de Groot et al. 2008; Kraneveld et al. 2011; Timmermans et al. 2018). Other important *C. albicans* adhesins are Hwp1 (hyphal wall protein), Eap1 (enhanced adhesion to polystyrene), and Csh1 (cell surface hydrophobicity), while the latter is also indicated as essential for *C. dubliniensis* cell surface hydrophobicity and virulence (Singleton et al. 2001, 2005; Li and Palecek 2003; Hazen 2004; Naglik et al. 2006; Ene and Bennett 2009). In the case of *C. albicans*, *C. parapsilosis* and *C. glabrata* it was shown that cell surface hydrophobicity correlates with adhesion to human buccal epithelial cells and denture acrylic surfaces (Panagoda et al. 2001; Luo and Samaranayake 2002). Also, it was reported that cells of *C. albicans* strain, isolated from patients with chronic periodontitis and diabetes that showed high surface hydrophobicity had a greater ability to adhere and invade human gingival fibroblasts (Sardi et al. 2012b).

The change in the morphological form and the formation of hyphae, on which surface typical adhesive proteins are abundantly exposed, may also increase the adhesiveness of *Candida* yeasts (Kumamoto and Vinces 2005; Tronchin et al. 2008; Mayer et al. 2013). Hyphal growth of *C. albicans* cells was previously considered as essential for penetration into gingival tissues in chronic periodontitis (Jarvensivu et al. 2004). The cigarette smoke condensate promoted the hyphae formation by *C. albicans*, increased expression of *HWP1* and *EAP1* genes, biofilm-forming ability, and fungal adhesion to human gingival fibroblasts, whereas e-cigarette vapor stimulated the binding of fungi to gingival epithelial cells and elongation of hyphae (Alanazi et al. 2014, 2019; Semlali et al. 2014).

In the case of *A. fumigatus*, the mechanisms of fungal adhesion are not sufficiently well understood. Initially, it was pointed out that some proteins of the conidia surface, including conidial hydrophobin RodA, 37-kDa allergen Asp f 2, and extracellular thaumatin domain protein (AfCalA) may bind to host ligands (Thau et al. 1994;

Banerjee et al. 1998; Upadhyay et al. 2009). However, the major role in the adhesion of these fungi to plastic, fibronectin, intact basal lamina, and the epithelium is generally assigned to surface-exposed negatively charged carbohydrates and galactosaminogalactan (GAG) (Wasylnka and Moore 2000; Gravelat et al. 2013; Rambach et al. 2015).

6.3.2 Fungal Hydrolytic Enzymes and Toxins

The hydrolytic activity is essential for fungal pathogens to obtain nutrients, degrade host tissues and proteins during the invasion, and evade the immune system response. Therefore fungi secrete extracellularly or present on their cell surface a variety of hydrolytic enzymes, including lipases, phospholipases, phosphatases, proteases, and hemolysins (Barrett-Bee et al. 1985; Robinson et al. 1990; Monod et al. 1994; Hube et al. 2000; Luo et al. 2001; Wartenberg et al. 2011).

One of the groups of yeast hydrolases that have been most thoroughly described regarding their involvement in the pathogenesis is secreted aspartic proteinases (Sap). Two catalytic Asp residues being a part of an Asp-Ser-Gly/Asp-Thr-Gly motif are responsible for the Sap hydrolytic activity at acidic pH as optimal (Koelsch et al. 2000; Rapala-Kozik et al. 2018). In the case of *C. albicans*, there are identified ten members of the *SAP* gene family (*SAP1–10*), three for *C. parapsilosis* (*SAPP1–3*), and four for *C. tropicalis* proteinases (*SAPT1–4*) (Zaugg et al. 2001; Hube and Naglik 2001; Singh et al. 2019). The broad substrate specificity of Saps ensures the participation of these proteins not only in the acquisition of nutrients, adhesion, and tissue damage but also additionally in many processes related to the interaction with host proteins, including the degradation of antimicrobial peptides, the modulation of inflammatory state or the deactivation of the complement system (Villar et al. 2007; Gropp et al. 2009; Bras et al. 2012, 2013; Kozik et al. 2015; Svoboda et al. 2015; Bochenska et al. 2015, 2016; Yu et al. 2016; Singh et al. 2019). The genome of *C. glabrata* contains at least 11 genes encoding yapsins (*YPS*)—extracellular glycosylphosphatidylinositol-linked aspartyl proteases also involved in fungal adhesion, virulence and cell wall maintenance (Kaur et al. 2007; Rasheed et al. 2018).

Other pathogenic fungi—*A. fumigatus* also produce aspartyl proteinases, including 38 kDa Pep1 protein (Reichard et al. 1994), 32 kDa serine protease Alp1 (Reichard et al. 1990) and metalloproteinase Mep1 (Shende et al. 2018). The latter two were identified as responsible for the degradation of the complement system components (Behnsen et al. 2010; Shende et al. 2018).

Apart from proteolytic activity, also other hydrolytic enzymes are essential for fungal virulence. In *C. albicans*, such enzymes also include the lipase family consisting of 10 members (*LIP1–10*) (Hube et al. 2000) and different phospholipases. Of the *Candida* family of phospholipases, the best described is the extracellular phospholipase Plb1—a lipolytic enzyme considered as an important virulence factor of *C. albicans* (Leidich et al. 1998; Hruskova-Heidingsfeldova 2008). Phospholipase activity was also detected for *A. fumigatus*, and assigned for proteins afPLB1, afPLB3, and phospholipase D (Shen et al. 2004; Li et al. 2012). Furthermore, for pathogenic representatives of different fungal genera, extracellular hemolytic activity is very important in virulence since it allows them to obtain iron from host erythrocytes in the infectious site (Manns et al. 1994; Luo et al. 2001; Kaveemongkonrat et al. 2019).

In addition to enzymatic proteins, also toxins play a significant role in the pathogenesis of fungal infections. *C. albicans* secreted cytolytic toxin—candidalysin, which is released from the Ece1 (extent of cell elongation) protein by the Golgi-located Kex2 protease, has been identified recently as one of the key virulence factors (Moyes et al. 2016). This secreted candidal toxin is involved in the direct damage of the epithelial cells by their permeabilization but can also activate pro-inflammatory host protection by mononuclear phagocytes (Naglik et al. 2019; König et al. 2020).

A wide variety of toxins are produced also by fungi other than *Candida* species, including *A. fumigatus*, for which several toxic secondary metabolites have been shown to contribute significantly to fungal pathogenicity (Rementeria et al. 2005). Among the best described mycotoxins of *A. fumigatus* are gliotoxin and fumagillin. Gliotoxin is an immunosuppressive and cytotoxic molecule highly affecting fungal virulence and responsible for the inhibition of the respiratory burst and phagocytosis, and the stimulation of apoptosis in leukocytes (Reeves et al. 2004; Sugui et al. 2007; Gayathri et al. 2020). Whereas fumagillin exhibits antiangiogenic properties through the irreversible inhibition of methionine aminopeptidase (MetAP) type 2 (Sin et al. 1997), it has also the ability to reduce the activity of neutrophils (Fallon et al. 2010) and is involved in the damage of the epithelial cells during infection (Guruceaga et al. 2018).

6.3.3 Fungal—Host Interactions

C. albicans Als3 protein enables effective invasion of the oral epithelium through binding to E-cadherin, cell adhesion molecule forming adherens junctions (Phan et al. 2007). Such interaction induces microfilament rearrangement, which activates hyphal endocytosis through the endothelial layer, allowing the invasion of tissues (Phan et al. 2007). It has been shown that the contact of fungal adhesins with gingival fibroblasts induces the production of proinflammatory factors such as tumor necrosis factor α (TNF-α), interleukin-1β (IL-1β), interleukin 13 (IL-13) by the activation of toll-like receptor 2 (TLR2) expressed on the cell surface (Pinheiro et al. 2018). Additionally, data presented by Sardi et al. indicated that gingival fibroblasts respond to fungal adhesion by producing nitric oxide (NO). A high concentration of NO in the microenvironment of periodontal disease may have a cytotoxic and cytostatic effect against infected cells and results in rapid tissue damage (Sardi et al. 2012b).

Adhesion to the surface of host proteins is a critical step in the initiation of biofilm formation (Phan et al. 2007). The appearance of such complex structures is a considerable challenge for the immune system, and the host cell response is closely dependent on the phase of biofilm development. At the initial stage of infection, access to the fungal cells is easier, and the immune system is effectively activated, mostly by recognizing the elements of the microbial cell wall or the secreted proteins. However, to date, several examples of the attenuation of the host cell's innate response to mature biofilm infections have been described. The main barrier that effectively hinders contact of the immune system with pathogen cells is the dense extracellular matrix (ECM) layer surrounding the mature biofilm (Sandai et al. 2016). Neutrophils, the first white blood cells recruited to the infection site, are equipped with several killing mechanisms, among which the most important in the context of fighting extensive fungal infections is the netosis process (Urban et al. 2006, 2009). During netosis, activated neutrophils release decondensed chromatin fibers, decorated with antimicrobial proteins derived from granules (NET) (Brinkmann et al. 2004). The research revealed that netosis is an effective mechanism to prevent invasion by large-size microorganisms that are relatively difficult to neutralize by phagocytosis (Branzk et al. 2014), including the filamentous forms of *Candida* and *Aspergillus* spp. Johnson et al. indicated that contact of neutrophils with *C. albicans* biofilm impairs NET release, making biofilm more resistant to neutrophil attack (Johnson et al. 2016). This inhibitory effect appears to be related to the presence of a high concentration of polysaccharides—the main components of ECM. It has been shown that *C. albicans* biofilms also resist attack by monocytes. Studies conducted on monocyte-like cell line THP-1 indicated the downregulation of TNF-α production, compared to planktonic cells (Katragkou et al. 2010). Inhibition of the production of this cytokine contributes to a significant reduction of phagocytosis. Taking into consideration that TNF-α has been shown to directly inhibit *C. albicans* biofilm formation by interaction with N,N′-diacetylchitobiose, this mechanism may represent an evolutionary adaptation of fungal biofilm to immune system evasion (Kernien et al. 2017; Rocha et al. 2017). Similar results, indicating a protective role of ECM, were obtained for *Aspergillus* biofilms. In this case, the ECM's most essential components are galactomannan and GAG (Loussert et al. 2010). The secreted enzyme Agd3 deacetylates these ECM components. Such modification is a critical step in forming *A. fumigatus* biofilms, ensuring the adhesion to anionic surfaces, including host tissues (Lee et al. 2016). Moreover, cell wall-bound GAG has been shown to influence the resistance of *Aspergillus* to neutrophil killing, probably by inhibition of NET binding to hyphae (Lee et al. 2015).

The main consequence of fungal proteolytic activity for a broad spectrum of host peptides and proteins, including lactoferrin, lactoperoxidase, cathepsin D, albumin, hemoglobin, LL37, histatin, contact system components, and the extracellular matrix proteins (Naglik et al. 2003; Rapala-Kozik et al. 2010, 2015) is their hydrolysis by Saps. Sap1–3 has been shown to degrade complement components preventing yeast opsonization and phagocytosis (Gropp et al. 2009). Saps are also the main fungal proteins involved in the penetration of the epithelium layer through the degradation of E-cadherin, mainly due to the activity of Sap5, which contributes to the destruction of host tissues (Villar et al. 2007). These enzymes might also activate the innate immune response. It has been indicated that Saps exhibit strong chemoattractant properties and drive the influx of neutrophils and macrophages to the place of infection and these proinflammatory properties are rather independent of proteolytic activity (Pietrella et al. 2010; Gabrielli et al. 2016). Through the activation of Akt/NF-κB pathway, Saps activate the expression of a broad spectrum of cytokines such as IL-1β, IL-6, and TNF-α (Pietrella et al. 2010).

Furthermore, Saps are involved in the activation of ROS generation and have the potential to induce NET formation (Hornbach et al. 2009; Zawrotniak et al. 2017). This mechanism may be of particular importance in the context of effective protection against oral infections, as it has been shown that neutrophils isolated from the peripheral blood and gingival crevice from individuals with periodontitis have significantly reduced phagocytic properties (Asif and Kothiwale 2010). In contrast to Saps, the role of *C. albicans* phospholipases in pathogenesis is not so well elucidated. Phospholipases facilitate the adhesion of yeast to the tissue surface and degrade the components of biological membranes, which are mainly composed of phospholipids (Sardi et al. 2010).

Also, cadidalysin, the main toxin produced by *C. albicans*, plays a critical role in developing oral infections (Moyes et al. 2016; Pellon et al. 2020). Moyes et al. demonstrated for the first time that at the early stages of infection, this peptide induces a strong inflammatory response in epithelial cells, including the production of IL-6, IL-1 α/β, GM-CSF (granulocyte-macrophage colony-stimulating factor), G-CSF (granulocyte colony-stimulating factor), CCL20, antimicrobial peptides (β-defensins), and damage-associated molecular patterns (IL-1α and S100A8/9) via p-MKP1/c-Fos pathway (Moyes et al. 2010, 2014; Verma et al. 2017). Accumulation of candidalysin during the progression of the infection causes direct tissue damage through the mechanism of intercalation, permeabilization, and calcium influx (Moyes et al. 2016). The rapid release of IL-1α and its synergistic role with EGFR ligands on the innate activation of human oral epithelial cells being the effect of candidalysin was also confirmed (Hanaoka and Domae 2020) and it was proved that candidalysin induces the release of antimicrobial peptides hBD2, hBD3 and LL37 and epithelial alarmins ATP, ROS/RNS and S100A8 during oral yeast infection (Ho et al. 2020). Moreover, candidalysin-induced production of IL-1α/β drives the activation of a specialized subpopulation of lymphocytes T_H17 expressing IL-17 (Verma et al. 2017). IL-17 production plays a crucial role in modulating the antifungal response (Mengesha and Conti 2017).

6.4 The Mixed-Species Biofilm Formation in the Periodontitis: Mechanism of Mutual Interactions Between Fungi and Bacteria

For many years, the common oral fungal colonizers were considered to be of minor importance in periodontal diseases. However, the observation that fungi can form multispecies biofilms with different types of bacteria has shed new light on the role of these microorganisms in periodontitis (Peters et al. 2017; Sultan et al. 2018; Delaney et al. 2018). The mixed-species microbial biofilm has a supragingival and subgingival location, where metabolic diversity encourages different species to aggregate and compete for space and nutrients (Aruni et al. 2015).

The formation of a polymicrobial community starts with the attachment of early colonizers to the tooth surface. The bacteria are represented by oral streptococci and *Actinomyces* species, which form the best environment for further appearance of bridging colonizers such as *Fusobacterium nucleatum.* Finally, the late colonizers—the red complex species: *P. gingivalis, T. forsythia*, and *Treponema denticola* find the best condition for existence, becoming the keystone species of periodontal disease (Könönen and Müller 2014; Mira et al. 2017; Manji et al. 2018; Valm 2019). Among them, *P. gingivalis* is also identified as a significant risk factor for developing cardiovascular disease (Oliveira et al. 2015), diabetes (Lamont et al. 2018), rheumatoid arthritis (Koziel et al. 2014), and Alzheimer's or Parkinson's diseases (Dominy et al. 2019; Olsen et al. 2020).

The biofilm cell mass is covered by an extracellular matrix that includes polysaccharides, lipids, and glycoproteins, not only of microbial but also of host origin. The biofilm matrix plays a unique role in protecting the formed microbial consortium against external factors, i.e. host defense factors, antioxidants, or antibiotics used in anti-biofilm therapies (Bowen et al. 2018).

During multispecies biofilm progression, identified fungal species can interact with bacteria, and the interactions may vary from synergism to competition or antagonism. Moreover, depending on the population diversity, opportunistic organisms, like yeasts belonging to *Candida* species, may switch from commensals to pathogens, equipped with an additional set of virulence factors. In periodontitis, the main synergistic microbial interactions were identified primarily for *C. albicans*, where fungal biofilm supports the milieu changes from aerobic to anaerobic (Fox et al. 2014), favoring the growth of Gram-negative obligate anaerobes. The microbial quorum-sensing molecules (QSM) or metabolic by-products may be useful in mutual pathogen communication. Moreover, the first colonizer's primary host tissue infection may predispose host cells to facilitate colonization by the subsequent pathogen (Basavaraju et al. 2016). On the other hand, the competition between the interacting microbes for nutrition or binding sites can restrict early facultative aerobes' growth, giving the late colonizer space for propagation.

The changes in the bacterial composition within mixed biofilm create a necessity for the lifestyle adaptation for fungi and modifying their interactions with emerging new bacterial partners. The different omic approaches supported the knowledge about the interactions and metabolic responses of microbial community members (Shokeen et al. 2021). These mutual microbial interactions can affect health or disease progression in the host organism. But the host activity can also impact the biofilm composition and the interplay between coexisting microbes by the immune system responses or divergences in the nutrient supply (Marsh and Zaura 2017; Lamont et al. 2018).

6.4.1 Synergistic Interactions

The mutual interaction of bacteria with fungi may be considered on two significant levels. The first one concerns the physical contact, where fungal filamentous cell morphotype (hyphae) serves as a structural platform for a mixed-species biofilm and plays a synergistic role. The second focuses on chemical and metabolic interactions and may also be antagonistic. In this case, the produced compounds modified the local environment's properties by changing pH and oxygen availability and enabling bacterial development in unfavorable conditions. It also includes the action of QSM that might regulate the microbial morphology and population abundance (Marsh and Zaura 2017; Diaz and Valm 2020).

The mechanical stabilization of biofilms is carried out using non-specific physical interactions and specific protein-protein contact that facilitate microbes' cooperation, especially in the teeth area, where pathogens are exposed to mechanical removal from the surface. Co-adhesion and co-aggregation ensure the proximity of microorganisms and enable collaboration within the mixed biofilm (Kolenbrander 2000). The important role of fungi in creating the specific interactions between microorganisms is related to the possible morphological changes of fungal cells, best documented for *Candida* yeasts, where formed elongated hyphae supported pathogens' interactions within mixed biofilm structure (Cui et al. 2013; Diaz et al. 2014; Janus et al. 2016). The best-known relationships occurred between yeasts and the primary colonizers of oral cavity—streptococci (Diaz et al. 2012; Xu et al. 2014b), especially *S. mutans* and *S. gordonii* (Gross et al. 2012; Metwalli et al. 2013), where the fungi promote streptococcal biofilm formation, while streptococci enhance the invasive property of *C. albicans* (Diaz et al. 2012; Xu et al. 2014a). The synergistic physical coexistence between *Streptococcus* sp. and *C. albicans* leads to intimate corn-cobb-like structures in the supragingival sites (Zijnge et al. 2010) where *S. gordonii* promotes hyphal development of *C. albicans* cells. The adhesive contact between the microbes occurs due to specific protein-protein interactions in which the streptococcal cell surface adhesins CshA, SspA, and SspB are involved (Holmes et al. 1996; Silverman et al. 2010; Xu et al. 2014a). Fungal adhesins responsible for the recognition of streptococci are Hwp1 and the members of the Als protein family (Klotz et al. 2007; Bamford et al. 2009; Silverman et al. 2010). Additionally, the interactions between microorganisms are mediated by salivary proline-rich proteins, which adsorbed on the surface of *S. gordonii* may be recognized as receptors by *C. albicans* cells (Holmes et al. 1995; Bamford et al. 2009; Silverman et al. 2010).

Another example of an interplay between bacteria and fungi is the interaction of internalin-family surface protein—InlJ of *P. gingivalis* and *C. albicans* Als3 in the dental pockets (Sztukowska et al. 2018). The same fungal adhesin can also form complexes with the hemagglutinin domains of RgpA and Kgp—the major *P. gingivalis* cysteine proteinases—gingipains. Moreover, gingipains interact also with another *C. albicans* surface mannoprotein—Mp65, overproduced during contact with this bacterium, and important for hyphal morphogenesis, membrane cell wall organization, and fungal biofilm formation. An interesting relation was also observed for fungal enolase, a cytosolic protein that appears on the fungal cell surface as "moonlighting protein" (Satala et al. 2020a, b). The binding affinity between RgpA and enolase was even threefold higher than between RgpA and Als3 (Bartnicka et al. 2019).

The adhesion of *P. gingivalis* to fungal cells was also dependent on the activity of bacterial peptidylarginine deiminase (PPAD)—an enzyme that can perform the citrullination of both bacterial and fungal surface proteins. The decrease in interactions between these microorganisms, observed for the bacterial mutant strain with PPAD depletion, demonstrated the significant role of PPAD-mediated modification in these interactions (Karkowska-Kuleta et al. 2018).

Gingipains produced by *P. gingivalis* activate the host immune responses, but the host cell's coinfection by bacteria and *C. albicans* cells leads to the alleviation of the inflammatory process, pointing to the possible protection of bacterial cells by fungal biofilm. Such a hypothesis was supported by the experiments performed using the mouse model, where the fungus initiated a sequential infection, and a reduction of mouse mortality was observed. On the other hand, mixed species biofilm assured more prolonged survival of bacteria within host tissues, suggesting the dual-species infection's chronic nature (Bartnicka et al. 2020).

The interacting microorganisms are secured by the extracellular polysaccharide matrix (EPM) formed by both types of microbes (Xu et al. 2008). For example, *S. mutans* accompanied by *C. albicans* induces glucosyltransferase B (GtfB) gene (Falsetta et al. 2014), encoding enzyme that catalyzes the production of bacterial α-glucans. Their deposition on *C. albicans* cell surface leads to the expansion of mutual EPM and enhancing microbes adhesion (O'Sullivan et al. 2000; Falsetta et al. 2014; Hwang et al. 2017). Moreover, the presence of EPM is also beneficial for fungi, as EPM structures might trap an antifungal drug—fluconazole, thus protecting *C. albicans* from its effects (Kim et al. 2018). The mixed biofilm of *C. albicans* and *S. gordonii* can also be stabilized through the extracellular release of bacterial DNA (Jack et al. 2015).

The relationship between yeasts and bacteria related to metabolism and its by-products is an essential aspect of microbial cooperation responsible for the maintenance of appropriate environmental conditions and the proper development of biofilm above and below the gum line (Lof et al. 2017). Some specific adaptations of *C. albicans* metabolism to co-existence with *S. gordonii* were correlated with an increased expression of genes involved in fungal arginine biosynthesis, probably required to overcome the effects of the oxidative stress generated by these bacteria (Dutton et al. 2016).

In gingival pockets and dental canals, several bacteria are either facultative or obligate anaerobes. Maintaining a low oxygen concentration in such a milieu is essential for their proper growth and development. *C. albicans* survives in both aerobic and anaerobic conditions but possesses the ability to lower the local oxygen concentration (Fox et al. 2014; van Leeuwen et al. 2016; Janus et al. 2017), thus creates the conditions optimal for the development of a mixed-species biofilm with *C. perfringens*, *B. fragilis*, *P. gingivalis*, and *C. difficile* (Tamai et al. 2011; van Leeuwen et al. 2016; Karkowska-Kuleta et al. 2018).

6.4.2 Antagonistic Interactions

The interactions within the mixed biofilms depend on the fast adaptation to the coexisting microorganisms' immediate needs and infection progression. The antagonistic relationships for microbes involved in periodontic diseases were again best documented for *Candida* species, especially for dominating *C. albicans* (Krüger et al. 2019).

For one of the first bacterial colonizers involved in these infections—streptococci, a synergistic collaboration with *C. albicans* was documented as contributing to plaque formation (Metwalli et al. 2013; Koo et al. 2018). But some contrary effects were also observed. The streptococci acidify the environment excreting carboxylic compounds, like lactate, pyruvate, or α-ketoglutarate. Nonetheless, low pH is unfavorable for fungal hyphae development. That finding was supported by observing the diminished formation of fungal hyphae in the *Galleria mellonella* larval tissue treated with *S. mutans* cell supernatant, resulting in the reduction of fungal cell pathogenicity. On the other hand, the cariogenic potential of *S. mutans* was attenuated in the biofilm formed with *C. albicans* cells, where higher pH was detected in comparison to one species, bacterial biofilm (Willems et al. 2016). *C. albicans* can use lactate as a carbon source for cell growth, thus alkalizing the environment, and maintaining a local pH at 5.3–5.5 (Vylkova et al. 2011; Krom et al. 2014; Danhof et al. 2016). Moreover, *C. albicans* is also able to generate ammonia, as a product of amino acid catabolism, increasing environmental pH and the hyphae formation (Vylkova et al. 2011). The alcalification of mixed biofilm milieu by fungi could be the way to counteract a microbial switch towards a cariogenic community and prevent the demineralization of teeth (Jenkinson et al. 1990).

Competition for access to limited nutrients may also affect the population dynamics within the biofilm. A good example is the heme competition between *P. gingivalis* and *C. albicans*, where at the restricted access to heme, the promotion of bacterial virulence was observed, correlated with the increase in expression of bacterial genes involved in heme utilization (Guo et al. 2020).

For antagonistic interactions, microbes also used the QSM production system (Barbosa et al. 2016). Such compounds include *S. mutans* signaling molecules like S-*trans*-2-decenoic acid, which is relevant in shaping multispecies fungal—bacteria biofilms, suppressing germ tube formation (Vílchez et al. 2010).

Another is the competence-stimulating peptide (CSP) produced during the early stages of *S. mutans* cell growth and fungal interaction with *S. gordonii*, influencing the fungal filamentation (Jarosz et al. 2009; Jack et al. 2015). A similar effect was also observed for mutanobactin A, that induced formation of yeast forms by *C. albicans*. Such inhibition of fungal cell filamentation may prevent activation of the immune system and proinflammatory cytokine production by macrophages (Joyner et al. 2010).

Other oral cavity bacteria—*Streptococcus sanguinis* presented antibacterial activity towards periodontal pathogens by the production of hydrogen peroxide and bacteriocin, which is localized within the *S. sanguinis* cells (Zhu and Kreth 2010). The hydrogen peroxide also influences fungal QSM—farnesol production and thus hyphae formation. On the other hand, the cell extract containing bacteriocin has antifungal activity and suppresses *C. albicans* and *C. tropicalis* cell growth (Ma et al. 2014).

C. albicans isolate from the root canal, and periapical infections are also accompanied with *Enterococcus faecalis* (Dahlén et al. 2012)—the opportunistic pathogenic bacterium also using bacteriocin (EntV), to inhibit fungal hyphal growth, biofilm formation, and virulence, without effect on cell viability (Cruz et al. 2013; Graham et al. 2017).

Another bacteria of aggressive periodontal disease—*A. actinomycetemcomitans* produces a different QSM—an autoinducer-2 (AI-2)—to inhibit *C. albicans* cell filamentation and suppress biofilm formation (Baker et al. 2017; Bachtiar et al. 2014). The genetic analysis of multispecies biofilm formed by *A. actinomycetemcomitans* showed the suppression of fungal hyphal-associated genes (*ALS3* and *HWP1*) without any effect on gene representing yeast form (*YWP1*) (Bachtiar and Bachtiar 2020). However, the action of AI-2 released by *S. gordonii* presented an opposite effect on *C. albicans* cells, suggesting that different bacteria produce AI-2 derivates with a miscellaneous activity (Lof et al. 2017).

Antagonistic effects were also observed for *C. albicans* and *C. dubliniensis* collaborating with "bridging" bacteria—*F. nucleatum*, which is involved in colonization succession of the oral polymicrobial community (Grimaudo and Nesbitt 1997; Jabra-Rizk et al. 1999; Signat et al. 2011). The mutual influence of microbes is mediated by direct interaction between their surface proteins, fusobacterial membrane protein RadD and *Candida* cell wall protein Flo9 (Wu et al. 2015; Bor et al. 2016). Moreover, the supposed effects of their reduced virulence towards the host do not result from the secretion of regulatory particles or metabolic products. But they have the source in resulted inhibition of hyphal morphogenesis and fungal cell growth. It was proposed that the observed mutual attenuation of virulence may promote a rather commensal lifestyle of both biofilm-forming species (Bor et al. 2016).

6.4.3 The Periodontal Cells in the Face of Mixed Infections

With emerging evidence that the development of periodontitis is associated with the presence of both bacterial and fungal pathogens, understanding of the biological consequences of the cross-kingdom interactions of microbes for host response is crucial.

The most attention has been devoted to the model of mixed infections caused by *P. gingivalis* and *C. albicans*. In 2011 Tamai et al. demonstrated that heat-killed *C. albicans*, but also mannoprotein-β-glucan complex constituting a component of the yeast cell wall, enhanced invasion of human gingival fibroblasts and epithelial cells by *P. gingivalis* (Tamai et al. 2011). The mechanism underlying this phenomenon is unclear, but the authors put forward two hypotheses. Firstly, pretreatment with *C. albicans* or cell wall components promotes the recruitment of clathrin and clathrin-mediated invasion of host cells by *P. gingivalis*, more efficiently than by *P. gingivalis* alone. On the other hand, *C. albicans* induces trafficking of cell receptors such as toll-like receptors (TLRs) to lipid rafts, membrane microdomains involved in the entry of *P. gingivalis* into the epithelial cells (Tsuda et al. 2008; Tamai et al. 2011). Additionally, Haverman et al. observed that the interaction between *P. gingivalis* and *Candida* spp. (*C. glabrata* and *C. kefyr*) also lead to the inhibition of oral epithelial cell migration more than either pathogen separately (Tsuda et al. 2008).

Studies conducted on the human macrophage cell line (THP-1) have shown that contact with biofilm formed by *P. gingivalis* and *C. albicans* was characterized by a different level of production of the main proinflammatory cytokines—IL-1β, TNF-α and IL-8 (Bartnicka et al. 2020). The macrophage response to the supernatant obtained from a mono-species bacterial biofilm resulted in lower IL-1β production, compared to the reactions of THP-1 to *C. albicans* biofilm. Unexpectedly, the contact of THP cells with mixed biofilm resulted in a significant increase in IL-1β. On the other hand, a 24-h incubation of THP-1 with mixed biofilm caused a dramatic decrease in TNF-α and IL-8 levels compared to the treatment with mono-species biofilm. The reason for the observed changes in the level of cytokines may result from the activity of proteolytic enzymes produced by *P. gingivalis*. They can effectively degrade the cytokines produced by the host cells, impairing the immune response (Stathopoulou et al. 2009; Bartnicka et al. 2020), especially considering that their activity could be attenuated during contact with a fungal partner (Bartnicka et al. 2020). Another explanation might be that *C. albicans* cells prevent bacteria from being recognized by the immune cell receptors. Moreover, mixed biofilm also resulted in a reduced neutrophil response, as evidenced by a significantly lowered elastase activity compared to bacterial biofilm (Bartnicka et al. 2020).

An example of bacteria frequently co-isolated with *C. albicans* in periodontal pockets is *F. nucleatum* (Bor et al. 2016). *In vitro* studies have shown that their interaction retains *C. albicans* in morphology less sensitive to RAW macrophage killing. Moreover, the co-infection of *C. albicans* with *F. nucleatum* has been shown to inhibit the bacteria-induced production of proinflammatory molecules such as chemokine MCP-1 (monocyte chemoattractant protein 1) and cytokine TNF-α. The immune response's attenuation was more effective when *C. albicans* was grown in yeast-like morphology (Bor et al. 2016).

6.5 New Trends in Prevention and Treatment of Periodontitis

The high incidence of periodontal diseases, the unsatisfactory effect of commonly used antibiotics, and the reports about the acquisition of drug resistance by pathogenic microorganisms present in the oral cavity necessitate the search for novel, alternative methods of preventing and treating periodontitis (Shlezinger et al. 2017; Cheng et al. 2017; Kinane et al. 2017). An additional problem in periodontal diseases caused by both bacterial and fungal pathogens is the accurate diagnosis and the correct identification of fungal microorganisms in subgingival sites and, consequently, the necessity for the modification of the conventional treatment. The diagnosis of fungal involvement in periodontal disease depends on the proper collection of a sample from the periodontal pockets after careful removal of the supragingival biofilm (Jewtuchowicz et al. 2008; Urzúa et al. 2008). Further diagnosis includes conventional fungal identification methods involving the observation of morphological changes or growth on differential media, however, a method based on the amplification of the internal transcribed spacer (ITS) is currently recommended (Persoon et al. 2017). In the case of the presence of fungi in the periodontal pockets during periodontitis, together with traditional treatment based on proper oral hygiene, conventional periodontal therapy, the use of topical antiseptics, and systemic antibiotics, the use of antifungals should be considered, including nystatin or fluconazole

(Sardi et al. 2011a). Oral administration of antifungal drugs from other classes might also be required (De-la-Torre et al. 2017).

However, in view of the emerging resistance of pathogens in the biofilm (Sardi et al. 2011a; Jarvensivu et al. 2004), attempts are made to develop therapeutics based on microbial virulence mechanisms and targeted at specific pathogenic species. One of the attractive targets are *P. gingivalis* cysteine proteases—Rgp and Kgp, which play a significant role in developing periodontal diseases. It was reported that extract of *Aspergillus oryzae* S-03 grown on fat-free soybean inhibits *P. gingivalis* cell growth through gingipain inhibitors, probably derived from the digestion of soybeans by the *A. oryzae* protease (Danshiitsoodol et al. 2014). Another natural secondary bioactive metabolite is terrein from *Aspergillus terreus*. It has been shown that this compound inhibits osteoclastogenesis and related excessive bone resorption, correlated with periodontitis, by blocking the main osteoclast regulator's activity—the activated cytoplasmic T cell nuclear factor 1. It was postulated that synthetic terrein, due to its low molecular mass and short half-life, might be used as an oral agent to prevent periodontal diseases (Nakagawa et al. 2020). There were also reports on the secretion of substances inhibiting bacterial and fungal growth by *Malassezia* and *Pichia* species. *M. globosa* spent medium has been shown to have antibacterial properties against *S. mutans* and *S. mitis* (Baraniya et al. 2020), while *Pichia* spent medium inhibits the growth of *Candida*, *Aspergillus*, and *Fusarium* (Mukherjee et al. 2014).

Another promising treatment in the fight against periodontitis is antisense therapy based on the peptide, which may block the synthesis of microbial proteins due to their strong affinity for mRNA. It was reported that the use of antisense peptides effectively inhibits the growth of two pathogens associated with periodontitis, *P. gingivalis*, and *A. actinomycetemcomitans* (Sugimoto et al. 2019). Similarly, the use of cerium-doped nanoparticles also inhibits the *P. gingivalis* and *F. nucleatum*. The acidic environment of the gingival pockets of patients with chronic periodontitis promotes the hydrolysis of nanoparticles, which releasing of ions modulate the microenvironment's pH, consequently changing them to unfavorable for bacterial growth (Li et al. 2019). In turn, polylactic-co-glycolic acid (PLGA) nanoparticles, coated with a peptide derived from *S. gordonii* significantly reduce the virulence of *P. gingivalis* in a mouse model of periodontitis and strongly inhibit the adhesion of *P. gingivalis* to streptococci, preventing mixed infections (Mahmoud et al. 2019).

Potential candidates for alternative therapy are also sought among natural substances. For example, it has been shown that alcoholic red propolis extract has antifungal activity against *Candida* species isolated from patients with chronic periodontitis (Siqueira et al. 2015), while cardamom extract is bactericidal against periodontal pathogens—*P. gingivalis*, *F. nucleatum*, *A. actinomycetemcomitans*, and *P. intermedia* (Souissi et al. 2020).

6.6 Conclusion and Perspectives

Dental biofilm, where diverse microbes form a compact structure of biofilm, could be, in states of dysbiosis, the primary source of dental caries and periodontal diseases, which can further implicate systemic diseases by translocation the included microbes by the bloodstream into a variety of host tissues.

Although biofilm-forming microbes sustain their repertoires of virulence factors, the interaction with the environment and surrounding microbes, often from different species, gives rise to the emergent new properties of the microbial community.

Here we highlighted the role of fungal species forming biofilm with periodontal bacteria, where physical, metabolic, and chemical interactions shape their properties. Although there is evidence of the presence of particular representatives of various fungal genera—including *Candida*, *Aspergillus*, and *Rhodotorula*—in the subgingival biofilm of patients with severe chronic periodontitis, the most data are available for *C. albicans*. Numerous studies show that fungi are not an infection bystander but also an active member of the biofilm consortium. Fungal dimorphism and extracellular matrix, varied in the composition, may serve as bacterial shelter from the adverse environment or camouflage from host recognition, where synergistic or antagonistic interactions are presented. The understanding of cross-kingdom interplay that underlies the biofilm functioning, especially during contact with host cells, raises the need to search for new infection models that will enable the analysis of such diverse interactions. To meet this challenge, the development of organoids and organ-on-chip technology is the priority. It will create an opportunity to develop improved patient treatment strategies without the risk of infection recurrence.

Acknowledgements This work was supported in part by the National Science Centre of Poland (grant N° 2019/33/B/NZ6/02284 to M.R.-K.).

References

Aas JA, Barbuto SM, Alpagot T et al (2007) Subgingival plaque microbiota in HIV positive patients. J Clin Periodontol 34:189–195. https://doi.org/10.1111/j.1600-051X.2006.01034.x

Alanazi H, Semlali A, Perraud L et al (2014) Cigarette smoke-exposed *Candida albicans* increased chitin production and modulated human fibroblast cell responses. Biomed Res Int 2014:1–11. https://doi.org/10.1155/2014/963156

Alanazi H, Semlali A, Chmielewski W, Rouabhia M (2019) E-cigarettes increase *Candida albicans* growth and modulate its interaction with gingival epithelial cells. Int J Environ Res Public Health 16:294. https://doi.org/10.3390/ijerph16020294

Aruni AW, Dou Y, Mishra A, Fletcher HM (2015) The biofilm community-rebels with a cause. Curr Oral Heal Rep 2:48–56. https://doi.org/10.1007/s40496-014-0044-5

Asif K, Kothiwale S (2010) Phagocytic activity of peripheral blood and crevicular phagocytes in health and periodontal disease. J Indian Soc Periodontol 14:8. https://doi.org/10.4103/0972-124X.65427

Aslani N, Janbabaei G, Abastabar M et al (2018) Identification of uncommon oral yeasts from cancer patients by MALDI-TOF mass spectrometry. BMC Infect Dis 18:24. https://doi.org/10.1186/s12879-017-2916-5

Babitha GA, Vangara A, Shobha P et al (2018) *Candida albicans*: can it be a periodontal pathogen? Int J Adv Res 6:1216–1225. https://doi.org/10.21474/IJAR01/6340

Bachtiar EW, Bachtiar BM (2020) Effect of cell-free spent media prepared from *Aggregatibacter actinomycetemcomitans* on the growth of *Candida albicans* and *Streptococcus mutans* in co-species biofilms. Eur J Oral Sci 128:395–404. https://doi.org/10.1111/eos.12725

Bachtiar EW, Bachtiar BM, Jarosz LM et al (2014) AI-2 of *Aggregatibacter actinomycetemcomitans* inhibits *Candida albicans* biofilm formation. Front Cell Infect Microbiol 4:94. https://doi.org/10.3389/fcimb.2014.00094

Baker JL, Bor B, Agnello M et al (2017) Ecology of the oral microbiome: beyond bacteria. Trends Microbiol 25:362–374. https://doi.org/10.1016/j.tim.2016.12.012

Bamford CV, D'Mello A, Nobbs AH et al (2009) *Streptococcus gordonii* modulates *Candida albicans* biofilm formation through intergeneric communication. Infect Immun 77:3696–3704. https://doi.org/10.1128/IAI.00438-09

Banerjee B, Greenberger PA, Fink JN, Kurup VP (1998) Immunological characterization of Asp f 2, a Major Allergen from *Aspergillus fumigatus* associated with allergic bronchopulmonary aspergillosis. Infect Immun 66:5175–5182. https://doi.org/10.1128/IAI.66.11.5175-5182.1998

Baraniya D, Chen T, Nahar A et al (2020) Supragingival mycobiome and inter-kingdom interactions in dental caries. J Oral Microbiol 12:1729305. https://doi.org/10.1080/20002297.2020.1729305

Barbosa JO, Rossoni RD, Vilela SFG et al (2016) *Streptococcus mutans* can modulate biofilm formation and attenuate the virulence of *Candida albicans*. PLoS One 11:e0150457. https://doi.org/10.1371/journal.pone.0150457

Barrett-Bee K, Hayes Y, Wilson RG, Ryley JF (1985) A comparison of phospholipase activity, cellular adherence and pathogenicity of yeasts. J Gen Microbiol 131:1217–1221. https://doi.org/10.1099/00221287-131-5-1217

Bartnicka D, Gonzalez-Gonzalez M, Sykut J et al (2020) *Candida albicans* shields the periodontal killer *Porphyromonas gingivalis* from recognition by the host immune system and supports the bacterial infection of gingival tissue. Int J Mol Sci 21:1984. https://doi.org/10.3390/ijms21061984

Bartnicka D, Karkowska-Kuleta J, Zawrotniak M et al (2019) Adhesive protein-mediated cross-talk between *Candida albicans* and *Porphyromonas gingivalis* in dual species biofilm protects the anaerobic bacterium in unfavorable oxic environment. Sci Rep 9:4376. https://doi.org/10.1038/s41598-019-40771-8

Basavaraju M, Sisnity VS, Palaparthy R, Addanki PK (2016) Quorum quenching: signal jamming in dental plaque biofilms. J Dent Sci 11:349–352. https://doi.org/10.1016/j.jds.2016.02.002

Bastos JA, Diniz CG, Bastos MG et al (2011) Identification of periodontal pathogens and severity of periodontitis in patients with and without chronic kidney disease. Arch Oral Biol 56:804–811. https://doi.org/10.1016/j.archoralbio.2010.12.006

Baumgardner DJ (2019) Oral fungal microbiota: to thrush and beyond. J Patient Cent Res Rev 6:252–261. https://doi.org/10.17294/2330-0698.1705

Baumgartner JC, Watts CM, Xia T (2000) Occurrence of *Candida albicans* in infections of endodontic origin. J Endod 26:695–698. https://doi.org/10.1097/00004770-200012000-00003

Behnsen J, Lessing F, Schindler S et al (2010) Secreted *Aspergillus fumigatus* protease Alp1 degrades human complement proteins C3, C4, and C5. Infect Immun 78:3585–3594. https://doi.org/10.1128/IAI.01353-09

Bochenska O, Rapala-Kozik M, Wolak N et al (2015) Inactivation of human kininogen-derived antimicrobial peptides by secreted aspartic proteases produced by the pathogenic yeast *Candida albicans*. Biol Chem 396:1369–1375. https://doi.org/10.1515/hsz-2015-0167

Bochenska O, Rapala-Kozik M, Wolak N et al (2016) The action of ten secreted aspartic proteases of pathogenic yeast *Candida albicans* on major human salivary antimicrobial peptide, histatin 5. Acta Biochim Pol 63:403–410. https://doi.org/10.18388/abp.2016_1318

Bor B, Cen L, Agnello M et al (2016) Morphological and physiological changes induced by contact-dependent interaction between *Candida albicans* and *Fusobacterium nucleatum*. Sci Rep 6:27956. https://doi.org/10.1038/srep27956

Bowen WH, Burne RA, Wu H, Koo H (2018) Oral biofilms: pathogens, matrix, and polymicrobial interactions in microenvironments. Trends Microbiol 26:229–242. https://doi.org/10.1016/j.tim.2017.09.008

Branzk N, Lubojemska A, Hardison SE et al (2014) Neutrophils sense microbe size and selectively release neutrophil extracellular traps in response to large pathogens. Nat Immunol 15:1017–1025. https://doi.org/10.1038/ni.2987

Bras G, Bochenska O, Rapala-Kozik M et al (2012) Extracellular aspartic protease SAP2 of *Candida albicans* yeast cleaves human kininogens and releases proinflammatory peptides, Met-Lys-bradykinin and des-Arg(9)-Met-Lys-bradykinin. Biol Chem 393:829–839. https://doi.org/10.1515/hsz-2012-0157

Bras G, Bochenska O, Rapala-Kozik M et al (2013) Release of biologically active kinin peptides, Met-Lys-bradykinin and Leu-Met-Lys-bradykinin from human kininogens by two major secreted aspartic proteases of *Candida parapsilosis*. Peptides 48:114–123. https://doi.org/10.1016/j.peptides.2013.08.003

Brinkmann V, Reichard U, Goosmann C et al (2004) Neutrophil extracellular traps kill bacteria. Science 303:1532–1535. https://doi.org/10.1126/science.1092385

Brusca MI, Rosa A, Albaina O et al (2010) The impact of oral contraceptives on women's periodontal health and the subgingival occurrence of aggressive periodontopathogens and *Candida species*. J Periodontol 81:1010–1008. https://doi.org/10.1902/jop.2010.090575

Canabarro A, Valle C, Farias MR et al (2013) Association of subgingival colonization of *Candida albicans* and other yeasts with severity of chronic periodontitis. J Periodontal Res 48:428–432. https://doi.org/10.1111/jre.12022

Casanova L, Hughes FJ, Preshaw PM (2014) Diabetes and periodontal disease: a two-way relationship. Br Dent J 217:433–437. https://doi.org/10.1038/sj.bdj.2014.907

Cheng L, Zhang K, Zhang N et al (2017) Developing a new generation of antimicrobial and bioactive dental resins. J Dent Res 96:855–863. https://doi.org/10.1177/0022034517709739

Cruz MR, Graham CE, Gagliano BC et al (2013) *Enterococcus faecalis* inhibits hyphal morphogenesis and virulence of *Candida albicans*. Infect Immun 81:189–200. https://doi.org/10.1128/IAI.00914-12

Cui L, Morris A, Ghedin E (2013) The human mycobiome in health and disease. Genome Med 5:63. https://doi.org/10.1186/gm467

Dahlén G, Blomqvist S, Almståhl A, Carlén A (2012) Virulence factors and antibiotic susceptibility in enterococci isolated from oral mucosal and deep infections. J Oral Microbiol 4:10855. https://doi.org/10.3402/jom.v4i0.10855

Danhof HA, Vylkova S, Vesely EM et al (2016) Robust extracellular pH modulation by *Candida albicans* during growth in carboxylic acids. MBio 7:e01646-16. https://doi.org/10.1128/mBio.01646-16

Danshiitsoodol N, Yamashita H, Noda M et al (2014) Aspergillus oryzae S-03 produces gingipain inhibitors as a virulence factor for *Porphyromonas gingivalis*. J Bacteriol Virol 44:152. https://doi.org/10.4167/jbv.2014.44.2.152

de Groot PWJ, Kraneveld EA, Yin QY et al (2008) The cell wall of the human pathogen *Candida glabrata*: differential incorporation of novel adhesin-like wall proteins. Eukaryot Cell 7:1951–1964. https://doi.org/10.1128/EC.00284-08

Delaney C, Kean R, Short B et al (2018) Fungi at the scene of the crime: innocent bystanders or accomplices in oral infections? Curr Clin Microbiol Rep 5:190–200. https://doi.org/10.1007/s40588-018-0100-3

De-la-Torre J, Ortiz-Samperio ME, Marcos-Arias C et al (2017) In vitro antifungal susceptibility of oral *Candida* isolates from patients suffering from caries and chronic periodontitis. Mycopath 182(5-6):471–485. https://doi.org/10.1007/s11046-017-0112-1

De-la-Torre J, Quindós G, Marcos-Arias C et al (2018) Oral *Candida* colonization in patients with chronic periodontitis. Is there any relationship? Rev Iberoam Micol 35:134–139. https://doi.org/10.1016/j.riam.2018.03.005

Dewhirst FE, Chen T, Izard J et al (2010) The human oral microbiome. J Bacteriol 192:5002–5017. https://doi.org/10.1128/JB.00542-10

Diaz PI, Valm AM (2020) Microbial interactions in oral communities mediate emergent biofilm properties. J Dent Res 99:18–25. https://doi.org/10.1177/0022034519880157

Diaz PI, Xie Z, Sobue T et al (2012) Synergistic interaction between *Candida albicans* and commensal oral streptococci in a novel *in vitro* mucosal model. Infect Immun 80:620–632. https://doi.org/10.1128/IAI.05896-11

Diaz PI, Strausbaugh LD, Dongari-Bagtzoglou A (2014) Fungal-bacterial interactions and their relevance to oral health: linking the clinic and the bench. Front Cell Infect Microbiol 4:101. https://doi.org/10.3389/fcimb.2014.00101

Dominy SS, Lynch C, Ermini F et al (2019) *Porphyromonas gingivalis* in Alzheimer's disease brains: evidence for disease causation and treatment with small-molecule inhibitors. Sci Adv 5:eaau3333. https://doi.org/10.1126/sciadv.aau3333

Dranginis AM, Rauceo JM, Coronado JE, Lipke PN (2007) A biochemical guide to yeast adhesins: glycoproteins for social and antisocial occasions. Microbiol Mol Biol Rev 71:282–294. https://doi.org/10.1128/MMBR.00037-06

Dutton LC, Paszkiewicz KH, Silverman RJ et al (2016) Transcriptional landscape of trans-kingdom communication between *Candida albicans* and *Streptococcus gordonii*. Mol Oral Microbiol 31:136–161. https://doi.org/10.1111/omi.12111

Egan MW, Spratt DA, Ng Y-L et al (2002) Prevalence of yeasts in saliva and root canals of teeth associated with apical periodontitis. Int Endod J 35:321–329. https://doi.org/10.1046/j.1365-2591.2002.00478.x

El-Kirat-Chatel S, Beaussart A, Derclaye S et al (2015) Force nanoscopy of hydrophobic interactions in the fungal pathogen *Candida glabrata*. ACS Nano 9:1648–1655. https://doi.org/10.1021/nn506370f

Ene IV, Bennett RJ (2009) Hwp1 and related adhesins contribute to both mating and biofilm formation in *Candida albicans*. Eukaryot Cell 8:1909–1913. https://doi.org/10.1128/EC.00245-09

Fallon JP, Reeves EP, Kavanagh K (2010) Inhibition of neutrophil function following exposure to the *Aspergillus fumigatus* toxin fumagillin. J Med Microbiol 59:625–633. https://doi.org/10.1099/jmm.0.018192-0

Falsetta ML, Klein MI, Colonne PM et al (2014) Symbiotic relationship between *Streptococcus mutans* and *Candida albicans* synergizes virulence of plaque biofilms *in vivo*. Infect Immun 82:1968–1981. https://doi.org/10.1128/IAI.00087-14

Fechney JM, Browne GV, Prabhu N et al (2019) Preliminary study of the oral mycobiome of children with and without dental caries. J Oral Microbiol 11:1536182. https://doi.org/10.1080/20002297.2018.1536182

Fox EP, Cowley ES, Nobile CJ et al (2014) Anaerobic bacteria grow within *Candida albicans* biofilms and induce biofilm formation in suspension cultures. Curr Biol 24:2411–2416. https://doi.org/10.1016/j.cub.2014.08.057

Gabrielli E, Sabbatini S, Roselletti E et al (2016) *In vivo* induction of neutrophil chemotaxis by secretory aspartyl proteinases of *Candida albicans*. Virulence 5594. https://doi.org/10.1080/21505594.2016.1184385

Gayathri L, Akbarsha MA, Ruckmani K (2020) In vitro study on aspects of molecular mechanisms underlying invasive aspergillosis caused by gliotoxin and fumagillin, alone and in combination. Sci Rep 10:14473. https://doi.org/10.1038/s41598-020-71367-2

Ghannoum MA, Jurevic RJ, Mukherjee PK et al (2010) Characterization of the oral fungal microbiome (mycobiome) in healthy individuals. PLoS Pathog 6:e1000713. https://doi.org/10.1371/journal.ppat.1000713

Ghogre P (2014) Endodontic mycology: a new perspective of root canal infection. RRJDS 1:43–50. e-ISSN: 2320-7949 and p-ISSN: 2322-0090

Glee PM, Sundstrom P, Hazen KC (1995) Expression of surface hydrophobic proteins by *Candida albicans in vivo*. Infect Immun 63:1373–1379. https://doi.org/10.1128/IAI.63.4.1373-1379.1995

Gomes C, Fidel S, Fidel R, de Moura Sarquis MI (2010) Isolation and taxonomy of filamentous fungi in endodontic infections. J Endod 36:626–629. https://doi.org/10.1016/j.joen.2010.01.016

Gomes CC, Pinto LCC, Victor FL et al (2015) *Aspergillus* in endodontic infection near the maxillary sinus. Braz J Otorhinolaryngol 81:527–532. https://doi.org/10.1016/j.bjorl.2015.07.013

Graham CE, Cruz MR, Garsin DA, Lorenz MC (2017) *Enterococcus faecalis* bacteriocin EntV inhibits hyphal morphogenesis, biofilm formation, and virulence of *Candida albicans*. Proc Natl Acad Sci U S A 114:4507–4512. https://doi.org/10.1073/pnas.1620432114

Gravelat FN, Beauvais A, Liu H et al (2013) *Aspergillus* galactosaminogalactan mediates adherence to host constituents and conceals hyphal β-glucan from the immune system. PLoS Pathog 9:e1003575. https://doi.org/10.1371/journal.ppat.1003575

Grimaudo NJ, Nesbitt WE (1997) Coaggregation of *Candida albicans* with oral *Fusobacterium species*. Oral Microbiol Immunol 12:168–173. https://doi.org/10.1111/j.1399-302x.1997.tb00374.x

Gropp K, Schild L, Schindler S et al (2009) The yeast *Candida albicans* evades human complement attack by secretion of aspartic proteases. Mol Immunol 47:465–475. https://doi.org/10.1016/j.molimm.2009.08.019

Gross EL, Beall CJ, Kutsch SR et al (2012) Beyond *Streptococcus mutans*: dental caries onset linked to multiple species by 16S rRNA community analysis. PLoS One 7:e47722. https://doi.org/10.1371/journal.pone.0047722

Guo Y, Wang Y, Wang Y et al (2020) Heme competition triggers an increase in the pathogenic potential of *Porphyromonas gingivalis* in *Porphyromonas gingivalis-Candida albicans* mixed biofilm. Front Microbiol 11:596459. https://doi.org/10.3389/fmicb.2020.596459

Guruceaga X, Ezpeleta G, Mayayo E et al (2018) A possible role for fumagillin in cellular damage during host infection by *Aspergillus fumigatus*. Virulence

9:1548–1561. https://doi.org/10.1080/21505594.2018.1526528

Hammad MM, Darwazeh AMG, Idrees MM (2013) The effect of glycemic control on *Candida* colonization of the tongue and the subgingival plaque in patients with type II diabetes and periodontitis. Oral Surg Oral Med Oral Pathol Oral Radiol 116:321–326. https://doi.org/10.1016/j.oooo.2013.05.013

Hanaoka M, Domae E (2020) IL-1α released from oral epithelial cells upon candidalysin exposure initiates an early innate epithelial response. Int Immunol 33:dxaa070. https://doi.org/10.1093/intimm/dxaa070

Hazen KC (2004) Relationship between expression of cell surface hydrophobicity protein 1 (CSH1p) and surface hydrophobicity properties of *Candida dubliniensis*. Curr Microbiol 48:447–451. https://doi.org/10.1007/s00284-003-4223-1

Hellstein JW, Marek CL (2019) Candidiasis: red and white manifestations in the oral cavity. Head Neck Pathol 13:25–32. https://doi.org/10.1007/s12105-019-01004-6

Ho J, Wickramasinghe DN, Nikou SA et al (2020) Candidalysin is a potent trigger of alarmin and antimicrobial peptide release in epithelial cells. Cells 9:699. https://doi.org/10.3390/cells9030699

Holmes AR, Cannon RD, Jenkinson HF (1995) Interactions of *Candida albicans* with bacteria and salivary molecules in oral biofilms. J Ind Microbiol 15:208–213. https://doi.org/10.1007/BF01569827

Holmes ARR, McNab R, Jenkinson HFF (1996) *Candida albicans* binding to the oral bacterium *Streptococcus gordonii* involves multiple adhesin-receptor interactions. Infect Immun 64:4680–4685

Hornbach A, Heyken A, Schild L et al (2009) The glycosylphosphatidylinositol-anchored protease Sap9 modulates the interaction of *Candida albicans* with human neutrophils. Infect Immun 77:5216–5224. https://doi.org/10.1128/IAI.00723-09

Hou Y, Wang X, Zhang C-X et al (2017) Risk factors of periodontal disease in maintenance hemodialysis patients. Medicine (Baltimore) 96:e7892. https://doi.org/10.1097/MD.0000000000007892

Hoyer LL (2001) The ALS gene family of *Candida albicans*. Trends Microbiol 9:176–180. https://doi.org/10.1016/s0966-842x(01)01984-9

Hoyer LL, Cota E (2016) *Candida albicans* agglutinin-like sequence (Als) family vignettes: a review of als protein structure and function. Front Microbiol 7:280. https://doi.org/10.3389/fmicb.2016.00280

Hruskova-Heidingsfeldova O (2008) Secreted proteins of *Candida albicans*. Front Biosci 13:7227. https://doi.org/10.2741/3224

Hube B, Naglik J (2001) *Candida albicans* proteinases: resolving the mystery of a gene family. Microbiology 147:1997–2005. https://doi.org/10.1099/00221287-147-8-1997

Hube B, Stehr F, Bossenz M et al (2000) Secreted lipases of *Candida albicans*: cloning, characterisation and expression analysis of a new gene family with at least ten members. Arch Microbiol 174:362–374. https://doi.org/10.1007/s002030000218

Hwang G, Liu Y, Kim D et al (2017) *Candida albicans* mannans mediate *Streptococcus mutans* exoenzyme GtfB binding to modulate cross-kingdom biofilm development in vivo. PLoS Pathog 13:1–25. https://doi.org/10.1371/journal.ppat.1006407

Jabra-Rizk MA, Falkler WA, Merz WG et al (1999) Coaggregation of *Candida dubliniensis* with *Fusobacterium nucleatum*. J Clin Microbiol 37:1464–1468

Jack AA, Daniels DE, Jepson MA et al (2015) *Streptococcus gordonii* comCDE (competence) operon modulates biofilm formation with *Candida albicans*. Microbiology 161:411–421. https://doi.org/10.1099/mic.0.000010

Janus MM, Willems HME, Krom BP (2016) *Candida albicans* in multispecies oral communities; a keystone commensal? In: Advances in experimental medicine and biology. Springer, Dordrecht, pp 13–20

Janus MM, Crielaard W, Volgenant CMC et al (2017) *Candida albicans* alters the bacterial microbiome of early *in vitro* oral biofilms. J Oral Microbiol 9:1–10. https://doi.org/10.1080/20002297.2016.1270613

Jarosz LM, Deng DM, van der Mei HC et al (2009) *Streptococcus mutans* competence-stimulating peptide inhibits *Candida albicans* hypha formation. Eukaryot Cell 8:1658–1664. https://doi.org/10.1128/EC.00070-09

Jarvensivu A, Hietanen J, Rautemaa R et al (2004) *Candida* yeasts in chronic periodontitis tissues and subgingival microbial biofilms *in vivo*. Oral Dis 10:106–112. https://doi.org/10.1046/j.1354-523X.2003.00978.x

Javed F, Klingspor L, Sundin U et al (2009) Periodontal conditions, oral *Candida albicans* and salivary proteins in type 2 diabetic subjects with emphasis on gender. BMC Oral Health 9:12. https://doi.org/10.1186/1472-6831-9-12

Jenkinson HF, Lala HC, Shepherd MG (1990) Coaggregation of *Streptococcus sanguis* and other streptococci with *Candida albicans*. Infect Immun 58:1429–1436

Jewtuchowicz VM, Mujica MT, Brusca MI et al (2008) Phenotypic and genotypic identification of *Candida dubliniensis* from subgingival sites in immunocompetent subjects in Argentina. Oral Microbiol Immunol 23:505–509. https://doi.org/10.1111/j.1399-302X.2008.00465.x

Johansson A, Dahlén G (2018) Bacterial virulence factors that contribute to periodontal pathogenesis. In: Pathogenesis of periodontal diseases. Springer, Cham, pp 31–49

Johnson CJ, Cabezas-Olcoz J, Kernien JF et al (2016) The extracellular matrix of *Candida albicans* biofilms impairs formation of neutrophil extracellular traps. PLOS Pathog 12:e1005884. https://doi.org/10.1371/journal.ppat.1005884

Joyner PM, Liu J, Zhang Z et al (2010) Mutanobactin A from the human oral pathogen *Streptococcus mutans* is a cross-kingdom regulator of the yeast-mycelium transition. Org Biomol Chem 8:5486–5489. https://doi.org/10.1039/c0ob00579g

Kamma JJ, Nakou M, Baehni PC (1999) Clinical and microbiological characteristics of smokers with early onset periodontitis. J Periodontal Res 34:25–33. https://doi.org/10.1111/j.1600-0765.1999.tb02218.x

Kapellas K, Singh A, Bertotti M et al (2019) Periodontal and chronic kidney disease association: a systematic review and meta-analysis. Nephrology 24:202–212. https://doi.org/10.1111/nep.13225

Karkowska-Kuleta J, Rapala-Kozik M, Kozik A (2009) Fungi pathogenic to humans: molecular bases of virulence of *Candida albicans*, *Cryptococcus neoformans* and *Aspergillus fumigatus*. Acta Biochim Pol 56:211–224

Karkowska-Kuleta J, Bartnicka D, Zawrotniak M et al (2018) The activity of bacterial peptidylarginine deiminase is important during formation of dual-species biofilm by periodontal pathogen *Porphyromonas gingivalis* and opportunistic fungus *Candida albicans*. Pathog Dis 76:fty033. https://doi.org/10.1093/femspd/fty033

Katragkou A, Kruhlak MJ, Simitsopoulou M et al (2010) Interactions between human phagocytes and *Candida albicans* biofilms alone and in combination with antifungal agents. J Infect Dis 201:1941–1949. https://doi.org/10.1086/652783

Kaur R, Ma B, Cormack BP (2007) A family of glycosylphosphatidylinositol-linked aspartyl proteases is required for virulence of *Candida glabrata*. Proc Natl Acad Sci 104:7628–7633. https://doi.org/10.1073/pnas.0611195104

Kaveemongkonrat S, Duangsonk K, Houbraken J et al (2019) Partial characteristics of hemolytic factors secreted from airborne *Aspergillus* and *Penicillium*, and an enhancement of hemolysis by *Aspergillus* micronesiensis CAMP-like factor via *Staphylococcus aureus*-sphingomyelinase. J Microbiol 57:1086–1094. https://doi.org/10.1007/s12275-019-9133-4

Kernien JF, Snarr BD, Sheppard DC, Nett JE (2017) The interface between fungal biofilms and innate immunity. Front Immunol 8:1968. https://doi.org/10.3389/fimmu.2017.01968

Kim D, Liu Y, Benhamou RI et al (2018) Bacterial-derived exopolysaccharides enhance antifungal drug tolerance in a cross-kingdom oral biofilm. ISME J 12:1427–1442. https://doi.org/10.1038/s41396-018-0113-1

Kinane DF, Stathopoulou PG, Papapanou PN (2017) Periodontal diseases. Nat Rev Dis Prim 3:17038. https://doi.org/10.1038/nrdp.2017.38

Klotz SA, Chasin BS, Powell B et al (2007) Polymicrobial bloodstream infections involving *Candida species*: analysis of patients and review of the literature. Diagn Microbiol Infect Dis 59:401–406. https://doi.org/10.1016/j.diagmicrobio.2007.07.001

Koelsch G, Tang J, Loy JA et al (2000) Enzymic characteristics of secreted aspartic proteases of *Candida albicans*. Biochim Biophys Acta Protein Struct Mol Enzymol 1480:117–131. https://doi.org/10.1016/S0167-4838(00)00068-6

Kolenbrander PE (2000) Oral microbial communities: biofilms, interactions, and genetic systems. Annu Rev Microbiol 54:413–437. https://doi.org/10.1146/annurev.micro.54.1.413

König A, Hube B, Kasper L (2020) The dual function of the fungal toxin candidalysin during *Candida albicans*-macrophage interaction and virulence. Toxins 12:469. https://doi.org/10.3390/toxins12080469

Könönen E, Müller H-P (2014) Microbiology of aggressive periodontitis. Periodontology 2000 65:46–78. https://doi.org/10.1111/prd.12016

Könönen E, Gursoy M, Gursoy UK (2019) Periodontitis: a multifaceted disease of tooth-supporting tissues. J Clin Med 8:1135. https://doi.org/10.3390/jcm8081135

Koo H, Andes DR, Krysan DJ (2018) *Candida–streptococcal* interactions in biofilm-associated oral diseases. PLOS Pathog 14:e1007342. https://doi.org/10.1371/journal.ppat.1007342

Koziel J, Mydel P, Potempa J (2014) The link between periodontal disease and rheumatoid arthritis: an updated review. Curr Rheumatol Rep 16:408. https://doi.org/10.1007/s11926-014-0408-9

Kozik A, Gogol M, Bochenska O et al (2015) Kinin release from human kininogen by 10 aspartic proteases produced by pathogenic yeast *Candida albicans*. BMC Microbiol 15:60. https://doi.org/10.1186/s12866-015-0394-8

Kraneveld EA, de Soet JJ, Deng DM et al (2011) Identification and differential gene expression of adhesin-like wall proteins in *Candida glabrata* biofilms. Mycopathologia 172:415–427. https://doi.org/10.1007/s11046-011-9446-2

Krom BP, Kidwai S, Ten Cate JM (2014) *Candida* and other fungal species: forgotten players of healthy oral microbiota. J Dent Res 93:445–451. https://doi.org/10.1177/0022034514521814

Krüger W, Vielreicher S, Kapitan M et al (2019) Fungal-Bacterial interactions in health and disease. Pathogens 8:70. https://doi.org/10.3390/pathogens8020070

Kucharíková S, Neirinck B, Sharma N et al (2015) *In vivo Candida glabrata* biofilm development on foreign bodies in a rat subcutaneous model. J Antimicrob Chemother 70:846–856. https://doi.org/10.1093/jac/dku447

Kumamoto CA, Vinces MD (2005) Contributions of hyphae and hypha-co-regulated genes to *Candida albicans* virulence. Cell Microbiol 7:1546–1554. https://doi.org/10.1111/j.1462-5822.2005.00616.x

Lamont RJ, Koo H, Hajishengallis G (2018) The oral microbiota: dynamic communities and host interactions. Nat Rev Microbiol 16:745–759. https://doi.org/10.1038/s41579-018-0089-x

Lee MJ, Liu H, Barker BM et al (2015) The fungal exopolysaccharide galactosaminogalactan mediates virulence by enhancing resistance to neutrophil extracellular traps. PLoS Pathog 11:e1005187. https://doi.org/10.1371/journal.ppat.1005187

Lee MJ, Geller AM, Bamford NC et al (2016) Deacetylation of fungal exopolysaccharide mediates adhesion and biofilm formation. MBio 7:e00252–e00216. https://doi.org/10.1128/mBio.00252-16
Leidich SD, Ibrahim AS, Fu Y et al (1998) Cloning and disruption of ca PLB1, a phospholipase b gene involved in the pathogenicity of *Candida albicans*. J Biol Chem 273:26078–26086. https://doi.org/10.1074/jbc.273.40.26078
Li F, Palecek SP (2003) EAP1, a *Candida albicans* gene involved in binding human epithelial cells. Eukaryot Cell 2:1266–1273. https://doi.org/10.1128/EC.2.6.1266-1273.2003
Li X, Gao M, Han X et al (2012) Disruption of the phospholipase D gene attenuates the virulence of *Aspergillus fumigatus*. Infect Immun 80:429–440. https://doi.org/10.1128/IAI.05830-11
Li X, Qi M, Li C et al (2019) Novel nanoparticles of cerium-doped zeolitic imidazolate frameworks with dual benefits of antibacterial and anti-inflammatory functions against periodontitis. J Mater Chem B 7:6955–6971. https://doi.org/10.1039/C9TB01743G
Lof M, Janus MM, Krom BP (2017) Metabolic interactions between bacteria and fungi in commensal oral biofilms. J Fungi 3:40. https://doi.org/10.3390/jof3030040
Loussert C, Schmitt C, Prevost M-C et al (2010) *In vivo* biofilm composition of *Aspergillus fumigatus*. Cell Microbiol 12:405–410. https://doi.org/10.1111/j.1462-5822.2009.01409.x
Luo G, Samaranayake LP (2002) *Candida glabrata*, an emerging fungal pathogen, exhibits superior relative cell surface hydrophobicity and adhesion to denture acrylic surfaces compared with *Candida albicans*. APMIS 110:601–610. https://doi.org/10.1034/j.1600-0463.2002.1100902.x
Luo G, Samaranayake LP, Yau JYY (2001) *Candida* species exhibit differential *in vitro* hemolytic activities. J Clin Microbiol 39:2971–2974. https://doi.org/10.1128/JCM.39.8.2971-2974.2001
Ma S, Li H, Yan C et al (2014) Antagonistic effect of protein extracts from *Streptococcus sanguinis* on pathogenic bacteria and fungi of the oral cavity. Exp Ther Med 7:1486–1494. https://doi.org/10.3892/etm.2014.1618
Mahmoud MY, Steinbach-Rankins JM, Demuth DR (2019) Functional assessment of peptide-modified PLGA nanoparticles against oral biofilms in a murine model of periodontitis. J Control Release 297:3–13. https://doi.org/10.1016/j.jconrel.2019.01.036
Manji F, Dahlen G, Fejerskov O (2018) Caries and periodontitis: contesting the conventional wisdom on their aetiology. Caries Res 52:548–564. https://doi.org/10.1159/000488948
Manns JM, Mosser DM, Buckley HR (1994) Production of a hemolytic factor by *Candida albicans*. Infect Immun 62:5154–5156. https://doi.org/10.1128/IAI.62.11.5154-5156.1994
Marsh PD (2006) Dental plaque as a biofilm and a microbial community – implications for health and disease. BMC Oral Health 6:S14. https://doi.org/10.1186/1472-6831-6-S1-S14
Marsh PD, Zaura E (2017) Dental biofilm: ecological interactions in health and disease. J Clin Periodontol 44:S12–S22. https://doi.org/10.1111/jcpe.12679
Masuoka J, Hazen KC (2004) Cell wall mannan and cell surface hydrophobicity in *Candida albicans* serotype a and b strains. Infect Immun 72:6230–6236. https://doi.org/10.1128/IAI.72.11.6230-6236.2004
Matić Petrović S, Cimbaljević M, Radunović M et al (2015) Detection and sampling methods for isolation of *Candida* spp. from oral cavities in diabetics and non-diabetics. Braz Oral Res 29:1–7. https://doi.org/10.1590/1807-3107BOR-2015.vol29.0077
Matic Petrovic S, Radunovic M, Barac M et al (2019) Subgingival areas as potential reservoirs of different *Candida* spp in type 2 diabetes patients and healthy subjects. PLoS One 14:e0210527. https://doi.org/10.1371/journal.pone.0210527
Mayer FL, Wilson D, Hube B (2013) *Candida albicans* pathogenicity mechanisms. Virulence 4:119–128. https://doi.org/10.4161/viru.22913
Melton JJ, Redding SW, Kirkpatrick WR et al (2010) Recovery of *Candida dubliniensis* and other *Candida* species from the oral cavity of subjects with periodontitis who had well-controlled and poorly controlled type 2 diabetes: a pilot study. Spec Care Dentist 30:230–234. https://doi.org/10.1111/j.1754-4505.2010.00159.x
Mengesha B, Conti H (2017) The role of IL-17 in protection against mucosal *Candida* infections. J Fungi 3:52. https://doi.org/10.3390/jof3040052
Metwalli KH, Khan SA, Krom BP, Jabra-Rizk MA (2013) *Streptococcus mutans*, *Candida albicans*, and the human mouth: a sticky situation. PLoS Pathog 9:e1003616. https://doi.org/10.1371/journal.ppat.1003616
Mira A, Simon-Soro A, Curtis MA (2017) Role of microbial communities in the pathogenesis of periodontal diseases and caries. J Clin Periodontol 44(Suppl 1):S23–S38. https://doi.org/10.1111/jcpe.12671
Monod M, Togni G, Hube B, Sanglard D (1994) Multiplicity of genes encoding secreted aspartic proteinases in *Candida* species. Mol Microbiol 13:357–368. https://doi.org/10.1111/j.1365-2958.1994.tb00429.x
Monteiro-da-Silva F, Araujo R, Sampaio-Maia B (2014) Interindividual variability and intraindividual stability of oral fungal microbiota over time. Med Mycol 52:498–505. https://doi.org/10.1093/mmy/myu027
Moyes DL, Runglall M, Murciano C et al (2010) A biphasic innate immune MAPK response discriminates between the yeast and hyphal forms of *Candida albicans* in epithelial cells. Cell Host Microbe 8:225–235. https://doi.org/10.1016/j.chom.2010.08.002
Moyes DL, Shen C, Murciano C et al (2014) Protection against epithelial damage during *Candida albicans* infection is mediated by PI3K/Akt and mammalian target of rapamycin signaling. J Infect Dis 209:1816–1826. https://doi.org/10.1093/infdis/jit824

Moyes DL, Wilson D, Richardson JP et al (2016) Candidalysin is a fungal peptide toxin critical for mucosal infection. Nature 532:64–68. https://doi.org/10.1038/nature17625

Mukherjee PK, Chandra J, Retuerto M et al (2014) Oral mycobiome analysis of HIV-infected patients: identification of *Pichia* as an antagonist of opportunistic fungi. PLoS Pathog 10:e1003996. https://doi.org/10.1371/journal.ppat.1003996

Murakami S, Mealey BL, Mariotti A, Chapple ILC (2018) Dental plaque-induced gingival conditions. J Periodontol 89:S17–S27. https://doi.org/10.1002/JPER.17-0095

Naglik JR, Challacombe SJ, Hube B (2003) *Candida albicans* secreted aspartyl proteinases in virulence and pathogenesis. Microbiol Mol Biol Rev 67:400–428. https://doi.org/10.1128/MMBR.67.3.400-428.2003

Naglik JR, Fostira F, Ruprai J et al (2006) *Candida albicans* HWP1 gene expression and host antibody responses in colonization and disease. J Med Microbiol 55:1323–1327. https://doi.org/10.1099/jmm.0.46737-0

Naglik JR, Gaffen SL, Hube B (2019) Candidalysin: discovery and function in *Candida albicans* infections. Curr Opin Microbiol 52:100–109. https://doi.org/10.1016/j.mib.2019.06.002

Nakagawa S, Omori K, Nakayama M et al (2020) The fungal metabolite (+)-terrein abrogates osteoclast differentiation via suppression of the RANKL signaling pathway through NFATc1. Int Immunopharmacol 83:106429. https://doi.org/10.1016/j.intimp.2020.106429

Nikawa H, Egusa H, Makihira S et al (2006) An *in vitro* evaluation of the adhesion of *Candida* species to oral and lung tissue cells. Mycoses 49:14–17. https://doi.org/10.1111/j.1439-0507.2005.01176.x

Norhammar A, Kjellström B, Habib N et al (2019) Undetected dysglycemia is an important risk factor for two common diseases, myocardial infarction and periodontitis: a report from the PAROKRANK study. Diabetes Care 42:1504–1511. https://doi.org/10.2337/dc19-0018

O'Connell LM, Santos R, Springer G et al (2020) Site-specific profiling of the dental mycobiome reveals strong taxonomic shifts during progression of early-childhood caries. Appl Environ Microbiol 86:e02825–e02819. https://doi.org/10.1128/AEM.02825-19

O'Donnell LE, Millhouse E, Sherry L et al (2015) Polymicrobial *Candida* biofilms: friends and foe in the oral cavity. FEMS Yeast Res 15:fov077. https://doi.org/10.1093/femsyr/fov077

O'Sullivan JM, Jenkinson HF, Cannon RD (2000) Adhesion of *Candida albicans* to oral streptococci is promoted by selective adsorption of salivary proteins to the streptococcal cell surface. Microbiology 146:41–48. https://doi.org/10.1099/00221287-146-1-41

Øilo M, Bakken V (2015) Biofilm and dental biomaterials. Materials (Basel) 8:2887–2900. https://doi.org/10.3390/ma8062887

Oliveira FAF, Forte CPF, Silva PG de B et al (2015) Molecular analysis of oral bacteria in heart valve of patients with cardiovascular disease by real-time polymerase chain reaction. Medicine (Baltimore) 94:e2067. https://doi.org/10.1097/MD.0000000000002067

Olsen I, Kell DB, Pretorius E (2020) Is *Porphyromonas gingivalis* involved in Parkinson's disease? Eur J Clin Microbiol Infect Dis 39:2013–2018. https://doi.org/10.1007/s10096-020-03944-2

Page RC, Kornman KS (1997) The pathogenesis of human periodontitis: an introduction. Periodontology 2000 14:9–11. https://doi.org/10.1111/j.1600-0757.1997.tb00189.x

Panagoda GJ, Ellepola ANB, Samaranayake LP (2001) Adhesion of *Candida parapsilosis* to epithelial and acrylic surfaces correlates with cell surface hydrophobicity. Mycoses 44:29–35. https://doi.org/10.1046/j.1439-0507.2001.00611.x

Pellon A, Sadeghi Nasab SD, Moyes DL (2020) New insights in *Candida albicans* innate immunity at the mucosa: toxins, epithelium, metabolism, and beyond. Front Cell Infect Microbiol 10:81. https://doi.org/10.3389/fcimb.2020.00081

Persoon IF, Buijs MJ, Özok et al (2017) The mycobiome of root canal infections is correlated to the bacteriome. Clin Oral Invest 21:1871–1881. https://doi.org/10.1007/s00784-016-1980-3

Peters BA, Wu J, Hayes RB, Ahn J (2017) The oral fungal mycobiome: characteristics and relation to periodontitis in a pilot study. BMC Microbiol 17:157. https://doi.org/10.1186/s12866-017-1064-9

Phan QT, Myers CL, Fu Y et al (2007) Als3 is a *Candida albicans* invasin that binds to cadherins and induces endocytosis by host cells. PLoS Biol 5:e64. https://doi.org/10.1371/journal.pbio.0050064

Pietrella D, Rachini A, Pandey N et al (2010) The inflammatory response induced by aspartic proteases of *Candida albicans* is independent of proteolytic activity. Infect Immun 78:4754–4762. https://doi.org/10.1128/IAI.00789-10

Pinheiro CR, Coelho AL, de Oliveira CE et al (2018) Recognition of *Candida albicans* by gingival fibroblasts: the role of TLR2, TLR4/CD14, and MyD88. Cytokine 106:67–75. https://doi.org/10.1016/j.cyto.2017.10.013

Rambach G, Blum G, Latgé J-P et al (2015) Identification of *Aspergillus fumigatus* surface components that mediate interaction of conidia and hyphae with human platelets. J Infect Dis 212:1140–1149. https://doi.org/10.1093/infdis/jiv191

Rapala-Kozik M, Karkowska-Kuleta J, Ryzanowska A et al (2010) Degradation of human kininogens with the release of kinin peptides by extracellular proteinases of *Candida* spp. Biol Chem 391:823–830. https://doi.org/10.1515/BC.2010.083

Rapala-Kozik M, Bochenska O, Zawrotniak M et al (2015) Inactivation of the antifungal and immunomodulatory properties of human cathelicidin LL-37 by aspartic proteases produced by the pathogenic yeast *Candida albicans*. Infect Immun 83:2518–2530. https://doi.org/10.1128/IAI.00023-15

Rapala-Kozik M, Bochenska O, Zajac D et al (2018) Extracellular proteinases of *Candida* species patho-

genic yeasts. Mol Oral Microbiol 33:113–124. https://doi.org/10.1111/omi.12206
Rasheed M, Battu A, Kaur R (2018) Aspartyl proteases in *Candida glabrata* are required for suppression of the host innate immune response. J Biol Chem 293:6410–6433. https://doi.org/10.1074/jbc.M117.813741
Reeves EP, Messina CG, Doyle S, Kavanagh K (2004) Correlation between gliotoxin production and virulence of *Aspergillus fumigatus* in *Galleria mellonella*. Mycopathologia 158:73–79. https://doi.org/10.1023/b:myco.0000038434.55764.16
Reichard U, Buttner S, Eiffert H et al (1990) Purification and characterisation of an extracellular serine proteinase from *Aspergillus fumigatus* and its detection in tissue. J Med Microbiol 33:243–251. https://doi.org/10.1099/00222615-33-4-243
Reichard U, Eiffert H, Rüchel R (1994) Purification and characterization of an extracellular aspartic proteinase from *Aspergillus fumigatus*. Med Mycol 32:427–436. https://doi.org/10.1080/02681219480000581
Rementeria A, López-Molina N, Ludwig A et al (2005) Genes and molecules involved in Aspergillus fumigatus virulence. Rev Iberoam Micol 22:1–23. https://doi.org/10.1016/s1130-1406(05)70001-2
Reynaud AH, Nygaard-Østby B, Bøygard G-K et al (2001) Yeasts in periodontal pockets. J Clin Periodontol 28:860–864. https://doi.org/10.1034/j.1600-051x.2001.028009860.x
Robinson BW, Venaille TJ, Mendis AH, McAleer R (1990) Allergens as proteases: an *Aspergillus fumigatus* proteinase directly induces human epithelial cell detachment. J Allergy Clin Immunol 86:726–731. https://doi.org/10.1016/s0091-6749(05)80176-9
Rocha FAC, Alves AMCV, Rocha MFG et al (2017) Tumor necrosis factor prevents *Candida albicans* biofilm formation. Sci Rep 7:1206. https://doi.org/10.1038/s41598-017-01400-4
Sandai D, Tabana YM, Ouweini AE, Ayodeji IO (2016) Resistance of *Candida albicans* biofilms to drugs and the host immune system. Jundishapur J Microbiol 9:e37385. https://doi.org/10.5812/jjm.37385
Santhana Krishnan G, Naik D, Uppoor A et al (2020) Candidal carriage in saliva and subgingival plaque among smokers and non-smokers with chronic periodontitis-a cross-sectional study. PeerJ 8:e8441. https://doi.org/10.7717/peerj.8441
Sardi JCO, Duque C, Mariano FS et al (2010) *Candida* spp. in periodontal disease: a brief review. J Oral Sci 52:177–185. https://doi.org/10.2334/josnusd.52.177
Sardi JC, Almeida AM, Mendes Giannini MJ (2011a) New antimicrobial therapies used against fungi present in subgingival sites-a brief review. Arch Oral Biol 56:951–959. https://doi.org/10.1016/j.archoralbio.2011.03.007
Sardi JCO, Duque C, Camargo GACG et al (2011b) Periodontal conditions and prevalence of putative periodontopathogens and *Candida* spp. in insulin-dependent type 2 diabetic and non-diabetic patients with chronic periodontitis-a pilot study. Arch Oral Biol 56:1098–1105. https://doi.org/10.1016/j.archoralbio.2011.03.017
Sardi JCO, Duque C, Höfling JF, Gonçalves RB (2012a) Genetic and phenotypic evaluation of *Candida albicans* strains isolated from subgingival biofilm of diabetic patients with chronic periodontitis. Med Mycol 50:467–475. https://doi.org/10.3109/13693786.2011.633233
Sardi JCO, Duque C, Mariano FS et al (2012b) Adhesion and invasion of *Candida albicans* from periodontal pockets of patients with chronic periodontitis and diabetes to gingival human fibroblasts. Med Mycol 50:43–49. https://doi.org/10.3109/13693786.2011.586133
Satala D, Karkowska-Kuleta J, Zelazna A et al (2020a) Moonlighting proteins at the candidal cell surface. Microorganisms 8:1046. https://doi.org/10.3390/microorganisms8071046
Satala D, Satala G, Karkowska-Kuleta J et al (2020b) Structural insights into the interactions of candidal enolase with human vitronectin, fibronectin and plasminogen. Int J Mol Sci 21:7843. https://doi.org/10.3390/ijms21217843
Schelenz S, Abdallah S, Gray G et al (2011) Epidemiology of oral yeast colonization and infection in patients with hematological malignancies, head neck and solid tumors. J Oral Pathol Med 40:83–89. https://doi.org/10.1111/j.1600-0714.2010.00937.x
Schwarz F, Becker K, Rahn S et al (2015) Real-time PCR analysis of fungal organisms and bacterial species at peri-implantitis sites. Int J Implant Dent 1:9. https://doi.org/10.1186/s40729-015-0010-6
Semlali A, Killer K, Alanazi H et al (2014) Cigarette smoke condensate increases *C. albicans* adhesion, growth, biofilm formation, and EAP1, HWP1 and SAP2 gene expression. BMC Microbiol 14:61. https://doi.org/10.1186/1471-2180-14-61
Shen D-K, Noodeh AD, Kazemi A et al (2004) Characterisation and expression of phospholipases B from the opportunistic fungus *Aspergillus fumigatus*. FEMS Microbiol Lett 239:87–93. https://doi.org/10.1016/j.femsle.2004.08.019
Shende R, Wong SSW, Rapole S et al (2018) *Aspergillus fumigatus* conidial metalloprotease Mep1p cleaves host complement proteins. J Biol Chem 293:15538–15555. https://doi.org/10.1074/jbc.RA117.001476
Shi M, Wei Y, Hu W et al (2018a) The subgingival microbiome of periodontal pockets with different probing depths in chronic and aggressive periodontitis: a pilot study. Front Cell Infect Microbiol 8:124. https://doi.org/10.3389/fcimb.2018.00124
Shi Q, Zhang B, Xing H et al (2018b) Patients with chronic obstructive pulmonary disease suffer from worse periodontal health-evidence from a meta-analysis. Front Physiol 9:33. https://doi.org/10.3389/fphys.2018.00033
Shlezinger M, Khalifa L, Houri-Haddad Y et al (2017) Phage therapy: a new horizon in the antibacterial treatment of oral pathogens. Curr Top Med Chem 17:1199–1211. https://doi.org/10.2174/1568026616666160930145649

Shokeen B, Dinis MDB, Haghighi F et al (2021) Omics and interspecies interaction. Periodontology 2000 (85):101–111. https://doi.org/10.1111/prd.12354

Signat B, Roques C, Poulet P, Duffaut D (2011) *Fusobacterium nucleatum* in periodontal health and disease. Curr Issues Mol Biol 13:25–36

Silva-Dias A, Miranda IM, Branco J et al (2015) Adhesion, biofilm formation, cell surface hydrophobicity, and antifungal planktonic susceptibility: relationship among *Candida* spp. Front Microbiol 6:205. https://doi.org/10.3389/fmicb.2015.00205

Silverman RJ, Nobbs AH, Vickerman MM et al (2010) Interaction of *Candida albicans* cell wall Als3 protein with *Streptococcus gordonii* SspB adhesin promotes development of mixed-species communities. Infect Immun 78:4644–4652. https://doi.org/10.1128/IAI.00685-10

Sin N, Meng L, Wang MQ, Wen JJ, Bornmann WG, Crews CM (1997) The anti-angiogenic agent fumagillin covalently binds and inhibits the methionine aminopeptidase, MetAP-2. Proc Natl Acad Sci USA 94:6099–6103. https://doi.org/10.1073/pnas.94.12.6099

Singh DK, Németh T, Papp A et al (2019) Functional characterization of secreted aspartyl proteases in *Candida parapsilosis*. mSphere 4:e00484–e00419. https://doi.org/10.1128/mSphere.00484-19

Singleton DR, Masuoka J, Hazen KC (2001) Cloning and analysis of a *Candida albicans* gene that affects cell surface hydrophobicity. J Bacteriol 183:3582–3588. https://doi.org/10.1128/JB.183.12.3582-3588.2001

Singleton DR, Fidel PL, Wozniak KL, Hazen KC (2005) Contribution of cell surface hydrophobicity protein 1 (Csh1p) to virulence of hydrophobic *Candida albicans* serotype A cells. FEMS Microbiol Lett 244:373–377. https://doi.org/10.1016/j.femsle.2005.02.010

Siqueira JF Jr, Sen BH (2004) Fungi in endodontic infections. Oral Surg Oral Med Oral Pathol Oral Radiol Endod 97:632–641. https://doi.org/10.1016/S1079210404000046

Siqueira JF Jr, Rôças IN, Lopes HP et al (2002) Fungal infection of the radicular dentin. J Endod 28:770–773. https://doi.org/10.1097/00004770-200211000-00006

Siqueira ABS, Rodriguez LRN de A, Santos RKB et al (2015) Antifungal activity of propolis against *Candida* species isolated from cases of chronic periodontitis. Braz Oral Res 29:1–6. https://doi.org/10.1590/1807--3107BOR-2015.vol29.0083

Souissi M, Azelmat J, Chaieb K, Grenier D (2020) Antibacterial and anti-inflammatory activities of cardamom (Elettaria cardamomum) extracts: potential therapeutic benefits for periodontal infections. Anaerobe 61:102089. https://doi.org/10.1016/j.anaerobe.2019.102089

Stathopoulou PG, Benakanakere MR, Galicia JC, Kinane DF (2009) The host cytokine response to *Porphyromonas gingivalis* is modified by gingipains. Oral Microbiol Immunol 24:11–17. https://doi.org/10.1111/j.1399-302X.2008.00467.x

Sugimoto S, Maeda H, Kitamatsu M et al (2019) Selective growth inhibition of *Porphyromonas gingivalis* and *Aggregatibacter actinomycetemcomitans* by antisense peptide nucleic acids. Mol Cell Probes 43:45–49. https://doi.org/10.1016/j.mcp.2018.11.006

Sugui JA, Pardo J, Chang YC et al (2007) Gliotoxin is a virulence factor of *Aspergillus fumigatus*: gliP deletion attenuates virulence in mice immunosuppressed with hydrocortisone. Eukaryot Cell 6:1562–1569. https://doi.org/10.1128/EC.00141-07

Sultan AS, Kong EF, Rizk AM, Jabra-Rizk MA (2018) The oral microbiome: a lesson in coexistence. PLoS Pathog 14:e1006719. https://doi.org/10.1371/journal.ppat.1006719

Svoboda E, Schneider AE, Sándor N et al (2015) Secreted aspartic protease 2 of *Candida albicans* inactivates factor H and the macrophage factor H-receptors CR3 (CD11b/CD18) and CR4 (CD11c/CD18). Immunol Lett 168:13–21. https://doi.org/10.1016/j.imlet.2015.08.009

Sztukowska MN, Dutton LC, Delaney C et al (2018) Community development between *Porphyromonas gingivalis* and *Candida albicans* mediated by InlJ and Als3. mBio 9:e00202-18. https://doi.org/10.1128/mBio.00202-18

Tamai R, Sugamata M, Kiyoura Y (2011) *Candida albicans* enhances invasion of human gingival epithelial cells and gingival fibroblasts by *Porphyromonas gingivalis*. Microb Pathog 51:250–254. https://doi.org/10.1016/j.micpath.2011.06.009

Taylor GW, Manz MC, Borgnakke WS (2004) Diabetes, periodontal diseases, dental caries, and tooth loss: a review of the literature. Compend Contin Educ Dent 25:179–184, 186–8, 190

Thau N, Monod M, Crestani B et al (1994) Rodletless mutants of *Aspergillus fumigatus*. Infect Immun 62:4380–4388. https://doi.org/10.1128/IAI.62.10.4380-4388.1994

Timmermans B, De Las PA, Castaño I, Van Dijck P (2018) Adhesins in *Candida glabrata*. J Fungi 4:60. https://doi.org/10.3390/jof4020060

Tomar SL, Asma S (2000) Smoking-attributable periodontitis in the united states: findings from NHANES III. J Periodontol 71:743–751. https://doi.org/10.1902/jop.2000.71.5.743

Tronchin G, Pihet M, Lopes-Bezerra LM, Bouchara J-P (2008) Adherence mechanisms in human pathogenic fungi. Med Mycol 46:749–772. https://doi.org/10.1080/13693780802206435

Tsuda K, Furuta N, Inaba H et al (2008) Functional analysis of α5β1 integrin and lipid rafts in invasion of epithelial cells by *Porphyromonas gingivalis* using fluorescent beads coated with bacterial membrane vesicles. Cell Struct Funct 33:123–132. https://doi.org/10.1247/csf.08012

Upadhyay SK, Mahajan L, Ramjee S et al (2009) Identification and characterization of a laminin-binding protein of *Aspergillus fumigatus*: extracellular thaumatin domain protein (AfCalAp). J Med Microbiol 58:714–722. https://doi.org/10.1099/jmm.0.005991-0

Urban CF, Reichard U, Brinkmann V, Zychlinsky A (2006) Neutrophil extracellular traps capture and kill *Candida albicans* yeast and hyphal forms. Cell Microbiol 8:668–676. https://doi.org/10.1111/j.1462-5822.2005.00659.x

Urban CF, Ermert D, Schmid M et al (2009) Neutrophil extracellular traps contain calprotectin, a cytosolic protein complex involved in host defense against *Candida albicans*. PLoS Pathog 5:e1000639. https://doi.org/10.1371/journal.ppat.1000639

Urzúa B, Hermosilla G, Gamonal J et al (2008) Yeast diversity in the oral microbiota of subjects with periodontitis: *Candida albicans* and *Candida dubliniensis* colonize the periodontal pockets. Med Mycol 46:783–793. https://doi.org/10.1080/13693780802060899

Valm AM (2019) The structure of dental plaque microbial communities in the transition from health to dental caries and periodontal disease. J Mol Biol 431:2957–2969. https://doi.org/10.1016/j.jmb.2019.05.016

van Leeuwen PT, van der Peet JM, Bikker FJ et al (2016) Interspecies interactions between *Clostridium difficile* and *Candida albicans*. mSphere 1:e00187–e00116. https://doi.org/10.1128/mSphere.00187-16

Verma AH, Richardson JP, Zhou C et al (2017) Oral epithelial cells orchestrate innate type 17 responses to *Candida albicans* through the virulence factor candidalysin. Sci Immunol 2:eaam8834. https://doi.org/10.1126/sciimmunol.aam8834

Verstrepen KJ, Klis FM (2006) Flocculation, adhesion and biofilm formation in yeasts. Mol Microbiol 60:5–15. https://doi.org/10.1111/j.1365-2958.2006.05072.x

Vílchez R, Lemme A, Ballhausen B et al (2010) *Streptococcus mutans* inhibits *Candida albicans* hyphal formation by the fatty acid signaling molecule trans-2-decenoic acid (SDSF). Chembiochem 11:1552–1562. https://doi.org/10.1002/cbic.201000086

Villar CC, Kashleva H, Nobile CJ et al (2007) Mucosal tissue invasion by *Candida albicans* is associated with E-cadherin degradation, mediated by transcription factor Rim101p and protease Sap5p. Infect Immun 75:2126–2135. https://doi.org/10.1128/IAI.00054-07

Vylkova S, Carman AJ, Danhof HA et al (2011) The fungal pathogen *Candida albicans* autoinduces hyphal morphogenesis by raising extracellular pH. MBio 2:e00055–e00011. https://doi.org/10.1128/mBio.00055-11

Wahid A, Chaudhry S, Ehsan A et al (2012) Bidirectional relationship between chronic kidney disease; periodontal disease: a review. Pakistan J Med Sci 29:211–215. https://doi.org/10.12669/pjms.291.2926

Waltimo TM, Sirén EK, Torkko HL et al (1997) Fungi in therapy-resistant apical periodontitis. Int Endod J 30:96–101. https://doi.org/10.1046/j.1365-2591.1997.00058.x

Waltimo TMT, Sen BH, Meurman JH et al (2003) Yeasts in apical periodontitis. Crit Rev Oral Biol Med 14:128–137. https://doi.org/10.1177/154411130301400206

Wartenberg D, Lapp K, Jacobsen ID et al (2011) Secretome analysis of *Aspergillus fumigatus* reveals Asp-hemolysin as a major secreted protein. Int J Med Microbiol 301:602–611. https://doi.org/10.1016/j.ijmm.2011.04.016

Wasylnka JA, Moore MM (2000) Adhesion of *Aspergillus* species to extracellular matrix proteins: evidence for involvement of negatively charged carbohydrates on the conidial surface. Infect Immun 68:3377–3384. https://doi.org/10.1128/IAI.68.6.3377-3384.2000

Willems HM, Kos K, Jabra-Rizk MA, Krom BP (2016) *Candida albicans* in oral biofilms could prevent caries. Pathog Dis 74:ftw039. https://doi.org/10.1093/femspd/ftw039

Wu T, Cen L, Kaplan C et al (2015) Cellular components mediating coadherence of *Candida albicans* and *Fusobacterium nucleatum*. J Dent Res 94:1432–1438. https://doi.org/10.1177/0022034515593706

Xu X-L, Lee RTH, Fang H-M et al (2008) Bacterial peptidoglycan triggers *Candida albicans* hyphal growth by directly activating the adenylyl cyclase Cyr1p. Cell Host Microbe 4:28–39. https://doi.org/10.1016/j.chom.2008.05.014

Xu H, Jenkinson HF, Dongari-Bagtzoglou A (2014a) Innocent until proven guilty: mechanisms and roles of *Streptococcus-Candida* interactions in oral health and disease. Mol Oral Microbiol 29:99–116. https://doi.org/10.1111/omi.12049

Xu H, Sobue T, Thompson A et al (2014b) *Streptococcal* co-infection augments *Candida* pathogenicity by amplifying the mucosal inflammatory response. Cell Microbiol 16:214–231. https://doi.org/10.1111/cmi.12216

Yu S, Li W, Liu X et al (2016) Distinct expression levels of ALS, LIP, and SAP genes in *Candida tropicalis* with diverse virulent activities. Front Microbiol 7:1175. https://doi.org/10.3389/fmicb.2016.01175

Zaugg C, Borg-von Zepelin M, Reichard U et al (2001) Secreted aspartic proteinase family of *Candida tropicalis*. Infect Immun 69:405–412. https://doi.org/10.1128/IAI.69.1.405-412.2001

Zawrotniak M, Bochenska O, Karkowska-Kuleta J et al (2017) Aspartic proteases and major cell wall components in *Candida albicans* trigger the release of neutrophil extracellular traps. Front Cell Infect Microbiol 7:414. https://doi.org/10.3389/fcimb.2017.00414

Zhao D, Khawaja AT, Jin L et al (2018) The directional and non-directional associations of periodontitis with chronic kidney disease: a systematic review and meta-analysis of observational studies. J Periodontal Res 53:682–704. https://doi.org/10.1111/jre.12565

Zhu L, Kreth J (2010) Role of *Streptococcus mutans* eukaryotic-type serine/threonine protein kinase in interspecies interactions with Streptococcus sanguinis. Arch Oral Biol 55:385–390. https://doi.org/10.1016/j.archoralbio.2010.03.012

Zijnge V, Van Leeuwen MBM, Degener JE et al (2010) Oral biofilm architecture on natural teeth. PLoS One 5:1–9. https://doi.org/10.1371/journal.pone.0009321

Herpesviruses in Periodontitis: An Umbrella Review

7

Aleksandar Jakovljevic, Miroslav Andric, Jelena Jacimovic, Jelena Milasin, and Javier Enrique Botero

Abstract

Background: Despite numerous studies indicating a high prevalence of herpesviruses in both apical and marginal periodontitis samples, their exact role in the pathogenesis of a periodontal disease is still unclear.

Objective: This umbrella review aimed to summarize data on herpesviruses detection in marginal periodontitis (MP) and apical periodontitis of endodontic origin (APEO) samples.

Methods: The study protocol has been drafted *a priori* and registered to the International Prospective Register of Systematic Reviews (PROSPERO) (CRD42020215922). The literature search was conducted using the following electronic databases: Clarivate Analytics' Web of Science, Scopus, PubMed and Cochrane Database of Systematic Reviews, from inception to October 2020, with no language restrictions. Systematic reviews with or without meta-analysis that evaluated the association between the occurrence of herpesviruses and different forms of periodontal diseases were included. Other types of studies, including narrative reviews, were excluded. Two reviewers independently performed a literature search, data extraction, and quality assessment of included studies. Any disagreements or doubts were resolved by a third reviewer. The quality of the reviews was assessed using the AMSTAR 2 tool (A measurement tool to assess systematic reviews).

Results: Six systematic reviews were included in the current review. One was graded as *high* quality, another one was graded as *moderate* quality, whereas the other four were graded as critically low-quality reviews. The presence of herpesviruses in subgingival samples was associated with an increased risk of MP, supported by the corresponding meta-analyses. Although the association was strong (OR > 3.0), the confidence intervals were wide, heterogeneity was significant, and stud-

A. Jakovljevic (✉)
Department of Pathophysiology, School of Dental Medicine, University of Belgrade, Belgrade, Serbia
e-mail: dr.sasuli@hotmail.com; a.jakovljevic@stomf.bg.ac.rs

M. Andric
Department of Oral Surgery, School of Dental Medicine, University of Belgrade, Belgrade, Serbia
e-mail: miroslav.andric@stomf.bg.ac.rs

J. Jacimovic
Central Library, School of Dental Medicine, University of Belgrade, Belgrade, Serbia
e-mail: jelena.jacimovic@stomf.bg.ac.rs

J. Milasin
Department of Human Genetics, School of Dental Medicine, University of Belgrade, Belgrade, Serbia
e-mail: jelena.milasin@stomf.bg.ac.rs

J. E. Botero
Facultad de Odontología, Universidad de Antioquia, Medellín, Colombia
e-mail: javier.botero@udea.edu.co

J. Santi-Rocca (ed.), *Periodontitis*, Advances in Experimental Medicine and Biology 1373,
https://doi.org/10.1007/978-3-030-96881-6_7

ies were of small sample size. In addition, publication bias was detected. Contrary, data from systematic reviews that assessed APEO and herpesviruses did not show any significant associations.

Conclusions: Low-quality studies with high uncertainty suggest a strong association between herpesviruses and MP, but not with APEO.

Keywords

marginal periodontitis · periapical periodontitis of endodontic origin · herpesviruses · herpes simplex virus · Epstein-Barr virus · human cytomegalovirus · systematic review

Abbreviations

AgP	Aggressive periodontitis
AMSTAR	Assessment of multiple systematic reviews
APEO	Apical periodontitis of endodontic origin
BZLF-1	BamHI fragment Z leftward open reading frame 1
CI	Confidence interval
CP	Chronic periodontitis
DEEP	DART-European E-theses portal
EBV	Epstein-Barr virus
EThOS	Opening access to UK theses
HCMV	Human cytomegalovirus
HSV	Herpes simplex virus
KJD	Korean journal database
MP	Marginal periodontitis
OR	Odds ratio
PE	Porphyromonas endodontalis
PRISMA	Preferred reporting items for systematic reviews and meta-analyses protocols
PROSPERO	International prospective register of systematic reviews
RSCI	Russian science citation index
SCIELO	SciELO citation index
WoS	Web of science core collection
ZEBRA	BamHI fragment Z EB replication activator

Highlights

- This umbrella review which includes six systematic reviews investigates the association between the occurrence of herpesviruses and the development of different forms of periodontal disease.
- AMSTAR 2 tool was employed for a

Patient Summary

Despite numerous studies pointing to a high prevalence of herpesviruses in both apical and marginal periodontitis samples, their exact role in the pathogenesis of a periodontal disease is still unclear. This umbrella aimed to critically assess the association between herpesviruses and the development of different forms of periodontal disease. Six systematic reviews were included in the final assessment based on the eligibility criteria. Based on the presented results herpesviruses may be implicated in the pathogenesis of marginal periodontitis. However, there is no evidence that herpesviruses contribute to the development of apical periodontitis of endodontic origin. Further investigations are necessary to elucidate the role of herpesviruses in the pathogenesis of periodontal diseases.

7.1 Introduction

Marginal periodontitis (MP) comprises a variety of clinical presentations, all of them resulting in progressive loss of tooth-supporting tissues. It is a widespread disease of complex etiopathogenesis, involving numerous interactions between oral microbiota and host responses. In contrast, apical periodontitis of endodontic origin (APEO) is strictly related to the infected root canal, but both conditions share common mechanisms of pathogenesis and disease progression.

Bacterial involvement in chronic and aggressive periodontitis and APEO is well recognized (Siqueira and Rôçac 2009; Hajishengallis and Lamont 2012). However, since the early 2000s, data suggesting that several viruses, in particular, those from the *Herpesviridae* family might contribute to the destruction of periodontal tissues, commonly seen in these conditions, have accumulated. Herpesviruses are ubiquitous pathogens capable of achieving life-time latency in infected hosts. Among human herpesviruses, Human cytomegalovirus (HCMV) and Epstein-Barr virus (EBV) are most commonly associated with MP and APEO (Slots et al. 2003; Slots 2010).

Despite numerous studies showing a high prevalence of herpesviruses in both APEO and MP samples, their exact role in the pathogenesis of the periodontal disease is still unclear. It was suggested that HCMV and EBV might promote the production of proinflammatory cytokines and initiate direct cytotoxic reactions in apical periodontal tissues, but also favor the overgrowth of pathogenic bacteria (Slots et al. 2003). A similar hypothesis was expressed concerning severe periodontitis, implicating synergistic action of herpesviruses and periodontopathic bacteria in the progression of this condition (Chen et al. 2020). On the other hand, since latency is one of the most striking features of herpesviral infection, it is essential to elucidate if their presence in periodontal tissues is just a consequence of persistent inflammation or actually affects the course of the disease. In line with this, it was suggested that the occurrence of herpesviruses might be merely a result of viral infection of inflammatory cells that infiltrate affected tissues (Ferreira et al. 2011). Therefore, it is significant not only to report detection rates of herpesviruses in periodontitis samples, but also to explain which mechanisms might contribute to their pathogenic effect. In line with this, several studies reported that the presence of herpesviruses is related to increased levels of proinflammatory cytokines and bone-resorption mediators (Hernádi et al. 2013; Jakovljevic et al. 2018b), which reinforces their role in the etiopathogenesis of periodontal disease.

As information from the literature has accumulated, several systematic reviews and meta-analyses have been published, aiming to synthesize data on herpesviruses in MP and APEO. Most of these studies focus on the detection of viruses in relation to clinical features of periodontitis (e.g., chronic or aggressive periodontitis and symptomatic or asymptomatic APEO) (Jakovljevic and Andric 2014; Gao et al. 2017; Li et al. 2017). Still, some reviews have analyzed data on herpesviruses in MP against healthy controls (Zhu et al. 2015; Alzahrani 2017), and in one study, the odds ratio (OR) for MP/APEO in HCMV-positive and HCMV-negative patients were evaluated (Botero et al. 2020).

Having in mind substantial differences in methodology and results of available systematic reviews and meta-analyses, it would be of interest to summarize their findings and analyze potential limitations. Quality of evidence obtained from a specific systematic review and/or meta-analysis depends on both the methodological quality of the review itself and the quality of original studies included in the analysis.

In this context, umbrella reviews aim to synthesize evidence from existing systematic reviews addressing the specific topic and also to emphasize consistent or contradictory findings, thus providing a better understanding of the subject in question. Therefore, this umbrella review aims to summarize data on herpesviruses detection in MP and APEO samples and the possible relationship between these viruses and clinical features of disease. In addition, it is significant to analyze the quality of available data and suggest potential improvements for future studies in this field.

7.2 Materials and Methods

The current umbrella review has been performed according to the Preferred Reporting Items for Systematic Reviews and Meta-Analyses Protocols (PRISMA) statement (Moher et al. 2015; Shamseer et al. 2015). The study protocol has been drafted *a priori* and registered to the International Prospective Register of Systematic Reviews (PROSPERO) (CRD42020215922).

7.2.1 Review Questions

Is there an association between the detection of herpesviruses and different forms of periodontitis?

7.2.2 Outcome Measures

Occurrence of herpesviruses in different forms of periodontal disease.

7.2.3 Selection Criteria

Systematic reviews with or without meta-analysis that evaluated the detection of herpesviruses and different forms of periodontitis were included. Case reports, clinical studies, laboratory studies, animal studies and narrative reviews were excluded.

7.2.4 Literature Search

Several international and national databases were searched to identify relevant systematic reviews and meta-analyses investigating the association of herpesviruses and different forms of periodontitis. The author team systematically and independently explored the following electronic databases: Clarivate Analytics' Web of Science (including Web of Science Core Collection – WoS, Korean Journal Database – KJD, Russian Science Citation Index – RSCI, SciELO Citation Index – SCIELO) [1980–2020], Scopus [1960–2020], PubMed [1964–2020] and Cochrane Database of Systematic Reviews (CDSR) [1996–2020] till October 2020, without language restrictions applied. Using preliminary searches of the mentioned databases, several retrieval strategies were developed and evaluated, while index terms, free keywords, and synonyms related to the main concepts of interest (herpesviruses and periodontitis) were identified as well. In the current review, key terms and the optimum search strategy differed according to the database being searched, using the combination of the most common free keywords and relevant controlled vocabulary (Medical Subject Headings – MeSH descriptors, https://www.ncbi.nlm.nih.gov/mesh), Boolean operators, truncation, and proximity operators. Complete search strategies applied for all databases are presented in detail in Table 7.1.

Furthermore, unpublished manuscripts, research reports, conference papers, doctoral dissertations, and other grey literature were searched in available repositories (e.g., OpenGrey, Networked Digital Library of Theses and Dissertations, Open Access Theses and Dissertations, DART-Europe E-theses Portal – DEEP, Opening access to UK theses – EThOS) and Google Scholar (first 100 returns). Additional searches of reference lists of the included reviews and relevant narrative reviews were also performed. For duplicate removal and further analysis, all records obtained were imported in the Rayyan QCRI environment (Ouzzani et al. 2016). To identify relevant systematic reviews and meta-analyses, two independent reviewers (A.J., J.J.) were involved in screening titles and abstracts and then reading the full text. Any disagreement between the two reviewers was resolved by consensus and discussion with the third reviewer (J.E.B.).

7.2.5 Data Extraction

A data extraction form, created and customized for the current review, included the following details: names of the authors and year published, name and country of the first author, name of the journal published, databases searched, period searched, the language used to restrict search results, number of included studies/samples (experimental/control group), design of included studies, used quality assessment tool, as well as objectives, analysis method, outcomes assessed and principal findings. Two reviewers (J.E.B., A.J.) extracted the data independently according to the inclusion and exclusion criteria, while any disagreement was resolved by consensus or discussing with a third reviewer (J.M.). If needed, the authors of the analyzed systematic reviews and meta-analyses were contacted for clarification or to obtain missing information.

7.2.6 Quality Assessment of Included Reviews Using the AMSTAR 2 Tool

The Assessment of Multiple Systematic Reviews 2 (AMSTAR 2) tool (Shea et al. 2017) was used to evaluate the quality of the included studies. Out of 16 AMSTAR 2 items, seven represent critical domains upon which the quality rating of individual systematic reviews depends. For most of the appraised domains in the AMSTAR 2 checklist, possible response options are *yes* or *no*, while some contain the third *partial yes*. Two independent reviewers (J.E.B., A.J.) conducted a quality assessment of the included systematic reviews, and any dis-

Table 7.1 Electronic databases and search strategy

Database (n)		Search strategy #1 AND #2 AND #3
WoS (n=181) SCIELO (n=5) RSCI (n=2) KJD (n=0)	#1	TS=((Herpesviridae OR HHV OR herpesvirus* OR Herpes Simplex OR HSV OR Chickenpox OR Herpes zoster OR Varicella-Zoster OR VZV OR Mononucleosis OR Epstein-Barr OR EBV OR Cytomegalovir* OR CMV OR HCMV OR "betaherpesvirus 6" OR HBLV OR KSHV) OR (Virus* AND (herpes OR Burkitt* OR EB OR Epstein-Barr OR Salivary Gland OR B-Lymphotropic OR Kaposi* Sarcoma)) OR (((Herpesvirus OR HHV) AND (1 OR 2 OR 3 OR 4 OR 5 OR 6 OR 7 OR 8)) OR HHV?) OR (((Herpes Simplex Virus OR HSV) AND (1 OR 2)) OR HSV1 OR HSV2)) (n=191,141)
	#2	TS=(periodontitis OR periodontosis OR periodontoses OR (periodontal NEAR/0 (disease$ OR abscess* OR pocket*))) (n=46,659)
	#3	TS=(review OR meta-analysis) OR DT=review (n=5,896,817)
Scopus (n=299)	#1	TITLE-ABS-KEY((Herpesviridae OR HHV OR herpesvirus OR "Herpes Simplex" OR HSV OR Chickenpox OR "Herpes zoster" OR Varicella-Zoster OR VZV OR Mononucleosis OR Epstein-Barr OR EBV OR Cytomegalovir* OR CMV OR HCMV OR {betaherpesvirus 6} OR HBLV OR KSHV) OR (Virus AND (herpes OR Burkitt* OR EB OR Epstein-Barr OR "Salivary Gland" OR B-Lymphotropic OR "Kaposi* Sarcoma")) OR ((Herpesvirus OR HHV) W/0 (1 OR 2 OR 3 OR 4 OR 5 OR 6 OR 7 OR 8)) OR ((("Herpes Simplex Virus" OR HSV) W/0 (1 OR 2)) OR HSV1 OR HSV2)) (n=284,940)
	#2	TITLE-ABS-KEY(periodontitis OR periodontosis OR periodontoses OR (periodontal W/0 (disease OR abscess OR pocket))) (n=89,597)
	#3	TITLE-ABS-KEY(review OR meta-analysis) OR DOCTYPE(review) (n=5,036,547)

(continued)

Table 7.1 (continued)

Database (n)		Search strategy #1 AND #2 AND #3
PubMed (n=52)	#1	"herpesviridae"[MeSH Terms] OR "herpesviridae"[All Fields] OR "HHV"[All Fields] OR "herpesvirus*"[All Fields] OR "Herpes Simplex"[All Fields] OR "HSV"[All Fields] OR ("chickenpox"[MeSH Terms] OR "chickenpox"[All Fields]) OR "Herpes zoster"[All Fields] OR "Varicella-Zoster"[All Fields] OR ("herpesvirus 3, human"[MeSH Terms] OR "human herpesvirus 3"[All Fields] OR "vzv"[All Fields]) OR ("infectious mononucleosis"[MeSH Terms] OR ("infectious"[All Fields] AND "mononucleosis"[All Fields]) OR "infectious mononucleosis"[All Fields] OR "mononucleosis"[All Fields]) OR "Epstein-Barr"[All Fields] OR ("herpesvirus 4, human"[MeSH Terms] OR "human herpesvirus 4"[All Fields] OR "ebv"[All Fields]) OR "cytomegalovir*"[All Fields] OR "Cytomegalovirus"[MeSH Terms] OR "CMV"[All Fields] OR "HCMV"[All Fields] OR ("betaherpesvirus"[All Fields] AND "6"[All Fields]) OR ("herpesvirus 6, human"[MeSH Terms] OR "human herpesvirus 6"[All Fields] OR "hblv"[All Fields]) OR "herpesvirus 7, human"[MeSH Terms] OR ("herpesvirus 8, human"[MeSH Terms] OR "human herpesvirus 8"[All Fields] OR "kshv"[All Fields]) OR ("virus*"[All Fields] AND ("herpes"[All Fields] OR "herps"[All Fields] OR "burkitt*"[All Fields] OR "EB"[All Fields] OR "Epstein-Barr"[All Fields] OR "Salivary Gland"[All Fields] OR "B-Lymphotropic"[All Fields] OR ("kaposi*"[All Fields] AND ("sarcoma"[MeSH Terms] OR "sarcoma"[All Fields] OR "sarcomas"[All Fields] OR "sarcoma s"[All Fields])))) OR (("herpesviridae"[MeSH Terms] OR "herpesviridae"[All Fields] OR "herpesvirus"[All Fields] OR "HHV"[All Fields]) AND (1[UID] OR 2[UID] OR 3[UID] OR 4[UID] OR 5[UID] OR 6[UID] OR 7[UID] OR 8[UID])) OR ((("Herpes Simplex Virus"[All Fields] OR "HSV"[All Fields]) AND (1[UID] OR 2[UID])) OR ("herpesvirus 1, human"[MeSH Terms] OR "human herpesvirus 1"[All Fields] OR "hsv1"[All Fields]) OR ("herpesvirus 2, human"[MeSH Terms] OR "human herpesvirus 2"[All Fields] OR "hsv2"[All Fields])) (n=203,876)
	#2	"periodontitis"[MeSH Terms] OR "periodontitis"[All Fields] OR "periodontitides"[All Fields] OR "aggressive periodontitis"[MeSH Terms] OR "Chronic Periodontitis"[Mesh] OR "Periapical Periodontitis"[Mesh] OR "Periodontal Abscess"[Mesh] OR "Periodontal Pocket"[Mesh] "periodontosis"[All Fields] OR "periodontoses"[All Fields] OR ("periodontal"[All Fields] AND ("disease*"[All Fields] OR "abscess*"[All Fields] OR "pocket*"[All Fields])) (n=54,459)
	#3	"systematic review"[Publication Type] OR "meta-analysis"[Publication Type] OR "review"[Title/Abstract] OR "meta-analysis"[Title/Abstract] (n=1,756,826)

(continued)

Table 7.1 (continued)

Database (n)		Search strategy #1 AND #2 AND #3
CDSR (n=0)	#1	((Herpesviridae OR HHV OR herpesvirus* OR Herpes Simplex OR HSV OR Chickenpox OR Herpes zoster OR Varicella-Zoster OR VZV OR Mononucleosis OR Epstein-Barr OR EBV OR Cytomegalovir* OR CMV OR HCMV OR "betaherpesvirus 6" OR HBLV OR KSHV) OR (Virus* AND (herpes OR Burkitt* OR EB OR Epstein-Barr OR Salivary Gland OR B-Lymphotropic OR Kaposi* Sarcoma)) OR (((Herpesvirus OR HHV) NEAR/0 (1 OR 2 OR 3 OR 4 OR 5 OR 6 OR 7 OR 8)) OR HHV?) OR (((Herpes Simplex Virus OR HSV) NEAR/0 (1 OR 2)) OR HSV1 OR HSV2)) OR ([mh "herpesviridae"] OR [mh "chickenpox"] OR [mh "herpesvirus 1, human"] OR [mh "herpesvirus 2, human"] OR [mh "herpesvirus 3, human"] OR [mh "herpesvirus 4, human"] OR [mh "herpesvirus 5, human"] OR [mh "herpesvirus 6, human"] OR [mh "herpesvirus 7, human"] OR [mh "herpesvirus 8, human"] OR [mh "Cytomegalovirus"]) (n=7,759)
	#2	(periodontitis OR periodontosis OR periodontoses OR (periodontal NEAR/0 (disease$ OR abscess* OR pocket*))) OR ([mh "periodontitis"] OR [mh "aggressive periodontitis"] OR [mh "Chronic Periodontitis"] OR [mh "Periapical Periodontitis"] OR [mh "Periodontal Abscess"] OR [mh "Periodontal Pocket"]) (n=5,363)
	#3	Cochrane Reviews (n=8,427)

n number of hits, *WoS* Web of Science Core Collection, *KJD* Korean Journal Database, *RSCI* Russian Science Citation Index, *SCIELO* SciELO Citation Index, *CDSR* Cochrane Database of Systematic Reviews, *TS* Topic (article title, abstract and keywords), *DT* Document type

agreement between the two reviewers was resolved by a third (J.M.). The final classification of each systematic review, as *high*, *moderate*, *low*, or *critically low-quality* review, was generated using the online AMSTAR 2 checklist available on the AMSTAR website (https://amstar.ca/Amstar_Checklist.php).

7.3 Results

7.3.1 Literature Search Process

Figure 7.1 shows a PRISMA flow diagram of the study selection process based on the presented exclusion criteria. A thorough search of the chosen databases and other relevant sources retrieved 539 references for potential inclusion in this umbrella review. In the next step, 146 duplicates were identified and removed from the database. After the title and abstract screening, 387 studies were determined for exclusion, while six studies were eligible for full-text assessment. Finally, six systematic reviews were included in the present umbrella review. Four systematic reviews were related only to MP (Zhu et al. 2015; Gao et al. 2017; Li et al. 2017; Alzahrani 2017), another one only to APEO (Jakovljevic and Andric 2014), and there was one related to both MP and APEO (Botero et al. 2020).

7.3.2 Characteristics of Included Studies

Table 7.2 shows the most important characteristics of the included systematic reviews. The reviews were published between 2014 and 2020 in the *Journal of Endodontics* (n = 1), *PLoS One* (n = 2), *The Saudi Journal for Dental*

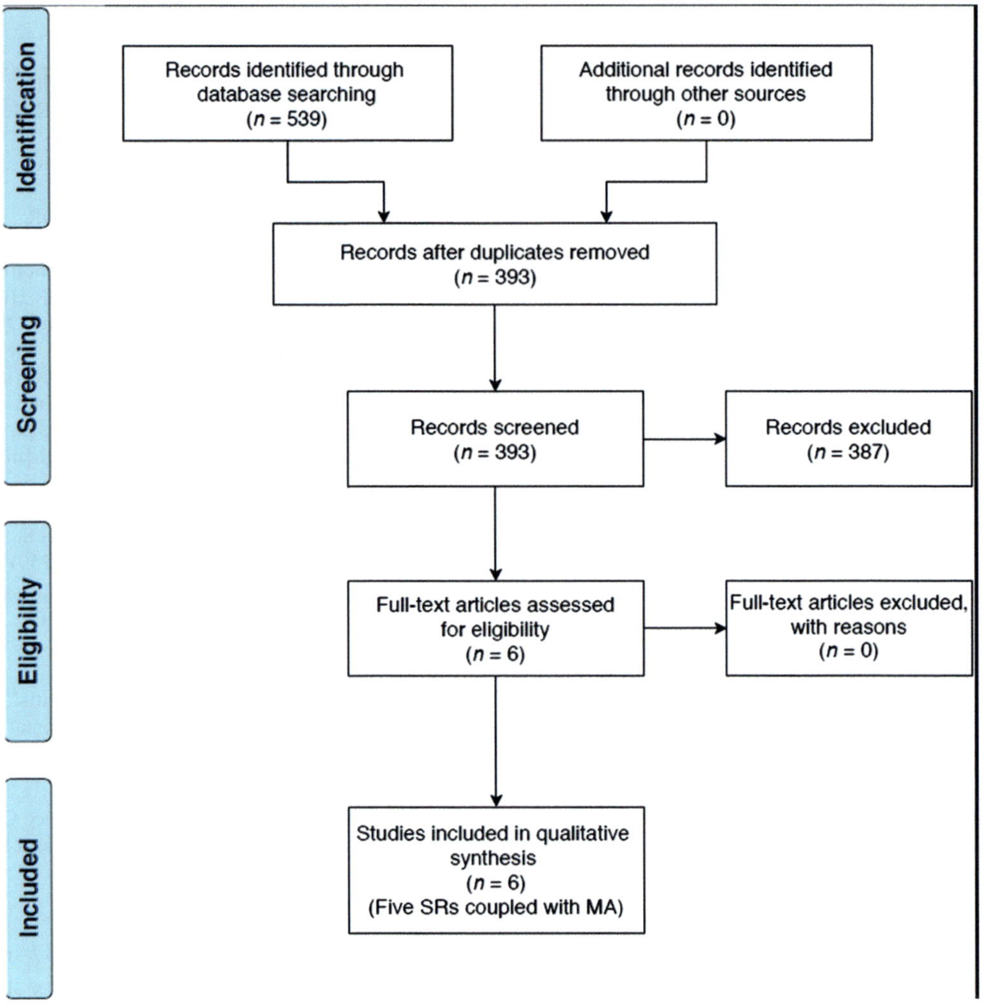

Fig. 7.1 Flow diagram

Research (n = 1), *Medicine (Baltimore)* (n = 1), and *Journal of Periodontal Research* (n = 1). The individual systematic reviews used PubMed, Web of Science, Scopus, Cochrane, Embase, Wanfang, and SciELO databases to identify relevant studies in the literature search process. The search period within the reviews ranged from inception to 2019. The literature electronic search was restricted to the English language, except in one review who also searched for papers in Spanish and Portuguese (Botero et al. 2020). The number of studies included in each systematic review ranged from 12 to 32. Quality assessments of the primary studies were performed in two out of six systematic reviews. One review (Li et al. 2017) used the Newcastle-Ottawa Quality Assessment tool, while the other one (Botero et al. 2020) used the National Institute of Health quality assessment tool.

7.3.3 Summary of Meta-analysis

Five out of six systematic reviews conducted a meta-analysis (Jakovljevic and Andric 2014; Zhu et al. 2015; Gao et al. 2017; Li et al. 2017; Botero et al. 2020). In these studies, the meta-analysis was conducted using the Mantel-Haenszel, random, or fixed-effects model to estimate effect size such as pooled OR and 95%

Table 7.2 Electronic databases and search strategy

No	Author, year	Name of the journal published	Database searched	Country of the first author	Search period	Language	Number of included studies/samples (experimental/control group)	Study design of included studies	Instrument of quality assessment
1	Jakovljevic and Andric (2014)	*Journal of Endodontics*	PubMed, Web of Science, Scopus, and Cochrane	Serbia	1996 to 2013	English	17/641 (508/133)	Cross-sectional studies	NP
2	Zhu et al. (2015)	*PLoS One*	PubMed, and Embase,	China	NS	English	12/923 (552/371)	Case-control	NP
3	Alzahrani (2017)	*The Saudi Journal for Dental Research*	Medline, EMBASE, Scopus, ISI Web of knowledge and Google-Scholar	Saudi Arabia	1970 to 2016	English	12/629 (330/229)	Case-control, cross-sectional, prospective	NP
4	Gao et al. (2017)	*Medicine (Baltimore).*	PubMed, Embase, and Wanfang	China	NS	NS	21/1559 (995/564)	Case-control	NP
5	Li et al. (2017)	*PLoS One*	PubMed and Embase	China	NS	English	12/664 (322/342)	Case-control	Newcastle-Ottawa Scale
6	Botero et al. (2020)	*Journal of Periodontal Research*	Medline, Scopus and SciELO	Colombia	Inception to 2019	English, Spanish and Portuguese	32/2702 (1543/1159)	Case-control, cross-sectional	NIH quality assessment tool

NS not specified, *NP* not performed

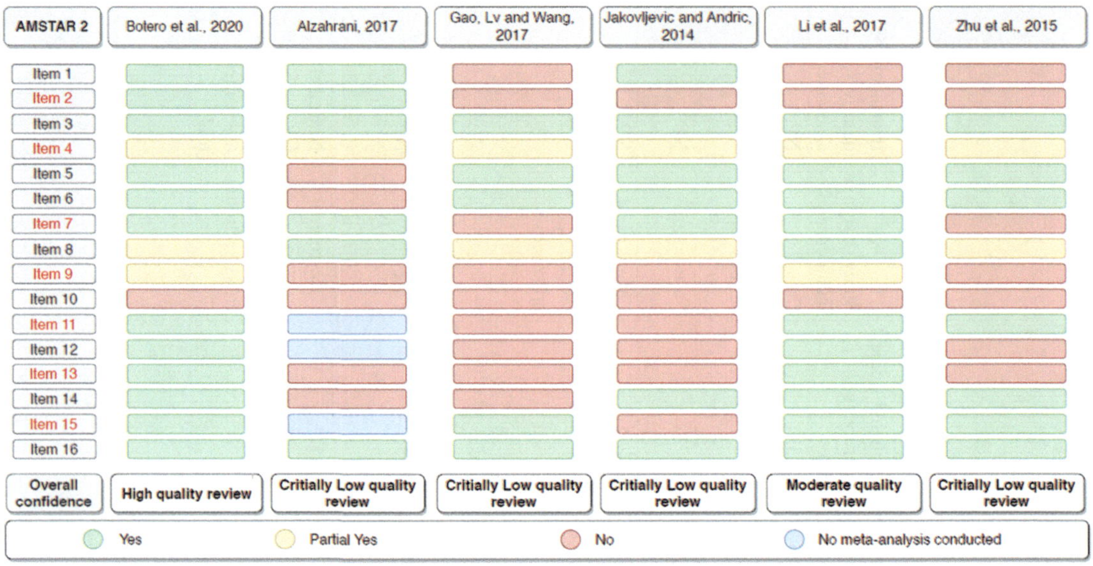

Fig. 7.2 Critical appraisal of the included systematic reviews. AMSTAR 2 – Assessing the Methodological Quality of Systematic Reviews 2; PICO – Population, Intervention, Comparison, Outcome; * – AMSTAR 2 critical domain; 1. Did the research questions and inclusion criteria for the review include the components of PICO?; 2. Did the report of the review contain an explicit statement that the review methods were established prior to the conduct of the review and did the report justify any significant deviations from the protocol?*; 3. Did the review authors explain their selection of the study designs for inclusion in the review?; 4. Did the review authors use a comprehensive literature search strategy?*; 5. Did the review authors perform study selection in duplicate?; 6. Did the review authors perform data extraction in duplicate?; 7. Did the review authors provide a list of excluded studies and justify the exclusions?*; 8. Did the review authors describe the included studies in adequate detail?; 9. Did the review authors use a satisfactory technique for assessing the risk of bias (RoB) in individual studies that were included in the review?*; 10. Did the review authors report on the sources of funding for the studies included in the review?; 11. If meta-analysis was performed did the review authors use appropriate methods for statistical combination of results?*; 12. If meta-analysis was performed, did the review authors assess the potential impact of RoB in individual studies on the results of the meta-analysis or other evidence synthesis?; 13. Did the review authors account for RoB in individual studies when interpreting/discussing the results of the review?*; 14. Did the review authors provide a satisfactory explanation for, and discussion of, any heterogeneity observed in the results of the review?; 15. If they performed quantitative synthesis did the review authors carry out an adequate investigation of publication bias (small study bias) and discuss its likely impact on the results of the review?*; 16. Did the review authors report any potential sources of conflict of interest, including any funding they received for conducting the review?

confidence intervals (CI). The heterogeneity was evaluated using I^2 statistics. The authors reported that a fixed or random effects model was selected based on a heterogeneity test (i.e., $I^2 > 50\%$ – random effect model, $I^2 < 50\%$ – fixed effect model). Publication bias was assessed in four (Zhu et al. 2015; Gao et al. 2017; Li et al. 2017; Botero et al. 2020) out of six systematic reviews using the Begg rank correlation test and/or Egger's asymmetry test evaluated via linear regression model.

7.3.4 Methodological Quality

The methodological quality of the systematic reviews included in this umbrella review is presented in Fig. 7.2. Among six reviews, one (Botero et al. 2020) was graded as *high* quality, another one (Li et al. 2017) was graded as *moderate* quality, whereas the other four were graded as critically low-quality (Jakovljevic and Andric 2014; Zhu et al. 2015; Gao et al. 2017; Alzahrani 2017). None of the six reviews addressed Item 10

(Did the review authors report on the sources of funding for the studies included in the review?).

7.3.5 Principal Findings

The principal findings of each study included in the systematic systematic review are reported in Table 7.3.

7.3.5.1 The Detection of Herpesviruses in Marginal Periodontitis

Five systematic reviews investigated the association between the detection of herpesviruses and different forms of periodontitis (Zhu et al. 2015; Alzahrani 2017; Gao et al. 2017; Li et al. 2017; Botero et al. 2020). EBV, HCMV, and Herpes simplex virus (HSV) were the most frequently detected viruses in MP.

Detection of EBV was significantly associated with an overall increased risk for MP (OR = 6.199, 95% CI = 3.119–12.319, $P < 0.001$) (Gao et al. 2017). Moreover, two systematic reviews revealed that the presence of EBV was significantly associated with an increased risk for aggressive periodontitis (AgP) (OR = 6.11, 95% CI = 2.13–17.51, $P = 0.0008$; OR = 8.361, 95% CI = 2.109–33.143, $P = 0.003$, respectively) (Li et al. 2017; Gao et al. 2017). Similarly, two systematic reviews discovered a significant association between increased risk for chronic periodontitis (CP) and detection of EBV (OR = 5.74, 95% CI = 2.53–13.00, $P < 0.001$; OR = 6.586, 95% CI = 3.042–14.262, $P < 0.001$, respectively) (Zhu et al. 2015; Gao et al. 2017). A significant heterogeneity was observed among all presented results ($I^2 = 74.9\%$, $P < 0.001$; $I^2 = 69.8\%$, $P < 0.001$; $I^2 = 74\%$, $P < 0.001$; $I^2 = 67\%$, $P = 0.001$; $I^2 = 69.9\%$, $P < 0.001$, respectively).

The overall MP risk increase was associated with HCMV (OR = 5.31; 95% CI = 3.15–8.97, $P < 0.001$) (Botero et al. 2020); moreover, the detection of HCMV was significantly associated with an increased risk for both, AgP (OR = 3.63, 95% CI = 2.15–6.13, $P = 0.009$) (Li et al. 2017), and CP (OR = 3.59, 95% CI = 1.41–9.16, $P = 0.007$) (Zhu et al. 2015), respectively. However, marked heterogeneity was observed among all presented results ($I^2 = 67\%$, $P < 0.00001$; $I^2 = 56.3\%$, $P = 0.009$; $I^2 = 74\%$, $P = 0.0001$, respectively). Significant publication bias was revealed for the detection of HCMV overall (Botero et al. 2020) and for the detection of EBV in AgP (Li et al. 2017).

Table 7.3 Principal findings of the included systematic reviews

No	Author, year	Objectives	Method of analysis	Outcome assessed	Principal findings
1	Jakovljevic and Andric (2014)	To analyze the evidence indicating that HCMV and EBV can contribute to the pathogenesis of periapical lesions	Qualitative and meta-analysis	Occurrence of HCMV and EBV in symptomatic compared to asymptomatic apical periodontitis lesions	No statistically significant relationship between the presence of HCMV or EBV mRNA transcripts (P = .083 and P = .306, respectively) and the clinical features of apical periodontitis. Values of the I^2 test indicated that these analyses were homogenous and moderately heterogeneous (I^2 = 0.0% and I^2 = 31, 14 %, respectively).

(continued)

Table 7.3 (continued)

No	Author, year	Objectives	Method of analysis	Outcome assessed	Principal findings
2	Zhu et al. (2015)	To clarify the association between detection of herpesviruses and risk of chronic periodontitis through conducting a meta-analysis based on relevant case-control studies.	Qualitative and meta-analysis	Detection rate of herpesviruses in chronic periodontitis compared to healthy control	EBV shows a significant association with chronic periodontitis compared to periodontally healthy group (OR = 5.74, 95% CI = 2.53–13.00, P<0.001). HCMV also shows a significant association with chronic periodontitis (OR = 3.59, 95% CI = 1.41–9.16, *P* = 0.007). The association between HSV and chronic periodontitis was inconclusive (OR = 2.81 95% CI = 0.95–8.27, *P* = 0.06).
3	Alzahrani (2017)	To elucidate the association between HHVs–HCMV, EBV and HSV, and risk of AgP and AP.	Qualitative	Association between HHVs and AgP and HHVs implication in the pathogenesis of AgP and AP	Human herpes virus (HSV, CMV and EBV) levels are increased and are found to be associated with AgP and AP as compared to healthy individuals.

(continued)

Table 7.3 (continued)

No	Author, year	Objectives	Method of analysis	Outcome assessed	Principal findings
4	Gao et al. (2017)	To establish whether the EBV is associated with periodontal diseases.	Qualitative and meta-analysis	Association between detection of EBV and different types of periodontitis	The association between overall increased risks of periodontitis and the detection of EBV was significant (OR=6.199, 95% CI=3.119–12.319, P<0.001). In the disease-type analysis, the pooled ORs for chronic periodontitis and aggressive periodontitis were 6.586 (95% CI=3.042–14.262, P<0.001) and 8.361 (95% CI=2.109–33.143, P= 0.003), respectively. In the subgroup analysis of ethnicity, our results suggested that high EBV-detecting frequencies were correlated with increased risks of periodontitis in Asians, Europeans, and Americans (P<0.001).

(continued)

Table 7.3 (continued)

No	Author, year	Objectives	Method of analysis	Outcome assessed	Principal findings
5	Li et al. (2017)	To clarify the association between occurrence of herpesviruses and risk of AgP by performing a meta-analysis based on available case – control studies (patients with AgP versus periodontally healthy individuals).	Qualitative and meta-analysis	Association between detection of herpesviruses and different types of periodontitis	The quantitative synthesis results for EBV showed significance (10 studies: $p = 0.0008$, OR = 6.11, 95% CI = 2.13±17.51); nevertheless, evidence of publication bias for EBV was considerable (EBV: Egger's test, $p < 0.001$). HCMV and HSV-1 showed significant association with AgP (12 studies for HCMV: $p = 0.009$, OR = 3.63, 95% CI = 2.15±6.13; 4 studies for HSV-1: $p < 0.001$, OR = 19.19, 95% CI = 4.16±79.06). Sensitivity analyses showed the results yielded consistency, and no significant publication bias was observed for HCMV. The association between HSV-2 and AgP was inconclusive (2 studies: $p = 0.20$, OR = 3.46, 95% CI = 0.51±23.51).
6	Botero et al. (2020)	To evaluate the association between HCMV and periodontitis, and apical periodontitis of endodontic origin.	Qualitative and meta-analysis	The odds ratio for periodontitis/apical periodontitis in HCMV-positive patients as compared to HCMV-negative patients	Increased odds for periodontitis when subgingival HCMV was detected (OR 5.31; 95% CI 3.15–8.97). Sensitivity analysis based on quality of the included studies, showed consistent results. In contrast, the odds ratio for apical periodontitis when HCMV was detected in apical lesions was not statistically significant (OR 3.65; 95% CI 0.49–27.10).

HHV human herpesviruses, *HCMV* human cytomegalovirus, *EBV* Epstein–Barr virus, *HSV* herpes simplex virus, *AgP* aggressive periodontitis, *AP* advanced periodontitis, *CI* confidence interval

Only one systematic review revealed a significant association between increased risk for AgP and detection of HSV-1 (OR = 3.63, 95% CI = 2.15–6.13, $P = 0.009$) (Li et al. 2017). On the other hand, due to the small number of primary studies included in the pooled study, the association between HSV and CP was inconclusive (Zhu et al. 2015).

7.3.5.2 The Detection of Herpesviruses in Apical Periodontitis of Endodontic Origin

Two systematic reviews investigated the association between the detection rates of herpesviruses and APEO (Jakovljevic and Andric 2014; Botero et al. 2020). Botero et al. (2020) did not find a significant association between HCMV and increased risk for APEO (OR = 3.65; 95% CI = 0.49–27.10, $P = 0.21$). Jakovljevic and Andric (2014) did not show significant difference in the detection rates of both investigated viruses [EBV (OR = 1.502; 95% CI = 0.699–3.275, $P = 0.306$) ($I^2 = 0\%$, $P = 0.991$;); HCMV (OR = 2.467; 95% CI = 0.888–6.854, $P = 0.083$) ($I^2 = 31.1\%$, $P = 0.202$;)] between symptomatic and asymptomatic APEO lesions. Heterogeneity was observed only in the case of HCMV detection in overall APEO cases ($I^2 = 83\%$, $P < 0.0001$) (Jakovljevic and Andric 2014).

7.4 Discussion

This umbrella review provides an overview on the association between the detection of herpesviruses and different forms of periodontitis. The presence of HCMV, HSV, and EBV in subgingival samples was associated with increased risk of MP, supported by the corresponding meta-analyses. Although the association was strong (OR > 3.0), the confidence intervals were wide, heterogeneity was notable, and studies had small sample size. In addition, publication bias was detected. On the other hand, data from systematic reviews that assessed APEO and herpesviruses did not show any significant associations.

Our analyses found five out of six systematic reviews with meta-analysis, of which four applied publication bias assessment (Begg rank correlation test and/or Egger's asymmetry test). All detected significant heterogeneity and small-study effects. The heterogeneity detected among the systematic reviews may be the result of real differences between studies or bias within studies included in the meta-analyses (Ioannidis et al. 2007; Egger et al. 1997). A combination of different study designs (case-control, cohort, cross-sectional) in the meta-analyses produces significant heterogeneity. In addition, the different prevalence of periodontitis in various geographical locations, disease, and control definitions could also account for heterogeneity. Consequently, the association presented in the meta-analyses should be interpreted carefully, as this evidence does not represent causality.

The association of herpesviruses with periodontal disease as co-destructive factors has been proposed based on the available biological studies. The alteration of the immunoinflammatory process within infected periodontal tissues in samples from human periodontal pockets and periapical tissues and its correlation with clinical parameters suggest plausibility (Slots et al. 2003; Contreras et al. 2014; Slots and Slots 2019). Our results point to a strong association between herpesviruses and MP, and this represents a supporting evidence of their contribution to periodontitis pathogenesis. However, more studies addressing other causality aspects, such as longitudinal studies and intervention studies, are necessary to help understand its role as risk factor.

Herpesviruses have been proposed as putative factors in APEO development. It was hypothesized that there is a bi-directional relationship between viruses and different endodontic bacterial species (Slots et al. 2003). Slots et al. (2003) assumed that endodontic bacteria are able to reactivate latent herpesviral infection, while herpesviruses induce local immune impairment and overgrowth of residual bacteria. Eventually, this bi-directional relationship, in a vicious cycle, ultimately leads to the destruction of periapical tissues. This umbrella review includes two systematic reviews related to the occurrence of herpesviruses in APEO. Jakovljevic and Andric (2014) included 17 studies for qualitative analysis, while pooled results of six were used to perform the meta-analysis. The authors revealed no significant difference in the occurrence of both EBV and HCMV between symptomatic and asymptomatic APEO lesions. Moreover, Botero and co-workers (2020) investigated the difference in the presence of HCMV between APEO lesions and healthy controls in six primary studies (three of six primary studies were included in the previous meta-analysis). They also revealed no significant differences in the occurrence of HCMV between investigated groups. The results in both systematic reviews imply that there is no clear evidence of herpesviral involvement in the pathogenesis of APEO. These results may be partially explained by the presence of significant methodological inadequacies in primary studies. This issue has been thoroughly elaborated in a recent publication (Jakovljevic et al. 2018a).

Although there is no clear evidence of herpesviral involvement in the pathogenesis of APEO, it must be stressed that several recent investigations potentially reinforced the previously proposed bi-directional model (Jakovljevic and Milasin 2021). Makino et al. (2018) showed that n-butyric acid produced by *Porphyromonas endodontalis* (PE) reactivated latent EBV infection in an *in vitro* model. The authors found that the promoter region of BamHI fragment Z leftward open reading frame 1 (BZLF-1) and BamHI fragment Z EB replication activator (ZEBRA) protein was expressed by Daudi cells (i.e., a Burkitt lymphoma cell line positive for the presence of EBV) in a dose-dependent manner after the treatment with PE culture supernatants, pointing to the induction of virus reactivation. The same group of authors strengthened the idea of herpesviral-bacterial synergy in APEO by showing that *Fusobacterium nucleatum*, another periodontopathic bacterium, also induced the expression of BZLF-1 mRNA and ZEBRA protein, two indicators of EBV reactivation (Himi et al. 2020). Also, it is important to emphasize that several molecular mechanisms of EBV induced bone resorption in APEO have been proposed (Jakovljevic et al. 2016). These studies dealt with ideas that the presence of EBV in APEO samples may induce periapical bone resorption via increased production of reactive oxygen species and bone resorption regulators (Jakovljevic et al. 2018b, 2020). Although there is evident progress in understanding the potential mechanism by which herpesviruses may contribute to the pathogenesis of APEO, it is still necessary to reinforce these hypotheses and provide more reliable data that will strengthen the presented results.

A limitation of this umbrella review is the quality of the included systematic reviews. Only one study was rated as high quality, one as moderate, and four as critically low-quality studies. Systematic reviews published before 2017 were low quality compared to the one published in 2020. Differences in protocol and quality assessments may explain these differences. In addition, the inclusion of primary studies of low quality or with different definitions of periodontal disease and control (e.g., periodontitis subjects vs. non-periodontitis subjects or healthy sites within periodontitis subjects) may account for the heterogeneous results. This creates real differences between studies that are related to biological aspects of the disease that, in the end, may not be comparable.

Besides its limitations, this umbrella review has several strengths. The authors developed and submitted for registration *a priori* protocol in the PROSPERO database. Moreover, a comprehensive literature search with no language restriction was performed in four electronic databases, including the grey literature. The literature search, data extraction process, and quality assessment of included studies were conducted by two independent reviewers, and disagreements were resolved by a third experienced reviewer. Finally, all included systematic reviews were critically appraised and assessed for quality using the AMSTAR 2 tool.

This systematic review did not attempt to perform meta-analysis as overlapping of many studies among systematic reviews was observed. This means that meta-analysis from individual systematic reviews frequently included the same studies. As a result, pooling the results of multiple systematic reviews would produce an increased statistical power to multiply redundant primary studies (Pieper et al. 2014). Overlapping analysis should be included in standardized umbrella review protocols.

7.5 Conclusion and Perspectives

Low-quality studies with high uncertainty suggest a strong association between herpesviruses and marginal periodontitis, but not with apical periodontitis of endodontic origin.

Acknowledgements This study was not funded. The authors deny any conflict of interest related to this study.

References

Alzahrani AA (2017) Association between human herpes virus and aggressive periodontitis: a systematic review. Saudi J Dent Res 8:97–104

Botero JE, Rodríguez-Medina C, Jaramillo-Echeverry A et al (2020) Association between human cytomegalovirus and periodontitis: a systematic review and meta-analysis. J Periodontal Res 55:551–558

Chen C, Feng P, Slots J (2020) Herpesvirus-bacteria synergistic interaction in periodontitis. Periodontology 2000 82:42–64

Contreras A, Botero JE, Slots J (2014) Biology and pathogenesis of cytomegalovirus in periodontal disease. Periodontology 2000 64:40–56

Egger M, Davey Smith G, Schneider M, Minder C (1997) Bias in meta-analysis detected by a simple, graphical test. BMJ 315:629–634

Ferreira DC, Ras IN, Paiva SSM et al (2011) Viral-bacterial associations in acute apical abscesses. Oral Surg Oral Med Oral Pathol Oral Radiol Endodontol 112:264–271

Gao Z, Lv J, Wang M (2017) Epstein-Barr virus is associated with periodontal diseases: A meta-analysis based on 21 case-control studies. Medicine (Baltimore) 96:e5980

Hajishengallis G, Lamont RJ (2012) Beyond the red complex and into more complexity: the polymicrobial synergy and dysbiosis (PSD) model of periodontal disease etiology. Mol Oral Microbiol 27:409–419

Hernádi K, Gyöngyösi E, Mészáros B et al (2013) Elevated tumor necrosis factor-alpha expression in periapical lesions infected by epstein-barr virus. J Endod 39:456–460

Himi K, Takeichi O, Imai K, Hatori K, Tamura T, Ogiso B (2020) Epstein-Barr virus reactivation by persistent apical periodontal pathogens. Int Endod J 53:492–505

Ioannidis JP, Patsopoulos NA, Evangelou E (2007) Uncertainty in heterogeneity estimates in meta-analyses. BMJ 335:914–916

Jakovljevic A, Andric M (2014) Human cytomegalovirus and Epstein-Barr virus in etiopathogenesis of apical periodontitis: a systematic review. J Endod 40:6–15

Jakovljevic A, Milasin J (2021) Herpesviral-bacterial synergy in the pathogenesis of apical periodontitis: new insights and future perspectives. J Dent Sci 16:793–794

Jakovljevic A, Andric M, Miletic M et al (2016) Epstein-Barr virus infection induces bone resorption in apical periodontitis via increased production of reactive oxygen species. Med Hypotheses 94:40–42

Jakovljevic A, Andric M, Knezevic A et al (2018a) Herpesviral infection in Periapical Periodontitis. Curr Oral Health Rep 5:255–263

Jakovljevic A, Andric M, Nikolic N et al (2018b) Levels of oxidative stress biomarkers and bone resorption regulators in apical periodontitis lesions infected by Epstein-Barr virus. Int Endod J 51:593–604

Jakovljevic A, Nikolic N, Carkic J et al (2020) Notch – a possible mediator between Epstein-Barr virus infection and bone resorption in apical periodontitis. Acta Odontol Scand 78:126–131

Li F, Zhu C, Deng FY, Wong M, Lu HX, Feng XP (2017) Herpesviruses in etiopathogenesis of aggressive periodontitis: a meta-analysis based on case-control studies. PLoS One 12:e0186373

Makino K, Takeichi O, Imai K et al (2018) Porphyromonas endodontalis reactivates latent Epstein-Barr virus. Int Endod J 51:1410–1419

Moher D, Shamseer L, Clarke M, Ghersi D, Liberati A, Petticrew M, Shekelle P, Stewart LA, PRISMA-P Group (2015) Preferred reporting items for systematic review and meta-analysis protocols (PRISMA-P) 2015 statement. Syst Rev 4:1

Ouzzani M, Hammady H, Fedorowicz Z, Elmagarmid A (2016) Rayyan: a web and mobile app for systematic reviews. Syst Rev 5:210

Pieper D, Antoine SL, Mathes T, Neugebauer EA, Eikermann M (2014) Systematic review finds overlapping reviews were not mentioned in every other overview. J Clin Epidemiol 67:368–375

Shamseer L, Moher D, Clarke M, Ghersi D, Liberati A, Petticrew M, Shekelle P, Stewart LA, PRISMA-P Group (2015) Preferred reporting items for systematic review and meta-analysis protocols (PRISMA-P) 2015: elaboration and explanation. Br Med J 349:g7647

Shea BJ, Reeves BC, Wells G et al (2017) AMSTAR 2: a critical appraisal tool for systematic reviews that include randomised or non-randomised studies of healthcare interventions, or both. Br Med J 358:j4008

Siqueira JF, Rôçac IN (2009) Critical review in oral biology and medicine: diversity of endodontic microbiota revisited. J Dent Res 88:969–981

Slots J (2010) Human viruses in periodontitis. Periodontology 2000 53:89–110

Slots J, Slots H (2019) Periodontal herpesvirus morbidity and treatment. Periodontology 2000 79:210–220

Slots J, Sabeti M, Simon JH (2003) Herpesviruses in periapical pathosis: An etiopathogenic relationship? Oral Surg Oral Med Oral Pathol Oral Radiol Endod 96:327–331

Zhu C, Li F, Wong MCM, Feng XP, Lu HX, Xu W (2015) Association between herpesviruses and chronic periodontitis: a meta-Analysis based on case-control studies. PLoS One 10:e0144319

Part II

Periodontitis Pathophysiology

8 Recent Updates on Microbial Biofilms in Periodontitis: An Analysis of *In Vitro* Biofilm Models

Maick Meneguzzo Prado, Nathalia Figueiredo, Andréa de Lima Pimenta, Tamires Szeremeske Miranda, Magda Feres, Luciene Cristina Figueiredo, Josiane de Almeida, and Bruno Bueno-Silva

Abstract

The development of oral biofilm models has been extremely important to study the specific role of most microbial species at the early stages of periodontitis. The current knowledge on monospecies or multispecies biofilms originates mainly from the observation of *in vitro* dynamic or static biofilm model systems, which were engineered to mimic clinical oral conditions. In the last few decades, mounting evidence has confirmed that biofilms are the major form of bacterial lifestyle, and more importantly, that microorganisms dwelling in sessile mixed-species aggregates display completely different phenotypes and physiological characteristics than when living in planktonic pure cultures. Interspecies interactions within these communities, mediated by chemical communication systems, have been shown to affect biofilm physiology and increase antimicrobial resistance by up to 1000 fold. These aspects reinforce the importance of developing multispecies biofilm models to better understand and control biofilms. Literature reports demonstrate that while monospecies models are still most commonly used in caries research, authors have used different multispecies models to study periodontal diseases.

M. M. Prado (✉)
Chemical and Food Engineering Department, Federal University of Santa Catarina, Florianópolis, SC, Brazil
e-mail: maickmeneguzzo@gmail.com

N. Figueiredo · T. S. Miranda · M. Feres
L. C. Figueiredo
Dental Research Division, Guarulhos University, Guarulhos, SP, Brazil
e-mail: nathaliaf.figueiredo@gmail.com; tami.szeremeske@gmail.com; mferes@ung.br; lucienedefigueiredo@gmail.com

A. d. L. Pimenta
Chemical and Food Engineering Department, Federal University of Santa Catarina, Florianópolis, SC, Brazil

Department of Biology, ERRMECe, Université de CergyPontoise, Maison Internationale de la Recherche, Pontoise, France
e-mail: andrea@intelab.ufsc.br

J. de Almeida
Faculty of Dentistry, Department of Endodontics, University of Southern Santa Catarina, Tubarão, Brazil
e-mail: dealmeidajosiane@hotmail.com

B. Bueno-Silva
Department of Biology, ERRMECe, Université de CergyPontoise, Maison Internationale de la Recherche, Pontoise, France
e-mail: brunobue@gmail.com

J. Santi-Rocca (ed.), *Periodontitis*, Advances in Experimental Medicine and Biology 1373,
https://doi.org/10.1007/978-3-030-96881-6_8

Periodontitis is a polymicrobial biofilm-dependent disease mainly associated with *Porphyromonas gingivalis*, *Tannerella forsythia*, and *Treponema denticola*. Interestingly, these species hardly adhere to substrates commonly used for biofilm formation, which makes multispecies models essential for an accurate analysis of periodontitis-related biofilms. The multispecies models currently available are generally composed of 6–10 species, but a more recent 34-species model was developed to better examine the dynamics within oral biofilms. The complexity of such polymicrobial biofilm models mimics more consistently the oral microbiome and different aspects of the oral environment. Collectively, the evidence on multispecies biofilm models described herein may support future studies on the use of antimicrobials for biofilm control as well as provide research opportunities to expand the current knowledge on interspecies interactions. The present manuscript reviews the most recent updates on *in vitro* biofilm model systems for periodontitis.

Keywords

Multispecies biofilm · Periodontitis · *In vitro* model

Abbreviations

AHLs	N-acyl homoserine lactones
AI	Auto Inducers
AIPs	autoinducing peptides
CBD	Calgary biofilm device
c-di-GMP	cyclic diguanylate
CF	cystic fibrosis
DFR	drip-flow reactor
DFS	Diffusible Signal Factor
ECM	extracellular matrix
EPS	exopolysaccharides
PAME	Palmitic Acid Methyl Ester
PQS	Pseudomonas Quinolone Signal
QS	quorum sensing
SAM	S-adenosylmethionine
sRNAs	small RNAs
TCS	two-component systems

Highlights

- The structure and composition of biofilms may significantly contribute to antimicrobial resistance.
- Periodontitis is a polymicrobial biofilm-dependent disease that has been studied through several biofilm model systems.
- Multispecies biofilm models are more accurate to investigate interspecies interactions within oral biofilms.
- Models that better mimic the oral environment are preferred for their similarities with clinical oral conditions.

Considerations for Practice

- Biofilms are the major form of bacterial lifestyle and provide additional survival and resistance-related benefits than the planktonic form.
- Biofilm models are relevant to understand microbial interactions and provide preliminary evidence for future clinical trials testing novel antimicrobial strategies.

Patient Summary

Bacteria can develop biofilms in the oral cavity to serve as an energy source, protection against antimicrobials, and cell-to-cell interactions. Dysbiosis of subgingival biofilms is considered the major etiological factor associated with the onset and progression of periodontitis. Several biofilm models have been proposed in the literature; more recently, a novel model composed of over 30 species was developed, including disease- and health-associated bacteria, to evaluate the preventive effects of antimicrobial strategies in periodontal and peri-implant diseases. The use of a bioreactor is also an interesting approach to develop multispecies biofilms as it manages to mimic more accurately the oral environment and its inherent variables.

8.1 Introduction

More than 700 bacterial species have been identified in the oral cavity (Paster et al. 2005; Utter et al. 2016). The analysis of microbiological cultures and molecular biology has indicated that microbial communities differ in composition according to the environment in which they are established (Simón-Soro et al. 2013). For instance, the ecological conditions of supra- and subgingival milieus are substantially different and thereby promote the establishment of a diverse range of microbial communities (Marsh and Devine 2011). Once the dental surfaces are entirely covered by commensal bacteria, other exogenous and often pathogenic species cannot thrive (Hajishengallis 2015). Nevertheless, if not removed through daily oral hygiene, undisrupted biofilms on the tooth surface may trigger a dysbiotic state resulting in an inflammatory reaction of periodontal tissues, known as gingivitis. In susceptible individuals, gingivitis may progress to periodontitis, which is a more complex condition of multifactorial pathogenesis characterized by major inflammatory reactions in the tooth support tissues, with significant loss of the alveolar bone (Knight et al. 2016).

Until now, the specific role that most species play in the beginning and progression of periodontitis remains unknown. Therefore, research using microcosms and oral biofilm models is extremely important (Crielaard and Cate 2010; Koopman et al. 2015). The current knowledge about monospecies or multispecies biofilm originates from observations of *in vitro* biofilm model systems, whether dynamic (in which the culture media circulate continuously over the substrate) or static (with no continuous flow of the culture media), with both advantages and disadvantages to consider (Røder et al. 2016).

The oldest static system for microbial growth is the agar plate, in which bacterial cultures are grown on a suitable solid medium. However, this method requires a large area and a high amount of culture media for biofilm formation and does not reproduce accurately the actual biofilm environment. To overcome these limitations, static systems have been significantly improved (Røder et al. 2016). The next generation of biofilm models was the microtiter plate system, which enabled the analysis of cell adhesion and multispecies biofilm formation at the bottom of the wells (Luo et al. 2020; Røder et al. 2016). A combination of the microtiter plate system with the Calgary Biofilm Device (CBD) resulted in a new biofilm model, which allows a better distribution of the biofilm and mimics the organization of bacterial cells within the oral environment. In this model, biofilms are formed around 96 pins that are adhered to the lid of the device and fit inside the 96 wells, once the plate is closed. The bottom of the well serves to channel the media flow through the pins, thereby creating a consistent shear force on all pins which results in equivalent biofilm formation around each pin (Ceri et al. 1999; Harrison et al. 2010). This system has been used in multispecies biofilm studies to quantify biomass and biofilm composition (Ren et al. 2014). Although they are still used, static models do not simulate the flow of fluids present in the oral microbiome and may be subject to nutrient depletion (Sim et al. 2016).

Some approaches used to simulate biofilm formation more consistently are continuous flow systems such as chemostats, constant depth film fermenters, flow cells, and artificial mouth biofilm models (Bradshaw et al. 1996a, b; Busscher and Van Der Mei 2006; Prado et al. 2020). Some of these systems have a cultivation chamber that can be observed directly under a microscope (Pamp et al. 2009). Other variations of continuous systems are the drip flow reactor (Goeres et al. 2009) and the flow systems using a standard well format, composed of microfluidic channels that allow for a higher transfer rate analysis (Benoit et al. 2010).

Several therapeutic strategies have attempted to mitigate the formation of bacterial biofilm complexes associated with the onset of chronic periodontitis, a condition with high prevalence rates (Dommisch and Kebschull 2014) and one of the main causes of tooth loss (Griffen et al. 2012). Nevertheless, although a wide variety of biofilm systems have been successfully developed to improve the experimental conditions under which oral biofilms are formed and matured, the ideal study model is far from being achieved. Thus, expanding our knowledge about the formation and development of pathogenic biofilms and intra- and interspecies interactions within biofilm communities is very much needed.

In this review, we discuss the most recent updates on *in vitro* biofilm model systems for periodontitis. Collectively, the evidence described herein may support future studies on the use of antimicrobials for biofilm control and provide further research opportunities to elucidate cell-to-cell interactions.

8.2 Biofilms

8.2.1 The History of a Lifestyle

Since Koch and Pasteur viewed microbes under a microscope and developed methods to isolate pure cultures and study microorganisms, microbiologists have produced mounting evidence on microbial biology, pathogenesis, and general metabolism by growing bacteria as single organisms, floating in growth media, isolated from each other. As a result, most of our current knowledge on how microbes live and infect hosts came from studies carried out with microorganisms grown under free-swimming (planktonic) conditions. Despite the enormous advances brought by these studies to the management of infectious diseases and improvement of human health, microbiologists were still faced with difficulties in eradicating infections or growing isolated cultures in the laboratory. These difficulties were the first indications that the single-cell, 'individualistic lifestyle" that constituted the basis of microbiologists working model for understanding bacterial metabolism and pathogenesis, did not fully correspond to microorganism's actual lifestyle "outside the lab".

Even though planktonic growth was adopted as the model of choice for studying microorganisms, sessile surface-associated bacteria had already been described since the early 1900s. In 1933, Henrici reported his surprise to identify bacteria growing attached to glass slides as microcolonies, formed by individuals displaying a different morphology from their free-living counterparts, which were "not removed by washing under tap water" and "were clearly surrounded by a sheath of gum which also served to fasten the colony to the glass". thus established a long-lasting field of research on what came to be known as "biofouling, (Henrici 1932).

The link between chronic infections and bacterial aggregates was established later, in 1973, when Høiby & Axelsen examined sputum from cystic fibrosis (CF) patients with chronic P. aeruginosa lung infection (Høiby and Axelsen 1973). Although the term "biofilm" had been used for the first time in 1975 to describe "microbial films that develop upon all objects submerged in natural waters" (Mack et al. 1975), it was only in 1981 that it appeared for the first time in a clinical report describing the "acquired pellicle which forms on all solid surfaces in the oral cavity" (Jendresen and Glantz 1981). Since then, several studies have demonstrated that biofilms represent the lifestyle of choice for most microorganisms, not only in nature but also in industrial and clinical environments, including symbionts and pathogens when colonizing their hosts (Geesey et al. 1977; Hall-Stoodley et al. 2004). It became then clear that microorganisms in these sessile mixed-species aggregates displayed completely different phenotypes and physiological characteristics than when living in planktonic pure cultures. As a result, microbiologists were led to reassess their working models to generate data that better reflected microbial metabolism, ecology, and pathogenesis, and consequently it became of paramount importance to understand the life of microorganisms inside this structure called "biofilm".

8.2.2 What Are Biofilms and Why They Are so Important?

Biofilms are generally defined as communities of sessile microorganisms, aggregated and/or attached to a surface, embedded in layers of slime mostly secreted by the microbes themselves. These communities thrive in any environment that provides minimal humidity and nutrients, generally in solid-liquid interfaces, although floating biofilms have also been described, composed of flocs of cell aggregates, not attached to any surface (Flemming and Wingender 2010; Nosyk et al. 2008). Biofilms are normal components of human, animal, and plant microbiomes, found ubiquitously in natural environments such as mouth; digestive, urinary and respiratory tracts; sludge, clean and wastewater pipes; cleaning sponges; and toothbrushes. There appear to be few limits to the ability of biofilms to colonize any environment.

Beneficial biofilms play an important role in maintaining health and homeostasis, – a healthy biofilm contributes to host and environmental health. Biofilms are also important to industrial processes such as wastewater treatment, biodegradation, and bioremediation, they teach our immune system to recognize self from non-self, and keep pathogens from colonizing host, contribute to oral health, prevent intestine disfunctions and infections in general (Morikawa 2006; Vestby et al. 2020). On the downside, imbalanced biofilms are very difficult to eliminate and may lead to the onset of chronic diseases, infections, biofouling, pipe clogging and corrosion, and contamination in food industry and indwelling devices in hospital patients, thereby causing major economic burden and loss of human lives (Vestby et al. 2020; Vishwakarma 2020). Clinical biofilms are responsible for approximately 80% of all chronic wounds (Sen et al. 2020) and, according to the National Institutes of Health (USA), more than 80% of all microbial infections are biofilm-related (Musk Jr. and Hergenrother 2006). When organized in biofilms, bacteria can survive antibiotic treatments at concentrations up to a thousand times higher than those necessary to kill their planktonic counterparts. The biofilm structure also protects colonizers against dehydration, acid attacks, salinity, and phagocytosis (Pereira et al. 2014). Therefore, it is not surprising that all antimicrobial-resistant bacteria listed by the World Health Organization as high priority pathogens for which new therapies are needed are biofilm-forming bacteria (World Health Organization 2017).

Both benefits and drawbacks of biofilms derive from their complex structure and physiology, which are related to the metabolic heterogeneity of their components due to adaptation to the community lifestyle.

8.2.3 Biofilm Structure

As previously mentioned, biofilms are composed of aggregates of microbial cells embedded in a slime secreted by the microorganisms themselves. This extracellular matrix (ECM) accounts for 90% of their dry mass and is composed of exopolysaccharides (EPS), extracellular DNA (eDNA), RNA, proteins, and lipids. The ECM plays a crucial role in biofilm formation and maintenance, and is responsible for its longstanding attachment to biotic and abiotic surfaces. It mediates cell aggregation, provides cohesion required to preserve biofilm's architecture, allows cell-to-cell communication, acts as a protective barrier against desiccation, confers tolerance to antimicrobial agents and resistance to host defenses during infection. It also enables the accumulation of nutrients from the environment (e.g., C, N, and P), recycles components from lysed cells, facilitates horizontal gene transfers between members of the community, and enables the enzymatic digestion of structural EPS, allowing the release of cells from the biofilm in the final steps of its lifecycle (Flemming and Wingender 2010).

Analysis of three of the best biofilm model organisms, *Pseudomonas aeruginosa, Escherichia coli,* and *Vibrio cholerae*, indicated that biofilm formation can be described as a three-stage process, involving (i) attachment of planktonic cells, (ii) maturation of the sessile community, and (iii) dispersal of cells to form a new biofilm (Davey and O'toole 2000), a process that can be described as "bacterial metastasis.

Biofilm formation begins when planktonic (free-swimming) bacteria are stimulated to adopt a sessile lifestyle and reversibly adhere to a surface in response to environmental stimuli such as temperature, pH, osmolality, and nutrient availability (Desai and Kenney 2019). The most efficient way for bacteria to approach surfaces is swimming by flagella-mediated motility, while non-motile bacteria must rely on passive transport, either via Brownian motion, flotation, or gravitational forces, to reach their target (Berne et al. 2018). After this initial approach, the balance between attractive and repulsive forces between bacteria and biotic or abiotic surfaces determines the outcome of the subsequent adhesion steps: Van der Walls interactions are usually attractive, hydrophobic interactions can be attractive or repulsive depending on the environment and bacterial surface, whereas electrostatic interactions are modulated by the ionic strength and pH of the surrounding media. Bacterial cellular appendages such as flagella and pili, LPS, and possibly membrane proteins, are also important at this early stage to establish an initial reversible adhesion. Besides having a crucial role in swimming motility, flagella are also used to explore the surface topology and establish physical contact acting as adhesins. Once a bacterial monolayer is formed over the surface, flagella production is downregulated and pili-mediated adhesion and twitching motility, which have been shown to be crucial to the establishment of more stable bacteria-surface interactions, are used to consolidate the initial attachment, allowing cells to reposition over the surface before more adhesins are produced (Zamani and Salehzadeh 2018). Irreversible adhesion initiates through interactions mediated by bacterial appendages and hydrogen bonds, and is reinforced by hydrophobic interactions promoted by the displacement of interfacial water between bacteria and the surface. Contact with the surface stimulates adhesin production and strengthens bacteria-surface irreversible interactions (Berne et al. 2018). Once individual cells achieve irreversible adhesion colonization of the surface and formation of multicellular biofilm begin, with bacteria aggregating in microcolonies through twitching motility. These microcolonies grow both by proliferation of previously attached bacteria and recruitment of planktonic cells from the environment through cell signaling.

As biofilm matures, EPS production is triggered forming the scaffold that supports biofilm's complex structure. This 3D architecture is not a random structure, it is genetically determined and constituted by towers crossed by water channels that are essential to keep the flow of liquid that provides nutrients and carry toxic metabolic waste away from the community (Desai and Kenney 2019). These tower structures are often described as a "network of mushrooms", in which the stalks are formed by bacterial clonal growth while the caps are formed by motile bacteria that use type IV pili mediated motility climb the stalks (Bjarnsholt et al. 2013). To build such an intricate architecture, free-swimming (planktonic) cells undergo a series of phenotypic and metabolic modifications during the biofilm's life cycle, most of which involving transcriptional regulation control in response to cell-to-cell communication.

8.2.4 Genetic Control of Biofilms

Biofilm formation is a fine-tuned developmental process that converts motile individual cells into sessile cooperative communities. As previously mentioned, the first step in this transition, initial attachment, is controlled by a number of environmental cues, including nutrient availability, osmolarity, pH, iron availability, oxygen tension, and temperature. These cues are perceived by bacterial sensors and trigger phenotypic changes through gene activation and inactivation, the most spectacular of which leads to flagella depolymerization and pili production (Davey and O'toole 2000; Kilmury and Burrows 2018).

The next steps in biofilm development rely upon even more spectacular phenotypic transitions to allow individual bacteria to adopt the cooperative behavior necessary for a community lifestyle. A central requirement to allow individual cells to cooperate, is their ability to communicate so it came as no surprise when biofilm maturation was reported to be regulated by cell signaling mechanisms such as Quorum-Sensing (QS), cyclic diguanylate (c-di-GMP), two-component systems (TCS), and small RNAs (sRNAs) (Yi et al. 2019). Amongst these four signaling systems involved in biofilm regulation,

Quorum Sensing (QS) is the most widely used, controlling most of the physiological changes during biofilm maturation.

QS is a cell-to-cell communication system that allows microorganisms to sense population density and coordinate their response in the community environment through synchronized changes in gene expression. QS was first reported in the 1970s to describe the mechanism that controls bioluminescence in marine bacteria such as Vibrio fischeri and Vibrio harveyi (Nealson and Hastings 1979). This chemical system allows cellular communication within and between species of bacteria, as well as interkingdom communication between microorganisms and their hosts, from fungi to plants and human cells (Khan et al. 2019). More recently, QS systems were described in viral species and reported to mediate communication between viruses and their hosts (Erez et al. 2017; Silpe and Bassler 2019).

Within biofilms, when community reaches a certain population density, QS triggers a coordinated expression of virulence-related genes, such as resistance to antibiotics, motility control, or EPS production, through a process that includes production, accumulation, detection, and response to chemical signals called Auto Inducers (AI). The most prevalent AIs that regulate QS are N-acyl homoserine lactones (AHLs) in Gram-negatives, autoinducing peptides (AIPs) in Gram-positives, and Autoinducer-2 (AI-2) in both Gram-negatives and Gram positives. Other signal molecules have also been described in the literature, including Autoinducer-3 (AI-3), Pseudomonas Quinolone Signal (PQS), Diffusible Signal Factor (DSF), S-adenosylmethionine (SAM), and hydroxyl-Palmitic Acid Methyl Ester (PAME) (Khan et al. 2019; Yi et al. 2019).

In classical QS mechanisms, the constitutive expression of signal molecules by microorganisms leads to a progressive accumulation of AIs within the biofilm. When these signaling molecules reach a critical concentration, they switch transcription of effectors genes on or off, through interaction and activation of specific receptors or response-regulators (Valen and Scheie 2018). This intricate and fine-tuned communication system coordinates biofilm maturation until its final step, signaling for disassembling the biofilm community.

Just as QS-controlled biofilm formation and maturation is important for keeping microbial communities in a protected environment, biofilm dispersion is vital to enable microorganisms to colonize new niches when nutrients become limited and metabolic wastes accumulate. This last step in biofilm life cycle is also controlled by cell-to-cell communication via QS, and it can be achieved by either inhibition of synthesis of the matrix compounds, matrix degradation, or disruption of non-covalent interactions between matrix components through upregulation of surfactant production (Irie et al. 2005; Solano et al. 2014).

As mentioned above, QS allows not only intraspecies but also interspecies and even interkingdom communication. Since biofilms in natural environments rarely exist as monospecies, being composed of microorganisms of different species and genera, the ability to exchange signals among community members is essential for biofilms evolutive success. These interspecies interactions have been shown to affect biofilm physiology, so in order to better understand and control biofilm-related diseases and hazards it is important to develop and study multispecies biofilm models.

8.3 Periodontal Disease

Periodontitis is a chronic inflammatory condition initiated by pathogenic microorganisms that compromises support and protective tissues of the tooth, such as gingiva, alveolar bone, cementum, and periodontal ligament (Socransky and Haffajee 1994; Socransky and Haffajee 2005). The irreversible destruction of the connective tissue and alveolar bone causes the main clinical signs associated with periodontitis: gingival inflammation, deep pockets, loss of clinical attachment, mobility and, possibly, tooth loss. A systematic revision of 72 studies with data from 291,170 individuals aged above 15 years from 37 countries, predicted that the global prevalence of severe periodontitis in 2010 was 10.8%, indicating that approximately 743 million people were affected worldwide (Kassebaum et al. 2014).

The etiology of periodontitis is related to an imbalance of the oral microbiome. In the last years, scientific studies using different microbiological techniques were carried out to identify the microorganisms present in the subgingival biofilm. Nowadays, it is well established that the orange and red complex species described in 1998 by Socransky et al. (1998) are the main periodontal pathogens. The species in the red complex (*Porphyromonas gingivalis*, *Tannerella forsythia,* and *Treponema denticola*) are considered the most pathogenic for their strong positive correlation with poor clinical parameters of periodontal disease, such as probing depth, clinical attachment level and gingival bleeding (Socransky and Haffajee 1994; Socransky and Haffajee 2005). The orange complex is composed of *Fusobacterium nucleatum* ssp., *Fusobacterium periodonticum*, *Prevotella intermedia*, *Prevotella nigrescens*, *Parvimonas micra*, *Eubacterium nodatum*, *Campylobacter rectus*, *Campylobacter showae*, *Campylobacter gracilis* and *Streptococcus constellatus*.

Scientific evidence shows that the progression of periodontitis occurs as a result of the host immune response to toxins and virulence factors released by pathogens. The host's response to the attack of microorganisms stimulates the recruitment of cells that actively participate in the inflammatory process and immune response (*e.g.*, polymorphonuclear neutrophils, monocytes, lymphocytes, and plasma cells). These cells, together with those present in the periodontium, such as endothelial cells, fibroblasts, osteoblasts, cementoblasts, osteocytes, and osteoclasts, produce a range of immunoinflammatory mediators, cytokines, and proteolytic enzymes, biomarkers that play an important role in the pathogenesis of periodontitis.

In the last 30 years, the classification of periodontitis was repeatedly revisited to comply with the most recent scientific evidence available. Most studies tried to determine whether the different phenotypes represented different diseases or variations of a single disease. Therefore, the 2017 World Workshop on the New Classification about Periodontal and Peri-Implants Diseases and Conditions reviewed the scientific evidence and adopted a new classification scheme. Disease forms described then as "chronic" and "aggressive" were allocated to the "periodontitis" category. A multidimensional system of stages and degrees was developed to better describe the different manifestations of periodontitis in individual cases. The stages (I, II, III, and IV) describe the severity and spread of the disease, whereas the degrees (A, B, and C) describe the likely rate of progression (Caton et al. 2018; Papapanou et al. 2018).

8.4 Biofilm Models

8.4.1 Monospecies Biofilms

Monospecies biofilm models are widely used for their ease to assess the most remarkable metabolic characteristics of microbial communities (Lemos et al. 2013) and to determine the effects of chemical and natural agents on specific microbial species (Bueno-Silva et al. 2017; Tomás et al. 2007; Veloo et al. 2012). However, monospecies biofilm models do not represent the actual condition of periodontal disease whose development results from strong interactions between several microbial species. Therefore, monospecies biofilm models are more adapted for caries research, in which the analysis of a single species seems to be more relevant. *Streptococcus mutans* is considered an important etiological agent (and secondary colonizer) of tooth decay due to its virulence factors, such as EPS production, aciduricity, and acidogenicity (Loesche 1986; Heller et al. 2016).

Several substrates have been described in the literature for the analysis of monospecies biofilms, including 96-well plates or Petri dishes. However, cultivating bacteria on the bottom of these devices raised the question of whether biofilms were formed by passive cell deposition due to gravity or were truly the result of an active process of biofilm formation. Thus, the development of a new device consisting of hydroxyapatite discs placed vertically in 24-well plates onto which biofilms were "actively" adhered, was considered a great advance to test the antibiofilm effects of antimicrobial drugs.

This model is comprised of saliva-coated hydroxyapatite (sHA) discs placed in a vertical position in 24-well plates. *Streptococcus mutans* cultures are inoculated into the wells along with culture media supplemented with sucrose. The choice for sucrose is based on previous evidence showing that it induces greater EPS production and pH drops than other sugars (Koo et al. 2010). In this model, biofilms are initially grown undisturbed for 24 h, and then it is possible to perform two 1-min treatments per day, until day 5, mimicking the use of a dental hygiene product. Usually, right after the experimental phase (5 days), the following biochemical data can be analyzed: total biomass, EPS content (extracellular and intracellular insoluble and soluble polysaccharides), protein content, and the number of viable cells. In addition, during the 5 days of biofilm formation, samples can also be analyzed under Scanning Electron Microscopy (Cunha et al. 2013) and/or Confocal Laser Scanning Microscopy. The latter allows to quantify biofilm thickness and determine the proportions of bacterial cells and EPS (Freires et al. 2015). At last, this monospecies biofilm model can also be used to evaluate gene expression in *S. mutans* (Bueno-Silva et al. 2013) and to test the antibiofilm effects of novel antimicrobials, such as natural products (Cunha et al. 2013; Freires et al. 2015) or novel dental materials incorporated with an antimicrobial agent (Bim-Júnior et al. 2020; Rodrigues et al. 2020).

Although they are relevant to caries research, monospecies biofilms rarely occur in the oral cavity, which is mostly composed of dynamic multispecies biofilms.

8.4.2 Multispecies Biofilms

Biofilms are highly resistant to antimicrobial agents, mechanical forces, nutrient deprivation, extreme pH fluctuations, and the immune system (Shaddox et al. 2010). Experiments show that mixed biofilms display a higher degree of pathogenicity when compared to monospecies cultures as a result of polymicrobial synergism between community members (Ebersole et al. 2017; Lamont et al. 2018).

Oral multispecies biofilm models have been used for several distinct purposes, namely: kinetics of biofilm formation (Sánchez et al. 2011); alternative strategies to prevent peri-implantitis (Pingueiro et al. 2019); and to compare the effectiveness of mouthwashes (de Miranda et al. 2020) and modified dental materials (Yang et al. 2020), among others.

A recent study (Shany-Kdoshim et al. 2019) reported the use of lethal photosensitization or antimicrobial photodynamic therapy (aPDT) as a complementary approach in the treatment of periodontitis. The authors determined the immediate and delayed effects of a non-cohesive treatment with a blue light on the viability of anaerobic multispecies biofilms *in vitro* and provided insights into possible mechanisms of action. Multispecies biofilms were formed with *Streptococcus sanguinis, Actinomyces naeslundii, Porphyromonas gingivalis,* and *Fusobacterium nucleatum,* and examined by flow cytometry and confocal microscopy (Shany-Kdoshim et al. 2019).

There has been a great effort of the scientific community to find and/or develop new antimicrobials capable of controlling subgingival multispecies biofilm formation. Hence, *in vitro* biofilm models are widely used for the initial evaluation of the antimicrobial activity of new compounds and/or formulations. For instance, Miranda et al. (2019) and Figueiredo et al. (2020) investigated the antimicrobial effects of the ethanolic extract of Brazilian red propolis (BRP) on multispecies biofilms. This biofilm model includes 34 species from all bacterial complexes described by Socransky et al. (1998). Biofilms are developed in the Calgary Biofilm Device (CBD), which contains a lid with 96 polystyrene pegs that fit the wells of a 96-well plate. The advantages of this model include the huge number of samples that may be tested per run; a vast number of bacterial strains that can be tested; and the "active" binding of bacteria onto vertically-positioned pegs (de Figueiredo et al. 2020; Miranda et al. 2019).

Bacterial growth curves are prepared for each species in the model. Then, 1x 10^4 CFU of each microorganism is placed over the CBD and incubated at 37 °C under anaerobic conditions. On day 3, the spent medium is replaced and biofilm cultures are kept at 37 °C under anaerobic conditions for an additional 4 days, to obtain seven-day-old biofilms. At this point, two possible treatment schemes can be tested: (1) two 1-min daily treatments from day 3 until day 7; and (2) 24-h treatment, to be initiated on day 6. The first therapeutic scheme was developed to mimic the topical use of dental hygiene products, such as toothpastes and/or mouthwashes, whereas the second scheme was based on the systemic use of common antimicrobials, such as amoxicillin and metronidazole (de Figueiredo et al. 2020; de Miranda et al. 2020).

This model also applies to different biomaterials for dental implants. Several strategies have been proposed to reduce or even eliminate bacterial load in peri-implant areas. Two major analyses can be performed using this model, namely: (1) determination of the pulsed electromagnetic field (PEMF) that can affect the cytoplasmatic membrane, leading to pore formation, leakage of cellular components, and cell death (Faveri et al. 2020); and (2) the chemical treatment of implant surfaces to prevent bacterial colonization of dental implants and peri-implantitis (Pingueiro et al. 2019).

8.4.3 Dynamic Oral Biofilm Models

Dynamic oral biofilm models can mimic the oral microbiome more consistently, allowing the control and simulation of different oral conditions, such as blood and salivary flow, gas tension, pH, and substrates available on the biofilm surface (Dezelic et al. 2009; McBain et al. 2005; Sissons et al. 1991). These models can be run in various systems, such as fed-batch bioreactors (De Winter et al. 2019; Prado et al. 2020), chemostats (Bradshaw et al. 1996a, b; McBain et al. 2005), constant depth film fermenters (Pratten et al. 2000), artificial mouth models (Sissons et al. 1991), flow systems (Benoit et al. 2010), among others. In the *in vitro* work proposed by Prado et al. (2020), 29 bacterial species formed periodontal biofilms in a bioreactor with agitation for 3 days. Figure 8.1 shows the model proposed by Prado et al. (2020), which can be compared with the static model using the CBD proposed by Figueiredo et al. (2020).

De Winter et al. (2019) also used a bioreactor to elucidate interspecies bacterial interactions. The two-species *in vitro* model with *Streptococcus gordonii* and *Porphyromonas gingivalis* allows existing Genome-Scale Metabolic Networks (GSMNs) to be selected and transformed into Dynamic Flow Balance Analysis (dFBA) models (De Winter et al. 2019). In both studies (Prado et al. 2020; De Winter et al. 2019), the bioreactor used in batch form allowed the control of parameters such as pH, anaerobiosis, temperature, and agitation, to mimic the oral environment.

The use of bioreactors in continuous format also allowed the formation of biofilms in aerobiosis. Using a multispecies model with 10 bacterial strains, researchers concluded that mixed cultures may protect obligate anaerobes from the toxic effects of oxygen, both in the *in vitro* formed biofilm and environmental growth (Bradshaw et al. 1996a, b). McBain et al. (2005) developed the Multiple Sorbarod Device (MSD), which works in continuous operation mode (in pulses) for growth of oral biofilms from human saliva. This device allowed the growth and viability of bacteria and the development of complex and stable salivary biofilms. It is considered a simple and reproducible tool for modeling oral bacterial ecosystems and for studying oral microecology (McBain et al. 2005). Another study tested a multispecies model produced in a constant deep film fermenter (CDFF) using bovine enamel discs for cell adhesion. Biofilms were fed with artificial saliva in continuous mode. The microstructure of the biofilms proved to be complex, with groups of bacteria developing over time, separated by channels (Pratten et al. 2000).

The drip-flow reactor (DFR) was tested in continuous format for 48 h and allowed the devel-

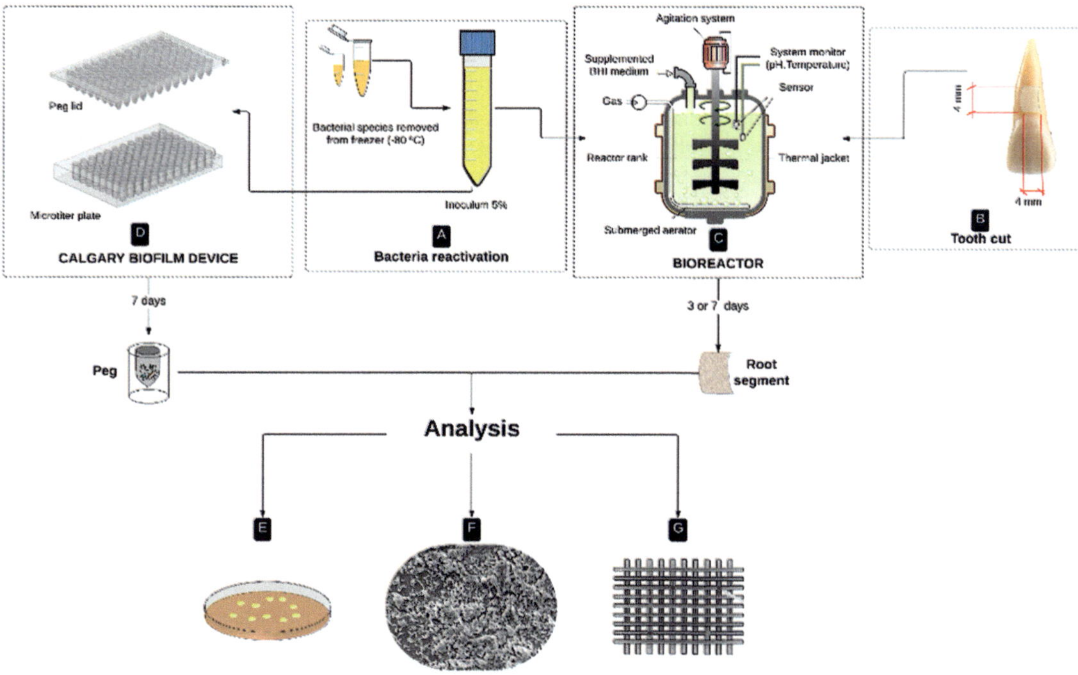

Fig. 8.1 Experimental design – (**a**) reactivation of bacterial species; (**b**) sectioning of tooth specimens; (**c**) cultivation in bioreactor (biofilm formation); (**d**) cultivation in the Calgary Biofilm Device; biofilm analysis: (**e**) CFU, (**f**) SEM, and (**g**) Checkerboard DNA-DNA Hybridization

opment of low shear monospecies biofilms of *Pseudomonas aeruginosa* (Goeres et al. 2009) and *Streptococcus mutans* (Adams et al. 2002). The advantages of this method include biofilm formation near the air-liquid interface, and the simulation of environments such as food processing in conveyor chains, catheters, lungs with cystic fibrosis, and the oral cavity. Because of their low shear, formed biofilms may have a smooth and loosely bound appearance. Microscopically, biofilms look similar to a leaf with few architectural details. Some advantages of this type of reactor are a relatively short time required for biofilm formation and the testing of novel drugs for their antibiofilm properties (Yuan et al. 2012).

The use of flow systems using a standard well format, composed of microfluidic channels and a pneumatic pump, allowed a higher continuous or intermittent flow rate analysis at a wide range of speeds for 96 individual *P. aeruginosa* biofilms. The plate used in these systems is compatible with most readers, allowing a fast and accurate analysis of biofilm viability. In addition, the effects of various antimicrobials on biofilm viability can be quickly determined (Benoit et al. 2010).

The artificial mouth of multiple stations developed by Sissons et al. (1991) uses saliva as an inoculum and artificial media for the development of multispecies biofilms. Importantly, this method produces biofilms that have a metabolism and pH typical of *in vivo* biofilms and which can be analyzed during biofilm development. This system also allows the manipulation of important environmental variables for pH control within the biofilm (Sissons et al. 1991). Subsequent studies in the literature examined the composition of formed biofilms through changes in media supplementation (Wong and Sissions 2001). Table 8.1 shows a summary of the dynamic biofilm models described in this review.

Lastly, microscopic observations of multispecies biofilms revealed the presence of physiologically distinct cells, commonly present in

Table 8.1 Dynamic biofilm models using different systems to mimic the oral flow

Dynamic biofilm models				
Model type	Number of bacterial species	Model	Biofilm formation time	References
Monospecies	1	Reactor Drip Flow	2 days	Adams et al. (2002) and Goeres et al. (2009)
Multispecies	Human saliva	Artificial Mouth	7 days	Sissons et al. (1991)
Multispecies	Human saliva	Constant Depth Film Fermenter	9 days	Pratten et al. (2000)
Multispecies	3	BioFlux Device	<1 day	Benoit et al. (2010)
Multispecies	6	Hydroxyapatite Discs	4 days	Sánchez et al. (2011)
Multispecies	6	Bioreactor	<1 day	De Winter et al. (2019)
Multispecies	32	Stirred Bioreactor	3 and 7 days	Prado et al. (2020)

subpopulations, which differed during biofilm development (Pamp et al. 2009). Interestingly, previous studies clarified that such cell subpopulations have different sensitivity to antimicrobial agents (Haagensen et al. 2007).

8.5 Conclusion and Perspectives

In vitro biofilm research under controlled conditions provides attractive insights into how bacteria can form highly complex multicellular structures. Dynamic biofilm models reproduce more accurately the interactions taking place within the microbial community and the arrangement of the biofilm structure. Therefore, they are especially relevant to the study of dental diseases such as caries and periodontitis. Despite this, dental practice should be based on clinical trials and/or systematic reviews. Patients and dental clinicians should be aware that *in vitro* models are excellent for initial studies, but no clinical procedures should be based only on such a limited level of evidence.

Future studies supported by sophisticated approaches and new system technologies will enable the use of other dynamic oral biofilm models and increase our understanding of the microbial lifestyle in multicellular communities. Faster and more accurate analyses will provide additional information on gene expression and global transcription in biofilm cells and may straightforward the discovery and development of novel antimicrobial drugs.

References

Adams H, Winston MT, Heersink J, Buckingham-Meyer K, Costerton JW, Stoodley P (2002) Development of a laboratory model to assess the removal of biofilm from interproximal spaces by powered tooth brushing. Am J Dent. https://pubmed.ncbi.nlm.nih.gov/12516676/. Accessed 3 Oct 2020

Benoit MR, Conant CG, Ionescu-Zanetti C, Schwartz M, Matin A (2010) New device for high-throughput viability screening of flow biofilms. Appl Environ Microbiol 76(13):4136–4142. https://doi.org/10.1128/AEM.03065-09

Berne C, Ellison CK, Ducret A, Brun YV (2018, outubro 1) Bacterial adhesion at the single-cell level. Nature Reviews Microbiology. Nature Publishing Group. https://doi.org/10.1038/s41579-018-0057-5

Bim-Júnior O, Gaglieri C, Bedran-Russo AK, Bueno-Silva B, Bannach G, Frem R et al (2020) MOF-based erodible system for on-demand release of bioactive flavonoid at the polymer-tissue Interface. ACS Biomat Sci Eng 6(8):4539–4550. https://doi.org/10.1021/acsbiomaterials.0c00564

Bjarnsholt T, Alhede M, Alhede M, Eickhardt-Sørensen SR, Moser C, Kühl M et al (2013, setembro) The in vivo biofilm. Trends in Microbiology. Trends Microbiol. https://doi.org/10.1016/j.tim.2013.06.002

Bradshaw DJ, Marsh PD, Allison C, Schilling KM (1996a) Effect of oxygen, inoculum composition and flow rate on development of mixed-culture oral biofilms. Microbiology 142(3):623–629. https://doi.org/10.1099/13500872-142-3-623

Bradshaw DU, Marsh PD, Schilling KM, Cummins D (1996b) A modified chemostat system to study the ecology of oral biofilms. J Appl Bacteriol 80(2):124–130. https://doi.org/10.1111/j.1365-2672.1996.tb03199.x

Bueno-Silva B, Koo H, Falsetta ML, Alencar SM, Ikegaki M, Rosalen PL (2013) Effect of neovestitol-vestitol containing Brazilian red propolis on accumulation of biofilm in vitro and development of dental caries in vivo. Biofouling 29(10):1233–1242. https://doi.org/10.1080/08927014.2013.834050

Bueno-Silva B, Marsola A, Ikegaki M, Alencar SM, Rosalen PL (2017) The effect of seasons on Brazilian red propolis and its botanical source: chemical composition and antibacterial activity. Nat Prod Res 31(11):1318–1324. https://doi.org/10.1080/14786419.2016.1239088

Busscher HJ, Van Der Mei HC (2006) Microbial adhesion in flow displacement systems. Clin Microbiol Rev 19(1):127–141. https://doi.org/10.1128/CMR.19.1.127-141.2006

Ceri H, Olson ME, Stremick C, Read RR, Morck D (1999) The calgary biofilm device. J Clin Microbiol 37(6):1771–1776

Crielaard RAMEW, Ten Cate JM (2010) Different response to amine fluoride by streptococcus mutans and polymicrobial biofilms in a novel high-throughput active attachment model. Res Output Contrib J:372–379. https://doi.org/10.1159/000316541

Da Cunha MG, Franchin M, Galvão LCDC, Bueno-Silva B, Ikegaki M, De Alencar SM, Rosalen PL (2013) Apolar bioactive fraction of Melipona scutellaris geopropolis on streptococcus mutans biofilm. Evid Based Complement Alternat Med 2013:256287–256287. https://doi.org/10.1155/2013/256287

Davey ME, O'toole GA (2000) Microbial biofilms: from ecology to molecular genetics. Microbiol Mol Biol Rev 64(4):847–867. https://doi.org/10.1128/mmbr.64.4.847-867.2000

de Figueiredo KA, da Silva HDP, Miranda SLF, Gonçalves FJDS, de Sousa AP, de Figueiredo LC et al (2020) Brazilian red propolis is as effective as amoxicillin in controlling red-complex of multispecies subgingival mature biofilm in vitro. Antibiotics 9(8):1–11. https://doi.org/10.3390/antibiotics9080432

de Miranda SLF, Damaceno JT, Faveri M, Figueiredo LC, Soares GMS, Bueno-Silva B (2020) In vitro antimicrobial effect of cetylpyridinium chloride on complex multispecies subgingival biofilm. Braz Dent J 31(2):103–108. https://doi.org/10.1590/0103-6440202002630

De Winter K, Ghesquière J, Teughels W, Waldherr S, Bernaerts K (2019) Towards the construction of a GSMN-based community model for an oral biofilm. IFAC-PapersOnLine 52(26):193–199. https://doi.org/10.1016/j.ifacol.2019.12.257

Desai SK, Kenney LJ (2019) Switching lifestyles is an in vivo adaptive strategy of bacterial pathogens. Front Cell Infect Microbiol 9(December):1–9. https://doi.org/10.3389/fcimb.2019.00421

Dezelic T, Guggenheim B, Schmidlin PR (2009) Multispecies biofilm formation on dental materials and an adhesive patch. Oral Health and Prev Dent 7(1):47–53. https://doi.org/10.3290/j.ohpd.a15267

Dommisch H, Kebschull M (2014) 23: chronic periodontitis I pocket dentistry. Carranza's clinical periodontology.https://pocketdentistry.com/23-chronic-periodontitis/. Accessed 1 Oct 2020

Ebersole JL, Dawson D, Emecen-Huja P, Nagarajan R, Howard K, Grady ME et al (2017) The periodontal war: microbes and immunity. Periodontology 2000 75(1):52–115. https://doi.org/10.1111/prd.12222

Erez Z, Steinberger-Levy I, Shamir M, Doron S, Stokar-Avihail A, Peleg Y et al (2017) Communication between viruses guides lysis-lysogeny decisions. Nature 541(7638):488–493. https://doi.org/10.1038/nature21049

Faveri M, Miquelleto DEC, Bueno-Silva B et al (2020) Antimicrobial effects of a pulsed electromagnetic field: an in vitro polymicrobial periodontal subgingival biofilm model. Biofouling 36(7):862–869. https://doi:10.1080/08927014.2020.1825694

Flemming HC, Wingender J (2010) The biofilm matrix. Nat Rev Microbiol 8(9):623–633. https://doi.org/10.1038/nrmicro2415

Freires IA, Bueno-Silva B, Galvão LCDC, Duarte MCT, Sartoratto A, Figueira GM et al (2015) The effect of essential oils and bioactive fractions on streptococcus mutans and Candida albicans biofilms: a confocal analysis. Evid Based Complement Alternat Med 2015:871316–871316. https://doi.org/10.1155/2015/871316

Caton GJ, Armitage G, Berglundh T, Chapple ILC, Jepsen S, Kornman SK et al (2018) A new classification scheme for periodontal and peri-implant diseases and conditions – introduction and key changes from the 1999 classification. J Clin Periodontol 45:S1–S8. https://doi.org/10.1111/jcpe.12935

Geesey GG, Richardson WT, Yeomans HG, Irvin RT, Costerton JW (1977) Microscopic examination of natural sessile bacterial populations from an alpine stream. Can J Microbiol 23(12):1733–1736. https://doi.org/10.1139/m77-249

Goeres DM, Hamilton MA, Beck NA, Buckingham-Meyer K, Hilyard JD, Loetterle LR et al (2009) A method for growing a biofilm under low shear at the air-liquid interface using the drip flow biofilm reactor. Nat Protoc 4(5):783–788. https://doi.org/10.1038/nprot.2009.59

Griffen AL, Beall CJ, Campbell JH, Firestone ND, Kumar PS, Yang ZK et al (2012) Distinct and complex bacterial profiles in human periodontitis and health revealed by 16S pyrosequencing. ISME J 6(6):1176–1185. https://doi.org/10.1038/ismej.2011.191

Haagensen JAJ, Klausen M, Ernst RK, Miller SI, Folkesson A, Tolker-Nielsen T, Molin S (2007) Differentiation and distribution of colistin- and sodium dodecyl sulfate-tolerant cells in Pseudomonas aeruginosa biofilms. J Bacteriol 189(1):28–37. https://doi.org/10.1128/JB.00720-06

Hajishengallis G (2015) Periodontitis: from microbial immune subversion to systemic inflammation. Nat Rev Immunol 15(1):30–44. https://doi.org/10.1038/nri3785.Periodontitis

Hall-Stoodley L, Costerton JW, Stoodley P (2004) Bacterial biofilms: from the natural environment to infectious diseases. Nat Rev Microbiol 2(2):95–108. https://doi.org/10.1038/nrmicro821

Harrison JJ, Stremick CA, Turner RJ, Allan ND, Olson ME, Ceri H (2010) Microtiter susceptibility testing of microbes growing on peg lids: a miniaturized biofilm model for high-throughput screening. Nat Protoc 5(7):1236–1254. https://doi.org/10.1038/nprot.2010.71

Henrici AT (1932) Studies of freshwater bacteria. https://doi.org/10.1128/jb.25.3.277-287.1933

Heller D, Helmerhorst EJ, Gower AC, Siqueira WL, Paster BJ, Oppenheim FG (2016) Microbial Diversity in the Early In Vivo-Formed Dental Biofilm. Appl Environ Microbiol. 82(6):1881–1888. Published Jan 8 2016. https://doi:10.1128/AEM.03984-15

Høiby N, Axelsen NH (1973) Identification and quantitation of precipitins against pseudomonas aeruginosa in patients with cystic fibrosis by means of crossed Immunoelectrophoresis with intermediate gel. Acta Patholog Microbiol Scand Section B Microbiol Immunol 81B(3):298–308. https://doi.org/10.1111/j.1699-0463.1973.tb02207.x

Irie Y, O'Toole GA, Yuk MH (2005) Pseudomonas aeruginosa rhamnolipids disperse Bordetella bronchiseptica biofilms. FEMS Microbiol Lett 250(2):237–243. https://doi.org/10.1016/j.femsle.2005.07.012

Jendresen MD, Glantz PO (1981) Clinical adhesiveness of selected dental materials: an in-vivo study. Acta Odontol Scand 39(1):39–45. https://doi.org/10.3109/00016358109162257

Kassebaum NJ, Bernabé E, Dahiya M, Bhandari B, Murray CJL, Marcenes W (2014, November 11) Global burden of severe periodontitis in 1990–2010: a systematic review and meta-regression. J Dent Res. SAGE. https://doi.org/10.1177/0022034514552491

Khan F, Javaid A, Kim Y-M (2019) Functional diversity of quorum sensing receptors in pathogenic bacteria: interspecies, Intraspecies and Interkingdom level. Curr Drug Targets 20(6):655–667. https://doi.org/10.2174/1389450120666181123123333

Kilmury SLN, Burrows LL (2018) The Pseudomonas aeruginosa pilSR two-component system regulates both twitching and swimming motilities. J Clin Microbiol 9(4):1–13

Knight ET, Liu J, Seymour GJ, Faggion CM, Cullinan MP (2016) Risk factors that may modify the innate and adaptive immune responses in periodontal diseases. Periodontology 2000 71(1):22–51. https://doi.org/10.1111/prd.12110

Koo H, Xiao J, Klein MI, Jeon JG (2010) Exopolysaccharides produced by Streptococcus mutans glucosyltransferases modulate the establishment of microcolonies within multispecies biofilms. J Bacteriol 192(12):3024–3032. https://doi.org/10.1128/JB.01649-09

Koopman JE, Röling WFM, Buijs MJ, Sissons CH, Cate JM, Keijser BJF (2015) Stability and resilience of oral microcosms toward acidification and Candida outgrowth by arginine supplementation, pp 422–433. https://doi.org/10.1007/s00248-014-0535-x

Lamont RJ, Koo H, Hajishengallis G (2018, December 1) The oral microbiota: dynamic communities and host interactions. Nat Rev Microbiol. Nature Publishing Group. https://doi.org/10.1038/s41579-018-0089-x

Lemos JA, Quivey RG, Koo H, Abranches J (2013) Streptococcus mutans: a new Gram-positive paradigm? Microbiology (United Kingdom) 159(PART3):436–445. https://doi.org/10.1099/mic.0.066134-0

Loesche WJ (1986) Role of Streptococcus mutans in human dental decay – PubMed. Microbiol Rev. https://pubmed.ncbi.nlm.nih.gov/3540569/. Accessed 3 Dec 2020

Luo W, Huang Y, Zhou X, Han Q, Peng X, Ren B et al (2020) The effect of disaggregated nano-hydroxyapatite on oral biofilm in vitro. Dent Mater 36(7):e207–e216. https://doi.org/10.1016/j.dental.2020.04.005

Mack WN, Mack JP, Ackerson AO (1975) Microbial film development in a trickling filter. Microb Ecol 2(3):215–226. https://doi.org/10.1007/BF02010441

Marsh PD, Devine DA (2011) How is the development of dental biofilms influenced by the host? J Clin Periodontol 38(SUPPL. 11):28–35. https://doi.org/10.1111/j.1600-051X.2010.01673.x

McBain AJ, Sissons C, Ledder RG, Sreenivasan PK, De Vizio W, Gilbert P (2005) Development and characterization of a simple perfused oral microcosm. J Appl Microbiol 98(3):624–634. https://doi.org/10.1111/j.1365-2672.2004.02483.x

Miranda SLF, Damasceno JT, Faveri M, Figueiredo L, da Silva HD, de Alencar SMA et al (2019) Brazilian red propolis reduces orange-complex periodontopathogens growing in multispecies biofilms. Biofouling 35(3):308–319. https://doi.org/10.1080/08927014.2019.1598976

Morikawa M (2006) Beneficial biofilm formation by industrial bacteria Bacillus subtilis and related species. J Biosci Bioeng 101(1):1–8. https://doi.org/10.1263/jbb.101.1

Musk D Jr, Hergenrother P (2006) Chemical countermeasures for the control of bacterial biofilms: effective compounds and promising targets. Curr Med Chem 13(18):2163–2177. https://doi.org/10.2174/092986706777935212

Nealson KH, Hastings JW (1979) Bacterial bioluminescence: its control and ecological significance. Microbiol Rev 43(4):496–518. https://doi.org/10.1128/mmbr.43.4.496-518.1979

Nosyk O, ter Haseborg E, Metzger U, Frimmel FH (2008) A standardized pre-treatment method of biofilm flocs for fluorescence microscopic characterization. J Microbiol Methods 75(3):449–456. https://doi.org/10.1016/j.mimet.2008.07.024

Pamp SJ, Sternberg C, Tolker-Nielsen T (2009) Insight into the microbial multicellular lifestyle via flow-cell technology and confocal microscopy. Cytometry A 75(2):90–103. https://doi.org/10.1002/cyto.a.20685

Papapanou PN, Sanz M, Buduneli N, Dietrich T, Feres M, Fine DH et al (2018) Periodontitis: consensus report of workgroup 2 of the 2017 world workshop on the classification of periodontal and Peri-implant diseases and conditions. In: Journal of clinical periodontology, vol 45. Blackwell Munksgaard, pp S162–S170. https://doi.org/10.1111/jcpe.12946

Paster BJ, Stokes LN, Olsen I, Dewhirst FE, Aas JA (2005) Defining the Normal bacterial Flora of the Oral cavity. J Clin Microbiol 43(11):5721–5732. https://doi.org/10.1128/JCM.43.11.5721

Pereira UA, Barbosa LCA, Maltha CRA, Demuner AJ, Masood MA, Pimenta AL (2014) γ-Alkylidene-γ-lactones and isobutylpyrrol-2(5H)-ones analogues to rubrolides as inhibitors of biofilm formation by Gram-positive and Gram-negative bacteria. Bioorganic Med Chem Lett 24(4):1052–1056. https://doi.org/10.1016/j.bmcl.2014.01.023

Pingueiro J, Piattelli A, Paiva J, De Figueiredo LC, Feres M, Shibli J, Bueno-Silva B (2019) Additive manufacturing of titanium alloy could modify the pathogenic microbial profile: an in vitro study. Braz Oral Res 33. https://doi.org/10.1590/1807-3107BOR-2019.VOL33.0065

Prado MM, Kovalski DJ, Torrez WB, Bueno-Silva B, Feres M, de Almeida J, Porto LM (2020) Development of a multispecies periodontal biofilm model within a stirred bioreactor. Biofouling 36(6):725–735. https://doi.org/10.1080/08927014.2020.1805600

Pratten J, Andrews CS, Craig DQM, Wilson M (2000) Structural studies of microcosm dental plaques grown under different nutritional conditions. FEMS Microbiol Lett 189(2):215–218. https://doi.org/10.1016/S0378-1097(00)00287-1

Ren D, Madsen JS, de la Cruz-Perera CI, Bergmark L, Sørensen SJ, Burmølle M (2014) High-throughput screening of multispecies biofilm formation and quantitative PCR-based assessment of individual species proportions, useful for exploring interspecific bacterial interactions. Microb Ecol 68(1):146–154. https://doi.org/10.1007/s00248-013-0315-z

Røder HL, Sørensen SJ, Burmølle M (2016) Studying bacterial multispecies biofilms: where to start? Trends Microbiol 24(6):503–513. https://doi.org/10.1016/j.tim.2016.02.019

Rodrigues MC, Rolim WR, Viana MM, Souza TR, Gonçalves F, Tanaka CJ et al (2020) Biogenic synthesis and antimicrobial activity of silica-coated silver nanoparticles for esthetic dental applications. J Dent 96:103327. https://doi.org/10.1016/j.jdent.2020.103327

Sánchez MC, Llama-Palacios A, Blanc V, León R, Herrera D, Sanz M (2011) Structure, viability, and bacterial kinetics of an in vitro biofilm model using six bacteria from the subgingival microbiota. J Periodontal Res 46(2):252–260. https://doi.org/10.1111/j.1600-0765.2010.01341.x

Sen CK, Mathew-Steiner SS, Das A, Sundaresan VB, Roy S (2020) Electroceutical management of bacterial biofilms and surgical infection. Antioxid Redox Signal 33(10):713–724. https://doi.org/10.1089/ars.2020.8086

Shaddox LM, Alfant B, Tobler J, Walker C (2010) Perpetuation of subgingival biofilms in an in vitro model. Mol Oral Microbiol 25(1):81–87. https://doi.org/10.1111/j.2041-1014.2009.00549.x

Shany-Kdoshim S, Polak D, Houri-Haddad Y, Feuerstein O (2019) Killing mechanism of bacteria within multi-species biofilm by blue light. J Oral Microbiol 11(1):1628577–1628577. https://doi.org/10.1080/20002297.2019.1628577

Silpe JE, Bassler BL (2019) A host-produced quorum-sensing autoinducer controls a phage lysis-lysogeny decision. Cell 176(1–2):268–280.e13. https://doi.org/10.1016/j.cell.2018.10.059

Sim CPC, Dashper SG, Reynolds EC (2016) Oral microbial biofilm models and their application to the testing of anticariogenic agents. J Dent 50:1–11. https://doi.org/10.1016/j.jdent.2016.04.010

Simón-Soro Á, Tomás I, Cabrera-Rubio R, Catalan MD, Nyvad B, Mira A (2013) Microbial geography of the oral cavity. J Dent Res 92(7):616–621. https://doi.org/10.1177/0022034513488119

Sissons CH, Cutress TW, Hoffman MP, Wakefield JSJ (1991) A multi-station dental plaque microcosm (artificial mouth) for the study of plaque growth, metabolism, pH, and mineralization. J Dent Res 70(11):1409–1416. https://doi.org/10.1177/00220345910700110301

Socransky, S. S., & Haffajee, A. D. (1994). Implications of periodontal microbiology for the treatment of periodontal infections. Compendium (Newtown, Pa.). Supplement (18):S684–5, 688. https://europepmc.org/article/med/8039206. Accessed 2 Dec 2020

Socransky SS, Haffajee AD, Cugini MA, Smith C, Kent RL (1998) Microbial complexes in subgingival plaque. J Clin Periodontol 25(2):134–144. https://doi.org/10.1111/j.1600-051X.1998.tb02419.x

Socransky SS, Haffajee AD (2005) Periodontal microbial ecology. Periodontology 2000. https://doi.org/10.1111/j.1600-0757.2005.00107.x

Solano C, Echeverz M, Lasa I (2014) Biofilm dispersion and quorum sensing. Curr Opin Microbiol. Elsevier Ltd. https://doi.org/10.1016/j.mib.2014.02.008

Tomás I, Tomás M, Álvarez M, Velasco D, Potel C, Limeres J, Diz P (2007) Susceptibility of oral obligate anaerobes to telithromycin, moxifloxacin, and a number of commonly used antibacterials. Oral Microbiol Immunol 22(5):298–303. https://doi.org/10.1111/j.1399-302X.2007.00359.x

Utter DR, Mark Welch JL, Borisy GG (2016) Individuality, stability, and variability of the plaque microbiome. Front Microbiol 7(APR). https://doi.org/10.3389/fmicb.2016.00564

Valen H, Scheie AA (2018) Biofilms and their properties. Eur J Oral Sci 126(S1):13–18. https://doi.org/10.1111/eos.12425

Veloo ACM, Seme K, Raangs E, Rurenga P, Singadji Z, Wekema-Mulder G, Van Winkelhoff AJ (2012) Antibiotic susceptibility profiles of oral pathogens. Int J Antimicrob Agents 40(5):450–454. https://doi.org/10.1016/j.ijantimicag.2012.07.004

Vestby LK, Grønseth T, Simm R, Nesse LL (2020) Bacterial biofilm and its role in the pathogenesis of

disease. Antibiotics 9(2). https://doi.org/10.3390/antibiotics9020059
Vishwakarma V (2020) Impact of environmental biofilms: industrial components and its remediation. J Basic Microbiol 60(3):198–206. https://doi.org/10.1002/jobm.201900569
Wong L, Sissions CH (2001) A comparison of human dental plaque microcosm biofilms grown in an undefined medium and a chemically defined artificial saliva. Arch Oral Biol 46(6):477–486. https://doi.org/10.1016/S0003-9969(01)00016-4
World Health Organization. (2017) Global priority list of antibiotic-resistant bacteria to guide research, discovery, and development of new antibiotics. WHO. http://www.who.int/medicines/publications/global-priority-list-antibiotic-resistant-bacteria/en/. Accessed 2 Dec 2020
Yang L, Teles F, Gong W, Dua SA, Martin L, Schoenfisch MH (2020) Antibacterial action of nitric oxide-releasing hyperbranched polymers against ex vivo dental biofilms. Dent Mater 36(5):635–644. https://doi.org/10.1016/j.dental.2020.03.012
Yi L, Li J, Liu B, Wang Y (2019) Advances in research on signal molecules regulating biofilms. World J Microbiol Biotechnol 35(8):1–8. https://doi.org/10.1007/s11274-019-2706-x
Yuan L, Yang X, Liang P, Wang L, Huang ZH, Wei J, Huang X (2012) Capacitive deionization coupled with microbial fuel cells to desalinate low-concentration salt water. Bioresour Technol 110:735–738. https://doi.org/10.1016/j.biortech.2012.01.137
Zamani H, Salehzadeh A (2018) Biofilm formation in uropathogenic Escherichia coli: association with adhesion factor genes. Turkish J Med Sci 48(1):162–167. https://doi.org/10.3906/sag-1707-3

Update on B Cell Response in Periodontitis

9

Julien Demoersman and Jacques Olivier Pers

Abstract

B cells have a central and dual role in the physio-pathological mechanisms of periodontitis. They take part in the elimination of the periodontal germs, the induction of tissue destructions and the regulation of the immune response. B cells play an essential role in the destruction of alveolar bone in periodontitis by immunomodulation, rather than by production of antibodies. In the periodontal cell network, B cells are in constant interaction with other immune cells and mesenchymal cells. Periodontitis is characterized by a cellular conversion from a dominant T-cell lesion to a dominant B-cell lesion, particularly enriched in plasma cells. This evolution results from abnormal interactions between B and T cells in periodontitis. Moreover, B cells are at the crossroads of the immune and the bone systems and are involved in the autoimmune mechanisms described in periodontitis. Different subsets of B cells are involved in periodontal destruction, in particular memory B cells, plasma cells and B1 cells. Effector memory B cells strongly express mRANKL in periodontitis and constitute the precursors of plasma cells. B1 cells are also involved in tissue destruction but also in the mechanisms of regulation, in particular via the natural secretion of IL-10 by $CD11b^+$ B1 cells which form a part of the B10 cells that regulate the inflammatory response. As such, periodontitis seems to be associated with a deficit in regulation. In peripheral blood, B cells can also be systemic markers of infection and periodontal inflammation: circulating memory B cells are increased in periodontitis while circulating $CD11b^+$ B1 cells are decreased. The study of B cells in periodontitis is constantly evolving for a better knowledge of the periodontitis setting, the evaluation and the follow-up of periodontitis, but also for the design of new therapeutic targets that may be promising in the management of severe periodontitis.

Keywords

B cells · Periodontitis · IL-10 · Bone resorption · Regulation

Abbreviations

APCs	antigen-presenting cells
DCs	dentritic cells
FDCs	follicular dendritic cells

J. Demoersman
Nantes, France
e-mail: julien.demoersman@gmail.com

J. O. Pers (✉)
LBAI, UMR1227, Univ Brest, Inserm, CHU de Brest, Brest, France
e-mail: pers@univ-brest.fr

J. Santi-Rocca (ed.), *Periodontitis*, Advances in Experimental Medicine and Biology 1373,
https://doi.org/10.1007/978-3-030-96881-6_9

ILC	sinnate lymphoid cells
iTreg	inductible regulatory T cells
lncRNAs	long non-coding RNAs
MAIT	Mucosa-Associated invariant
NK	Natural Killer
nTreg	natural regulatory T cells
Pg	*Porphyromonas gingivalis*
RANKL	receptor activator of nuclear factor B-Ligand
Tfh	T-follicular helper
Treg	regulatory T cells

Highlights

- B cells play a central role in immune cell orchestration and modulation leading to periodontal destructions
- B cells are a source of secreted osteoclastogenic factor of activated T cells that can induce osteoclastogenesis
- B cell depletion leads to an improvement of the clinical periodontal parameters

Considerations for Practice

- B cells as biomarkers of periodontitis
- B cells as a potential target for new therapeutics in periodontitis

Patient Summary

B cells have a central and a dual role in periodontitis (destructive and protective). They are activated in the presence of periodontal germs. Their activation leads to an increase of memory B cells, in particular implied in periodontal bone loss. This increase is visible in the blood of the patients with severe periodontitis. In the same way, B cells with regulatory functions are decreased in blood. These two sub-populations of B cells could be markers of periodontitis evolution. B cells could also represent a promising therapeutic target to treat severe forms of periodontitis.

9.1 Introduction

B cell implications in periodontal diseases have been studied since the 1970s. For 40 years, periodontitis has been characterized by a cellular conversion: T cells are predominant in gingivitis and B cells are predominant in periodontitis (Seymour et al. 1979; Page and Schroeder 1976).

B cells have a complex role, both protective and destructive (Berglundh et al. 2007; Zouali 2017) but essential in the evolution of periodontitis (Hajishengallis and Korostoff 2017; Ebersole et al. 2001). The absence or a deficiency of B cells in animal studies may lead to an absence of tissue destruction (Baker et al. 2009; Oliver-Bell et al. 2015; Harada et al. 2006; Klausen et al. 1989). In humans, it has been recently shown that an anti-CD20 immunotherapy, responsible for a partial depletion of B cells, leads to an improvement of the clinical periodontal parameters (Coat et al. 2015).

B cells appears to be involved at different levels in the development of periodontal diseases, particularly at the bone/immune system interface, where the immune system and bone metabolism interact (Dar et al. 2018). Similarly, B cells are involved in the autoimmune component of periodontitis (Kaur et al. 2017).

B cells could be an important player in the bidirectional relationships with some systemic diseases, and especially with autoimmune or immunoinflammatory components. Therefore, the modulation of B cells, and more particularly some B cell subsets, could constitute a new therapeutic approach to treat some forms of periodontitis, especially the most severe, actually resistant to conventional treatments or associated with systemic diseases.

9.2 B Cells in Periodontitis

B cells are part of the humoral component of the adaptive immune system. They are specialized in antibody secretion, in the mediation of the immuno-inflammatory response by cytokine secretions and, in antigen presentation to naïve T cells (Youinou 2007). They also have regulatory functions through the secretion of IL-10. These cells are then called B10 or Breg cells. B cells are also involved in the extracellular matrix destruction (MMPs), the modulation of the bone balance (RANKL, RANK, OPG), and even act through phagocytosis as osteoclast-like cells (Gao et al. 2012; Kawai et al. 2006; Ali et al. 2015; Horowitz et al. 2010; Parra et al. 2012; Zamboni et al. 2012; Pugliese et al. 2012) (Fig. 9.1).

Two main populations can be distinguished among B cells in mice, i.e. B1 and B2 cells. These two subsets are differentiated at an early stage of otongenesis, and present different tissue circulation and functions. B1 and B2 contribute to innate and adaptive immunity respectively (Montecino-Rodriguez and Dorshkind 2012; Prieto and Felippe 2017). B2 cells can be subdivided as follows: plasmablasts (the immature precursor of plasma cells), memory B cells, marginal zone (MZ) B cells (located in the marginal zone of the spleen), and immature B cells (transitional and pre-naive B cells), and naïve B cells.

9.3 B Cells in Periodontal Cellular Network

B cells and plasma cells are part of a periodontal cellular network where all cells interact with each other (immune and mesenchymal cells). Their proportion evolves throughout the progression of periodontitis and the degree of tissue destruction is correlated with the predominance of plasma cells (Carvel and Carr 1982; Younes et al. 2009). The composition of the cell infiltrate, inside the periodontal cellular network, is dependent on the biofilm, notably *Porphyromonas gingivalis (Pg)*, but also on the cytokinic environment (Johannessen et al. 1990; Mallison 3rd et al.

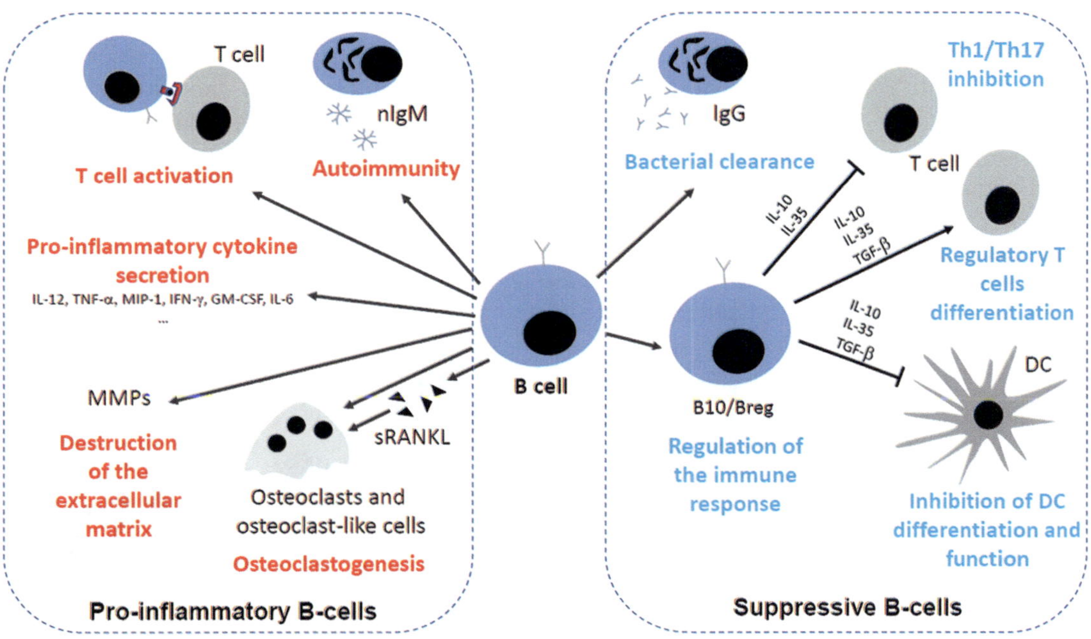

Fig. 9.1 **B cells have a dual role in periodontitis, both protective and destructive**. Proinflammatory B cells secrete sRANKL and induce ostoclastogenesis. They are a source of MMPs and pro-inflammatory cytokines promoting Th17 and plasma cell differentiation (IL-6) or macrophage activation (IFN-γ and GM-CSF). They participate in autoimmune mechanisms by secreting natural immoglobulins and activate T cells acting as antigen presenting cells. Regulatory B cells or B10, characterized by IL-10 secretion, inhibit Th1 and Th17 and promote regulatory T cell differentiation. B cells are not only responsible for bacterial clearance but are also important modulators of the immune response.
IgG Immunoglobulin G, *nIgM* natural immunoglobulin M, *RANK* receptor activator of nuclear factor-kappa B, *sRANKL* soluble RANK Ligand, *IL-* interleukin, *MMPs* matrix metalloproteinase, *MIP* Macrophage inflammatory protein, *TNF-α* Tumor necrosis factor, *DC* Dendritic cell

1991; Gurses et al. 1996). However, the histological aspects of the periodontal tissues do not permit to distinguish the different types of periodontitis (Johnson et al. 1980).

Inside the periodontal tissues, cells are organized in specific aggregates (Thorbert-Mros et al. 2015; Jotwani et al. 2001; Okada et al. 1983) (Fig. 9.2). A rich region of T cells (T cell zone), monocytes/macrophages, and dentritic cells (DCs), in contact with T cells, is described just subjacent to the pocket or the sulcular epithelium. In the central area of lamina propria, located away from the microbial agent, plasma cells are organized in aggregates, separated from the T cell zone, located under the lamina propria, and the B cell zone (rich in B cells not yet differentiated into plasma cells) (Thorbert-Mros et al. 2015; Jotwani et al. 2001; Okada et al. 1983). Plasma cell aggregates also contain some DCs which are mainly intraepithelial (Cirrincione et al. 2002; Lins et al. 2008). This intraepithelial zone is also composed of innate lymphoid cells (ILCs) (Natural Killer (NK), and γδ-T cells) and Th22 cells (Parisi et al. 2017; Dutzan et al. 2016). Unconventional T cells are composed with Mucosa-Associated invariant T (MAIT) cells defined by their innate-like characteristics and broad antimicrobial responsiveness (Sobkowiak et al. 2019; Figueredo et al. 2019). These cells are in contact with the biofilm, and their reactions will participate in the development of the cytokine environment and tissue destruction, notably by secreting high amount of IL-17A (Sobkowiak et al. 2019).

The T cell zone is composed with activated $CD4^+$ T cells (Th), in particular Th1 and Th17 cells (Dutzan et al. 2016; Stashenko et al. 2007). In healthy gingiva, T cells are mainly Th cells, followed by $CD8^+$ T cells and a small proportion of γδ-T cells. A part of $CD4^+$ T cells are regulatory T cells (Tregs) expressing FOXP3 and a majority are memory or naïve T cells (Dutzan et al. 2016). However, it is important to point out that all these studies do not exclude intravascular naive cellularity and that naive cells are not necessarily interstitial cells (inside the tissue ground substance) but present in the vessels or in the blood and extravasated after excision of the specimens.

In B cell infiltrates, the majority of B cells in healthy gingiva and in gingivitis are memory B cells, whereas in periodontitis the majority of B cells are differentiated in plasma cells (Mahanonda et al. 2016). In addition, a low population of naïve B cells is observed within the gingiva (Mahanonda et al. 2016), regardless of its

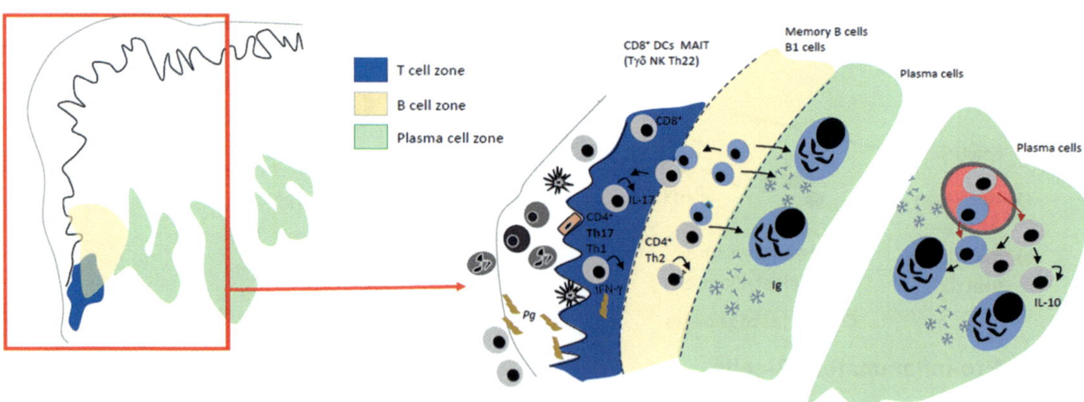

Fig. 9.2 The periodontal cell network is organized in specific aggregates. A T cell zone (Th1 and Th17) is visible under the lamina propria in close relation with the MAIT and DCs. This zone interacts with an underlying B cell zone, probably composed of memory B cells and B1 cells. At greater depth, different aggregates of plasma cells can be distinguished, associated with some DCs, macrophages and osteoclasts. Plasma cells can be derived from both B1 cells and local memory B cells, but also from circulating B1 cells and memory B cells attracted by chemokines in inflammatory periodontal tissues. *MAIT* Mucosa-Associated Invariant T, *DCs* Dendritic cells, *Pg: Porphyromonas gingivalis*, *Ig* Immunoglobulin

pathological state. The migration of B cells is connected with mesenchyme environment of connective tissues. CXCL13-secreting cells, responsible for B cell chemotaxis in inflammatory periodontal tissues are increased in the connective tissue underlying the pocket epithelium when compared to gingivitis and are correlated with B cell numbers. CXCL13-secreting cells are distributed in B cell zones and associated or not with follicular dendritic cells (FDCs) (Nakajima et al. 2008).

Some studies suggest that these intercellular interactions between B cells and their mesenchymal environment could be attributed to an attempt to form a germinal center (Lins et al. 2008), but to date, no reticular network of FDCs, or formation of a lymphoid structure with an active germinal center, has been identified in gingival biopsies (Nakajima et al. 2008). The formation of ectopic tertiary lymphoid organs in periodontal tissues remains a hypothesis to be explored.

Thus, healthy gums contain memory B cells, which could have an immunological sentinel role with the ability to differentiate locally into plasma cells in periodontitis. It is likely that the B cell zones in periodontitis are partly populated by these local or resident memory B cells.

In addition, the presence of $CD5^+$ B cells in the connective tissue, under the pocket epithelium, has been described since 1992 in severe periodontitis (Afar et al. 1992; Sugawara et al. 1992; Aramaki et al. 1998; Berglundh et al. 2002). It is more important inside B cell infiltrate in periodontitis (Thorbert-Mros et al. 2015). $CD5^+$ B cells are associated with a loss of collagen structure and a self-reactive function, producing anti-collagen antibodies (Sugawara et al. 1992). Plasma cells infiltrates are associated with macrophages, close to osteoclasts, and the number of fibroblasts decreases with the decrease in collagen (Sugawara et al. 1992). The CD5 expression was correlated with B1 cells in these studies and it was thus suggested that autoreactive B cells were important in patients with advanced periodontitis (Berglundh et al. 2007; Berglundh et al. 2002). However, CD5, initially described as a B1 marker, is now considered as a marker of activity or immaturity (Lee et al. 2015). Mechanisms of autoimmunity are involved in periodontal destruction, but the blood characteristics highlighted for periodontitis, particularly severe ones, rather suggest an immunoinflammatory pathology (Demoersman et al. 2018; Schmidt et al. 2014).

9.4 Abnormal Interactions Between B and T Cells in Periodontitis

The T cell/B cell ratio is lower in active lesions than in inactive ones, suggesting an abnormal regulation of B cells by Th cells (Reinhardt et al. 1988). When the resolution is not achieved, the activation of T and B cells is crucial for the control of chronic inflammation through constitutive cytokine secretion and modulation of osteoclastogenesis (Figueredo et al. 2019). All lymphocytes, whatever their origin, have a similar kinetics response to periodontal bacteria (Donaldson et al. 1984).

When inflammation is not resolved, antigen-presenting cells (APCs) are activated by bacterial products and interact with naïve Th cells (Th0) in secondary lymphoid organs, driving their differentiation into several subsets, such as Th1, Th2, Th9, Th17, T-follicular helper (Tfh), and regulatory T (Treg) cells (Campbell et al. 2016). Overexpression of the Th17/Treg axis is seen in disease initiation (Ebersole et al. 2014; Moutsopoulos et al. 2012; Cardoso et al. 2009), notably via the production of IL17-A (Korn et al. 2009; Dutzan and Abusleme 2019). However, Th17 might not be the main source of IL-17A in periodontal tissues: MAIT and Treg cells also produce IL-17A, suggesting an important role of these cells in the pathogenesis of periodontal disease (Parachuru et al. 2018; Okui et al. 2012; Sobkowiak et al. 2019). Interestingly, activated B cells seem able to product IL-17 (Bermejo et al. 2013).

A gingival Th17 response is also induced by *Pg* via the secretion of IL-1, IL-6 and IL-23, by epithelial cells. Gingipains cleave IL-12 and thus direct the differentiation of naïve T cells towards the Th17 phenotype. This secretion leads indirectly to the stimulation of bone loss via IL-17, without the direct involvement of DCs (Bittner-Eddy et al. 2016; Moutsopoulos et al. 2012; Korn et al. 2009; Song et al. 2017; Theoleyre et al. 2004).

Furthermore, Th17 cells drive, unlike Th1 cells, the differentiation of B cells into plasma cells by secreting IL-21 and IL-17 (Mitsdoerffer et al. 2010). The interaction of Th17 cells with B cells, and notably the secretion of IL-21, like for Tfh cells (Mitsdoerffer et al. 2010), is responsible for the formation of germinal centers in either peripheral lymph nodes or tertiary lymph organs that can result in clonal activation of B cells, which produce antibodies to recognize bacterial components; however, production of autoantibodies to collagen, fibronectin and laminin can contribute to local destruction of the gingival tissue. (Jones et al. 2016; Mitsdoerffer et al. 2010; Figueredo et al. 2019).

Activated Th1, Th2, and Th17 cells can produce a variety of pro-inflammatory cytokines, such as IL-1, IL-17E (IL-25) and IL-17, that activate other immune cells such as DCs, neutrophils, and B cells. Activation of both T cells and subsequently B cells can cause the production of the receptor activator of nuclear factor B -Ligand (RANKL), leading to alveolar bone resorption by osteoclasts, and tooth loss (Campbell et al. 2016; Dar et al. 2018).

Natural (nTreg) and inductible (iTreg) cells are regulatory T cell subsets characterized by the expression of FoxP3 and the secretion of IL-10 and TGF-β (Gao et al. 2017). nTreg cells inhibit the proliferation of effectors T cells and B cells through direct cell contact (Xu et al. 2016). They thus play a role in the regulation of autoimmunity. However, they cannot inhibit the proliferation of highly activated effector T cells or the production of pro-inflammatory cytokines (Peterson 2012). In chronic periodontitis, a high proportion of circulating nTreg cells has been reported compared to healthy subjects (Sabarish et al. 2016).

iTreg cells occurs at the periphery under the induction of IL-10 (Tr1, secreting IL-10) and TGF-β (Th3 cells) from activated naïve T cells. Th3 suppresses B cell responses mainly via the TGF-β receptor signaling pathway, without causing cell apoptosis. Thus Th3 inhibits B cell differentiation into plasma cells, antibody production and promote B cell orientation towards a phenotype of B10 cells, a subset of B cells with a regulatory function producing IL-10. Moreover, B10 cells can notably promote the conversion of naïve T cells into Tr1, and naïve B cells can induce directly a specific Treg-of-B cell subset (Chien and Chiang 2017). Tr1 cells subsequently induce an expansion of B10 cells, which themselves allow the recruitment of nTreg cells (Zheng et al. 2014). Th3 cells are more particularly located at the mucous membranes and are able to control immune responses in link with the biofilm.

In periodontitis, the percentage of Treg cells is increased and associated with higher proportions of B cells relative to T cells (Nakajima et al. 2005). Treg cells thus play a regulatory role in periodontitis by blocking B cells and the proliferation of Th1, Th17 and Th2 cells, but not the production of Th2 cytokines (Faustino et al. 2012). Th2 and Treg cells cooperate to suppress bone resorption. Finally, a lack of Treg cells, or an inability of those present to reduce local inflammatory responses by other immune cells may play a role in the chronic inflammation associated with periodontitis (Campbell et al. 2016; Figueredo et al. 2019), and favors the differentiation of B cells into plasma cells.

Similarly to T cells, B cells may indeed be a source of secreted osteoclastogenic factor of activated T cells (SOFAT) that can induce osteoclastogenesis in a RANKL-independent manner (Figueredo et al. 2019). SOFAT has been described increased in periodontal tissues, compared to healthy gingiva (Jarry et al. 2013). It usually secreted by activated T cells, but B cells and plasma cells also exhibited strong staining for SOFAT in diseased periodontal tissue (Jarry et al. 2016). Therefore, SOFAT might have an important role in periodontal disease by activating RANKL-related osteoclastogenesis.

9.5 Plasma Cells and B Cells Are Predominant in Periodontal Lesions

In 1976, Page and Schroeder described the histological development of the gingiva according to the different stages of periodontitis (Page and Schroeder 1976). The advanced lesion, the ultimate stage in the progression of periodontal disease, is then described as a lesion with a large infiltration of plasma cells. In 1974, and again in 1978, Mackler et al. showed that bacteria LPS was able to stimulate B cells in healthy patients and in patients with periodontitis. However only the interaction between B cells and T cells in patients with periodontitis resulted in the terminal differentiation of B cells into plasma cells (Mackler et al. 1977, 1978). Plasma cells secrete immunoglobulins, pro- and anti-inflammatory cytokines, sRANKL and MMP and their presence has notably been associated with the persistence of *Pg* (Kim et al. 2010; Mahanonda et al. 2016). Antibodies reactive to *Pg* are locally produced in gingival lesions (Mizutani et al. 2014). Viral infections also seem to play a role in the evolution of periodontitis, notably via Epstein-Barr virus-infected plasma cells (Olivieri et al. 2020). Similarly, long non-coding RNAs (lncRNAs) could have potential functions in periodontitis lesions, notably by regulating the immune reaction. A dysregulation of multiple IncRNAs in plasma cells seems associated with the development of periodontitis (Wu et al. 2020).

The origin of plasma cells in gingiva remains a point to be clarified. Plasma cells could be issued from a local activation of resident gingival B1 or memory B cells, initially described in the healthy gum, or from B cells migrating to the gingiva. B1 cells and switched or unswitched memory B cells can be recruited in peripheral tissues and differentiated in plasma cells secreting specific or natural antibodies (Donati et al. 2009). Their cytokine release could also be different depending on their origins. Indeed, the release of cytokines linked to the differentiation of B cells into plasma cells (IL-4, IL-5 and IL-6) is increased in periodontitis (Fujihashi et al. 1993b).

Plasma cell survival is notably dependent on BAFF/BMCA/APRIL interactions. APRIL binding to BCMA and BAFF binding to BAFF-R block cell apoptosis (Sakai and Akkoyunlu 2017). Recently, APRIL and BAFF have been described as increased in periodontitis compared to healthy gums, underlining an important activation of B cells and a maintenance of plasma cells (Abe et al. 2015; Gumus et al. 2014). The secretion of APRIL takes place mainly in the epithelium and BAFF in the connective tissue and both can be induced by bacterial LPS (Abe et al. 2015; Sakai and Akkoyunlu 2017). Notably, DCs promote the survival of B cells via BAFF and APRIL secretions and their differentiation into plasma cells via IL-6 secretion (Na et al. 2016; Craxton et al. 2003; Jego et al. 2003; Garcia-Marquez et al. 2013). DCs induce an increase in bone loss by promoting the survival of B cells and plasma cell-expressing RANKL.

9.5.1 B1 Cells in Periodontitis

B1 cells originate from the fetal liver and circulate homeostatically between the peripheral tissues and the peritoneum in mice (Griffin et al. 2011; Rothstein et al. 2013; Geherin et al. 2016; Ansel et al. 2002). B1 cells would represent 2–4% of the circulating B cells (Rothstein et al. 2013). Once in the peripheral tissues, B1 cells are responsible, according to their phenotype, for the secretion of natural antibodies, notably implicated in autoimmune reactions (Panda and Ding 2015; Deng et al. 2016). They can also have regulatory, phagocytosis and osteoclastic-like functions (Griffin et al. 2012; Pugliese et al. 2012). The human phenotype ($CD20^{+}CD27^{+}CD43^{+}CD70^{-}CD69^{-}$) of B1 cells has been described (Griffin et al. 2011; Rothstein et al. 2013). B1 cells were initially distinguished as B1a ($CD5^{+}$) and B1b ($CD5^{-}$) in mice, but this distinction does not apply in humans. CD5 would be a typical marker of immature B cells and would distinguish more between anergic ($CD5^{+}$) and active ($CD5^{-}$) forms (Lee et al. 2015; Werner-Favre et al. 1989). CD5 plays a major role in down regulation of BCR response in the B1a cell subset. It has a pivotal role in maintaining anergy in autoreactive B cells: anergic B1a cells are CD5$^+$ (Garaud et al. 2009; Hippen et al. 2000). CD5 expression is not a specific characteristic of B1 cells (Pers et al. 1999; Lee et al. 2015). The CD5 expression on B cells can be for some populations "native" and for others "acquired" by activation (Youinou and Renaudineau 2007). Long considered as an autoimmune marker, CD5 on B cells may play a paradoxical role, both in the prevention and induction of autoimmunity (Youinou and Renaudineau 2007).

The expression of CD11b, a factor that interacts with ICAM-1 on the surface of endothelial cells and within the extracellular matrix, confers migratory capacities to B1 cells (Kawai et al. 2005; Geherin et al. 2016) and distinguishes between the orchestrator B1 cells (CD11b$^+$) and the natural antibody secretory B1 cells (CD11b$^-$) (Fig. 9.3). The orchestrator CD11b$^+$ B1 cells play a role in regulating T cells via a spontaneous secretion of IL-10 (Griffin and Rothstein 2012).

B1 cells have a particular sensitivity to TLR signals, and express different TLRs than other B cell subsets (Popi et al. 2016; Popi 2015; Sindhava and Bondada 2012). In contrast to other B cells, activation by TLRs signals can activate B1 cells which will be able to get out of their state of anergy and will proliferate and differentiate into natural antibody-secreting mature plasma cells, on a T cell-independent manner. B1 cells thus control the bacterial load while awaiting the adaptive response (Genestier et al. 2007). Consequently, T cell activation or high doses of TLR ligands can overcome the anergic state of B1 cells and allow their activation during infection (Sindhava and Bondada 2012). Activated CD11b$^-$CD5$^+$ B1 cells are less sensitive to IgM-induced apoptosis, and this mechanism may facilitate their proliferation, while inducing apoptosis in non-activated CD11b$^-$CD5$^-$ B1 cells

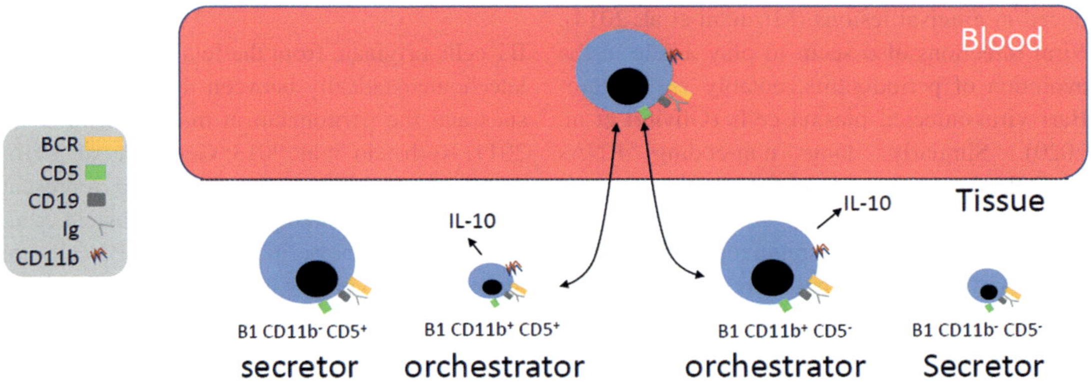

Fig. 9.3 RANKL is expressed by B cells and T cells. RANK-RANKL-OPG triade modulates osteocalastogenesis, angiogenesis and periodontal pocket formation after interaction with endothelial cells, epithelial cells and osteoclastic precursor cells. RANKL inhibits neoangiogenesis by inhibiting OPG, which in turn inhibits TRAIL, an inducer of endothelial cell apoptosis. Under inflammatory conditions the production of OPG by endothelial cells is increased. Interactions between B cells acting as Lymphoid tissue inductors (LTi), in inflamed tissues, and stromal cells, acting as Lymphoid Tissue organizers (LTo), result in the formation of a lymphoid tissue. RANKL acts as an amplifier of the interactions between these cell populations and promotes the growth of primary lymphoid tissues. It may have a similar role in the establishment of tertiary lymphoid organs. The interaction between B cells, through the release of LTαβ and resident stromal cells leads to the production of homeostatic chemokines (CXCL13, CXCL12, CCL21 and CCL19) which could allow cell recruitment and formation of a tertiary lymphoid organ. The action of RANKL on RANK on the surface of the LTi induces the production of LTαβ which will stimulate the productions of LTo, including RANKL. MAIT and other immune cells may also have a role of LTi in inducing the formation of tertiary lymphoid organs by the expression of various proinflammatory cytokines (IL-17, IL-22, IL-23) and LTαβ. *TNF* Tumor Necrosis Factor, *LTαβ* Lymphotoxin αβ, *CXCL* chemokine (C-X-C motif) ligand, *CCL* chemokine ligands, *MAIT* Mucosa-Associated invariant T, *OPG* osteopotegerin, *TRAIL* tumor necrosis factor (TNF)-related apoptosis-inducing ligand, *EC* endothelial cells

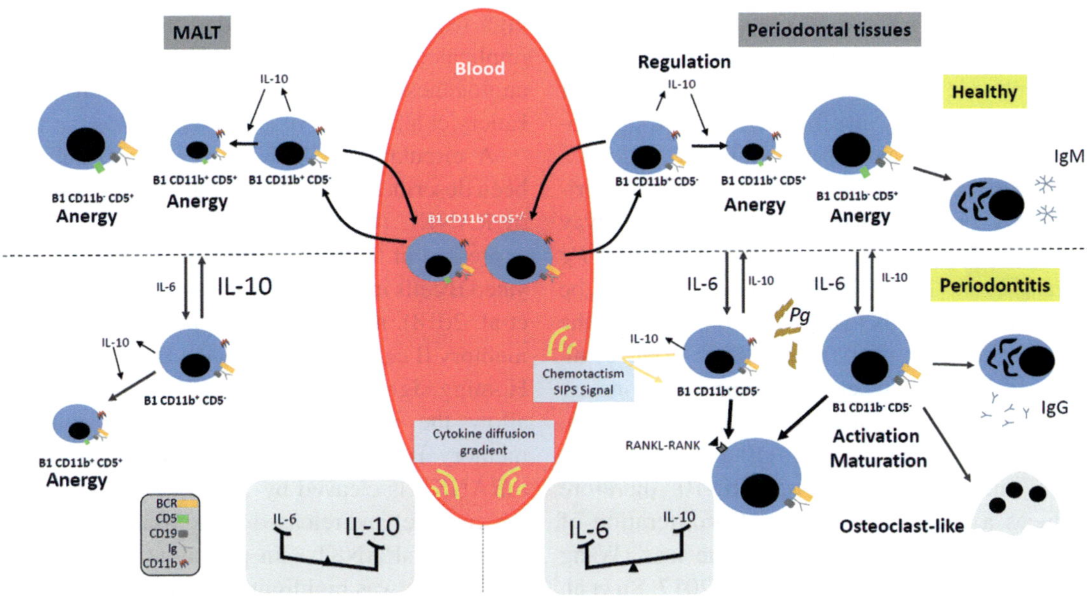

Fig. 9.4 Two sub-populations of B1 cells in humans. B1 CD11b⁻ cells are natural antibody secretors (B1 secretors) and B1 CD11b⁺ cells are natural IL-10 secretors (orchestrators). CD11b⁺ B1 cells have a high migration capacity and both populations can express or not CD5. Orchestrators are mainly CD5⁻ and secretors CD5⁺. IL: interleukin

(Pers et al. 2002). Its systemic diffusion could be at the origin of a decrease in CD11b⁺ B1 cells in peripheral blood by blocking their renewal at the level of the peritoneum or in secondary lymphoid organs. The IL-10/IL-6 ratio influences the phenotype of local and circulating B1 cells and the balance between their regulatory and protective properties and their destructive capacities.

The expression of CD5 by the different subsets of B1 would be variable and also dependent on the cytokine environment, in particular IL-10 and IL-6 (Garaud et al. 2009; Youinou and Renaudineau 2007) (Fig. 9.4). IL-6 could be a regulator of CD5 expression and B1 migration. This proinflammatory cytokine is highly expressed in periodontal inflammatory tissues and has antagonistic effects on IL-10. It is responsible for a decrease in membranous CD5 and the differentiation of B cells into plasma cells (Fujihashi et al. 1993a; Garaud et al. 2009; Hunter and Jones 2015; Seymour and Gemmell 2001).

Since Griffin et al. (Griffin et al. 2011; Rothstein et al. 2013), there is no study on the involvement of B1 cells in periodontal tissues, except for their role as B10 cells.

9.6 Regulatory B Cells (Breg/B10) in Periodontitis

B10 cells are a heterogeneous subpopulation of B cells with a high potential to modulate the immune response against infectious or tumor agents (Goode et al. 2013; Rosser and Mauri 2015; Lykken et al. 2015; Rincon-Arevalo et al. 2016; Tedder 2015; Mauri and Menon 2017). They are characterized by their ability to secrete IL-10 and are involved in autoimmune diseases (Goode et al. 2013; Vadasz et al. 2013). Their role in the regulation of the periodontal immune response, and in the reduction of bone loss in periodontitis has been shown recently (Yu et al. 2017; Wang et al. 2017; Dai et al. 2017; Hu et al. 2017; Hetta et al. 2020; Gu and Han 2020; Shi et al. 2020). The level of IL-10 is significantly higher in inflammatory gingival tissue than in healthy tissue and the proportion of IL-10 producing B cells represents a small proportion of the gingival B cell population (Dai et al. 2017). B10 cells are increased in periodontitis, but their number is probably insufficient to inhibit the production of proinflammatory cytokines.

The secretion of IL-10 by B10 cells in periodontitis can be induced by the TLR pathway (TLR2/4 and 9), in the presence of LPS from *Pg* (Wang et al. 2017). This secretion can also be induced by Th cells (Yu et al. 2017), and by the cytokine environment (IL-21, Tim1) (Hu et al. 2017; Su et al.). Th cells and their secretions are responsible for an increase in IL-10 secretion, and a decrease in RANKL/OPG ratio, TNF-α, IL-1 and IL-17 secretions. IL-10 therefore induces a decrease in the local proliferation of Th17 cells and a decrease in bone loss (Wang et al. 2017; Yu et al. 2017; Hu et al. 2017; Su et al. 2008; Shi et al. 2020). This immunomodulatory effect of B10 cells during periodontitis is independent of antibody production (Shi et al. 2020). TLR signaling induces both pro-inflammatory and anti-inflammatory pathways in activated B cells and it is thus conceivable that promoting the optimal B10 function requires selective activation of TLR signaling and availability of co-stimulatory molecules (Gu and Han 2020).

9.7 B Cells and Bone Resorption

B cells are central players in the concept of osteo-immunology defining the cross interactions between bone metabolism and immune system (Fig. 9.5) (Choi et al. 2001; Arron and Choi 2000; Kumar and Roger 2019; AlQranei and Chellaiah 2020). The triad RANK-RANKL-OPG is common to both pathways (Dougall et al. 1999; Kong et al. 1999), and B cells, plasma cells and T cells are described as the main sources of sRANKL in periodontal tissues (Kawai et al. 2006; Mahanonda et al. 2016). B cells produce even more RANKL than T cells and other lymphocytes during periodontal diseases and memory B cells are the most prone to secrete RANKL (AlQranei and Chellaiah 2020; Han et al. 2019). Membrane expression is low on plasma cells while RANKL's intracytoplasmic expression is high (Mahanonda et al. 2016). RANKL is also implicated in periodontal pocket formation and angiogenesis in periodontitis (Sojod et al. 2017; Emery et al. 1998; Pritzker et al. 2004).

A circulating RANKL$^+$ memory B cell has been described in rheumatoid arthritis as a source of RANKL (Ota et al. 2016) and RANKL are expressed at the membrane (mRANKL) of activated B cells in severe periodontitis (Demoersman et al. 2018). mRANKL is expressed in switched memory B cells activated via the BCR and CD40. Homing signals, such as CXCR3 and CXCL10, allow the migration of mRANKL$^+$ switched memory B cells to inflammatory tissues where mRANKL is cleaved by TACE (TNF-α converting enzyme) and released in the tissues as a soluble form (sRANKL). In periodontitis, a similar mechanism was highlighted: adoptive transfer of pathogen-sensitized memory B cells in rats with periodontitis increased the extent of alveolar bone resorption compared to control rats with periodontitis (Kanzaki et al. 2016; Kawai et al. 2006). In the same way, memory B cell cultures from annimals with periodontitis, promote in a RANKL-dependent manner the differentiation into osteoclasts whereas the other subsets of B cells do not (Han et al. 2019).

B cells are inducers of osteoclatogenesis when activated, but they may also have a physiological role in bone balance. They are also secretors of OPG (Titanji et al. 2014; Perlot and Penninger 2012; Okamoto et al. 2017; Weitzmann 2017) and their expansion and their expression of RANKL, especially by B1 cells, is stimulated by signals, such as IL-33, produced as the result of mechanical, traumatic or infectious gingival stimulations (Malcolm et al. 2015; Laperine et al. 2016; Tada et al. 2016).

The RANKL-RANK-OPG triad is therefore not only an important factor in bone metabolism, but also in the organization of the immune response. It is involved in the formation of lymphoid organs, in particular in communications between immune cells and mesenchymal cells (Hess et al. 2012; Mueller and Hess 2012). These mechanisms are reproduced in the formation of tertiary lymphoid organs with an essential role for B cells as lymphoid tissue organizers cells.

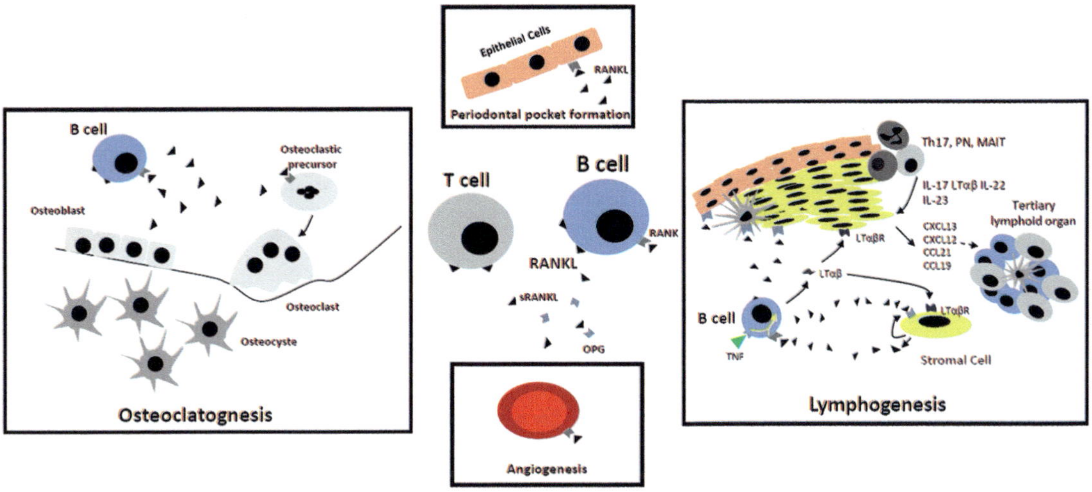

Fig. 9.5 **IL-6 and IL-10 are both increased in periodontitis and have antagonistic roles.** The ratio IL-10/IL-6 could explain the local phenotypic orientations of the different B1 cells, but also at the systemic level. A diffusion of IL-6 from the periodontal tissues to the other mucosa-associated lymphoid tissue (MALT) could explain the decrease of B1 CD11b$^+$ cells, observed in severe periodontitis. IL-6 and IL-10 regulate the expression of CD5 and the maintenance of B1 cells in state of anergy within tissues. B1 cells migrate to inflamed tissues due to by SIPS signal and chemokines. *IL* interleukin, *Ig* Immunoglobulin, *S1P* sphingosine 1 phosphate

9.8 Circulating B Cells and Periodontitis

Memory B cells subsets (unswitched, switched, and IgD/CD27 double negative) have probably different implications in periodontitis (Fig. 9.6). Unswitched memory B cells have not met antigen but could have a role against encapsulated bacteria such as *Pg* (Weller and Descatoire 2015; Weller et al. 2004). Switched forms have met antigen in germinal centers and are therefore specific to antigen whereas IgD/CD27 double negative forms are the result of multiple circulations of memory B cells between germinal centers and inflammatory tissues. The increase in switched forms may be a marker of an increase in the affinity of the immune response against the antigen (Claes et al. 2016; Wu et al. 2011; Moura et al. 2017) and interestingly, the increase in switched memory B cells has been highlighted in severe periodontitis (Demoersman et al. 2018). All subsets of memory B cells were previously shown as increased in the chronic forms of periodontitis (Schmidt et al. 2014). This increase of memory B cells could be a sign of displacement of lymphopoiesis in the secondary lymphoid organs observed in chronic inflammatory conditions (Jones et al. 2016; Cain et al. 2009). However, this mechanism would be disrupted in the aggressive forms because of the local progression of the disease that would be limited by the systemic response in the chronic forms, which are nonexistent in the aggressive forms.

Similarly, a decrease in circulating CD11b$^+$ B1 cells has been shown in severe periodontitis (Demoersman et al. 2018). They could be associated with an acquired or constitutive decrease of local regulation. In inflammatory conditions, B1 cells seem to have a stronger propensity to enter in inflammatory tissues than memory B cells, following specific homing signals (Geherin et al. 2016). This decrease could be a biomarker of severe periodontitis, or of periodontal risk (Griffin et al. 2011; Rothstein et al. 2013; Geherin et al. 2016; Ansel et al. 2002).

RANKL expression concerns all the circulating B cells, and seems to be more important on the activated forms (Demoersman et al. 2018).

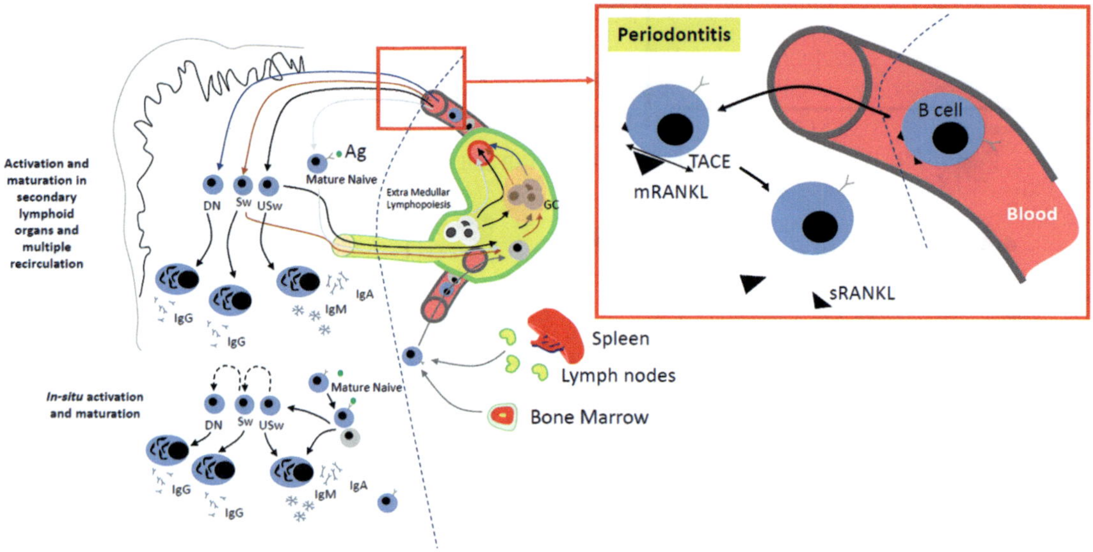

Fig. 9.6 Memory B cells have a preponderant role in periodontitis. Memory B cells are the main source of RANKL and can be activated locally and in the lymph nodes. Antigen presenting cells (APCs) interacts with T cells which activate naïve B cells against a periodontal Ag. A germinal center is formed and will lead to the production of switched (SW) Memory B cells with a high affinity towards the antigen. SW memory B cells migrate following homing signals in periodontal tissues where some of them differentiate into plasma cells. They present a RANKL$^+$ phenotype in blood and periodontal tissues. A cleavage of mRANKL by TACE leads to the release of sRANKL in the periodontal tissues. Multiple recirculation of SW memory B cells between periodontal tissues and secondary lymphoid organs allows to improve the affinity of the memory populations towards Ags. This mechanism is at the origin of IgD$^-$CD27$^-$ double negative (DN) memory populations. The unswitched (USw) memory populations has not met the Ag and also circulate between the secondary lymphoid organs and the periodontal tissues. As in chronic inflammatory situations, extra medullary lymphopoiesis could be observed in the lymph nodes.
The memory B cells residing in the gum could be locally activated in periodontitis. The local increase of their affinity and the formation of DN memory B cells could be the sign of the presence of transitory ectopic germinal center within the periodontal tissues. *TACE* TNF alpha converting enzyme, *Ig* Immunoglobulin, *Ag* Antigen

9.9 Discussion

B cells have a central role in the establishment of periodontitis. In healthy gums or gums presenting a subclinical inflammation. B1 and memory B cells could form a resident population directly activated locally in order to maintain homeostasis. The activation of B cells is carried out in the presence of bacterial antigens and according to the cytokine environment induced by an infection or mechanical stimulus. They could also have a role in bone adaptation under physiological conditions.

Their activation leads to the organization of the local immune response, notably by the secretion of RANKL, and cytokines. The migration of APC in the lymph nodes leads to the activation of memory B cells specific to periodontal antigens that will later migrate in the periodontal tissues. Memory B cells will participate, organize and regulate the local immune response, via the production of RANKL, cytokines and antibodies. Circulating B1 cells will also have this dual role, independently of the antigenic specificities. They will participate, organize and regulate the local immune response, via the production of RANKL, cytokines and antibodies.

Dysregulation of these mechanisms could be responsible for the development of a severe and rapidly progressive disease. The insufficiency of a systemic response to periodontitis leads to a lack of specificity of the immune response, a defect of local regulation and to the rapid and extensive development of severe periodontal lesions. These lesions could be characterized by important local autoimmune mechanisms and the establishment of tertiary lymphoid organs. Further studies are required to determine the genetic or epigenetic characteristics and factors involved in this dysfunctional systemic response.

In slow-progressive forms, periodontal infection would be responsible for a displacement of lymphopoiesis towards structures in periphery and the increase of switched memory B cells, potentially expressing RANKL, organizing and regulating the local immune response, while directly participating in tissue destruction. The mechanisms observed in periodontitis would thus be similar to those observed in certain rheumatic pathologies.

B cell subsets could be blood biomarkers of the type of periodontal damage and of the type of systemic response. In addition to allowing the follow-up of the evolution of periodontitis, B cell distribution would inform on the progression of the periodontal lesions.

From these observations, different therapeutic strategies could be proposed to treat periodontitis, in order to rebalance the immune T cell/B cell ratio, by targeting the B cells. In addition, treatments leading to a depletion of mature B cells, participating in tissue destruction, such as anti-CD20 antibodies, would make it possible to reduce the populations of B cells favoring tissue destruction, while maintaining the immature forms that are the source of B10. Therefore, interventional strategies targeting TLR signaling and immune regulatory T/B cells may be a promising approach to rebalance the immune response and alleviate bone loss in periodontal disease (Gu and Han 2020).

These therapeutic strategies must be studied because periodontitis is a local and systemic disease, involved with other systemic diseases and responsible for a significant loss of quality of life for the patients.

References

Abe T, AlSarhan M, Benakanakere MR, Maekawa T, Kinane DF, Cancro MP, Korostoff JM, Hajishengallis G (2015) The B cell-stimulatory cytokines BLyS and APRIL are elevated in human periodontitis and are required for B cell-dependent bone loss in experimental murine periodontitis. J Immunol 195(4):1427–1435. https://doi.org/10.4049/jimmunol.1500496

Afar B, Engel D, Clark EA (1992) Activated lymphocyte subsets in adult periodontitis. J Periodontal Res 27(2):126–133

Ali MF, Driscoll CB, Walters PR, Limper AH, Carmona EM (2015) Beta-glucan-activated human B lymphocytes participate in innate immune responses by releasing proinflammatory cytokines and stimulating neutrophil chemotaxis. J Immunol 195(11):5318–5326. https://doi.org/10.4049/jimmunol.1500559

AlQranei MS, Chellaiah MA (2020) Osteoclastogenesis in periodontal diseases: possible mediators and mechanisms. J Oral Biosci 62(2):123–130. https://doi.org/10.1016/j.job.2020.02.002

Ansel KM, Harris RB, Cyster JG (2002) CXCL13 is required for B1 cell homing, natural antibody production, and body cavity immunity. Immunity 16(1):67–76

Aramaki M, Nagasawa T, Koseki T, Ishikawa I (1998) Presence of activated B-1 cells in chronic inflamed gingival tissue. J Clin Immunol 18(6):421–429

Arron JR, Choi Y (2000) Bone versus immune system. Nature 408(6812):535–536. https://doi.org/10.1038/35046196

Baker PJ, Boutaugh NR, Tiffany M, Roopenian DC (2009) B cell IgD deletion prevents alveolar bone loss following murine oral infection. Interdiscip Perspect Infect Dis 2009:864359. https://doi.org/10.1155/2009/864359

Berglundh T, Liljenberg B, Tarkowski A, Lindhe J (2002) The presence of local and circulating autoreactive B cells in patients with advanced periodontitis. J Clin Periodontol 29(4):281–286

Berglundh T, Donati M, Zitzmann N (2007) B cells in periodontitis: friends or enemies? Periodontol 45:51–66. https://doi.org/10.1111/j.1600-0757.2007.00223.x

Bermejo DA, Jackson SW, Gorosito-Serran M, Acosta-Rodriguez EV, Amezcua-Vesely MC, Sather BD, Singh AK, Khim S, Mucci J, Liggitt D, Campetella O, Oukka M, Gruppi A, Rawlings DJ (2013) Trypanosoma cruzi trans-sialidase initiates a program independent of the transcription factors RORgammat and Ahr that leads to IL-17 production by activated B cells. Nat Immunol 14(5):514–522. https://doi.org/10.1038/ni.2569

Bittner-Eddy PD, Fischer LA, Kaplan DH, Thieu K, Costalonga M (2016) Mucosal langerhans cells promote differentiation of Th17 cells in a murine model of periodontitis but are not required for porphyromo-

nas gingivalis-driven alveolar bone destruction. J Immunol 197(4):1435–1446. https://doi.org/10.4049/jimmunol.1502693
Cain D, Kondo M, Chen H, Kelsoe G (2009) Effects of acute and chronic inflammation on B-cell development and differentiation. J Invest Dermatol 129(2):266–277. https://doi.org/10.1038/jid.2008.286
Campbell L, Millhouse E, Malcolm J, Culshaw S (2016) T cells, teeth and tissue destruction - what do T cells do in periodontal disease? Mol Oral Microbiol 31(6):445–456. https://doi.org/10.1111/omi.12144
Cardoso CR, Garlet GP, Crippa GE, Rosa AL, Junior WM, Rossi MA, Silva JS (2009) Evidence of the presence of T helper type 17 cells in chronic lesions of human periodontal disease. Oral Microbiol Immunol 24(1):1–6. https://doi.org/10.1111/j.1399-302X.2008.00463.x
Carvel RI, Carr RF (1982) Clinico-pathologic correlation of 385 lesions of marginal destructive periodontitis. J Periodontol 53(5):328–333. https://doi.org/10.1902/jop.1982.53.5.328
Chien CH, Chiang BL (2017) Regulatory T cells induced by B cells: a novel subpopulation of regulatory T cells. J Biomed Sci 24(1):86. https://doi.org/10.1186/s12929-017-0391-3
Choi Y, Woo KM, Ko SH, Lee YJ, Park SJ, Kim HM, Kwon BS (2001) Osteoclastogenesis is enhanced by activated B cells but suppressed by activated CD8(+) T cells. Eur J Immunol 31(7):2179–2188. https://doi.org/10.1002/1521-4141(200107)31:7<2179::aid-immu2179>3.0.co;2-x. PMID: 11449372.
Cirrincione C, Pimpinelli N, Orlando L, Romagnoli P (2002) Lamina propria dendritic cells express activation markers and contact lymphocytes in chronic periodontitis. J Periodontol 73(1):45–52. https://doi.org/10.1902/jop.2002.73.1.45
Claes N, Fraussen J, Vanheusden M, Hellings N, Stinissen P, Van Wijmeersch B, Hupperts R, Somers V (2016) Age-associated B cells with proinflammatory characteristics are expanded in a proportion of multiple sclerosis patients. J Immunol 197(12):4576–4583. https://doi.org/10.4049/jimmunol.1502448
Coat J, Demoersman J, Beuzit S, Cornec D, Devauchelle-Pensec V, Saraux A, Pers JO (2015) Anti-B lymphocyte immunotherapy is associated with improvement of periodontal status in subjects with rheumatoid arthritis. J Clin Periodontol 42(9):817–823. https://doi.org/10.1111/jcpe.12433
Craxton A, Magaletti D, Ryan EJ, Clark EA (2003) Macrophage- and dendritic cell--dependent regulation of human B-cell proliferation requires the TNF family ligand BAFF. Blood 101(11):4464–4471. https://doi.org/10.1182/blood-2002-10-3123
Dai J, Bi L, Lin J, Qi F (2017) Evaluation of interleukin-10 producing CD19+ B cells in human gingival tissue. Arch Oral Biol 84:112–117. https://doi.org/10.1016/j.archoralbio.2017.09.009
Dar HY, Azam Z, Anupam R, Mondal RK, Srivastava RK (2018) Osteoimmunology: the nexus between bone and immune system. Front Biosci (Landmark Ed) 23:464–492
Demoersman J, Pochard P, Framery C, Simon Q, Boisrame S, Soueidan A, Pers JO (2018) B cell subset distribution is altered in patients with severe periodontitis. PLoS One 13(2):e0192986. https://doi.org/10.1371/journal.pone.0192986
Deng J, Wang X, Chen Q, Sun X, Xiao F, Ko KH, Zhang M, Lu L (2016) B1a cells play a pathogenic role in the development of autoimmune arthritis. Oncotarget 7(15):19299–19311. https://doi.org/10.18632/oncotarget.8244
Donaldson SL, Ranney RR, Tew JG (1984) B-lymphocyte blastogenesis in response to periodontitis-associated bacteria. Kinetics and proportion of total response. J Periodontol 55(6):359–363. https://doi.org/10.1902/jop.1984.55.6.359
Donati M, Liljenberg B, Zitzmann NU, Berglundh T (2009) B-1a cells and plasma cells in periodontitis lesions. J Periodontal Res 44(5):683–688. https://doi.org/10.1111/j.1600-0765.2008.01178.x
Dougall WC, Glaccum M, Charrier K, Rohrbach K, Brasel K, De Smedt T, Daro E, Smith J, Tometsko ME, Maliszewski CR, Armstrong A, Shen V, Bain S, Cosman D, Anderson D, Morrissey PJ, Peschon JJ, Schuh J (1999) RANK is essential for osteoclast and lymph node development. Genes Dev 13(18):2412–2424
Dutzan N, Abusleme L (2019) T helper 17 cells as pathogenic drivers of periodontitis. Adv Exp Med Biol 1197:107–117. https://doi.org/10.1007/978-3-030-28524-1_9
Dutzan N, Konkel JE, Greenwell-Wild T, Moutsopoulos NM (2016) Characterization of the human immune cell network at the gingival barrier. Mucosal Immunol 9(5):1163–1172. https://doi.org/10.1038/mi.2015.136
Ebersole JL, Cappelli D, Holt SC (2001) Periodontal diseases: to protect or not to protect is the question? Acta Odontol Scand 59(3):161–166
Ebersole JL, Kirakodu S, Novak MJ, Stromberg AJ, Shen S, Orraca L, Gonzalez-Martinez J, Burgos A, Gonzalez OA (2014) Cytokine gene expression profiles during initiation, progression and resolution of periodontitis. J Clin Periodontol 41(9):853–861. https://doi.org/10.1111/jcpe.12286
Emery JG, McDonnell P, Burke MB, Deen KC, Lyn S, Silverman C, Dul E, Appelbaum ER, Eichman C, DiPrinzio R, Dodds RA, James IE, Rosenberg M, Lee JC, Young PR (1998) Osteoprotegerin is a receptor for the cytotoxic ligand TRAIL. J Biol Chem 273(23):14363–14367
Faustino L, Mucida D, Keller AC, Demengeot J, Bortoluci K, Sardinha LR, Carla Takenaka M, Basso AS, Faria AM, Russo M (2012) Regulatory T cells accumulate in the lung allergic inflammation and efficiently suppress T-cell proliferation but not Th2 cytokine production. Clin Dev Immunol 2012:721817. https://doi.org/10.1155/2012/721817
Figueredo CM, Lira-Junior R, Love RM (2019) T and B cells in periodontal disease: new functions in a complex scenario. Int J Mol Sci 20(16). https://doi.org/10.3390/ijms20163949

Fujihashi K, Beagley KW, Kono Y, Aicher WK, Yamamoto M, DiFabio S, Xu-Amano J, McGhee JR, Kiyono H (1993a) Gingival mononuclear cells from chronic inflammatory periodontal tissues produce interleukin (IL)-5 and IL-6 but not IL-2 and IL-4. Am J Pathol 142(4):1239–1250

Fujihashi K, Kono Y, Beagley KW, Yamamoto M, McGhee JR, Mestecky J, Kiyono H (1993b) Cytokines and periodontal disease: immunopathological role of interleukins for B cell responses in chronic inflamed gingival tissues. J Periodontol 64(5 Suppl):400–406

Gao J, Ma X, Gu W, Fu M, An J, Xing Y, Gao T, Li W, Liu Y (2012) Novel functions of murine B1 cells: active phagocytic and microbicidal abilities. Eur J Immunol 42(4):982–992. https://doi.org/10.1002/eji.201141519

Gao L, Zhao Y, Wang P, Zhang L, Zhang C, Chen Q, Zhao C (2017) Detection of Th17/Treg cells and related factors in gingival tissues and peripheral blood of rats with experimental periodontitis. Iran J Basic Med Sci 20(3):294–300. https://doi.org/10.22038/ijbms.2017.8359

Garaud S, Le Dantec C, Jousse-Joulin S, Hanrotel-Saliou C, Saraux A, Mageed RA, Youinou P, Renaudineau Y (2009) IL-6 modulates CD5 expression in B cells from patients with lupus by regulating DNA methylation. J Immunol 182(9):5623–5632. https://doi.org/10.4049/jimmunol.0802412

Garcia-Marquez M, Shimabukuro-Vornhagen A, von Bergwelt-Baildon M (2013) Complex interactions between B cells and dendritic cells. Blood 121(12):2367–2368. https://doi.org/10.1182/blood-2012-12-468017

Geherin SA, Gomez D, Glabman RA, Ruthel G, Hamann A, Debes GF (2016) IL-10+ innate-like B cells are part of the skin immune system and require alpha-4beta1 integrin to migrate between the peritoneum and inflamed skin. J Immunol 196(6):2514–2525. https://doi.org/10.4049/jimmunol.1403246

Genestier L, Taillardet M, Mondiere P, Gheit H, Bella C, Defrance T (2007) TLR agonists selectively promote terminal plasma cell differentiation of B cell subsets specialized in thymus-independent responses. J Immunol 178(12):7779–7786

Goode I, Xu H, Ildstad ST (2013) Regulatory B cells: the new “it” cell. Transplant Proc. https://doi.org/10.1016/j.transproceed.2013.08.075

Griffin DO, Rothstein TL (2012) Human “orchestrator” CD11b(+) B1 cells spontaneously secrete interleukin-10 and regulate T-cell activity. Mol Med 18:1003–1008. https://doi.org/10.2119/molmed.2012.00203

Griffin DO, Holodick NE, Rothstein TL (2011) Human B1 cells in umbilical cord and adult peripheral blood express the novel phenotype CD20+CD27+CD43+CD70−. J Exp Med 208(1):67–80. https://doi.org/10.1084/jem.20101499

Griffin DO, Quach T, Batliwalla F, Andreopoulos D, Holodick NE, Rothstein TL (2012) Human CD11b+ B1 cells are not monocytes: a reply to “Gene profiling of CD11b+ and CD11b- B1 cell subsets reveals potential cell sorting artifacts”. J Exp Med 209(3):434–436. https://doi.org/10.1084/jem.20120403

Gu Y, Han X (2020) Toll-like receptor signaling and immune regulatory lymphocytes in periodontal disease. Int J Mol Sci 21(9). https://doi.org/10.3390/ijms21093329

Gumus P, Nizam N, Lappin DF, Buduneli N (2014) Saliva and serum levels of B-cell activating factors and tumor necrosis factor-alpha in patients with periodontitis. J Periodontol 85(2):270–280. https://doi.org/10.1902/jop.2013.130117

Gurses N, Unlu F, Hekimgil M, Keskinoglu A (1996) Immunohistochemical characterization of lymphocyte subsets in chronic adult periodontitis. J Nihon Univ Sch Dent 38(2):94–101

Hajishengallis G, Korostoff JM (2017) Revisiting the Page & Schroeder model: the good, the bad and the unknowns in the periodontal host response 40 years later. Periodontol 75(1):116–151. https://doi.org/10.1111/prd.12181

Han Y, Jin Y, Miao Y, Shi T, Lin X (2019) Improved RANKL expression and osteoclastogenesis induction of CD27+CD38- memory B cells: a link between B cells and alveolar bone damage in periodontitis. J Periodontal Res 54(1):73–80. https://doi.org/10.1111/jre.12606

Harada Y, Han X, Yamashita K, Kawai T, Eastcott JW, Smith DJ, Taubman MA (2006) Effect of adoptive transfer of antigen-specific B cells on periodontal bone resorption. J Periodontal Res 41(2):101–107. https://doi.org/10.1111/j.1600-0765.2005.00839.x

Hess E, Duheron V, Decossas M, Lezot F, Berdal A, Chea S, Golub R, Bosisio MR, Bridal SL, Choi Y, Yagita H, Mueller CG (2012) RANKL induces organized lymph node growth by stromal cell proliferation. J Immunol 188(3):1245–1254. https://doi.org/10.4049/jimmunol.1101513

Hetta HF, Mwafey IM, Batiha GE, Alomar SY, Mohamed NA, Ibrahim MA, Elkady A, Meshaal AK, Alrefai H, Khodeer DM, Zahran AM (2020) CD19(+) CD24(hi) CD38(hi) regulatory B cells and memory B cells in periodontitis: association with pro-inflammatory and anti-inflammatory cytokines. Vaccines (Basel) 8(2). https://doi.org/10.3390/vaccines8020340

Hippen KL, Tze LE, Behrens TW (2000) CD5 maintains tolerance in anergic B cells. J Exp Med 191(5):883–890

Horowitz MC, Fretz JA, Lorenzo JA (2010) How B cells influence bone biology in health and disease. Bone 47(3):472–479. https://doi.org/10.1016/j.bone.2010.06.011

Hu Y, Yu P, Yu X, Hu X, Kawai T, Han X (2017) IL-21/anti-Tim1/CD40 ligand promotes B10 activity in vitro and alleviates bone loss in experimental periodontitis in vivo. Biochim Biophys Acta 1863 9:2149–2157. https://doi.org/10.1016/j.bbadis.2017.06.001

Hunter CA, Jones SA (2015) IL-6 as a keystone cytokine in health and disease. Nat Immunol 16(5):448–457. https://doi.org/10.1038/ni.3153

Jarry CR, Duarte PM, Freitas FF, de Macedo CG, Clemente-Napimoga JT, Saba-Chujfi E, Passador-

Santos F, de Araujo VC, Napimoga MH (2013) Secreted osteoclastogenic factor of activated T cells (SOFAT), a novel osteoclast activator, in chronic periodontitis. Hum Immunol 74(7):861–866. https://doi.org/10.1016/j.humimm.2013.04.013

Jarry CR, Martinez EF, Peruzzo DC, Carregaro V, Sacramento LA, Araujo VC, Weitzmann MN, Napimoga MH (2016) Expression of SOFAT by T- and B-lineage cells may contribute to bone loss. Mol Med Rep 13(5):4252–4258. https://doi.org/10.3892/mmr.2016.5045

Jego G, Palucka AK, Blanck JP, Chalouni C, Pascual V, Banchereau J (2003) Plasmacytoid dendritic cells induce plasma cell differentiation through type I interferon and interleukin 6. Immunity 19(2):225–234

Johannessen AC, Nilsen R, Kristoffersen T, Knudsen GE (1990) Variation in the composition of gingival inflammatory cell infiltrates. J Clin Periodontol 17(5):298–305

Johnson RJ, Matthews JL, Stone MJ, Hurt WC, Newman JT (1980) Immunopathology of periodontal disease. I. immunologic profiles in periodontitis and juvenile periodontitis. The. J Periodontol 51(12):705–712. https://doi.org/10.1902/jop.1980.51.12.705

Jones GW, Hill DG, Jones SA (2016) Understanding immune cells in tertiary lymphoid organ development: it is all starting to come together. Front Immunol 7. https://doi.org/10.3389/fimmu.2016.00401

Jotwani R, Palucka AK, Al-Quotub M, Nouri-Shirazi M, Kim J, Bell D, Banchereau J, Cutler CW (2001) Mature dendritic cells infiltrate the T cell-rich region of oral mucosa in chronic periodontitis: in situ, in vivo, and in vitro studies. J Immunol 167(8):4693–4700

Kanzaki H, Makihira S, Suzuki M, Ishii T, Movila A, Hirschfeld J, Mawardi H, Lin X, Han X, Taubman MA, Kawai T (2016) Soluble RANKL cleaved from activated lymphocytes by TNF-alpha-converting enzyme contributes to osteoclastogenesis in periodontitis. J Immunol 197(10):3871–3883. https://doi.org/10.4049/jimmunol.1601114

Kaur G, Mohindra K, Singla S (2017) Autoimmunity-basics and link with periodontal disease. Autoimmun Rev 16(1):64–71. https://doi.org/10.1016/j.autrev.2016.09.013

Kawai K, Tsuno NH, Matsuhashi M, Kitayama J, Osada T, Yamada J, Tsuchiya T, Yoneyama S, Watanabe T, Takahashi K, Nagawa H (2005) CD11b-mediated migratory property of peripheral blood B cells. J Allergy Clin Immunol 116(1):192–197. https://doi.org/10.1016/j.jaci.2005.03.021

Kawai T, Matsuyama T, Hosokawa Y, Makihira S, Seki M, Karimbux NY, Goncalves RB, Valverde P, Dibart S, Li YP, Miranda LA, Ernst CW, Izumi Y, Taubman MA (2006) B and T lymphocytes are the primary sources of RANKL in the bone resorptive lesion of periodontal disease. Am J Pathol 169(3):987–998. https://doi.org/10.2353/ajpath.2006.060180

Kim YC, Ko Y, Hong SD, Kim KY, Lee YH, Chae C, Choi Y (2010) Presence of Porphyromonas gingivalis and plasma cell dominance in gingival tissues with periodontitis. Oral Dis 16(4):375–381. https://doi.org/10.1111/j.1601-0825.2009.01649.x

Klausen B, Hougen HP, Fiehn NE (1989) Increased periodontal bone loss in temporarily B lymphocyte-deficient rats. J Periodontal Res 24(6):384–390

Kong YY, Yoshida H, Sarosi I, Tan HL, Timms E, Capparelli C, Morony S, Oliveira-dos-Santos AJ, Van G, Itie A, Khoo W, Wakeham A, Dunstan CR, Lacey DL, Mak TW, Boyle WJ, Penninger JM (1999) OPGL is a key regulator of osteoclastogenesis, lymphocyte development and lymph-node organogenesis. Nature 397(6717):315–323. https://doi.org/10.1038/16852

Korn T, Bettelli E, Oukka M, Kuchroo VK (2009) IL-17 and Th17 Cells. Annu Rev Immunol 27:485–517. https://doi.org/10.1146/annurev.immunol.021908.132710

Kumar G, Roger PM (2019) From crosstalk between immune and bone cells to bone erosion in infection. Int J Mol Sci 20(20). https://doi.org/10.3390/ijms20205154

Laperine O, Cloitre A, Caillon J, Huck O, Bugueno IM, Pilet P, Sourice S, Le Tilly E, Palmer G, Davideau JL, Geoffroy V, Guicheux J, Beck-Cormier S, Lesclous P (2016) Interleukin-33 and RANK-L interplay in the alveolar bone loss associated to periodontitis. PLoS One 11(12):e0168080. https://doi.org/10.1371/journal.pone.0168080

Lee J, Sim JH, Kim IJ (2015) Peripheral immature B cells: modulators of autoimmunity. Int J Rheum Dis 18(2):200–207. https://doi.org/10.1111/1756-185X.12432

Lins RD, Figueiredo CR, Queiroz LM, da Silveira EJ, Freitas Rde A (2008) Immunohistochemical evaluation of the inflammatory response in periodontal disease. Braz Dent J 19(1):9–14

Lykken JM, Candando KM, Tedder TF (2015) Regulatory B10 cell development and function. Int Immunol 27(10):471–477. https://doi.org/10.1093/intimm/dxv046

Mackler BF, Frostad KB, Robertson PB, Levy BM (1977) Immunoglobulin bearing lymphocytes and plasma cells in human periodontal disease. J Periodontal Res 12(1):37–45

Mackler BF, Waldrop TC, Schur P, Robertson PB, Levy BM (1978) IgG subclasses in human periodontal disease. I. Distribution and incidence of IgG subclass bearing lymphocytes and plasma cells. J Periodontal Res 13(2):109–119

Mahanonda R, Champaiboon C, Subbalekha K, Sa-Ard-Iam N, Rattanathammatada W, Thawanaphong S, Rerkyen P, Yoshimura F, Nagano K, Lang NP, Pichyangkul S (2016) Human memory B cells in healthy gingiva, gingivitis, and periodontitis. J Immunol 197(3):715–725. https://doi.org/10.4049/jimmunol.1600540

Malcolm J, Awang RA, Oliver-Bell J, Butcher JP, Campbell L, Adrados Planell A, Lappin DF, Fukada SY, Nile CJ, Liew FY, Culshaw S (2015) IL-33 exacerbates periodontal disease through induction of RANKL. J Dent Res 94(7):968–975. https://doi.org/10.1177/0022034515577815

Mallison SM 3rd, Smith JP, Schenkein HA, Tew JG (1991) Accumulation of plasma cells in inflamed sites: effects of antigen, nonspecific microbial activators, and chronic inflammation. Infect Immun 59(11):4019–4025

Mauri C, Menon M (2017) Human regulatory B cells in health and disease: therapeutic potential. J Clin Invest 127(3):772–779. https://doi.org/10.1172/JCI85113

Mitsdoerffer M, Lee Y, Jager A, Kim HJ, Korn T, Kolls JK, Cantor H, Bettelli E, Kuchroo VK (2010) Proinflammatory T helper type 17 cells are effective B-cell helpers. Proc Natl Acad Sci U S A 107(32):14292–14297. https://doi.org/10.1073/pnas.1009234107

Mizutani Y, Tsuge S, Takeda H, Hasegawa Y, Shiogama K, Onouchi T, Inada K, Sawasaki T, Tsutsumi Y (2014) In situ visualization of plasma cells producing antibodies reactive to Porphyromonas gingivalis in periodontitis: the application of the enzyme-labeled antigen method. Mol Oral Microbiol 29(4):156–173. https://doi.org/10.1111/omi.12052

Montecino-Rodriguez E, Dorshkind K (2012) B-1 B cell development in the fetus and adult. Immunity 36(1):13–21. https://doi.org/10.1016/j.immuni.2011.11.017

Moura RA, Quaresma C, Vieira AR, Goncalves MJ, Polido-Pereira J, Romao VC, Martins N, Canhao H, Fonseca JE (2017) B-cell phenotype and IgD-CD27-memory B cells are affected by TNF-inhibitors and tocilizumab treatment in rheumatoid arthritis. PLoS One 12(9):e0182927. https://doi.org/10.1371/journal.pone.0182927

Moutsopoulos NM, Kling HM, Angelov N, Jin W, Palmer RJ, Nares S, Osorio M, Wahl SM (2012) Porphyromonas gingivalis promotes Th17 inducing pathways in chronic periodontitis. J Autoimmun 39(4):294–303. https://doi.org/10.1016/j.jaut.2012.03.003

Mueller CG, Hess E (2012) Emerging Functions of RANKL in Lymphoid Tissues. Front Immunol 3:261. https://doi.org/10.3389/fimmu.2012.00261

Na H, Cho M, Chung Y (2016) Regulation of Th2 cell immunity by dendritic cells. Immune Netw 16(1):1–12. https://doi.org/10.4110/in.2016.16.1.1

Nakajima T, Ueki-Maruyama K, Oda T, Ohsawa Y, Ito H, Seymour GJ, Yamazaki K (2005) Regulatory T-cells infiltrate periodontal disease tissues. J Dent Res 84(7):639–643. https://doi.org/10.1177/154405910508400711

Nakajima T, Amanuma R, Ueki-Maruyama K, Oda T, Honda T, Ito H, Yamazaki K (2008) CXCL13 expression and follicular dendritic cells in relation to B-cell infiltration in periodontal disease tissues. J Periodontal Res 43(6):635–641. https://doi.org/10.1111/j.1600-0765.2008.01042.x

Okada H, Kida T, Yamagami H (1983) Identification and distribution of immunocompetent cells in inflamed gingiva of human chronic periodontitis. Infect Immun 41(1):365–374

Okamoto K, Nakashima T, Shinohara M, Negishi-Koga T, Komatsu N, Terashima A, Sawa S, Nitta T, Takayanagi H (2017) Osteoimmunology: the conceptual framework unifying the immune and skeletal systems. Physiol Rev 97(4):1295–1349. https://doi.org/10.1152/physrev.00036.2016

Okui T, Aoki Y, Ito H, Honda T, Yamazaki K (2012) The presence of IL-17+/FOXP3+ double-positive cells in periodontitis. J Dent Res 91(6):574–579. https://doi.org/10.1177/0022034512446341

Oliver-Bell J, Butcher JP, Malcolm J, MacLeod MK, Adrados Planell A, Campbell L, Nibbs RJ, Garside P, McInnes IB, Culshaw S (2015) Periodontitis in the absence of B cells and specific anti-bacterial antibody. Mol Oral Microbiol 30(2):160–169. https://doi.org/10.1111/omi.12082

Olivieri CV, Raybaud H, Tonoyan L, Abid S, Marsault R, Chevalier M, Doglio A, Vincent-Bugnas S (2020) Epstein-Barr virus-infected plasma cells in periodontitis lesions. Microb Pathog 143:104128. https://doi.org/10.1016/j.micpath.2020.104128

Ota Y, Niiro H, Ota S, Ueki N, Tsuzuki H, Nakayama T, Mishima K, Higashioka K, Jabbarzadeh-Tabrizi S, Mitoma H, Akahoshi M, Arinobu Y, Kukita A, Yamada H, Tsukamoto H, Akashi K (2016) Generation mechanism of RANKL(+) effector memory B cells: relevance to the pathogenesis of rheumatoid arthritis. Arthritis Res Ther 18:67. https://doi.org/10.1186/s13075-016-0957-6

Page RC, Schroeder HE (1976) Pathogenesis of inflammatory periodontal disease. A summary of current work. Lab Investig 34(3):235–249

Panda S, Ding JL (2015) Natural antibodies bridge innate and adaptive immunity. J Immunol 194(1):13–20. https://doi.org/10.4049/jimmunol.1400844

Parachuru VPB, Coates DE, Milne TJ, Rich AM, Seymour GJ (2018) FoxP3(+) regulatory T cells, interleukin 17 and mast cells in chronic inflammatory periodontal disease. J Periodontal Res 53(4):622–635. https://doi.org/10.1111/jre.12552

Parisi L, Bassani B, Tremolati M, Gini E, Farronato G, Bruno A (2017) Natural killer cells in the orchestration of chronic inflammatory diseases. J Immunol Res 2017:4218254. https://doi.org/10.1155/2017/4218254

Parra D, Rieger AM, Li J, Zhang YA, Randall LM, Hunter CA, Barreda DR, Sunyer JO (2012) Pivotal advance: peritoneal cavity B-1 B cells have phagocytic and microbicidal capacities and present phagocytosed antigen to CD4+ T cells. J Leukoc Biol 91(4):525–536. https://doi.org/10.1189/jlb.0711372

Perlot T, Penninger JM (2012) Development and function of murine B cells lacking RANK. J Immunol 188(3):1201–1205. https://doi.org/10.4049/jimmunol.1102063

Pers JO, Jamin C, Predine-Hug F, Lydyard P, Youinou P (1999) The role of CD5-expressing B cells in health and disease (review). Int J Mol Med 3(3):239–245

Pers JO, Berthou C, Porakishvili N, Burdjanadze M, Le Calvez G, Abgrall JF, Lydyard PM, Youinou P, Jamin C (2002) CD5-induced apoptosis of B cells in some patients with chronic lymphocytic leukemia. Leukemia 16(1):44–52. https://doi.org/10.1038/sj.leu.2402327

Peterson RA (2012) Regulatory T-cells: diverse phenotypes integral to immune homeostasis and suppression. Toxicol Pathol 40(2):186–204. https://doi.org/10.1177/0192623311430693

Popi AF (2015) B-1 phagocytes: the myeloid face of B-1 cells. Ann N Y Acad Sci 1362:86–97. https://doi.org/10.1111/nyas.12814

Popi AF, Longo-Maugeri IM, Mariano M (2016) An overview of B-1 cells as antigen-presenting cells. Front Immunol 7:138. https://doi.org/10.3389/fimmu.2016.00138

Prieto JMB, Felippe MJB (2017) Development, phenotype, and function of non-conventional B cells. Comp Immunol Microbiol Infect Dis 54:38–44. https://doi.org/10.1016/j.cimid.2017.08.002

Pritzker LB, Scatena M, Giachelli CM (2004) The role of osteoprotegerin and tumor necrosis factor-related apoptosis-inducing ligand in human microvascular endothelial cell survival. Mol Biol Cell 15(6):2834–2841. https://doi.org/10.1091/mbc.E04-01-0059

Pugliese LS, Goncalves TO, Popi AF, Mariano M, Pesquero JB, Lopes JD (2012) B-1 lymphocytes differentiate into functional osteoclast-like cells. Immunobiology 217(3):336–344. https://doi.org/10.1016/j.imbio.2011.07.014

Reinhardt RA, Bolton RW, McDonald TL, DuBois LM, Kaldahl WB (1988) In situ lymphocyte subpopulations from active versus stable periodontal sites. J Periodontol 59(10):656–670. https://doi.org/10.1902/jop.1988.59.10.656

Rincon-Arevalo H, Sanchez-Parra CC, Castano D, Yassin L, Vasquez G (2016) Regulatory B cells and mechanisms. Int Rev Immunol 35(2):156–176. https://doi.org/10.3109/08830185.2015.1015719

Rosser EC, Mauri C (2015) Regulatory B cells: origin, phenotype, and function. Immunity 42(4):607–612. https://doi.org/10.1016/j.immuni.2015.04.005

Rothstein TL, Griffin DO, Holodick NE, Quach TD, Kaku H (2013) Human B-1 cells take the stage. Ann N Y Acad Sci 1285:97–114. https://doi.org/10.1111/nyas.12137

Sabarish R, Rao SR, Lavu V (2016) Natural T regulatory cells (n Treg) in the peripheral blood of healthy subjects and subjects with chronic periodontitis – a pilot study. J Clin Diagn Res 10(3):ZC36-39. https://doi.org/10.7860/JCDR/2016/15449.7446

Sakai J, Akkoyunlu M (2017) The role of BAFF system molecules in host response to pathogens. Clin Microbiol Rev 30(4):991–1014. https://doi.org/10.1128/CMR.00046-17

Schmidt J, Jentsch H, Stingu CS, Sack U (2014) General immune status and oral microbiology in patients with different forms of periodontitis and healthy control subjects. PLoS One 9(10):e109187. https://doi.org/10.1371/journal.pone.0109187

Seymour GJ, Gemmell E (2001) Cytokines in periodontal disease: where to from here? Acta Odontol Scand 59(3):167–173

Seymour GJ, Powell RN, Davies WI (1979) Conversion of a stable T-cell lesion to a progressive B-cell lesion in the pathogenesis of chronic inflammatory periodontal disease: an hypothesis. J Clin Periodontol 6(5):267–277

Shi T, Jin Y, Miao Y, Wang Y, Zhou Y, Lin X (2020) IL-10 secreting B cells regulate periodontal immune response during periodontitis. Odontology 108(3):350–357. https://doi.org/10.1007/s10266-019-00470-2

Sindhava VJ, Bondada S (2012) Multiple regulatory mechanisms control B-1 B cell activation. Front Immunol 3:372. https://doi.org/10.3389/fimmu.2012.00372

Sobkowiak MJ, Davanian H, Heymann R, Gibbs A, Emgard J, Dias J, Aleman S, Kruger-Weiner C, Moll M, Tjernlund A, Leeansyah E, Sallberg Chen M, Sandberg JK (2019) Tissue-resident MAIT cell populations in human oral mucosa exhibit an activated profile and produce IL-17. Eur J Immunol 49(1):133–143. https://doi.org/10.1002/eji.201847759

Sojod B, Chateau D, Mueller CG, Babajko S, Berdal A, Lezot F, Castaneda B (2017) RANK/RANKL/OPG signalization implication in periodontitis: new evidence from a RANK transgenic mouse model. Front Physiol 8:338. https://doi.org/10.3389/fphys.2017.00338

Song L, Dong G, Guo L, Graves DT (2017) The function of dendritic cells in modulating the host response. Mol Oral Microbiol. https://doi.org/10.1111/omi.12195

Stashenko P, Goncalves RB, Lipkin B, Ficarelli A, Sasaki H, Campos-Neto A (2007) Th1 immune response promotes severe bone resorption caused by Porphyromonas gingivalis. Am J Pathol 170(1):203–213. https://doi.org/10.2353/ajpath.2007.060597

Su EW, Lin JY, Kane LP (2008) TIM-1 and TIM-3 proteins in immune regulation. Cytokine 44(1):9–13. https://doi.org/10.1016/j.cyto.2008.06.013

Sugawara M, Yamashita K, Yoshie H, Hara K (1992) Detection of, and anti-collagen antibody produced by, CD5-positive B cells in inflamed gingival tissues. J Periodontal Res 27(5):489–498

Tada H, Matsuyama T, Nishioka T, Hagiwara M, Kiyoura Y, Shimauchi H, Matsushita K (2016) Porphyromonas gingivalis gingipain-dependently enhances IL-33 production in human gingival epithelial cells. PLoS One 11(4):e0152794. https://doi.org/10.1371/journal.pone.0152794

Tedder TF (2015) B10 cells: a functionally defined regulatory B cell subset. J Immunol 194(4):1395–1401. https://doi.org/10.4049/jimmunol.1401329

Theoleyre S, Wittrant Y, Tat SK, Fortun Y, Redini F, Heymann D (2004) The molecular triad OPG/RANK/RANKL: involvement in the orchestration of pathophysiological bone remodeling. Cytokine Growth Factor Rev 15(6):457–475. https://doi.org/10.1016/j.cytogfr.2004.06.004

Thorbert-Mros S, Larsson L, Berglundh T (2015) Cellular composition of long-standing gingivitis and periodontitis lesions. J Periodontal Res 50(4):535–543. https://doi.org/10.1111/jre.12236

Titanji K, Vunnava A, Sheth AN, Delille C, Lennox JL, Sanford SE, Foster A, Knezevic A, Easley KA, Weitzmann MN, Ofotokun I (2014) Dysregulated B

cell expression of RANKL and OPG correlates with loss of bone mineral density in HIV infection. PLoS Pathog 10(10):e1004497. https://doi.org/10.1371/journal.ppat.1004497
Vadasz Z, Haj T, Kessel A, Toubi E (2013) B-regulatory cells in autoimmunity and immune mediated inflammation. FEBS Lett 587(13):2074–2078. https://doi.org/10.1016/j.febslet.2013.05.023
Wang Y, Yu X, Lin J, Hu Y, Zhao Q, Kawai T, Taubman MA, Han X (2017) B10 cells alleviate periodontal bone loss in experimental periodontitis. Infect Immun. https://doi.org/10.1128/IAI.00335-17
Weitzmann MN (2017) Bone and the immune system. Toxicol Pathol 45(7):911–924. https://doi.org/10.1177/0192623317735316
Weller S, Descatoire M (2015) IgM+IgD+CD27+ B cells in human: an essential role in the protection against encapsulated bacteria. Med Sci (Paris) 31(6–7):647–653. https://doi.org/10.1051/medsci/20153106018
Weller S, Braun MC, Tan BK, Rosenwald A, Cordier C, Conley ME, Plebani A, Kumararatne DS, Bonnet D, Tournilhac O, Tchernia G, Steiniger B, Staudt LM, Casanova JL, Reynaud CA, Weill JC (2004) Human blood IgM "memory" B cells are circulating splenic marginal zone B cells harboring a prediversified immunoglobulin repertoire. Blood 104(12):3647–3654. https://doi.org/10.1182/blood-2004-01-0346
Werner-Favre C, Vischer TL, Wohlwend D, Zubler RH (1989) Cell surface antigen CD5 is a marker for activated human B cells. Eur J Immunol 19(7):1209–1213. https://doi.org/10.1002/eji.1830190709
Wu YC, Kipling D, Dunn-Walters DK (2011) The relationship between CD27 negative and positive B cell populations in human peripheral blood. Front Immunol 2:81. https://doi.org/10.3389/fimmu.2011.00081
Wu D, Zhou P, Cao F, Lin Z, Liang D, Zhou X (2020) Expression profiling and cell type classification analysis in periodontitis reveal dysregulation of multiple lncRNAs in plasma cells. Front Genet 11:382. https://doi.org/10.3389/fgene.2020.00382
Xu A, Liu Y, Chen W, Wang J, Xue Y, Huang F, Rong L, Lin J, Liu D, Yan M, Li QZ, Li B, Song J, Olsen N, Zheng SG (2016) TGF-beta-induced regulatory T cells directly suppress B cell responses through a noncytotoxic mechanism. J Immunol 196(9):3631–3641. https://doi.org/10.4049/jimmunol.1501740
Youinou P (2007) B cell conducts the lymphocyte orchestra. J Autoimmun 28(2-3):143–151. https://doi.org/10.1016/j.jaut.2007.02.011
Youinou P, Renaudineau Y (2007) The paradox of CD5-expressing B cells in systemic lupus erythematosus. Autoimmun Rev 7(2):149–154. https://doi.org/10.1016/j.autrev.2007.02.016
Younes R, Ghorra C, Khalife S, Igondjo-Tchen-Changotade S, Yousfi M, Willig C, Senni K, Godeau G, Naaman N (2009) Pertinent cell population to characterize periodontal disease. Tissue Cell 41(2):141–150. https://doi.org/10.1016/j.tice.2008.09.003
Yu P, Hu Y, Liu Z, Kawai T, Taubman MA, Li W, Han X (2017) Local induction of B cell interleukin-10 competency alleviates inflammation and bone loss in ligature-induced experimental periodontitis in mice. Infect Immun 85(1). https://doi.org/10.1128/IAI.00645-16
Zamboni DS, Popi AF, Osugui L, Perez KR, Longo-Maugéri IM, Mariano M (2012) Could a B-1 cell derived phagocyte "be one" of the peritoneal macrophages during LPS-driven inflammation? PLoS One 7(3). https://doi.org/10.1371/journal.pone.0034570
Zheng M, Xing C, Xiao H, Ma N, Wang X, Han G, Chen G, Hou C, Shen B, Li Y, Wang R (2014) Interaction of CD5 and CD72 is involved in regulatory T and B cell homeostasis. Immunol Investig 43(7):705–716. https://doi.org/10.3109/08820139.2014.917096
Zouali M (2017) The emerging roles of B cells as partners and targets in periodontitis. Autoimmunity 50(1):61–70. https://doi.org/10.1080/08916934.2016.1261841

10 Polarization Profiles of T Lymphocytes and Macrophages Responses in Periodontitis

Franco Cavalla and Marcela Hernández

Abstract

Periodontitis is a multifactorial, chronic inflammatory disease affecting the supporting structures of teeth triggered by the complex interactions between a dysbiotic bacterial biofilm and the host's immune response that results in the characteristic loss of periodontal attachment and alveolar bone. The differential phenotypic presentations of periodontitis emerge from inter-individual differences in immune response regulatory mechanisms. The monocyte-macrophage system has a crucial role in innate immunity and the initiation of the T and B lymphocyte adaptive immune responses. Macrophages involve a heterogeneous cell population that shows wide plasticity and differentiation dynamics. In response to the inflammatory milieu, they can skew at the time of TLR ligation to predominant M1 –pro-inflammatory- or M2 –anti-inflammatory/healing- functional phenotypes. The perpetuation of inflammation by M1 macrophages leads to the recruitment of the adaptive immune response, promoting Th1, Th17, and Th22 differentiation, which are directly associated with periodontal breakdown. In contrast, M2 macrophages induce Th2 and Treg responses which are associated with periodontal homeostasis. In this article, we review the recent advances comprising the role of macrophages and lymphocyte polarization profiles and their reprogramming as potential therapeutic strategies. For this purpose, we reviewed the available literature targeting periodontitis, macrophage, and lymphocyte subpopulations with an emphasis in the later 5 years. The active reprogramming of macrophages and lymphocytes polarization cross-talk opens a promising area for therapeutic development.

Keywords

Periodontitis · Immunity · Inflammation · Macrophages · Lymphocytes · Polarization

Abbreviations

A. actinomycetemcomitans	Aggregatibacter actinomycetemcomitans
ADP	Adenosine diphosphate

F. Cavalla
Conservative Dentistry Department, Faculty of Dentistry, University of Chile, Santiago, Chile
e-mail: icavalla@odontologia.uchile.cl

M. Hernández (✉)
Laboratory of Periodontal Biology and Department of Oral Pathology and Medicine, Faculty of Dentistry, University of Chile, Santiago, Chile
e-mail: mhernandezrios@odontologia.uchile.cl

J. Santi-Rocca (ed.), *Periodontitis*, Advances in Experimental Medicine and Biology 1373,
https://doi.org/10.1007/978-3-030-96881-6_10

AhR	Aryl hydrocarbon receptor
Akt	Protein kinase B
AMP	Adenosine monophosphate
APC	Antigen-presenting cells
ARG-1	Arginase 1
ATP	Adenosine triphosphate
C5a	Complement component 5a
C5aR	Complement component 5a receptor
CCL	CC chemokine ligand
CCR	CC chemokine receptor
CD	Cluster of differentiation
CTLA-4	Cytotoxic T-lymphocyte-associated protein 4
DAMPs	Endogenous danger-associated molecular patterns
DNA	Deoxyribonucleic acid
Foxp3	Forkhead box P3
GITR	Glucocorticoid-induced tumour necrosis factor receptor family related protein
HLA-DR	Human Leukocyte Antigen – DR isotype
IFN-γ	Interferon-gamma
IL	Interleukin
IL-1RA	Interleukin −1 receptor antagonist
iNOS	Inducible nitric oxide synthase
IRF	Interferon regulatory factor
iTregs	Induced Tregs
JMJD3	Jumonji domain-containing protein D3
JNK	c-Jun N-terminal Kinase
LPS	Lipopolysaccharide
M-CSF	Macrophage colony stimulating factor
MHC	Major Histocompatibility Complex
MMPs	Matrix metalloproteinases
MSCs	Mesenchymal stem cells
NF-κB	Nuclear factor kappa-light-chain-enhancer of activated B cells
NK	Natural killer
NKT	Natural killer T cells
nTregs	Natural Tregs
OPG	Osteoprotegerin
P. gingivalis	Porphyromonas gingivalis
PAMPs	Pathogen-associated molecular patterns
PDL	Programmed Death-ligand
PPARγ	Peroxisome proliferator-activated receptors
PRRs	Pattern recognition receptors particularly
RANKL	Receptor activator of NF-kappa B ligand
RORγT	Retinoic acid-related orphan receptor gamma t
STAT1	Signal transducer and activator of transcription 1
T2DM	Type 2 diabetes mellitus
T-bet	T-box expressed in T cells
TCR	T cell receptor
TET	Ten-eleven translocation
TGF-β	Transforming growth factor
Th	T helper
TLR	Toll-like receptors
TM	Memory T cell subsets
TNF-α	Tumor necrosis factor-alpha
TRAILDR5	Tumor necrosis factor related apoptosis-inducing ligand-death receptor 5
Tregs	Regulatory T-cells
VEGF	Vascular endothelial growth factor

Highlights

- A dysregulated immune response is the ultimate determinant of periodontal tissue destruction and alveolar bone resorption.
- Active periodontal lesions associate with the predominance of M1 macrophages and Th1/Th17 lymphocytes over M2 macrophages and Th2/Treg lymphocyte subpopulations.
- Th22 and activated T memory lymphocytes can further exert pro-inflammatory and bone resorptive effects.
- Intervention strategies aiming at an active reprogramming of macrophage and/or lymphocyte subpopulations represent a promising therapeutic approach

Considerations for Practice

- An intimate understanding of the cellular processes underlying the clinical evolution of periodontitis is essential to guide therapeutic interventions.
- The development of molecular biology, material science, and nanotechnology promise the arrival of novel interventions for periodontitis in the future

Patient Summary

Periodontitis is the inflammatory destruction of tooth-supporting tissues, leading to loosening tooth. It initiates as a response to bacterial plaque, but its evolution is the result of the deregulation of the host's immune system, which is influenced by genes and the environment. In this sense, periodontitis shares similar disease mechanisms with other non-communicable diseases, such as diabetes and cardiovascular disease. Periodontitis requires long-term therapeutic management by a competent professional. This review discusses the detailed immune mechanisms that cause the disease and serves as an update for the interested clinician to assess cutting-edge evidence in the field.

10.1 Introduction

Periodontitis is a multifactorial, chronic inflammatory disease that affects the supporting structures of teeth, – gingiva, alveolar bone, periodontal ligament, and cementum- that can lead to tooth loss and systemic inflammation. The disease results from the complex interactions between a dysbiotic bacterial biofilm and the disrupted host's immune response with the loss of periodontal homeostasis (Almubarak et al. 2020; Garaicoa-Pazmino et al. 2019).

Periodontitis is triggered by an infectious challenge to periodontal tissues. Extensive evidence demonstrates that the presence of bacteria is the necessary stimulus for disease initiation (Hajishengallis 2015), though the pathological changes that occur in the periodontal tissues are almost entirely caused by the immune and inflammatory mechanisms of infection control (Cavalla et al. 2021). These immune mechanisms are in turn modulated by a plethora of host and environmental factors that create a complex interplay between defense mechanisms, tissue homeostasis, and infecting microorganisms. Accordingly, periodontitis is a chronic non-communicable disease, associated with a dysbiotic microbiota, but ultimately caused by the host's immune-inflammatory response (Jepsen et al. 2020).

The immune response, independently of its nature, degree, extension, and regulation will always be incapable to eradicate the infecting periodontal microorganisms located in the subgingival periodontal biofilm, as they are inaccessible to the microbicide effector mechanisms of immunity. The differential phenotypic presentations of the diseases are not a result of a differential efficiency of immunity in terms of infection eradication, but they emerge from inter-individual differences in immune response regulatory mechanisms (Nibali et al. 2019; Cavalla et al. 2018). In this context, the understanding of the polarization and balance of macrophages, different T cell subtypes, and their interactions during the immune cell response is essential to explain the fate of periodontitis.

In this article, we review the recent advances comprising the role of macrophages and lymphocyte polarization profiles and their crosstalk in periodontitis pathogenesis. For this purpose, we reviewed the available literature targeting periodontitis, macrophage, and lymphocyte subpopulations with an emphasis in the later 5 years.

10.2 Macrophage Polarization Profiles in Periodontitis

The monocyte-macrophage system has a central role in innate immunity and the initiation of the adaptive immune response. Macrophages, as resident cells or monocyte-derived cells recruited upon inflammation, can effectively recognize microbes using a repertoire of pattern recognition receptors (PRRs), which include Toll-like receptors (TLRs) that recognize distinct classes of pathogen-associated molecular patterns (PAMPs) from microbes or endogenous danger-associated molecular patterns (DAMPs) released from necrotic or dying host cells, leading to macrophage activation, phagocytosis, antigen presentation, activation of adaptive immunity, and release of inflammatory mediators (Holden et al. 2014; Sima and Glogauer 2013).

Macrophages involve a heterogeneous cell population that shows wide plasticity and differentiation dynamics (Sima and Glogauer 2013). Macrophages will polarize to predominant M1 or M2 functional phenotypes at the time of TLR ligation in response to the cytokine milieu. M1 macrophages play an important role in fighting intracellular pathogens and differentiate as a result from the classical activation pathway triggered mainly by IFN-γ; as result, they release pro-inflammatory cytokines, such as TNF-α, IL-1β, IL-6, IL-12, and IL-23, as well as display increased antigen presentation capabilities (Derlindati et al. 2015). On the other hand, as a result of the exposure to IL-4/IL-13 (M2a), immune complexes (M2b) or IL-10 (M2c), M2 macrophages are involved in immune regulation and tissue repair by secreting the anti-inflammatory cytokines, and IL-10, IL-1RA, and transforming growth factor (TGF)-β; instead they are not efficient in antigen presentation and microbial killing (Ortiz et al. 2015), though in the presence of pathogens they can rapidly switch to the functional M1 phenotype. Macrophage functional phenotypes are not only based on their differential cytokine production, but also their chemokine production, NO metabolism, phagocytosis, and transcriptional profiles (Derlindati et al. 2015). Because *in vivo* M1 and M2 macrophage profiles represent the extremes of a continuum of intermediate cells, human studies have been troubled by the availability of valid phenotypic markers (Ambarus et al. 2012; Barros et al. 2013; Ortiz et al. 2015). To date, flow cytometry and immunofluorescence/immunohistochemistry are the most common methodologies for immune cell subtyping in tissues (Li et al. 2020b), but the former has the advantage to permit the analysis of multiple markers on a single cell.

Macrophages are tissue-resident professional phagocytes and antigen-presenting cells (APC), which differentiate from circulating peripheral blood monocytes. Human monocytes are classified as classical, non-classical, or intermediate monocytes similar to macrophages. Whist classical monocytes can differentiate to macrophages at the sites of injury, non-classical monocytes play a role in vasculature surveillance. They can also accumulate at sites of chronic bacterial infection and produce predominantly anti-inflammatory cytokines. A third set corresponds to intermediate monocytes, which are central players of inflammation and are also referred to as hyperinflammatory monocytes. (Almubarak et al. 2020).

The perpetuation of inflammation by macrophages leads to the recruitment of the adaptive immune response. M1 macrophages promote T helper 1 and T helper 17 responses, known to be tightly associated with periodontal tissue destruction, through the secretion of IL-12 and IL-23, respectively (Hernandez et al. 2011; Alvarez et al. 2019); M2 macrophages, on the other hand, interact with Th2 and Treg cell responses, via IL-4 and transforming growth factor-beta (TGF-β), respectively, which are regarded to be periodontally protective responses. Accordingly, the temporal imbalances in macrophage functional phenotypes may lead to pathologic changes (Derlindati et al. 2015; Sima and Glogauer 2013).

Clinical human studies identifying macrophage subpopulations based on more than one surface marker combined with a pan-macrophage marker are limited. A recent study reported larger populations of intermediate ($CD14^+$ $CD16^+$) and non-classical ($CD14^-$ $CD16^+$) monocytes in periodontitis individuals compared to healthy ones, showing higher HLA-DR and PDL-1 expression, suggestive of pro-inflammatory profile and activation status, especially in the former, which are the most pro-inflammatory subtype of monocytes. Also, M1 macrophages ($CD80^+$ $MHCII^+$) were higher and M2 macrophages ($CD206^+$ $CD163^+$) were lower in periodontitis-affected gingival tissues compared to healthy controls, supporting a disruption in macrophage homeostasis towards a pro-inflammatory phenotype. M1 macrophages in periodontitis individuals also overexpressed PDL-1, which delivers inhibitory signals that regulate the balance among T-cell activation and tolerance, and CD47, a glycoprotein of the immunoglobulin family that plays a critical role in self-recognition and impairment of phagocytosis (Almubarak et al. 2020). Also, high numbers of M2 macrophages have been reported in healthy tissues compared to periodontitis (Li et al. 2020a). Similar results, as well as higher intermediate monocyte subpopulations expressing high levels of PDL-1, were identified in periodontitis versus healthy sites in individuals with type 2 diabetes mellitus (T2DM). Also, elevated levels of the pro-inflammatory transcription factors, STAT1 and IRF1, and repressed of the anti-inflammatory factor JMJD3 were identified in circulating lymphocytes (Almubarak et al. 2020). Thus, macrophage-mediated regulation of chronic inflammation and especially M1 polarization might play a role in the pathogenesis of periodontitis, diabetic complications, and may explain the higher susceptibility of these individuals to periodontitis.

In line with the previous findings, a study in periapical periodontitis, which shares similar pathogenic mechanisms with periodontitis that lead to the destruction of the peri-radicular periodontal tissues, reported a higher macrophage M1/M2 ratio ($CD14^+$ $CD64^+$ $CD80^+$ / $CD14^+$ $CD163^+$ $CD206^+$) in symptomatic apical lesions compared to asymptomatic apical lesions and healthy periodontal ligament controls, along with significantly higher M1 cytokines, such as IL-6 and IL-23. Interestingly, CD163 protein was expressed with lower intensity in the symptomatic lesions. In tissue macrophages, CD163 transduces signals that lead to immunomodulation through IL-10 and IL-1RA and the inhibition of pro-inflammatory cytokines (Alvarado-Vazquez et al. 2017). The latter suggests that macrophages might reprogram themselves from the M2 phenotype towards M1 and trigger clinical symptoms and periodontal destruction (Veloso et al. 2020). It is proposed that periodontal homeostasis might be associated with tissue remodeling by adult tissue-resident macrophages, whilst periodontitis might be associated with tissue infiltration by monocyte-derived cells that would primarily assist in host defense (Epelman et al. 2014). Higher total and polarized ($iNOS^+$/M1 and $CD206^+$/M2) macrophages counts were reported in gingivitis, compared to periodontitis and healthy tissue samples (Garaicoa-Pazmino et al. 2019). Accordingly, the periodontally destructive phenotype might be determined predominantly by an M1/M2 macrophages imbalance rather than the quantity of total or polarized macrophages.

In vitro studies in monocytic human cell lines (THP-1) demonstrated that oral commensal bacteria primarily elicit an M2 phenotype, while the challenge of macrophages that have been driven towards an M1 phenotype with INF-γ and lipopolysaccharide (LPS) with commensal bacteria appears to provide some modulation of the inflammatory nature of the M1 cells. On the contrary, keystone periodontal pathogens *P. gingivalis* and especially *A. actinomycetemcomitans*, induced polarization predominantly towards an M1 phenotype. The induction of M1 cells, particularly through ligation of TLRs, triggers signaling cascades through the inflammatory master switch transcription factor NF-κB (Huang et al. 2016). A relevant role of ten-eleven translocation (TET)-1 proteins, which demethylates DNA, has been proposed in *P. gingivalis* LPS/IFN-γ-induced M1 macrophage polarization through the NF-κB pathway in THP-1 monocytic cells (Huang et al. 2019). Functionally, naïve, M1 and M2 murine macrophages (RAW 264.7 cell line) have shown different abilities to phagocyte and clear *P. gingivalis*. Among them, only M1 macrophages were able to produce a respiratory burst and kill the bacteria from the phagosome after 24 hours and induced high levels of TNF-α, IL-12, and iNOS. Also, gingipain extracts plus *P. gingivalis*-LPS-treated RAW264.7 in macrophages enhanced the expression of IL-12 and IL-23, while gingipain alone increased IL-12, IL-23, iNOS, TNF-α, IL-1β, and IL-6, but not IL-10; though only *P. gingivalis*-LPS increased CD86$^+$ M1 over CD206$^+$ M2 macrophages. M1 phenotype differentiation to *P. gingivalis* challenge has been associated with the engagement of C5a (Hou et al. 2017), which can mediate C5aR-TLR crosstalk to disengage bacterial clearance, and suppress of α-ketoglutarate metabolic pathway (Yu et al. 2018). Additionally, experimental *P. gingivalis*-LPS injection in mice supports that M1 macrophage infiltration of gingiva and reactive oxygen species synthesis are modulated by uncoupling protein-2 (Yan et al. 2020), whereas exposure to high levels of bacteria resulted in cell apoptosis (Lam et al. 2016).

Employing classically activated (M1) and alternatively activated (M2) murine bone-marrow-derived macrophages, immunologic activation of macrophages before *P. gingivalis* challenge induced the expression of TLR4 and CD14, although the cytokine profiles were predominantly anti-inflammatory (IL-10) in M1-skewed macrophages and proinflammatory (IL-6 and TNF-α) in M2 skewed macrophages (Papadopoulos et al. 2017). Similarly, *P. gingivalis*-LPS stimulation during the polarization phase of THP-1 cells resulted in lower TNF-α, IL-1β upon secondary LPS treatment (Belfield et al. 2017). Altogether, these differences might rely on specific monocyte/macrophage cells and protocols of macrophage differentiation, priming, and bacterial exposure used. Though it can be generally accepted that *P. gingivalis* elicits a predominantly M1 macrophage response, the inflammatory milieu encountered by macrophages before pathogen interaction might impact the quality and quantity of the response (Papadopoulos et al. 2017).

Osteoclasts are mainly derived from monocyte/macrophages by their exposure to M-CSF and RANKL. M2 (CD206$^+$) subpopulations were suggested to have a stronger osteoclastogenenic potential than M1 (CD86$^+$) subpopulations, based on their enhanced osteoclast differentiation, acting reings formation and bone-resorptive activity in mouse bone marrow cells, due to low expression of the transcription factor IRF5, which induces M1 differentiation (Yang et al. 2019). Likewise, the addition of M1 macrophages significantly suppressed RANKL-induced osteoclastogenesis in RAW264.7 cells or bone marrow cells, compared to non-stimulated (M0) and M2 macrophages; Similarly, adoptive-transfer of M1 macrophages *in vivo* using a ligature-induced periodontitis mouse model resulted in lower alveolar bone loss and TRAP$^+$ osteoclasts when compared to M0 or M2 macrophage transfer, which might be mediated by the secretion of soluble IFN-γ and IL-12 (Yamaguchi et al. 2016). The M1 transcription factors IRF5, IRF8, and STAT1were able to inhibit osteoclast differentiation, while the M2 factors, IRF4 and PPARγ, are known to increase osteoclast differentiation (Yang et al. 2019). Conversely, another study showed that M1 macrophage infiltration was associated with periodontal inflammation and bone destruction in experimentally-induced periodontitis. Moreover, M1 macrophage-derived conditioned medium suppressed osteoblastogenesis while increasing osteoclastogenesis, probably via a paracrine effect (Zhu et al. 2019). M2 cells, on the other hand, promoted osteogenic differentiation, as well as increased cementoblastic differentiation when bone marrow mononuclear cells secreting IL-10 and vascular endothelial growth factor (VEGF) were co-cultured with periodontal ligament stem cells, showing enhanced alkaline phosphatase activity, via Akt and c-Jun N-terminal Kinase (JNK) signaling pathways (Li et al. 2020b). Despite these contradictions, the cited studies support that macrophage subtypes might impact the osteoclastogenic potential and, that it might be reversed by phenotype switching. Whether M1/M2 skewed macrophages themselves can better differentiate to osteoclasts is controversial, but the paracrine effects of M1 macrophages appear to promote osteoclastogenesis.

10.3 Lymphocyte Subpopulations in Periodontitis

The recognition of different lineages of Th lymphocytes characterized by distinctive cytokine secretory patterns was a keystone in the understanding of immunological responses and the advancement of the current paradigms of inflammatory disease etiology. The original Th1 and Th2 subtypes of T helper cells were identified in the late 1980s, followed by the rapid recognition of a wide variety of new subtypes, including Th17, Th22, and Treg. All these subtypes have been detected and investigated in periodontal tissues (Li et al. 2020a).

Th1 cells are a subpopulation of T helper cells characterized by the expression of the transcription factor T-bet under the influence of IL-12. The archetypical cytokine product of Th1 lymphocytes is IFN-γ and has been heavily implicated in the activation of macrophage microbicide mechanisms. The Th1 preponderance hypothesis in periodontitis is supported by experimental evidence of the overexpression of IFN-γ in mononuclear cells isolated from periodontitis patients, and by the relative abundance of IFN- γ in gingival crevicular fluid and saliva of periodontitis patients (Medara et al. 2021a; Koidou et al. 2020). Th1 cytokines intensify the activation of neutrophils and macrophages, generating a proinflammatory loop that augments local levels of tissue destructive mediators. The Th1 lineage is responsible for the production of various cytokines that maintain and amplify inflammation, directly or indirectly promoting the recruitment and perpetuity of active immune cells in periodontal tissues. Both IL-1β and TNFα are characteristic secreted products of Th1 lymphocytes (Cavalla et al. 2021; Araujo-Pires et al. 2014). This upregulation in inflammation produces enhanced periodontal destruction. It is worth observing that IFN-γ- mediated Th1 responses are essential to avoid pathogen dissemination and that experimental murine evidence suggests that IFN-γ deficient mice suffer from widespread and often lethal systemic infection after being orally challenged with periodontopathic bacteria (Garlet et al. 2008).

Th2 cells are characterized by the expression of the transcription factor GATA-3 and the secretion of a wide array of cytokines, including IL-4, IL-5, IL-9, IL-13, and IL-25, being IL-4 the main positive feedback signal for Th2 lineage commitment (Cavalla et al. 2021). Th2 mediators seem to cooperate with anti-inflammatory cytokines, given the similarity between the properties of IL-10 and IL-4. Also, there is some evidence linking Th2 polarization with a B-cell cooperative axis, where IL-4 would support the local production of antibodies against periodontopathic bacteria (Pradeep et al. 2008; Zhu and Zhu 2020). Also, IL-4 seems to prime macrophages for M2 polarization, acting as a key mediator to drive the immune response towards a regulatory response (Lam et al. 2016).

Th17 cells are a subpopulation of T helper cells characterized by the expression of the transcription factor RORγT under the influence of IL-21, and able to secrete IL-17 (Bunte and Beikler 2019). Th17 cells are implicated in inflammatory osteolytic conditions, probably via RANKL upregulation and their capacity to enhance neutrophil and macrophage activity (Cavalla et al. 2021). Both Th1 and Th17 subtypes seem to collaborate in sustaining a proinflammatory loop characterized by a strong influx and activation of neutrophil microbicidal activities, secretion of MMPs, and osteoclastogenesis, collectively responsible for periodontal destruction (Monasterio et al. 2018; Chen et al. 2016).

Regulatory T-cells (or Tregs) are a heterogeneous population of $CD4^+$ $CD25^+$ $FoxP3^+$ T helper cells. At least two distinct Treg subtypes exist, thymic-derived natural Tregs (nTregs) and peripherally induced Tregs (iTregs). Both subtypes are characterized by high levels of secretion of IL-10, and of TGF-β which are responsible for their immune and inflammatory suppressor functions (Haribhai et al. 2016). Despite these classical functions, Tregs might eventually promote the gain of Th17-related functions, such as the production of IL-17A, in a chronic inflammatory milieu, particularly in the presence of sustained IL-6 and TGF-β stimuli (Alvarez et al. 2020; Kimura and Kishimoto 2010). There is plenty of evidence that supports the notion of multiple immune and inflammatory paths leading to periodontal destruction by a relative increase of RANKL over OPG. On the other hand, emerging evidence points to a previously unimagined Th2/Treg cooperation to drive periodontal stabilization. Th2 and Treg secretion products are known to inhibit the production of RANKL and other proinflammatory destructive cytokines, moreover, Th2/Treg secreted cytokines may directly or indirectly upregulate OPG, inhibiting lesion progression (Cavalla et al. 2021).

Tregs have a battery of suppressive mechanisms that may function in four distinctive ways: (Tonetti et al. 2013) modulation of antigen-presenting cell (APC) maturation or function, (2) suppression by killing targeted cells, (3) suppression by metabolic disruption, and (4) suppression by inhibitory cytokines (Deng et al. 2019; Barbi et al. 2014). An example of APC inhibition is the Treg expression of CTLA-4, the archetypal inhibitory receptor relative to the T cell costimulatory molecule CD28. While CD28 signaling promotes T cell activation, CTLA-4 suppresses the T cell response by interacting with costimulatory receptors expressed at the APC surface. This contact inhibition leads to anergy by the downregulation and sequestration of costimulatory receptors. Tregs can also induce apoptosis in their target cells through cell contact-dependent cytolysis or via the tumor necrosis factor-related apoptosis-inducing ligand-death receptor 5 pathway (TRAILDR5). Tregs also mediate suppressive metabolic disturbance of T cells by utilization of local IL-2, which limits T cell proliferation. Yet another suppressive mechanism is the expression of cell surface enzymes, CD39 and CD73, which catalyze extracellular ATP hydrolysis to ADP, AMP, and adenosine. Adenosine signals may inhibit APCs as well as activated T lymphocytes by elevation of intracellular cAMP. All these suppressor mechanisms are cooperative and work in conjunction with the secretion of inhibitory cytokines (such as IL-10, IL-35, and TGF-β) (Ferreira et al. 2019; Raffin et al. 2020).

Many studies using animal models of periodontitis have confirmed the importance of Treg suppressive functions during the late stages of the disease. For instance, the inhibition of Tregs with anti-GITR in an *A. actinomycetemcomitans*-induced model of periodontitis showed increased alveolar bone loss associated with the reduction of IL-10, CTLA-4, and TGF-β levels (Garlet et al. 2010). In another murine study, Tregs seemed to cooperate with Th2 cells, where the coexistence and expression of IL-4, Foxp3, and IL-10 correlated with attenuation of osteolysis. In this model, IL-4 induced CCL22 expression that modulated CCR4-dependent Treg migration. Specifically, experimental periodontitis in IL-4 knockout mice shows an almost total reduction of Tregs and CCL22 production/expression. Therefore, Treg functionality is needed to sustain a controlled immune response that might avoid the disease progression or reactivation (Alvarez et al. 2018).

Abundant experimental evidence from animal models has demonstrated that periodontal destruction is strongly associated with a relative predominance of Th1 and Th17 cytokines, whereas Th2 and Treg cytokine products are associated with lesion stability and regression (Vernal et al. 2014; Monasterio et al. 2019b; Araujo-Pires et al. 2015). Similarly, studies from human tissue samples and gingival crevicular fluid have demonstrated a relative abundance of Th1 and Th17 cells and secreted products in progressing periodontal lesions, while Th2 and Treg cells and secreted products predominate in healthy tissues or post periodontal treatment (Bi et al. 2019; Chen et al. 2016; Zhao et al. 2011).

Recent evidence suggests that CCL22 may be the molecular connection between Th2 and Tregs cell subtypes. IL-4 is capable of inducing CCL22 production and secretion, which stimulates Treg migration in a CCR4/CCL22-dependent axis (Araujo-Pires et al. 2015). The transit from an active periodontal lesion to an inactive phenotype requires the synergic contribution of Th2 and Treg cell subtypes. Characteristic Th2/Treg secretion products, such as IL4, IL10, TGF-β are abundant in inactive lesions. Moreover, a low RANKL/OPG ratio seems to be the direct consequence of the relative abundance of Th2/Treg secretion products (Araujo-Pires et al. 2015; Francisconi et al. 2016). It is also significant to notice that other cell populations may collaborate to create a suppressive environment characteristic of stable lesions. For example, mesenchymal stem cells (MSCs) are known to produce IL10 and TGF-β and could be activated to secrete these immune-suppressive cytokines by the molecular environment generated by the Th2/Treg collaborative axis (Araujo-Pires et al. 2014).

More recently, Th22 cells (CD4+ IL-22+ AhR+) have been typified by the production and secretion of IL-22, which can exert pro-inflammatory effects by a synergistic action with classic pro-inflammatory mediators, such as TNF-α and IL-17 (Diaz-Zuniga et al. 2017). IL-22 can upregulate RANKL expression and therefore induce osteoclastogenesis. Murine animal models have demonstrated that higher levels of IL-22, AhR, and RANKL, as well as IL-1β, IL-6, IL-12, IL-17, IL-23, and TNF-α, were expressed in periodontal lesions of infected mice compared with periodontal tissues of sham-infected and non-infected controls. Likewise, high RANKL localization was observed in periodontal tissues of infected mice; however, few RANKL was observed in controls. Furthermore, higher recognition of $CD4^+$ $IL\text{-}22^+$ AhR^+ T lymphocytes was evident in periodontal lesions and draining cervical lymph nodes of infected mice compared with non-infected controls. Finally, the overexpression of IL-22 and RANKL showed a significant positive correlation with bone resorption in experimental periodontal lesions (Monasterio et al. 2019a).

Memory T cell subsets (TM) are also present in gingival tissue and are believed to play a role in the immune surveillance of the gingival mucosal barrier. Two memory T cell populations; a $CD69^-$ recirculating population and a $CD69^+$ gingiva-resident memory T cell population have been identified in healthy gingiva. Among them, $CD4^+$ T cells with transitional memory phenotype ($CD45RA^-$ $CCR7^-$ $CD28^+$ $CD95^+$) constitute the major subset. The invasion of bacteria and the presence of bacterial antigens in deep compartments of the gingival connective tissue has the potential to activate these memory T cells and trigger deleterious inflammatory and immune responses capable of disrupting periodontal homeostasis (Mahanonda et al. 2018; Medara et al. 2021b).

Emerging evidence also suggests a role of myeloid-derived suppressor cells (MDSC) in periodontitis. MDSC represents a heterogeneous population of myeloid progenitor and immature myeloid cells with potent immune-suppressive activity, which seems to depend on the inflammatory microenvironment, particularly induced by PGE2, IL-1 beta, and IL-6 (Valero-Monroy et al. 2016). MDSC can induce the generation of Th17 and Treg cells in other chronic inflammatory diseases (Nagaraj et al. 2013). Experimental periodontitis induced by *P. gingivalis* promoted the expansion of these immune-suppressive cells, as suggested by the differential expression of the granulocyte and monocyte markers Ly6G and Ly6C, respectively. On the one hand, MDSC exerted an inhibitory effect on the proliferative activity of CD4+ T cells. On the other, CD11b + Ly6G+ Ly6C++ MDSC could differentiate into TRAP+ osteoclasts. Consecutively, MDSC may contribute to the disease by immune suppression but also through bone-resorbing activity (Su et al. 2017). Further studies are needed to elucidate the emerging roles of these cells in periodontitis.

Periodontitis is linked to other chronic/non-communicable diseases (e.g., diabetes mellitus, cardiovascular diseases, obesity), probably by the systemic effects of bacteria and their products in distant body compartments (Sanz et al. 2020; Polak and Shapira 2018; Baeza et al. 2020). Recent evidence points to an effect of periodontal treatment over the polarization pattern of T helper cells in peripheral blood. Interestingly, it has been demonstrated that the majority of $Foxp3^+$ and $IL\text{-}17^+$ cells were $TCR\alpha\beta^+$ $CD4^+$ (true T helper cells), whereas the majority of $IFN\text{-}\gamma^+$ and $IL\text{-}4^+$ cells were $TCR\alpha\beta^+$ $CD4^-$ followed by $TCR\alpha\beta^+$ $CD4^+$ cells. Likely, these $TCR\alpha\beta^+$ CD4 cells are $CD8^+$ T cells, $CD4^+$ $TCR\alpha\beta^-$ cells are NK and NKT cells, and $TCR\alpha\beta^-$ $CD4^-$ cells are γδ T cells, macrophages, basophils, or B cells (Medara et al. 2021b). This is an important point since it has been long been assumed that $IFN\text{-}\gamma^+$, $IL\text{-}4^+$, $IL\text{-}17^+$, or $IL\text{-}10^+$ expression equals Th1, Th2, Th17, and Treg cells, while it now appears that non-canonical cytokine-producing cells may also contribute to peripheral immune dysregulation in periodontitis.

10.4 Immune Reprogramming Intervention Strategies

Overall, the pathogenesis of periodontitis is associated with a higher ratio of proinflammatory M1 (classically activated) macrophages to anti-inflammatory M2 (alternatively activated). Likewise, the pattern of Th response seems to be a critical determinant of periodontal destruction or stability, since such cell types not only regulate the overall immune response pattern but also may directly interfere in the RANKL/OPG balance, which is essential for bone homeostasis (Fig. 10.1). Accordingly, intervention strategies aiming at an active reprogramming of immune polarization phenotypes to promote homeostasis in chronic inflammatory destructive diseases, such as periodontitis, open a promising area for therapeutic development (Alvarez et al. 2019).

Cranberry concentrates (100 μg/mL) down-regulated proinflammatory cytokine expression (IL-8 and IL-6) and M1 ($CD68^+$ and $CCR7^+$) polarization in LPS-stimulated THP-1 macrophages *in vitro,* whereas anti-inflammatory IL-10 and M2 ($CD68^+$ and $CD206^+$) polarization were significantly upregulated after 24 hours of exposure (Galarraga-Vinueza et al. 2020). Also, promising results have been recently reported in experimentally induced periodontitis. C-C motif

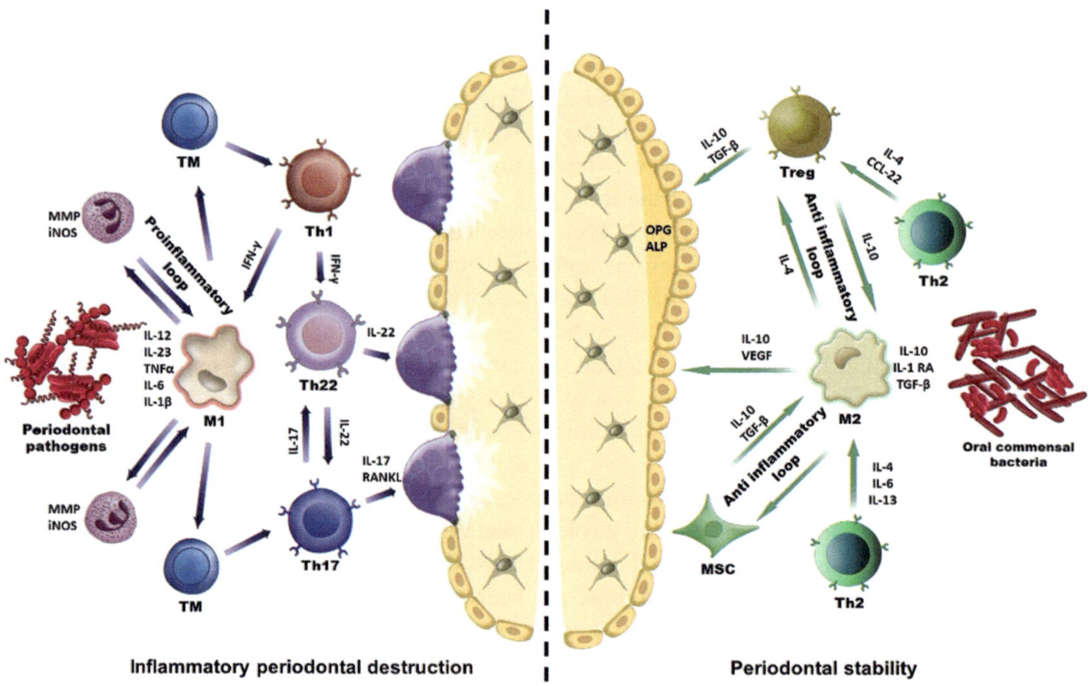

Fig. 10.1 Polarization profiles of T lymphocytes and macrophages responses in periodontitis. Periodontitis is triggered by the complex interactions established between a dysbiotic bacterial biofilm and the host's immune response that results in periodontal supporting tissue loss. In response to periodontal pathogens macrophages polarize predominantly to M1 –pro-inflammatory- functional phenotype, promoting Th1- and Th17 differentiation subtypes, which are directly associated with periodontal destruction. Also, Th22 and activated T memory (TM) lymphocytes contribute to periodontal tissue breakdown through pro-inflammatory and pro-resorptive effects. In contrast, the periodontal milieu in the presence of oral commensal bacteria associates with M2 macrophages as well as Th2 and Treg responses, which are associated with periodontal homeostasis and stability. Noteworthy, IL10 and TGF-β can also be produced by resident mesenchymal stem cells (MSCs) and contribute to local immunosuppression

chemokine ligand 2 (CCL2) was shown to increase the number of M2 macrophages, decreasing TNF-α secretion. CCL2 controlled-release nanoparticles enhanced chemotaxis of RAW264.7 cells and effectively inhibited inflammatory alveolar bone loss in *P. gingivalis*- and ligature-induced periodontitis models. Furthermore, local administration of CCL2-releasing nanoparticles increased the number of M2 polarized macrophages in the *P. gingivalis*-induced periodontitis model. Accordingly, a significant increase in IL-1RA and a decrease in RANKL mRNA expression were observed (Zhuang et al. 2019). Recently, a combination of IL-4, IL-13, and IL-10 was reported to differentiate macrophages into a subset of M2 polarization cells that express much higher levels of CD206, ARG-1 to promote tissue regeneration, and PDL-2 to suppress T cell proliferation. After testing with a mixed lymphocyte reaction assay, this subset of macrophages increased IL-2 secretion and suppressed IL-17, IL-6, and TNF-α secretion. Moreover, injection of these M2 macrophages to *P. gingivalis*-induced periodontitis mice model increased the ratio of regulatory T cells ($CD4^+$ /$Foxp3^+$) and reduced osteoclast differentiation, showing a striking immune homeostatic potential (Miao et al. 2020). These data indicate that manipulation of the macrophage equilibrium could be a viable option in the treatment and/or prevention of periodontitis in susceptible individuals. Also, the macrophage-based nanosystems exhibit great potential for controlled drug delivery in periodontitis treatment, but more studies are needed.

Experimental approaches have already demonstrated the feasibility of harnessing the protective potential of Treg cells by selective recruitment to inflamed periodontal sites using a nano polymer capable of controlled CCL22 liberation (Glowacki et al. 2013). Alternative approaches using specific monoclonal antibodies against key mediators of bone destruction have also demonstrated their utility in the prevention of bone fractures in postmenopausal women and are, at least potentially, a plausible alternative to modulate periodontal destruction (Deeks 2018; Sobacchi et al. 2019). The greatest challenge to transition from experimental procedures to therapeutic interventions is the development of adequate vehicles to deliver bioactive molecules, avoiding unwanted systemic effects while achieving desired outcomes in the periodontal tissues.

10.5 Conclusion and Perspectives

The crosstalk between macrophages and T lymphocytes polarization profiles plays a determinant role in periodontitis. Active periodontal lesions are associated with the predominance of M1 macrophages and Th1/Th17 lymphocytes over M2 macrophages and Th2/Treg lymphocyte subpopulations. The possibility of safe local immune modulation, harnessing the body's natural regulatory capacity to suppress destructive immune responses represents an attractive perspective to be explored as a future approach to revolutionize the clinical management of periodontitis.

Acknowledgments FONDECYT grants #1160741, #1200098 and #1200098; IADR regional development program, Latin American region, and Oral Reconstruction Foundation Grant 42004.

References

Almubarak A, Tanagala KKK, Papapanou PN, Lalla E, Momen-Heravi F (2020) Disruption of monocyte and macrophage homeostasis in periodontitis. Front Immunol 11:330. https://doi.org/10.3389/fimmu.2020.00330

Alvarado-Vazquez PA, Bernal L, Paige CA, Grosick RL, Moracho Vilrriales C, Ferreira DW, Ulecia-Moron C, Romero-Sandoval EA (2017) Macrophage-specific nanotechnology-driven CD163 overexpression in human macrophages results in an M2 phenotype under inflammatory conditions. Immunobiology 222(8–9):900–912. S0171-2985(17)30093-1 [pii]

Alvarez C, Monasterio G, Cavalla F, Cordova LA, Hernandez M, Heymann D, Garlet GP, Sorsa T, Parnanen P, Lee HM, Golub LM, Vernal R, Kantarci A (2019) Osteoimmunology of oral and maxillofacial diseases: translational applications based on biological mechanisms. Front Immunol 10:1664. https://doi.org/10.3389/fimmu.2019.01664

Alvarez C, Rojas C, Rojas L, Cafferata EA, Monasterio G, Vernal R (2018) Regulatory T lymphocytes in

periodontitis: a translational view. Mediat Inflamm 2018:7806912. https://doi.org/10.1155/2018/7806912
Alvarez C, Suliman S, Almarhoumi R, Vega ME, Rojas C, Monasterio G, Galindo M, Vernal R, Kantarci A (2020) Regulatory T cell phenotype and anti-osteoclastogenic function in experimental periodontitis. Sci Rep 10(1):19018. https://doi.org/10.1038/s41598-020-76038-w
Ambarus CA, Krausz S, van Eijk M, Hamann J, Radstake TR, Reedquist KA, Tak PP, Baeten DL (2012) Systematic validation of specific phenotypic markers for in vitro polarized human macrophages. J Immunol Methods 375(1–2):196–206. https://doi.org/10.1016/j.jim.2011.10.013. S0022–1759(11)00294–8 [pii]
Araujo-Pires AC, Biguetti CC, Repeke CE, Rodini Cde O, Campanelli AP, Trombone AP, Letra A, Silva RM, Garlet GP (2014) Mesenchymal stem cells as active prohealing and immunosuppressive agents in periapical environment: evidence from human and experimental periapical lesions. J Endod 40(10):1560–1565. https://doi.org/10.1016/j.joen.2014.02.012
Araujo-Pires AC, Vieira AE, Francisconi CF, Biguetti CC, Glowacki A, Yoshizawa S, Campanelli AP, Trombone AP, Sfeir CS, Little SR, Garlet GP (2015) IL-4/CCL22/CCR4 axis controls regulatory T-cell migration that suppresses inflammatory bone loss in murine experimental periodontitis. J Bone Miner Res 30(3):412–422. https://doi.org/10.1002/jbmr.2376
Baeza M, Morales A, Cisterna C, Cavalla F, Jara G, Isamitt Y, Pino P, Gamonal J (2020) Effect of periodontal treatment in patients with periodontitis and diabetes: systematic review and meta-analysis. J Appl Oral Sci 28:e20190248. https://doi.org/10.1590/1678-7757-2019-0248
Barbi J, Pardoll D, Pan F (2014) Treg functional stability and its responsiveness to the microenvironment. Immunol Rev 259(1):115–139. https://doi.org/10.1111/imr.12172
Barros MH, Hauck F, Dreyer JH, Kempkes B, Niedobitek G (2013) Macrophage polarisation: an immunohistochemical approach for identifying M1 and M2 macrophages. PLoS One 8(11):e80908. https://doi.org/10.1371/journal.pone.0080908
Belfield LA, Bennett JH, Abate W, Jackson SK (2017) Exposure to Porphyromonas gingivalis LPS during macrophage polarisation leads to diminished inflammatory cytokine production. Arch Oral Biol 81:41–47. https://doi.org/10.1016/j.archoralbio.2017.04.021
Bi CS, Sun LJ, Qu HL, Chen F, Tian BM, Chen FM (2019) The relationship between T-helper cell polarization and the RANKL/OPG ratio in gingival tissues from chronic periodontitis patients. Clin Exp Dent Res 5(4):377–388. https://doi.org/10.1002/cre2.192
Bunte K, Beikler T (2019) Th17 cells and the IL-23/IL-17 Axis in the pathogenesis of periodontitis and immune-mediated inflammatory diseases. Int J Mol Sci 20(14):doi:10.3390/ijms20143394
Cavalla F, Biguetti CC, Melchiades JL, Tabanez AP, Azevedo MCS, Trombone APF, Faveri M, Feres M, Garlet GP (2018) Genetic association with subgingival bacterial colonization in chronic periodontitis. Genes (Basel) 9(6):doi:10.3390/genes9060271
Cavalla F, Letra A, Silva RM, Garlet GP (2021) Determinants of periodontal/periapical lesion stability and progression. J Dent Res 100(1):29–36. https://doi.org/10.1177/0022034520952341
Chen XT, Chen LL, Tan JY, Shi DH, Ke T, Lei LH (2016) Th17 and Th1 lymphocytes are correlated with chronic periodontitis. Immunol Investig 45(3):243–254. https://doi.org/10.3109/08820139.2016.1138967
Deeks ED (2018) Denosumab: a review in postmenopausal osteoporosis. Drugs Aging 35(2):163–173. https://doi.org/10.1007/s40266-018-0525-7
Deng G, Song X, Fujimoto S, Piccirillo CA, Nagai Y, Greene MI (2019) Foxp3 post-translational modifications and Treg suppressive activity. Front Immunol 10:2486. https://doi.org/10.3389/fimmu.2019.02486
Derlindati E, Dei Cas A, Montanini B, Spigoni V, Curella V, Aldigeri R, Ardigo D, Zavaroni I, Bonadonna RC (2015) Transcriptomic analysis of human polarized macrophages: more than one role of alternative activation? PLoS One 10(3):e0119751. https://doi.org/10.1371/journal.pone.0119751
Diaz-Zuniga J, Melgar-Rodriguez S, Monasterio G, Pujol M, Rojas L, Alvarez C, Carvajal P, Vernal R (2017) Differential human Th22-lymphocyte response triggered by Aggregatibacter actinomycetemcomitans serotypes. Arch Oral Biol 78:26–33. https://doi.org/10.1016/j.archoralbio.2017.02.008
Epelman S, Lavine KJ, Randolph GJ (2014) Origin and functions of tissue macrophages. Immunity 41(1):21–35. https://doi.org/10.1016/j.immuni.2014.06.013
Ferreira LMR, Muller YD, Bluestone JA, Tang Q (2019) Next-generation regulatory T cell therapy. Nat Rev Drug Discov 18(10):749–769. https://doi.org/10.1038/s41573-019-0041-4
Francisconi CF, Vieira AE, Biguetti CC, Glowacki AJ, Trombone AP, Letra A, Menezes Silva R, Sfeir CS, Little SR, Garlet GP (2016) Characterization of the protective role of regulatory T cells in experimental periapical lesion development and their Chemoattraction manipulation as a therapeutic tool. J Endod 42(1):120–126. https://doi.org/10.1016/j.joen.2015.09.022
Galarraga-Vinueza ME, Dohle E, Ramanauskaite A, Al-Maawi S, Obreja K, Magini R, Sader R, Ghanaati S, Schwarz F (2020) Anti-inflammatory and macrophage polarization effects of Cranberry Proanthocyanidins (PACs) for periodontal and peri-implant disease therapy. J Periodontal Res 55(6):821–829. https://doi.org/10.1111/jre.12773
Garaicoa-Pazmino C, Fretwurst T, Squarize CH, Berglundh T, Giannobile WV, Larsson L, Castilho RM (2019) Characterization of macrophage polarization in periodontal disease. J Clin Periodontol 46(8):830–839. https://doi.org/10.1111/jcpe.13156
Garlet GP, Cardoso CR, Campanelli AP, Garlet TP, Avila-Campos MJ, Cunha FQ, Silva JS (2008) The essential role of IFN-gamma in the control of lethal Aggregatibacter actinomycetemcomitans infection

in mice. Microbes Infect 10(5):489–496. https://doi.org/10.1016/j.micinf.2008.01.010
Garlet GP, Cardoso CR, Mariano FS, Claudino M, de Assis GF, Campanelli AP, Avila-Campos MJ, Silva JS (2010) Regulatory T cells attenuate experimental periodontitis progression in mice. J Clin Periodontol 37(7):591–600. https://doi.org/10.1111/j.1600-051X.2010.01586.x
Glowacki AJ, Yoshizawa S, Jhunjhunwala S, Vieira AE, Garlet GP, Sfeir C, Little SR (2013) Prevention of inflammation-mediated bone loss in murine and canine periodontal disease via recruitment of regulatory lymphocytes. Proc Natl Acad Sci U S A 110(46):18525–18530. https://doi.org/10.1073/pnas.1302829110
Hajishengallis G (2015) Periodontitis: from microbial immune subversion to systemic inflammation. Nat Rev Immunol 15(1):30–44. https://doi.org/10.1038/nri3785
Haribhai D, Chatila TA, Williams CB (2016) Immunotherapy with iTreg and nTreg cells in a murine model of inflammatory bowel disease. Methods Mol Biol 1422:197–211. https://doi.org/10.1007/978-1-4939-3603-8_19
Hernandez M, Dutzan N, Garcia-Sesnich J, Abusleme L, Dezerega A, Silva N, Gonzalez FE, Vernal R, Sorsa T, Gamonal J (2011) Host-pathogen interactions in progressive chronic periodontitis. J Dent Res 90(10):1164–1170. doi:0022034511401405 [pii]
Holden JA, Attard TJ, Laughton KM, Mansell A, O'Brien-Simpson NM, Reynolds EC (2014) Porphyromonas gingivalis lipopolysaccharide weakly activates M1 and M2 polarized mouse macrophages but induces inflammatory cytokines. Infect Immun 82(10):4190–4203. https://doi.org/10.1128/IAI.02325-14
Hou Y, Yu H, Liu X, Li G, Pan J, Zheng C, Yu W (2017) Gingipain of Porphyromonas gingivalis manipulates M1 macrophage polarization through C5a pathway. In Vitro Cell Dev Biol Anim 53(7):593–603. https://doi.org/10.1007/s11626-017-0164-z
Huang CB, Alimova Y, Ebersole JL (2016) Macrophage polarization in response to oral commensals and pathogens. Pathog Dis 74(3). https://doi.org/10.1093/femspd/ftw011
Huang Y, Tian C, Li Q, Xu Q (2019) TET1 knockdown inhibits Porphyromonas gingivalis LPS/IFN-gamma-induced M1 macrophage polarization through the NF-kappaB pathway in THP-1 cells. Int J Mol Sci 20(8):doi:10.3390/ijms20082023
Jepsen S, Suvan J, Deschner J (2020) The association of periodontal diseases with metabolic syndrome and obesity. Periodontology 2000 83(1):125–153. https://doi.org/10.1111/prd.12326
Kimura A, Kishimoto T (2010) IL-6: regulator of Treg/Th17 balance. Eur J Immunol 40(7):1830–1835. https://doi.org/10.1002/eji.201040391
Koidou VP, Cavalli N, Hagi-Pavli E, Nibali L, Donos N (2020) Expression of inflammatory biomarkers and growth factors in gingival crevicular fluid at different healing intervals following non-surgical periodontal treatment: a systematic review. J Periodontal Res 55(6):801–809. https://doi.org/10.1111/jre.12795
Lam RS, O'Brien-Simpson NM, Holden JA, Lenzo JC, Fong SB, Reynolds EC (2016) Unprimed, M1 and M2 macrophages differentially interact with Porphyromonas gingivalis. PLoS One 11(7):e0158629. https://doi.org/10.1371/journal.pone.0158629
Li W, Zhang Z, Wang ZM (2020a) Differential immune cell infiltrations between healthy periodontal and chronic periodontitis tissues. BMC Oral Health 20(1):293. https://doi.org/10.1186/s12903-020-01287-0
Li Y, Wang X, Wang S, Zhu C, Guo J, Li K, Li A (2020b) Complement 3 mediates periodontal destruction in patients with type 2 diabetes by regulating macrophage polarization in periodontal tissues. Cell Prolif 53(10):e12886. https://doi.org/10.1111/cpr.12886
Mahanonda R, Champaiboon C, Subbalekha K, Sa-Ard-Iam N, Yongyuth A, Isaraphithakkul B, Rerkyen P, Charatkulangkun O, Pichyangkul S (2018) Memory T cell subsets in healthy gingiva and periodontitis tissues. J Periodontol 89(9):1121–1130. https://doi.org/10.1002/JPER.17-0674
Medara N, Lenzo JC, Walsh KA, Holden JA, Reynolds EC, Darby IB, O'Brien-Simpson NM (2021a) Peripheral memory T-cell profile is modified in patients undergoing periodontal management. J Clin Periodontol 48(2):249–262. https://doi.org/10.1111/jcpe.13399
Medara N, Lenzo JC, Walsh KA, O'Brien-Simpson NM, Reynolds EC, Darby IB (2021b) Peripheral T helper cell profiles during management of periodontitis. J Clin Periodontol 48(1):76–90. https://doi.org/10.1111/jcpe.13389
Miao Y, He L, Qi X, Lin X (2020) Injecting immunosuppressive M2 macrophages alleviates the symptoms of periodontitis in mice. Front Mol Biosci 7:603817. https://doi.org/10.3389/fmolb.2020.603817
Monasterio G, Budini V, Fernandez B, Castillo F, Rojas C, Alvarez C, Cafferata EA, Vicencio E, Cortes BI, Cortez C, Vernal R (2019a) IL-22-expressing CD4(+) AhR(+) T lymphocytes are associated with RANKL-mediated alveolar bone resorption during experimental periodontitis. J Periodontal Res 54(5):513–524. https://doi.org/10.1111/jre.12654
Monasterio G, Castillo F, Ibarra JP, Guevara J, Rojas L, Alvarez C, Fernandez B, Agucro A, Betancur D, Vernal R (2018) Alveolar bone resorption and Th1/Th17-associated immune response triggered during Aggregatibacter actinomycetemcomitans-induced experimental periodontitis are serotype-dependent. J Periodontol 89(10):1249–1261. https://doi.org/10.1002/JPER.17-0563
Monasterio G, Fernandez B, Castillo F, Rojas C, Cafferata EA, Rojas L, Alvarez C, Fernandez A, Hernandez M, Bravo D, Vernal R (2019b) Capsular-defective Porphyromonas gingivalis mutant strains induce less alveolar bone resorption than W50 wild-type strain due to a decreased Th1/Th17 immune response and less osteoclast activity. J Periodontol 90(5):522–534. https://doi.org/10.1002/JPER.18-0079
Nagaraj S, Youn JI, Gabrilovich DI (2013) Reciprocal relationship between myeloid-derived suppressor

cells and T cells. J Immunol 191(1):17–23. https://doi.org/10.4049/jimmunol.1300654
Nibali L, Bayliss-Chapman J, Almofareh SA, Zhou Y, Divaris K, Vieira AR (2019) What is the heritability of periodontitis? A systematic review. J Dent Res 98(6):632–641. https://doi.org/10.1177/0022034519842510
Ortiz MC, Lefimil C, Rodas PI, Vernal R, Lopez M, Acuna-Castillo C, Imarai M, Escobar A (2015) Neisseria gonorrhoeae modulates immunity by polarizing human macrophages to a M2 profile. PLoS One 10(6):e0130713. https://doi.org/10.1371/journal.pone.0130713
Papadopoulos G, Shaik-Dasthagirisaheb YB, Huang N, Viglianti GA, Henderson AJ, Kantarci A, Gibson FC 3rd (2017) Immunologic environment influences macrophage response to Porphyromonas gingivalis. Mol Oral Microbiol 32(3):250–261. https://doi.org/10.1111/omi.12168
Polak D, Shapira L (2018) An update on the evidence for pathogenic mechanisms that may link periodontitis and diabetes. J Clin Periodontol 45(2):150–166. https://doi.org/10.1111/jcpe.12803
Pradeep AR, Roopa Y, Swati PP (2008) Interleukin-4, a T-helper 2 cell cytokine, is associated with the remission of periodontal disease. J Periodontal Res 43(6):712–716. https://doi.org/10.1111/j.1600-0765.2007.01079.x
Raffin C, Vo LT, Bluestone JA (2020) Treg cell-based therapies: challenges and perspectives. Nat Rev Immunol 20(3):158–172. https://doi.org/10.1038/s41577-019-0232-6
Sanz M, Marco Del Castillo A, Jepsen S, Gonzalez-Juanatey JR, D'Aiuto F, Bouchard P, Chapple I, Dietrich T, Gotsman I, Graziani F, Herrera D, Loos B, Madianos P, Michel JB, Perel P, Pieske B, Shapira L, Shechter M, Tonetti M, Vlachopoulos C, Wimmer G (2020) Periodontitis and cardiovascular diseases: consensus report. J Clin Periodontol 47(3):268–288. https://doi.org/10.1111/jcpe.13189
Sima C, Glogauer M (2013) Macrophage subsets and osteoimmunology: tuning of the immunological recognition and effector systems that maintain alveolar bone. Periodontology 2000 63(1):80–101. https://doi.org/10.1111/prd.12032
Sobacchi C, Menale C, Villa A (2019) The RANKL-RANK Axis: a bone to thymus round trip. Front Immunol 10:629. https://doi.org/10.3389/fimmu.2019.00629
Su L, Xu Q, Zhang P, Michalek SM, Katz J (2017) Phenotype and function of myeloid-derived suppressor cells induced by Porphyromonas gingivalis infection. Infect Immun 85(8). https://doi.org/10.1128/IAI.00213-17
Tonetti MS, Van Dyke TE, working group 1 of the joint EFPAAPW (2013) Periodontitis and atherosclerotic cardiovascular disease: consensus report of the Joint EFP/AAP Workshop on Periodontitis and Systemic Diseases. J Periodontol 84(Suppl 4S):S24–S29. https://doi.org/10.1902/jop.2013.1340019
Valero-Monroy O, Garcia-Cervantes G, Marquez-Corrales LF, Leija-Montoya AG, Sandoval-Basilio J, Martinez-Coronilla G, Isiordia-Espinoza MA, Serafin-Higuera N (2016) Myeloid derived suppressor cell: a new player in periodontal disease? Med Hypotheses 95:35–38. https://doi.org/10.1016/j.mehy.2016.08.010
Veloso P, Fernandez A, Terraza-Aguirre C, Alvarez C, Vernal R, Escobar A, Hernandez M (2020) Macrophages skew towards M1 profile through reduced CD163 expression in symptomatic apical periodontitis. Clin Oral Investig 24(12):4571–4581. https://doi.org/10.1007/s00784-020-03324-2
Vernal R, Diaz-Zuniga J, Melgar-Rodriguez S, Pujol M, Diaz-Guerra E, Silva A, Sanz M, Garcia-Sanz JA (2014) Activation of RANKL-induced osteoclasts and memory T lymphocytes by Porphyromonas gingivalis is serotype dependant. J Clin Periodontol 41(5):451–459. https://doi.org/10.1111/jcpe.12236
Yamaguchi T, Movila A, Kataoka S, Wisitrasameewong W, Ruiz Torruella M, Murakoshi M, Murakami S, Kawai T (2016) Proinflammatory M1 macrophages inhibit RANKL-induced osteoclastogenesis. Infect Immun 84(10):2802–2812. https://doi.org/10.1128/IAI.00461-16
Yan X, Yuan Z, Bian Y, Jin L, Mao Z, Lei J, Chen N (2020) Uncoupling protein-2 regulates M1 macrophage infiltration of gingiva with periodontitis. Cent Eur J Immunol 45(1):9–21. https://doi.org/10.5114/ceji.2020.94664
Yang J, Park OJ, Kim J, Kwon Y, Yun CH, Han SH (2019) Modulation of macrophage subtypes by IRF5 determines osteoclastogenic potential. J Cell Physiol 234(12):23033–23042. https://doi.org/10.1002/jcp.28863
Yu S, Ding L, Liang D, Luo L (2018) Porphyromonas gingivalis inhibits M2 activation of macrophages by suppressing alpha-ketoglutarate production in mice. Mol Oral Microbiol 33(5):388–395. https://doi.org/10.1111/omi.12241
Zhao L, Zhou Y, Xu Y, Sun Y, Li L, Chen W (2011) Effect of non-surgical periodontal therapy on the levels of Th17/Th1/Th2 cytokines and their transcription factors in Chinese chronic periodontitis patients. J Clin Periodontol 38(6):509–516. https://doi.org/10.1111/j.1600-051X.2011.01712.x
Zhu LF, Li L, Wang XQ, Pan L, Mei YM, Fu YW, Xu Y (2019) M1 macrophages regulate TLR4/AP1 via paracrine to promote alveolar bone destruction in periodontitis. Oral Dis 25(8):1972–1982. https://doi.org/10.1111/odi.13167
Zhu X, Zhu J (2020) CD4 T helper cell subsets and related human immunological disorders. Int J Mol Sci 21(21):doi:10.3390/ijms21218011
Zhuang Z, Yoshizawa-Smith S, Glowacki A, Maltos K, Pacheco C, Shehabeldin M, Mulkeen M, Myers N, Chong R, Verdelis K, Garlet GP, Little S, Sfeir C (2019) Induction of M2 macrophages prevents bone loss in murine periodontitis models. J Dent Res 98(2):200–208. https://doi.org/10.1177/0022034518805984

11 Complementary Experimental Methods in Genetics Open Up New Avenues of Research to Elucidate the Pathogenesis of Periodontitis

Arne S. Schaefer

Abstract

A complex disease such as periodontitis is the sum of environmental and genetic effects. The personal genetic constitution interacts with the effects of internal and external risk factors like smoking, oral hygiene, malnutrition, emotional stress, and age. Accordingly, individuals who live in the same environmental context and share comparable lifestyle habits have different disease risks. Genetic research offers the identification of DNA sequence variants that have a causal role in disease etiology and allows the identification of disease relevant immune and metabolic pathways that contribute to disease susceptibility and pathogenesis in specific situations. Real advances have been made in genetic medical research in the last years. Starting from candidate gene association studies, new approaches were employed that have expanded the study design of genomewide association studies to genomewide meta-analyses and gene x environment interaction studies. Cost efficient whole-exome and whole-genome sequencing of patients with rare severe forms of periodontitis has the potential to identify genes and pathways with a direct role in the pathogenesis of common forms. In parallel, animal models were developed that use genetically highly diverse mouse lines to identify risk genes of human diseases. This chapter presents the main studies and the identified susceptibility genes that have clear statistical evidence. In addition, it describes pioneering studies that used advanced methods in experimental dental research, opening up new avenues of research. Although the knowledge of the genetic architecture of periodontitis is still in its infancy, genetic research is building the basis for future works with the potential to advance dental medicine in ways that will determine the various causes of periodontal diseases. This knowledge may eventually allow making predictions about disease risk for individual patients and leading to diagnosis and treatments that do not treat the symptoms but heal the disease.

Keywords

Complex disease · Susceptibility · Genetic risk · Candidate gene association studies · Genomewide association studies · Meta-analyses · Gene x environment interaction studies

A. S. Schaefer (✉)
Genetics of Oral Inflammatory Diseases Group, Department of Periodontology, Oral Medicine and Oral Surgery, Institute for Dental and Craniofacial Sciences, Charité – University Medicine Berlin, Corporate Member of Freie Universität Berlin, Humboldt-Universität zu Berlin, and Berlin Institute of Health, Berlin, Germany
e-mail: arne.schaefer@charite.de

J. Santi-Rocca (ed.), *Periodontitis*, Advances in Experimental Medicine and Biology 1373,
https://doi.org/10.1007/978-3-030-96881-6_11

Abbreviations

GWAS Genomewide association study
RIL Recombinant inbred line
WES Whole exome sequencing

Highlights

- GWAS identified *SIGLEC5*, *DEFA1A3* and *ATP6V1C1* as risk loci of periodontitis.
- Complementary study concepts that integrate common and rare severe early-onset disease manifestations in human patients and mouse models will identify the causative factors of periodontitis.
- Different disease manifestations have different causations.

Considerations for Practice

- Moderate late-onset and severe early-onset forms of periodontitis have different causations, which are largely found in environmental and age related or genetic factors, respectively.
- Genetic testing cannot currently predict the individual risk to develop periodontitis.
- Complementary study concepts have the potential to identify the genetic risk factors of periodontitis. This knowledge will eventually contribute to the elucidation of the causative factors underlying the disease, leading to improved diagnosis and treatment options.

Patient Summary

Periodontitis has no certain known causative factors such as specific bacteria. The lack of known causality results in almost the same treatment options for all periodontal diseases. Advances were made in the last years in genetic medical research and although in its infancy, susceptibility genes with specific roles in wound healing and immune defense were identified. This chapter highlights important studies that identified these genes and introduces pioneering studies that used advanced experimental methods opening up new avenues of research. These may allow leading to enhanced understanding of the different causation of periodontal diseases, improved diagnosis and treatments.

11.1 Meaning of Genetic Studies for Periodontal Diseases

11.1.1 The Causative Factors of Periodontal Diseases Seem Elusive

Periodontitis is an inflammatory oral disease that affects a large proportion of the world's population (Eke et al. 2012; Marcenes et al. 2013). It is supposed to develop as a result of disruption of the microbial homeostasis (Lamont et al. 2018; Darveau 2010), a complex network of interactions between the microbial community of the oral mucosa, host epithelial and immune cells (Clemente et al. 2012). The oral microbiome is shaped and stabilized by the host immune system and the local environment (Yost et al. 2017). Factors such as smoking and nutrient availability but also changes and by-products of host immune responses during life and ageing can propagate outgrowth and over-representation of microorganisms that are able to adapt to altered environmental conditions. These microbial groups may selectively expand, thereby creating an imbalance of the community, with increases in the production of virulence factors beyond a threshold that can instigate periodontitis. Likewise, in periodontitis, species or genera that dominate disease-associated polymicrobial communities are also found in health but at reduced relative abundance (Abusleme et al. 2013). The current concept that underlies the etiology of periodontal diseases describes that there is not a single microbial composition representing a periodontal state associated with health or disease. Instead, the life history of the host in interaction with the host genome shapes the composition and quantities of the individual's oral microbiome (Nibali et al. 2009; Nibali et al. 2014).

The lack of a specific causative microbial factor for periodontal diseases is unlike other diseases that involve microorganisms, e.g. like pneumonia, where specific bacteria are known to be the cause. Moreover, the effects of internal and external factors like age, smoking, malnutrition and emotional stress (Kinane et al. 2006) differ largely between individual cases who live in the same environmental context and share comparable lifestyle habits. The causative factors of periodontal diseases seem elusive. Accordingly, no medication exists to cure periodontitis and no general accepted laboratory tests are available to advance diagnosis. This is why clinicians resort to diagnose the clinical presentations of periodontal diseases by the symptoms, subdividing different stages of periodontal diseases according to the extent and progression of tissue destruction. However, many patients meet criteria for multiple groups depending on the measurement time point, and symptoms can blend into one another, blurring boundaries. This has raised scepticism concerning whether the diagnostic categories correspond to real natural entities. Furthermore, the knowledge gap between the clinical symptoms and the underlying causes results in almost the same treatment options offered for all periodontal diseases, irrespective of the causation. Quantifying this problem is the startling statistic of tooth implants lost several years after periodontitis therapy.

11.1.2 The Genetic Architecture Determines the Susceptibility to Periodontal Diseases

A complex disease such as periodontitis is the sum of genetic and environmental effects. Genetic research offers the identification of DNA sequence variants that are directly involved in disease pathogenesis. These variants point to genes that have a causal role in the etiology. The knowledge of these genes and variants, and their relationship with environmental factors, allows the identification of immune regulatory or metabolic pathways that are involved in the etiology of periodontitis and gives direct insight into the molecular mechanisms that contribute to disease susceptibility and pathogenesis. In this respect, knowledge of the genetic etiology has the potential to guide stratification of patients into subgroups by the causation and not the symptoms of the disease.

The genetic contributions to the variability of the phenotype in a population is estimated by the heritability. The heritability for periodontal bone loss is estimated at 0.4–0.5 and increases with younger age of disease onset and severity (Nibali et al. 2019). It is important to consider that periodontitis is usually the result of long-term gingivitis. This is implied because, for example, bleeding on probing for 25 years results in an almost 50% increased tooth loss (Lang et al. 2009). However, gingivitis showed no heritability (Michalowicz et al. 2000; Nibali et al. 2019). In other words, genetic factors do not determine the susceptibility to gingivitis. Instead, it can be extrapolated that it is the ability to confine gingival inflammation by endowing appropriate barrier integrity, efficient wound healing, and immune defense, which in the end causes alveolar bone loss. Moreover, the increased heritability at a younger age of disease onset and severity implies that specifically the age of onset and the rate of progression of periodontal bone loss is determined by genetic factors. This entails that the genetic variation, which results from combinations of segregating alleles in the genomes of a population, plays a central role in determining time of disease-onset, severity, and progression, resulting in the different clinical manifestations of periodontal diseases. In this context it would not be very speculative to hypothesize that microbial factors and subsequent gingivitis initiate the pathogenesis of periodontitis, but genetic variation paves the way to disease. In other words, it is considered for complex a disease that the susceptibility is determined by the individual genetic architecture (Timpson et al. 2018; Yong et al. 2020), the characteristics of genetic variation that are responsible for heritable phenotypic variability. The genetic architecture depends on the number and allele frequencies of genetic variants affecting a disease, the magnitude of their effects and their interactions with each other and the environment. Therefore, gene–gene as well as gene–environmental interactions play a fundamental role in the disease etiology. From this follows the objective of genetic research that is to identify the genes that are involved in the pathogenesis and to estimate the genetic effects that the potential risk variants exert in interaction with specific environmental factors.

In the long-term, this knowledge will allow using an individual's genetic profile to guide decisions made about the diagnosis, treatment, and prevention of disease. Genetic medical research, in a specific molecular way, gives direct insight into disease relevant differences between individuals. Likewise, the practice of medicine emerging from genetic research is termed personalized medicine. It has the potential to transform our understanding in ways that may allow making predictions about disease risk for individual patients that can help to choose a prevention or treatment plan that is right for them. The stratification of patients by different causes of diseases allows in some instances the possibility of picking the right drug at the right dose for the right person instead of the same treatment options offered for all patients and all clinical presentations. Although personalized medicine is still in its infancy, as a former director of the National Institute of Health in the USA, Francis Collins put it, this may be the biggest revolution in medicine in a very long time.

11.2 Current Strategies to Identify the Genetic Basis of Periodontal Diseases

11.2.1 Candidate Gene Studies

The study of the genetic variation in candidate genes that were selected based on literature review had long been the only strategy for the identification of disease genes. These studies required an a priori hypothesis on the involvement of the selected gene in the disease risk (Wilkening et al. 2009). In principle, there are two different selection strategies for a candidate gene. It can be asked whether specific loci within a specific biological pathway are involved in the increase of the genetic risk of the disease, or effects of a risk variant from other diseases are tested. Both approaches can give answers whether a selected candidate gene carries genetic variants that increase the risk of PD. The hypotheses for the selection of the genes of interest are depending on the current knowledge of the genetic etiology, which has usually been very incomplete, and many disease relevant genes will not be selected because their functions are unknown or lie within pathways that have not been implicated with the disease.

Corresponding to the former understanding of PD as a disease caused by bacteria, most studies focused on genes that were selected for their roles in the immune system and a few investigated genes that are involved in tissue destructive processes or metabolism (Laine et al. 2010; Loos et al. 2005). Most of the candidate genes for any human disease turned out not to be right. Moreover, most studies also were limited by insufficient sample sizes and incomplete analysis of the spectrum of genetic variants at the selected candidate genes, and often missed replications of the results. Therefore, these studies were largely unable to draw unambiguous conclusions, even when the results were negative.

11.2.2 Genomewide Association Studies

The hypothesis-free approach of genome-wide association studies (GWAS) got us past the candidate gene approach. It allows the simultaneous testing of millions of genetic variants across the entire genome, in principle considering all genes of the entire genome as candidates. A GWAS searches across the entire genome using DNA sequence variants as markers of a large sample of patients with a specific disease and a large sample of healthy individuals who are otherwise well matched. It finds genetic variations associated with a particular disease. These are positions in the genome that must be involved in disease risk because of consistent differences between both groups. However, the associated variants themselves may not directly cause the disease. They may just be "tagging along" with the actual causal variants. For this reason, researchers often need to take additional steps, to identify the exact genetic change involved in the disease. GWAS were particularly useful in finding genetic variations and identified almost all of the common susceptibility variants for almost any complex human disease and trait. As of 2020–09, the GWAS Catalog (https://www.ebi.ac.uk/gwas/home) contains 197,708 genetic associations that contribute to human traits and diseases, described in 4694 publications.

11.2.2.1 Statistical Limitations of GWAS

Nevertheless, genome-wide testing of genetic variants for associations with a disease is problematic. For each common genetic variant of the genome, a GWAS tests the hypothesis, that it has an effect on disease. The probability of detecting an effect, if there is a true effect present to detect, is expressed by the statistical power of the GWAS. The statistical power is affected mainly by the effect size and the sample size used to detect it. Large effects are easier to detect than small effects, while large samples offer greater test sensitivity than small samples. In Fig. 11.1, the power calculation of the minimum sample size required for detecting an association as a function of allele frequency and effect size is summarized. For example, for detecting the association of a variant with an effect size of odds ratio = 1.3 and an allele frequency of 20% in the general population, a sample size of 1500 is required.

In particular, there is the problem of multiple testing. Allele frequencies in independently sampled populations are susceptible to random gene drifts, which arise without selection pressure across populations. Given random drifts, it is possible by chance alone that many alleles are more frequent in one sample than in another sample, e.g. consisting of patients and controls, respectively. The more often an association test is

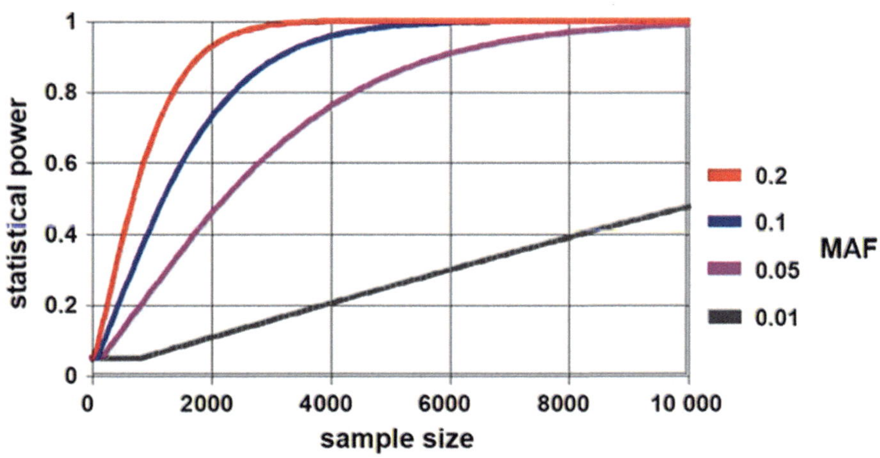

Fig. 11.1 Statistical power in relation to sample size, allele frequency, and OR
To identify a genetic risk variant with a minor allele frequency (MAF) of, for example, 20% in the general population (corresponding to 0.2 in the graph), ~1500 cases and 1500 controls are required to achieve the necessary statistical power of 0.8. For rare alleles, with a MAF of 1% (corresponding to 0.01 in the graph), more than 20,000 each of cases and controls are necessary. The statistical power was calculated for an average OR of 1.3 and an equal ratio of controls to cases. (Dupont and Plummer 1998)

performed, and GWAS test each of the millions of common variants of the genome for association with the disease individually, the probability of statistical errors rises. To give an example, when one million single nucleotide polymorphisms are independently tested, one would expect to find over 50,000 disease-associated variants with a P value of <.05, simply by chance. To avoid this error, the P value obtained from the chi-square statistic must be corrected for multiple testing. This is addressed by setting a genome-wide significance threshold by correcting for the number of tests performed (Balding 2006). The current standard for declaring statistical significance at the genome-wide level is a combined P value (including 'initial discovery' genome-wide association study and replication cohorts) of $<5 \times 10^{-8}$ (Manolio 2010). This approach reduces the number of false positive results but also markedly decreases the power to detect single nucleotide polymorphisms associated with disease. Thus, if the sample size or effect sizes of the risk alleles are limited, many true associations cannot be separated from chance. This requires strategies to increase the sample size or to conceive additional study concepts.

11.2.2.2 Alternative Study Concepts to Increase Testing Power of GWAS

Two different study concepts are constructive to achieve the necessary statistical power for genome-wide association studies (GWAS). Studies can take advantage of existing extremely large population-based cohorts with both genetic data and proxy phenotypes of the disease available. Most individuals of these samples have a moderate phenotype because severe phenotypes are rare in the general population, these studies make a trade off between the power provided by large sample sizes and the power provided by the high heritability of severe disease forms. Furthermore, this study concept does not allow the identification of risk alleles that modify different disease stages in subgroups of patients, because they have no refined clinical data. Alternatively, studies can take advantage of the increased heritability of cases with severe and early-onset disease phenotypes. These studies make a reverse trade off between heritability and sample size, because there are logistical and financial constraints to recruit cases with rare phenotypes and detailed phenotype information. By taking advantage of the refined information of clinical phenotypes, these studies may allow stratification of subgroups of cases by clinical presentations to identify alleles that modify specific clinical presentations.

For periodontal diseases, difficulties in generating appropriate case-control samples have been the major cause for the slow progress in identification of the genetic risk loci of periodontitis compared with other complex human diseases. Consequently, and despite an unclear number of candidate-gene association studies, susceptibility genes of periodontal diseases with strong statistical evidence are scarce.

11.2.2.3 SIGLEC5

The largest GWAS that has been performed for periodontal diseases to date combined periodontitis cases from seven studies (N cases = 17,353, N controls = 28,210) from the Gene-Lifestyle Interactions in Dental Endpoints (GLIDE) consortium with cases who self-reported loose teeth (N cases = 18,979, N controls = 442,052) from the UK Biobank (Fig. 11.2) (Shungin et al. 2019). The single association that passed the genome wide significance threshold of $P = 5 \times 10^{-8}$ located to the gene sialic acid binding Ig like lectin 5 (*SIGLEC5*), with an intronic variant (rs12461706) showing the smallest P-value of association with $P = 3.9 \times 10^{-9}$. The frequency of the effect allele was 40% and the effect estimates in GLIDE and the UK Biobank were consistent at this lead locus with an odds ratio = 1.05 for periodontitis (GLIDE consortium) and 1.06 for loose teeth (UK biobank).

Interestingly, the association at *SIGLEC5* was earlier identified in a GWAS that used cases with ≥30% bone loss at more than 2 teeth under 35 years of age (Munz et al. 2017). According to the current classification of periodontal diseases and conditions (Caton et al. 2018), that diagnoses periodontal diseases according to severity (by staging) and rate of progression (by grading),

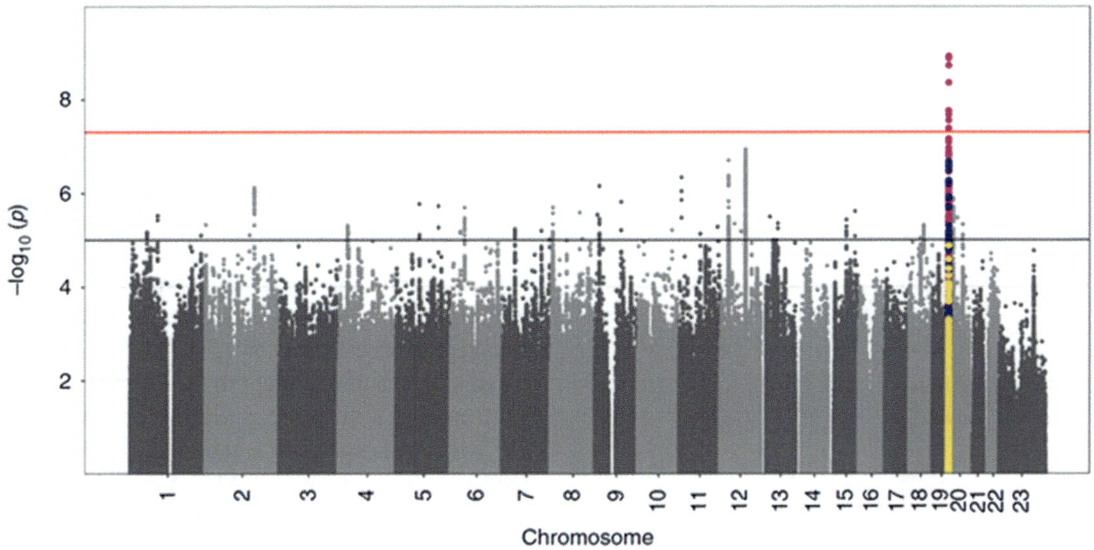

Fig. 11.2 Manhattan plot of the large meta-analysis of GWAS data from the GLIDE consortium and the UK biobank that included 36,332 cases and 470,262 controls. The red line indicates the threshold for genome-wide significance at ($P = 5 \times 10^{-8}$), and the black line indicates a suggestive threshold for association at ($P = 1 \times 10^{-5}$). The single-locus meeting the genomewide significance threshold is colored in magenta. P-values for the same locus are shown in yellow for periodontitis in GLIDE, and blue for loose teeth in UKB. The association peak at chromosome 19 locates to the gene *SIGLEC5* (rs12461706, $P = 3.9 \times 10^{-9}$). (Taken from Shungin et al. 2019)

these cases were classified as the most severe phenotypes (Stages III and IV) with a rapid rate of progression (Grade C). The increased power provided by the high heritability of this severe phenotype of comparatively early age of disease onset compared to a more moderate and later onset phenotype is impressively indicated by the P-values of the associations that were observed in both studies. Whereas the GWAS that included 1119 stage III-IV Grade C cases and 7668 controls showed a P-value of 1.1×10^{-8}, the GWAS that used ~30 times more cases of a more moderate late onset phenotype, observed a P-value that was only marginally smaller, with $P = 3.9 \times 10^{-9}$. The increased heritability of early onset cases was also reflected by the considerably higher effect size of the associated allele with an odds ratio = 1.34 (95% confidence interval = 1.19–1.50), indicating enrichment of the effect allele in the case sample.

The results from these two GWAS indicated that the locus at *SIGLEC5* is an important risk gene that plays a role in the etiology of severe early-onset as well as moderate common forms of periodontitis. The role of *SIGLEC5* in the context of oral inflammatory diseases is currently unknown. However, sialoside-based pattern recognition by members of the SIGLEC family seem to discriminate infections from aseptic tissue injuries and attenuate innate immunity (Chen et al. 2009). Recognition of pathogens-associated molecular patterns (PAMPs) by Toll-like receptors (TLRs), NOD-like receptors (NLRs) and RIG-I-like receptors (RLR) plays a critical role in protecting host against pathogens. In addition, TLR and NLR also recognize danger-associated molecular patterns (DAMPs) to initiate limited innate immune responses. While innate immune response to DAMPs may be important for tissue repairs and wound healing, it is normally well controlled to avoid autoimmune destruction. In the mouse model, it was shown that CD24–Siglec10/G interaction selectively dampened host response to DAMPs but not PAMPs, thus discriminating infections from aseptic tissue injuries (Fig. 11.3). In conclusion, it is possible that SIGLEC5 has a role in wound healing. Future studies will aim to identify the causal vari-

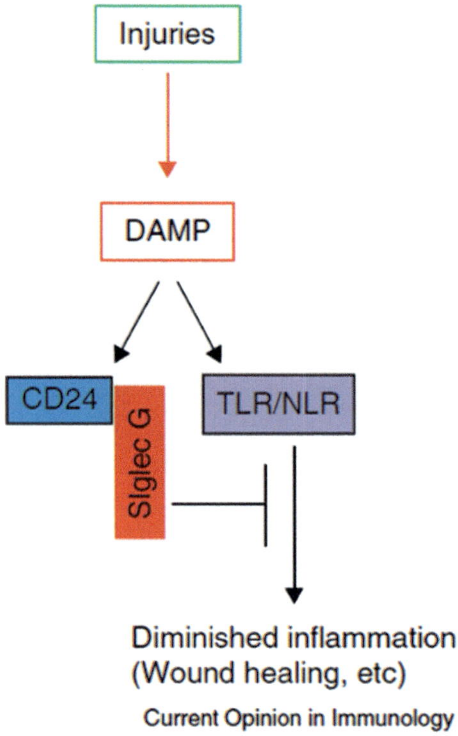

Fig. 11.3 Sialoside-based pattern recognition discriminates infections from aseptic tissue injuries
CD24 forms trimolecular complex with DAMPs and SIGLEC10 that inhibits activation of TLR/NLR. Second, pathogen-encoded sialidases prevent CD24 from interacting with Siglec 10. As a result, DAMPs and PAMPs become indistinguishable during infection. In particular, since CD24–Siglec 10 interaction selectively dampens host response to DAMPs but not PAMPs, this sialoside-based pattern recognition may serve as a foundation to discriminate PAMPs from DAMPs, which is essential for efficient wound healing. (Taken from Liu et al. 2011)

ant of the associations and to elucidate the regulatory mechanism that is affected by the effect alleles. Additionally, a detailed characterization of the function of SIGLEC5 for oral immunity or barrier regeneration will give profound insight into the etiology of periodontitis.

11.2.2.4 Alpha Defensin Genes

The GWAS that used Stages III - IV Grade C patients also identified a second genetic region associated with periodontal diseases at genomewide significance level. Interestingly, an earlier GWAS on chronic periodontitis reported a suggestive association of this region with more common periodontitis forms and a later age of disease onset (Teumer et al. 2013). The associated variants located to the antimicrobial peptide genes defensin alpha 1 (*DEFA1*), −3 and −4. These genes belong to the family of alpha defensins that cluster on chromosome 8 and play a role in host defense against microorganisms (Fig. 11.4). The genes *DEFA1* and *DEFA3* are highly copy variable and differ only by a single base substitution in the coding sequence. They seem to be interchangeable occupants of a 19-kb-long copy-variable repeat unit, with both *DEFA1* and *DEFA3* gene numbers showing variation (Khan et al. 2013). For this reason, the composite designation *DEFA1A3* was suggested (Aldred et al. 2005).

Noteworthy, in the sample of early onset cases and controls alone, the p-value of association was not genomewide significant for the GWAS lead variant rs2978951, with P = 4.49×10^{-7} (effect allele frequency 41%, odds ratio = 1.31). However, after increasing the sample size by pooling 1116 early onset PD cases and 7654 controls from Germany, the Netherlands, and Turkey with the GWAS sample that previously found suggestive association of this region with periodontitis (2211 German cases with chronic periodontitis and 1817 controls), the p-value passed the genomewide significance threshold with P = 2.06×10^{-8}. However, this combined sample of early-onset and late-onset cases showed reduced heritably (odds ratio = 1.25).

11.2.2.5 ATP6V1C1

Another study combined several GWAS data sets from stages III - IV grade C patients and more moderate, late-onset periodontitis cases (Munz et al. 2019). This study identified two novel loci that showed genome-wide significant associations with PD. One association is located 13 kilo basepairs (kb) upstream to the gene *ATP6V1C1* (chr8, rs16870060-G, P = 3.69×10^{-9}, OR = 1.36, 95% CI = [1.23–1.51]). This gene encodes a component of vacuolar ATPase (V-ATPase), a multisubunit enzyme that mediates acidification of intracellular compartments of eukaryotic cells. V-ATPase is comprised of a cytosolic V1 domain and a transmembrane V0 domain. *ATP6V1C1* is one of two genes that encode the V1 domain C subunit pro-

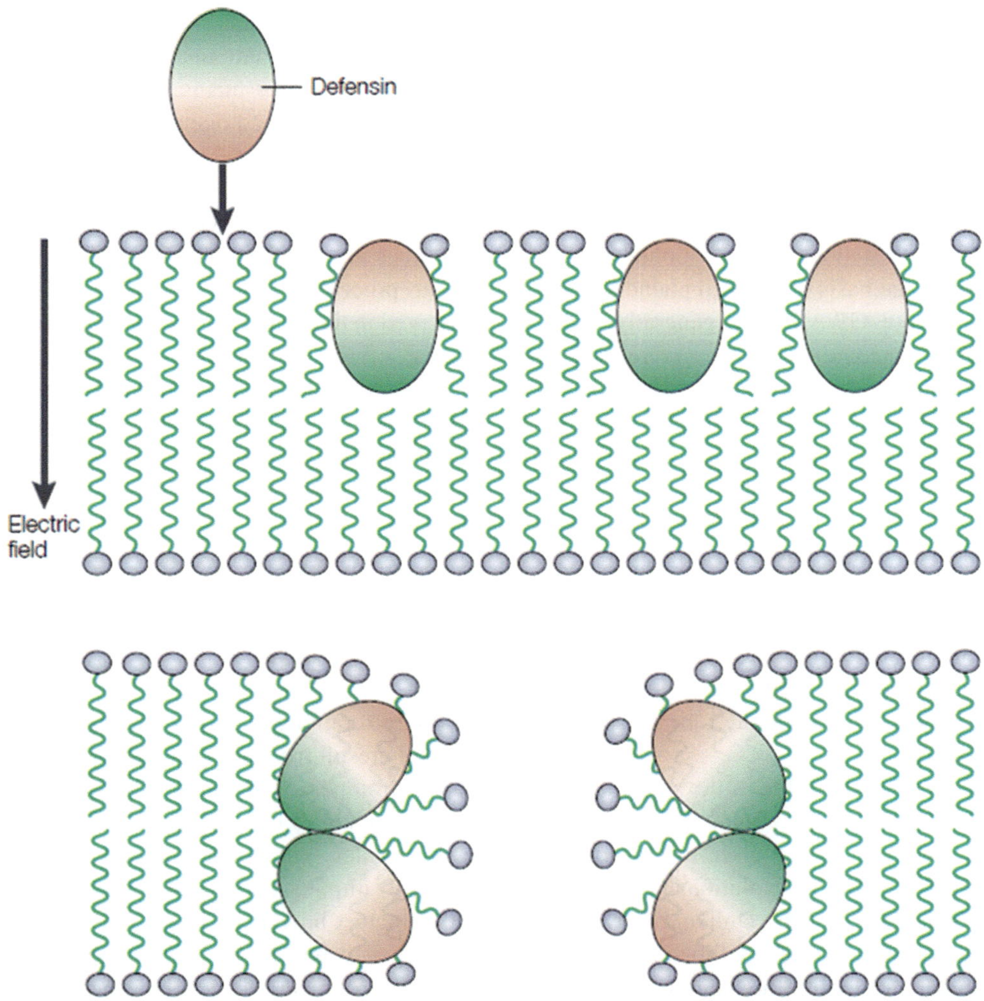

Fig. 11.4 A model of action of defensins
Most defensins are amphipathic molecules that have positively and negatively charged amino-acid side chains. This allows them to interact with bacterial membranes. Electrostatic attraction and the transmembrane bioelectric field pull the peptides towards and into the membrane (upper panel). As peptides accumulate, the membrane is strained and the peptides transition into another arrangement (lower panel) that lowers the strain but results in the formation of membrane pores. The pores result in leakages of the bacteria. (Taken from Ganz 2003)

teins and is highly expressed in osteoclasts. The expression is highly induced by RANKL [receptor activator for NF-kappaB (nuclear factor kappaB) ligand] during osteoclast differentiation. Osteoclasts are bone resorbing cells acting as key mediators of alveolar bone loss in periodontal diseases. In mice, knockdown of *Atp6v1c1* expression severely impaired osteoclast acidification activity and bone resorption, and it was shown that *Atp6v1c1* is an essential component of the osteoclast proton pump and an important regulator in osteoclast activation (Feng et al. 2009). Moreover, downregulation of the gene *TCIRG1* (also known as Atp6i), encoding the ATPase H+ transporting V0 subunit a3, in mice strongly reduced *Porphyromonas gingivalis* infection-stimulated bone loss and gingival inflammation (Jiang et al. 2013). Taken together, this strongly implies an important role of V-ATPase in alveolar bone loss during oral inflammation.

The second variant was located in the intron of the long intergenic non protein coding RNA (lncRNA) LOC107984137 on chromosome 16, downstream of the gene SHISA9 (rs729876-T, $P = 9.77 \times 10{-}9$, OR = 1.24, 95% CI = [1.15–1.34]). To date, the function of this lncRNA is not known.

In addition to the described studies, GWAS were performed that focused on common less severe manifestations of periodontal diseases, selected on the basis of detailed phenotype information. These GWAS were generally small in size and detected no variants that were associated at a genome-wide significance level. However, these GWAS reported several variants as suggestive genetic susceptibility variants of common forms of periodontitis (Divaris et al. 2013; Hong et al. 2015; Munz et al. 2017; Sanders et al. 2016; Shaffer et al. 2014; Shimizu et al. 2015; Teumer et al. 2013). The reason for the comparably low success of these studies is very likely found in the reduced statistical power as a result of both the limited sample size and the lower heritability of the late onset more moderate forms of periodontal diseases.

In the context of the little success to find genetic risk variants for common late-onset manifestations of periodontal diseases, it is important to bring to mind the evolutionary meaning of periodontitis for health. Unlike other diseases like pneumonia, diabetes mellitus, or atherosclerosis that are often life threatening, tooth loss as end point of untreated periodontitis generally results in the resolution of inflammation leading to healing. According to an emerging concept of evolutionary medicine (Nesse and Stein 2012), periodontal bone loss might have evolved because it serves as useful response to long-lasting inflammation. Such regulatory response can be compared, to some degree, with fever or cough, unpleasant symptoms that turn on when they are needed to restore health. Tooth loss as a response to gingivitis may provide a selective advantage in situations where the cost of losing teeth to resolve inflammation, which could otherwise lead to cost-intensive immune reactions and increased risk of bacteremia and sepsis, are lower than keeping them. This view would explain both the increased frequency and reduced heritability of periodontitis in the elderly, that can be explained by 'evolutionary trade-offs', cost-benefit decisions when limited amounts of resources do not allow to increase or maintain several protective traits, for example pathogen defense and efficient wound healing, at once (Garland 2014). Evolutionary trade-offs are useful to save costs for the maintenance of many vital functions when resources are limited. They imply for many different processes in humans and can also be observed in the young, for example between children's growth and immune function (McDade et al. 2008). For periodontitis, the concept of evolutionary trade-offs is novel and has not been addressed experimentally, e.g., in mouse models. The interrelation with diabetes might involve this situation.

11.2.3 Gene x Environment Interactions

Adding to complexity, some forms of periodontitis likely result from an evolutionary mismatch to the environment. For instance, lifestyle changes can contribute to disruptions of the microbiome in particular vulnerable individuals. This is strongly implicated in modern epidemics of inflammatory bowel diseases or obesity. Different genotypes respond differently to exposure of the same environmental factor. However, the effects of the genetic variants become manifest only in specific situations, thus exposing the mismatch to an environmental factor. Thus, understanding genotype x environment interaction is important for an improved understanding of the etiology of PD and for the identification of specific risk groups and finding appropriate therapies. Genotype x environment interactions can be tested using genomewide genotype data sets, if the exposure of the test persons to the environmental factors was documented. Applying a logistic regression model to the case-only design provides a clear advantage of statistical power for gene x environment interaction studies compared to the case-control design (Piegorsch et al. 1994). Verification that the interacting alleles and environmental factor are uncorrelated in the general population is subsequently tested in a sample of healthy controls. A requisite of the case-only design is that the disease is rare, with a prevalence <5% in the general population, making stages III - IV Grade C patients particularly appropriate.

Smoking is a well-established environmental risk factor for periodontitis but only a portion of smokers develops periodontitis. Specifically, among adults ≥30 years of age 19% of the current smokers but only 6% of the non-smokers had severe chronic PD (Eke et al. 2015). The percentage of cases was even higher according to EFP (Tonetti et al. 2005) definitions (26% of the current smokers and 7% of the non-smokers had severe chronic PD). A recent study investigated whether the relative disease risk associated with smoking is modified by common genetic variants in a genomewide gene x smoking interaction analysis using stage III-IV grade C cases. It found various suggestive associations and notably, the most illustrative located to the gene Sclerostin (*SOST*) (Freitag-Wolf et al. 2019), which is a potent suppressor of bone formation. SOST is primarily produced by osteocytes and inhibits osteoblast proliferation and stimulates osteoblast apoptosis (Winkler et al. 2003; van Bezooijen et al. 2004; Sutherland et al. 2004). Mice overexpressing SOST are osteopenic whereas mice lacking SOST exhibit increased Wnt signaling, bone density, and bone strength. The gene x smoking interaction study of Freitag-Wolf et al. also showed that cells stimulated with tobacco smoke extract significantly transcribed more *SOST in vitro*. This might provide a molecular link that potentially connects the effects of tobacco smoking with increased alveolar bone loss.

11.2.4 Mouse Models

Not alone smoking but other environmental factors, too, increase the disease risk for periodontitis. For example, dietary patterns influence the oral microbiome and overexposure to fermentable carbohydrates and saturated fatty acids (SFAs) influence the structure and composition of dental biofilms and promote a microbial shift towards pathogenic organisms (reviewed in Lamont et al. 2018). However, the immune response to excess SFA strongly varies between individuals living in the same nutritional environment. This variation is explained to a significant extent by the genetic variability. The understanding of how this variation translates into different clinical manifestations has potential to guide personalized medicine. In the absence of available human analysis populations that allow directly testing the interactions of human genes and susceptibility variants with a specific environmental factor, mouse models are particularly suitable as tools for the identification of risk genes in complex human diseases. For example, the aspect of excess dietary fat as a risk factor for periodontitis was demonstrated in a mouse model that showed that alveolar bone loss is significantly increased in individuals living on high fat diet (HFD) compared to normal chow and this observation was independent of infection with periodontal pathogens (Fig. 11.5; Blasco-Baque et al. 2016). These findings are in agreement with other studies that showed alveolar bone loss in rodent animals after HFD consumption (Cavagni et al. 2013; Fujita and Maki 2015).

The majority of genes in mice have orthologous in the human genome and comparative mapping showed that mouse models have the potential to recapitulate human conditions and can be used for the identification of risk genes in complex human diseases (Iraqi et al. 2008). Whereas common inbred mouse lines show a generally decreased genetic diversity that may impede observation of phenotypic effects caused by genetic variability present in the diverse mouse genomes, genetically highly diverse sets of recombinant inbred mouse lines (RIL) are good tools for the identification of risk genes in complex human diseases. For example, the Collaborative Cross (CC) is a unique, genetically highly diverse set of 75 RILs, derived from crosses that were created by full reciprocal 8-way matings of eight divergent strains of mice (Collaborative Cross 2012; Churchill et al. 2004). Three founders of the CC are wild-derived, representing the subspecies *Mus musculus castaneus*, *musculus* and *domesticus*, and contribute a large

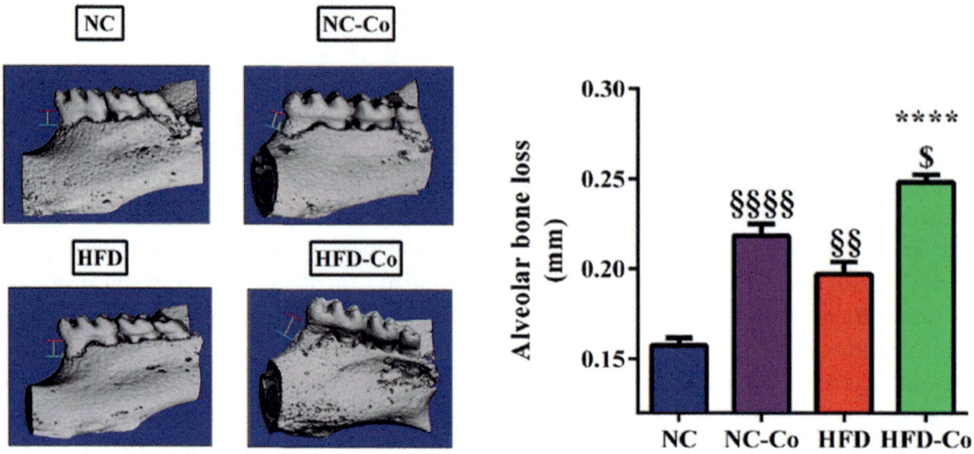

Fig. 11.5 HFD without oral bacterial colonisation induces periodontitis
Mice were colonised with the oral bacterial pathogens *Porphyromonas gingivalis*, *Fusobacterium nucleatum* and *Prevotella intermedia* or by vehicle solution for 1 month and then randomised into four groups: normal chow (NC, blue bar, n = 6), normal chow colonised (NC-Co, purple bar, n = 6), high-fat diet (HFD red bar, n = 7) and high-fat diet colonised (HFD-Co, green bar, n = 10; §p < 0.05; §§p < 0.001, §§§§p < 0.0001 when compared to NC and $p < 0.05 when compared to NC-Co.). (From Blasco-Baque et al. 2016, modified)

number of sequence variants that do not segregate among classical strains, which descended from *Mus musculus domesticus*. Thus, each RIL of the CC is regarded homozygous but a high genetic variation exists between the different lines. Their advantage is that they allow an analysis of molecular interactions in a context that is the most relevant to the clinical trait, namely, multiple genetic perturbations (as in a natural population), instead of an individual genetic perturbation as in a transgenic mouse (Civelek and Lusis 2014). Loci that confer susceptibility to alveolar bone loss upon exposure to an environmental factor can be mapped across the entire murine genome and the human orthologous regions are subsequently searched for disease-associated variants. To narrow down the list of potential candidates even further, expression of the genes within a mapped genomic region can be compared between susceptible and resistant mice, and only those genes with a significant difference in expression would be selected for association tests in human case-control samples.

A recent study employed this approach and identified three differently expressed genes in the mouse genome, which carry variants for increased susceptibility to alveolar bone loss in response to *Porphyromonas gingivalis* infection (Shusterman et al. 2017). All variants within 200 kb upstream and downstream of the three protein coding human orthologues of the differently expressed genes of the mapped trait loci were included in the subsequent explorative association analysis using the genotypes of the recent large genome-wide association study on aggressive periodontitis (Munz et al. 2017). Two genetic regions suggested association with periodontitis. These were located at the platelet factor 4/ pro-platelet basic protein/C-X-C motif chemokine ligand 5 (PF4/ PPBP/ CXCL5) gene cluster and at the gene UDP glucuronosyltransferase family 2 member A1 complex locus (UGT2A1). The best associated single nucleotide polymorphism of each of the two regions was subsequently validated using the genotypes of the European- American genome-wide association study on chronic periodontitis described above (Divaris et al. 2013). In this sample, a confined genetic region downstream of the *PF4/PPBP/CXCL5* gene cluster indicated association with late onset periodontitis. The best associated single nucleotide polymorphism, rs1595009, was subsequently replicated in a sample of cases of German descent with chronic periodontitis (399 of whom were < 60 years of age at disease onset) and 1633 controls. The combined estimate of association from the German aggressive periodontitis sample, the European-American chronic periodontitis sample, and the German chronic periodontitis sample was $P = 2.9 \times 10^{-5}$.

11.2.5 Genetic Susceptibility Variants That Are Shared Between Different Diseases

Similar to the concept of candidate genes, see above, GWAS data sets from epidemiologically related diseases can be combined to increase statistical power and to thus identify risk factors that are shared between both diseases. This can be an efficient approach because pleiotropy, which occurs when one gene influences two or more seemingly unrelated phenotypic traits, is common. GWASs frequently showed that the same genetic variants can be significantly associated with multiple diseases and traits when the phenotypes are measured on different individuals so that no environmental associations confound the results (Bulik-Sullivan et al. 2015; Sivakumaran et al. 2011). Evidence implies that at some loci, the same causal variants are driving the observations of associations across diseases and analytical methods that estimate genetic correlations from GWAS data have provided evidence for widespread pleiotropy at the level of genes (63%) and SNPs (31%) (Watanabe et al. 2019). Pleiotropy may inform reasons for comorbidity between traits, pointing to underlying shared genetic mechanisms, and may aid in establishing the direction of causality between traits.

For example, coronary artery disease (CAD) has a well-established genetic basis and a strong inflammatory component. An association between CAD and periodontitis was reported in several clinical and observational studies, and may be independent of smoking (Lockhart et al. 2012) and adiposity (Shungin et al. 2015). To identify shared genetic risk variants a meta-analysis of GWAS data sets for CAD and periodontitis was performed. To this end, genotype data from the CAD GWAS meta-analysis from the "Coronary Artery Disease Genome-wide Replication and Meta-analysis plus The Coronary Artery Disease" (CARDIoGRAMplusC4D) consortium (60,801 cases vs 123,504 controls) (Howson et al. 2017) and the GWASs of stage III/IV grade C and moderate late-onset periodontitis cases (4415 cases vs 5935) was performed (Munz et al. 2018). Two SNPs at the known CAD risk loci *ADAMTS7* (rs11634042) and *VAMP8* (rs1561198) were identified.

The association of VAMP8 (vesicle-associated membrane protein 8) illustrates how the discovery of risk genes opens up new avenues of medical research. VAMP8 has a role in mucin secretion in various mucosal interfaces such as airway epithelial cells (Jones et al. 2012) and intestinal goblet cells (Cornick et al. 2017). In these barrier tissues, the epithelium is covered by a mucin layer that microorganisms have to overcome to invade the mucosa. It was shown that certain microorganism are able of mucin proteolysis, which evokes mucin hypersecretion eventually leading to mucus depletion. This process results in direct contact of epithelial cells and microorganisms, which may cause host cell death and epithelial damage, enabling microbial tissue invasion.

Various studies showed that ablation of VAMP8 leads to impaired mucus secretion, increased adherence of microorganism to epithelial cells and induction of an aggressive proinflammatory response an enhanced cell death (Cornick et al. 2017). It was concluded that VAMP8 orchestrates the mechanism controlling mucus release in response to pathogenic infection. Future studies will be showing if VAMP8 has the same role in the oral mucosa and how this is related to the increased risk for coronary heart disease. The increased disease risk of patients with VAMP8 susceptibility variants may result from impaired barrier integrity that makes them susceptible to microbial tissue invasion. This would be an entirely different causation compared to e.g. an impaired would healing capacity, which is implied by susceptibility variants at *SIGLEC5*. Understanding of the impaired pathogenic processes has the potential to find different treatment options that address the causes of the disease and not the symptoms.

11.2.6 Severe Early-Onset Diseases Are Tools to Identify Susceptibility Genes of Common Disease Phenotypes

Variants of large effects are inconsistent with most diagnoses of common diseases, since they result in more severe and/or early-onset diagnoses (Wray et al. 2010). This explains why the heritability declines with increasing age of onset and decreasing severity. Additionally, many genetic variants show effects only in specific situations, adding to complexity. In consequence, for highly polygenic traits such as common complex diseases, millions of individuals are required to identify all susceptibility variants of the effect sizes that exist in nature and survive natural selection at genome-wide significance ($P < 5 \times 10^{-8}$), (Watanabe et al. 2019). An alternative to increasing the size of clinical analysis samples such extensively is taking advantage of the opposite scenario. Literature shows that risk genes detected by GWAS harbor rare high-effect susceptibility variants, which are independent (not in linkage disequilibrium) of the associations with common diseases and cause severe early-onset disease manifestations (Do et al. 2015; Flannick et al. 2014; Fuchsberger et al. 2016; Luo et al. 2017). Therefore, rare disorders that are distinguished from common forms by age-of-onset and severity can be used to identify not only rare variants with large effect sizes but are tools to identify candidate susceptibility genes of common disease forms. These candidate genes can be tested for associations in available GWAS genotype data sets (Antonarakis and Beckmann 2006). The reduced number of multiple testing increases the statistical power of the currently available samples.

GWAS cannot identify rare variants that cause rare early-onset forms of disease because they are not covered by the genotyping arrays. Instead, the DNA of the patients and, ideally, of the core family including the parents and siblings, need to be sequenced to identify the causative variants. When an entire genome is being sequenced, the process is called 'whole-genome sequencing.' An alternative to whole-genome sequencing is the targeted sequencing of parts of a genome. Most often, this involves just sequencing the protein-coding regions, which reside within DNA segments called 'exons' and reflect the currently 'best understood' part of most genomes. For example, all of the exons in the human genome (the human 'exome') correspond to ~1.5% of the total human genome. Methods are now available to isolate and sequence the exons to generate a 'whole-exome sequence' (WES) of a genome.

Several WES studies have been performed for periodontal diseases to date. In a consanguineous Turkish family with very early EO-PD (first diagnosis in early childhood), a homozygous missense mutation of the gene *CTSC*, the causative gene of the Papillon-Lefèvre syndrome, was identified in the affected family members (Molitor et al. 2019). This confirmed results from an earlier study that identified two variants within *CTSC* to be specific to nonsyndromic extreme EO-PD (Hewitt et al. 2004). In another WES study of two families with severe EO-PD cases (formerly classified as aggressive periodontitis, currently PD grade C) from Japan, a heterozygous missense mutation in *NOD2* was described (Sudo et al. 2017), that was previously reported to contribute to inflammatory bowel disease (Rivas et al. 2011). Another WES-study from Brazil sequenced two families including a child and parent affected with PD grade C cases (29 and 32 years of age at the year of enrollment) and an unaffected parent and sibling. This study proposed various putative deleterious mutations in the genes *EEFSEC*, *ELN*, *GPRC6A*, *KRT25*, and *ZNF* (Sudo et al. 2017). In summary, to date, some WES studies have been conducted with few EA-PD patients. However, these studies rediscovered, among others, the known EO-PD risk gene *CTSC*.

11.3 Conclusion and Perspectives

Substantial advances have been made in genetic research, although progress has been much slower in dentistry compared to other fields of medicine. Starting from candidate gene association studies, new approaches were employed that have expanded the study design of GWAS to genomewide meta-analyses and gene x environment interaction studies. Technical advances have enabled cost efficient whole-genome and whole-exome sequencing of patients with rare forms of periodontitis with very severe and rapid progression and early age of onset, that has the potential to identify genes and pathways with a direct role in pathogenesis. In parallel, animal models were developed that use genetically highly diverse mouse lines to identify risk genes of human diseases. These experimental and technical advances are real improvements that will strongly contribute to a more fundamental understanding of the manifold causes of periodontal diseases.

The knowledge of genetic loci that are associated with periodontal diseases is not synonymous, however, with the presence of meaning. The challenge of our current times and into the future is to leverage the wealth of genomic data to derive true biological meaning from GWAS-implicated disease risk loci. The bottleneck in our understanding of risk loci found by GWAS is likely to be due to a lack of disease-focused functional biological studies downstream of GWAS locus discovery, and in medical genetic research, the main priority now is to translate genome-based information into biological relevant knowledge. Various experimental 'wet lab' approaches have been established to test the functions of potential causal variants at these regions. Now, an increased emphasis on the functional dissection of already identified GWAS loci is most likely to benefit knowledge of pathophysiology. These methods are laborious, require time, and compared to GWAS analyses, different experimental and analytical expertise. This is the reason why there will not be real improvements until we bridge the gap that separates dental medicine from natural sciences, a project that is just getting started. In human genetics, the requirements of GWAS' fundamentally transformed the ways in which research was being performed. The necessities to generate large sample sizes, which involved many different groups, high costs for sample collection and processing, as well as expensive technology and quickly developing data analyses methods, transformed single research groups into international consortia that worked highly cooperative on the same projects, exchanging data and methods, and incorporated both clinicians and natural scientists of various specialisations. Similar developments will be required for dentistry to transform our understanding in ways that eventually lead to diagnosis and treatments that do not describe and treat the symptoms but heal the disease.

References

Abusleme L, Dupuy AK, Dutzan N, Silva N, Burleson JA, Strausbaugh LD, Gamonal J, Diaz PI (2013) The subgingival microbiome in health and periodontitis and its relationship with community biomass and inflammation. ISME J 7(5):1016–1025. https://doi.org/10.1038/ismej.2012.174

Aldred PM, Hollox EJ, Armour JA (2005) Copy number polymorphism and expression level variation of the human alpha-defensin genes DEFA1 and DEFA3. Hum Mol Genet 14(14):2045–2052. https://doi.org/10.1093/hmg/ddi209

Antonarakis SE, Beckmann JS (2006) Mendelian disorders deserve more attention. Nat Rev Genet 7(4):277–282. https://doi.org/10.1038/nrg1826

Balding DJ (2006) A tutorial on statistical methods for population association studies. Nat Rev Genet 7(10):781–791. https://doi.org/10.1038/nrg1916

Blasco-Baque V, Garidou L, Pomie C, Escoula Q, Loubieres P, Le Gall-David S, Lemaitre M, Nicolas S, Klopp P, Waget A, Azalbert V, Colom A, Bonnaure-Mallet M, Kemoun P, Serino M, Burcelin R (2016) Periodontitis induced by Porphyromonas gingivalis drives periodontal microbiota dysbiosis and insulin resistance via an impaired adaptive immune response. Gut. https://doi.org/10.1136/gutjnl-2015-309897

Bulik-Sullivan B, Finucane HK, Anttila V, Gusev A, Day FR, Loh PR, ReproGen C, Psychiatric Genomics C, Genetic Consortium for Anorexia Nervosa of the Wellcome Trust Case Control C, Duncan L, Perry JR, Patterson N, Robinson EB, Daly MJ, Price AL, Neale BM (2015) An atlas of genetic correlations across human diseases and traits. Nat Genet 47(11):1236–1241. https://doi.org/10.1038/ng.3406

Caton JG, Armitage G, Berglundh T, Chapple ILC, Jepsen S, Kornman KS, Mealey BL, Papapanou PN, Sanz M, Tonetti MS (2018) A new classification scheme for periodontal and peri-implant diseases and conditions – introduction and key changes from the 1999 classification. J Periodontol 89(Suppl 1):S1–S8. https://doi.org/10.1002/JPER.18-0157

Cavagni J, Wagner TP, Gaio EJ, Rego RO, Torres IL, Rosing CK (2013) Obesity may increase the occurrence of spontaneous periodontal disease in Wistar rats. Arch Oral Biol 58(8):1034–1039. https://doi.org/10.1016/j.archoralbio.2013.03.006

Chen GY, Tang J, Zheng P, Liu Y (2009) CD24 and Siglec-10 selectively repress tissue damage-induced immune responses. Science 323(5922):1722–1725. https://doi.org/10.1126/science.1168988

Churchill GA, Airey DC, Allayee H, Angel JM, Attie AD, Beatty J, Beavis WD, Belknap JK, Bennett B, Berrettini W, Bleich A, Bogue M, Broman KW, Buck KJ, Buckler E, Burmeister M, Chesler EJ, Cheverud JM, Clapcote S, Cook MN, Cox RD, Crabbe JC, Crusio WE, Darvasi A, Deschepper CF, Doerge RW, Farber CR, Forejt J, Gaile D, Garlow SJ, Geiger H, Gershenfeld H, Gordon T, Gu J, Gu W, de Haan G, Hayes NL, Heller C, Himmelbauer H, Hitzemann R, Hunter K, Hsu HC, Iraqi FA, Ivandic B, Jacob HJ, Jansen RC, Jepsen KJ, Johnson DK, Johnson TE, Kempermann G, Kendziorski C, Kotb M, Kooy RF, Llamas B, Lammert F, Lassalle JM, Lowenstein PR, Lu L, Lusis A, Manly KF, Marcucio R, Matthews D, Medrano JF, Miller DR, Mittleman G, Mock BA, Mogil JS, Montagutelli X, Morahan G, Morris DG, Mott R, Nadeau JH, Nagase H, Nowakowski RS, O'Hara BF, Osadchuk AV, Page GP, Paigen B, Paigen K, Palmer AA, Pan HJ, Peltonen-Palotie L, Peirce J, Pomp D, Pravenec M, Prows DR, Qi Z, Reeves RH, Roder J, Rosen GD, Schadt EE, Schalkwyk LC, Seltzer Z, Shimomura K, Shou S, Sillanpaa MJ, Siracusa LD, Snoeck HW, Spearow JL, Svenson K, Tarantino LM, Threadgill D, Toth LA, Valdar W, de Villena FP, Warden C, Whatley S, Williams RW, Wiltshire T, Yi N, Zhang D, Zhang M, Zou F (2004) The Collaborative Cross, a community resource for the genetic analysis of complex traits. Nat Genet 36(11):1133–1137. doi:ng1104-1133 [pii]. https://doi.org/10.1038/ng1104-1133

Civelek M, Lusis AJ (2014) Systems genetics approaches to understand complex traits. Nat Rev Genet 15(1):34–48. https://doi.org/10.1038/nrg3575

Clemente JC, Ursell LK, Parfrey LW, Knight R (2012) The impact of the gut microbiota on human health: an integrative view. Cell 148(6):1258–1270. https://doi.org/10.1016/j.cell.2012.01.035

Collaborative Cross C (2012) The genome architecture of the Collaborative Cross mouse genetic reference population. Genetics 190(2):389–401. https://doi.org/10.1534/genetics.111.132639

Cornick S, Moreau F, Gaisano HY, Chadee K (2017) Entamoeba histolytica-induced mucin exocytosis is mediated by VAMP8 and is critical in mucosal innate host defense. mBio 8 (5). https://doi.org/10.1128/mBio.01323-17

Darveau RP (2010) Periodontitis: a polymicrobial disruption of host homeostasis. Nat Rev Microbiol 8(7):481–490. https://doi.org/10.1038/nrmicro2337

Divaris K, Monda KL, North KE, Olshan AF, Reynolds LM, Hsueh WC, Lange EM, Moss K, Barros SP, Weyant RJ, Liu Y, Newman AB, Beck JD, Offenbacher S (2013) Exploring the genetic basis of chronic periodontitis: a genome-wide association study. Hum Mol Genet https://doi.org/10.1093/hmg/ddt065. doi:ddt065 [pii]

Do R, Stitziel NO, Won HH, Jorgensen AB, Duga S, Angelica Merlini P, Kiezun A, Farrall M, Goel A, Zuk O et al (2015) Exome sequencing identifies rare ldlr and apoa5 alleles conferring risk for myocardial infarction. Nature 518(7537):102–106

Dupont WD, Plummer WD Jr (1998) Power and sample size calculations for studies involving linear regression. Control Clin Trials 19(6):589–601

Eke PI, Dye BA, Wei L, Slade GD, Thornton-Evans GO, Borgnakke WS, Taylor GW, Page RC, Beck JD, Genco RJ (2015) Update on Prevalence of Periodontitis in Adults in the United States: NHANES 2009 to 2012. J Periodontol 86(5):611–622. https://doi.org/10.1902/jop.2015.140520

Eke PI, Dye BA, Wei L, Thornton-Evans GO, Genco RJ (2012) Prevalence of periodontitis in adults in the United States: 2009 and 2010. J Dent Res 91(10):914–920. https://doi.org/10.1177/0022034512457373. doi:0022034512457373 [pii]

Feng S, Deng L, Chen W, Shao J, Xu G, Li YP (2009) Atp6v1c1 is an essential component of the osteoclast proton pump and in F-actin ring formation in osteoclasts. Biochem J 417(1):195–203. https://doi.org/10.1042/BJ20081073

Flannick J, Thorleifsson G, Beer NL, Jacobs SB, Grarup N, Burtt NP, Mahajan A, Fuchsberger C, Atzmon G, Benediktsson R et al (2014) Loss-of-function mutations in slc30a8 protect against type 2 diabetes. Nat Genet 46(4):357–363

Freitag-Wolf S, Munz M, Wiehe R, Junge O, Graetz C, Jockel-Schneider Y, Staufenbiel I, Bruckmann C, Lieb W, Franke A, Loos BG, Jepsen S, Dommisch H, Schaefer AS (2019) Smoking Modifies the Genetic Risk for Early-Onset Periodontitis. J Dent Res 98(12):1332–1339. https://doi.org/10.1177/0022034519875443

Fuchsberger C, Flannick J, Teslovich TM, Mahajan A, Agarwala V, Gaulton KJ, Ma C, Fontanillas P, Moutsianas L, McCarthy DJ et al (2016) The genetic architecture of type 2 diabetes. Nature 536(7614):41–47

Fujita Y, Maki K (2015) High-fat diet-induced obesity triggers alveolar bone loss and spontaneous periodontal disease in growing mice. BMC Obes 3:1. https://doi.org/10.1186/s40608-016-0082-8

Ganz T (2003) Defensins: antimicrobial peptides of innate immunity. Nat Rev Immunol 3(9):710–720. https://doi.org/10.1038/nri1180

Garland T Jr (2014) Trade-offs. Curr Biol 24(2):R60–R61. https://doi.org/10.1016/j.cub.2013.11.036

Hewitt C, McCormick D, Linden G, Turk D, Stern I, Wallace I, Southern L, Zhang L, Howard R, Bullon P,

Wong M, Widmer R, Gaffar KA, Awawdeh L, Briggs J, Yaghmai R, Jabs EW, Hoeger P, Bleck O, Rudiger SG, Petersilka G, Battino M, Brett P, Hattab F, Al-Hamed M, Sloan P, Toomes C, Dixon M, James J, Read AP, Thakker N (2004) The role of cathepsin C in Papillon-Lefevre syndrome, prepubertal periodontitis, and aggressive periodontitis. Hum Mutat 23:222–228. https://doi.org/10.1002/humu.10314

Hong KW, Shin MS, Ahn YB, Lee HJ, Kim HD (2015) Genomewide association study on chronic periodontitis in korean population: Results from the yangpyeong health cohort. J Clin Periodontol 42(8):703–710. https://doi.org/10.1111/jcpe.12437

Howson JMM, Zhao W, Barnes DR, Ho WK, Young R, Paul DS, Waite LL, Freitag DF, Fauman EB, Salfati EL, Sun BB, Eicher JD, Johnson AD, Sheu WHH, Nielsen SF, Lin WY, Surendran P, Malarstig A, Wilk JB, Tybjaerg-Hansen A, Rasmussen KL, Kamstrup PR, Deloukas P, Erdmann J, Kathiresan S, Samani NJ, Schunkert H, Watkins H, CardioGramplusC4D, Do R, Rader DJ, Johnson JA, Hazen SL, Quyyumi AA, Spertus JA, Pepine CJ, Franceschini N, Justice A, Reiner AP, Buyske S, Hindorff LA, Carty CL, North KE, Kooperberg C, Boerwinkle E, Young K, Graff M, Peters U, Absher D, Hsiung CA, Lee WJ, Taylor KD, Chen YH, Lee IT, Guo X, Chung RH, Hung YJ, Rotter JI, Juang JJ, Quertermous T, Wang TD, Rasheed A, Frossard P, Alam DS, Majumder AAS, Di Angelantonio E, Chowdhury R, Epic CVD, Chen YI, Nordestgaard BG, Assimes TL, Danesh J, Butterworth AS, Saleheen D (2017) Fifteen new risk loci for coronary artery disease highlight arterial-wall-specific mechanisms. Nat Genet 49(7):1113–1119. https://doi.org/10.1038/ng.3874

Iraqi FA, Churchill G, Mott R (2008) The Collaborative Cross, developing a resource for mammalian systems genetics: a status report of the Wellcome Trust cohort. Mamm Genome 19(6):379–381. https://doi.org/10.1007/s00335-008-9113-1

Jiang H, Chen W, Zhu G, Zhang L, Tucker B, Hao L, Feng S, Ci H, Ma J, Wang L, Stashenko P, Li YP (2013) RNAi-mediated silencing of Atp6i and Atp6i haploinsufficiency prevents both bone loss and inflammation in a mouse model of periodontal disease. PLoS One 8(4):e58599. https://doi.org/10.1371/journal.pone.0058599

Jones LC, Moussa L, Fulcher ML, Zhu Y, Hudson EJ, O'Neal WK, Randell SH, Lazarowski ER, Boucher RC, Kreda SM (2012) VAMP8 is a vesicle SNARE that regulates mucin secretion in airway goblet cells. J Physiol 590(3):545–562. https://doi.org/10.1113/jphysiol.2011.222091

Khan FF, Carpenter D, Mitchell L, Mansouri O, Black HA, Tyson J, Armour JA (2013) Accurate measurement of gene copy number for human alpha-defensin DEFA1A3. BMC Genomics 14:719. https://doi.org/10.1186/1471-2164-14-719

Kinane DF, Peterson M, Stathopoulou PG (2006) Environmental and other modifying factors of the periodontal diseases. Periodontol 2000 40:107–119. https://doi.org/10.1111/j.1600-0757.2005.00136.x

Laine ML, Loos BG, Crielaard W (2010) Gene polymorphisms in chronic periodontitis. Int J Dent 2010:324719. https://doi.org/10.1155/2010/324719

Lamont RJ, Koo H, Hajishengallis G (2018) The oral microbiota: dynamic communities and host interactions. Nat Rev Microbiol 16(12):745–759. https://doi.org/10.1038/s41579-018-0089-x

Lang NP, Schätzle MA, Löe H (2009) Gingivitis as a risk factor in periodontal disease. J Clin Periodontol 36(Suppl 10):3–8. https://doi.org/10.1111/j.1600-051X.2009.01415.x

Liu Y, Chen GY, Zheng P (2011) Sialoside-based pattern recognitions discriminating infections from tissue injuries. Curr Opin Immunol 23(1):41–45. https://doi.org/10.1016/j.coi.2010.10.004

Lockhart PB, Bolger AF, Papapanou PN, Osinbowale O, Trevisan M, Levison ME, Taubert KA, Newburger JW, Gornik HL, Gewitz MH, Wilson WR, Smith SC Jr, Baddour LM (2012) Periodontal disease and atherosclerotic vascular disease: does the evidence support an independent association?: a scientific statement from the American Heart Association. Circulation 125(20):2520–2544. doi:CIR.0b013e31825719f3 [pii]. https://doi.org/10.1161/CIR.0b013e31825719f3

Loos BG, John RP, Laine ML (2005) Identification of genetic risk factors for periodontitis and possible mechanisms of action. J Clin Periodontol 32(Suppl 6):159–179

Luo Y, de Lange KM, Jostins L, Moutsianas L, Randall J, Kennedy NA, Lamb CA, McCarthy S, Ahmad T, Edwards C et al (2017) Exploring the genetic architecture of inflammatory bowel disease by whole-genome sequencing identifies association at adcy7. Nat Genet 49(2):186–192

Manolio TA (2010) Genomewide association studies and assessment of the risk of disease. N Engl J Med 363(2):166–176. https://doi.org/10.1056/NEJMra0905980

Marcenes W, Kassebaum NJ, Bernabe E, Flaxman A, Naghavi M, Lopez A, Murray CJ (2013) Global burden of oral conditions in 1990-2010: a systematic analysis. J Dent Res 92(7):592–597. https://doi.org/10.1177/0022034513490168. doi:0022034513490168 [pii]

McDade TW, Reyes-Garcia V, Tanner S, Huanca T, Leonard WR (2008) Maintenance versus growth: investigating the costs of immune activation among children in lowland Bolivia. Am J Phys Anthropol 136(4):478–484. https://doi.org/10.1002/ajpa.20831

Michalowicz BS, Diehl SR, Gunsolley JC, Sparks BS, Brooks CN, Koertge TE, Califano JV, Burmeister JA, Schenkein HA (2000) Evidence of a substantial genetic basis for risk of adult periodontitis. J Periodontol 71(11):1699–1707. https://doi.org/10.1902/jop.2000.71.11.1699

Molitor A, Prud'homme T, Miao Z, Conrad S, Bloch-Zupan A, Pichot A, Hanauer A, Isidor B, Bahram S, Carapito R (2019) Exome sequencing identifies a novel missense variant in ctsc causing nonsyndromic aggressive periodontitis. J Hum Genet 64(7):689–694

Munz M, Richter GM, Loos BG, Jepsen S, Divaris K, Offenbacher S, Teumer A, Holtfreter B, Kocher T, Bruckmann C, Jockel-Schneider Y, Graetz C, Ahmad I, Staufenbiel I, van der Velde N, Uitterlinden AG, de Groot L, Wellmann J, Berger K, Krone B, Hoffmann P, Laudes M, Lieb W, Franke A, Erdmann J, Dommisch H, Schaefer AS (2019) Meta-analysis of genome-wide association studies of aggressive and chronic periodontitis identifies two novel risk loci. Eur J Hum Genet 27(1):102–113. https://doi.org/10.1038/s41431-018-0265-5

Munz M, Richter GM, Loos BG, Jepsen S, Divaris K, Offenbacher S, Teumer A, Holtfreter B, Kocher T, Bruckmann C, Jockel-Schneider Y, Graetz C, Munoz L, Bhandari A, Tennstedt S, Staufenbiel I, van der Velde N, Uitterlinden AG, de Groot L, Wellmann J, Berger K, Krone B, Hoffmann P, Laudes M, Lieb W, Franke A, Dommisch H, Erdmann J, Schaefer AS (2018) Genome-wide association meta-analysis of coronary artery disease and periodontitis reveals a novel shared risk locus. Sci Rep 8(1):13678. https://doi.org/10.1038/s41598-018-31980-8

Munz M, Willenborg C, Richter GM, Jockel-Schneider Y, Graetz C, Staufenbiel I, Wellmann J, Berger K, Krone B, Hoffmann P, Velde NV, Uitterlinden AG, Groot L, Sawalha A, Direskeneli H, Saruhan-Direskeneli G, Guzeldemir-Akcakanat E, Keceli G, Laudes M, Noack B, Teumer A, Holtfreter B, Kocher T, Eickholz P, Meyle J, Doerfer C, Bruckmann C, Lieb W, Franke A, Schreiber S, Nohutcu RM, Erdmann J, Loos BG, Jepsen S, Dommisch H, Schaefer AS (2017) A genome-wide association study identifies nucleotide variants at SIGLEC5 and DEFA1A3 as risk loci for periodontitis. Hum Mol Genet. https://doi.org/10.1093/hmg/ddx151

Nesse RM, Stein DJ (2012) Towards a genuinely medical model for psychiatric nosology. BMC Med 10:5. https://doi.org/10.1186/1741-7015-10-5

Nibali L, Bayliss-Chapman J, Almofareh SA, Zhou Y, Divaris K, Vieira AR (2019) What is the heritability of periodontitis? A systematic review. J Dent Res 98(6):632–641. https://doi.org/10.1177/0022034519842510

Nibali L, Donos N, Henderson B (2009) Periodontal infectogenomics. J Med Microbiol 58(Pt 10):1269–1274. doi:jmm.0.012021-0 [pii]. https://doi.org/10.1099/jmm.0.012021-0

Nibali L, Henderson B, Sadiq ST, Donos N (2014) Genetic dysbiosis: the role of microbial insults in chronic inflammatory diseases. J Oral Microbiol 6. https://doi.org/10.3402/jom.v6.22962 22962 [pii]

Piegorsch WW, Weinberg CR, Taylor JA (1994) Non-hierarchical logistic models and case-only designs for assessing susceptibility in population-based case-control studies. Stat Med 13(2):153–162

Rivas MA, Beaudoin M, Gardet A, Stevens C, Sharma Y, Zhang CK, Boucher G, Ripke S, Ellinghaus D, Burtt N et al (2011) Deep resequencing of gwas loci identifies independent rare variants associated with inflammatory bowel disease. Nat Genet 43(11):1066–1073

Sanders AE, Sofer T, Wong Q, Kerr KF, Agler C, Shaffer JR, Beck JD, Offenbacher S, Salazar CR, North KE et al (2016) Chronic periodontitis genome-wide association study in the hispanic community health study/study of latinos. J Dent Res 96(1):64–72. https://doi.org/10.1177/0022034516664509

Shaffer JR, Polk DE, Wang X, Feingold E, Weeks DE, Lee MK, Cuenco KT, Weyant RJ, Crout RJ, McNeil DW et al (2014) Genome-wide association study of periodontal health measured by probing depth in adults ages 18–49 years. G3 (Bethesda) 4(2):307–314

Shimizu S, Momozawa Y, Takahashi A, Nagasawa T, Ashikawa K, Terada Y, Izumi Y, Kobayashi H, Tsuji M, Kubo M et al (2015) A genome-wide association study of periodontitis in a japanese population. J Dent Res 94(4):555–561

Shungin D, Cornelis MC, Divaris K, Holtfreter B, Shaffer JR, Yu YH, Barros SP, Beck JD, Biffar R, Boerwinkle EA, Crout RJ, Ganna A, Hallmans G, Hindy G, Hu FB, Kraft P, McNeil DW, Melander O, Moss KL, North KE, Orho-Melander M, Pedersen NL, Ridker PM, Rimm EB, Rose LM, Rukh G, Teumer A, Weyant RJ, Chasman DI, Joshipura K, Kocher T, Magnusson PK, Marazita ML, Nilsson P, Offenbacher S, Davey Smith G, Lundberg P, Palmer TM, Timpson NJ, Johansson I, Franks PW (2015) Using genetics to test the causal relationship of total adiposity and periodontitis: Mendelian randomization analyses in the Gene-Lifestyle Interactions and Dental Endpoints (GLIDE) Consortium. Int J Epidemiol 44(2):638–650. https://doi.org/10.1093/ije/dyv075

Shungin D, Haworth S, Divaris K, Agler CS, Kamatani Y, Keun Lee M, Grinde K, Hindy G, Alaraudanjoki V, Pesonen P, Teumer A, Holtfreter B, Sakaue S, Hirata J, Yu YH, Ridker PM, Giulianini F, Chasman DI, Magnusson PKE, Sudo T, Okada Y, Volker U, Kocher T, Anttonen V, Laitala ML, Orho-Melander M, Sofer T, Shaffer JR, Vieira A, Marazita ML, Kubo M, Furuichi Y, North KE, Offenbacher S, Ingelsson E, Franks PW, Timpson NJ, Johansson I (2019) Genome-wide analysis of dental caries and periodontitis combining clinical and self-reported data. Nat Commun 10(1):2773. https://doi.org/10.1038/s41467-019-10630-1

Shusterman A, Munz M, Richter G, Jepsen S, Lieb W, Krone B, Hoffman P, Laudes M, Wellmann J, Berger K, Kocher T, Offenbacher S, Divaris K, Franke A, Schreiber S, Dommisch H, Weiss E, Schaefer AS, Houri-Haddad Y, Iraqi FA (2017) The PF4/PPBP/CXCL5 gene cluster is associated with periodontitis. J Dent Res:22034517706311. https://doi.org/10.1177/0022034517706311

Sivakumaran S, Agakov F, Theodoratou E, Prendergast JG, Zgaga L, Manolio T, Rudan I, McKeigue P, Wilson JF, Campbell H (2011) Abundant pleiotropy in human complex diseases and traits. Am J Hum Genet 89(5):607–618. https://doi.org/10.1016/j.ajhg.2011.10.004. S0002-9297(11)00438-1 [pii]

Sudo T, Okada Y, Ozaki K, Urayama K, Kanai M, Kobayashi H, Gokyu M, Izumi Y, Tanaka T (2017) Association of nod2 mutations with aggressive periodontitis. J Dent Res 96(10):1100–1105

Sutherland MK, Geoghegan JC, Yu C, Winkler DG, Latham JA (2004) Unique regulation of SOST, the sclerosteosis gene, by BMPs and steroid hormones in human osteoblasts. Bone 35(2):448–454. https://doi.org/10.1016/j.bone.2004.04.019

Teumer A, Holtfreter B, Volker U, Petersmann A, Nauck M, Biffar R, Volzke H, Kroemer HK, Meisel P, Homuth G et al (2013) Genome-wide association study of chronic periodontitis in a general german population. J Clin Periodontol 40(11):977–985

Timpson NJ, Greenwood CMT, Soranzo N, Lawson DJ, Richards JB (2018) Genetic architecture: the shape of the genetic contribution to human traits and disease. Nat Rev Genet 19(2):110–124. https://doi.org/10.1038/nrg.2017.101

Tonetti MS, Claffey N, European Workshop in Periodontology group C (2005) Advances in the progression of periodontitis and proposal of definitions of a periodontitis case and disease progression for use in risk factor research. Group C consensus report of the 5th European Workshop in Periodontology. J Clin Periodontol 32(Suppl 6):210–213. https://doi.org/10.1111/j.1600-051X.2005.00822.x

van Bezooijen RL, Roelen BA, Visser A, van der Wee-Pals L, de Wilt E, Karperien M, Hamersma H, Papapoulos SE, ten Dijke P, Lowik CW (2004) Sclerostin is an osteocyte-expressed negative regulator of bone formation, but not a classical BMP antagonist. J Exp Med 199(6):805–814. https://doi.org/10.1084/jem.20031454

Watanabe K, Stringer S, Frei O, Umicevic Mirkov M, de Leeuw C, Polderman TJC, van der Sluis S, Andreassen OA, Neale BM, Posthuma D (2019) A global overview of pleiotropy and genetic architecture in complex traits. Nat Genet 51(9):1339–1348. https://doi.org/10.1038/s41588-019-0481-0

Wilkening S, Chen B, Bermejo JL, Canzian F (2009) Is there still a need for candidate gene approaches in the era of genome-wide association studies? Genomics

Winkler DG, Sutherland MK, Geoghegan JC, Yu C, Hayes T, Skonier JE, Shpektor D, Jonas M, Kovacevich BR, Staehling-Hampton K, Appleby M, Brunkow ME, Latham JA (2003) Osteocyte control of bone formation via sclerostin, a novel BMP antagonist. EMBO J 22(23):6267–6276. https://doi.org/10.1093/emboj/cdg599

Wray NR, Yang J, Goddard ME, Visscher PM (2010) The genetic interpretation of area under the ROC curve in genomic profiling. PLoS Genet 6:e1000864. https://doi.org/10.1371/journal.pgen.1000864

Yong SY, Raben TG, Lello L, Hsu SDH (2020) Genetic architecture of complex traits and disease risk predictors. Sci Rep 10(1):12055. https://doi.org/10.1038/s41598-020-68881-8

Yost S, Duran-Pinedo AE, Krishnan K, Frias-Lopez J (2017) Potassium is a key signal in host-microbiome dysbiosis in periodontitis. PLoS Pathog 13(6):e1006457. https://doi.org/10.1371/journal.ppat.1006457

Part III

Periodontitis-Associated Conditions

Update on the Bidirectional Link Between Diabetes and Periodontitis

12

Leila Salhi and Michèle Reners

Abstract

Aim: To provide an update on the evidence on the bidirectional relationship between periodontitis and diabetes.

Methods: This narrative review was focused on recent studies between 2015 and 2020. The literature search was performed on PubMed. The inclusion criteria were systematic reviews, consensus reports and controlled trials assessing the effect of diabetes on periodontitis, the effect of periodontitis on diabetes, and the influence of periodontitis non-surgical treatment on diabetes.

Results: Data concerning the influence of periodontitis on diabetes, and the influence of diabetes on periodontitis were summarized in descriptive tables.

Conclusion: The control of hyperglycemia in the prevention of periodontitis and the control of periodontitis systemic inflammation in the prevention of diabetes, should be take into account in the treatment planning of both diseases.

Keywords

Periodontitis · Diabetes · Periodontitis non-surgical treatment

Abbreviations

AGEs	Advanced Glycation Endproducts)
CI	Confidence Interval
EFP	European Federation of Periodontology
HR	Hazard Ratio
PNST	Periodontal Non-Surgical Treatment
RR	Relative Risk (or Risk Ratio)
T1DM	Type 1 Diabetes Mellitus
T2DM	Type 2 Diabetes Mellitus
VEGF	Vascular Endothelial Growth Factors
WHO	World Health Organization
ORCA	European Organization for Caries Research
RAGE	Receptor for Advanced Glycation Endproducts
OPG	Osteoprotegerin Ratio
PMNs	Polymorphonuclear Leucocytes

L. Salhi (✉)
Department of Periodontology, Buccal Surgery and Implantology, University of Liège, Liège, Belgium
e-mail: l.salhi@chuliege.be

M. Reners
Private Practice, Parolìège, Liège, Belgium
e-mail: mreners@gmail.com

J. Santi-Rocca (ed.), *Periodontitis*, Advances in Experimental Medicine and Biology 1373,
https://doi.org/10.1007/978-3-030-96881-6_12

Highlights

Periodontitis and diabetes presented bidirectional influence:

- People with diabetes have an increased risk for periodontitis prevalence and tooth loss;
- People with periodontitis have an increased risk for systemic inflammation and diabetes onset.

The treatment of periodontitis reduces HBA1c, CRP, TNF-α, IL-6 level, contributing to the control of diabetes of glycemia and prevent further complications.

Considerations For Practice

- All diabetic patients should be screened for periodontal disease and should receive information about how to prevent and to treat periodontal disease
- It might be relevant, for patients with poor response after periodontal treatment, to propose the screening of diabetes by their doctor.
- The periodontal non-surgical treatment lead, after 3 months, to a significant decrease of HBA1c

Patient Summary

A good oral health contributes to a better control of diabetes.

A well-controlled diabetes contributes to oral health and avoid periodontitis. Prevention is the best way to keep diseases away.

Regularly visit your doctor and your dental practitioner for screening diabetes and periodontal disease.

12.1 Introduction

According to the world health organization (WHO), diabetes is a chronic metabolic disease that affects more than 422 million of adults worldwide (Tonetti et al. 2013), including 59.3 million of European people aged from 20–79 years (Atlas 2019). This global epidemic disease occurs when the pancreas fails to produce enough insulin or, when the body becomes resistant to insulin. Both of these conditions are characterized by hyperglycemia in the blood with a decrease of glucose storage in the liver, but also a limited use of glucose by other cells and organs as retina, heart and kidney. Additionally, hyperglycemia induces the expression of the vascular endothelial growth factors (VEGF), advanced glycation endproducts (AGEs), and oxidative stress (Nardi et al. 2020) that negatively impair the microvascularisation and increase the risk of diabetes complications. The adverse events involved in the disease concern microvascular complications (retinopathy, nephropathy and neuropathy), macrovascular complications (coronary artery disease, peripheral arterial disease and stroke)(Skyler et al. 2017). Additionally, during pregnancy, gestational diabetes can provoke eclampsia with preterm birth and baby with low birth weight (Tonetti et al. 2013; Atlas 2019). Besides, diabetes has been widely associated with the increase of sensibility to infection (foot ulcers) (Alavi et al. 2014; Everett and Mathioudakis 2018) and periodontal disease (Preshaw and Bissett 2013, 2019).

Periodontitis is a chronic inflammatory disease characterized by gram negative bacteria organized in a dental plaque biofilm within microbial dysbiosis leads to a chronic non-resolving and destructive inflammatory response (Jepsen et al. 2017; Meyle and Chapple 2015) that induces local inflammation and progressive destruction of the supporting tissues around teeth (Meyle and Chapple 2015; Schenkein 2006; Haffajee and Socransky 1994; Amano 2010a, b). Approximately 50% of the adult population suffers from this chronic disease (Eke et al. 2016), and severe periodontitis affects 11.2% of the population (Kassebaum et al. 2014). Moreover, due to mastication and tooth brushing, the periodontal pathogens or/and their sub-products (SPs) are able to disseminate (Forner et al. 2006) from the pocket depth to the blood circulation, inducing a systemic inflammation with an extra-oral metastatic infection (Van Dyke and van Winkelhoff 2013; Loos 2005; Chapple et al. 2013; Szulc et al. 2015; Tonetti et al. 2013; Mealey 1999; van Winkelhoff and Slots 1999). This pathway contributes to the relationship between periodontitis and chronic non-communicable diseases as cardiovascular diseases and diabetes (Chapple et al. 2013; Tonetti et al. 2013; Sanz et al. 2018; Salhi et al. 2019; Salhi et al. 2020).

Periodontitis and diabetes are two associated diseases with a well-described bidirectional influence (Preshaw and Bissett 2019; Sanz et al. 2018; Preshaw et al. 2012; Kumar et al. 2014; Taylor 2001). This review aims to support the bidirectional influence of these two diseases.

12.2 Method

This narrative review was focused on recent studies between 2015 and 2020. The literature search was performed on PubMed, and inclusion criteria were systematic reviews, consensus reports and controlled trials assessing the relationship between periodontitis and diabetes, and the influence of periodontitis non-surgical treatment on diabetes.

12.3 Results

This narrative review aims to highlight the updated literature on the bidirectional link between periodontitis and diabetes. Three questions were asked: (1) what is the effect of diabetes on periodontitis?, (2) what is the effect of periodontitis on diabetes?, and (3) what is the effect of periodontal non-surgical treatment on diabetes? The above-described search were summarized respectively in Tables 12.1, 12.2 and 12.3.

Table 12.1 Effect of diabetes on periodontitis

Authors	Study design	Conclusion
Dicembrini et al. (2020)	Systematic review on 5 studies	T1DM is a relevant risk factor for the development of PD. The prevalence of PD in type 1 diabetes was 18.5%
Nardi et al. (2020)	Systematic review on 8 studies	The microvasculature of the periodontium is affected in diabetic patients towards the overexpression of VEGF
Herrmann et al. (2019)	Controlled study on 20 patients with chronic periodontitis T2DM + CP patients with type 2 diabetes mellitus and chronic periodontitis; CP: Systemically healthy participants with chronic periodontit Patients with type 2 diabetes Systemically healthy patients	Overexpression of PMN's in patients with T2DM
Sanz et al. (2018)	Consensus report	Poor glycaemic control in diabetes is associated with poorer periodontal status and outcomes
Nascimento et al. (2018)	Systematic review and meta-analysis on 13 studies	Diabetes increased the risk of incidence or progression of periodontitis by 86% (RR 1.86 [95% CI 1.3–2.8])
Chapple et al. (2017)	Consensus report	Hyperglycaemia drives oxidative stress and advanced glycation end-pro-ducts that can trigger a hyper inflammatory state and periodontitis

T1DM type 1 diabetes mellitus, *T2DM* type 2 diabetes mellitus, *PD* periodontal disease, *VEGF* vascular endothelial growth factors, *PMN's* polymorphonuclear leucocytes

Table 12.2 Effect of periodontitis on diabetes

Authors	Study design	Conclusion
Jensen et al. (2020)	Systematic review and meta-analysis on 23 studies	Risk markers for periodontal disease, PI, GI, BOP, PD and CAL were found to be more pronounced among children and adolescents with T1DM compared to healthy control
Rapone et al. (2020)	Systematic review and meta-analysis on 10 studies	PI and GI parameters affects the impacts of uncontrolled diabetes
Nguyen et al. (2020)	Systematic review on 14 studies	Higher risks for diabetic ($p < 0.001$) Retinopathy (odds ratios -OD: 2.8–8.7) Neuropathy (OD: 3.2–6.6) Nephropathy (OD: 1.9–8.5) Cardiovascular complications (OD: 1.28–17.7)
Graziani et al. (2018)	Systematic review and meta-analysis on 20 studies	Periodontitis has a significant impact on the control, the incidence and the complications of diabetes. The hazar ratio (HR) to develop diabetes in periodontitis compromised patient is 1.29; 95% CI, 1.11–1.46, $p < .0001$)
ZiuKaite et al. (2018)	Systematic review and meta-analysis on 27 studies	Periodontitis influence both the prevalence and odds of having diabetes
Sanz et al. (2018)	Consensus report	Periodontitis is associated with dysglycaemia and increased insulin resistance in people with diabetes, as well as increased risk for incident diabetes and diabetes complications
Abariga et al. (2016)	Systematic review and meta-analysis on 10 studies	periodontitis is associated with a significant increased risk for gestational diabetes mellitus compared to women without periodontitis

PI plaque index, *GI* gingival index, *BOP* bleeding on probing, *PD* pocket depth and *CAL* clinical attachment loss

Table 12.3 Effect of periodontal non-surgical treatment on diabetes

Authors	Study design	Conclusion
Baeza et al. (2020)	Systematic review and meta-analysis on 9 studies	Scaling and root planing reduces the blood concentration of HbA1c and CRP ($p < 0.01$).
El-Makaky et al. (2019)	RCT on 88 patients with chronic periodontitis and uncontrolled diabetes	Non-surgical periodontal treatment reduces the level of HbA1c ($p < 0.001$)
Cao et al. (2019)	Systematic review and meta-analysis on 14 studies	Periodontal treatment induces the decrease of the HbA1c% in T2DM
Garde et al. (2019)	Systematic review and meta-analysis on 7 studies	Periodontal therapy induces the changes of total cholesterol and triglycerides levels in patients with type 2 DM
Nishioka et al. (2019)	RCT on 74 patients with borderline T2DM	After periodontal therapy, lower BOP (%) is associated with significant improvements in fasting serum insulin
Lima et al. (2019)	Systematic review on 15 studies	Periodontal therapy has beneficial effects on the level of IL-6
Sanz et al. (2018)	Consensus report	Periodontal therapy improves serum HbA1C levels
D'Aiuto et al. 2018	RCT on 264 patients with T2DM	After periodontal treatment, the HbA1c was 0·6% (95% CI 0·3–0·9; $p < 0.0001$) lower in the test group than in control
Kocher et al.2019	Prospective controlled study on Periodontitis patients With normal HbA1c Prediabetes Unknown diabetes	Non-surgical periodontal treatment leads to the decrease of HbA1c in pre-diabetic patients from 5.9% (95% CI, 5.9% to 6.0%) to 5.4% (95% CI, 5.3% to 5.5%)
Tsobgny-Tsague et al. (2018)	RCT on 30 patients with T2DM	Non-surgical periodontal treatment induced a reduction of HbA1c levels from 9.7 ± 1.6% at baseline to 6.7 ± 2.0% 3 months after NSPT ($p < 0.001$)

(continued)

Table 12.3 (continued)

Authors	Study design	Conclusion
Madianos et al. (2017)	Meta-analysis of 7 RCT	Reduction in HbA1c at 3–4 months was reported ranging from −0.27% (95% CI: −0.46, −0.07, $p = 0.007$) to −1.03% (95% CI: 0.36, −1.70, $p = 0.003$)
Mauri-Obradors et al. (2017)	RCT on 90 patients with T2DM	Non-surgical periodontal treatment induced a reduction of HbA1c levels ($p < 0.05$)
Mizuno et al. (2017)	RCT on 40 patients with T2DM	Non-surgical periodontal treatment improved systemic oxidative stress
Hasuike et al. (2017)	Systematic review and meta-analysis on 13 studies	Periodontal treatment in diabetic patients reduces the level of HbA1c
Simpson et al. (2015)	Systematic review on 34 studies	Periodontal therapy leads to the reduction in HbA1c ($p = 0.003$).
Artese et al. (2015)	Systematic review on 9 studies	Periodontal treatment leads to a significant decrease of both TNF- α and CRP level ($p < 0.001$)
Kaur et al. (2015)	RCT on 100 patients with T2DM.	Non-surgical periodontal therapy improved glycemic control, with the decrease of HbA1c decreased by 10.8%

RCT randomized controlled trial, *HbA1* haemoglobin A1c or glycated haemoglobin, *TNF* tumor necrosis factor, *CRP* C reactive protein

12.3.1 Effect of Diabetes on Periodontitis

The effect of diabetes on periodontitis is summarized in Table 12.1. Recent systematic reviews (Nguyen et al. 2020; Dicembrini et al. 2020; Nascimento et al. 2018), controlled prospective trials (Herrmann et al. 2020) and consensus reports (Sanz et al. 2018; Chapple et al. 2017) highlighted that diabetes impairs periodontitis. Diabetes significantly increased the incidence (Dicembrini et al. 2020; Nascimento et al. 2018) and the progression of periodontitis with a relative risk (RR) of 1.86 ([95% CI 1.3–2.8]) (Nascimento et al. 2018). Indeed, diabetic patients harbored significantly worst periodontal status with lower outcomes than healthy patients (Sanz et al. 2018; Dicembrini et al. 2020). Furthermore, the pathophysiological processes involved in the relationship, were mechanisms that trigger hyper inflammatory state and periodontitis, as the impairment of microvasculature (Nascimento et al. 2018), the over-expression of PMN'S (Herrmann et al. 2020), oxidative stress and the advanced glycation end-products (Chapple et al. 2017).

12.3.2 Effect of Periodontitis on Diabetes

The effect of periodontitis on diabetes is summarized in Table 12.2. Recent systematic reviews with meta-analysis (Jensen et al. 2021; Rapone et al. 2020; Nguyen et al. 2020; Graziani et al. 2018; Ziukaite et al. 2018; Abariga and Whitcomb 2016), consensus report (Sanz et al. 2018) confirm that periodontitis impairs diabetes. Periodontitis, with associated risk markers (Jensen et al. 2021; Rapone et al. 2020) as plaque index, probing depth, gingival index, impairs both the incidence (Graziani et al. 2018; Ziukaite et al. 2018) and the complications of diabetes (Sanz et al. 2018; Nguyen et al. 2020; Graziani et al. 2018). Patients with periodontitis present significant higher risk of developing diabetes with a hazard ratio (HR) of 1.29; 95% CI, 1.11–1.46, $p < .0001$ (Graziani et al. 2018). The dysregulation of glycemia and the increased insulin were the mechanisms involved in the relation of periodontitis on diabetes (Sanz et al. 2018).

12.3.3 Effect of Periodontal Non-surgical Treatment on Diabetes

The effect of periodontal non-surgical treatment on diabetes is summarized in Table 12.3. Recent systematic reviews with meta-analysis (Baeza et al. 2020; Cao et al. 2019; Lima et al. 2019; Garde et al. 2019; Madianos and Koromantzos 2018; Hasuike et al. 2017; Simpson et al. 2015; Artese et al. 2015), randomized controlled trials (Kaur et al. 2015; Mizuno et al. 2017; Mauri-Obradors et al. 2018; Tsobgny-Tsague et al. 2018; D'Aiuto et al. 2018; Nishioka et al. 2019; El-Makaky and Shalaby 2020), controlled prospective trials (Kocher et al. 2019) and consensus report (Sanz et al. 2018) confirm the positive effect of periodontal non-surgical treatment (PNST) on diabetic outcomes. Among assessed parameters, PNST significantly decreased the blood concentration of glycated haemoglobin (HbA1c) (Sanz et al. 2018; Baeza et al. 2020; Cao et al. 2019; Madianos and Koromantzos 2018; Hasuike et al. 2017; Simpson et al. 2015; Kaur et al. 2015; Tsobgny-Tsague et al. 2018; D'Aiuto et al. 2018; El-Makaky and Shalaby 2020; Kocher et al. 2019; Mauri-Obradors et al. 2015), total cholesterol, triglycerides (Garde et al. 2019) and insulin (Nishioka et al. 2019). Furthermore, the blood level of biomarkers related to inflammation (IL-6,TNF- α and CRP) (Baeza et al. 2020; Lima et al. 2019; Artese et al. 2010) and oxidative stress (Mizuno et al. 2017) were also significantly decreased after PNST.

12.4 Discussion

Periodontitis and diabetes are two associated diseases with a well-described bidirectional influence (Preshaw and Bissett 2019; Sanz et al. 2018; Preshaw et al. 2012; Kumar et al. 2014; Taylor 2001). Nonetheless, distinction must be made between the tree categories of diabetes (Atlas 2019). The type 1 diabetes mellitus (T1DM) occurs most frequently in children and young people with genetic and environmental factors. However, the increase of overweight and obesity in young people raise the prevalence of diabetes type 2 diabetes in this population. T1DM is characterized by an autoimmune reaction that leads to the destruction of the insulin-producing beta cells of the pancreas (pancreatic islet cells) (Paschou et al. 2018). Indeed, the lymphocytes T recognize the B cells of the pancreas as non-self-cells, and therefore destroy them. As a result, a little quantity or the absence of insulin is secreted (Paschou et al. 2018; Wang et al. 2019; Smith et al. 2017). Therefore, in order to maintain the appropriate blood level of glucose and to avoid the complications of diabetes, a daily injection of insulin is needed.

On the contrary, the type 2 diabetes mellitus (T2DM) occurs most frequently in elderly population, with obesity being a major risk factor in addition to complex genetic and environmental etiology (Chan et al. 1994; Colditz et al. 1995). Furthermore, the increment of overweight, obesity and sedentary lifestyle in young people increase the apparition of the disease in this population. T2DM is characterized by inadequate use of insulin from the body enhancing insulin resistance and leading to elevated levels of glucose in blood (Atlas 2019).

Gestational diabetes happens to women with hyperglycemia during the beginning of pregnancy, but also to women presenting insufficient insulin secretion to overcome the diminished action of insulin due to hormone production by the placenta (Atlas 2019). Nevertheless, once hyperglycemia occurs, diabetic people present the same complications as the micro- and the macro-vascular complications (Skyler et al. 2017), the sensibility to infection (Alavi et al. 2014; Everett and Mathioudakis 2018), as well as periodontitis (Preshaw and Bissett 2013, 2019), but with different rate of progression (Skyler et al. 2017). Regarding patients with T2DM, they have higher risk to suffer from severe form of periodontitis (Genco and Borgnakke 2013; Albert et al. 2012; Lalla et al. 2006) (Table 12.1), and in the other way, patients with periodontitis presented a higher risk to develop T2DM (Demmer et al. 2008; Chapple et al. 2013; Borgnakke et al. 2013) (Table 12.2).

Diabetic patients present an increased risk of the incidence and the progression of periodontitis by 86% (Nascimento et al. 2018). As described in 2017 (Chapple et al. 2017) by the consensus report of the joint workshop the European Federation of Periodontology (EFP) and the European Organization for Caries Research (ORCA), the absence of the control of glycemia leads to hyperglycemia that enhance the expression of AGE-RAGE complex (Advanced Glycation Endproducts- Receptor for Advanced Glycation Endproducts) in periodontal tissues. Its expression triggers an hyper inflammatory state and promotes periodontitis. Subsequently, an elevation of proinflammatory mediators (PIM) such as TNF-α, IL-1βIL-6, NF-kβ, osteoprotegerin ratio (OPG), oxidative stress (ROS) and immune dysfunction as with the overexpression of the polymorphonuclear leucocytes (PMNs) (Herrmann et al. 2020) occur that contribute to the destruction of connective tissue surrounding the teeth, and the progression of periodontitis (Sanz et al. 2018; Nascimento et al. 2018; Herrmann et al. 2020; Chapple et al. 2017; Polak and Shapira 2018). Additionally, the recent systematic review of Nardi et al. (Chapple et al. 2017) supported that in diabetic patient, the microvasculature of periodontium is affected by the overexpression of vascular endothelial growth factor (VEGF) that enhances periodontitis progression. As emphasized recently by the consensus report by the International Diabetes Federation and the European Federation of Periodontology (Sanz et al. 2018) poor glycaemic control is associated with poorer periodontal status and outcomes.

Furthermore, recent systematic reviews (Sanz et al. 2018; Nguyen et al. 2020; Graziani et al. 2018; Ziukaite et al. 2018) describe the influence of periodontitis on both the prevalence of diabetes and the risk to develop further complications. Indeed, in patients affected by the disease, the presence of bacteria or their metabolites (LPS) induce the host secretion inside the bloodstream of pro-inflammatory cytokines (TNF-α, IL-6, CRP, oxygen radicals) involved in the increase of systemic inflammatory state and in the impairment of insulin signaling and resistance (Sanz et al. 2018; Graziani et al. 2018; Polak and Shapira 2018). Therefore, as described in the recent consensus report and guidelines of the joint workshop on periodontal diseases and diabetes by the International Diabetes Federation (IDF) and the European Federation of Periodontology (EFP) (Sanz et al. 2018), periodontitis is associated with dysglycaemia and insulin resistance by increasing the HBA1c, and as a result, augmenting the risk of diabetes development and its complications. Likewise, in the update of Polak et al. 2017 (Polak and Shapira 2018), it was underlined that periodontitis systemic inflammation contributes to the under-controlled diabetes and its complications. Additionally, the periodontitis chronicity elicits cytokine sensitivity, contributing to the increase of diabetes (Sanz et al. 2018; Graziani et al. 2018; Polak and Shapira 2018) and, according to the recent systematic review of Nguyen et al. (Nguyen et al. 2020), contribute to the long-term complications of diabetes, as retinopathy, neuropathy, nephropathy and cardiovascular adverse events. Besides, as well-described, the periodontal non-surgical treatment leads to significant decrease of blood biomarkers related to diabetes, as HBA1c, CRP, TNF-α, IL-6 (Cao et al. 2019; Garde et al. 2019; Madianos and Koromantzos 2018; El-Makaky and Shalaby 2020; Kocher et al. 2019). These clinical observations support both the negative effect of periodontitis on diabetes and the positive effect of periodontitis treatment on diabetes (Tables 12.2 and 12.3).

12.5 Conclusion

This review supports the bidirectional link between diabetes and periodontitis. Furthermore, it is relevant to bring in light the role of the prevention and the control of hyperglycemia in the prevention of periodontitis, and the control of periodontitis systemic inflammation in the prevention of diabetes.

References

Abariga SA, Whitcomb BW (2016) Periodontitis and gestational diabetes mellitus: a systematic review and meta-analysis of observational studies. BMC Pregnancy Childbirth 16(1):344. https://doi.org/10.1186/s12884-016-1145-z

Alavi A, Sibbald RG, Mayer D, Goodman L, Botros M, Armstrong DG, Woo K, Boeni T, Ayello EA, Kirsner RS (2014) Diabetic foot ulcers: part I. Pathophysiology and prevention. J Am Acad Dermatol 70 (1):1 e1–18; quiz 19–20. https://doi.org/10.1016/j.jaad.2013.06.055

Albert DA, Ward A, Allweiss P, Graves DT, Knowler WC, Kunzel C, Leibel RL, Novak KF, Oates TW, Papapanou PN, Schmidt AM, Taylor GW, Lamster IB, Lalla E (2012) Diabetes and oral disease: implications for health professionals. Ann N Y Acad Sci 1255:1–15. https://doi.org/10.1111/j.1749-6632.2011.06460.x

Amano A (2010a) Host-parasite interactions in periodontitis: microbial pathogenicity and innate immunity. Periodontology 2000 54(1):9–14. https://doi.org/10.1111/j.1600-0757.2010.00376.x

Amano A (2010b) Host-parasite interactions in periodontitis: subgingival infection and host sensing. Periodontology 2000 52(1):7–11. https://doi.org/10.1111/j.1600-0757.2009.00328.x

Artese HP, Foz AM, Rabelo Mde S, Gomes GH, Orlandi M, Suvan J, D'Aiuto F, Romito GA (2015) Periodontal therapy and systemic inflammation in type 2 diabetes mellitus: a meta-analysis. PLoS One 10(5):e0128344. https://doi.org/10.1371/journal.pone.0128344

Artese HP, Sousa CO, Luiz RR, Sansone C, Torres MC (2010) Effect of non-surgical periodontal treatment on chronic kidney disease patients. Braz Oral Res 24(4):449–454. https://doi.org/10.1590/s1806-83242010000400013

Atlas ID (2019) 2019. International Diabetes Federation; 2019. Online version of IDF Diabetes Atlas: www.diabetesatlas.org

Baeza M, Morales A, Cisterna C, Cavalla F, Jara G, Isamitt Y, Pino P, Gamonal J (2020) Effect of periodontal treatment in patients with periodontitis and diabetes: systematic review and meta-analysis. J Appl Oral Sci 28:e20190248. https://doi.org/10.1590/1678-7757-2019-0248

Borgnakke WS, Ylostalo PV, Taylor GW, Genco RJ (2013) Effect of periodontal disease on diabetes: systematic review of epidemiologic observational evidence. J Periodontol 84(4 Suppl):S135–S152. https://doi.org/10.1902/jop.2013.1340013

Cao R, Li Q, Wu Q, Yao M, Chen Y, Zhou H (2019) Effect of non-surgical periodontal therapy on glycemic control of type 2 diabetes mellitus: a systematic review and Bayesian network meta-analysis. BMC Oral Health 19(1):176. https://doi.org/10.1186/s12903-019-0829-y

Chan JM, Rimm EB, Colditz GA, Stampfer MJ, Willett WC (1994) Obesity, fat distribution, and weight gain as risk factors for clinical diabetes in men. Diabetes Care 17(9):961–969. https://doi.org/10.2337/diacare.17.9.961

Chapple IL, Bouchard P, Cagetti MG, Campus G, Carra MC, Cocco F, Nibali L, Hujoel P, Laine ML, Lingstrom P, Manton DJ, Montero E, Pitts N, Range H, Schlueter N, Teughels W, Twetman S, Van Loveren C, Van der Weijden F, Vieira AR, Schulte AG (2017) Interaction of lifestyle, behaviour or systemic diseases with dental caries and periodontal diseases: consensus report of group 2 of the joint EFP/ORCA workshop on the boundaries between caries and periodontal diseases. J Clin Periodontol 44(Suppl 18):S39–S51. https://doi.org/10.1111/jcpe.12685

Chapple IL, Genco R, Working group 2 of joint EFPAAPw (2013) Diabetes and periodontal diseases: consensus report of the Joint EFP/AAP Workshop on Periodontitis and Systemic Diseases. J Clin Periodontol 40(Suppl 14):S106–S112. https://doi.org/10.1111/jcpe.12077

Colditz GA, Willett WC, Rotnitzky A, Manson JE (1995) Weight gain as a risk factor for clinical diabetes mellitus in women. Ann Intern Med 122(7):481–486. https://doi.org/10.7326/0003-4819-122-7-199504010-00001

D'Aiuto F, Gkranias N, Bhowruth D, Khan T, Orlandi M, Suvan J, Masi S, Tsakos G, Hurel S, Hingorani AD, Donos N, Deanfield JE (2018) Systemic effects of periodontitis treatment in patients with type 2 diabetes: a 12 month, single-centre, investigator-masked, randomised trial. Lancet Diabetes Endocrinol 6(12):954–965. https://doi.org/10.1016/s2213-8587(18)30038-x

Demmer RT, Kocher T, Schwahn C, Volzke H, Jacobs DR Jr, Desvarieux M (2008) Refining exposure definitions for studies of periodontal disease and systemic disease associations. Community Dent Oral Epidemiol 36(6):493–502. https://doi.org/10.1111/j.1600-0528.2008.00435.x

Dicembrini I, Serni L, Monami M, Caliri M, Barbato L, Cairo F, Mannucci E (2020) Type 1 diabetes and periodontitis: prevalence and periodontal destruction-a systematic review. Acta Diabetol 57(12):1405–1412. https://doi.org/10.1007/s00592-020-01531-7

Eke PI, Wei L, Borgnakke WS, Thornton-Evans G, Zhang X, Lu H, McGuire LC, Genco RJ (2016) Periodontitis prevalence in adults >/= 65 years of age, in the USA. Periodontology 2000 72(1):76–95. https://doi.org/10.1111/prd.12145

El-Makaky Y, Shalaby HK (2020) The effects of non-surgical periodontal therapy on glycemic control in diabetic patients: a randomized controlled trial. Oral Dis 26(4):822–829. https://doi.org/10.1111/odi.13256

Everett E, Mathioudakis N (2018) Update on management of diabetic foot ulcers. Ann N Y Acad Sci 1411(1):153–165. https://doi.org/10.1111/nyas.13569

Forner L, Larsen T, Kilian M, Holmstrup P (2006) Incidence of bacteremia after chewing, tooth brushing and scaling in individuals with periodontal inflammation. J Clin Periodontol 33(6):401–407. https://doi.org/10.1111/j.1600-051X.2006.00924.x

Garde S, Akhter R, Nguyen MA, Chow CK, Eberhard J (2019) Periodontal therapy for improving lipid profiles in patients with type 2 diabetes mellitus: a systematic review and meta-analysis. Int J Mol Sci 20(15). https://doi.org/10.3390/ijms20153826

Genco RJ, Borgnakke WS (2013) Risk factors for periodontal disease. Periodontology 2000 62(1):59–94. https://doi.org/10.1111/j.1600-0757.2012.00457.x

Graziani F, Gennai S, Solini A, Petrini M (2018) A systematic review and meta-analysis of epidemiologic observational evidence on the effect of periodontitis on diabetes an update of the EFP-AAP review. J Clin Periodontol 45(2):167–187. https://doi.org/10.1111/jcpe.12837

Haffajee AD, Socransky SS (1994) Microbial etiological agents of destructive periodontal diseases. Periodontology 2000 5:78–111. https://doi.org/10.1111/j.1600-0757.1994.tb00020.x

Hasuike A, Iguchi S, Suzuki D, Kawano E, Sato S (2017) Systematic review and assessment of systematic reviews examining the effect of periodontal treatment on glycemic control in patients with diabetes. Med Oral Patol Oral Cir Bucal 22(2):e167–e176. https://doi.org/10.4317/medoral.21555

Herrmann JM, Sonnenschein SK, Groeger SE, Ewald N, Arneth B, Meyle J (2020) Refractory neutrophil activation in type 2 diabetics with chronic periodontitis. J Periodontal Res 55(2):315–323. https://doi.org/10.1111/jre.12717

Jensen E, Allen G, Bednarz J, Couper J, Pena A (2021) Periodontal risk markers in children and adolescents with type 1 diabetes: a systematic review and meta-analysis. Diabetes Metab Res Rev 37(1):e3368. https://doi.org/10.1002/dmrr.3368

Jepsen S, Blanco J, Buchalla W, Carvalho JC, Dietrich T, Dorfer C, Eaton KA, Figuero E, Frencken JE, Graziani F, Higham SM, Kocher T, Maltz M, Ortiz-Vigon A, Schmoeckel J, Sculean A, Tenuta LM, van der Veen MH, Machiulskiene V (2017) Prevention and control of dental caries and periodontal diseases at individual and population level: consensus report of group 3 of joint EFP/ORCA workshop on the boundaries between caries and periodontal diseases. J Clin Periodontol 44(Suppl 18):S85–S93. https://doi.org/10.1111/jcpe.12687

Kassebaum NJ, Bernabe E, Dahiya M, Bhandari B, Murray CJ, Marcenes W (2014) Global burden of severe tooth loss: a systematic review and meta-analysis. J Dent Res 93(7 Suppl):20S–28S. https://doi.org/10.1177/0022034514537828

Kaur PK, Narula SC, Rajput R, R KS, Tewari S (2015) Periodontal and glycemic effects of nonsurgical periodontal therapy in patients with type 2 diabetes stratified by baseline HbA1c. J Oral Sci 57(3):201–211. https://doi.org/10.2334/josnusd.57.201

Kocher T, Holtfreter B, Petersmann A, Eickholz P, Hoffmann T, Kaner D, Kim TS, Meyle J, Schlagenhauf U, Doering S, Gravemeier M, Prior K, Rathmann W, Harks I, Ehmke B, Koch R (2019) Effect of periodontal treatment on HbA1c among patients with prediabetes. J Dent Res 98(2):171–179. https://doi.org/10.1177/0022034518804185

Kumar M, Mishra L, Mohanty R, Nayak R (2014) Diabetes and gum disease: the diabolic duo. Diabetes Metab Syndr 8(4):255–258. https://doi.org/10.1016/j.dsx.2014.09.022

Lalla E, Cheng B, Lal S, Tucker S, Greenberg E, Goland R, Lamster IB (2006) Periodontal changes in children and adolescents with diabetes: a case-control study. Diabetes Care 29(2):295–299. https://doi.org/10.2337/diacare.29.02.06.dc05-1355

Lima RPE, Belem FV, Abreu LG, Cunha FA, Cota LOM, da Costa JE, Costa FO (2019) Effect of periodontal therapy on serum levels of IL-6 in type 2 diabetics: a systematic review. Int J Periodontics Restorative Dent 39(1):e1–e10. https://doi.org/10.11607/prd.3866

Loos BG (2005) Systemic markers of inflammation in periodontitis. J Periodontol 76(11 Suppl):2106–2115. https://doi.org/10.1902/jop.2005.76.11-S.2106

Madianos PN, Koromantzos PA (2018) An update of the evidence on the potential impact of periodontal therapy on diabetes outcomes. J Clin Periodontol 45(2):188–195. https://doi.org/10.1111/jcpe.12836

Mauri-Obradors E, Jane-Salas E, Sabater-Recolons Mdel M, Vinas M, Lopez-Lopez J (2015) Effect of nonsurgical periodontal treatment on glycosylated hemoglobin in diabetic patients: a systematic review. Odontology 103(3):301–313. https://doi.org/10.1007/s10266-014-0165-2

Mauri-Obradors E, Merlos A, Estrugo-Devesa A, Jane-Salas E, Lopez-Lopez J, Vinas M (2018) Benefits of non-surgical periodontal treatment in patients with type 2 diabetes mellitus and chronic periodontitis: a randomized controlled trial. J Clin Periodontol 45(3):345–353. https://doi.org/10.1111/jcpe.12858

Mealey BL (1999) Influence of periodontal infections on systemic health. Periodontology 2000 21(1):197–209

Meyle J, Chapple I (2015) Molecular aspects of the pathogenesis of periodontitis. Periodontology 2000 69(1):7–17. https://doi.org/10.1111/prd.12104

Mizuno H, Ekuni D, Maruyama T, Kataoka K, Yoneda T, Fukuhara D, Sugiura Y, Tomofuji T, Wada J, Morita M (2017) The effects of non-surgical periodontal treatment on glycemic control, oxidative stress balance and quality of life in patients with type 2 diabetes: a randomized clinical trial. PLoS One 12(11):e0188171. https://doi.org/10.1371/journal.pone.0188171

Nardi GM, Ferrara E, Converti I, Cesarano F, Scacco S, Grassi R, Gnoni A, Grassi FR, Rapone B (2020) Does diabetes induce the vascular endothelial growth factor (VEGF) expression in periodontal tissues? A systematic review. Int J Environ Res Public Health 17(8). https://doi.org/10.3390/ijerph17082765

Nascimento GG, Leite FRM, Vestergaard P, Scheutz F, Lopez R (2018) Does diabetes increase the risk of periodontitis? A systematic review and meta-regression analysis of longitudinal prospective studies. Acta Diabetol 55(7):653–667. https://doi.org/10.1007/s00592-018-1120-4

Nguyen ATM, Akhter R, Garde S, Scott C, Twigg SM, Colagiuri S, Ajwani S, Eberhard J (2020) The association of periodontal disease with the complications of diabetes mellitus. A systematic review. Diabetes Res Clin Pract 165:108244. https://doi.org/10.1016/j.diabres.2020.108244

Nishioka S, Maruyama K, Tanigawa T, Miyoshi N, Eguchi E, Nishida W, Osawa H, Saito I (2019) Effect of non-surgical periodontal therapy on insulin resistance and insulin sensitivity among individuals with borderline diabetes: a randomized controlled trial. J Dent 85:18–24. https://doi.org/10.1016/j.jdent.2019.04.005

Paschou SA, Papadopoulou-Marketou N, Chrousos GP, Kanaka-Gantenbein C (2018) On type 1 diabetes mellitus pathogenesis. Endocr Connect 7(1):R38–R46. https://doi.org/10.1530/EC-17-0347

Polak D, Shapira L (2018) An update on the evidence for pathogenic mechanisms that may link periodontitis and diabetes. J Clin Periodontol 45(2):150–166. https://doi.org/10.1111/jcpe.12803

Preshaw PM, Alba AL, Herrera D, Jepsen S, Konstantinidis A, Makrilakis K, Taylor R (2012) Periodontitis and diabetes: a two-way relationship. Diabetologia 55(1):21–31. https://doi.org/10.1007/s00125-011-2342-y

Preshaw PM, Bissett SM (2013) Periodontitis: oral complication of diabetes. Endocrinol Metab Clin N Am 42(4):849–867. https://doi.org/10.1016/j.ecl.2013.05.012

Preshaw PM, Bissett SM (2019) Periodontitis and diabetes. Br Dent J 227(7):577–584. https://doi.org/10.1038/s41415-019-0794-5

Rapone B, Corsalini M, Converti I, Loverro MT, Gnoni A, Trerotoli P, Ferrara E (2020) Does periodontal inflammation affect type 1 diabetes in childhood and adolescence? A meta-analysis. Front Endocrinol (Lausanne) 11:278. https://doi.org/10.3389/fendo.2020.00278

Salhi L, Rompen E, Sakalihasan N, Laleman I, Teughels W, Michel JB, Lambert F (2019) Can periodontitis influence the progression of abdominal aortic aneurysm? A systematic review. Angiology 70(6):479–491. https://doi.org/10.1177/0003319718821243

Salhi L, Sakalihasan N, Okroglic AG, Labropoulos N, Seidel L, Albert A, Teughels W, Defraigne JO, Lambert F (2020) Further evidence on the relationship between abdominal aortic aneurysm and periodontitis: a cross-sectional study. J Periodontol 91(11):1453–1464. https://doi.org/10.1002/JPER.19-0671

Sanz M, Ceriello A, Buysschaert M, Chapple I, Demmer RT, Graziani F, Herrera D, Jepsen S, Lione L, Madianos P, Mathur M, Montanya E, Shapira L, Tonetti M, Vegh D (2018) Scientific evidence on the links between periodontal diseases and diabetes: consensus report and guidelines of the joint workshop on periodontal diseases and diabetes by the International Diabetes Federation and the European Federation of Periodontology. J Clin Periodontol 45(2):138–149. https://doi.org/10.1111/jcpe.12808

Schenkein HA (2006) Host responses in maintaining periodontal health and determining periodontal disease. Periodontology 2000(40):77–93. https://doi.org/10.1111/j.1600-0757.2005.00144.x

Simpson TC, Weldon JC, Worthington HV, Needleman I, Wild SH, Moles DR, Stevenson B, Furness S, Iheozor-Ejiofor Z (2015) Treatment of periodontal disease for glycaemic control in people with diabetes mellitus. Cochrane Database Syst Rev 11:CD004714. https://doi.org/10.1002/14651858.CD004714.pub3

Skyler JS, Bakris GL, Bonifacio E, Darsow T, Eckel RH, Groop L, Groop PH, Handelsman Y, Insel RA, Mathieu C, McElvaine AT, Palmer JP, Pugliese A, Schatz DA, Sosenko JM, Wilding JP, Ratner RE (2017) Differentiation of diabetes by pathophysiology, natural history, and prognosis. Diabetes 66(2):241–255. https://doi.org/10.2337/db16-0806

Smith MJ, Simmons KM, Cambier JC (2017) B cells in type 1 diabetes mellitus and diabetic kidney disease. Nat Rev Nephrol 13(11):712–720. https://doi.org/10.1038/nrneph.2017.138

Szulc M, Kustrzycki W, Janczak D, Michalowska D, Baczynska D, Radwan-Oczko M (2015) Presence of periodontopathic bacteria DNA in atheromatous plaques from coronary and carotid arteries. Biomed Res Int 2015:825397. https://doi.org/10.1155/2015/825397

Taylor GW (2001) Bidirectional interrelationships between diabetes and periodontal diseases: an epidemiologic perspective. Ann Periodontol 6(1):99–112. https://doi.org/10.1902/annals.2001.6.1.99

Tonetti MS, Van Dyke TE, working group 1 of the joint EFPAAPw (2013) Periodontitis and atherosclerotic cardiovascular disease: consensus report of the Joint EFP/AAPWorkshop on Periodontitis and Systemic Diseases. J Periodontol 84(Suppl 4S):S24–S29. https://doi.org/10.1902/jop.2013.1340019

Tsobgny-Tsague NF, Lontchi-Yimagou E, Nana ARN, Tankeu AT, Katte JC, Dehayem MY, Bengondo CM, Sobngwi E (2018) Effects of nonsurgical periodontal treatment on glycated haemoglobin on type 2 diabetes patients (PARODIA 1 study): a randomized controlled trial in a sub-Saharan Africa population. BMC Oral Health 18(1):28. https://doi.org/10.1186/s12903-018-0479-5

Van Dyke TE, van Winkelhoff AJ (2013) Infection and inflammatory mechanisms. J Periodontol 84(4 Suppl):S1–S7. https://doi.org/10.1902/jop.2013.1340018

van Winkelhoff AJ, Slots J (1999) Actinobacillus actinomycetemcomitans and Porphyromonas gingivalis in nonoral infections. Periodontology 2000 20:122–135. https://doi.org/10.1111/j.1600-0757.1999.tb00160.x

Wang P, Xu YY, Lv TT, Guan SY, Li XM, Li XP, Pan HF (2019) Subclinical atherosclerosis in patients with type 1 diabetes mellitus: a systematic review and meta-analysis. Angiology 70(2):141–159. https://doi.org/10.1177/0003319718787366

Ziukaite L, Slot DE, Van der Weijden FA (2018) Prevalence of diabetes mellitus in people clinically diagnosed with periodontitis: a systematic review and meta-analysis of epidemiologic studies. J Clin Periodontol 45(6):650–662. https://doi.org/10.1111/jcpe.12839

Periodontitis as a Risk Factor for Alzheimer's Disease: The Experimental Journey So Far, with Hope of Therapy

13

Alice Harding, Shalini Kanagasingam, Richard Welbury, and Sim K. Singhrao

Abstract

Periodontitis and Alzheimer's disease (AD) exist globally within the adult population. Given that the risk of AD incidence doubles within 10 years from the time of periodontal disease diagnosis, there is a window of opportunity for slowing down or preventing AD by risk-reduction-based intervention. Literature appraisal on the shared risk factors of these diseases suggests a shift to a healthy lifestyle would be beneficial. Generalised (chronic) periodontitis with an established dysbiotic polymicrobial aetiology affects the tooth supporting tissues with eventual tooth loss. The cause of AD remains unknown, however two neurohistopathological lesions – amyloid-beta plaques and neurofibrillary tangles, together with the clinical history, provide AD diagnosis at autopsy. Historically, prominence was given to the two hallmark lesions but now emphasis is placed on cerebral inflammation and what triggers it. Low socioeconomic status promotes poor lifestyles that compromise oral and personal hygiene along with reliance on poor dietary intake. Taken together with advancing age and a declining immune protection, these risk factors may negatively impact on periodontitis and AD. These factors also provide a tangible solution to controlling pathogenic bacteria indigenous to the oral and gastrointestinal tract microbioes in vulnerable subjects. The focus here is on *Porphyromonas gingivalis,* one of several important bacterial pathogens associated with both periodontitis and AD. Recent research has enabled advances in our knowledge of the armoury of *P. gingivalis* via reproduction of all clinical and neuropathological hallmark lesions of AD and chronic periodontal disease in vitro and in vivo experimental models, thus paving the way for better future management.

Keywords

Alzheimer's disease · Dysbiosis · Inflammation · *Porphyromonas gingivalis* · Periodontitis · Therapy

Abbreviations

AD	Alzheimer's disease
AGE	advanced glycation end products

A. Harding · S. Kanagasingam · R. Welbury
S. K. Singhrao (✉)
Brain and Behavior Centre, Faculty of Clinical and Biomedical Sciences, School of Dentistry, University of Central Lancashire, Preston, UK
e-mail: AHarding7@uclan.ac.uk; SKanagasingam1@uclan.ac.uk; RWelbury@uclan.ac.uk; sksinghrao@uclan.ac.uk

J. Santi-Rocca (ed.), *Periodontitis*, Advances in Experimental Medicine and Biology 1373,
https://doi.org/10.1007/978-3-030-96881-6_13

APOE ϵ4	*apolipoprotein E, allele 4*
$ApoE^{-/-}$	apolipoprotein knockout
APP	amyloid precursor protein
APP-Tg	APP-transgenic
ARUK	Alzheimer's Research UK
Aβ	Amyloid-beta
C9/C1q and C1s	complement component/ sub-complement
CR1	complement receptor 1
DNA	deoxyribonucleic acid
ELISA	enzyme-linked immunoassay
GI	gastrointestinal tract
GWAS	Genome-Wide-Association Studies
IL	interleukins
IL-1R	interleukin-1 receptor
LPS	lipopolysaccharide
NFTs	neurofibrillary tangles
NF-ϰB	nuclear factor kappaB
NIH	National Institutes of Health
NSAIDs	Non-Steroidal Inflammatory Drugs
OMVs	outer membrane vesicles
PG	prostaglandin
PHF	paired helical filaments
RT-PCR	reverse transcriptase-polymperase chain reaction
TLRs	toll-like receptors
TNFα	tumour necrosis factor-α

Highlights

- Periodontal disease has been associated with the onset and progression of Alzheimer's Disease, which causes irreversible cognitive and functional decline
- This chapter summarises the diverse range of evidence which implicates periodontal disease and its key pathogen, *P. gingivalis*, as a significant risk factor for AD
- The potential for prevention is high as the dental team can play an important role in preventing and/or delaying the development of AD by proactive management of periodontitis
- A drug COR388 targeting the *P. gingivalis* protease enzyme is a novel treatment modality and may offer promising results, considering the number of failed AD therapeutic clinical trials

Considerations for Practice

- To deliver personalised preventative advice to patients, not only with regards to oral health, but also to healthy lifestyle choices
- To maintain and stabilise periodontal disease in a bid to reduce or delay AD progression or development
- To risk assess each patient and deliver prevention and care accordingly in order to enable 'successful ageing' and reduce disease development in their patient base

Patient Summary

Dentists understand that maintaining a healthy mouth contributes to maintaining a healthy body. Research suggests that periodontal disease can lead to the development and progression of AD. Patients should be aware of all modifiable risk factors which have been linked to AD, including their oral health and the lifestyle choices they make. The onus comes down to the dental profession to deliver this preventative advice to their patients. It is imperative to maintain gingival health and make healthy lifestyle choices to not only to reduce their risk of AD but also to delay the onset, in susceptible individuals.

13.1 Introduction

There is a growing body of evidence that supports oral (generalised periodontitis) and gastrointestinal (GI) tract dysbiosis as having a negative impact on mental health. Briefly, the enteric nervous system supplies the brain with "psychobiotics" (serotonin, dopamine and Gamma aminobutyric acid) courtesy of healthy gut bacteria (Strandwitz 2018; Strandwitz et al. 2019). Dysbiosis of the gut bacteria leads to a deficit of these neurotransmitters and causes anxiety and depression (Clapp et al. 2017). Depression, in this context, may either be a risk factor or an early sign of dementia. Alzheimer's disease (AD) being the most common example of dementia.

Shared features of generalised periodontitis and AD include a progressive, inflammatory disease pathway, and shared disease related risk factors including ageing, infection, immunosuppression, genetic predisposition and socioeconomic factors. This chapter will explain the dental hypothesis surrounding the associations made between periodontal disease and AD. The associations between generalised periodontal disease and AD are evidenced by experimental research data and describe how the hypothesis that "periodontitis is a risk factor for AD" came about. The evidence to support this hypothesis is substantial, whilst studies supporting a causal relationship are in progress. The aim is to highlight the risk factor involvement of generalised periodontal disease on AD development with a specific focus on *Porphyromonas gingivalis,* because this microbe is at the heart of current research, which suggests potential for much desired therapy.

13.2 Periodontal Disease

Periodontitis is an oral disease presenting with a polymicrobial dysbiosis of the subgingival microbiome, which, if left untreated, leads to tooth loss. *P. gingivalis* is considered a keystone pathogen of periodontal disease alongside its companion bacteria (*Treponema denticola, Tannerella forsythia*) of the red complex in adult periodontitis (Socransky et al. 1998; Holt and Ebersole 2005; Hajishengallis et al. 2012). Around 50% of all humans in middle age (50 years and over) appear to fall victim to periodontitis (Eke et al. 2015). In the previous classification, the adult form of periodontitis was known as "chronic periodontitis". However, a new classification (Caton et al. 2018; Dietrich et al. 2019) refers to the formerly known "chronic" periodontitis as generalised periodontitis stages I–IV, grades A–C. This is difficult to integrate into a story that appeals to dental and non-dental professionals so we will refer to "chronic periodontitis" as "generalised periodontitis" to be consistent with the new changes throughout this chapter.

The nature of periodontal disease is episodic with characteristic recurrent periods of active disease progression followed by periods of quiescence in individuals who are unable to prevent commensals (healthy microbiome) converting into pathogens (pathobiome) (Dioguardi et al. 2020). Periodontal disease affects the tooth supporting tissues, and interaction of specific bacteria and consequently the host's immune responses play a pivotal role (Haffajee et al. 1988). The host's response to the pathogenic bacteria and their virulence factors such as lipopolysaccharide (LPS), proteases such as gingipains, fimbriae, hemagglutinins, and outer membrane vesicles (OMVs) determine the severity and extent of periodontal disease (Kinane and Marshall 2001; How et al. 2016). Bacterial LPS is located in the outer membrane of Gram-negative bacteria and is a potent stimulator of host innate immune signal transduction pathways (Beutler 2000). The acute bacterial challenge stimulates the junctional pocket epithelium to produce inflammatory mediators to protect against tissue damage via an acute phase receptor-mediated cytokine production and neuropeptide release, resulting in vasodilation of local vessels. Gingival bleeding, swelling and redness together with the presence of neutrophils and macrophages within the inflamed gingival tissues indicate clinical signs of inflammation (Hasturk et al. 2012). In susceptible individuals, the acute phase responses fail to clear infection and chronic inflammatory lesions develop. The subgingival sulcus serves as a niche enabling a cyclic chronic inflammatory process, which facilitates recurrent bacteraemias, enabling micro-organisms and their virulence factors to access the systemic circulation (Forner et al. 2006; Lockhart et al. 2008; Bahrani-Mougeot et al. 2008). Each time we brush or chew on a periodontally-affected tooth, bacteraemia consisting of a spectrum of oral bacteria occurs. In any one day this bacteraemia can last for a total of 3 h (Bahrani-Mougeot et al. 2008; Tomás et al. 2012). In addition, viruses, bacteriophages, and yeasts within the periodontal pocket may follow the bacteria into the blood stream along with inflammatory mediators from the inflamed periodontal tissues (Olsen and Singhrao 2015; Li et al. 2020). From the blood stream these bacteria may be carried to wherever they lodge, potentially providing a nidus for further organ specific inflammatory pathologies. Thus, periodontal disease has been associated with a number of inflammatory pathologies, including cardiovascular disease, diabetes and AD (Makiura et al. 2008; Bale et al. 2017; Demmer et al. 2020; Stein et al. 2007; Poole et al. 2013; Dominy et al. 2019). Indeed, the cause of death in AD cases can result from cerebrovascular diseases such as stroke and pneumonia suggesting a greater risk of dementia for individuals who have suffered multiple co-morbidities in their lives (van Oijen et al. 2007).

13.3 Alzheimer's Disease (AD)

There are two main forms of AD; familial or early-onset form, which involves mutated genes such as amyloid precursor protein (APP) and presenilin 1 and presenilin 2. In the sporadic or late-onset form of AD, Genome-Wide-Association Studies (GWAS) have identified a number of susceptibility genes expressed by the brain cells. Of these, the *apolipoprotein E, allele 4* (*APOE* ϵ4) is firmly established as the second strongest risk factor after advancing age (Corder et al. 1993; Saunders et al. 1993). AD is a leading neurodegenerative disease with clinical signs of deteriorating memory, which together with its two neuropathological hallmark lesions, amyloid-beta (Aβ) and neurofibrillary tangles (NFTs), complete its post-mortem diagnosis (Hyman et al. 2012). Although not pathognomonic of AD, other lesions which present in AD pathophysiology include neuronal and synaptic loss and gliosis (intracerebral inflammation) (Dugger and Dickson 2017). The origins and the roles of Aβ plaques and NFTs are quite distinct but they both lead to neurodegeneration within the associated regions of the cerebral cortex and medial temporal lobes. The hippocampus, by contrast, typically contains abundant intra-neuronal NFTs composed of abnormally phosphorylated tau protein representing destabilized microtubules that are non–membrane-bound masses of abnormal straight and/or paired helical filaments (PHF) (Grundke-Iqbal et al. 1986; Goedert et al. 2006). The NFTs were thought to first appear in the entorhinal cortex leading to the hippocampus, but this view has changed identifying early involvement in the subcortical nuclei such as the locus coeruleus in the pons (Braak and Braak 1991; Braak et al. 2011).

The AD plaque is composed of an insoluble form of Aβ and continues to form the basis of the influential "Amyloid Cascade Hypothesis" (Hardy and Selkoe 2002). Molecular cloning methodologies identified Aβ as a cleavage fragment of a single membrane-spanning receptor-like glycoprotein known as amyloid precursor protein (APP) that is inserted, in part, in the plasma membrane (Kang et al. 1987). APP also occurs in internal vesicular membranes, including the Golgi apparatus and endosomes (Choi et al. 2012) as explained by Singhrao and Olsen (2019) in intracellular bacterial infections such as *P. gingivalis*. APP proteolytic cleavage involves three proteases, namely α-, β- and γ-secretases, the β- secretase acts with γ-secretases to release Aβ which becomes insoluble and fibrillar in appearance (Hook et al. 2008) giving rise to the neuritic/senile or Aβ plaques of AD. Activated microglia and astrocytes (gliosis) accompany these plaques and contribute to intracerebral inflammation (Perry et al. 2010; Boche et al. 2013; Heneka et al. 2015).

13.4 The Emerging Association Between Periodontal Disease and Alzheimer's Disease Provides a Rationale for Therapy

Researchers have shown that the genetic variant of the *APOE ϵ4* increases a person's risk of developing AD (Genin et al. 2011). An indirect effect of inheriting the *APOE* ϵ4 gene variant with respect to patients with periodontal disease may be the increased susceptibility of the host to microbial infections (de Bont et al. 1999; Moretti et al. 2005; Watts et al. 2008). This genotype has also been linked to increased inflammatory burdens in terms of cytokines (systemic circulation and brain) in response to LPS (Tsoi et al. 2007; Watts et al. 2008; Hubacek et al. 2010).

Previous research has indicated that those individuals from low socioeconomic status and low levels of education show poorer cognitive functioning in later life, and are at greater risk of developing AD (Brayne and Calloway 1990; Chen and Miller 2013; Russ et al. 2013; Marden et al. 2017). Low socioeconomic status groups have also shown an increased likelihood of engaging in habits (poor dietary choices) and lifestyle choices that are detrimental to their oral and general health. These behavioural traits (for example, smoking and drinking excessive amounts of alcohol) increase the risk for an individual to develop both periodontitis, and AD (Singh et al. 2013). A poor diet contributes to making the immune system less robust in its ability to fight infection. Smoking tobacco has a detrimental impact on general health, increasing the likelihood of developing cardiovascular and cerebrovascular disease, but more importantly increases the severity to which periodontal disease progresses due to decreased oxygen and an increased inflammatory burden (Kamma et al. 1999; Grossi et al. 1995).

It is important to point out that not everyone who develops clinical AD is comorbid with periodontal disease (Stein et al. 2007; Farhad et al. 2014). The reason for this observation remains unknown. However, a plausible explanation could be that the more virulent strains of *P. gingivalis* may develop in poorly managed oral health and in cases of periodontitis that eventually lead to sporadic AD. Figure 13.1 illustrates the bidirectional relationship between periodontitis and AD.

13.5 The Experimental Journey Towards Risk Factor Identification Between Periodontitis and Alzheimer's Disease

The journey began in 1994 with the Japanese elderly population based epidemiological study which introduced the concept of fewer teeth in later life as a risk factor for dementia (Kondo et al. 1994). For a significant length of time this concept was left unexplored. Later Chalmers et al. (2002) showed that cognitively impaired institutionalised patients exhibited more retained roots, carious teeth and missing teeth due to poor oral health compared to community-dwelling older adults. A systematic review by Cerutti-Kopplin et al. (2016) concluded that there was evidence to support that tooth loss increased the risk of cognitive impairment and dementia.

It was Stein et al. (2007) who investigated a potential association between the history of oral disease and dementia development in female subjects of a religious order. The study included analysis of the previously collected data from 10 annual cognitive assessments of 144 Milwaukee participants in the "Nun Study". Participants with fewer teeth directly correlated with an increased risk of prevalence and incidence of dementia (Stein et al. 2007). It also concluded that missing up to nine teeth carried the highest risk for developing late-onset AD with an odds ratio of 2.2 (95% CI 1.1, 4.5), proposing tooth loss due to periodontitis was doubling the risk factor for AD development. Together with other studies, this helped to consolidate the idea that potential neglect of oral health could have a detrimental effect on brain health (Stein et al. 2007).

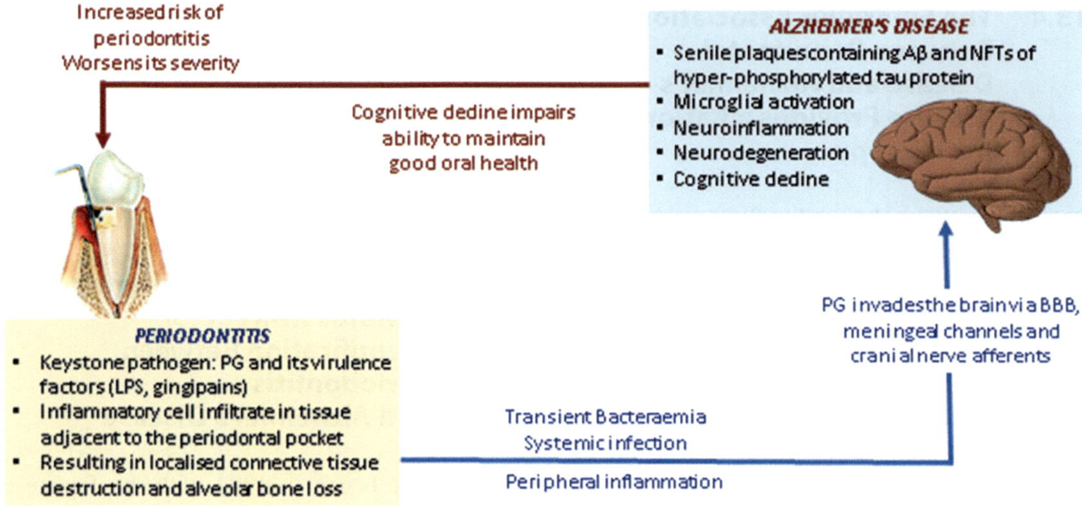

Fig. 13.1 The bidirectional relationship between periodontitis and AD

13.6 Does Peripheral inflammation Induce Cerebral Inflammation?

In establishing the events leading from poor oral health to AD manifestation, Noble et al. (2009) used a cross-sectional methodology to investigate an association between cognitive impairment and *P. gingivalis* via serum markers of its infection. It was found that those subjects who were cognitively impaired had higher mean *P. gingivalis* IgG levels compared to those who were cognitively healthy. Differences in individuals' humoral responses supported the association of *P. gingivalis* with cognitive impairment. Also, Holmes et al. (2009) had proposed that peripheral inflammation may be a key determinant of the cognitive decline associated with AD progression. This prompted Kamer et al. (2009) to establish if inflammatory markers such as TNF-α cytokine and antibodies to periodontal bacteria could discriminate between AD and non-AD subjects using case-control methodology. Sparks Stein et al. (2012) also investigated whether or not serum antibodies to periodontal pathogens were a risk factor for AD. They used a case-control methodology with information gathered from a larger cohort study. They found that antibodies to certain periodontal pathogens over 10 years were significantly higher at baseline in those who went on to develop AD (Sparks Stein et al. 2012). A large scale Taiwanese insurance based epidemiological survey also reported that individuals with chronic periodontitis for at least 10 years had a 70% higher risk of developing AD compared to individuals without periodontal disease (Chen et al. 2017). This reinforced the postulated 10-year lag phase from the time of periodontal disease diagnosis to developing AD and lends credence to the bidirectional relationship between the periodontitis-AD brain risk-axis.

The report by Poole et al. (2013) analysed human AD and non-AD post-mortem brains using immunolabelling and immunoblotting and demonstrated the LPS of *P. gingivalis* exclusively in the AD brains and not the controls. This finding was consolidated by Dominy et al. (2019) who used more advanced molecular techniques identifying *P. gingivalis* deoxyribonucleic acid (DNA), LPS and gingipain antigens within the brain tissue of AD subjects. These two studies (Poole et al. 2013; Dominy et al. 2019) provided the rationale and the impetus for developing periodontal disease models for AD to better understand the potential of a bacterial aetiology for this neurodegenerative disease occurring via the peripheral (systemic) system.

The first mouse model used genetically modified apolipoprotein E knock-out (ApoE$^{-/-}$) mice, which were orally infected with *P. gingivalis* to initiate experimental chronic periodontitis (Velsko et al. 2014). Poole et al. (2015) showed that within 3 months of *P. gingivalis* oral infection, bacterial DNA and LPS had spread to the $ApoE^{-/-}$ mice brains. In addition, there was evidence of intracerebral inflammatory pathology and complement activation once the bacterium had entered the hippocampus, compromising the health of neurons (Poole et al. 2015). Due to the genetic weaknesses of the $ApoE^{-/-}$ *P. gingivalis*-mono-infection model, demonstrating the Aβ plaque hallmark of AD (intracerebral pathology) was not possible.

The next stage saw researchers introducing genetic mutations to their periodontitis mouse model. They induced *P. gingivalis* mono-infection into the APP-transgenic (APP-Tg) model carrying the Swedish and Indiana mutations (Ishida et al. 2017). The aim was to assess the role of chronic periodontitis in the development of AD hallmark pathology. However, Ishida et al. (2017) concluded that their results reflected an overall susceptibility to AD rather than having contributed to the overall Aβ hallmark pathology. In retrospect this interpretation appears to be plausible, as APP in the familial form of AD, harbours a missense mutation, (where a single nucleotide is changed with the substitution of a different amino acid) (Rossor et al. 1993), suggesting that the bacterial and/or host's β-secretase equivalent digestive enzymes (cathepsin B and gingipains) (Dominy et al. 2019; Hook et al. 2008) may be unable to cleave APP in significant amounts to notice Aβ contribution in this model.

Moving away from genetically modified mice, Ilieviski et al. (2018) chose a wild-type mouse as their model of periodontitis for reproducing AD pathophysiology. They confirmed the spread of *P. gingivalis* from the oral niche to the brain in their wild-type mice and observed glial cell activation, the same as reported by Poole et al. (2015), however in addition, the infection reproduced the cardinal neuropathological hallmark lesions (Aβ plaques and phosphorylated tau protein at serine396 position) of AD for the first time. This is corroborated by the fact that *P. gingivalis* has its own β-secretase (in the form of gingipains) activity to cleave APP (Dominy et al. 2019). Both bacterial and host β-secretase (in the form of neuronal cathepsin B) enhances the overall APP cleavage (Hook et al. 2008).

P. gingivalis gingipains and its LPS are also known to activate kinases such as glycogen synthase kinase-3 (GSK-3) which subsequently phosphorylates neuronal tau at two sites (serine 396 and Threonine231), involved with the NFT AD lesion (Haditsch et al. 2020; Ilieviski et al. 2018; Bahar and Singhrao 2021).

Further studies have examined the mechanistic links for cognition with *P. gingivalis*. The first of these functional testing studies was reported by Ishida et al. (2017) in their APP-Tg mice orally infected with *P. gingivalis*. They demonstrated that cognitive function was significantly impaired in periodontitis-induced APP-Tg mice compared to the sham-infected group when tested using a water maze (Morris 1984). An explanation for the greater cognitive deficit in the infected APP-Tg group was an increased inflammatory mediator (cytokine) burden following induction of experimental periodontitis.

Subsequent investigations performed similar behavioural tests on orally mono-infected *P. gingivalis* mice and supported the mechanism of cognitive deficit due to inflammation resulting from infection in which ageing was also a factor (Ding et al. 2018). These mouse model based studies are a proof of concept that periodontitis detrimentally impacts cognition via release of inflammatory mediators into the blood stream.

13.6.1 Systemic LPS and Its Effect on Cognition

Wu et al. (2017) were the first to report that chronic exposure to LPS from *P gingivalis* elicited AD-like phenotypes in middle-aged (12 months old) mice. The phenotypes included learning and memory deficits, intracellular Aβ in neurons, and microglia-induced neuroinflammation in the hippocampus. The suggested mechanism is that Aβ is cleaved indirectly by the action of cathepsin B on the parent APP. APP is initially stimulated by the interaction of the cytokine interleukin (IL)-1β with its cognate receptor (IL-1R) on neurons. Exposure to LPS from *P. gingivalis* led to a significant increase in microglia implying their activation and subsequent secretion of cytokines and neurons. This indicates that the memory deficit seen following systemic exposure to LPS of *P. gingivalis* in middle-aged mice is depended on cathepsin B and gingipains.

Following on from this, Zhang et al. (2018) conducted functional testing, but specifically analysing the effect of *P. gingivalis* LPS on cognitive function. Behavioural changes were assessed with the open field test, Morris water maze, and passive avoidance test. Using immunohistochemistry, they assessed for activation of astrocytes and microglia within the cerebral cortex and hippocampus; and assessed for pro-inflammatory cytokine expression of Interleukins (IL) IL-1β, IL-6, IL-8, Tumour necrosis factor-α (TNFα), toll-like receptors (TLRs) and CD-14 using reverse transcriptase-polymerase chain reaction (RT-PCR), enzyme-linked immunoassay (ELISA) and western blotting. The mice infected with *P. gingivalis*-LPS showed impairment in spatial learning and memory, along with activation of microglia and astrocytes within both the cerebral cortex and hippocampus. In addition, there was up-regulation of pro-inflammatory cytokines and activation of the TLR4/nuclear factor kappaB (NF-κB) signaling pathway. These findings suggested that *P. gingivalis*-LPS can lead to impaired memory and learning, which would appear to be mediated by the TLR4 signalling pathway. This demonstrates the potential for periodontal pathogen endotoxin as a risk factor for cognitive disorders.

13.6.2 Systemic Aβ; Does It Contribute to the Intracerebral Burden of This Hallmark in AD?

It has recently been suggested that periodontitis increases the levels of peripheral Aβ from gingival tissue and animal models and inflammatory cell sources (Leira et al. 2019; Nie et al. 2019). This has led to the proposal that in the human scenario, there is a potential for this peripheral source of Aβ to gain access into the brain thereby adding to the Aβ pool in the AD brain. Whether this would be a plausible mechanism in the human AD brain is not known. However, Zeng et al. (2020), in their *P. gingivalis* infection mouse model, identified advanced glycation end products (AGE) as a plausible receptor for Aβ in cerebral endothelial cells in mice. This implies that AGE products-receptor are a plausible mediator of cerebrovascular-related Aβ accumulation in the brain; supporting the hypothesis that patients harboring the generalized form of periodontal disease may be at risk of developing AD via multiple pathways involving periodontitis.

Limitations to the study included the intraperitoneal administration of *P. gingival*-LPS (rather than oral infection of whole *P. gingivalis* as with other research); the route of administration may have had some bearing on the passage into the cerebral tissues. Also, all mice were of the same age, and therefore the effect of ageing was not assessed by this study. However, it provided useful information with regard to the potential effect of periodontal pathogen endotoxin on cognitive function, and that inflammation, which plays an important role in cognitive impairment, was mediated by the TLR4 signaling pathway, giving an indication of possible underlying mechanisms.

These studies have demonstrated major advances that have been made experimentally to investigate the relationship between periodontal pathogens and their associated products, cognitive impairment, and AD pathophysiology. This research has been fundamental in the positive progress of the development of potential therapeutic agents against AD.

13.6.3 Evidence Supporting the Inflammatory Burden of Periodontitis Affecting Memory

An observational cohort study by de Rolim et al. (2014) involved 29 participants with clinically mild AD. Intervention involved a complete evaluation performed by a dental surgeon, and included a clinical questionnaire; research diagnostic criteria for temporomandibular disorders; the McGill pain questionnaire; oral health impact profile; decayed, missing and filled teeth index; and complete periodontal examination before and after the trial, which involved any dental treatment deemed necessary based on the findings of the initial evaluation. The dental treatment most frequently performed included periodontal treatment (scaling), extractions, and application of topical Nystatin (anti-fungal). The study found an increase in orofacial pain, followed by improvement in the mandibular function and periodontal indices in patients with mild AD after treatment. These improvements were maintained until the last evaluation after 6 months, and this was followed by a reduction in the functional cognitive impairment. The limitation of this study is that it lacked a bigger cohort and appropriate (non-AD) controls. The second observational cohort study was performed by Ide et al. (2016) who set out to test the hypothesis that circulating inflammatory cytokines due to periodontal disease bacteria were linked to greater rates of cognitive decline in clinical AD cases. The study recruited 59 participants with mild to moderate AD in which cognition and circulating inflammatory markers were tested. Fifty-two of the participants were followed-up at 6 months when they all underwent repeat assessment of their initial biomarkers for any changes. The findings of the study concluded that the presence of periodontitis at baseline was associated with a six-fold increase in the rate of cognitive decline in participants over the 6-month follow-up period. Periodontitis at baseline was also associated with a relative increase in the pro-inflammatory state over the 6-month follow-up. The authors concluded that periodontitis is associated with an increase in cognitive decline in AD. This study upheld the view linking cognitive decline with the body's inflammatory responses. The major weakness of the study was the absence of participants with intact cognition as controls. A more recent report by Li et al. (2020) also supports a potential link with inflammatory cell counts as part of systemic inflammation due to periodontal disease and the risk of cognitive decline in the US elderly population. Table 13.1 summarizes the key evidence which supports the association between periodontitis and AD.

13.7 AD Brain Amyloid-β: A Potentially Shared Pool from Peripheral and Intracerebral Sources Towards a Blood Biomarker for Clinical AD

It has recently been suggested that periodontitis increases the levels of peripheral Aβ within gingival tissues and human serum (Kamer et al. 2015; Gil-Montoya et al. 2017; Nie et al. 2019; Leira et al. 2020). Blood-based biomarkers are very appealing as obtaining cerebrospinal fluid from patients is not ideal. Ashton et al. (2021) recently suggested that blood pTau231 has the potential to indicate early amyloid formation. This residue (Threonine231) phosphorylation in tau protein is also a site that *P. gingivalis* infection contributes to (Haditsch et al. 2020), and further supports a causative role for periodontitis in AD development.

13.7.1 The Inflammatory Hypothesis – A Potentially Shared Pathway

Inflammation is the inevitable consequence of an infectious episode in the body. This is better appreciated in the case of generalised periodontitis (due to polymicrobial aetiology). The existence of an intact blood-brain barrier avoids any impact of peripheral inflammation on the brain. Details of the blood-brain barrier will not be described here, as the reader is encouraged to read Singhrao and Olsen (2018) for related information. However, reports (Marques et al. 2013; Montagne et al. 2015, 2016; Halliday et al. 2016) have suggested that during ageing and manifestation of AD, the blood-brain barrier becomes defective. This may be the result of a compromised immune protection coupled with poor

Table 13.1 Key references supporting scientific evidence for the link between periodontitis and AD

Type of Study	Study	Relevant Methodology	Outcome of interest
Postmortem human brains	Poole et al. (2013)	Matched 10 AD cases for tissue from Brains for Dementia Research alongside 10 non-AD age-related controls with similar or greater postmortem interval	LPS from periodontal bacteria can enter the AD brain but were not found in controls
	Emery et al. (2017)	Pilot study using 16S ribosomal New Generation Sequencing to assess the bacterial component of the microbiome in frozen and fixed post-mortem tissue from AD and control temporal cortex	AD brains had increased bacterial loads compared to controls. Species associated with skin, nasopharyngeal and the oral cavity were most consistently detected
	Riviere et al. (2002)	Molecular and immunological techniques to detect oral Treponema in postmortem human brains. 16 donors with AD (55–87 years of age, eightmales) and 18 controls	AD brains more likely to have Treponema than controls as well as increased number of Treponema species. This study supports the hypothesis that oral Treponema reached the brain via the trigeminal nerve
	Dominy et al. (2019)	Tissue microarrays containing sex- and age-matched brain tissue cores from the middle temporal gyrus of AD patients and controls were used for immunohistochemical analysis	Identification of P. gingivalis DNA in AD brains. Gingipains were also identified in the brain of AD patients, and levels correlated with tau and ubiquitin pathology
Inflammatory biomarkers	Sparks Stein et al. (2012)	Serum samples from 158 participants. Antibody levels were compared between controls and subjects with AD. Median time from baseline assessment to diagnosis for AD was 9.6 years, Mean length of follow-up for the controls was 12.5 years	Elevated antibodies to periodontal bacteria in subjects years before cognitive impairment, suggesting that periodontal disease could potentially contribute to the risk of AD onset/progression
	Kamer et al. (2015)	18 with AD and 16 cognitively normal	Plasma TNF-α and antibodies against periodontal bacteria were elevated in AD patients compared with NL and independently associated with AD. The number of positive IgG to periodontal bacteria incremented the TNF-α classification of clinical AD and NL. This study shows that TNF-α and elevated numbers of antibodies against periodontal bacteria associate with AD and contribute to the AD diagnosis
	Li et al. (2020)	766 participants aged above 60 years and who had periodontal and cognitive examinations in the National Health and Nutrition Examination Survey (NHANES) 2001–2002 in the U.S.	Participants with increased periodontal health obtained higher digit symbol substitution test (DSST) scores. White blood cell count acted as a mediator in the association between bleeding on probing and DSST we well as periodontal inflamed surface area and DSST. This supports the role of systemic inflammatory factors as a mediator of the association between periodontal inflammation and cognitive function

(continued)

Table 13.1 (continued)

Type of Study	Study	Relevant Methodology	Outcome of interest
Epidemiological studies	Tzeng et al. (2016)	2207 CP and gingivitis patients were selected from the National Health Insurance Research Database of Taiwan, with 6621 controls matched for sex and age. After adjusting for confounding factors, Cox proportional hazards analysis was used to compare the risk of developing dementia during the 10-year follow-up period	Patients with CP and gingivitis had 2 times higher risk of developing AD
	Chen et al. (2017)	Retrospective matched-cohort study (National Health Insurance Research Database) in Taiwan 9291 patients with CP matched to 18,672 patients without CP. Regression analysis to determine risk of AD	10-year CP exposure was associated with 1.7 fold increase in risk of developing AD
	Lin et al. (2020)	Taiwan's National Insurance database was used to evaluate associations between dental health and AD; 209,112 new cases of AD were matched 1:4 with 836,448 dementia-free controls	Oral health care was associated with lower odds of developing AD
	Beydoun et al. (2020)	IgG titers against P. gingivalis and other periodontal pathogens	Evidence for an association between periodontal pathogens and AD, which was stronger for older adults

oral hygiene and increased circulating inflammatory mediators. These mediators are implicated with age-related defects (cardiovascular health) and with possible microbial components in vulnerable subjects. It is therefore plausible to suggest that pathogenic bacteria from chronic infections and their endotoxins do spread to other anatomical sites (Makiura et al. 2008; Bale et al. 2017; Poole et al. 2013; Rokad et al. 2017; Dominy et al. 2019; Demmer et al. 2020). This invariably impacts on glial cell activation. Subsequent exposure of oral microbial debris or their secondary products to the brain causes the already reactive microglial cells to ferociously up-regulate cytokine secretion. Based on the reviews by Olsen et al. (2017); Olsen and Singhrao (2019, 2020), there is an increasing body of evidence to support complement as a potential shared pathway for both periodontitis and AD. *P. gingivalis* has developed impressively successful strategies for immune evasion in the periodontal pocket (Hajishengallis 2011). This is because *P. gingivalis* has to evade immune-mediated killing to survive and yet requires inflammation in order to obtain nutrients to flourish. This may be the reason for how this bacterium has resolved this predicament whilst also benefiting its companion species within the microbial community as a whole, and particularly under inflammatory environments to compete for dominance (Hajishengallis 2011; Singhrao et al. 2015). An imbalance in complement activity may influence dysbiosis of the host's microbiome. We will not describe the complement system in detail as it is a subject of its own and the reader is directed to excellent reviews (Morgan and Harris 2015; Olsen et al. 2017).

Complement evasion in periodontitis nearly always involves gingipains, the digestive enzymes specific to *P. gingivalis*. These enzymes break down the host's protective barriers enabling the pathogen to penetrate periodontal tissues. In doing so, they switch on the host's immune responses. Gingipains have dual functionality and can exert dose-dependent biphasic effects on complement activation. At low concentrations they can activate complement, which have the advantage of eliminating complement sensitive commensals which may compete with *P. gingivalis* for niche space and nutrients. In addition, *P. gingivalis* is relatively resistant to complement-mediated opsonisation and killing, and is intrinsically resistant to the lytic action of complement. Conversely at high concentration, *P. gingivalis* is likely to be established, and so gingipains can inhibit bactericidal activity of complement thus preventing opsonisation of complement-sensitive bacteria promoting a mixed species biofilm. Sustained complement activation is a potent driver of inflammation in the body, including the brain. Inappropriate complement activity also plays a part in AD pathogenesis.

GWAS has identified four possible genes linking complement to AD. These include Complement and sub-complement 1 s (C1s), Complement receptor 1 (CR1), Clusterin and Complement component 9 (C9). The concern is that the brain does not have a traditional lymphatic drainage system, meaning an efficient complement system (which also acts as a pathway for clearing debris) is essential for the clearance of damaged cerebral tissue. Defective complement genes may disable phagocytic activity of local microglia resulting in ineffective clearance of proteins like Aβ and NFTs. Two clearance systems, unique to the brain, have been described. Firstly, the glymphatic system, a form of brain cleansing system that works better during night sleep (Iliff et al. 2013; Nedergaard 2013). Hence adequate sleep is implicated for better brain function. The second system is the intramural periarterial drainage pathway. This clears solutes from the brain to the periphery along arterial smooth muscle cell basement membranes. This works more efficiently during the daytime. For more information on these clearance pathways in relation to bacterial products see Singhrao and Olsen (2018).

It is becoming clear that the susceptibility gene *APOE ϵ4* also exerts its own risk for developing AD via inflammation mediated by the classical complement pathway linked to deregulating C1q to keep the classical complement pathway activated (Yin et al. 2019). This has an impact on further inflammatory responses via cytokine liberation by activated monocytes/macrophages/microglia (Ihara and Yamamoto 2016). Interestingly, a combined action of inappropriate complement activity can instruct microglia to over-prune synapses and this has implications in loss of synapses and memory (Vasek et al. 2016).

It has been demonstrated that susceptibility genes, such as *APOE*, can exert adverse effects on the host in combination with environmental factors, which are often controllable or modifiable. Hence controlling the environmental factors may exert a protective effect on cognitive functioning. This appears to be the case in the human AD interventional study carried out by Ide et al. (2016). The inflammatory burden is often described as the mechanism of risk; and certainly the impact of any lifestyle modification which has an anti-inflammatory effect will be significant for reducing both AD and periodontitis.

13.8 Therapeutic Advances – Are We Nearly There Yet?

13.8.1 Periodontal Therapy

Current therapies for periodontal treatment include non-surgical instrumentation to disrupt the biofilm and remove calculus, root surface instrumentatio and, removal of plaque retention factors (such as overhangs on restorations). Non-surgical therapy must be combined with changing the patient's behaviour and improving oral hygiene measures for it to succeed. More complex and advanced periodontal therapy include full mouth disinfection, local antimicrobial application, systemic antibiotic therapy, host modulation therapy, and advanced surgical techniques such as flap surgery, soft tissue grafts, bone grafting, guided tissue regeneration and tissue stimulating proteins. By managing periodontal disease, we can keep the numbers of pathogenic bacteria in the periodontal pocket low. This should prevent their spread.

13.8.2 AD Treatment Based on Cholinesterase Inhibitors

For AD, potential treatments were developed based on the rationale that neurodegeneration led to neurotransmitter deficit, in particular of the cholinergic system. Cholinesterase inhibitors approved for treatment of mild to moderate AD include tacrine (First Horizon Pharmaceuticals), which causes liver toxicity and is therefore rarely prescribed, donepezil (Pfizer) (also approved for severe AD); rivastigmine (Novartis) and galantamine (Janssen). Certainly, cholinesterase inhibitors are currently still the main drug administered to AD patients. Unfortunately, this treatment is proving to be inadequate as cholinesterase inhibitors do not slow the progression of AD symptoms (Massoud and Gauthier 2010).

13.8.3 AD Treatment Based on Targeting Aβ

According to the Amyloid Cascade Hypothesis (Hardy and Selkoe 2002) the Aβ plaque lesions of AD were considered to be the cause of AD. Although this has subsequently been shown to be untrue, this hypothesis still remains influential in the field of AD research because of its link with the Aβ diagnostic burden and disease progression. Therapies were based on two main aspects of the amyloid cascade hypothesis. Firstly, the idea of inhibiting α-, β- and γ-secretases was explored (Nunan and Small 2000). This proved not to be successful, but formed the basis for exploration of anti-Aβ immunotherapy to solubilise aggregated Aβ, thus reducing it in the AD brain. Two anti-Aβ drugs, gantenerumab and solanezumab, were designed to help remove excess Aβ in the brain. For a more comprehensive read on the anti-Aβ based therapies the reader should consult (Kozin et al. 2018; van Dyck 2018). What is clear is that altering the course of AD by protease inhibitors of the amyloid cascade and/or mopping up the already aggregated Aβ in the form of plaques has not been successful in humans. This has opened the field of therapy to other related strategies.

13.8.4 AD Treatment Based on Targeting Tau Protein

NFT formation and its spread in the AD brain correlate with symptom severity and neurodegeneration, thus providing the rationale for developing therapies that target tau in its abnormal state. This formed the basis for developing inhibitors for kinases that phosphorylate tau and/or immunotherapy to control aggregation of abnormally phosphorylated tau, or to stabilise microtubule assembly within neurons. Of these, free cytosolic hyperphosphorylated tau fragments are implicated in the spreading of tau pathology. Researchers have shown that extracts of cytosolic tau from AD brains reproduce tau pathology when injected into mice brains. Therapeutically, these approaches also have their challenges in terms of being toxic and/or a lack of efficacy. Drugs based on tau toxicity are also being developed but this is not our niche. However, the interested reader is directed to some comprehensive references on this subject (Congdon and Sigurdsson 2018; Sayas 2020).

13.8.5 AD Treatment Based on Targeting Neuroinflammation

Based on the observation of neuroinflammation within the pathogenesis of AD, it became of great interest to assess the impact of Non-Steroidal Inflammatory Drugs (NSAIDs) on the development and progresion of AD. Development of specific NSAIDs along with discovery of NSAID targets in diseased brains led to a multitude of animal and human trials. Many NSAIDs have been found to cross the blood-brain barrier (Parepally et al. 2006) and cyclo-oxygenase (COX)-mediated prostaglandin (PG) synthesis has led to an increase in the understanding of healthy and diseased brain function. However, it has been shown that neuroinflammation and neurodegenerative disease are complex and NSAID efficacy on pathology and behaviour vary according to the specific disease, the region of the brain, and study design (Moore et al. 2010).

Trials assessing the effect of COX-2 inhibitors on the cognitive decline seen in AD, showed initial evidence that indomethacin might have beneficial effects (Rogers et al. 1993). Unfortunately, large scale clinical trials assessing cognitive outcomes following NSAID administration have been disappointing (de Jong et al. 2008; Pasqualetti et al. 2009) and the general consensus was that NSAID treatment becomes ineffective once memory declines and the associated pathologies have already developed (McGeer and McGeer 2007; Townsend and Practico 2005).

13.8.6 AD Treatment Based on Targeting *P. gingivalis*

The research by Dominy et al. (2019) highlighted the potential importance of gingipains which have been shown to mediate the toxicity of *P. gingivalis* in epithelial and endothelial cells (Stathopoulou et al. 2009; Sheets et al. 2005; Kinane et al. 2012). They may also be implicated as narrow spectrum virulence targets (Guo et al. 2010; Travis and Potempa 2000; Clatworthy et al. 2007). Blocking gingipain activity with short peptide analogues reduces the virulence of *P. gingivalis* (Kadowaki et al. 2004).

Dominy et al. (2019) hypothesised that *P. gingivalis* acts in AD pathogenesis via gingipain secretion to promote neuronal damage. This was supported with experimental evidence demonstrating that gingipain immunoreactivity in AD human brain was significantly greater than in non-AD brains; and correlated with the presence of gingipains to the increase in tau load and tau NFTs. As a result of this evidence, a potent, selective, brain-penetrating, small-molecule gingipain inhibitor was developed which, *in vivo,* shows promise with the potential to act as a disease modifier for AD (Ryder 2020).

Based on the ground-breaking research from Poole et al. (2013) and Dominy et al. (2019), the US pharmaceutical company have developed COR388 (Cortexyme), a novel virulence factor inhibitor, that targets gingipains from *P. gingivalis*. This drug is in phase 2/3 of clinical trials with promising results (Ryder 2020). It is thought to be able to block brain infiltration of *P. gingivalis* along with the subsequent downstream pathology of AD, including Aβ production, neuroinflammation and neurodegeneration. If effective, it is unclear how this drug will be administered on daily basis. However, speculation suggests it may be an adjunctive management.

13.8.7 AD Treatment Based on Managing Lifestyles

The Lancet Commission on Dementia Prevention, Intervention and Care have recognised elements of lifestyle such as smoking, obesity, excessive alcohol consumption, physical inactivity, depression and low social contact as potentially modifiable risk factors for dementia (Ngandu et al. 2015). Thus, making better lifestyle choices could prevent or delay dementia. This was supported by findings from the large-scale randomised controlled trial; the Finnish Geriatric Intervention Study to Prevent Cognitive Impairment and Disability (FINGER) which investigated multi-domain intervention in the at-risk segment of the general elderly population. The intervention group, which received risk reducing modifications to lifestyle patterns including diet, exercise, cognitive training and vascular risk monitoring, showed maintenance of, or improved cognitive functioning and a reduction in cognitive decline after the 2-year follow-up. This was the first robust large-scale trial which showed that modifying lifestyle factors could be beneficial to cognition (Ngandu et al. 2015). Although this may not be directly extrapolated to AD patients, the study population, being 60–77 year olds, could be in a pre-dementia state. Therefore, further research involving multi-modal intervention, designed to track participants over longer periods; assessing the level of impact of each factor on disease progression would be useful. Potential therapeutic effects can certainly be observed in pre-symptomatic and pre-dementia stages.

In summary, the importance of lifestyle factors is being increasingly recognized and investigated. Table 13.2 highlights current approaches to treating AD, aimed at maintaining mental function, managing behavioural symptoms, and

Table 13.2 Therapeutic advances; are we nearly there yet?

AD treatment modalities	Mechanism of action
Disease-Modifying Biologics	Monoclonal antibodies targeting the two hallmarks of Alzheimer's disease, toxic amyloid and the malfunctioning tau protein, belong to this class e.g. Aducanumab, Solanezumab and Gantenerumab
Disease-Modifying Small Molecules	Earlier small molecule AD treatments (cholinesterase inhibitors and memantine) did not alter the course of the disease but could only stabilise some symptoms. Newer small molecule therapies are aimed at reducing inflammation, nurturing the growth of synapse and inhibiting the activity of toxic enzymes e.g. curcumin, omega-3 fatty acids, tau-active agents (methylene blue)
Symptom-Reducing Small Molecules	Drugs given to control some of the symptoms of dementia, such as aggression and delusions e.g. Nabilone or dronabinol (components of marijuana); lemborexant, piromelatine and zolpidem (sleep-enhancing agents); escitalopram and mirtazapine (antidepressants); brexpiprazole and pimavanserine (antipsychotics)
Non-medication management	Non-pharmacologic trials are testing light therapy, acupuncture, transcranial direct current stimulation, transcranial magnetic stimulation, electroconvulsive therapy and deep brain stimulation Additional treatment strategies; nutritious diet, sufficient sleep, psychosocial therapies and exercise
Infection-based treatment	COR388, a bacterial protease inhibitor which targets gingipain produced by *P. gingivalis* which may reduce neuro-inflammation and hippocampal degeneration

At the time of publication, there are over 120 therapeutic agents undergoing clinical trials for the treatment of AD. The vast majority are disease-modifying agents aimed at altering the underlying biology of AD. The other categories of agents are intended to enhance cognition and manage neuropsychiatric symptoms. Infection-based treatment utilising the drug COR388 by Cortexyme targeting gingipains, the neurotoxic protease of *P. gingivalis* is currently undergoing clinical trials. Alternative therapies have also been introduced including light therapy and lifestyle modifications. The diverse AD therapeutic modalities have been summarised below. (Adaikkan et al. 2019; Cummings et al. 2020, Alzheimers.org, gaintrial.com)

slowing down the onset of symptoms of the disease.

Finally, poor sleep has been implicated with AD risk (Sprecher et al. 2017). Adequate sleep in advancing age is necessary for retaining memory and for effective functioning of the glymphatic system (Iliff et al. 2013; Nedergaard 2013). In 2019, The World Health Organisation has recognised poor sleep as a risk factor for general health and made recommendations of minimum hours of sleep per night for specific age groups.

The question as to whether we are there yet remains open, despite great efforts being made towards developing fundamental treatments and a push to adopting better lifestyles. In the absence of an adequate treatment for AD, the World Dementia Council (https://worlddementiacouncil.org/), National Institutes of Health (NIH) and Alzheimer's Research UK (ARUK) (www.alzheimersresearchuk.org/ARUK), lend their support towards prevention by modifiable risk factor identification and implementation during life to stay cognitively healthy for a happier and longer life.

13.9 Conclusions

We set out to understand how one common, clinical condition such as generalised periodontitis becomes a potential risk factor for the development of AD and discussed the factors related to oral microbial infection and the behavioural component. The GAIN trials will demonstrate if this is a causative relationship.

The onus is now on both the individual and the dental profession and emphasises the key roles every individual and dentist have in the prevention or delay of AD development. This gives added importance to the need for adequate control and management of periodontal disease, as well as the prevention message that patients receive. Oral health monitoring should complement other lifestyle management commissions such as those conducted in the FINGER trial. Periodontal disease, its pathogens and their virulence factors, have now been linked to many systemic diseases, and our role in maintaining not only the dental health, but now the general health of our patient base has become a key factor in enabling our patients to age 'successfully' and enjoy a healthy life.

References

Adaikkan C, Middleton SJ, Marco A et al (2019) Gamma entrainment binds higher-order brain regions and offers neuroprotection. Neuron. https://doi.org/10.1016/j.neuron.2019.04.011

Ashton NJ, Pascoal TA, Karikari TK et al (2021) Plasma p-tau231: a new biomarker for incipient Alzheimer's disease pathology. Acta Neuropathol. https://doi.org/10.1007/s00401-021-02275-6

Bahar B, Singhrao SK (2021) An evaluation of the molecular mode of action of *trans*-resveratrol in the *Porphyromonas gingivalis* lipopolysaccharide challenged neuronal cell model. Mol Biol Rep 48(1):147–156

Bahrani-Mougeot FK, Paster BJ, Coleman S et al (2008) Diverse and novel oral bacterial species in blood following dental procedures. J Clin Microbiol 46:2129–2132

Bale BF, Doneen AL, Vigerust DJ (2017) High-risk periodontal pathogens contribute to the pathogenesis of atherosclerosis. Postgrad Med J 93:215–220

Beutler B (2000) Endotoxin, toll-like receptor 4, and the afferent limb of innate immunity. Curr Opin Microbiol 3(1):23–28

Beydoun MA, Beydoun HA, Hossain S et al (2020) Clinical and bacterial markers of periodontitis and their association with incident all-cause and Alzheimer's disease dementia in a large national survey. J Alzheimers Dis 75(1):157–172

Boche D, Perry VH, Nicoll JAR (2013) Activation patterns of microglia and their identification in the human brain. Neuropathol Appl Neurobiol 39:3–18

Braak H, Braak E (1991) Neuropathological stageing of Alzheimer-related changes. Acta Neuropathol 82:239–259

Braak H, Thal DR, Ghebremedhin E et al (2011) Stages of the pathologic process in Alzheimer disease: age categories from 1 to 100 years. J Neuropathol Exp Neurol 70:960–969

Brayne C, Calloway P (1990) The association of education and socioeconomic status with the mini mental state examination and the clinical diagnosis of dementia in elderly people. Age Ageing 19:91–96

Caton JG, Armitage G, Berglundh T et al (2018) A new classification scheme for periodontal and peri-implant diseases and conditions – introduction and key changes from the 1999 classification. J Clin Periodontol 45(20):S1–S8. https://doi.org/10.1111/jcpe.12935

Cerutti-Kopplin D, Feine J, Padilha DM et al (2016) Tooth loss increases the risk of diminished cognitive function: a systematic review and meta-analysis. JDR Clin Trans Res 1(1):10–19

Chalmers JM, Hodge C, Fuss JM et al (2002) The prevalence and experience of oral diseases in Adelaide nursing home residents. Aust Dent J 47(2):123–130

Chen E, Miller GE (2013) Socioeconomic status and health: mediating and moderating factors. Annu Rev Clin Psychol 9:723–749

Chen C-K, Wu Y-T, Chang Y-C (2017) Association between chronic periodontitis and the risk of Alzheimer's disease: a retrospective, population based, matched-cohort study. Alzheimers Res Ther 9:56. https://doi.org/10.1186/s13195-017-0282-6

Choi RW, Cheng Z, Schekman R (2012) Amyloid precursor protein (APP) traffics from the cell surface via endosomes for amyloid β (Aβ) production in the trans-Golgi network. Proc Natl Acad Sci U S A 109(30):E2077–E2082

Clapp M, Aurora N, Herrera L et al (2017) Gut microbiota's effect on mental health: the gut-brain axis. Clin Pract 7(4):987. https://doi.org/10.4081/cp.2017.987

Clatworthy AE, Pierson E, Hung DT (2007) Targeting virulence: a new paradigm for antimicrobial therapy. Nat Chem Biol 3:541–548

Congdon EE, Sigurdsson EM (2018) Tau-targeting therapies for Alzheimer disease. Nat Rev Neurol 14(7):399–415

Corder EH, Saunders AM, Strittmatter WJ et al (1993) Gene dose of apolipoprotein E type 4 allele and the risk of Alzheimer's disease in late onset families. Science 261(5123):921–923

Cummings J, Lee G, Ritter A et al (2020) Alzheimer's disease drug development pipeline. Alzheimers Dement (N Y) 6(1):e12050. https://doi.org/10.1002/trc2.12050

de Bont N, Netea MG, Demacker PN (1999) Apolipoprotein E knock-out mice are highly susceptible to endotoxemia and *Klebsiella pneumoniae* infection. J Lipid Res 40:680–685

de Jong D, Jansen R, Hoefnagels W et al (2008) No effect of one-year treatment with indomethacin on Alzheimer's disease progression: a randomized controlled trial. PLoS One 3(1):e1475. https://doi.org/10.1371/journal.pone.0001475

de Rolim TS, Fabri GM, Nitrini R et al (2014) Evaluation of patients with Alzheimer's disease before and after dental treatment. Arq Neuropsiquiatr 72(12):919–924

Demmer RT, Norby FL, Lakshminarayan K et al (2020) Periodontal disease and incident dementia: The Atherosclerosis Risk in Communities Study (ARIC). Neurology 95(12):e1660–e1671. https://doi.org/10.1212/WNL.0000000000010312

Dietrich T, Ower P, Tank M et al (2019) Periodontal diagnosis in the context of the 2017 classification system of periodontal diseases and conditions - implementation in clinical practice. Br Dent J 226(1):16–22

Ding Y, Ren J, Yu H et al (2018) *Porphyromonas gingivalis*, a periodontitis causing bacterium, induces memory impairment and age-dependent neuroinflammation in mice. Immun Ageing 215:6. https://doi.org/10.1186/s12979-017-0110-7

Dioguardi M, Crincoli V, Laino L et al (2020) The role of periodontitis and periodontal bacteria in the onset and progression of Alzheimer's disease: a systematic review. J Clin Med 9(2):495. https://doi.org/10.3390/jcm9020495

Dominy SS, Lynch C, Ermini F et al (2019) Porphorymonas gingivalis in Alzheimer's disease brains: evidence for disease causation and treatment with small molecule inhibitors. Sci Adv 5(1). https://doi.org/10.1126/sciadv.aau3333

Dugger BN, Dickson DW (2017) Pathology of neurodegenerative diseases. Cold Spring Harb Perspect Biol 9(7):a028035. https://doi.org/10.1101/cshperspect.a028035

Eke PI, Dye BA, Wei L et al (2015) Update on prevalence of periodontitis in adults in the United States: NHANES 2009 to 2012. J Periodontol 86(5):611–622. https://doi.org/10.1902/jop.2015.140520

Emery DC, Shoemark DK, Batstone TE et al (2017) 16S rRNA next generation sequencing analysis shows bacteria in Alzheimer's post-mortem brain. Front Aging Neurosci 9:195. https://doi.org/10.3389/fnagi.2017.00195

Farhad SZ, Amini S, Khalilian A et al (2014) The effect of chronic periodontitis on serum levels of tumor necrosis factor-alpha in Alzheimer disease. Dent Res J (Isfahan) 11(5):549–552

Forner L, Larsen T, Kilian M et al (2006) Incidence of bacteremia after chewing, tooth brushing, and scaling in individuals with periodontal inflammation. J Clin Periodontol 33:401–407

Genin E, Hannequin D, Wallon D et al (2011) APOE and Alzheimer disease: a major gene with semi-dominant inheritance. Mol Psychiatry 16:903–907

Gil-Montoya JA, Barrios R, Santana S et al (2017) Association between periodontitis and amyloid β peptide in elderly people with and without cognitive impairment. J Periodontol 88:1051–1058

Goedert M, Klug A, Crowther RA (2006) Tau protein, the paired helical filament and Alzheimer's disease. J Alzheimers Dis 9:195–207

Grossi SG, Genco RI, Machtei EE et al (1995) Assessment of risk for periodontal disease II. Risk indicators for alveolar bone loss. J Periodontol 66:23–29

Grundke-Iqbal I, Iqbal K, Quinlan M et al (1986) Microtubule-associated protein tau. A component of Alzheimer paired helical filaments. J Biol Chem 261:6084–6089

Guo Y, Nguyen KA, Potempa J (2010) Dichotomy of gingipains action as virulence factors: from cleaving substrates with the precision of a surgeon's knife to a meat chopper-like brutal degradation of proteins. Periodontol 2000 54:15–44

Haditsch U, Roth T, Rodriguez L et al (2020) Alzheimer's disease-like neurodegeneration in *Porphyromonas gingivalis* infected neurons with persistent expression of active gingipains. J Alzheimers Dis 75:1361–1376

Haffajee AD, Socransky SS, Dzink JL et al (1988) Clinical, microbiological and immunological features of subjects with destructive periodontal diseases. J Clin Periodontol 15:240–246

Hajishengallis G (2011) Immune evasion strategies of *Porphyromonas gingivalis*. J Oral Biosci 53(3):233–240

Hajishengallis G, Darveau RP, Curtis MA (2012) The keystone-pathogen hypothesis. Nat Rev Microbiol 10(10):717–725

Halliday MR, Rege SV, Ma Q et al (2016) Accelerated pericyte degeneration and blood-brain barrier breakdown in apolipoprotein E4 carriers with Alzheimer's disease. J Cereb Blood Flow Metab 36:216–227

Hardy J, Selkoe DJ (2002) The amyloid hypothesis of Alzheimer's disease: progress and problems on the road to therapeutics. Science 297(5580):353–356

Hasturk H, Kantarci A, Van Dyke TE (2012) Oral inflammatory diseases and systemic inflammation: role of the macrophage. Front Immunol 3:1–11

Heneka MT, Carson MJ, El Khoury J et al (2015) Neuroinflammation in Alzheimer's disease. Lancet Neurol 14(4):388–405

Holmes C, Cunningham C, Zotova E, Woolford J, Dean C, Kerr S, Culliford D, Perry VH (2009) Systemic inflammation and disease progression in Alzheimer's disease. Neurology 73:768–774

Holt SC, Ebersole JL (2005) *Porphyromonas gingivalis*, *Treponema denticola*, and *Tannerella forsythia*: the "red complex", a prototype polybacterial pathogenic consortium in periodontitis. Periodontol 2000 38:72–122

Hook V, Schechter I, Demuth HU et al (2008) Alternative pathways for production of beta-amyloid peptides of Alzheimer's disease. J Biol Chem 389:993–1006

How KY, Song KP, Chan KG (2016) *Porphyromonas gingivalis*: an overview of periodontopathic pathogen below the gum line. Front Microbiol 7:53. https://doi.org/10.3389/fmicb.2016.00053

Hubacek JA, Peasey A, Pikhart H et al (2010) APOE polymorphism and its effect on plasma C-reactive protein levels in a large general population sample. Hum Immunol 71(3):304–308

Hyman BT, Phelps CH, Beach TG et al (2012) National Institute on Aging–Alzheimer's association guidelines for the Neuropathologic assessment of Alzheimer's disease. Alzheimer's & Dement 8(1):1–13

Ide M, Harris M, Stevens A et al (2016) Periodontitis and cognitive decline in Alzheimer's disease. PLoS One 11(3):e0151081. https://doi.org/10.1371/journal.pone.0151081

Ihara M, Yamamoto Y (2016) Emerging evidence for pathogenesis of sporadic cerebral small vessel disease. Stroke 47:554–560

Ilieviski V, Zuchowska PK, Green SJ et al (2018) Chronic oral application of a periodontal pathogen results in brain inflammation, neurodegeneration and amyloid beta production in wild type mice. PLoS One. https://doi.org/10.1371/journal.pone.02049418

Iliff JJ, Lee H, Yu M et al (2013) Brain wide pathway for waste clearance captured by contrast-enhanced MRI. J Clin Invest 123:1299–1309

Ishida N, Ishihara Y, Ishida K et al (2017) Periodontitis induced by bacterial infection exacerbates features of Alzheimer's disease in transgenic mice. NPJ

Aging Mech Dis 6(3):15. https://doi.org/10.1038/s41514-017-0015-x
Kadowaki T, Baba A, Abe N et al (2004) Suppression of pathogenicity of *Porphyromonas gingivalis* by newly developed gingipain inhibitors. Mol Pharmacol 66:1599–1606
Kamer AR, Craig RG, Pirraglia E et al (2009) TNFα and antibodies to periodontal bacteria discriminate between Alzheimer's disease patients and normal subjects. J Immunol 216(1–2):92–97
Kamer AR, Pirraglia E, Tsui W et al (2015) Periodontal disease associates with higher brain amyloid load in normal elderly. Neurobiol Aging 36(2):627–633
Kamma JJ, Nakou M, Baehni PC (1999) Clinical and microbiological characteristics of smokers with early onset periodontitis. J Periodontal Res 34:25–33
Kang J, Lemaire HG, Unterbeck A et al (1987) The precursor of Alzheimer's disease amyloid A4protein resembles a cell-surface receptor. Nature 325:733–736
Kinane DF, Marshall GJ (2001) Periodontal manifestations of systemic disease. Aust Dent J 46:2–12
Kinane JA, Benakanakere MR, Zhao J et al (2012) *Porphyromonas gingivalis* influences actin degradation within epithelial cells during invasion and apoptosis. Cell Microbiol 14:1085–1096
Kondo K, Niino M, Shido K (1994) A case-control study of Alzheimer's disease in Japan-significance of lifestyles. Dementia 5:314–326
Kozin SA, Barykin EP, Mitkevich VA et al (2018) Anti-amyloid therapy of Alzheimer's disease: current state and prospects. Biochemistry (Mosc) 83(9):1057–1067
Leira Y, Iglesias-Rey R, Gómez-Lado N et al (2019) *Porphyromonas gingivalis* lipopolysaccharide-induced periodontitis and serum amyloid-beta peptides. Arch Oral Biol 99:120–125
Leira Y, Carballo Á, Orlandi M et al (2020) Periodontitis and systemic markers of neurodegeneration: a case-control study. J Clin Periodontol 47(5):561–571
Li A, Chen Y, van der Sluis LWM et al (2020) Mediation analysis of white blood cell count on the association between periodontal inflammation and digit symbol substitution test scoring cognitive function among older U.S. Adults. J Gerontol A Biol Sci Med Sci 76:glaa223. https://doi.org/10.1093/gerona/glaa223
Lin JW, Chang CH, Caffrey JL (2020) Feature article: examining the association between oral health status and dementia: a nationwide nested case-controlled study. Exp Biol Med (Maywood) 245(3):231–244
Lockhart PB, Brennan MT, Sasser HC et al (2008) Bacteremia associated with tooth brushing and dental extraction. Circulation 117:3118–3125
Makiura N, Ojima M, Kou Y et al (2008) Relationship of *Porphyromonas gingivalis* with glycemic level in patients with type 2 diabetes following periodontal treatment. Oral Microbiol Immunol 23(4):348–351
Marden JR, Tchetgen EJ, Kawachi I et al (2017) Contribution of socioeconomic status at three life course periods to late life memory function and decline: early and late predictors of dementia risk. Am J Epidemiol 186:805–814
Marques F, Sousa JC, Sousa N et al (2013) Blood-brain-barriers in aging and in Alzheimer's disease. Mol Neurodegener 8:38. https://doi.org/10.1186/1750-1326-8-38
Massoud F, Gauthier S (2010) Update on the pharmacological treatment of Alzheimer's disease. Curr Neuropharmacol 8(1):69–80
McGeer PL, McGeer EG (2007) NSAIDs and Alzheimer disease: epidemiological, animal model and clinical studies. Neurobiol Aging 28:639–647
Montagne A, Barnes SR, Sweeney MD (2015) Blood-brain barrier breakdown in the aging human hippocampus. Neuron 85:296–302
Montagne A, Nation DA, Pa J et al (2016) Brain imaging of neurovascular dysfunction in Alzheimer's disease. Acta Neuropathol (Berl) 131:687–707
Moore AH, Bigbee MJ, Boynton GE et al (2010) Non-steroidal anti-inflammatory drugs in Alzheimer's disease and Parkinson's disease: reconsidering the role of neuroinflammation. Pharmaceuticals (Basel) 3(6):1812–1841
Moretti EW, Morris RW, Podgoreanu M et al (2005) APOE polymorphism is associated with risk of severe sepsis in surgical patients. Perioperative Genetics and Safety Outcomes Study (PEGASUS) Investigative Team. Crit Care Med 33(11):2521–2526
Morgan BP, Harris CL (2015) Complement, a target for therapy in inflammatory and degenerative diseases. Nat Rev Drug Discov 14(12):857–877
Morris R (1984) Developments of a water-maze procedure for studying spatial learning in the rat. J Neurosci Methods 11:47–60
Nedergaard M (2013) Neuroscience. Garbage truck of the brain. Science 340(6140):1529–1530
Ngandu T, Lehtisalo J, Solomon A et al (2015) A 2 year multidomain intervention of diet, exercise, cognitive training, and vascular risk monitoring versus control to prevent cognitive decline in at-risk elderly people (FINGER): a randomised controlled trial. Lancet 385(9984):2255–2263
Nie R, Wu Z, Ni J et al (2019) *Porphyromonas gingivalis* infection induces amyloid-β accumulation in monocytes/macrophages. J Alzheimers Dis 72:479–494
Noble JM, Borrell LN, Papapnou PN et al (2009) Periodontitis is associated with cognitive impairment among older adults: analysis of NHSNES-III. J Neurol Neurosurg Psychiatr 80:1206–1211
Nunan J, Small DH (2000) Regulation of APP cleavage by alpha-, beta- and gamma-secretases. FEBS Lett 483(1):6–10
Olsen I, Singhrao SK (2015) Can oral infection be a risk factor for Alzheimer's disease? J Oral Microbiol 7:29143
Olsen I, Singhrao SK (2019) Is there a link between genetic defects in the complement cascade and *Porphyromonas gingivalis* in Alzheimer's disease? J Oral Microbiol 12:167648

Olsen I, Singhrao SK (2020) Interaction between genetic factors, *Porphyromonas gingivalis* and microglia to promote Alzheimer's disease. J Oral Microbiol 12(1):1820834

Olsen I, Lambris JD, Hajishengallis G (2017) *Porphyromonas gingivalis* disturbs host-commensal homeostasis by changing complement function. J Oral Microbiol 9(1):1340085

Parepally JMR, Mandula H, Smith QR (2006) Brain uptake of nonsteroidal anti-inflammatory drugs: ibuprofen, flurbiprofen, and indomethacin. Pharm Res 23:873–881

Pasqualetti P, Bonomini C, Dal Forno G et al (2009) A randomized controlled study on effects of ibuprofen on cognitive progression of Alzheimer's disease. Aging Clin Exp Res 21:102–110

Perry VH, Nicoll JA, Holmes C (2010) Microglia in neurodegenerative disease. Nat Rev Neurol 6:193–201

Poole S, Singhrao SK, Kesavalu L et al (2013) Determining the presence of periodontopathic virulence factors in short-term postmortem Alzheimer's disease brain tissue. J Alzheimers Dis 36:665–677

Poole S, Singhrao SK, Chukkapalli S et al (2015) Active infection of *Porphorymonas gingivalis* and infection-induced complement activation in $ApoE^{-/-}$ mice brains. J Alzheimers Dis 43(1):67–80

Riviere GR, Riviere KH, Smith KS (2002) Molecular and immunological evidence of oral *Treponema* in the human brain and their association with Alzheimer's disease. Oral Microbiol Immunol 17:113–118

Rogers J, Kirby LC, Hempelman SR et al (1993) Clinical trial of indomethacin in Alzheimer's disease. Neurology 43:1609–1611

Rokad F, Moseley R, Hardy RS et al (2017) Cerebral oxidative stress and microvasculature defects in TNF-α expressing transgenic and *Porphyromonas gingivalis*-infected $ApoE^{-/-}$ mice. J Alzheimers Dis 60:359–369

Rossor MN, Newman S, Frackowiak RS et al (1993) Alzheimer's disease families with amyloid precursor protein mutations. Ann N Y Acad Sci 695:198–202

Russ TC, Stamatakis E, Hamer M et al (2013) Socioeconomic status as a risk factor for dementia death: individual participant meta-analysis of 86,508 men and women from UK. Br J Psychiatry 203:10–17

Ryder MI (2020) *Porphyromonas gingivalis* and Alzheimer disease: recent findings and potential therapies. J Periodontol 91(Suppl 1):S45–S49. https://doi.org/10.1002/JPER.20-0104

Saunders AM, Strittmatter WJ, Schmechel D et al (1993) Association of apolipoprotein E allele epsilon 4 with late-onset familial and sporadic Alzheimer's disease. Neurology 43(8):1467–1472

Sayas CL (2020) Chapter 10: Tau-based therapies for Alzheimer's disease: promising novel neuroprotective approaches. In Neuroprotection in Autism, Schizophrenia and Alzheimer's Disease. Elsevier Inc., pp 245–272. https://doi.org/10.1016/B978-0-12-814037-6.00005-7

Sheets SM, Potempa J, Travis J et al (2005) Gingipains from *Porphyromonas gingivalis* W83 induce cell adhesion molecule cleavage and apoptosis in endothelial cells. Infect Immun 73:1543–1552

Singh A, Rouxel P, Watt RG et al (2013) Social inequalities in clustering of oral health related behaviors in a national sample of British adults. Prev Med 57:102–106

Singhrao SK, Olsen I (2018) Are *Porphyromonas gingivalis* outer membrane vesicles microbullets for sporadic Alzheimer's disease manifestation? J Alzheimers Dis Rep 2(1):219–228

Singhrao SK, Olsen I (2019) Assessing the role of *Porphyromonas gingivalis* in periodontitis to determine a causative relationship with Alzheimer's disease. J Oral Microbiol 11(1):1563405

Singhrao SK, Harding A, Poole S, et al (2015) Porphyromonas gingivalis periodontal infection and its putative links with Alzheimer's disease. Mediators Inflamm 2015:137357. https://doi.org/10.1155/2015/137357

Socransky SS, Haffajee AD, Cugini MA et al (1998) Microbial complexes in subgingival plaque. J Clin Periodontol 25(2):134–144

Sparks Stein P, Steffen MJ, Smith C et al (2012) Serum antibodies to periodontal pathogens are a risk factor for Alzheimer's disease. Alzheimer's Dementia 8(3):196–203

Sprecher KE, Koscik RL, Carlsson CM et al (2017) Poor sleep is associated with CSF biomarkers of amyloid pathology in cognitively normal adults. Neurology 89:445–453

Stathopoulou PG, Galicia JC, Benakanakere MR et al (2009) *Porphyromonas gingivalis* induce apoptosis in human gingival epithelial cells through a gingipain-dependent mechanism. BMC Microbiol 9:107

Stein PS, Desrosiers M, Donegan SJ et al (2007) Tooth loss, dementia and neuropathy in the Nun study. J Am Dent Assoc 138(10):1314–1322

Strandwitz P (2018) Neurotransmitter modulation by the gut microbiota. Brain Res 1693(Pt B):128–133. https://doi.org/10.1016/j.brainres.2018.03.015

Strandwitz P, Kim KH, Terekhova D et al (2019) GABA-modulating bacteria of the human gut microbiota. Nat Microbiol 4(3):396–403. https://doi.org/10.1038/s41564-018-0307-3

Tomás I, Diz P, Tobias A et al (2012) Periodontal health status and bacteraemia from daily oral activities: systematic review/meta-analysis. J Clin Periodontol 39:213–228

Townsend KP, Practico D (2005) Novel therapeutic opportunities for Alzheimer's disease: focus on nonsteroidal anti-inflammatory drugs. FASEB J 19(12):1592–1601

Travis J, Potempa J (2000) Bacterial proteinases as targets for the development of second-generation antibiotics. Biochim Biophys Acta 1477:35–50

Tsoi LM, Wong KY, Liu YM et al (2007) Apoprotein E isoform-dependent expression and secretion of pro-inflammatory cytokines TNF-α and IL-6 in macrophages. Arch Biochem Biophys 460:33–40

Tzeng NS, Chung CH, Yeh CB et al (2016) Are chronic periodontitis and gingivitis associated with dementia?

A nationwide, retrospective, matched-cohort study in Taiwan. Neuroepidemiology 47:82–93

van Dyck CH (2018) Anti-amyloid-β monoclonal antibodies for Alzheimer's disease: pitfalls and promise. Biol Psychiatry 83(4):311–319

van Oijen M, de Jong FJ, Witteman JC et al (2007) Atherosclerosis and risk for dementia. Ann Neurol 61:403–410

Vasek MJ, Garber C, Dorsey D et al (2016) A complement-microglial axis drives synapse loss during virus-induced memory impairment. Nature 534(7608):538–543

Velsko IM, Chukkapalli SS, Rivera MF et al (2014) Active invasion of oral and aortic tissues by *Porphyromonas gingivalis* in mice causally links periodontitis and atherosclerosis. PLoS One 9:e97811. https://doi.org/10.1371/journal.pone.0097811

Watts A, Crimmins EM, Gatz M (2008) Inflammation as a potential mediator for the association between periodontal disease and Alzheimer's disease. Neuropsychiatr Dis Treat 4:865–876

Wu Z, Ni J, Liu Y et al (2017) Cathepsin B plays a critical role in inducing Alzheimer's disease-like phenotypes following chronic systemic exposure to lipopolysaccharide from *Porphyromonas gingivalis* in mice. Brain Behav Immun 65:350–361

Yin C, Ackermann S, Ma Z et al (2019) ApoE attenuates unresolvable inflammation by complex formation with activated C1q. Nat Med 25(3):496–506

Zeng F, Liu Y, Huang W et al (2020) Receptor for advanced glycation end products up-regulation in cerebral endothelial cells mediates cerebrovascular-related amyloid β accumulation after *Porphyromonas gingivalis* infection. J Neurochem. https://doi.org/10.1111/jnc.15096

Zhang J, Yu C, Zhang X et al (2018) *Porphyromonas gingivalis* lipopolysaccharide induces cognitive dysfunction, mediated by neuronal inflammation via activation of the TLR4 signalling pathway in C57BL/6 mice. J Neuroinflammation 15(1):37. https://doi.org/10.1186/s12974-017-1052-x

14 Cardiovascular Diseases and Periodontitis

Peter Riis Hansen and Palle Holmstrup

Abstract

Periodontitis is a chronic inflammatory disease of the tooth-supporting connective tissue and alveolar bone that is initiated by a bacterial biofilm in periodontal pockets. It affects about half of adults in the Western world, and is associated with a range of systemic comorbidities, e.g., cardiovascular disease (CVD), diabetes and rheumatoid arthritis, and these diseases share overlapping systemic and target tissue inflammatory mechanisms. Indeed, mounting evidence has indicated that their association is causal and built on the presence of systemic low-grade inflammation (LGI). Prior research linking periodontitis to CVD has mainly been derived from experimental studies, observational data, and small interventional trials with surrogate markers of CVD, e.g., endothelial dysfunction. However, recent data from randomised studies have demonstrated that intensive treatment of periodontitis can reduce blood pressure in patients with hypertension in conjunction with reduction of systemic inflammatory markers. Furthermore, targeted anti-inflammatory therapy has been shown to reduce recurrent events in patients with established CVD and LGI. Along this line, the concept of residual inflammatory risk has emerged as an independent new risk factor for atherothrombotic CVD. The present review summarizes translational evidence indicating that periodontitis is a risk factor for CVD dependent on LGI, and we conclude that treatment of periodontitis is likely to contribute importantly to reduction of residual inflammatory risk.

Keywords

Cardiovascular disease · Myocardial infarction · Periodontitis · Inflammation · Comorbidity

P. R. Hansen (✉)
Department of Cardiology, Herlev-Gentofte University Hospital, The Capital Region of Denmark, Hellerup, Denmark

Department of Clinical Medicine, University of Copenhagen, Copenhagen, Denmark
e-mail: peter.riis.hansen@regionh.dk

P. Holmstrup
Department of Odontology, Faculty of Health and Medical Sciences, University of Copenhagen, Copenhagen, Denmark
e-mail: pah@sund.ku.dk

Abbreviations

CI	Confidence interval
CVD	Cardiovascular disease
HbA1c	Glycated hemoglobin
hs-CRP	high sensitivity C-reactive protein

J. Santi-Rocca (ed.), *Periodontitis*, Advances in Experimental Medicine and Biology 1373,
https://doi.org/10.1007/978-3-030-96881-6_14

IL	Interleukin
IRR	Adjusted incidence rate ratio
LDL	Low-density lipoprotein
LGI	Systemic low-grade systemic inflammation
MI	Myocardial infarction
MMP	Metalloproteinase
NLRP3	Nucleotide-binding oligomerization domain-like receptor family, pyrin domain containing 3
OR	Odds ratio
RA	Rheumatoid arthritis
RR	Relative risk
SCORE	Systematic coronary risk evaluation
Th	T helper cell
TNF	Tumor necrosis factor-α-2

Highlights

- Periodontitis is associated with other inflammatory diseases, e.g., cardiovascular disease, diabetes, and rheumatoid arthritis
- Periodontitis increases systemic low-grade inflammation
- Periodontitis contributes to residual inflammatory risk of cardiovascular disease
- Periodontal treatment may reduce risk of cardiovascular disease

Considerations for Practice

- Patients with cardiovascular disease should receive treatment for periodontitis if needed
- Periodontal treatment may improve glycemic control in patients with type 2 diabetes
- Periodontal treatment may reduce blood pressure and risk of atherosclerotic disease

Patient Summary

Inflammation in the mouth as seen in periodontitis can worsen other diseases with inflammatory background including, for example, cardiovascular diseases, diabetes, and rheumatoid arthritis. Markers of inflammation in blood tests are often increased in these diseases which entails increased risk of cardiovascular disease. Also, treatment of periodontitis may improve blood pressure and diabetes control and reduce the risk of cardiovascular disease. Therefore, it is vital to take care of your oral health and receive regular dental check-ups. Not only does this lower the risk of periodontitis, but it may also reduce the risk of other major diseases, e.g., cardiovascular disease.

14.1 Introduction

During the last decades, interaction of oral inflammation with prevalent non-communicable diseases has been widely explored. The most common oral inflammatory disease is periodontitis, which is a chronic inflammatory tissue-degrading disorder affecting the tooth-supporting structures (Pihlstrom et al. 2005). Periodontitis has been linked with numerous highly prevalent medical diseases, including, for example, cardiovascular diseases (CVD), diabetes, and rheumatoid arthritis (RA) (Lalla and Papapanou 2011; Lockhart et al. 2012; Belstrom et al. 2012; Potempa et al. 2017; Sanz et al. 2020; Herrera et al. 2020). Indeed, accumulating evidence has suggested that a causal relationship exists between periodontitis and its systemic comorbidities including CVD that this is based on a shared state of chronic low-grade inflammation (LGI) (Hansen et al. 2016; Holmstrup et al. 2017; Schenkein et al. 2020). However, more definite proof of causality, e.g., with positive results of well-powered randomized controlled trials of effects of periodontal treatment on periodontitis comorbidities has been lacking. This evidence gap is explained, in part, by realization that randomized trials with hard clinical comorbidity endpoints, e.g., myocardial infarction (MI), stroke, and death in studies with CVD, usually require long-term follow-up of thousands of patients. Also, there are problems with blinding of periodontal treatment and leaving periodontitis untreated in control groups of such studies. Consequently, definitive randomised trials are unlikely ever to be performed in this area of research and to guide public health authorities and patients, the field has remained largely dependent on aggregation of the available evidence accumulated from animal models, clinical studies with surrogate outcomes, and epidemiological work. However, recent data from rigorously conducted albeit small-scale studies have more firmly suggested that periodontal treatment can lower blood pressure in patients with hypertension (Zhou et al. 2017; Czesnikiewicz-Guzik et al. 2019) and improve glycemic control in patients with type 2 diabetes (Sgolastra et al. 2013; D'Aiuto et al. 2018). At the same time, large clinical trials have unequivocally established that atherosclerosis, the main contributor to CVD, is an inflammatory disease and that residual inflammatory risk in CVD can be treated by targeted anti-inflammatory therapy (Ridker et al. 2017; Tardif et al. 2019). Accordingly, although proof of a causal relationship of periodontitis and its comorbidities including CVD will probably remain an elusive goal, the sum of available evidence would seem to have reached a critical mass in terms of establishing such relationship. To further support this view and interpretation of the available data, we here provide an updated overview of evidence linking periodontitis and CVD, with emphasis on the importance of LGI and the associated residual inflammatory risk of CVD.

14.2 Periodontitis

The surface of teeth is not constantly sloughed and renewed, as is the case for other inner of outer surfaces of the human organism. Accordingly, bacterial accumulation on dental surfaces along the gingival margin is easily established in the form of biofilms (plaques). If these are not mechanically removed, the bacteria invade the gingival crevice (Fig. 14.1a, b), and gingivitis, which is a reversible inflammation process limited to the gingival soft tissue, ensues as primary response to the bacterial accumulation. Gingivitis affects 50–90% of the adult population (Pihlstrom et al. 2005; Al Qahtani et al. 2017) is characterized by gingival redness and edematous swelling. In addition, bleeding often occurs during oral hygiene procedures and chewing, but the dental biofilm can be removed with adequate oral hygiene, and the inflammatory lesion may then resolve without irreversible damage to the tooth-supporting tissues. However, in absence of biofilm removal, there may be a progressive overgrowth of opportunistic pathogenic bacteria that accumulate at the tooth surface below the gingival margin and over time form calculus. Depending on individual factors, gingivitis may thereby progress to periodontitis (Fig. 14.1c), characterized by deepening of the gingival crevice and formation of a gingival pocket with an anaerobic environment (Pihlstrom et al. 2005). The associated dysbiosis of the subgingival biofilm leads to a non-resolving and destructive inflammatory response typified by deepened periodontal pockets with ulcerated epithelial linings, irreversible degradation of periodontal tissues and loss of alveolar bone whereby affected teeth may loosen, change position, and, ultimately, be lost (Bartold and Van Dyke 2013). Moreover, the ulcerated epithelial lining of deepened periodontal pockets (Fig. 14.1c) enables bacteria in the pockets to penetrate the gingival soft tissue and microcirculation, which results in bacteremia during everyday procedures like chewing, tooth brushing and flossing (Forner et al. 2006a).

Periodontitis is the most common non-transmissible inflammatory disease in humans worldwide (Kassebaum et al. 2014) and in its mildest forms has a prevalence of around 50% in adults and > 60% in individuals over 65 years of age (White et al. 2012; Eke et al. 2016). Severe periodontitis affects 7–8% of the global adult population (Kassebaum et al. 2014) and is a

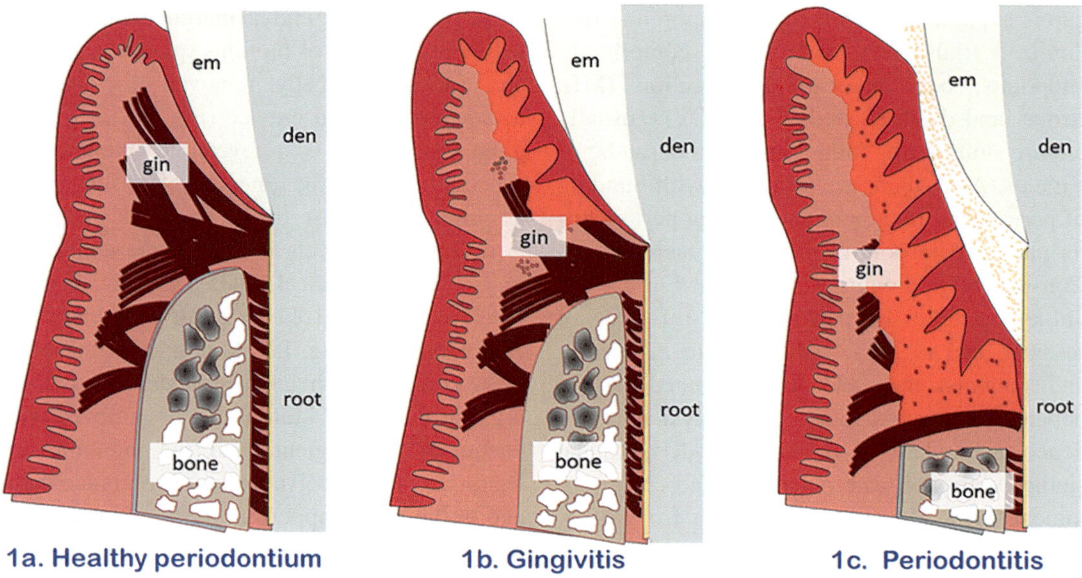

Fig. 14.1 Schematic drawing of the periodontal soft and hard tissue close to the gingival margin. em: Enamel; den: Dentin; bone: Alveolar bone; gin; Gingival soft tissue with epithelial attachment to enamel; root: Root covered by cementum with periodontal ligament attaching to root to the alveolar bone; in gingivitis the gingiva is inflamed with edema. In periodontitis the gingival connective tissue, the periodontal ligament and the supporting alveolar bone is undergoing degradation. There is a bacterial biofilm in the gingival pocket, and through ulcerations in the pocket epithelium bacteria have direct access to the connective tissue and the circulation

major cause of tooth loss, nutritional compromise, altered speech, low self-esteem and poor quality of life (Al-Harthi et al. 2013; Buset et al. 2016; Billings et al. 2018). Individual susceptibility to periodontitis varies greatly and is determined by a host of factors including oral hygiene, lifestyle, tobacco habits, medical diseases and genetics. For example, a twin study has suggested that genetic variance may account for as much as half of the variance in periodontitis (Michalowicz et al. 2000). Peritonitis usually only results in vague clinical symptoms, e.g., slight periodontal edema, redness and bleeding, and is often unnoticed by the patient until loosening and changed position of the teeth become notable. Professional scaling with removal of biofilm and calculus is necessary to stop progression of the disease, and surgical periodontal treatment with reduction of periodontal pocket depths may be required in subjects with deep pockets (Fig. 14.2a, b).

14.3 Cardiovascular Disease

Atherosclerosis, the major cause of CVD, e.g., MI, stroke, peripheral artery disease and cardiovascular mortality, is a chronic inflammatory disease. The earliest hallmark the disease is endothelial cell activation and dysfunction which is linked with CVD risk factor-driven oxidative stress and reduced levels of bioactive nitric oxide in the arterial wall, with subsequent subintimal accumulation of modified cholesterol particles, and local recruitment and activation of leukocytes of both innate and adaptive arms of the immune system. Further processes includes formation of fatty streaks characterized by subintimal accumulation of cholesterol-laden neointimal macrophages (foam cells), vascular smooth muscle cell proliferation, complex intra-lesion inflammatory reactions, e.g., activation of diverse cytokine networks, Toll-like receptor-NFκB signaling, and nucleotide-binding oligomerization domain-like receptor family, pyrin domain containing 3 (NLRP3) inflammasome activation by cho-

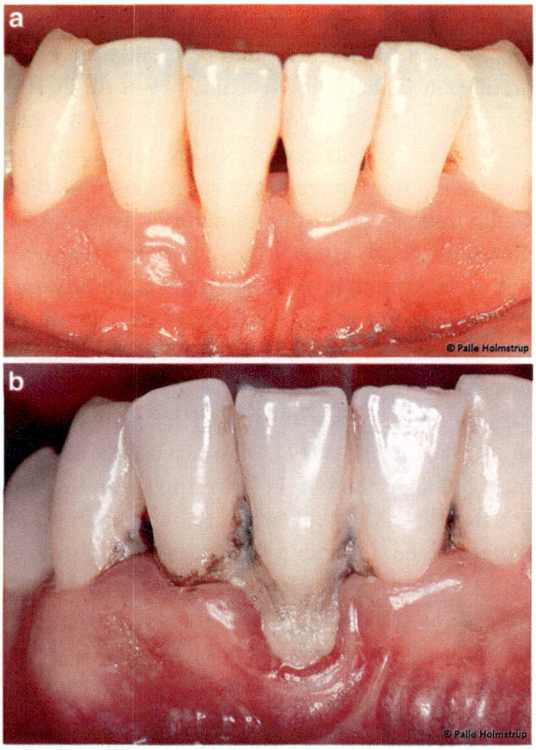

Fig. 14.2 (**a**) Periodontitis with plaque covering dental calculus at the exposed root surface. Interdental papillae are lost, and the gingival margin shows irregular edematous condition with retraction (**b**) Same individual as shown in Fig. 14.2a after periodontal treatment. There is no longer inflammation, but the periodontal tissue degradation is irreversible

lesterol crystals, vessel wall matrix remodeling, neo-angiogenesis and eventual progression to a mature atherosclerotic lesion with a lipid-rich necrotic lipid core, luminal fibrous cap and potential arterial lumen compromise (Hansson 2005; Libby 2012b). Along the course of this evolution, lesions may become unstable and display fibrous cap rupture or superficial erosion associated with increased local and systemic inflammation, activation of the coagulation system, and explosive local platelet activation and aggregation leading to thrombosis, arterial occlusion and clinical events, e.g., MI (Crea and Libby 2017).

14.3.1 Residual Inflammatory Risk

Considerable overlap and redundancy exist of inflammatory mechanisms irrespective of the initiating factors, and systemic inflammatory activation by infectious processes and microbial products can modulate atherosclerosis, most likely indirectly by contributing to a systemic inflammatory response that elicits a local 'echo' in atherosclerotic lesions, which again may lead to plaque destabilization and atherothrombotic manifestations (Libby et al. 2018). Indeed, it is well-established that patients with acute infections, e.g., influenza and pneumonia are at increased short-term risk of MI and the aggregated infectious burden throughout the lifetime has been proposed as a new risk factor (and potential treatment target) for atherosclerosis (Rosenfeld and Campbell 2011; Musher et al. 2019).

More than two decades ago, the importance of LGI in atherothrombotic CVD was suggested by studies demonstrating that circulating levels of inflammatory biomarkers, e.g., C-reactive protein (CRP) or interleukin (IL)-6 were strong independent predictors of future MI (Ridker et al. 2000; Ridker 2002). Since then the significance of inflammation as a marker and maker of CVD and, in fact, of most other chronic noncommunicable diseases that represent leading causes of disability, frailty and premature death worldwide, e.g., diabetes, cancer, chronic kidney disease, and autoimmune and neurodegenerative disorders, has been documented (Ferrucci and Fabbri 2018; Furman et al. 2019). Specifically, the magnitude of the increased risk of CVD associated with increased CRP levels is remarkably similar that of increased low-density lipoprotein (LDL) cholesterol levels and both these risk markers contribute independently to CVD prediction (Emerging Risk Factors et al. 2010). Recently, the causal role of inflammation in atherothrombosis was confirmed by results of the large randomized placebo-controlled Canakinumab Anti-inflammatory Thrombosis Outcome Study (CANTOS), where treatment with canakinumab, a monoclonal antibody targeting the proximal inflammatory cytokine IL-1β, reduced a combined CVD endpoint in patients with prior MI and LGI determined by plasma high sensitivity (hs)-CRP levels >2 mg/dl (Ridker et al. 2017). These patients already received aggressive lipid-lowering statin therapy and lipid levels were not affected by canakinumab. Notably, the favorable effect of canakinumab was limited to individuals who achieved reductions of hs-CRP levels and in these subjects, cardiovascular mortality and all-cause death were reduced by 31% ($p < 0.0001$) (Ridker et al. 2018). Furthermore, in another recent large, randomized placebo-controlled trial, colchicine, an anti-inflammatory agent that inhibits NLRP3 inflammasome activation and tubulin polymerization, reduced recurrent CVD in patients with recent MI (Tardif et al. 2019). In the wake of these and other results spanning two decades of research, the concept of residual inflammatory risk for CVD, namely LGI, has gained clinical momentum in the treatment of atherothrombotic CVD, with recognition of the potential for anti-inflammatory therapy to reduce CVD and the necessity of identifying and treating unrecognized but prevalent sources of LGI to reduce CVD (Czesnikiewicz-Guzik et al. 2020). Since weight loss, physical exercise, and smoking cessation all reduce inflammation, the residual inflammatory risk concept is readily useful to promote general lifestyle improvements, but other forms of anti-inflammatory therapy also seem posed to contribute to the therapeutic armamentarium against CVD. However, markers of residual inflammatory risk have not yet been uniformly adopted in CVD risk prediction algorithms. For example, inflammatory markers are not among entry data in the established European Society of Cardiology systematic coronary risk evaluation (SCORE) for assessment of global CVD risk in primary prevention, whereas hs-CRP levels contribute to some risk scores used elsewhere, e.g., the Reynolds Risk Score employed for CVD risk prediction in some US states (Conroy et al. 2003; Ridker 2007).

14.3.2 Increased Risk of CVD in Other Chronic Inflammatory Diseases

Increased risk of CVD observed in periodontitis is also manifested in 'traditional' chronic inflammatory diseases, e.g., RA, psoriasis, and inflammatory bowel disease, and can be viewed within the framework of LGI and residual inflammatory risk (Ahlehoff et al. 2013; Kristensen et al. 2013; Hansildaar et al. 2021). Although in the latter conditions, e.g., RA and psoriasis, target tissue and systemic inflammation is probably predominantly triggered by autoimmune mechanisms (not infection), much scientific work supporting a causal role of these diseases in CVD parallels studies done in periodontitis (Hansen 2018). In fact, bi-directional interrelationships have been established between periodontitis and other chronic inflammatory diseases, e.g., with periodontitis increasing the risk of RA, and RA increasing the risk of periodontitis (Potempa et al. 2017; Lorenzo-Pouso et al. 2020; Zhang et al. 2020). Evidence has also suggested that systemic anti-inflammatory treatment, e.g., with tumor necrosis factor-α (TNF) inhibitors or methotrexate, may reduce the risk of CVD in RA and psoriasis albeit that large, randomized trials powered for CVD outcomes are awaited (Yang et al. 2016; Hansen 2018; Hansildaar et al. 2021). Indeed, the European League Against Rheumatism (EULAR) recommendations for cardiovascular risk management in patients with RA and other forms of inflammatory arthritis advocates that rheumatologic disease activity should be controlled optimally to lower the risk of CVD, and that algorithms for cardiovascular risk prediction, e.g., SCORE, should be adopted for patients with RA by introducing a multiplication factor of 1.5 (Agca et al. 2017). The European Society of Cardiology also advices that this multiplication factor should be used for risk stratification in patients with RA, and potentially also employed on an individual basis in individuals with other immune-inflammatory diseases dependent on disease activity/severity (Piepoli et al. 2016). In recognition of the contribution of RA and systemic lupus erythematosus to the risk of CVD, the QRISK algorithm used for estimation of cardiovascular risk in the UK also adds the presence of these chronic inflammatory disease to CVD risk calculation (Hippisley-Cox et al. 2017).

14.4 Observational Studies Linking CVD with Periodontitis

A large body of observational studies have found positive associations between CVD and periodontitis (Holmstrup et al. 2017). For example, in the Swedish 'PAROKRANK' study which included 805 patients <75 years of age with first-time MI and 805 matched controls, periodontitis verified by radiographically-rated alveolar bone loss, was more prevalent in patients with MI than controls (43% vs. 33%; $p < 0.001$) (Ryden et al. 2016). There was an increased risk of MI (odds ratio [OR] 1.49; 95% confidence interval [CI] 1.21–1.81) among the patients with periodontitis, which remained significant (OR 1.28; 95% CI 1.03–1.60) after adjustment for co-variables (smoking, diabetes, and socioeconomic factors). These findings from the largest and most well-conducted case-control study to date clearly support an independent association between periodontitis and AMI (Ryden et al. 2016). In a Danish nationwide cohort study, we also found increased risk of MI (adjusted incidence rate ratio [IRR] 1.16, 95% CI 1.04–1.30), ischemic stroke (IRR 1.51, 95% CI 1.38–1.65), cardiovascular mortality (IRR 2.02, 95% CI 1.87–2.18), and all-cause mortality (IRR 2.70, 95% CI 2.60–2.81) in patients with periodontitis compared to controls (Hansen et al. 2016).

Comparable results have been reported in several systematic reviews and meta-analyses. In a recent systematic review and meta-analysis including 30 longitudinal cohort studies the risk of CVD (with coronary artery disease, MI and stroke as the main endpoints) was significantly higher in patients with periodontitis compared to periodontally healthy subjects (relative risk [RR] 1.20, 95% CI 1.14–1.26) (Larvin et al. 2020). CVD risk did not differ between clinical or self-reported periodontitis (RR 0.97, 95% CI 0.87–1.07), it was higher in men (RR 1.16, 95% CI 1.08–1.25) and severe periodontitis (RR 1.25, 95% CI 1.15–1.35), and among all types of CVD, the increase in risk of stroke was highest (RR 1.24, 95% CI 1.12–1.38). Another systematic review with meta-analysis of periodontitis as a risk factor for stroke included 10 studies and found an association between periodontitis and (ischemic and hemorrhagic) stroke with increased risk of stroke in cohort studies (RR 1.88. 95% CI 1.55–2.29; $p < 0.00001$) and for ischemic stroke events in case-control studies (RR 2.72, 95% CI 2.00–3.71; $p < 0.00001$) (Fagundes et al. 2019). Along this line, the most recent systematic review of a total of 48 cohorts and 5.71 million participants showed that periodontitis was linked with increased risk of all-cause mortality (RR 1.46, 95% CI 1.15–1.85), cardiovascular mortality (RR 1.47, 95% CI 1.14–1.90), coronary artery disease (RR 2.58, 95% CI 2.20–3.03), cerebrovascular disease (RR 3.11, 95% CI 2.42–3.98), and cancer (RR 1.38, 95% CI 1.24–1.53) (Romandini et al. 2021). Edentulism, the ultimate sequela of periodontitis, has also been associated with comparably increased risk of adverse cardiovascular endpoints and death (Watt et al. 2012; Lee et al. 2019; Romandini et al. 2021). In addition, periodontitis has been linked with other manifestations of atherosclerotic CVD, e.g., peripheral artery disease, aortic aneurysms, erectile dysfunction, atrial fibrillation, and heart failure, as well as measures of subclinical atherosclerotic disease including endothelial dysfunction, e.g., determined by attenuated brachial artery flow-mediated vasodilation, augmented arterial stiffness, inflammation in the major arteries determined by positron emission tomography, increased carotid intima-media thickness, and carotid artery atherosclerotic plaques and calcifications, respectively (Amar et al. 2003; Fifer et al. 2011; Orlandi et al. 2014; Gurav 2014; Frohlich et al. 2016; Zeng et al. 2016; Kaschwich et al. 2019; Bizzarro and Loos 2019; Yu et al. 2020; AlSakr et al. 2020; Hassan et al. 2021; Darnaud et al. 2021) .

14.5 Periodontitis and Risk Factors for CVD

14.5.1 Hypertension

The overall global prevalence of hypertension in adults is around 30–45% and increases to >60% in subjects aged >60 years, with hypertension being a leading cause of premature death and risk factor for CVD, e.g., MI, stroke, heart failure, peripheral artery disease and chronic kidney disease (Williams et al. 2018). Hypertension has long been associated with periodontitis and in a recent meta-analysis of 40 studies, moderate-severe periodontitis (OR 1.22; 95 CI 1.10–1.35) and severe periodontitis (OR 1.49; 95% CI 1.09–1.25) were linked with hypertension, as was also documented in prospective studies where periodontitis increased the risk of hypertension (OR 1.68; 95% CI 0.85–3.35) (Munoz Aguilera et al. 2020). In fact, it might be considered whether dentists can contribute to screening for hypertension, since patients usually visit dentists more frequently than their general physician for preventive checkups (Engstrom et al. 2011). Also, sublingual varices are associated with hypertension and may be used as indicator for blood pressure screening (Hedstrom et al. 2015).

A causal role of periodontitis in hypertension was suggested in a randomized trial (n = 97) of Chinese never-smoking patients with prehypertension and moderate-severe periodontitis, where intensive periodontal treatment for 4 consecutive weeks significantly reduced systolic and diastolic blood pressure after 6 months compared to controls that only received supragingival scaling (Zhou et al. 2017). These results were recently supported by a Mendelian randomization study where single nucleotide polymorphisms linked with periodontitis were significantly associated with increased blood pressure (Czesnikiewicz-Guzik et al. 2019). Even more importantly, these investigators performed a randomized study of effects of a single session of intensive periodontal treatment in patients with hypertension and moderate-to-severe periodontitis (n = 101). In the intensive periodontal treatment group, mean reductions of systolic blood pressure by 11.1 mmHg (95% CI 6.5–15.8 mmHg; $p < 0.001$) and diastolic blood pressure by 8.3 mmHg (95% CI 3.98–12.6; $p < 0.001$) were observed after 2 months compared to the control group. Systolic blood pressure reduction was correlated to measures of periodontal status improvement and intensive periodontal treatment also ameliorated endothelial function and reduced circulating levels of both inflammatory markers, e.g., interferon-γ, IL-17A, TNF, and share of circulating activated CD8 cells, respectively (Czesnikiewicz-Guzik et al. 2019). Accordingly, this proof of concept study clearly suggests that periodontal treatment can have favorable effects on hypertension and that reduction of LGI is a likely causal mediator. Along this line, a recent study with mediation analysis suggested that CRP levels was a mediator in the link between periodontitis and hypertension (Munoz Aguilera et al. 2020). Results in larger cohorts of patients with hypertension are awaited to establish the role of periodontal treatment as a novel nonpharmacological treatment of hypertension, and periodontitis is currently not mentioned in guidelines on management of hypertension (Williams et al. 2018).

14.5.2 Diabetes

Diabetes (type 1 and type 2) is an established risk factor for periodontitis and the link between the two diseases is bidirectional as periodontitis also adversely affects glycemic control and contributes to microvascular diabetic complications (Lalla and Papapanou 2011). Accordingly, dentists are well-posed for screening for diabetes and in a recent Danish study of patients with no history of diabetes that attended dental treatment, 3.1% and 27.1% were identified with diabetes and prediabetes, respectively (Holm et al. 2016). Type 2 diabetes accounts for >85% of diabetes prevalence and meta-analyses of randomized but usually underpowered studies of effects of periodontal treatment on glycemic control in patients with type 2 diabetes have consistently found reductions of HbA1c (glycated hemoglobin) levels (Teshome and Yitayeh 2016). Also, meta-analyses have indicated that periodontal therapy reduces circulating levels of CRP and TNF in patients with type 2 diabetes (Artese et al. 2015). In a recent rigorously conducted randomized trial, mean HbA1c in patients with type 2 diabetes and moderate-to-severe periodontitis ($n = 264$) receiving multi-phased intensive periodontal treatment were 0.6% (95% CI 0.3–0.9%; $p < 0.0001$) lower than in the control group after 2 months and was positively correlated with the change of periodontal probing depth (D'Aiuto et al. 2018). Also, in the intensive periodontal therapy group, endothelium-dependent vasodilation was increased, plasma levels of CRP and TNF were decreased, and quality of life and a calculated 10-year CVD risk score were improved (D'Aiuto et al. 2018). A decrease in HbA1c levels of 0.6% is similar to reductions observed with an added glucose-lowering drug and the potential clearly importance is clearly underlined by historic data suggesting that each 1% reduction of HbA1c is associated with reductions of 21% (95% CI 15–27%; $p < 0001$) for deaths related to diabetes, 14% (95% CI 8–21%; $p < 0.0001$) for MI, and 37% (95% CI 22–41%; $p < 0.0001$) for microvascular complications, respectively (Stratton et al. 2000). Based on the aggregated evidence, the UK National Health Service recently concluded that prevention, early diagnosis, and treatment of periodontitis is required in subjects with diabetes and that all individuals with periodontitis should have periodontal treatment to help prevent type 2 diabetes (NHS 2019).

14.5.3 Other Risk Factors for CVD

Periodontitis is associated with several other risk factors for CVD. For example, smoking increases the risk of periodontitis and exerts strong negative effects on the response to periodontal treatment (Pihlstrom et al. 2005). Also, the disease is linked to an atherogenic lipid profile with lowered high-density lipoprotein (HDL) levels, and elevations of LDL cholesterol, and triglyceride concentrations (Nepomuceno et al. 2017). Moreover, periodontitis is associated with obesity and low cardiorespiratory fitness, and physical activity has been tied with reduced prevalence of periodontitis (Keller et al. 2015; Ferreira et al. 2019). Low socioeconomic status confers increased risk of CVD that may be of magnitude equivalent to traditional risk factors (Schultz et al. 2018). In this regard, periodontitis is inversely related to socioeconomic status, and a strong negative educational gradient is reported in oral health-compromising behaviors (Borrell and Crawford 2012; Singh et al. 2013). In addition, periodontitis has been linked with other less traditional risk factors for CVD, e.g., chronic kidney disease, depression, non-alcoholic fatty liver disease, obstructive sleep apnea, and polycystic ovary syndrome (Al-Jewair et al. 2015; Alakhali et al. 2018; Deschamps-Lenhardt et al. 2019; Machado et al. 2020; Zheng et al. 2021). Notably, all above-mentioned risk factors for CVD are associated with LGI, many are interdependent, some may be unmeasured confounders in epidemiological studies, and heterogeneity is often apparent in studies of their linkage with periodontitis.

14.6 Pathogenic Mechanisms Linking CVD with Periodontitis

Numerous mechanisms can contribute to the association between CVD and periodontitis and add to the discussion of whether periodontitis should be considered as an independent risk factor for CVD. These mechanisms generally include (1) shared clinical, environmental, and genetic risk factors, (2) overlap of inflammatory pathways with increased systemic levels of inflammatory mediators, i.e. LGI, and (3) translocation of bacteria from periodontal pockets into the circulation with subsequent colonization of atheromatous plaques contributing to plaque instability and atherothrombotic events.

14.6.1 Shared Clinical, Environmental, and Genetic risk Factors

As indicated above, periodontitis has been associated with a range of clinical and environmental risk factors for CVD, e.g., hypertension, diabetes, smoking, dyslipidemia, obesity, low cardiorespiratory fitness, low socio-economic status, chronic kidney disease, depression, non-alcoholic fatty liver disease, obstructive sleep apnea, and polycystic ovary syndrome, and well as with other chronic inflammatory diseases that confer increased risk of CVD, e.g., RA, psoriasis, and inflammatory bowel disease. While these shared risk factors can confound determination of any independent causal role of periodontitis in CVD, it is notable that all shared risk factors are characterized by LGI. Also, evidence has suggested shared genetic susceptibilities for periodontitis and CVD, especially coronary artery disease (Aarabi et al. 2017). For example, significant associations have been reported between periodontitis and established coronary artery disease risk loci CDKN2B-AS1, VAMP3, and VAMP8, but exact mechanisms connecting these loci with the two diseases remain to be identified (Schaefer et al. 2011; Munz et al. 2018). However, neither shared gene variants or concurrent clinical and environmental risk factors appear to fully explain the observed association between periodontitis and CVD.

14.6.2 Overlap of Inflammatory Pathways

Local immuno-inflammatory responses leading to alveolar bone destruction in periodontitis are coupled with LGI as is the case for CVD and all other periodontitis comorbidities as well (Holmstrup et al. 2017). Periodontitis is a characterized by a hyperreactive inflammatory response in periodontal tissues with an extensive orchestra of inflammatory players and mediators, e.g., neutrophils, monocyte-macrophages, T cells with potential T helper cell (Th)17 polarization, B cells, complement, chemokine and cytokine networks, lipid mediators, toll-like receptors, citrullination, the NLRP3-inflammasome, and metalloproteinases (MMPs), which contribute to breakdown of tissue matrix, microvascular damage, and exudation and release of inflammatory cells and mediators to both gingival pocket fluid and the systemic circulation (Engstrom et al. 2011; Moutsopoulos et al. 2012; Cekici et al. 2014; Loos and Van Dyke 2020; Aral et al. 2020) . Many of these local inflammatory reactions are recapitulated in atherosclerotic plaque development (Hansson 2005; Libby 2012b; Crea and Libby 2017). In agreement with the notion that spill-over of inflammatory mediators from the inflamed periodontium promotes systemic inflammation, CRP and other circulating markers of LGI increase with the severity of periodontitis in a dose-dependent manner (Noack et al. 2001). These mediators also include, for example, TNF, IL-6, and IL-17, and increased levels of these can activate endothelial cells leading to endothelial dysfunction, i.e. the initiating event in atherogenesis, and contribute to other inflammatory processes in vascular lesions ultimately leading to clinical atherothrombotic manifestations (Hansson 2005; Huang and Vita 2006; Libby 2012a; Schenkein and Loos 2013; Crea and Libby 2017; Schenkein et al. 2020). Indeed, periodontitis is associated with endothelial dysfunction and as discussed earlier, plasma levels of CRP and IL-6 are strong independent predictors of future MI (Ridker et al. 2000; Ridker 2002; Amar et al. 2003; Gurav 2014). It remains unknown, however, if LGI and its sequelae, e.g., endothelial dysfunction, in periodontitis is mainly elicited by spill-over of inflammatory mediators from inflamed periodontal tissue to the circulation, or rather a consequence of translocation of periodontal bacteria and bacterial products from inflamed periodontium to the blood (see below). Along this line, bacteremia occurs more frequently in patients with periodontitis than in orally healthy controls, and root scaling is associated with increases in plasma IL-6 levels (Forner et al. 2006a, b). Illustratively, a mild systemic inflammatory response generated by administration of a vaccine containing bacterial lipopolysaccharide in healthy subjects also induces endothelial dysfunction and increased plasma levels of IL-6 (Hingorani et al. 2000).

14.6.3 Transfer of Bacteria from Periodontal Pockets to the Circulation with Subsequent Colonization of Atheromatous Plaques

From its onset, periodontitis causes ulcers in the gingival, tooth-surrounding epithelial barrier that normally protects the organism from bacterial invasion through the periodontal pockets. Consequently, routine procedures such as chewing and tooth brushing cause translocation of viable bacteria from the periodontium to the circulation (Forner et al. 2006a). Also, we recently showed that blood donors with periodontitis display increased risk of having viable bacteria in their freshly drawn blood donations compared to controls without periodontitis (Damgaard et al. 2020). Once in the circulation, bacteria may spread in the organism and contribute to progression of atherosclerosis, either directly by uptake in atherosclerotic lesions potentially leading to increased local inflammation and plaque instability, or indirectly by promoting LGI (Rosenfeld and Campbell 2011; Aarabi et al. 2015). Many studies with use of sensitive techniques have demonstrated the presence of DNA from oral bacteria in atherosclerotic plaques specimens, but it is possible that such DNA represents bacteria that are passive bystanders or nonviable after engulfment elsewhere by leukocytes that subsequently homed in on the plaques ((Joshi et al. 2021). However, with use of advanced cultivation techniques a few investigators have reported viable periodontal bacteria in human atherosclerotic plaques (Kozarov et al. 2005; Rafferty et al. 2011).

In addition to their obvious potential for activation of the immunoinflammatory response, selected oral keystone pathogens have shown experimental effects of potential relevance to atherothrombotic CVD. For example, *Porphyromonas gingivalis* is unique among prokaryotic species by containing a peptidylarginine deiminase responsible for protein citrullination (McGraw et al. 1999). This posttranslational modification may break the immune tolerance and lead to an antibody response against citrullinated proteins, which has been implicated as a causal link between periodontitis and RA (Potempa et al. 2017). Antibodies to citrullinated proteins are highly prevalent in subjects with RA where they are linked with endothelial dysfunction and CVD, but these antibodies also occur more frequently in subjects without RA at increased risk of coronary artery disease (Cambridge et al. 2013; Potempa et al. 2017). Furthermore, in vitro experiments have suggested that virulence factors from *P. gingivalis* can induce upregulation of endothelial cell adhesion molecules, vascular smooth muscle cell proliferation, and platelet activation (Lourbakos et al. 2001; Nakamura et al. 2008). Interestingly, pore-forming leukotoxin A, the major virulence factor of *Aggretibacter actinomycetemcomitans*, can also indirectly elicit protein citrullination by causing cytotoxic loss of membrane integrity in neutrophils, which triggers calcium-dependent activation of neutrophil peptidylaginine deiminases with consequent hypercitrullination of intracellular proteins that are then released from the dying cells (Konig et al. 2016). Leukotoxin A can also increase endothelial cell activation in vitro (Dietmann et al. 2013). At present, however, it remains unclear if systemic spread of periodontal bacteria affects the risk of CVD in patients with periodontitis.

14.7 Results from Experimental Models

A range of experimental investigations have examined mechanisms that may contribute to the link between periodontitis and CVD, with studies frequently done in apolipoprotein E-deficient mice, a dyslipidemic model that reiterates many facets of human atherosclerosis (Kolovou et al. 2008). In these mice, infection with *P. gingivalis* accelerates progression of atherosclerosis and is associated with localization of *P. gingivalis* in atherosclerotic lesions and other tissues, increased lipid levels, and systemic T helper cell (Th) 17 skewing of the immune system (Li et al. 2002; Maekawa et al. 2011; Cai et al. 2014). In this model, augmented atherosclerosis is also observed following infection with the oral spirochete *Treponema denticola*, and infection with *A. actinomycetemcomitans* is linked with a more severely proatherogenic lipid profile and increased MMP-9 expression in arterial lesions (Tuomainen et al. 2008; Chukkapalli et al. 2014). In addition, *P. gingivalis* increases vascular injury in other mice models and recurrent *P. gingivalis* bacteremia induces coronary and aortic atherosclerosis in both normocholesterolemic and hypercholesterolemic pigs (Brodala et al. 2005; Aoyama et al. 2011; Kobayashi et al. 2012). Also, in rats, ligature-induced experimental periodontitis promotes endothelial dysfunction and vascular inflammation (Brito et al. 2013). Intriguingly, in a mice model of MI, inoculation with *P. gingivalis* was reported to result in higher mortality rates due to cardiac rupture, with increased oxidative stress and MMP-9 expression in the post-MI myocardium (Shiheido et al. 2016). Moreover, immunization of mice with *P. gingivalis* lysate is associated with activation of Th1 immune responses, aortic inflammation and increased elevation of blood pressure and endothelial dysfunction in response to the vasopressor angiotensin II, thus providing a link between periodontitis and hypertension (Czesnikiewicz-Guzik et al. 2019). However, the extent to which the increase in cardiovascular injury observed in experimental studies is specific for periodontal pathogens or applicable to all infectious agents is unclear and the huge differences between such models and human disease limits the applicability of results.

14.8 Clinical Interventional Studies

Several interventional studies have been performed to examine the effect of periodontal treatment (e.g., tooth scaling) on CVD but have usually been small and underpowered. It is reasonably well-established, however, that treatment of periodontitis improves the CVD risk profile by lowering LGI reflected by reductions in circulating levels of biomarkers, e.g., hs-CRP and IL-6, and ameliorated lipid levels, e.g., reduced total cholesterol concentrations, especially in patients with established CVD and/or diabetes (D'Aiuto et al. 2013; Teeuw et al. 2014). Furthermore, periodontal treatment improves endothelial dysfunction although the initial inflammatory response to invasive treatment is accompanied by a short-term worsening of endothelial function and increase in hs-CRP and IL-6 levels, likely caused by bacteremia induced by the invasive procedure (Forner et al. 2006a; b; Tonetti et al. 2007; D'Aiuto et al. 2013; Orlandi et al. 2014; Teeuw et al. 2014). Levels of hs-CRP and IL-6 normalize within 7 days and hereafter endothelial function steadily improves over the next months and becomes significantly better than in controls (Tonetti et al. 2007; D'Aiuto et al. 2013; Teeuw et al. 2014; Orlandi et al. 2014; Graziani et al. 2015). A meta-analysis also demonstrated that periodontal treatment reduced carotid intima-media thickness (Orlandi et al. 2014). Along this line, in a prospective study of elderly (mean age 68 ± 8 years) New Yorkers, improvement of periodontal status assessed by periodontal pocket depth, gingival crevicular fluid IL-1β levels, and subgingival dental biofilm burden of etiologic bacteria, e.g., *P. gingivalis* and *A. actinomycetemcomitans*, respectively, was associated with decreased progression of carotid intima-media thickness over a 3-year follow-up period (Desvarieux et al. 2013).

As discussed earlier, no well-powered study of effects of periodontal treatment on hard CVD endpoints, e.g., MI, stroke, and cardiovascular death are available, and none are likely to be performed (Offenbacher et al. 2009; Lopez et al. 2012). However, register-based studies of large cohorts have consistently linked periodontal treatment with improved CVD outcomes and suggested that a dose-response relationship exists between periodontitis severity and CVD risk. For example, a population-based study from the National Health Insurance Research Database in Taiwan examined major adverse cardiovascular events in patients with mild or severe periodontitis ($n = 13{,}573$ in each group) that were propensity-matched for sex, age, diabetes, hypertension, and hyperlipidemia (Chou et al. 2015). After 10 years of follow-up, severe periodontitis was linked with increased risk of adverse cardiovascular events compared to milder periodontitis among patients >60 years of age (IRR 1.25, 95% CI 1.08–1.46) (Chou et al. 2015). Another nationwide population-based study from the same database with 7 years of follow-up examined a random sample of one million persons and found that individuals who received periodontal treatment with scaling experienced fewer MIs and strokes that those who did not, and that increased frequency of tooth scaling was correlated with increased risk reduction (Chen et al. 2012). Also, in a recent study of healthy Koreans ≥40 years of age (n = 247.696) who underwent an oral health screening program and had no history of major cardiovascular events, the risk of these events (MI, stroke, cardiovascular death, and heart failure) after a median follow-up of 9.5 years was higher for subjects with periodontal disease, a higher number of dental caries, or more tooth loss (Park et al. 2019). Moreover, ≥1 tooth brushing per day or ≥ 1 regular dental visit for professional cleaning per year reduced cardiovascular risk by 9% and 14%, respectively (Park et al. 2019). Because of the transient acute phase response with increased endothelial dysfunction observed initially after periodontal treatment, results of a prior US Medicaid database study indicating increased risk of MI and stroke in the first 4 weeks after invasive dental treatment raised some concerns (Minassian et al. 2010). However, in a recent larger study from Taiwan, invasive dental treatment did not appear to be associated with early risk of MI or stroke (Chen et al. 2019). Register-based studies are, of course, subject to inherent limitations, e.g., diverse biases and unmeasured confounders frequently including smoking, socioeconomic status, physical fitness, alcohol consumption, diet, and other factors that may contribute to healthy user effects associated with good oral hygiene. Notably, in view of the totality of available evidence the current European Society of Cardiology Guidelines for Prevention of Cardiovascular Diseases cites the link between periodontitis and CVD, with a cautious note that even if treatment or prevention of periodontitis improves, the effect on clinical prognosis remains unclear (Piepoli et al. 2016).

14.9 Conclusion

The role of LGI in mediating the risk of CVD has been amply documented and it is now well-established that CVD risk is effectively reduced by treatment of residual inflammation. Therefore, identification of previously less noticed sources of LGI can provide new and unique opportunities to lower CVD risk. Periodontitis is the most prevalent inflammatory disease and in later years our understanding of the mediating role of LGI in the link between periodontitis and CVD has developed considerably. It is now evident that periodontal health should be a key target for interventions aimed at reducing LGI and residual inflammatory risk of CVD. That improved oral health benefits CVD and other major noncommunicable diseases involving LGI is an exciting new perspective that is likely to have major impact on public health.

References

Aarabi G, Eberhard J, Reissmann DR, Heydecke G, Seedorf U (2015) Interaction between periodontal disease and atherosclerotic vascular disease-fact or fiction? Atherosclerosis 241(2):555–560. https://doi.org/10.1016/j.atherosclerosis.2015.04.819

Aarabi G, Zeller T, Seedorf H, Reissmann DR, Heydecke G, Schaefer AS, Seedorf U (2017) Genetic susceptibility contributing to periodontal and cardiovascular disease. J Dent Res 96(6):610–617. https://doi.org/10.1177/0022034517699786

Agca R, Heslinga SC, Rollefstad S, Heslinga M, McInnes IB, Peters MJ, Kvien TK, Dougados M, Radner H, Atzeni F, Primdahl J, Sodergren A, Wallberg Jonsson S, van Rompay J, Zabalan C, Pedersen TR, Jacobsson L, de Vlam K, Gonzalez-Gay MA, Semb AG, Kitas GD, Smulders YM, Szekanecz Z, Sattar N, Symmons DP, Nurmohamed MT (2017) Eular recommendations for cardiovascular disease risk management in patients with rheumatoid arthritis and other forms of inflammatory joint disorders: 2015/2016 update. Ann Rheum Dis 76(1):17–28. https://doi.org/10.1136/annrheumdis-2016-209775

Ahlehoff O, Skov L, Gislason G, Lindhardsen J, Kristensen SL, Iversen L, Lasthein S, Gniadecki R, Dam TN, Torp-Pedersen C, Hansen PR (2013) Cardiovascular disease event rates in patients with severe psoriasis treated with systemic anti-inflammatory drugs: a danish real-world cohort study. J Intern Med 273(2):197–204. https://doi.org/10.1111/j.1365-2796.2012.02593.x

Al Qahtani NA, Joseph B, Deepthi A, Vijayakumari BK (2017) Prevalence of chronic periodontitis and its risk determinants among female patients in the aseer region of ksa. J Taibah Univ Med Sci 12(3):241–248. https://doi.org/10.1016/j.jtumed.2016.11.012

Alakhali MS, Al-Maweri SA, Al-Shamiri HM, Al-Haddad K, Halboub E (2018) The potential association between periodontitis and non-alcoholic fatty liver disease: a systematic review. Clin Oral Investig 22(9):2965–2974. https://doi.org/10.1007/s00784-018-2726-1

Al-Harthi LS, Cullinan MP, Leichter JW, Thomson WM (2013) The impact of periodontitis on oral health-related quality of life: a review of the evidence from observational studies. Aust Dent J 58(3):274–277; quiz 384. https://doi.org/10.1111/adj.12076

Al-Jewair TS, Al-Jasser R, Almas K (2015) Periodontitis and obstructive sleep apnea's bidirectional relationship: a systematic review and meta-analysis. Sleep Breath 19(4):1111–1120. https://doi.org/10.1007/s11325-015-1160-8

AlSakr A, Blanchard S, Wong P, Thyvalikakath T, Hamada Y (2020) Association between intracranial carotid artery calcifications and periodontitis. Cone-beam computed tomography (CBCT) study. J Periodontol. https://doi.org/10.1002/JPER.20-0607

Amar S, Gokce N, Morgan S, Loukideli M, Van Dyke TE, Vita JA (2003) Periodontal disease is associated with brachial artery endothelial dysfunction and systemic inflammation. Arterioscler Thromb Vasc Biol 23(7):1245–1249. https://doi.org/10.1161/01.ATV.0000078603.90302.4A

Aoyama N, Suzuki J, Wang D, Ogawa M, Kobayashi N, Hanatani T, Takeuchi Y, Izumi Y, Isobe M (2011) Porphyromonas gingivalis promotes murine abdominal aortic aneurysms via matrix metalloproteinase-2 induction. J Periodontal Res 46(2):176–183. https://doi.org/10.1111/j.1600-0765.2010.01326.x

Aral K, Milward MR, Kapila Y, Berdeli A, Cooper PR (2020) Inflammasomes and their regulation in periodontal disease: a review. J Periodontal Res 55(4):473–487. https://doi.org/10.1111/jre.12733

Artese HP, Foz AM, Rabelo Mde S, Gomes GH, Orlandi M, Suvan J, D'Aiuto F, Romito GA (2015) Periodontal therapy and systemic inflammation in type 2 diabetes mellitus: a meta-analysis. PLoS One 10(5):e0128344. https://doi.org/10.1371/journal.pone.0128344

Bartold PM, Van Dyke TE (2013) Periodontitis: a host-mediated disruption of microbial homeostasis. Unlearning learned concepts. Periodontol 2000 62(1):203–217. https://doi.org/10.1111/j.1600-0757.2012.00450.x

Belstrom D, Damgaard C, Nielsen CH, Holmstrup P (2012) Does a causal relation between cardiovascular disease and periodontitis exist? Microbes Infect 14(5):411–418. https://doi.org/10.1016/j.micinf.2011.12.004

Billings M, Holtfreter B, Papapanou PN, Mitnik GL, Kocher T, Dye BA (2018) Age-dependent distribution of periodontitis in two countries: findings from NHANES 2009 to 2014 and ship-trend 2008 to 2012. J Periodontol 89(Suppl 1):S140–S158. https://doi.org/10.1002/JPER.17-0670

Bizzarro S, Loos BG (2019) The link between periodontitis and erectile dysfunction: a review. Br Dent J 227(7):599–603. https://doi.org/10.1038/s41415-019-0724-6

Borrell LN, Crawford ND (2012) Socioeconomic position indicators and periodontitis: examining the evidence. Periodontol 2000 58(1):69–83. https://doi.org/10.1111/j.1600-0757.2011.00416.x

Brito LC, DalBo S, Striechen TM, Farias JM, Olchanheski LR Jr, Mendes RT, Vellosa JC, Favero GM, Sordi R, Assreuy J, Santos FA, Fernandes D (2013) Experimental periodontitis promotes transient vascular inflammation and endothelial dysfunction. Arch Oral Biol 58(9):1187–1198. https://doi.org/10.1016/j.archoralbio.2013.03.009

Brodala N, Merricks EP, Bellinger DA, Damrongsri D, Offenbacher S, Beck J, Madianos P, Sotres D, Chang YL, Koch G, Nichols TC (2005) Porphyromonas gingivalis bacteremia induces coronary and aortic atherosclerosis in normocholesterolemic and hypercholesterolemic pigs. Arterioscler Thromb Vasc Biol 25(7):1446–1451. https://doi.org/10.1161/01.ATV.0000167525.69400.9c

Buset SL, Walter C, Friedmann A, Weiger R, Borgnakke WS, Zitzmann NU (2016) Are periodontal diseases really silent? A systematic review of their effect on quality of life. J Clin Periodontol 43(4):333–344. https://doi.org/10.1111/jcpe.12517

Cai Y, Kobayashi R, Hashizume-Takizawa T, Kurita-Ochiai T (2014) Porphyromonas gingivalis infection enhances Th17 responses for development of atherosclerosis. Arch Oral Biol 59(11):1183–1191. https://doi.org/10.1016/j.archoralbio.2014.07.012

Cambridge G, Acharya J, Cooper JA, Edwards JC, Humphries SE (2013) Antibodies to citrullinated peptides and risk of coronary heart disease. Atherosclerosis 228(1):243–246. https://doi.org/10.1016/j.atherosclerosis.2013.02.009

Cekici A, Kantarci A, Hasturk H, Van Dyke TE (2014) Inflammatory and immune pathways in the pathogenesis of periodontal disease. Periodontol 2000 64(1):57–80. https://doi.org/10.1111/prd.12002

Chen ZY, Chiang CH, Huang CC, Chung CM, Chan WL, Huang PH, Lin SJ, Chen JW, Leu HB (2012) The association of tooth scaling and decreased cardiovascular disease: a nationwide population-based study. Am J Med 125(6):568–575. https://doi.org/10.1016/j.amjmed.2011.10.034

Chen TT, D'Aiuto F, Yeh YC, Lai MS, Chien KL, Tu YK (2019) Risk of myocardial infarction and ischemic stroke after dental treatments. J Dent Res 98(2):157–163. https://doi.org/10.1177/0022034518805745

Chou SH, Tung YC, Lin YS, Wu LS, Lin CP, Liou EJ, Chang CJ, Kung S, Chu PH (2015) Major adverse cardiovascular events in treated periodontitis: a population-based follow-up study from Taiwan. PLoS One 10(6):e0130807. https://doi.org/10.1371/journal.pone.0130807

Chukkapalli SS, Rivera MF, Velsko IM, Lee JY, Chen H, Zheng D, Bhattacharyya I, Gangula PR, Lucas AR, Kesavalu L (2014) Invasion of oral and aortic tissues by oral spirochete treponema denticola in ApoE(-/-) mice causally links periodontal disease and atherosclerosis. Infect Immun 82(5):1959–1967. https://doi.org/10.1128/IAI.01511-14

Conroy RM, Pyorala K, Fitzgerald AP, Sans S, Menotti A, De Backer G, De Bacquer D, Ducimetiere P, Jousilahti P, Keil U, Njolstad I, Oganov RG, Thomsen T, Tunstall-Pedoe H, Tverdal A, Wedel H, Whincup P, Wilhelmsen L, Graham IM, group Sp (2003) Estimation of ten-year risk of fatal cardiovascular disease in europe: the score project. Eur Heart J 24(11):987–1003. https://doi.org/10.1016/s0195-668x(03)00114-3

Crea F, Libby P (2017) Acute coronary syndromes: the way forward from mechanisms to precision treatment. Circulation 136(12):1155–1166. https://doi.org/10.1161/CIRCULATIONAHA.117.029870

Czesnikiewicz-Guzik M, Osmenda G, Siedlinski M, Nosalski R, Pelka P, Nowakowski D, Wilk G, Mikolajczyk TP, Schramm-Luc A, Furtak A, Matusik P, Koziol J, Drozdz M, Munoz-Aguilera E, Tomaszewski M, Evangelou E, Caulfield M, Grodzicki T, D'Aiuto F, Guzik TJ (2019) Causal association between periodontitis and hypertension: evidence from mendelian randomization and a randomized controlled trial of non-surgical periodontal therapy. Eur Heart J 40(42):3459–3470. https://doi.org/10.1093/eurheartj/ehz646

Czesnikiewicz-Guzik M, D'Aiuto F, Deanfield JE (2020) Understanding residual inflammatory risk sheds new light on the clinical importance of periodontitis in cardiovascular disease. Eur Heart J 41(7):818–819. https://doi.org/10.1093/eurheartj/ehaa107

D'Aiuto F, Orlandi M, Gunsolley JC (2013) Evidence that periodontal treatment improves biomarkers and cvd outcomes. J Periodontol 84(4 Suppl):S85–S105. https://doi.org/10.1902/jop.2013.134007

D'Aiuto F, Gkranias N, Bhowruth D, Khan T, Orlandi M, Suvan J, Masi S, Tsakos G, Hurel S, Hingorani AD, Donos N, Deanfield JE (2018) Systemic effects of periodontitis treatment in patients with type 2 diabetes: a 12 month, single-Centre, investigator-masked, randomised trial. Lancet Diabetes Endocrinol 6(12):954–965. https://doi.org/10.1016/s2213-8587(18)30038-x

Damgaard C, Sækmose SG, Nilsson M, Nielsen CH, Holmstrup P (2020) Periodontitis increases risk of viable bacteria in freshly drawn blood donations. Blood Transfus 19:376–383

Darnaud C, Courtet A, Schmitt A, Boutouyrie P, Bouchard P, Carra MC (2021) Association between periodontitis and pulse wave velocity: a systematic review and meta-analysis. Clin Oral Investig 25(2):393–405. https://doi.org/10.1007/s00784-020-03718-2

Deschamps-Lenhardt S, Martin-Cabezas R, Hannedouche T, Huck O (2019) Association between periodontitis and chronic kidney disease: systematic review and meta-analysis. Oral Dis 25(2):385–402. https://doi.org/10.1111/odi.12834

Desvarieux M, Demmer RT, Jacobs DR, Papapanou PN, Sacco RL, Rundek T (2013) Changes in clinical and microbiological periodontal profiles relate to progression of carotid intima-media thickness: the oral infections and vascular disease epidemiology study. J Am Heart Assoc 2(6):e000254. https://doi.org/10.1161/JAHA.113.000254

Dietmann A, Millonig A, Combes V, Couraud PO, Kachlany SC, Grau GE (2013) Effects of aggregatibacter actinomycetemcomitans leukotoxin on endothelial cells. Microb Pathog 61–62:43–50. https://doi.org/10.1016/j.micpath.2013.05.001

Eke PI, Wei L, Thornton-Evans GO, Borrell LN, Borgnakke WS, Dye B, Genco RJ (2016) Risk indicators for periodontitis in us adults: NHANES 2009 to 2012. J Periodontol 87(10):1174–1185. https://doi.org/10.1902/jop.2016.160013

Emerging Risk Factors C, Kaptoge S, Di Angelantonio E, Lowe G, Pepys MB, Thompson SG, Collins R, Danesh J (2010) C-reactive protein concentration and risk of coronary heart disease, stroke, and mortality: an individual participant meta-analysis. Lancet 375(9709):132–140. https://doi.org/10.1016/S0140-6736(09)61717-7

Engstrom S, Berne C, Gahnberg L, Svardsudd K (2011) Efficacy of screening for high blood pressure in dental health care. BMC Public Health 11:194. https://doi.org/10.1186/1471-2458-11-194

Fagundes NCF, Almeida A, Vilhena KFB, Magno MB, Maia LC, Lima RR (2019) Periodontitis as a risk factor for stroke: a systematic review and meta-analysis. Vasc Health Risk Manag 15:519–532. https://doi.org/10.2147/VHRM.S204097

Ferreira RO, Correa MG, Magno MB, Almeida A, Fagundes NCF, Rosing CK, Maia LC, Lima RR (2019) Physical activity reduces the prevalence of periodontal disease: systematic review and meta-analysis. Front Physiol 10:234. https://doi.org/10.3389/fphys.2019.00234

Ferrucci L, Fabbri E (2018) Inflammageing: chronic inflammation in ageing, cardiovascular disease, and frailty. Nat Rev Cardiol 15(9):505–522. https://doi.org/10.1038/s41569-018-0064-2

Fifer KM, Qadir S, Subramanian S, Vijayakumar J, Figueroa AL, Truong QA, Hoffmann U, Brady TJ, Tawakol A (2011) Positron emission tomography measurement of periodontal 18F-fluorodeoxyglucose uptake is associated with histologically determined carotid plaque inflammation. J Am Coll Cardiol 57(8):971–976. https://doi.org/10.1016/j.jacc.2010.09.056

Forner L, Larsen T, Kilian M, Holmstrup P (2006a) Incidence of bacteremia after chewing, tooth brushing and scaling in individuals with periodontal inflammation. J Clin Periodontol 33(6):401–407. https://doi.org/10.1111/j.1600-051X.2006.00924.x

Forner L, Nielsen CH, Bendtzen K, Larsen T, Holmstrup P (2006b) Increased plasma levels of IL-6 in bacteremic periodontis patients after scaling. J Clin Periodontol 33(10):724–729. https://doi.org/10.1111/j.1600-051X.2006.00964.x

Frohlich H, Herrmann K, Franke J, Karimi A, Tager T, Cebola R, Katus HA, Zugck C, Frankenstein L (2016) Periodontitis in chronic heart failure. Tex Heart Inst J 43(4):297–304. https://doi.org/10.14503/THIJ-15-5200

Furman D, Campisi J, Verdin E, Carrera-Bastos P, Targ S, Franceschi C, Ferrucci L, Gilroy DW, Fasano A, Miller GW, Miller AH, Mantovani A, Weyand CM, Barzilai N, Goronzy JJ, Rando TA, Effros RB, Lucia A, Kleinstreuer N, Slavich GM (2019) Chronic inflammation in the etiology of disease across the life span. Nat Med 25(12):1822–1832. https://doi.org/10.1038/s41591-019-0675-0

Graziani F, Cei S, Orlandi M, Gennai S, Gabriele M, Filice N, Nisi M, D'Aiuto F (2015) Acute-phase response following full-mouth versus quadrant non-surgical periodontal treatment: a randomized clinical trial. J Clin Periodontol 42(9):843–852. https://doi.org/10.1111/jcpe.12451

Gurav AN (2014) The implication of periodontitis in vascular endothelial dysfunction. Eur J Clin Investig 44(10):1000–1009. https://doi.org/10.1111/eci.12322

Hansen GM, Egeberg A, Holmstrup P, Hansen PR (2016) Relation of periodontitis to risk of cardiovascular and all-cause mortality (from a danish nationwide cohort study). Am J Cardiol 118(4):489–493. https://doi.org/10.1016/j.amjcard.2016.05.036

Hansen PR (2018) Chronic inflammatory diseases and atherosclerotic cardiovascular disease: Innocent bystanders of partners in crime? Curr Pharm Des 24:281–290. https://doi.org/10.2174/1381612824666180110102341

Hansildaar R, Vedder D, Baniaamam M, Tausche AK, Gerritsen M, Nurmohamed MT (2021) Cardiovascular risk in inflammatory arthritis: rheumatoid arthritis and gout. Lancet Rheumatol 3(1):e58–e70. https://doi.org/10.1016/S2665-9913(20)30221-6

Hansson GK (2005) Inflammation, atherosclerosis, and coronary artery disease. N Engl J Med 352(16):1685–1695. https://doi.org/10.1056/NEJMra043430

Hassan A, Lip GYH, Harris RV (2021) Atrial fibrillation and cardiac arrhythmia associated with acute dental infection: a systematic literature review and case report. Int J Clin Pract 75(5):e13875. https://doi.org/10.1111/ijcp.13875

Hedstrom L, Albrektsson M, Bergh H (2015) Is there a connection between sublingual varices and hypertension? BMC Oral Health 15:78. https://doi.org/10.1186/s12903-015-0054-2

Herrera D, Molina A, Buhlin K, Klinge B (2020) Periodontal diseases and association with atherosclerotic disease. Periodontol 2000 83(1):66–89. https://doi.org/10.1111/prd.12302

Hingorani AD, Cross J, Kharbanda RK, Mullen MJ, Bhagat K, Taylor M, Donald AE, Palacios M, Griffin GE, Deanfield JE, MacAllister RJ, Vallance P (2000) Acute systemic inflammation impairs endothelium-dependent dilatation in humans. Circulation 102(9):994–999. https://doi.org/10.1161/01.cir.102.9.994

Hippisley-Cox J, Coupland C, Brindle P (2017) Development and validation of qrisk3 risk prediction algorithms to estimate future risk of cardiovascular disease: prospective cohort study. BMJ 357:j2099. https://doi.org/10.1136/bmj.j2099

Holm NC, Belstrøm DB, Østergaard JA, Schou S, Holmstrup P, Grauballe MB (2016) Identification of individuals with undiagnosed diabetes and pre-diabetes in a Danish cohort attending dental treatment. J Periodontol, 87:395–401. https://doi.org/10.1902/jop.2016.150266

Holmstrup P, Damgaard C, Olsen I, Klinge B, Flyvbjerg A, Nielsen CH, Hansen PR (2017) Comorbidity of periodontal disease: two sides of the same coin? An introduction for the clinician. J Oral Microbiol 9(1):1332710. https://doi.org/10.1080/20002297.2017.1332710

Huang AL, Vita JA (2006) Effects of systemic inflammation on endothelium-dependent vasodilation. Trends Cardiovasc Med 16(1):15–20. https://doi.org/10.1016/j.tcm.2005.10.002

Joshi C, Bapat R, Anderson W, Dawson D, Hijazi K, Cherukara G (2021) Detection of periodontal microorganisms in coronary atheromatous plaque specimens of myocardial infarction patients: a systematic review and meta-analysis. Trends Cardiovasc Med 31(1):69–82. https://doi.org/10.1016/j.tcm.2019.12.005

Kaschwich M, Behrendt CA, Heydecke G, Bayer A, Debus ES, Seedorf U, Aarabi G (2019) The association of periodontitis and peripheral arterial occlusive disease-a systematic review. Int J Mol Sci 20(12). https://doi.org/10.3390/ijms20122936

Kassebaum NJ, Bernabe E, Dahiya M, Bhandari B, Murray CJ, Marcenes W (2014) Global burden of severe periodontitis in 1990-2010: a systematic review and meta-regression. J Dent Res 93(11):1045–1053. https://doi.org/10.1177/0022034514552491

Keller A, Rohde JF, Raymond K, Heitmann BL (2015) Association between periodontal disease and overweight and obesity: a systematic review. J Periodontol 86(6):766–776. https://doi.org/10.1902/jop.2015.140589

Kobayashi N, Suzuki J, Ogawa M, Aoyama N, Hanatani T, Hirata Y, Nagai R, Izumi Y, Isobe M (2012) Porphyromonas gingivalis accelerates neointimal formation after arterial injury. J Vasc Res 49(5):417–424. https://doi.org/10.1159/000339583

Kolovou G, Anagnostopoulou K, Mikhailidis DP, Cokkinos DV (2008) Apolipoprotein e knockout

models. Curr Pharm Des 14(4):338–351. https://doi.org/10.2174/138161208783497769
Konig MF, Abusleme L, Reinholdt J, Palmer RJ, Teles RP, Sampson K, Rosen A, Nigrovic PA, Sokolove J, Giles JT, Moutsopoulos NM, Andrade F (2016) Aggregatibacter actinomycetemcomitans-induced hypercitrullination links periodontal infection to autoimmunity in rheumatoid arthritis. Sci Transl Med 8(369):369ra176. https://doi.org/10.1126/scitranslmed.aaj1921
Kozarov EV, Dorn BR, Shelburne CE, Dunn WA Jr, Progulske-Fox A (2005) Human atherosclerotic plaque contains viable invasive actinobacillus actinomycetemcomitans and porphyromonas gingivalis. Arterioscler Thromb Vasc Biol 25(3):e17–e18. https://doi.org/10.1161/01.ATV.0000155018.67835.1a
Kristensen SL, Ahlehoff O, Lindhardsen J, Erichsen R, Jensen GV, Torp-Pedersen C, Nielsen OH, Gislason GH, Hansen PR (2013) Disease activity in inflammatory bowel disease is associated with increased risk of myocardial infarction, stroke and cardiovascular death--a danish nationwide cohort study. PLoS One 8(2):e56944. https://doi.org/10.1371/journal.pone.0056944
Lalla E, Papapanou PN (2011) Diabetes mellitus and periodontitis: a tale of two common interrelated diseases. Nat Rev Endocrinol 7(12):738–748. https://doi.org/10.1038/nrendo.2011.106
Larvin H, Kang J, Aggarwal VR, Pavitt S, Wu J (2020) Risk of incident cardiovascular disease in people with periodontal disease: a systematic review and meta-analysis. Clin Exp Dent Res. https://doi.org/10.1002/cre2.336
Lee HJ, Choi EK, Park JB, Han KD, Oh S (2019) Tooth loss predicts myocardial infarction, heart failure, stroke, and death. J Dent Res 98(2):164–170. https://doi.org/10.1177/0022034518814829
Li L, Messas E, Batista EL Jr, Levine RA, Amar S (2002) Porphyromonas gingivalis infection accelerates the progression of atherosclerosis in a heterozygous apolipoprotein e-deficient murine model. Circulation 105(7):861–867. https://doi.org/10.1161/hc0702.104178
Libby P (2012a) Counterregulation rules in atherothrombosis. J Am Coll Cardiol 59(16):1438–1440. https://doi.org/10.1016/j.jacc.2012.01.023
Libby P (2012b) Inflammation in atherosclerosis. Arterioscler Thromb Vasc Biol 32(9):2045–2051. https://doi.org/10.1161/ATVBAHA.108.179705
Libby P, Loscalzo J, Ridker PM, Farkouh ME, Hsue PY, Fuster V, Hasan AA, Amar S (2018) Inflammation, immunity, and infection in atherothrombosis: JACC review topic of the week. J Am Coll Cardiol 72(17):2071–2081. https://doi.org/10.1016/j.jacc.2018.08.1043
Lockhart PB, Bolger AF, Papapanou PN, Osinbowale O, Trevisan M, Levison ME, Taubert KA, Newburger JW, Gornik HL, Gewitz MH, Wilson WR, Smith SC Jr, Baddour LM, American Heart Association Rheumatic Fever E, Kawasaki Disease Committee of the Council on Cardiovascular Disease in the Young CoE, Prevention CoPVD, Council on Clinical C (2012) Periodontal disease and atherosclerotic vascular disease: does the evidence support an independent association?: a scientific statement from the american heart association. Circulation 125(20):2520–2544. https://doi.org/10.1161/CIR.0b013e31825719f3
Loos BG, Van Dyke TE (2020) The role of inflammation and genetics in periodontal disease. Periodontol 2000 83(1):26–39. https://doi.org/10.1111/prd.12297
Lopez NJ, Quintero A, Casanova PA, Ibieta CI, Baelum V, Lopez R (2012) Effects of periodontal therapy on systemic markers of inflammation in patients with metabolic syndrome: a controlled clinical trial. J Periodontol 83(3):267–278. https://doi.org/10.1902/jop.2011.110227
Lorenzo-Pouso AI, Castelo-Baz P, Rodriguez-Zorrilla S, Perez-Sayans M, Vega P (2020) Association between periodontal disease and inflammatory bowel disease: a systematic review and meta-analysis. Acta Odontol Scand:1–15. https://doi.org/10.1080/00016357.2020.1859132
Lourbakos A, Yuan YP, Jenkins AL, Travis J, Andrade-Gordon P, Santulli R, Potempa J, Pike RN (2001) Activation of protease-activated receptors by gingipains from porphyromonas gingivalis leads to platelet aggregation: a new trait in microbial pathogenicity. Blood 97(12):3790–3797. https://doi.org/10.1182/blood.v97.12.3790
Machado V, Escalda C, Proenca L, Mendes JJ, Botelho J (2020) Is there a bidirectional association between polycystic ovarian syndrome and periodontitis? A systematic review and meta-analysis. J Clin Med 9(6). https://doi.org/10.3390/jcm9061961
Maekawa T, Takahashi N, Tabeta K, Aoki Y, Miyashita H, Miyauchi S, Miyazawa H, Nakajima T, Yamazaki K (2011) Chronic oral infection with porphyromonas gingivalis accelerates atheroma formation by shifting the lipid profile. PLoS One 6(5):e20240. https://doi.org/10.1371/journal.pone.0020240
McGraw WT, Potempa J, Farley D, Travis J (1999) Purification, characterization, and sequence analysis of a potential virulence factor from porphyromonas gingivalis, peptidylarginine deiminase. Infect Immun 67(7):3248–3256
Michalowicz BS, Diehl SR, Gunsolley JC, Sparks BS, Brooks CN, Koertge TE, Califano JV, Burmeister JA, Schenkein HA (2000) Evidence of a substantial genetic basis for risk of adult periodontitis. J Periodontol 71(11):1699–1707. https://doi.org/10.1902/jop.2000.71.11.1699
Minassian C, D'Aiuto F, Hingorani AD, Smeeth L (2010) Invasive dental treatment and risk for vascular events: a self-controlled case series. Ann Intern Med 153(8):499–506. https://doi.org/10.7326/0003-4819-153-8-201010190-00006
Moutsopoulos NM, Kling HM, Angelov N, Jin W, Palmer RJ, Nares S, Osorio M, Wahl SM (2012) Porphyromonas gingivalis promotes th17 inducing pathways in chronic periodontitis. J Autoimmun

39(4):294–303. https://doi.org/10.1016/j.jaut.2012.03.003
Munoz Aguilera E, Suvan J, Buti J, Czesnikiewicz-Guzik M, Barbosa Ribeiro A, Orlandi M, Guzik TJ, Hingorani AD, Nart J, D'Aiuto F (2020) Periodontitis is associated with hypertension: a systematic review and meta-analysis. Cardiovasc Res 116(1):28–39. https://doi.org/10.1093/cvr/cvz201
Munz M, Richter GM, Loos BG, Jepsen S, Divaris K, Offenbacher S, Teumer A, Holtfreter B, Kocher T, Bruckmann C, Jockel-Schneider Y, Graetz C, Munoz L, Bhandari A, Tennstedt S, Staufenbiel I, van der Velde N, Uitterlinden AG, de Groot L, Wellmann J, Berger K, Krone B, Hoffmann P, Laudes M, Lieb W, Franke A, Dommisch H, Erdmann J, Schaefer AS (2018) Genome-wide association meta-analysis of coronary artery disease and periodontitis reveals a novel shared risk locus. Sci Rep 8(1):13678. https://doi.org/10.1038/s41598-018-31980-8
Musher DM, Abers MS, Corrales-Medina VF (2019) Acute infection and myocardial infarction. N Engl J Med 380(2):171–176. https://doi.org/10.1056/NEJMra1808137
Nakamura N, Yoshida M, Umeda M, Huang Y, Kitajima S, Inoue Y, Ishikawa I, Iwai T (2008) Extended exposure of lipopolysaccharide fraction from porphyromonas gingivalis facilitates mononuclear cell adhesion to vascular endothelium via toll-like receptor-2 dependent mechanism. Atherosclerosis 196(1):59–67. https://doi.org/10.1016/j.atherosclerosis.2007.01.039
Nepomuceno R, Pigossi SC, Finoti LS, Orrico SRP, Cirelli JA, Barros SP, Offenbacher S, Scarel-Caminaga RM (2017) Serum lipid levels in patients with periodontal disease: a meta-analysis and meta-regression. J Clin Periodontol 44(12):1192–1207. https://doi.org/10.1111/jcpe.12792
NHS (2019) Commissioning standard: dental care for people with diabetes. NHS England and NHS Improvement
Noack B, Genco RJ, Trevisan M, Grossi S, Zambon JJ, De Nardin E (2001) Periodontal infections contribute to elevated systemic C-reactive protein level. J Periodontol 72(9):1221–1227. https://doi.org/10.1902/jop.2000.72.9.1221
Offenbacher S, Beck JD, Moss K, Mendoza L, Paquette DW, Barrow DA, Couper DJ, Stewart DD, Falkner KL, Graham SP, Grossi S, Gunsolley JC, Madden T, Maupome G, Trevisan M, Van Dyke TE, Genco RJ (2009) Results from the periodontitis and vascular events (PAVE) study: a pilot multicentered, randomized, controlled trial to study effects of periodontal therapy in a secondary prevention model of cardiovascular disease. J Periodontol 80(2):190–201. https://doi.org/10.1902/jop.2009.080007
Orlandi M, Suvan J, Petrie A, Donos N, Masi S, Hingorani A, Deanfield J, D'Aiuto F (2014) Association between periodontal disease and its treatment, flow-mediated dilatation and carotid intima-media thickness: a systematic review and meta-analysis. Atherosclerosis 236(1):39–46. https://doi.org/10.1016/j.atherosclerosis.2014.06.002
Park SY, Kim SH, Kang SH, Yoon CH, Lee HJ, Yun PY, Youn TJ, Chae IH (2019) Improved oral hygiene care attenuates the cardiovascular risk of oral health disease: a population-based study from Korea. Eur Heart J 40(14):1138–1145. https://doi.org/10.1093/eurheartj/ehy836
Piepoli MF, Hoes AW, Agewall S, Albus C, Brotons C, Catapano AL, Cooney MT, Corra U, Cosyns B, Deaton C, Graham I, Hall MS, Hobbs FDR, Lochen ML, Lollgen H, Marques-Vidal P, Perk J, Prescott E, Redon J, Richter DJ, Sattar N, Smulders Y, Tiberi M, van der Worp HB, van Dis I, Verschuren WMM, Binno S, Group ESCSD (2016) 2016 european guidelines on cardiovascular disease prevention in clinical practice: the sixth joint task force of the European Society of Cardiology and other societies on cardiovascular disease prevention in clinical practice (constituted by representatives of 10 societies and by invited experts) developed with the special contribution of the european association for cardiovascular prevention & rehabilitation (eacpr). Eur Heart J 37(29):2315–2381. https://doi.org/10.1093/eurheartj/ehw106
Pihlstrom BL, Michalowicz BS, Johnson NW (2005) Periodontal diseases. Lancet 366(9499):1809–1820. https://doi.org/10.1016/S0140-6736(05)67728-8
Potempa J, Mydel P, Koziel J (2017) The case for periodontitis in the pathogenesis of rheumatoid arthritis. Nat Rev Rheumatol 13(10):606–620. https://doi.org/10.1038/nrrheum.2017.132
Rafferty B, Jonsson D, Kalachikov S, Demmer RT, Nowygrod R, Elkind MS, Bush H Jr, Kozarov E (2011) Impact of monocytic cells on recovery of uncultivable bacteria from atherosclerotic lesions. J Intern Med 270(3):273–280. https://doi.org/10.1111/j.1365-2796.2011.02373.x
Ridker PM (2002) Inflammatory biomarkers, statins, and the risk of stroke: cracking a clinical conundrum. Circulation 105(22):2583–2585. https://doi.org/10.1161/01.cir.0000017822.82512.62
Ridker PM (2007) Inflammatory biomarkers and risks of myocardial infarction, stroke, diabetes, and total mortality: implications for longevity. Nutr Rev 65(12 Pt 2):S253–S259. https://doi.org/10.1111/j.1753-4887.2007.tb00372.x
Ridker PM, Rifai N, Stampfer MJ, Hennekens CH (2000) Plasma concentration of interleukin-6 and the risk of future myocardial infarction among apparently healthy men. Circulation 101(15):1767–1772
Ridker PM, Everett BM, Thuren T, MacFadyen JG, Chang WH, Ballantyne C, Fonseca F, Nicolau J, Koenig W, Anker SD, Kastelein JJP, Cornel JH, Pais P, Pella D, Genest J, Cifkova R, Lorenzatti A, Forster T, Kobalava Z, Vida-Simiti L, Flather M, Shimokawa H, Ogawa H, Dellborg M, Rossi PRF, Troquay RPT, Libby P, Glynn RJ, Group CT (2017) Antiinflammatory therapy with canakinumab for atherosclerotic disease. N Engl J Med 377(12):1119–1131. https://doi.org/10.1056/NEJMoa1707914

Ridker PM, MacFadyen JG, Everett BM, Libby P, Thuren T, Glynn RJ, Group CT (2018) Relationship of c-reactive protein reduction to cardiovascular event reduction following treatment with canakinumab: a secondary analysis from the CANTOS randomised controlled trial. Lancet 391(10118):319–328. https://doi.org/10.1016/S0140-6736(17)32814-3

Romandini M, Baima G, Antonoglou G, Bueno J, Figuero E, Sanz M (2021) Periodontitis, edentulism, and risk of mortality: a systematic review with meta-analyses. J Dent Res 100(1):37–49. https://doi.org/10.1177/0022034520952401

Rosenfeld ME, Campbell LA (2011) Pathogens and atherosclerosis: update on the potential contribution of multiple infectious organisms to the pathogenesis of atherosclerosis. Thromb Haemost 106(5):858–867. https://doi.org/10.1160/TH11-06-0392

Ryden L, Buhlin K, Ekstrand E, de Faire U, Gustafsson A, Holmer J, Kjellstrom B, Lindahl B, Norhammar A, Nygren A, Nasman P, Rathnayake N, Svenungsson E, Klinge B (2016) Periodontitis increases the risk of a first myocardial infarction: a report from the PAROKRANK study. Circulation 133(6):576–583. https://doi.org/10.1161/CIRCULATIONAHA.115.020324

Sanz M, Del Castillo AM, Jepsen S, Gonzalez-Juanatey JR, D'Aiuto F, Bouchard P, Chapple I, Dietrich T, Gotsman I, Graziani F, Herrera D, Loos B, Madianos P, Michel JB, Perel P, Pieske B, Shapira L, Shechter M, Tonetti M, Vlachopoulos C, Wimmer G (2020) Periodontitis and cardiovascular diseases. Consensus report. Glob Heart 15(1):1. https://doi.org/10.5334/gh.400

Schaefer AS, Richter GM, Dommisch H, Reinartz M, Nothnagel M, Noack B, Laine ML, Folwaczny M, Groessner-Schreiber B, Loos BG, Jepsen S, Schreiber S (2011) CDKN2BAS is associated with periodontitis in different european populations and is activated by bacterial infection. J Med Genet 48(1):38–47. https://doi.org/10.1136/jmg.2010.078998

Schenkein HA, Loos BG (2013) Inflammatory mechanisms linking periodontal diseases to cardiovascular diseases. J Clin Periodontol 40(Suppl 14):S51–S69. https://doi.org/10.1111/jcpe.12060

Schenkein HA, Papapanou PN, Genco R, Sanz M (2020) Mechanisms underlying the association between periodontitis and atherosclerotic disease. Periodontol 2000 83(1):90–106. https://doi.org/10.1111/prd.12304

Schultz WM, Kelli HM, Lisko JC, Varghese T, Shen J, Sandesara P, Quyyumi AA, Taylor HA, Gulati M, Harold JG, Mieres JH, Ferdinand KC, Mensah GA, Sperling LS (2018) Socioeconomic status and cardiovascular outcomes: challenges and interventions. Circulation 137(20):2166–2178. https://doi.org/10.1161/CIRCULATIONAHA.117.029652

Sgolastra F, Severino M, Pietropaoli D, Gatto R, Monaco A (2013) Effectiveness of periodontal treatment to improve metabolic control in patients with chronic periodontitis and type 2 diabetes: a meta-analysis of randomized clinical trials. J Periodontol 84(7):958–973. https://doi.org/10.1902/jop.2012.120377

Shiheido Y, Maejima Y, Suzuki JI, Aoyama N, Kaneko M, Watanabe R, Sakamaki Y, Wakayama K, Ikeda Y, Akazawa H, Ichinose S, Komuro I, Izumi Y, Isobe M (2016) Porphyromonas gingivalis, a periodontal pathogen, enhances myocardial vulnerability, thereby promoting post-infarct cardiac rupture. J Mol Cell Cardiol 99:123–137. https://doi.org/10.1016/j.yjmcc.2016.03.017

Singh A, Rouxel P, Watt RG, Tsakos G (2013) Social inequalities in clustering of oral health related behaviors in a national sample of British adults. Prev Med 57(2):102–106. https://doi.org/10.1016/j.ypmed.2013.04.018

Stratton IM, Adler AI, Neil HA, Matthews DR, Manley SE, Cull CA, Hadden D, Turner RC, Holman RR (2000) Association of glycaemia with macrovascular and microvascular complications of type 2 diabetes (UKPDS 35): prospective observational study. BMJ 321(7258):405–412. https://doi.org/10.1136/bmj.321.7258.405

Tardif JC, Kouz S, Waters DD, Bertrand OF, Diaz R, Maggioni AP, Pinto FJ, Ibrahim R, Gamra H, Kiwan GS, Berry C, Lopez-Sendon J, Ostadal P, Koenig W, Angoulvant D, Gregoire JC, Lavoie MA, Dube MP, Rhainds D, Provencher M, Blondeau L, Orfanos A, L'Allier PL, Guertin MC, Roubille F (2019) Efficacy and safety of low-dose colchicine after myocardial infarction. N Engl J Med 381(26):2497–2505. https://doi.org/10.1056/NEJMoa1912388

Teeuw WJ, Slot DE, Susanto H, Gerdes VE, Abbas F, D'Aiuto F, Kastelein JJ, Loos BG (2014) Treatment of periodontitis improves the atherosclerotic profile: a systematic review and meta-analysis. J Clin Periodontol 41(1):70–79. https://doi.org/10.1111/jcpe.12171

Teshome A, Yitayeh A (2016) The effect of periodontal therapy on glycemic control and fasting plasma glucose level in type 2 diabetic patients: systematic review and meta-analysis. BMC Oral Health 17(1):31. https://doi.org/10.1186/s12903-016-0249-1

Tonetti MS, D'Aiuto F, Nibali L, Donald A, Storry C, Parkar M, Suvan J, Hingorani AD, Vallance P, Deanfield J (2007) Treatment of periodontitis and endothelial function. N Engl J Med 356(9):911–920. https://doi.org/10.1056/NEJMoa063186

Tuomainen AM, Jauhiainen M, Kovanen PT, Metso J, Paju S, Pussinen PJ (2008) Aggregatibacter actinomycetemcomitans induces MMP-9 expression and proatherogenic lipoprotein profile in apoE-deficient mice. Microb Pathog 44(2):111–117. https://doi.org/10.1016/j.micpath.2007.08.011

Watt RG, Tsakos G, de Oliveira C, Hamer M (2012) Tooth loss and cardiovascular disease mortality risk-results from the Scottish Health Survey. PLoS One 7(2):e30797. https://doi.org/10.1371/journal.pone.0030797

White DA, Tsakos G, Pitts NB, Fuller E, Douglas GV, Murray JJ, Steele JG (2012) Adult dental health survey 2009: common oral health conditions and their impact on the population. Br Dent J 213(11):567–572. https://doi.org/10.1038/sj.bdj.2012.1088

Williams B, Mancia G, Spiering W, Agabiti Rosei E, Azizi M, Burnier M, Clement DL, Coca A, de Simone G, Dominiczak A, Kahan T, Mahfoud F, Redon J, Ruilope L, Zanchetti A, Kerins M, Kjeldsen SE, Kreutz R, Laurent S, Lip GYH, McManus R, Narkiewicz K, Ruschitzka F, Schmieder RE, Shlyakhto E, Tsioufis C, Aboyans V, Desormais I, Group ESCSD (2018) 2018 esc/esh guidelines for the management of arterial hypertension. Eur Heart J 39(33):3021–3104. https://doi.org/10.1093/eurheartj/ehy339

Yang ZS, Lin NN, Li L, Li Y (2016) The effect of TNF inhibitors on cardiovascular events in psoriasis and psoriatic arthritis: an updated meta-analysis. Clin Rev Allergy Immunol 51(2):240–247. https://doi.org/10.1007/s12016-016-8560-9

Yu X, Feng X, Xia L, Cao S, Wei X (2020) Risk of aortic aneurysm in patients with psoriasis: a systematic review and meta-analysis of cohort studies. Clin Cardiol 43(11):1266–1272. https://doi.org/10.1002/clc.23438

Zeng XT, Leng WD, Lam YY, Yan BP, Wei XM, Weng H, Kwong JS (2016) Periodontal disease and carotid atherosclerosis: a meta-analysis of 17,330 participants. Int J Cardiol 203:1044–1051. https://doi.org/10.1016/j.ijcard.2015.11.092

Zhang X, Gu H, Xie S, Su Y (2020) Periodontitis in patients with psoriasis: a systematic review and meta-analysis. Oral Dis. https://doi.org/10.1111/odi.13617

Zheng DX, Kang XN, Wang YX, Huang YN, Pang CF, Chen YX, Kuang ZL, Peng Y (2021) Periodontal disease and emotional disorders: a meta-analysis. J Clin Periodontol 48(2):180–204. https://doi.org/10.1111/jcpe.13395

Zhou QB, Xia WH, Ren J, Yu BB, Tong XZ, Chen YB, Chen S, Feng L, Dai J, Tao J, Yang JY (2017) Effect of intensive periodontal therapy on blood pressure and endothelial microparticles in patients with prehypertension and periodontitis: a randomized controlled trial. J Periodontol 88(8):711–722. https://doi.org/10.1902/jop.2017.160447

Part IV

Therapeutic Management of Periodontitis

Update on the Role of Cytokines as Oral Biomarkers in the Diagnosis of Periodontitis

15

Triana Blanco-Pintos, Alba Regueira-Iglesias, Carlos Balsa-Castro, and Inmaculada Tomás

Abstract

Periodontitis is one of the world's most common chronic human diseases and has a significant impact on oral health. Recent evidence has revealed a link between periodontitis and certain severe systemic conditions. Moreover, periodontal patients remain so for life, even following successful therapy, requiring ongoing supportive care to prevent the disease's recurrence. The first challenge in treating the condition is ensuring a timely and accurate diagnosis since the loss of periodontal bone and soft tissue is progressive and largely irreversible. Although current clinical and radiographic parameters are the best available for identifying and monitoring the disease, the scientific community has a particular interest in finding quantifiable biomarkers in oral fluids that can improve early detection rates of periodontitis and evaluations of its severity. It is widely accepted that periodontitis is associated with polymicrobial dysbiosis and a chronic inflammatory immune response in the host. This response causes the generation of mediators like cytokines. Higher concentrations of cytokines are involved in inflammation and disease progression, acting as a network of biological redundancy. Most of the cytokines investigated concerning the periodontitis pathogenesis are proinflammatory. Of all of them, interleukin (IL) 1beta has been studied the most, followed by tumor necrosis factor (TNF) alpha and IL6. In contrast, only a few papers have evaluated antiinflammatory cytokines, with the most researched being IL4 and IL10. Several systemic reviews have concluded that the specific cytokines present in patients with periodontitis have a distinctive profile, which may indicate their possible discriminatory potential. In this chapter, the focus is on analyzing studies that investigate the accuracy of diagnoses of periodontitis based on the cytokines present in gingival crevicular fluid and saliva. The findings of our research group are also described.

Keywords

Biomarkers · Cytokines · Accuracy · Diagnosis · Gingival crevicular fluid · Saliva · Periodontitis

T. Blanco-Pintos · A. Regueira-Iglesias
C. Balsa-Castro · I. Tomás (✉)
Oral Sciences Research Group, Special Needs Unit, Department of Surgery and Medical-surgical Specialties, School of Medicine and Dentistry, Health Research Institute of Santiago (IDIS), Universidade de Santiago de Compostela, Galicia, Spain
e-mail: triana.blanco.pintos@usc.es; albaregueira.iglesias@usc.es; cbalsa@coitt.es; inmaculada.tomas@usc.es

J. Santi-Rocca (ed.), *Periodontitis*, Advances in Experimental Medicine and Biology 1373,
https://doi.org/10.1007/978-3-030-96881-6_15

Abbreviations

ACC	accuracy (of a test)
AUC	area under the curve
BL	bone loss
BOP	bleeding on probing
CAL	clinical attachment loss
ELISA	enzyme linked immunosorbent assay
GCF	gingival crevicular fluid
GMCSF	granulocyte-macrophage colonystimulating factor
IFN	interferon
IL	interleukin
MIP	macrophage inflammatory protein
ml	milliliter
MMP	matrix metalloproteinase
PPD	probing pocket depth
ra	receptor antagonist
TNF	tumor necrosis factor
USA	United States of America

Highlights

- Biomarkers in oral fluids could play a crucial role in both the early detection of periodontitis and evaluation of its severity.
- The most studied cytokine biomarkers in gingival crevicular fluid are interleukin 1alpha and interleukin 1beta, both of which have outstanding potential for enabling periodontitis to be distinguished from periodontal health.
- The most studied cytokine biomarkers in saliva are interleukin 1beta and interleukin 6, both of which have outstanding-to-acceptable potential when it comes to distinguishing periodontitis from periodontal health
- The combination of interleukin 1alpha, interleukin 1beta and interleukin 6 with another molecule seems to enhance the discriminatory capability of these cytokines individually. This is not overcome in models with more than two biomarkers.
- The smoking/non-smoking status affects the diagnostic accuracy of cytokines in oral fluids when it comes to detecting periodontitis
- More high-quality evidence is required on the accuracy of cytokines in oral fluids for diagnosis of periodontitis

Considerations for Practice

- Determining the presence of biomarkers in oral fluids could improve the diagnosis and monitoring of periodontitis.
- Assuming a 45% periodontitis prevalence, a test of interleukin 1beta in the gingival crevicular fluid would be highly effective for detecting the presence or absence of periodontitis, although there is an increase in false positives in smokers.
- Assuming a 45% prevalence of periodontitis, a salivary test based on a combination of two biomarkers, such as the interleukins 1beta and 6, interleukin 6 and metalloproteinase 8, or interleukin 1beta and metalloproteinase 8, would be effective for detecting the presence or absence of periodontitis.
- More high-quality evidence is required on the effectiveness of tests of cytokines in oral fluids for diagnosis of periodontitis.

Patient Summary

Periodontitis is an oral disease that requires a timely and accurate diagnosis since the periodontal bone and soft tissue loss it induces is incremental and largely irreversible. A patient suffering from periodontitis continues to do so for life, even after successful therapy, and needs ongoing supportive care to prevent the disease's recurrence. More sensitive and specific tools based on the presence in oral fluids of quantifiable biomarkers like cytokines could supplement or, in some cases, replace the conventional clinical measurements used to diagnose and monitor periodontitis.

15.1 Epidemiology of Periodontitis, its Clinical Characteristics, and its Impact on Oral and Systemic Health

Periodontitis is recognized as one of the most prevalent chronic diseases in human beings globally (Dentino et al. 2013; Kassebaum et al. 2014). The condition affects over 50% of the adult population and 11% suffer from its most severe forms (Dentino et al. 2013; Tonetti et al. 2015b). In the United States of America, the prevalence of periodontitis is 47% in adults older than 30 years, equating to about 65 million people (Eke et al. 2016b). In Europe, this figure increases to 70–85% in those older than 60 (König et al. 2010).

Periodontitis is the most serious of the periodontal diseases and is characterized by the destruction of the tooth-supporting structures (Tonetti et al. 2005). It usually begins as gingivitis, which is considered to be a reversible condition (Kinane et al. 2017). However, if the inflammatory disease is maintained, the accumulation and maturation of bacterial plaque can lead to periodontitis in susceptible individuals. Periodontitis and gingivitis have typical signs and symptoms, such as gum inflammation with redness and bleeding, excess tartar, halitosis and increased tooth sensitivity. However, periodontal pockets, dental mobility and, ultimately, tooth loss are hallmarks of periodontitis. Like gingivitis, periodontitis is usually painless, enabling both these diseases to progress to severe forms before they are detected (Kinane et al. 2017).

Unlike gingivitis, periodontitis is a chronic disease that, even following successful therapy, requires periodic reassessment and ongoing supportive care to prevent the recurrence of the disease (Chapple et al. 2018). The condition is treated mechanically (scaling and root planing and/or surgery), as well as with the removal or reduction of risk factors and appropriate periodontal maintenance. New treatment modalities being actively explored include antimicrobial therapy, host-modulation therapy, laser therapy, and tissue engineering for tissue repair and regeneration (Sanz and Teughels 2008; Slots 2017).

According to the literature, periodontitis is not a "silent" problem. Indeed, it has been demonstrated that periodontal patients have a poorer perception of their oral health and a worse quality of life than healthy individuals (Al-Harthi et al. 2013). Advanced periodontitis can also compromise different activities of daily living, including mastication or even speech due to pronunciation difficulties (Borges et al. 2013; Durham et al. 2013; Meusel et al. 2015). In addition, as the disease often leads to tooth loss, it is associated with poor nutrition caused by the reduced diversity of the food consumed (Iwasaki et al. 2015; Sheiham et al. 2001). This effect has been correlated with morbidity and mortality in the elderly (Saunders et al. 2008). Smile aesthetics can also be impacted in advanced periodontitis due to tooth loss or gum recessions, leading to low self-esteem and negative repercussions in social relationships (Ferreira et al. 2017; Jansson et al. 2014). Moreover, as people are living to older ages and tend to have a higher quality of life for longer (Eke et al. 2016a), it is becoming increasingly essential to prevent or treat the disease (Santuchi et al. 2016; Shanbhag et al. 2012). The field of periodontal medicine emerged when scientific evidence revealed a relationship between periodontitis and certain systemic conditions like cardiovascular disease (Aarabi et al. 2017), diabetes (Badiger et al. 2019) and rheumatoid arthritis (de Pablo et al. 2009). In recent years, numerous epidemiological associations with respiratory illnesses (Hobbins et al. 2017), adverse pregnancy outcomes (Zi et al. 2015) and even Alzheimer's disease have also been identified (Abbayya et al. 2015).

15.2 Limitations of Clinical Parameters for Diagnosing Periodontitis

In periodontics, the first treatment challenge is ensuring a timely and accurate diagnosis, as the loss of periodontal bone and soft tissue is incremental and largely irreversible (Kinane et al. 2017). Traditional clinical measures are informative for evaluating the severity of periodontitis and the response to therapy. These include: the presence of plaque or poor oral hygiene; the degree of gingival inflammation and bleeding on probing (BOP); the probing pocket depth (PPD) and suppuration; the clinical attachment loss (CAL); and the radiographic bone loss (BL) (Chatzistavrianou and Blair 2017). These clinical and radiographic parameters are the best available for diagnosing and monitoring the health-disease state in the majority of patients, probably because they respond favorably to the fundamental principles of periodontal care (Tonetti et al. 2018). Nevertheless, they are neither sensitive nor specific enough to reflect the current biological activity of the disease or predict its future course (Buduneli and Kinane 2011).

The BOP is still the best negative predictor of periodontitis activity, with its absence foretelling a lack of tissue destruction, although it does produce too many false positives (low sensitivity value) (Buduneli and Kinane 2011). Meanwhile, evaluations of the CAL measure past episodes of bone destruction and require a 2–3 mm threshold change before a site can be recognized as having significantly broken down (Taba et al. 2005). An accurate diagnosis (at specific sites and in the patient overall) requires all of these clinical parameters to be recorded at six locations per tooth (whether affected or not), which is both time-consuming and dependent on the professional's clinical experience (error-prone measures). Furthermore, this tedious process is often poorly tolerated by patients, given the need to repeat it regularly during appointments to monitor the disease course (Taylor 2014). Consequently, researchers are striving to identify faster and more sensitive and specific tools based on the quantifiable biomarkers present in oral fluids, with the goal being to supplement or, in some cases, replace the need for conventional clinical measurements (AlMoharib et al. 2014; AlRowis et al. 2014; Ghallab 2018).

15.3 The Importance of Biomarkers in Oral Fluids: Diagnostic Accuracy Studies

In 1998, the National Institutes of Health Biomarkers Definitions Working Group defined a biomarker as "a characteristic that is objectively measured and evaluated as an indicator of normal biological processes, pathogenic processes, or pharmacologic responses to a therapeutic intervention" (Biomarkers Definitions Working Group 2001). In dentistry, gingival crevicular fluid (GCF) is easy to collect, and its composition results from the interplay between the bacterial biofilm and the cells of the periodontium. GCF is the liquid sample that most accurately reflects the pathophysiological condition of the gingival sulcus. Consequently, it is regarded as the most promising medium for the detection of the molecular biomarkers associated with periodontitis (Barros et al. 2016; Ghallab 2018). However, it is important to note that very low GCF volumes can have a significant effect on the concentrations of GCF biomarkers (Ghallab 2018). Saliva can also reflect the state of periodontal health and is considered to be a useful tool for screening and monitoring periodontitis because of how easy it is to collect (Miller et al. 2010) and its capacity to reflect the inflammatory status of the whole mouth (Ghallab 2018).

The complex etiopathogenesis of periodontal diseases has been studied for many years (Darveau 2010; Yucel-Lindberg and Båge 2013). A large number of molecules play a part in the different disease states, but the focus of this chapter is on the fundamental role of inflammatory mediators, including cytokines (Preshaw and Taylor 2011). Numerous articles have discussed the measurement of cytokines in GCF and saliva, confirming a distinct cytokine profile for patients with periodontitis (Jaedicke et al. 2016; Stadler et al. 2016). Nevertheless, on its own, the mere existence of this profile does not equate to a diagnosis. Indeed, investigations of the diagnostic capability of a biomarker require a specific accuracy study to produce estimates of test performance (e.g. sensitivity and specificity) (McInnes et al. 2018).

An "ideal" oral biomarker of periodontitis must be able to: diagnose the presence of the condition; reflect the severity of the disease; predict its progression; and monitor the response to treatment (Ji and Choi 2015). There are three types of accuracy study: (1) diagnostic, which focus on analyzing the capability of an index test (in this case, oral biomarkers) to distinguish patients who have a target condition from those who do not (in this case, periodontitis *versus* periodontal health); (2) prognostic, in which the information from the index test is used to identify patients who will have a later event, such as disease progression (i.e., the event has not occurred at the time the test was undertaken); and (3) predictive, where a test is used to identify patients who will or will not benefit from treatment (Bossuyt and Leeflang 2008). Depending on its purpose, an index text can be used for triage, as an add-on, or as a replacement for an existing test (Gopalakrishna et al. 2016).

15.4 Host Response in Periodontitis: Role of Cytokines

The primary hallmark of periodontitis, namely the destruction of periodontal tissue, is widely accepted to be the result of a chronic inflammatory host response caused by a polymicrobial dysbiosis (Eke et al. 2016a; Michalowicz et al. 2000; Zaura et al. 2017). This response is characterized by the infiltration of the gingival tissue by neutrophils, macrophages, and lymphocytes, as well as by the generation of high concentrations of mediators (Preshaw and Taylor 2011). These mediators not only act as initiators and regulators of the immune response but are also involved in the damage caused to tissue, leading to clinical disease (Ebersole et al. 2013a; Korte and Kinney 2016). The extent of this immune reaction can determine the progression and severity of periodontitis (Jaedicke et al. 2016).

Cytokines (in Greek: "Cyto" = cell; "Kinos" = movement) are low-molecular-weight soluble proteins that play an essential role in homeostasis (Stadler et al. 2016). They are produced by different cell types and are involved in the initiation and progression of immune-inflammatory processes acting as "messengers" transmitting signals to other cells (Preshaw and Taylor 2011). In the acute inflammation phase, cytokines are released by epithelial cells, fibroblasts, and phagocytes; in the adaptive immunity process, they are produced by lymphocytes (Ara et al. 2009; Cekici et al. 2014).

The cytokine production in the organism is extremely regulated, and in healthy people, their concentrations are measured in picomolar/ml (Stenken and Poschenrieder 2015). Higher concentrations of cytokines are associated with inflammation and disease progression (Stenken and Poschenrieder 2015). Some cytokines are among the most critical proinflammatory mediators in stimulating the activation of osteoclasts and potentiating the progression of periodontitis, and include interleukin (IL) 1beta, IL6 and IL17 (not IL17E), granulocyte-macrophage colony-stimulating factor (GMCSF), and tumor necrosis factor (TNF) alpha (Yucel-Lindberg and Båge 2013). Antiinflammatory cytokines, on the other hand, play a significant role in the regulation of the T-cell subsets that act at many levels. A number of these mediators, such as IL13 and interferon (IFN) gamma, have an inhibitory impact on osteoclastogenesis (Tomás et al. 2017). However, some cytokines can have both proinflammatory and antiinflammatory effects, depending on their function (Garlet 2010). They also act as a network, and so different versions carry out the same roles, meaning that, in the absence of a specific cytokine, another would take its place; consequently, the response would continue to be activated by another pathway. This vital mechanism is known as biological redundancy (Salvi and Lang 2005; Seymour and Gemmell 2001).

Given the inflammatory nature of periodontitis, most of the cytokines investigated in relation to its pathogenesis are proinflammatory. Of all of them, IL1beta has been studied the most, followed by TNFalpha and IL6 (Stadler et al. 2016). Unlike the position with proinflammatory cytokines, only a few papers have focused on the role of antiinflammatory mediators, with the most researched being IL4 and IL10 (Stadler et al. 2016). Several authors have reported that cytokines are not specific enough to detect periodontitis (Zhang et al. 2009), and that their levels in oral fluids can be affected by local or systemic factors like smoking, alcohol consumption and stress (Pussinen et al. 2007; Zhang et al. 2009). If correct, these biomarkers may be of limited use for predicting periodontitis (Kurdukar et al. 2015). Interestingly, however, the authors of several systematic reviews have concluded that there is a distinctive profile of the specific cytokines that are present in patients with periodontitis (Finoti et al. 2017; Stadler et al. 2016), which may be an indicator of their possible discriminatory potential. In this chapter, the focus is on analyzing studies of the diagnostic accuracy of cytokines for the detection of the disease, with the results obtained by our research group also described.

15.5 Diagnostic Accuracy of Cytokines in the Gingival Crevicular Fluid: Single and Multiple Biomarkers

Our review of the literature found little evidence of the diagnostic potential of single cytokines present in the GCF for detecting periodontitis. Indeed, most studies focus on the discriminatory capacity of this pathology in relation to periodontal health. The most researched cytokines are IL1alpha and IL1beta followed by IL17A (Table 15.1). Similarly, very few authors have investigated the diagnostic accuracy of the presence of multiple cytokines in the GCF, or cytokines in combination with other biomarkers (Table 15.2). In most GCF studies, cytokines were quantified by applying ELISA or multiparametric cytometry techniques, and predictive models were obtained by applying logistic regression or linear discriminant analysis techniques.

The promising diagnostic capability of IL1 identified in previous studies (Baeza et al. 2016) was later corroborated by the findings published by Tomás et al. (2017) on their series of patients with periodontitis and a group of periodontally healthy controls. In this particular research, the area under the curve (AUC) values were above 0.960 for both types of IL1, which was interpreted as evidence of their outstanding potential to distinguish periodontitis from periodontal health (Hosmer et al. 2013). Specifically, the study revealed sensitivity and specificity values of 94.5% and 91.9%, respectively, for IL1alpha and 93.2% and 94.6%, respectively, for IL1beta. This predictive capability of IL1alpha and IL1beta was improved further by incorporating an antiinflammatory cytokine (respectively, IFNgamma and IL10) in the predictive model: the accuracy (ACC) values increased from about 93% to 95%, with sensitivities and specificities of 93.2% and 97.3% for IL1alpha, and 94.5% and 94.6% for IL1beta (Tomás et al. 2017). In contrast, and somewhat curiously, the combination of IL1alpha and IL1beta with IL2 (in the form of a ratio) did not improve the diagnostic accuracy of these proinflammatory cytokines individually, with the ACC values ranging from 68.5% to 89.8% (Arias-Bujanda et al. 2018).

In relation to predictive models with more than two biomarkers, and unlike studies published in the early 1990s (Kitamura et al. 1991), Huang et al. (2020) were able to produce a predictive model of periodontitis and periodontal health involving two cytokines (IL1beta and IL8), a matrix metalloproteinase (MMP13), an osteoprotein, and an osteoactivin. Although this multibiomarker model had sensitivity and specificity values of 96.0%, this was not a significant improvement on the diagnostic potential of the two-cytokine models evaluated previously (Tomás et al. 2017).

In contrast, the great potential of IL1alpha and IL1beta to distinguish between periodontitis from periodontal health appears to be significantly reduced when their capacity to identify active periodontal sites in periodontitis patients is assessed. In this sense, Kitamura et al. (1991) obtained that the ACC values fell to 56.6% for

Table 15.1 Characteristics of diagnostic accuracy studies on single cytokines in the gingival crevicular fluid, including performance values

Study	Index test	No. CC/No. TC	Type CC/Type TC	Threshold	AUC	ACC (%)	SENS/SPEC (%)	PPV/NPV (%)	LR+/LR-	DOR
Baeza et al. (2016)	IL1	31/31	H/CP	5.95 (pg/ml)	0.920	87.1	80.6/93.5	92.6/82.9	12.5/0.2	60.4
Kitamura et al. (1991)	IL1α	13/13	CP_Non-CAL/CP_CAL	NS (NS)	NS	56.6	15.4/96.3	80.0/54.2	4.2/0.9	4.7
Tomás et al. (2017)	IL1α	74/73	H/M-S-Ge-CP	NS (pg/ml)	0.973	93.2	94.5/91.9	92.0/94.4	11.7/0.1	195.5
Arias-Bujanda et al. (2018)	IL1α	61/32	H/M-S-Ge-CP (non-smokers)	65,644 (pg/ml)	0.959	93.5	87.5/96.7	93.3/93.7	26.6/0.1	206.5
Arias-Bujanda et al. (2018)	IL1α	13/41	H/M-S-Ge-CP (smokers)	46,099 (pg/ml)	0.966	91.7	98.8/69.2	91.1/94.7	3.2/0.0	184.5
Kitamura et al. (1991)	IL1β	13/13	CP_Non-CAL/CP_CAL	NS (NS)	NS	60.4	23.1/96.3	85.7/56.5	6.2/0.8	7.8
Tomás et al. (2017)	IL1β	74/73	H/M-S-Ge-CP	NS (pg/ml)	0.963	93.9	93.2/94.6	94.4/93.3	17.2/0.1	238.0
Arias-Bujanda et al. (2018)	IL1β	61/32	H/M-S-Ge-CP (non-smokers)	5827 (pg/ml)	0.944	94.6	90.6/96.7	93.5/95.2	27.6/0.0	285.1
Arias-Bujanda et al. (2018)	IL1β	13/41	H/M-S-Ge-CP (smokers)	4732 (pg/ml)	0.968	94.4	97.6/84.6	95.2/91.7	6.3/0.0	220.0
Tomás et al. (2017)	IL17A	74/73	H/M-S-Ge-CP	NS (pg/ml)	0.937	89.1	89.0/89.2	89.0/89.2	8.2/0.1	67.0
Arias-Bujanda et al. (2018)	IL17A	61/32	H/M-S-Ge-CP (non-smokers)	17.1 (pg/ml)	0.914	88.2	78.1/93.4	86.2/89.1	11.9/0.2	50.8
Arias-Bujanda et al. (2018)	IL17A	13/41	H/M-S-Ge-CP (smokers)	11.0 (pg/ml)	0.940	92.6	95.1/84.6	95.1/84.6	6.1/0.0	107.2
Baeza et al. (2016)	IL6	31/31	H/CP	1.68 (pg/ml)	0.930	85.5	83.9/87.1	86.7/84.4	6.5/0.2	35.1
Baeza et al. (2016)	TNFα	31/31	H/CP	1.26 (pg/ml)	0.640	67.7	45.2/90.3	82.4/62.2	4.7/0.6	7.7

The performance values of each study were recalculated on the basis of the original classification tables, which were either provided directly in the paper or were calculated
IL interleukin, *TNF* tumor necrosis factor, *No*. number, *CC* control condition, *TC* target condition, *H* periodontally healthy, *CP* chronic periodontitis, *CAL* clinical attachment loss, *M* moderate, *S* severe, *Ge* generalised, *pg/ml* picograms/milliliter, *NS* not specified, *AUC* area under the curve, *ACC* accuracy, *SEN*, *S* sensitivity *SPEC* specificity, *PPV* positive predictive value, *NPV* negative predictive value, *LR* likelihood ratio, *DOR* diagnostic odds ratio

Table 15.2 Characteristics of diagnostic accuracy studies on combination of biomarkers with at least one cytokine in gingival crevicular fluid, including performance values

Study	Index test	No. CC/No. TC	Type CC/Type TC	AUC	ACC (%)	SENS/SPEC (%)	PPV/NPV (%)	LR+/LR-	DOR
Tomás et al. (2017)	IL1α, IFNγ	74/73	H/M-S-Ge-CP	0.986	95.2	93.2/97.3	97.1/93.5	34.4/0.0	489.6
Arias-Bujanda et al. (2018)	IL1α, IL2 (ratio)	61/32	H/M-S-Ge-CP (non-smokers)	0.911	87.8	84.9/90.5	89.9/85.9	8.9/0.1	53.9
Arias-Bujanda et al. (2018)	IL1α, IL2 (ratio)	13/41	H/M-S-Ge-CP (smokers)	0.878	68.5	99.3/37.8	61.3/98.2	1.5/0.0	88.8
Kitamura et al. (1991)	IL1α, IL1β, aCOL, tCOL, lCOL, LPS, PGE2	13/13	CP_Non-CAL/ CP_CAL	NS	79.2	61.5/96.3	94.1/72.2	16.6/0.3	41.6
Tomás et al. (2017)	IL1β, IL10	74/73	H/M-S-Ge-CP	0.971	94.6	94.5/94.6	94.5/94.6	17.4/0.0	301.8
Arias-Bujanda et al. (2018)	IL1β, IL2 (ratio)	61/32	H/M-S-Ge-CP (non-smokers)	0.886	83.7	80.8/86.5	85.5/82.1	5.9/0.2	26.9
Arias-Bujanda et al. (2018)	IL1β, IL2 (ratio)	13/41	H/M-S-Ge-CP (smokers)	0.906	89.8	94.5/85.1	86.3/94.0	6.3/0.0	98.7
Huang et al. (2020)	IL1β, IL8, MMP13, OPG, OA	12/12	H/S-Ge-P	NS	96.0	96.0/96.0	96.0/96.0	24.0/0.0	576.0
Tomás et al. (2017)	IL17A, IFNγ	74/73	H/M-S-Ge-CP	0.974	92.5	90.4/94.6	94.3/90.9	16.7/0.1	165.0
Arias-Bujanda et al. (2018)	IL17A, IL2 (ratio)	61/32	H/M-S-Ge-CP (non-smokers)	0.857	82.3	87.7/77.0	79.0/86.4	3.8/0.1	23.8
Arias-Bujanda et al. (2018)	IL17A, IL2 (ratio)	13/41	H/M-S-Ge-CP (smokers)	0.955	84.1	99.3/68.9	76.0/99.0	3.1/0.0	323.7

The performance values of each study were recalculated on the basis of the original classification tables, which were either provided directly in the paper or were calculated
IL interleukin, *IFN* interferon, *a* active, *t* total, *l* latent, *COL* collagenase, *LPS* lipopolysaccharide, *PG* prostaglandin, *MMP* matrix metalloproteinase, *OPG* osteoprotegerin, *OA* osteoactivin, *No.* number, *CC* control condition, *TC* target condition, *H* periodontally healthy, *M* moderate, *S* severe, *Ge* generalised, *CP* chronic periodontitis, *CAL* clinical attachment loss, *P* periodontitis, *AUC* area under a curve, *NS* not specified, *ACC* accuracy, *SEN S* sensitivity, *SPEC* specificity, *PPV* positive predictive value, *NPV* negative predictive value, *LR* likelihood ratio, *DOR* diagnostic odds ratio

IL1alpha and 60.4% for IL1beta due to low sensitivity values (15.4% and 23.1%, respectively).

The Tomás's series also revealed for the first time the high discriminatory power of IL17A in GCF (AUC value of 0.937): for periodontitis, the sensitivity was 89.0%, and for periodontal health, the specificity was 89.2%. As with IL1, this discriminatory potential of IL17A was enhanced when it was combined with an antiinflammatory cytokine (IFNgamma): the AUC value increased to 0.974 and the sensitivity and specificity to 90.4% and 94.6%, respectively (Tomás et al. 2017). In relation to other cytokines, Baeza et al. (2016) demonstrated the greater diagnostic accuracy of IL6 over TNFalpha (85.5% *vs* 67.7%), with the first biomarker having a sensitivity of 83.9% and the second only 45.2%.

Of the various methodological factors affecting the diagnostic accuracy of a biomarker in the GCF, the smoking status is of particular interest. Indeed, several studies have revealed that cytokines are present at different levels in the oral fluids of smokers and non-smokers with periodontitis (Javed et al. 2020; Rathnayake et al. 2013; Toker et al. 2012; Tymkiw et al. 2011), which could have a significant effect on the accuracy of cytokine models for the diagnosis of periodontitis. However, despite the inclusion of both non-smokers and smokers in some of the series included in the present chapter, only one study – published by our research group – took this variable into account, with adjustments made to its predictive models and performance measures (Tomás et al. 2017); later, we also produced specific models for use in non-smokers and smokers (Arias-Bujanda et al. 2018). If the smoking status is taken into account in models using individual cytokines, the accuracy of IL1alpha and IL1beta is similar in non-smokers and smokers (IL1alpha: 93.5% and 91.7%; IL1beta: 94.6% and 94.4%); with IL17A, the accuracy is lower for non-smokers (88.2% *vs* 92.6% in smokers). However, these three proinflammatory cytokines are less able to detect periodontitis in non-smokers (sensitivities of 87.5%, 90.6% and 78.1%, respectively, *vs* 98.8%, 97.6% and 95.1%); meanwhile, in smokers, they are less able to detect periodontal health (specificities of 69.2%, 84.6% and 84.6%, respectively, *vs* 96.7%, 96.7% and 93.4%) (Arias-Bujanda et al. 2018).

15.6 Cytokines in Saliva for the Diagnosis of Periodontitis: Single or Multiple Biomarkers

Like the literature on cytokines in GCF, there is little research on using salivary biomarkers to diagnose periodontitis and distinguish it from other periodontal conditions (periodontal health or gingivitis): IL1beta is the most studied, followed by IL6 (Table 15.3). As with GCF studies, cytokines in saliva were usually quantified using ELISA or multiparametric cytometry techniques, and predictive models were frequently obtained using logistic regression or linear discriminant analysis techniques.

The capacity of salivary IL1beta to distinguish a periodontally healthy subject from one with the disease varies, with AUC values varying from 0.960 (Sánchez et al. 2013) to 0.689 (Arias-Bujanda et al. 2020). Meanwhile, sensitivity ranges from 88.0% (Ebersole et al. 2013b) to 53.8% (Ramseier et al. 2009) and specificity from 96.8% (Sánchez et al. 2013) to 51.9% (Wu et al. 2018). The discriminatory potential of IL1beta in saliva could therefore be interpreted as varying from outstanding to acceptable (Hosmer et al. 2013). This discrepancy may be due to the studies' dissimilarities in relation to the different control and periodontal patient groups analyzed. Indeed, it is acknowledged in the diagnostic accuracy field that the spectrum of clinical conditions directly affects estimations of sensitivity and specificity (Reistma et al. 2009).

Another observation is the fact that all of the three studies that describe AUC values $\geq$0.950 for IL1beta in saliva (Afacan et al. 2018; Ebersole et al. 2013b; Sánchez et al. 2013) used fewer controls than periodontitis patients ($\leq$30 individuals). It is interesting that the specificity values detected in the three studies were not only very high, but also higher than the sensitivity values (Afacan et al. 2018; Ebersole et al. 2013b; Sánchez et al. 2013). This is probably due to the

Table 15.3 Characteristics of diagnostic accuracy studies on single cytokines in saliva, including performance values

Study	Index test	No. CC/No. TC	Type CC/Type TC	Threshold	AUC	ACC (%)	SENS/SPEC (%)	PPV/NPV (%)	LR+/LR-	DOR
Ramseier et al. (2009)	IL1β	40/39	H_G/Mi-M-S-CP	235.8 (pg/ml)	0.720	54.4	53.8/55.0	53.8/55.0	1.2/0.8	1.4
Ebersole et al. (2013b)	IL1β	30/50	H/CP	17.8 (pg/ml)	0.950	90.0	88.0/93.3	95.7/82.4	13.2/0.1	102.7
Sánchez et al. (2013)	IL1β	15/59	H/Mi-M-S-CP	212.0 (pg/ml)	0.960	81.9	78.0/96.8	98.9/53.6	24.2/0.2	106.2
Ebersole et al. (2015)	IL1β	108/101	H_G/P	24.0 (pg/ml)	0.830	75.6	75.2/75.9	74.5/76.6	3.1/0.3	9.6
Afacan et al. (2018)	IL1β	20/40	H/CP_Ge-AgP	445.7 (pg/mg)	0.950	83.3	80.0/90.0	94.1/69.2	8.0/0.2	36.0
Wu et al. (2018)	IL1β	27/30	Non-P/P	71.5 (pg/ml)	NS	68.4	83.3/51.9	65.8/73.7	1.7/0.3	5.4
Arias-Bujanda et al. (2020)	IL1β	40/36	H/Non-TP (non-smokers)	84.76 (pg/ml)	0.830	77.6	72.2/82.5	78.8/76.7	4.1/0.3	12.2
Arias-Bujanda et al. (2020)	IL1β	22/43	H/Non-TP (smokers)	42.78 (pg/ml)	0.689	70.8	67.4/77.3	85.3/54.8	2.9/0.4	7.0
Ramseier et al. (2009)	IL6	40/39	H_G/Mi-M-S-CP	22.4 (pg/ml)	0.710	59.5	59.0/60.0	59.0/60.0	1.5/0.7	2.2
Ebersole et al. (2013b)	IL6	30/50	H/CP	7.5 (pg/ml)	0.950	91.3	88.0/96.7	97.8/82.9	26.4/0.1	212.7
Ebersole et al. (2015)	IL6	108/101	H_G/P	5.11 (pg/ml)	0.849	78.5	78.2/78.7	77.5/79.4	3.7/0.3	13.3
Wu et al. (2018)	IL6	27/30	Non-P/P	3.7 (pg/ml)	NS	50.9	53.3/48.1	53.3/48.1	1.0/1.0	1.1
Al-Sabbagh et al. (2012)	MIP1α	40/40	H/M-S-Ge-CP	1.12 (pg/ml)	0.940	93.8	95.0/92.5	92.7/94.9	12.7/0.1	234.3
Ebersole et al. (2015)	MIP1α	108/101	H_G/P	3.28 (pg/ml)	0.723	67.0	66.3/67.6	65.7/68.2	2.0/0.5	4.1
Ebersole et al. (2013b)	TNFα	30/50	H/CP	NS (pg/ml)	0.630	65.0	60.0/73.3	78.9/52.4	2.3/0.5	4.1
Wu et al. (2018)	TNFα	27/30	Non-P/P	8.10 (pg/ml)	NS	36.8	23.3/51.9	35.0/37.8	0.5/1.5	0.3
Ebersole et al. (2013b)	IFNα	30/50	H/CP	NS (pg/ml)	0.750	70.8	54.0/98.4	98.2/56.6	32.9/0.5	70.4
Wu et al. (2018)	IL1ra	27/30	Non-P/P	8046.1 (pg/ml)	NS	36.8	23.3/51.9	35.0/37.8	0.5/1.5	0.3
Wu et al. (2018)	IL8	27/30	Non-P/P	478.0 (pg/ml)	NS	50.9	50.0/51.9	53.6/48.3	1.0/1.0	1.1
Ramseier et al. (2009)	IL10	40/39	H_G/Mi-M-S-CP	520.9 (pg/ml)	0.680	54.4	53.8/55.0	53.8/55.0	1.2/0.8	1.4

The performance values of each study were recalculated on the basis of the original classification tables, which were either provided directly in the paper or were calculated
IL interleukin, *MIP* macrophage inflammatory protein, *TNF* tumor necrosis factor, *IFN* interferon, *ra* receptor antagonist, *No.* number, *CC* control condition, *TC* target condition, *H* periodontally healthy, *G* gingivitis, *Mi* mild, *M* moderate, *S* severe, *CP* chronic periodontitis, *P* periodontitis, *Ge* generalised, *Ag* aggressive, *T* treated, *pg/ml* picograms/milliliter, *NS* not specified, *AUC* area under a curve, *ACC* accuracy, *SEN S* sensitivity, *SPEC* specificity, *PPV* positive predictive value, *NPV* negative predictive value, *LR* likelihood ratio, *DOR* diagnostic odds ratio

small sample size, which directly affects the AUC's discriminatory capacity. Confirming this, the research group of Ebersole et al. (2015) followed up their 2013 study by re-evaluating their sample of more than 100 clinically similar controls and periodontitis patients. In this second paper, the salivary IL1beta AUC values were much lower than those initially identified by the authors in their first investigation (0.830 *vs* 0.950) (Ebersole et al. 2013b; Ebersole et al. 2015).

Also, of the three studies in which the IL1beta AUC values are ≥0.950 (Afacan et al. 2018; Ebersole et al. 2013b, Sánchez et al. 2013), the samples in two of them include only non-smokers (Afacan et al. 2018; Sánchez et al. 2013); in other investigations with lower AUC values (≤0.800), groups of both non-smokers and smokers were examined (Isaza-Guzmán et al. 2017; Ramseier et al. 2009). Given the possible influence of smoking, Arias-Bujanda et al. (2020) conducted a study which found that IL1beta salivary levels have an excellent capacity to distinguish between untreated periodontitis and periodontal health, although this potential was notably reduced in patients who smoke (AUC values of 0.830 vs 0.689, respectively: sensitivity and specificity values of 72.2% and 82.5% vs 67.4% and 77.3%). The capacity of salivary IL1beta to discriminate untreated from treated periodontitis was acceptable, although, somewhat curiously, to a slightly lesser extent in non-smokers (AUC values of 0.671 and 0.708, respectively, in smokers) (Arias-Bujanda et al. 2020).

As with IL1beta, the findings on the predictive potential of salivary IL6 for distinguishing periodontitis from other clinical conditions (periodontal health or gingivitis) are fairly heterogeneous: while the work of some researchers produced very low accuracy scores (below 60.0%) (Ramseier et al. 2009; Wu et al. 2018), that of others identified accuracies of 78.5%, and even 91.3% (Ebersole et al. 2013b, Ebersole et al. 2015). The promising results of Al-Sabbagh et al. (2012) on macrophage inflammatory protein (MIP) 1alpha deserve to be highlighted: their AUC and accuracy values were 0.940 and 93.8%, respectively (sensitivity of 95.0% and specificity of 92.5%). Disappointingly, however, these results could not be replicated by later investigations (Ebersole et al. 2015). In relation to the diagnostic capability of other cytokines, IFNalpha has been found to produce the best accuracy scores (70.8%) (Ebersole et al. 2013b); the results for TNFalpha are mixed (65.0% and 36.8%) (Ebersole et al. 2013b; Wu et al. 2018), while the least accurate cytokines have been identified as IL1 receptor antagonist (ra), IL8 and IL10 (36.8%, 50.9% and 54.4%, respectively) (Ramseier et al. 2009; Wu et al. 2018).

The most studied combinations of salivary biomarkers for use to distinguish periodontitis from periodontal health or gingivitis are IL1beta with IL6 and MMP8, followed by IL6 with MMP8 (Table 15.4). The results published thus far come from studies carried out by the same research group, which produced better classification parameters in their 2013 series than in that from 2015 (Ebersole et al. 2013b, Ebersole et al. 2015). The best two-biomarker model was found to be that involving IL1beta and IL6, which produced an AUC of 0.983 (accuracy, sensitivity and specificity of 95.0%, 94.0% and 96.7%, respectively); the next most accurate came from IL6 and MMP8, with an AUC of 0.975 (accuracy of 95.7%, sensitivity of 94.0% and specificity of 98.4%), and IL1beta and MMP8 (AUC of 0.963; accuracy, sensitivity and specificity of 91.3%, 88.0% and 96.7%) (Ebersole et al. 2013b). These three predictive models can all be described as having an outstanding discrimination capacity (Hosmer et al. 2013). In contrast, the model with MIPalpha and MMP8 produced the worst results for distinguishing periodontitis from gingivitis (accuracy of 63.9%, sensitivity of 64.4%, specificity of 62.8%) (Ebersole et al. 2015).

The most studied salivary multibiomarker model for predicting periodontitis is that formed by IL1beta, IL6 and MMP8, to which MIP1alpha was added (Table 15.5). Interestingly, in general terms, none of the more than two-molecule combinations studied thus far have produced diagnostic accuracy results that are substantially better than those from models using two salivary biomarkers (accuracy from 95.0% to 72.9%, sensitivity from 94.0% to 60.0%, and specificity from 96.7% to 73.8%) (Ebersole et al. 2013b; Ebersole et al. 2015; Wu et al. 2018).

Table 15.4 Characteristics of diagnostic accuracy studies on combination of two biomarkers with at least one cytokine in saliva, including performance values

Study	Index test	No. CC/No. TC	Type CC/Type TC	AUC	ACC (%)	SENS/SPEC (%)	PPV/NPV (%)	LR+/LR-	DOR
Ebersole et al. (2013b)	IL1β, IL6	30/50	H/CP	0.983	95.0	94.0/96.7	97.9/90.6	28.2/0.0	454.3
Ebersole et al. (2015)	IL1β, IL6	65/101	H/P	NS	79.5	81.2/76.9	84.5/72.5	3.5/0.2	14.3
Ebersole et al. (2015)	IL1β, IL6	43/101	G/P	NS	77.1	77.2/76.7	88.6/58.9	3.3/0.2	11.1
Ebersole et al. (2015)	IL1β, MIP1α	65/101	H/P	NS	77.7	79.2/75.4	83.3/70.0	3.2/0.2	11.6
Ebersole et al. (2015)	IL1β, MIP1α	43/101	G/P	NS	76.4	75.2/79.1	89.4/57.6	3.5/0.3	11.4
Ebersole et al. (2013b)	IL1β, MMP8	30/50	H/CP	0.963	91.3	88.0/96.7	97.8/82.9	26.4/0.1	212.6
Ebersole et al. (2015)	IL1β, MMP8	65/101	H/P	NS	77.7	78.2/76.9	84.0/69.4	3.3/0.2	11.9
Ebersole et al. (2015)	IL1β, MMP8	43/101	G/P	NS	75.0	76.2/72.1	86.5/56.4	2.7/0.3	8.2
Ebersole et al. (2015)	IL6, MIP1α	65/101	H/P	NS	78.3	79.2/76.9	84.2/70.4	3.4/0.2	12.6
Ebersole et al. (2015)	IL6, MIP1α	43/101	G/P	NS	77.8	81.2/69.8	86.3/61.2	2.6/0.2	9.9
Ebersole et al. (2013b)	IL6, MMP8	30/50	H/CP	0.975	95.7	94.0/98.4	98.9/90.9	57.3/0.0	940.0
Ebersole et al. (2015)	IL6, MMP8	65/101	H/P	NS	77.7	81.2/72.3	82.0/71.2	2.9/0.2	11.2
Ebersole et al. (2015)	IL6, MMP8	43/101	G/P	NS	76.4	78.2/72.1	86.8/58.5	2.8/0.3	9.2
Ebersole et al. (2015)	MIP1α, MMP8	65/101	H/P	NS	72.9	70.3/76.9	82.6/62.5	3.0/0.3	7.8
Ebersole et al. (2015)	MIP1α, MMP8	43/101	G/P	NS	63.9	64.4/62.8	80.2/42.9	1.7/0.5	3.0

The performance values of each study were recalculated on the basis of the original classification tables, which were either provided directly in the paper or were calculated
IL interleukin, *MIP* macrophage inflammatory protein, *MMP* matrix metalloproteinase, *No.* number, *CC* control condition, *TC* target condition, *H* periodontally healthy, *CP* chronic periodontitis, *P* periodontitis, *G* gingivitis, *AUC* area under the curve, *NS* not specified, *ACC* accuracy, *SENS* sensitivity, *SPEC* specificity, *PPV* positive predictive value, *NPV* negative predictive value, *LR* likelihood ratio, *DOR* diagnostic odds ratio

Table 15.5 Characteristics of diagnostic accuracy studies on combination of more than two biomarkers with at least one cytokine in saliva, including performance values

Study	Index test	No. CC/No. TC	Type CC/Type TC	AUC	ACC (%)	SENS/SPEC (%)	PPV/NPV (%)	LR+/LR-	DOR
Ebersole et al. (2015)	IL1β, IL6, MIP1α	65/101	H/P	NS	79.5	80.2/78.5	85.3/71.8	3.7/0.2	14.7
Ebersole et al. (2015)	IL1β, IL6, MIP1α	43/101	G/P	NS	78.5	78.2/79.1	89.8/60.7	3.7/0.2	13.5
Ebersole et al. (2013b)	IL1β, IL6, MMP8	30/50	H/CP	0.984	95.0	94.0/96.7	97.9/90.6	28.2/0.0	454.3
Ebersole et al. (2015)	IL1β, IL6, MMP8	65/101	H/P	NS	77.7	78.2/76.9	84.0/69.4	3.3/0.2	11.9
Ebersole et al. (2015)	IL1β, IL6, MMP8	43/101	G/P	NS	75.7	76.2/74.4	87.5/57.1	2.9/0.3	9.3
Wu et al. (2018)	IL1β, IL1ra, MMP9	27/30	Non-P/P	0.853	80.7	73.3/88.9	88.0/75.0	6.6/0.3	22.0
Wu et al. (2018)	IL1β, IL1ra, TNFα	27/30	Non-P/P	0.838	78.9	70.0/88.9	87.5/72.7	6.3/0.3	18.6
Wu et al. (2018)	IL1β, TNFα, MMP9	27/30	Non-P/P	0.826	77.2	60.0/96.3	94.7/68.4	16.2/0.4	39.0
Ebersole et al. (2015)	IL1β, IL6, MIP1α, MMP8	65/101	H/P	NS	78.3	78.2/78.5	84.9/69.9	3.6/0.2	13.0
Ebersole et al. (2015)	IL1β, IL6, MIP1α, MMP8	43/101	G/P	NS	78.5	78.2/79.1	89.8/60.7	3.7/0.2	13.5
Nagarajan et al. (2015)	IL1β, IL6, MIP1α, MMP8	40/40	G/P	NS	77.5	77.5/77.5	77.5/77.5	3.4/0.2	11.8
Wu et al. (2018)	IL1β, IL1ra, TNFα, MMP9	27/30	Non-P/P	0.880	82.5	76.7/88.9	88.5/77.4	6.9/0.2	26.2
Ebersole et al. (2015)	IL6, MIP1α, MMP8	65/101	H/P	NS	77.1	79.2/73.8	82.5/69.6	3.0/0.2	10.7
Ebersole et al. (2015)	IL6, MIP1α, MMP8	43/101	G/P	NS	72.9	72.3/74.4	86.9/53.3	2.8/0.3	7.5
Wu et al. (2018)	IL1ra, TNFα, MMP9	27/30	Non-P/P	0.821	78.9	80.0/77.8	80.0/77.8	3.6/0.2	14.0
Inönü et al. (2020)	IL17, Del1, LFA1	90/90	H_G/ CP_Ge-AgP	0.893	83.3	83.3/83.3	83.3/83.3	5.0/0.2	25.0

The performance values of each study were recalculated on the basis of the original classification tables, which were either provided directly in the paper or were calculated
IL interleukin, *MIP* macrophage inflammatory protein, *MMP* matrix metalloproteinase, *ra* receptor antagonist, *TNF* tumor necrosis factor, *Del* developmental endothelial locus, *LFA* lymphocyte function-associated antigen, *No.* number, *CC* control condition, *TC* Target condition, *H* periodontally healthy, *P* periodontitis, *G* gingivitis, *CP* chronic periodontitis, *Ge* generalised, *Ag* aggressive, *AUC* area under the curve, *NS* not specified, *ACC* accuracy, *SENS* sensitivity, *SPEC* specificity, *PPV* positive predictive value, *NPV* negative predictive value, *LR* likelihood ratio, *DOR* diagnostic odds ratio

15.7 Cytokines for the Diagnosis of Periodontitis: Implications for Practice

Tonetti et al. (2018) have recently reiterated that biomarkers play a crucial role in the early detection of periodontitis and evaluations of its severity. Indeed, this view is so accepted that a proposed classification framework has newly introduced validated biomarkers into the case-definition system. Since the biological phenotype of periodontitis is not adequately reflected in the clinical phenotype (Srivastava et al. 2017), diagnoses would be improved if the biological phenotype could be evaluated in clinical practice through the determination of GCF biomarkers.

In addition, a salivary diagnostic tool could be a non-invasive, sensitive, and specific test for use as an adjunct to periodontal patient care and maintenance, as well as for screening large populations (Ghallab 2018; Miller et al. 2010). However, as the natural progression of periodontitis substantially complicates the identification of biomarkers in GCF (Kinane et al. 2017), it is reasonable to assume that this problem will be even worse in saliva. Clearly, therefore, although the advantages of salivary samples are obvious (Miller et al. 2010), their limitations should also be taken into account (Al-Tarawneh et al. 2011; Giannobile 2012; Miller et al. 2010; Srivastava et al. 2017).

Practically, our findings highlight that one of the main requirements when designing predictive biomarker studies of periodontitis is differentiating between subjects based on their smoking habit. The results of our investigations revealed that smokers had lower diagnostic thresholds for IL1beta in both GCF and saliva than non-smokers. The precise definition of diagnostic thresholds represents a first step in the design and construction of periodontitis diagnostic kits for use in clinical practice (Arias-Bujanda et al. 2018; Arias-Bujanda et al. 2020).

An ideal diagnostic test should, of course, have sensitivity and specificity values approaching 100% (Buduneli and Kinane 2011), which has not been possible to date in evaluations of biomarkers for use to diagnose periodontitis. Nevertheless, a first-line test, also known as a triage test, may still be clinically valuable in these circumstances, depending on the steps to be taken after testing (Leeflang 2014). In this sense, it is essential to analyze the predictive values associated with a diagnostic test considering the prevalence of a disease. If we focus on IL1beta, which is the most researched cytokine, and with an assumed periodontitis prevalence of 45% (Eke et al. 2016b; Tonetti et al. 2015a), our predictive results on this biomarker in GCF suggest that would, theoretically, be a highly effective test for detecting periodontitis and periodontal health (Arias-Bujanda et al. 2018). Therefore, if we consider that IL1beta test in GCF was used as a first-line test to decide who should be referred for additional testing (Leeflang 2014), this would mean that the test would be unable to detect disease during initial any screening in only 7.6% of periodontitis patients and only 8.1% of patients would have to undergo unnecessary periodontal testing. In smokers, testing for the presence of IL1beta in GCF would be associated with a reduction in false negatives (2.3%) and, in contrast, a rise in false positives (16.2%) (Fig. 15.1).

Nonetheless, due to the enormous ethiopathogenic complexity of periodontitis, it is very unlikely that a single salivary biomarker can be used to diagnose and predict its evolution (Zhang et al. 2009). In fact, if the salivary IL1beta test was used as a first-line screening tool to determine who should be referred for subsequent evaluation, about 70.8–77.1% of such patients would be identified, with the accuracy of the results depending on their smoking status (Fig. 15.2).

Accordingly, the use of combinations of more than one salivary biomarker may not only produce a more accurate assessment of a patient's periodontal status, but could also be useful for predicting the disease's progression (Ghallab 2018). From the point of view of developing a kit with clinical value, the best combination would be one with the lowest number of biomarkers possible; that is, two biomarkers that are associated with high diagnostic precision. Again, assuming a periodontitis prevalence of 45%, and with the focus on combinations of two of the most researched salivary biomarkers – IL1beta

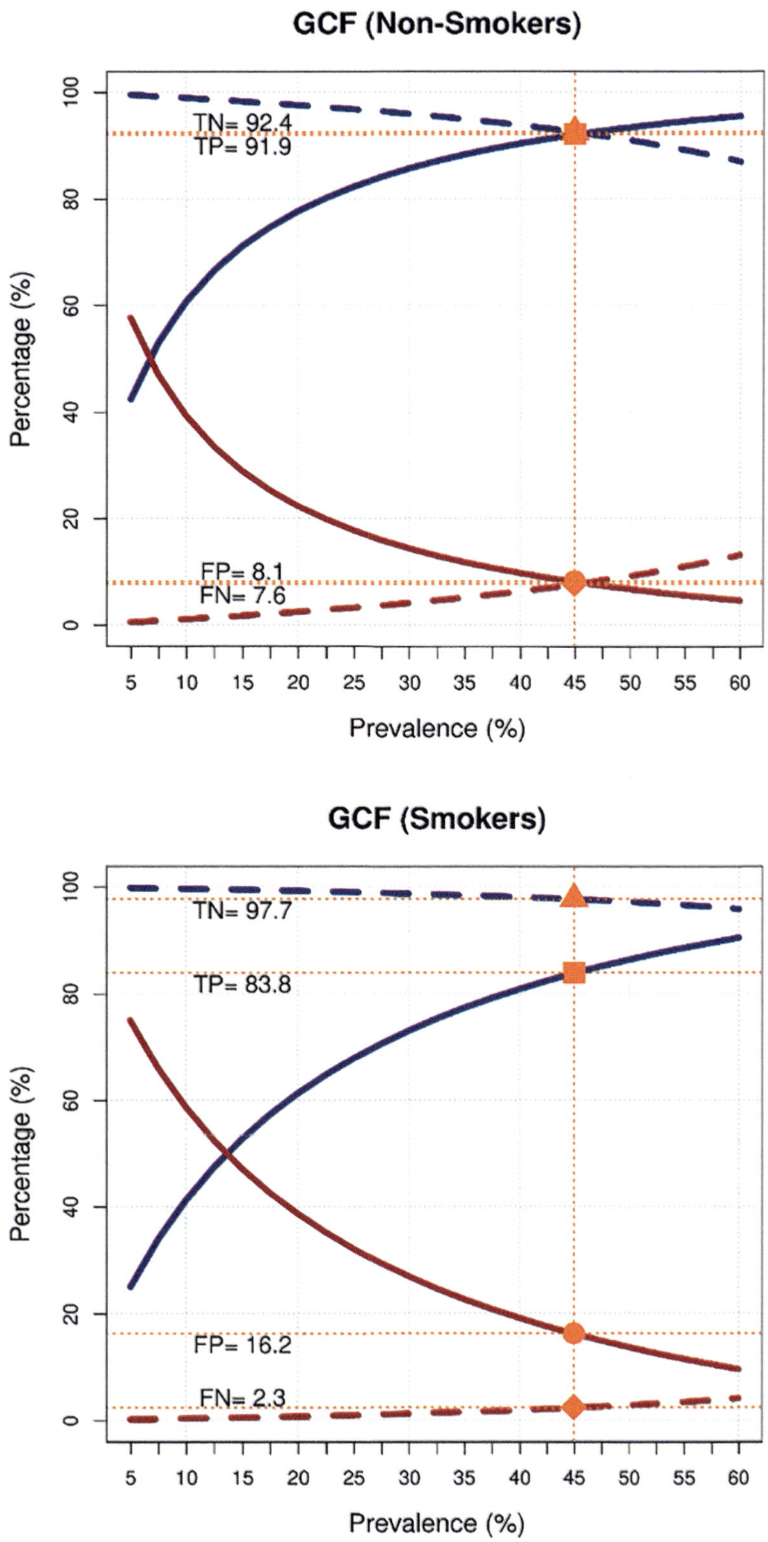

Fig. 15.1 Predictive percentages of IL1beta biomarker in gingival crevicular fluid for different prevalence values of periodontitis in non-smokers and smokers Discontinuous lines indicate the percentage of negative tests for the different prevalences of periodontitis. The continuous lines indicate the percentage of positive tests for the different prevalences of periodontitis. *TP* true positive, test is positive (indicates periodontitis and patient has periodontitis), FP: false positive, test is positive (indicates periodontitis but patient does not have periodontitis), *TN* true negative, test is negative (indicates periodontitis not present and patient does not have periodontitis), *FN* false negative, test is negative (indicates periodontitis not present but patient has periodontitis)

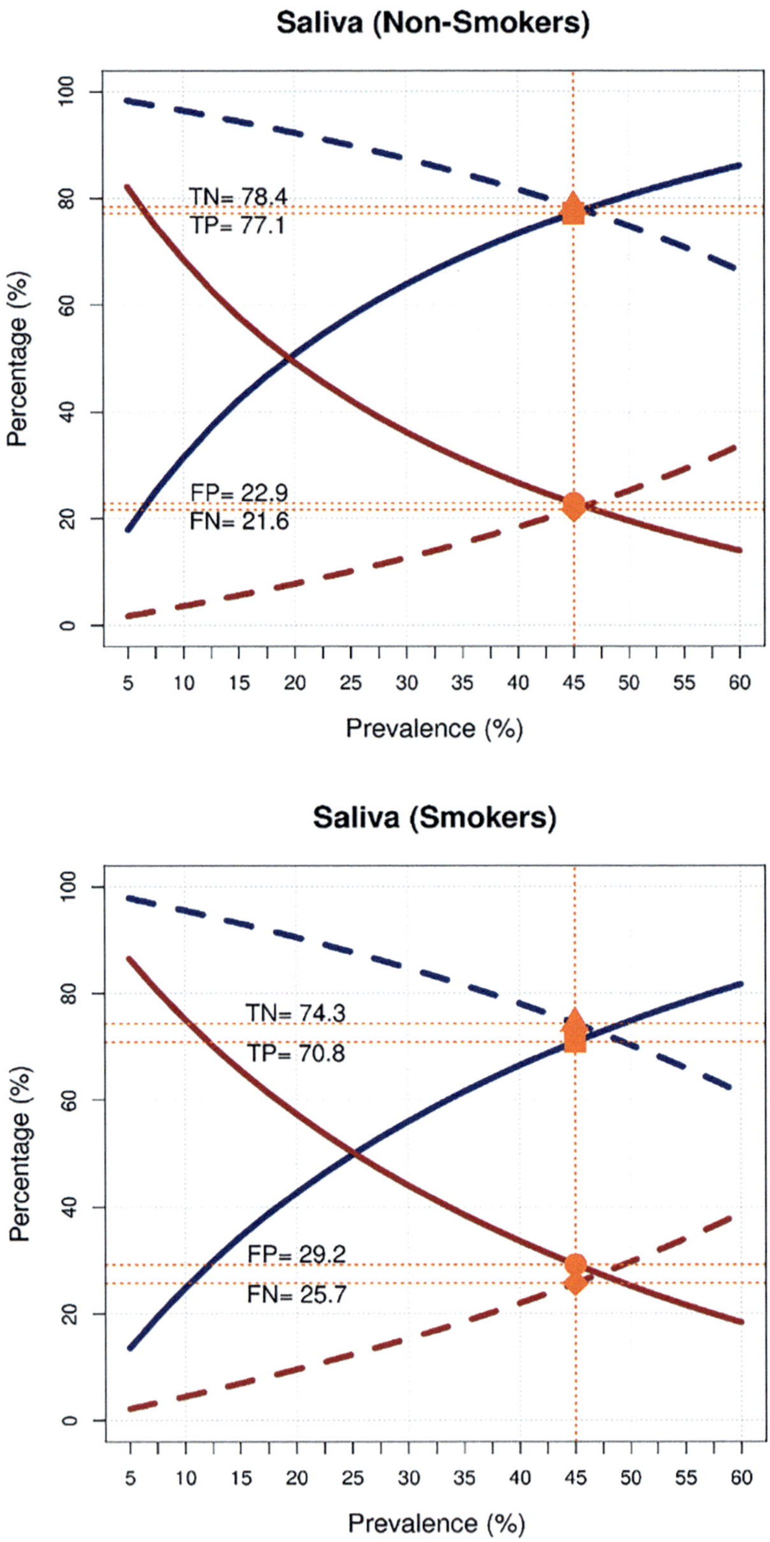

Fig. 15.2 Predictive percentages of IL1beta biomarker in saliva for different prevalence values of periodontitis in non-smokers and smokers
Discontinuous lines indicate the percentage of negative tests for the different prevalences of periodontitis. The continuous lines indicate the percentage of positive tests for the different prevalences of periodontitis. *TP* true positive, test is positive (indicates periodontitis and patient has periodontitis), *FP* false positive, test is positive (indicates periodontitis but patient does not have periodontitis), *TN* true negative, test is negative (indicates periodontitis not present and patient does not have periodontitis), *FN* false negative, test is negative (indicates periodontitis not present but patient has periodontitis)

and IL6; IL6 and MMP8; or IL1beta and MMP8 (Ebersole et al. 2013b; Ebersole et al. 2015) – 79.3–81.2% of patients with periodontitis would be identified; in terms of the total number of negative tests, 83.0–86.5% would detect patients with the non-periodontitis condition.

15.8 Conclusions and Perspectives

The most studied cytokine biomarkers in GCF are IL1alpha and IL1beta, which show outstanding potential when it comes to discriminating periodontitis from periodontal health; the most studied in saliva are IL1beta and IL6, whose potential is outstanding to acceptable. The combination of IL1alpha, IL1beta and IL6 with another molecule seems to improve the discriminatory capability of these cytokines when used individually, which is not overcome by models with more than two biomarkers. Assuming a periodontitis prevalence of 45%, a test of IL1beta in GCF would be highly effective at identifying both periodontitis and non-periodontitis; in saliva, a test based on a combination of two biomarkers, such as IL1beta and IL6, IL6 and MMP 8, or IL1beta and MMP 8, would be required. A patient's smoking/non-smoking status affects the diagnostic accuracy of cytokines in oral fluids and must be considered when determining precise diagnostic thresholds. More generally, additional high-quality research is needed on the accuracy of cytokines in oral fluids, as well as on the effectiveness of cytokine tests for diagnosis of periodontitis.

References

Aarabi G, Zeller T, Seedorf H et al (2017) Genetic susceptibility contributing to periodontal and cardiovascular disease. J Dent Res 96(6):610–617. https://doi.org/10.1177/0022034517699786

Abbayya K, Puthanakar NY, Naduwinmani S et al (2015) Association between periodontitis and Alzheimer's disease. N Am J Med Sci 7(6):241–246. https://doi.org/10.4103/1947-2714.159325

Afacan B, Öztürk VÖ, Emingil G et al (2018) Alarm antiprotease trappin-2 negatively correlates with proinflammatory cytokines in patients with periodontitis. J Periodontol 89(1):58–66. https://doi.org/10.1902/jop.2017.170245

Al-Harthi LS, Cullinan MP, Leichter JW et al (2013) The impact of periodontitis on oral health-related quality of life: a review of the evidence from observational studies. Aust Dent J 58(3):274–277. https://doi.org/10.1111/adj.12076

AlMoharib HS, AlMubarak A, AlRowis R et al (2014) Oral fluid based biomarkers in periodontal disease – part 1. Saliva J Int Oral Health 6(4):95–103

AlRowis R, AlMoharib HS, AlMubarak A et al (2014) Oral fluid-based biomarkers in periodontal disease – part 2. Gingival crevicular fluid. J Int Oral Health 6(5):126–135

Al-Sabbagh M, Alladah A, Lin Y et al (2012) Bone remodeling-associated salivary biomarker MIP-1α distinguishes periodontal disease from health. J Periodontal Res 47(3):389–395. https://doi.org/10.1111/j.1600-0765.2011.01445.x

Al-Tarawneh SK, Border MB, Dibble CF et al (2011) Defining salivary biomarkers using mass spectrometry-based proteomics: a systematic review. OMICS 15(6):353–361. https://doi.org/10.1089/omi.2010.0134

Ara T, Kurata K, Hirai K et al (2009) Human gingival fibroblasts are critical in sustaining inflammation in periodontal disease. J Periodontal Res 44(1):21–27. https://doi.org/10.1111/j.1600-0765.2007.01041.x

Arias-Bujanda N, Regueira-Iglesias A, Alonso-Sampedro M et al (2018) Cytokine thresholds in gingival crevicular fluid with potential diagnosis of chronic periodontitis differentiating by smoking status. Sci Rep 8(1):18003–18012. https://doi.org/10.1038/s41598-018-35920-4

Arias-Bujanda N, Regueira-Iglesias A, Blanco-Pintos T et al (2020) Diagnostic accuracy of IL1β in saliva: the development of predictive models for estimating the probability of the occurrence of periodontitis in non-smokers and smokers. J Clin Periodontol 47(6):702–714. https://doi.org/10.1111/jcpe.13285

Badiger AB, Gowda TM, Chandra K et al (2019) Bilateral interrelationship of diabetes and periodontium. Curr Diabetes Rev 15(5):357–362. https://doi.org/10.2174/1573399815666190115144534

Baeza M, Garrido M, Hernandez-Rios P et al (2016) Diagnostic accuracy for apical and chronic periodontitis biomarkers in gingival crevicular fluid: an exploratory study. J Clin Periodontol 43(1):34–45. https://doi.org/10.1111/jcpe.12479

Barros SP, Williams R, Offenbacher S et al (2016) Gingival crevicular fluid as a source of biomarkers for periodontitis. Periodontol 2000 70(1):53–64. https://doi.org/10.1111/prd.12107

Biomarkers Definitions Working Group (2001) Biomarkers and surrogate endpoints: preferred definitions and conceptual framework. Clin Pharmacol Ther 69(3):89–95. https://doi.org/10.1067/mcp.2001.113989

Borges TF, Regalo SC, Taba M et al (2013) Changes in masticatory performance and quality of life in individuals with chronic periodontitis. J Periodontol 84(3):325–331. https://doi.org/10.1902/jop.2012.120069

Bossuyt PM, Leeflang M (2008) Chapter 6: developing criteria for including studies. In: Deeks JJ, Bossuyt PM, Gatsonis C (eds) Cochrane handbook for systematic reviews of diagnostic test accuracy. The Cochrane Collaboration, pp 1–7

Buduneli N, Kinane DF (2011) Host-derived diagnostic markers related to soft tissue destruction and bone degradation in periodontitis. J Clin Periodontol 38(S11):85–105. https://doi.org/10.1111/j.1600-051X.2010.01670.x

Cekici A, Kantarci A, Hasturk H et al (2014) Inflammatory and immune pathways in the pathogenesis of periodontal disease. Periodontol 2000 64(1):57–80. https://doi.org/10.1111/prd.12002

Chapple ILC, Mealey BL, Van Dyke TE et al (2018) Periodontal health and gingival diseases and conditions on an intact and a reduced periodontium: consensus report of workgroup 1 of the 2017 world workshop on the classification of periodontal and peri-implant diseases and conditions. J Clin Periodontol 45(S20):S68–S77. https://doi.org/10.1111/jcpe.12940

Chatzistavrianou D, Blair F (2017) Diagnosis and management of chronic and aggressive periodontitis part 1: periodontal assessment and diagnosis. Dent Update 44(4):306–315. https://doi.org/10.12968/denu.2017.44.4.306

Darveau RP (2010) Periodontitis: a polymicrobial disruption of host homeostasis. Nat Rev Microbiol 8(7):481–490. https://doi.org/10.1038/nrmicro2337

de Pablo P, Chapple ILC, Buckley CD et al (2009) Periodontitis in systemic rheumatic diseases. Nat Rev Rheumatol 5(4):218–224. https://doi.org/10.1038/nrrheum.2009.28

Dentino A, Lee S, Mailhot J et al (2013) Principles of periodontology. Periodontol 2000 61(1):16–53. https://doi.org/10.1111/j.1600-0757.2011.00397.x

Durham J, Fraser HM, McCracken GI et al (2013) Impact of periodontitis on oral health-related quality of life. J Dent 41(4):370–376. https://doi.org/10.1016/j.jdent.2013.01.008

Ebersole JL, Dawson DR, Morford LA et al (2013a) Periodontal disease immunology: 'double indemnity' in protecting the host. Periodontol 2000 62(1):163–202. https://doi.org/10.1111/prd.12005

Ebersole JL, Schuster JL, Stevens J et al (2013b) Patterns of salivary analytes provide diagnostic capacity for distinguishing chronic adult periodontitis from health. J Clin Immunol 33(1):271–279. https://doi.org/10.1007/s10875-012-9771-3

Ebersole JL, Nagarajan R, Akers D et al (2015) Targeted salivary biomarkers for discrimination of periodontal health and disease(s). Front Cell Infect Microbiol 5:62. https://doi.org/10.3389/fcimb.2015.00062

Eke PI, Wei L, Borgnakke WS et al (2016a) Periodontitis prevalence in adults $\geq$ 65 years of age, in the USA. Periodontol 2000 72(1):76–95. https://doi.org/10.1111/prd.12145

Eke PI, Zhang X, Lu H et al (2016b) Predicting periodontitis at state and local levels in the United States. J Dent Res 95(5):515–522. https://doi.org/10.1177/0022034516629112

Ferreira MC, Dias-Pereira AC, Branco-de-Almeida LS et al (2017) Impact of periodontal disease on quality of life: a systematic review. J Periodontal Res 52(4):651–665. https://doi.org/10.1111/jre.12436

Finoti LS, Nepomuceno R, Pigossi SC et al (2017) Association between interleukin-8 levels and chronic periodontal disease: a PRISMA-compliant systematic review and meta-analysis. Medicine 96(22):e6932. https://doi.org/10.1097/MD.0000000000006932

Garlet GP (2010) Destructive and protective roles of cytokines in periodontitis: a re-appraisal from host defense and tissue destruction viewpoints. J Dent Res 89(12):1349–1363. https://doi.org/10.1177/0022034510376402

Ghallab NA (2018) Diagnostic potential and future directions of biomarkers in gingival crevicular fluid and saliva of periodontal diseases: review of the current evidence. Arch Oral Biol 87:115–124. https://doi.org/10.1016/j.archoralbio.2017.12.022

Giannobile WV (2012) Salivary diagnostics for periodontal diseases. J Am Dent Assoc 143(10 suppl):6S–11S. https://doi.org/10.14219/jada.archive.2012.0341

Gopalakrishna G, Langendam MW, Scholten RJ et al (2016) Defining the clinical pathway in cochrane diagnostic test accuracy reviews. BMC Med Res Methodol 16(1):153. https://doi.org/10.1186/s12874-016-0252-x

Hobbins S, Chapple I, Sapey E et al (2017) Is periodontitis a comorbidity of COPD or can associations be explained by shared risk factors/behaviors? Int J Chron Obstruct Pulmon Dis 12:1339–1349. https://doi.org/10.2147/COPD.S127802

Hosmer DW, Lemeshow S, Sturdivant RX (2013) Applied logistic regression, 3rd edn. John Wiley & Sons. https://doi.org/10.1002/9781118548387

Huang W, Wu J, Mao Y et al (2020) Developing a periodontal disease antibody array for the prediction of severe periodontal disease using machine learning classifiers. J Periodontol 91(2):232–243. https://doi.org/10.1002/JPER.19-0173

Inönü E, Kayis SA, Eskan MA et al (2020) Salivary Del-1, IL-17, and LFA-1 levels in periodontal health and disease. J Periodontal Res 55(4):511–518. https://doi.org/10.1111/jre.12738

Isaza-Guzmán DM, Medina-Piedrahíta VM, Gutiérrez-Henao C et al (2017) Salivary levels of NLRP3 inflammasome-related proteins as potential biomarkers of periodontal clinical status. J Periodontol 88(12):1329–1338. https://doi.org/10.1902/jop.2017.170244

Iwasaki M, Sato M, Minagawa K et al (2015) Longitudinal relationship between metabolic syndrome and periodontal disease among Japanese adults aged $\geq$70 tears: the Niigata Study. J Periodontol 86(4):491–498. https://doi.org/10.1902/jop.2015.140398

Jaedicke KM, Preshaw PM, Taylor JJ (2016) Salivary cytokines as biomarkers of periodontal diseases. Periodontol 2000 70(1):164–183. https://doi.org/10.1111/prd.12117

Jansson H, Wahlin Å, Johansson V et al (2014) Impact of periodontal disease experience on oral health-related quality of life. J Periodontol 85(3):438–445. https://doi.org/10.1902/jop.2013.130188

Javed F, Al-Zawawi AS, Allemailem KS et al (2020) Periodontal conditions and whole salivary IL-17A and −23 levels among young adult cannabis sativa (Marijuana)-smokers, heavy cigarette-smokers and non-smokers. Int J Environ Res Public Health 17(20):7435. https://doi.org/10.3390/ijerph17207435

Ji S, Choi Y (2015) Point-of-care diagnosis of periodontitis using saliva: technically feasible but still a challenge. Front Cell Infect Microbiol 5:65. https://doi.org/10.3389/fcimb.2015.00065

Kassebaum NJ, Bernabé E, Dahiya M et al (2014) Global burden of severe periodontitis in 1990-2010: a systematic review and meta-regression. J Dent Res 93(11):1045–1053. https://doi.org/10.1177/0022034514552491

Kinane DF, Stathopoulou PG, Papapanou PN (2017) Periodontal diseases. Nat Rev Dis Primers 3(1):17038. https://doi.org/10.1038/nrdp.2017.38

Kitamura M, Kohyama A, Ebisu S et al (1991) Distinction between active and inactive lesion by subgingival plaque bacteria and gingival crevicular fluid. Dent Jpn 28:151–154.

König J, Holtfreter B, Kocher T (2010) Periodontal health in Europe: future trends based on treatment needs and the provision of periodontal services – position paper 1. Eur J Dent Educ 14(S1):4–24. https://doi.org/10.1111/j.1600-0579.2010.00620.x

Korte DL, Kinney J (2016) Personalized medicine: an update of salivary biomarkers for periodontal diseases. Periodontol 2000 70(1):26–37. https://doi.org/10.1111/prd.12103

Kurdukar PA, Kurdukar AA, Mahale SA et al (2015) Biomarkers in gingival crevicular fluid IOSR JDMS 14(10):104–109. https://www.iosrjournals.org/iosr-jdms/papers/Vol14-issue10/Version-9/U014109104109.pdf

Leeflang MM (2014) Systematic reviews and meta-analyses of diagnostic test accuracy. Clin Microbiol Infect 20(2):105–113. https://doi.org/10.1111/1469-0691.12474

McInnes MDF, Moher D, Thombs BD et al (2018) Preferred reporting items for a systematic review and meta-analysis of diagnostic test accuracy studies: the PRISMA-DTA statement. JAMA 319(4):388–396. https://doi.org/10.1001/jama.2017.19163

Meusel DR, Ramacciato JC, Motta RHL et al (2015) Impact of the severity of chronic periodontal disease on quality of life. J Oral Sci 57(2):87–94. https://doi.org/10.2334/josnusd.57.87

Michalowicz BS, Diehl SR, Gunsolley JC et al (2000) Evidence of a substantial genetic basis for risk of adult periodontitis. J Periodontol 71(11):1699–1707. https://doi.org/10.1902/jop.2000.71.11.1699

Miller CS, Foley JD, Bailey AL et al (2010) Current developments in salivary diagnostics. Biomark Med 4(1):171–189. https://doi.org/10.2217/bmm.09.68

Nagarajan R, Miller CS, Dawson D et al (2015) Patient-specific variations in biomarkers across gingivitis and periodontitis. PLoS One 10(9):e0136792. https://doi.org/10.1371/journal.pone.0136792

Preshaw PM, Taylor JJ (2011) How has research into cytokine interactions and their role in driving immune responses impacted our understanding of periodontitis? J Clin Periodontol 38:60–84. https://doi.org/10.1111/j.1600-051X.2010.01671.x

Pussinen P, Paju S, Mantyla P et al (2007) Serum microbial- and host-derived markers of periodontal diseases: a review. Curr Med Chem 14(22):2402–2412. https://doi.org/10.2174/092986707781745604

Ramseier CA, Kinney JS, Herr AE et al (2009) Identification of pathogen and host-response markers correlated with periodontal disease. J Periodontol 80(3):436–446. https://doi.org/10.1902/jop.2009.080480

Rathnayake N, Åkerman S, Klinge B et al (2013) Salivary biomarkers of oral health – a cross-sectional study. J Clin Periodontol 40(2):140–147. https://doi.org/10.1111/jcpe.12038

Reistma JB, Rutjes AWS, Whiting P et al (2009) Chapter 9: assessing methodological quality. In: Deeks JJ, Bossuyt PM, Gatsonis C (eds) Cochrane handbook for systematic reviews of diagnostic test accuracy, 1st edn. The Cochrane Collaboration, pp 1–29

Salvi GE, Lang NP (2005) Host response modulation in the management of periodontal diseases. J Clin Periodontol 32(S6):108–129. https://doi.org/10.1111/j.1600-051x.2005.00785.x

Sánchez GA, Miozza VA, Delgado A et al (2013) Salivary IL-1β and PGE2 as biomarkers of periodontal status, before and after periodontal treatment. J Clin Periodontol 40(12):1112–1117. https://doi.org/10.1111/jcpe.12164

Santuchi CC, Cortelli JR, Cortelli SC et al (2016) Scaling and root planing per quadrant *versus* one-stage full-mouth disinfection: assessment of the impact of chronic periodontitis treatment on quality of life—a clinical randomized, controlled trial. J Periodontol 87(2):114–123. https://doi.org/10.1902/jop.2015.150105

Sanz M, Teughels W (2008) Innovations in non-surgical periodontal therapy: consensus report of the sixth European Workshop on Periodontology. J Clin Periodontol 35(S8):3–7. https://doi.org/10.1111/j.1600-051X.2008.01256.x

Saunders MJ, Stattmiller SP, Kirk KM (2008) Oral health issues in the nutrition of institutionalized elders. J Nutr Elder 26(3–4):39–58. https://doi.org/10.1300/j052v26n03_03

Seymour GJ, Gemmell E (2001) Cytokines in periodontal disease: where to from here? Acta Odontol Scand 59(3):167–173. https://doi.org/10.1080/000163501750266765
Shanbhag S, Dahiya M, Croucher R (2012) The impact of periodontal therapy on oral health-related quality of life in adults: a systematic review. J Clin Periodontol 39(8):725–735. https://doi.org/10.1111/j.1600-051X.2012.01910.x
Sheiham A, Steele JG, Marcenes W et al (2001) Prevalence of impacts of dental and oral disorders and their effects on eating among older people; a national survey in Great Britain. Community Dent Oral Epidemiol 29(3):195–203. https://doi.org/10.1034/j.1600-0528.2001.290305.x
Slots J (2017) Periodontitis: facts, fallacies and the future. Periodontol 2000 75(1):7–23. https://doi.org/10.1111/prd.12221
Srivastava N, Nayak PA, Rana S (2017) Point of care- a novel approach to periodontal diagnosis-a review. J Clin Diagn Res 11(8):ZE01–ZE06. https://doi.org/10.7860/JCDR/2017/26626.10411
Stadler AF, Angst PDM, Arce RM et al (2016) Gingival crevicular fluid levels of cytokines/chemokines in chronic periodontitis: a meta-analysis. J Clin Periodontol 43(9):727–745. https://doi.org/10.1111/jcpe.12557
Stenken JA, Poschenrieder AJ (2015) Bioanalytical chemistry of cytokines – a review. Anal Chim Acta 853:95–115. https://doi.org/10.1016/j.aca.2014.10.009
Taba M, Kinney J, Kim AS et al (2005) Diagnostic biomarkers for oral and periodontal diseases. Dent Clin North Am 49(3):551–571. https://doi.org/10.1016/j.cden.2005.03.009
Taylor JJ (2014) Protein biomarkers of periodontitis in saliva. ISRN Inflamm 2014:593151. https://doi.org/10.1155/2014/593151
Toker H, Akpınar A, Aydın H et al (2012) Influence of smoking on interleukin-1beta level, oxidant status and antioxidant status in gingival crevicular fluid from chronic periodontitis patients before and after periodontal treatment. J Periodontal Res 47(5):572–577. https://doi.org/10.1111/j.1600-0765.2012.01468.x
Tomás I, Arias-Bujanda N, Alonso-Sampedro M et al (2017) Cytokine-based predictive models to estimate the probability of chronic periodontitis: development of diagnostic nomograms. Sci Rep 7(1):11580. https://doi.org/10.1038/s41598-017-06674-2
Tonetti MS, Claffey N, European Workshop in Periodontology group C (2005) Advances in the progression of periodontitis and proposal of definitions of a periodontitis case and disease progression for use in risk factor research. Group C consensus report of the 5th European Workshop in Periodontology. J Clin Periodontol 32(S6):210–213. https://doi.org/10.1111/j.1600-051X.2005.00822.x
Tonetti MS, Chapple ILC, Jepsen S et al (2015a) Primary and secondary prevention of periodontal and peri-implant diseases: introduction to, and objectives of the 11th European Workshop on Periodontology consensus conference. J Clin Periodontol 42(S16):S1–S4. https://doi.org/10.1111/jcpe.12382
Tonetti MS, Eickholz P, Loos BG et al (2015b) Principles in prevention of periodontal diseases: consensus report of group 1 of the 11th European Workshop on Periodontology on effective prevention of periodontal and peri-implant diseases. J Clin Periodontol 42(S16):S5–S11. https://doi.org/10.1111/jcpe.12368
Tonetti MS, Greenwell H, Kornman KS (2018) Staging and grading of periodontitis: framework and proposal of a new classification and case definition. J Clin Periodontol 89(S1):S159–S172. https://doi.org/10.1002/JPER.18-0006
Tymkiw KD, Thunell DH, Johnson GK et al (2011) Influence of smoking on gingival crevicular fluid cytokines in severe chronic periodontitis. J Clin Periodontol 38(3):219–228. https://doi.org/10.1111/j.1600-051X.2010.01684.x
Wu Y, Ning L, Tu Y et al (2018) Salivary biomarker combination prediction model for the diagnosis of periodontitis in a Taiwanese population. J Formos Med Assoc 117(9):841–848. https://doi.org/10.1016/j.jfma.2017.10.004
Yucel-Lindberg T, Båge T (2013) Inflammatory mediators in the pathogenesis of periodontitis. Expert Rev Mol Med 15:e7. https://doi.org/10.1017/erm.2013.8
Zaura E, Brandt BW, Prodan A et al (2017) On the ecosystemic network of saliva in healthy young adults. ISME J 11(5):1218–1231. https://doi.org/10.1038/ismej.2016.199
Zhang L, Henson BS, Camargo PM et al (2009) The clinical value of salivary biomarkers for periodontal disease. Periodontol 2000 51(1):25–37. https://doi.org/10.1111/j.1600-0757.2009.00315.x
Zi MY, Longo PL, Bueno-Silva B et al (2015) Mechanisms involved in the association between periodontitis and complications in pregnancy. Front Public Health 2:290. https://doi.org/10.3389/fpubh.2014.00290

Non-surgical Periodontal Treatment: SRP and Innovative Therapeutic Approaches

16

Alexia Vinel, Antoine Al Halabi, Sébastien Roumi, Hélène Le Neindre, Pierre Millavet, Marion Simon, Constance Cuny, Jean-Sébastien Barthet, Pierre Barthet, and Sara Laurencin-Dalicieux

Abstract

Periodontitis is a major public health problem, that can have local and systemic consequences ranging from tooth loss to the aggravation of other chronic diseases. The consequences of which have an impact on patient's overall general health and quality of life. Periodontal treatments include a large range of techniques and concepts from plaque control to periodontal debridement, surgery and regeneration. Regardless of the treatment proposed, it always begins with the same first essential simple step that is etiological therapy which includes oral hygiene management and the control of periodontal risk factors. The aim of this first step, presented in this chapter, consists mainly in reducing oral bacterial load and inflammation by the means of daily oral hygiene methods and sub-gingival biofilm disruption. Although understanding of the pathogenesis and molecular and cellular mechanisms involved in periodontitis has increased, treatment wise, non-surgical debridement remains the keystone of every periodontal treatment and supportive periodontal therapy. Once risk factors are monitored and plaque control mastered by the patient, root instrumentation can be performed with hand or power-driven instruments. However effective, sub-gingival biofilm disruption has some limits and can be improved with adjunctive therapies such as antiseptics, antibiotics, air polishing or other emerging devices and therapies. Unfortunately, the lack of clear clinical guidelines, concerning these adjunctive therapies, still remains, thus pointing out the necessity of more standardized clinical studies. Also, if some patients can return to a healthy periodontal state, most periodontal patients will remain at periodontal risk for life. Proper assessment of the patient's periodontal risk will help establish correct monitoring of patients successfully treated for their periodontal disease.

Keywords

Periodontal treatment · Periodontitis · Non-surgical therapy · Sub-gingival debride-

A. Vinel
Department of Periodontology, Faculty of Dental Surgery, University of Toulouse III, Toulouse, France

I2MC, INSERM UMR 1297, University of Toulouse III, Toulouse, France
e-mail: vinel.a@chu-toulouse.fr

A. Al Halabi · S. Roumi · H. Le Neindre · P. Millavet
M. Simon · C. Cuny · J.-S. Barthet · P. Barthet
Department of Periodontology, Faculty of Dental Surgery, University of Toulouse III, Toulouse, France

S. Laurencin-Dalicieux (✉)
Department of Periodontology, Faculty of Dental Surgery, University of Toulouse III, Toulouse, France

INSERM UMR 1295, CERPOP, University of Toulouse III, Toulouse, France
e-mail: laurencin.s@chu-toulouse.fr

J. Santi-Rocca (ed.), *Periodontitis*, Advances in Experimental Medicine and Biology 1373,
https://doi.org/10.1007/978-3-030-96881-6_16

ment · Adjunctive therapy · Plaque control · Scaling and root planing

Abbreviations

AAP	American academy of periodontology
AMP	antimicrobial peptide
BOP	aleeding on probing
CAL	clinical attachment level
CAP	cold atmospheric `plasma
CHX	chlorhexidine
CPC	cetilpyridinium chloride
EFP	European federation of periodontology
FMD	full mouth disinfection
HMT	host modulation therapy
MMP	matrix metalloproteinase
OH	oral hygiene
OHI	oral hygiene instructions
PDT	photodynamic therapy
PPD	probing pocket depth
SAP	self-assembling peptides
SDD	subantimicrobial dose doxycycline
SRP	scaling and root planing

Highlights

- Although of great interest, heterogeneity in the studies does not allow to recommend photodynamic therapy as a complement of sub-gingival mechanical debridement.
- Emerging antiseptics like phytotherapy, propolis and boric acid-based gels seem to be promising as adjunctive therapies.
- Cold Atmospheric Plasma applied directly or indirectly could become an interesting adjunctive therapy and even a future alternative to scaling root planing.

Considerations for Practice

- Plaque control is an essential part of periodontal treatment and cannot be avoided.
- Non-surgical sub-gingival debridement remains the keystone of periodontal treatment. Hand and ultrasonic instruments are as effective regarding clinical parameters as well as quadrant-wise and full-mouth procedure approaches.
- Chlorhexidine mouthwashes potentiate mechanical debridement outcomes.

Patient Summary

As a result of excessive bacterial load, periodontitis-related inflammation destructs tooth supporting tissues ultimately resulting in tooth loss. Periodontal treatment aims at controlling bacteria load and reducing inflammation to stop the destruction process. For this purpose, the first treatment step requires the disorganization of the bacterial biofilm colonizing sub-gingival root surfaces by the means of non-surgical mechanical instrumentation. This procedure can only be performed once oral hygiene methods are perfectly controlled to avoid the recolonization of treated surfaces by oral bacteria. To improve mechanical treatment, pharmacological adjuvants can be prescribed like antiseptic mouthwashes and toothpastes, as well as antibiotics.

16.1 Introduction

Periodontal diseases are considered like the 6th most common non-communicable, chronic inflammatory disease in humans. They result from an imbalance between host resistance and a dysbiotic oral microbiota environment. This disruption triggers a chain of immune responses leading to a progressive, more or less severe yet irreversible tissular destruction and *in fine*, tooth loss. Periodontitis is a major public health problem due to its high prevalence ranging from 11.2% to 50% according to its severity (Billings et al. 2018; Kassebaum et al. 2014).

The latest classification on Periodontal and Peri-implant Diseases and Conditions (2017 European Federation of Periodontology and American Academy of Periodontology Consensus Report), introduces the notion of periodontal health as well as peri-implant conditions and states (health, mucositis, peri-implantitis) and reviews a great variety of gingival diseases and conditions not associated with dental plaque but based on their primary etiology. Concerning periodontitis, three forms of periodontitis are identified: necrotizing periodontitis, periodontitis as a manifestation of systemic pathology, and the former "chronic" and "aggressive" forms grouped under a single denomination of "periodontitis". Besides that, the classification of periodontitis is accompanied by a pluri-dimensional characterization according to the severity and complexity of the disease (stages I–IV) and the speed of progression, therapeutic possibilities and impact on the general health of the patient (grades A, B, C) (Papapanou et al. 2018).

Periodontal treatment aims at stabilizing the clinical attachment level, limiting tooth loss and improving oral-health related quality of life (Loos and Needleman 2020). Also, the difference is made between a patient with gingivitis who can return to a healthy periodontal state and a periodontal patient who remains at periodontal risk for life. This distinction was made in order to highlight the need for periodontal supportive therapy and global monitoring of patients successfully treated for their periodontal disease.

The concept of non-surgical periodontal therapy emerged in the 80s. While surgical approaches had been considered inescapable to treat periodontitis so far, randomized controlled clinical trials showed that scaling and root planing associated with oral hygiene instructions were viable options in terms of pocket reduction and attachment levels maintenance (Lang et al. 2019). Even before oral hygiene instructions delivery, patients must be educated about their diagnosis, the etiologies and risk factors of periodontitis as well as treatment modalities and the benefits/risk balance associated. Recent guidelines about the treatment of stage I–III periodontitis indicate that the first step of periodontal therapy aims at changing patients behaviors regarding oral hygiene and risk factors control whereas the second step consists in reducing inflammation and stopping disease progression by disrupting subgingival bacterial biofilm (Sanz et al. 2020).

16.2 Prerequisites for Non-surgical Periodontal Therapy

16.2.1 Oral hygiene Instructions

The control of bacterial load, pillar of periodontal therapy, essentially relies on patient's daily oral hygiene (OH), strongly linked to periodontal inflammation (Chapple et al. 2018; Lang and Bartold 2018). Indeed, insufficient OH combined with uncontrolled periodontal risk factors is associated with treatment failure and disease relapse (Carra et al. 2020; Graziani et al. 2017). OH is based on patient's motivation and manual skills and can be technically difficult in particular in inter-proximal spaces especially since it must be carried on a long term basis (Newton and Asimakopoulou 2015). For those reasons, oral hygiene instructions (OHI) and patient motivation by an oral health professional (dental hygienist, dental surgeon or dentist, oral health specialist) are necessary for efficient self-performed plaque control and patient OH education may be improved by behavioral interventions based on cognitive constructs and motivational interviewing principles. OHI must be delivered and reinforced at all stages of periodontal treatment (Carra et al. 2020) (Fig. 16.1a, b).

At least twice a day, removal of buccal, lingual and occlusal plaque must be performed with a tooth brush and is more effective using power-driven toothbrushes than manual ones (Van der Weijden and Slot 2015). However, tooth brushing does not allow to correctly remove interdental

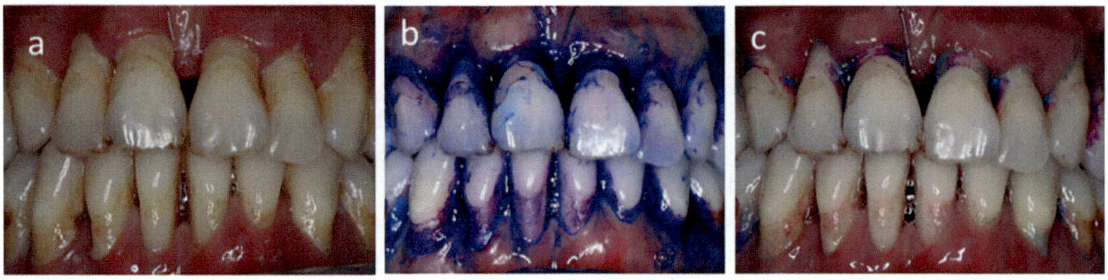

Fig. 16.1 Risk factor control (dental plaque) (**a**) Before Oral Hygiene Instructions; (**b**) After plaque coloration; (**c**) After tooth brushing

plaque from proximal sites (Fig. 16.1c). Interdental cleaning is therefore critical for maintaining OH and interdental brushes are favored as primary choice to execute it (Chapple et al. 2015).

16.2.2 Risk Factor Control

16.2.2.1 General Risk Factors

Smoking and diabetes are two established risk factors of periodontitis and their control is of major importance in periodontal treatment (Papapanou et al. 2018). Dental practitioners must alert their patients about the harmful effects of smoking and encourage them to quit even more when they are suffering from periodontitis. To this aim, smoking cessation may require the help from a tobacco addiction specialist but it has been shown that brief interventions about tobacco use cessation delivered in the dental practice are effective to improve periodontal treatment results (Ramseier et al. 2020).

As for metabolic diseases, diabetes control seems to improve periodontal treatment outcomes and although scientific evidence is insufficient to strongly recommend it, increasing physical activity, dietary counselling and weight loss interventions may participate in ameliorating periodontal therapy (Ramseier et al. 2020; Sanz et al. 2020).

16.2.2.2 Aggravating Local Risk Factors

Local factors have an impact on the periodontal micro-environment and either increase periodontal vulnerability or potentiate dental plaque negative effects. Aggravating local factors include anatomical factors such as dental malposition and radicular proximities, malocclusions, variations in root anatomy, and insufficient keratinized tissue around teeth as well as gingival recessions. Iatrogenic factors are also of importance with inadequate prosthetic or conservative restorations as well as behavioral factors such as mouth breathing, labial or lingual piercings and nail biting. All of the previously cited local factors have an impact on plaque control or directly attack periodontal tissues (Ercoli and Caton 2018; Jepsen et al. 2018). It is of importance to correct or control aggravating local factors when possible to allow optimal oral hygiene. For this purpose, professional mechanical plaque and calculus removal must be realized prior to root instrumentation (Sanz et al. 2020).

16.3 Mechanical Instrumentation

Only once the patient has acquired efficient oral hygiene methods will non-surgical periodontal treatment be performed. It consists in subgingival plaque and calcified deposit removal and its aim is to make root surface biologically compatible to promote periodontal wound healing (Suvan et al. 2020). It was initially considered that the complete removal of contaminated root cementum and reshaping of root surface was necessary to treat periodontitis. The evolution of scientific knowledge about periodontal healing has shifted this approach into a more conservative one and the terms of root planing have progressively been replaced by root debridement (Ciantar 2014;

Graziani et al. 2017). Non-surgical debridement is performed blindly with no way to really quantify biofilm removal; for this reason, treatment success is assessed by clinical signs, mainly reduction of pocket probing depth (PPD), clinical attachment gain and elimination of bleeding on probing (BOP) which vouches for the resolution of inflammation (Corbet et al. 1993; Suvan et al. 2020).

Non-surgical periodontal treatment may be performed with hand or power-driven instruments.

16.3.1 Hand Instrumentation

There is a large offer of hand instruments including curettes, scalers and hoes, and with clinical experience and operative skills improvement, the number of instruments can be reduced (Fig. 16.2). Scalers exhibit two cutting edges on both sides of the working part that can be flat or curved and form a 90 degrees angle with the handle; they have a pointed tip and are mostly used to eliminate supra-gingival calculus in inter-dental regions. Curettes were considered the gold standard for root planing as they allow to remove mineralized calculus and granulation tissue but also to reshape root surface. Curettes have usually one cutting edge, oriented at a 60 or 70 degree angle with the handle and a blunt end to avoid injuring healthy tissues; they must be operated in tension from the end of periodontal pockets towards the crown. Hoes are thin and round designed instruments with a blade oriented at a 99 degree angle with the handle; they are also used in tension towards the coronal part of the tooth. Most hand instruments are made of stainless steel and must be sharpened on a regular basis to allow precise and effective work. Massively used in the 1970s and 1980s, hand instruments have been progressively replaced by powered-driven instruments since the 1990s.

16.3.2 Sonic and Ultrasonic Instruments

16.3.2.1 Sonic Instruments

Sonic instruments operate using compressed air that triggers the metal blade to vibrate, with a 2–8 KHz frequency. The insert has an orbital oscillation and is oriented at 90 ° to the handpiece, allowing it to work parallel to the tooth's major axis. The amplitude motion of the insert is determined by air pressure and cannot be adjusted by the operator and its end is rather thin, leading to potential soft tissue injuries. For these reasons, sonic instruments are preferentially used for supra-gingival biofilm removal (Arabaci et al. 2007) (Table 16.1).

16.3.2.2 Ultrasonic Instruments

The effects of ultrasonic instruments result from the mechanical action of the insert associated with the cavitation of irrigation water. They

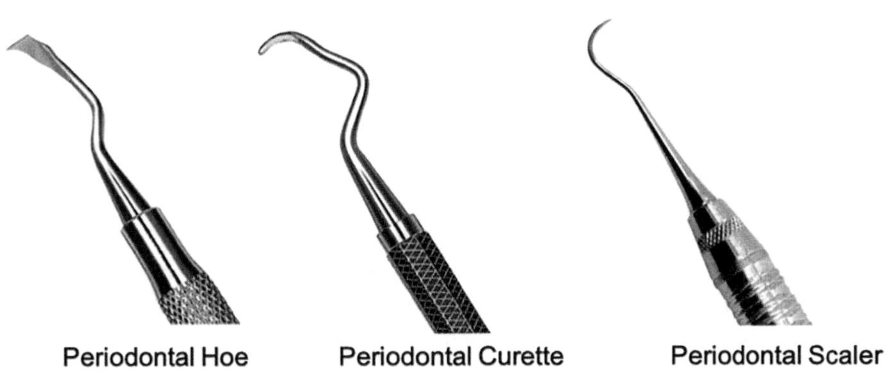

Fig. 16.2 Periodontal hand-instruments

Table 16.1 Comparison of different power-driven instrument technologies

Power-driven instruments	Sonic	Magnetostrictive (Ultrasonic)	Piezoelectric (Ultrasonic)
Transducer (converts energy to vibration)	Pressurized air	Metallic alloy blade	Quartz pastille
Optimal frequency range	2–8 KHz	20–45 kHz	20–35 KHz
Type of movement of the tip of the insert	Orbital oscillation	Elliptique	Linear
Active surface of the insert	Whole tip	Whole tip	Lateral surfaces of the tip

function in an inaudible frequency ranging from 25 to 40 kHz and more, depending on the technology they rely on, magnetostrictive or piezoelectric. Magnetoscriction is the ability of a ferromagnetic material to deform under the influence of a magnetic field. In power-driven instruments, a thin metallic blade placed in the handpiece is exposed to an alternative magnetic field; consecutive dimensional variations result in longitudinal vibrations producing high frequency oscillation of the insert (20–45 kHz). The insert describes an elliptical movement and kinetic energy is distributed in several plans with a large distribution on dental surfaces, allowing the operator to exert little pressure on teeth. In piezoelectricity, a quartz pastille placed in the handpiece exhibits dimensional changes following its exposure to an alternative current, inducing in turn longitudinal vibrations (Lea et al. 2009; Lea et al. 2003, 2004). The high frequency oscillations of the tip (20–35 kHz) obtained are linear and parallel to the instrument's major axis and in consequence perpendicular to the root surface, generating a hammering effect (Laleman et al. 2017). Piezoelectric instruments are operated with a swiping motion on the root surface (Table 16.1).

Ultrasonic instrumentation is associated with a continuous irrigation system that has several roles including insert cooling, debris flushing and cavitation. The latter consists in the formation of gas or steam bubbles that expand and implode as a result of vibrations to form micro-bubbles causing shock waves. Following every oscillation cycle at the tip of the instrument, a micro-bubble ring is formed enhancing biofilm disruption and bacterial dispersion effects of ultrasounds (Vyas et al. 2017).

16.3.2.3 Hand vs. Power-Driven Instrumentation

Hand and power-driven instruments have been compared in numerous studies in the context of non-surgical periodontal treatment. Most studies and meta-analysis do not find any statistically significant difference between the two approaches regarding their efficiency based on plaque and calculus removal or clinical parameters (clinical attachment gain, PPD and BOP reduction) (Krishna and De Stefano 2016; Suvan et al. 2020). A recent study also showed an increase of serum C-reactive protein, interleukin-6 and tumor necrosis factor (markers of inflammation) levels at day 1 followed by a return to baseline at day 7 with no difference between hand and ultrasonic instrumentations (Johnston et al. 2020).

However ultrasonic instruments allow continuous irrigation and flushing while operating, reduce operating time and operator and patient fatigue. They are more efficient than manual instruments to access limited access zones (furcation and deep infra-bony defects) thanks to cavitation effect (Maritato et al. 2018). Also, ultrasonic devices cause less tooth surface alterations in morphology and roughness and remove less cementum than manual curettes.

16.3.3 Full Mouth vs. Quadrant-Wise Scaling

Introduced by Quirynen in 1995, global disinfection appears as a new treatment modality of non-surgical periodontal therapy (Quirynen et al. 1995). Called full mouth disinfection (FMD), it consists in associating mechanical biofilm disorganization and chemical disinfection of the totality of the oral cavity in a 24-h interval. This approach is opposed to the classical non-surgical treatment in which root debridement is realized in four different sessions. The underlying idea is to avoid the recolonization of debrided periodontal pockets by periodontopathogen bacteria still present in the oral cavity (*Aggregatibacter actinomicetemcomitans (Aa), Porphyromonas gingivalis (Pg), Prevotella intermedia (Pi)*). Several studies showed the presence of the latter at different sites in the oral cavity such as the dorsal surface of the tongue, tonsils, saliva and mucosal surfaces, called ecological niches (Beikler et al. 2004; Umeda et al. 2004; Van Winkelhoff et al. 1988). The approach proposed by Quirynen is the following: (1) periodontal pockets instrumentation; (2) tongue's dorsum brushing with 1% chlorhexidine (CHX) gel during 1 min; (3) 0,2% CHX mouthwash during 1 min with 10 s gargle to reach the tonsils; (4) sub-gingival irrigation with 1% CHX gel repeated 3 times during 10 min, operation is renewed 8 days after root debridement; (5) 0,2% CHX mouthwash daily for 14 days following treatment and (6) tongue's dorsum brushing with 1% CHX gel twice daily (Quirynen et al. 1995). In 1999, the same team introduced a modification: home mouthwash duration was extended from 14 days to 2 months (Mongardini et al. 1999).

25 years after Quirynen's publication, 3 therapeutic approaches can be identified, quadrant-wise debridement with or without antiseptics (CHX) (Fig. 16.3), full mouth debridement without antiseptic and full mouth debridement with antiseptics (CHX) (Fig. 16.4). Since the first publication in 1995, several authors have compared the FMD concept to the quadrant-wise approach. The results are almost unanimous, no technique has showed significant statistical superiority and both are as effective in terms of PPD reduction, clinical attachment gain, BOP reduction at 1, 3, 6 and 9 months (Apatzidou and Kinane 2004; Jervøe-Storm et al. 2006; Swierkot et al. 2009; Wennström et al. 2005). At the microbiological level, FMD exhibits a slight advantage compared to the quadrant-wise approach. It is associated with a reduction of spirochetes, mobile bacteria as well as *Pg* and *Pi* concentrations. FMD doesn't seem to have any effect on *Aa* thanks to its ability to penetrate periodontal tissues and the use of antibiotics or a surgical approach are required to reduce its concentration (Quirynen et al. 2000; Roman-Torres et al. 2018). It is also associated with a reduction of the presence of viruses (HCMV, EBV, HSV) and protozoans (*Trichomonas tenax*) (Dubar et al. 2020; Grenier et al. 2009). Two Cochrane meta-analysis published in 2008 and 2015 conclude in the absence of significant difference between the three aforementioned approaches regarding clinical parameters (Eberhard et al. 2008, 2015). This is in adequation with the results from a systematic review published in 2008 (Lang et al. 2008). The authors of the last 3 reviews recommend that the choice between the 3 compared techniques should be made according to patients' preference, health state and tolerance regarding time spent on the dental chair on the one hand and to the practitioners and practice organization on the other hand. It is important to note that FMD is associated with transient bacteriemia and body temperature increase; on the other hand, quadrant-

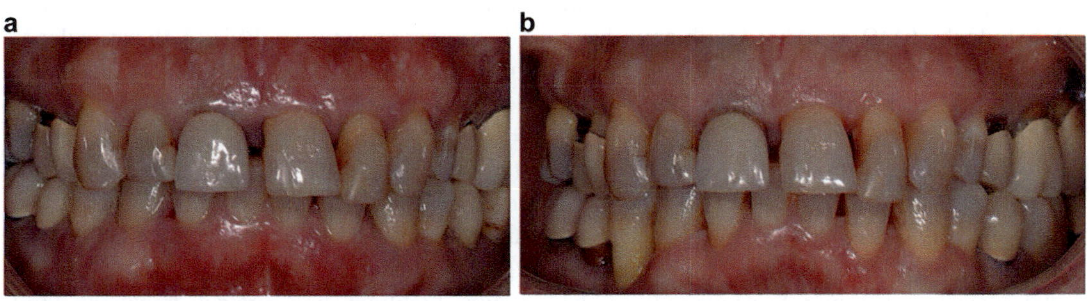

Fig. 16.3 Full-mouth disinfection therapy (**a**) Initial situation; (**b**) after FMD without antiseptics

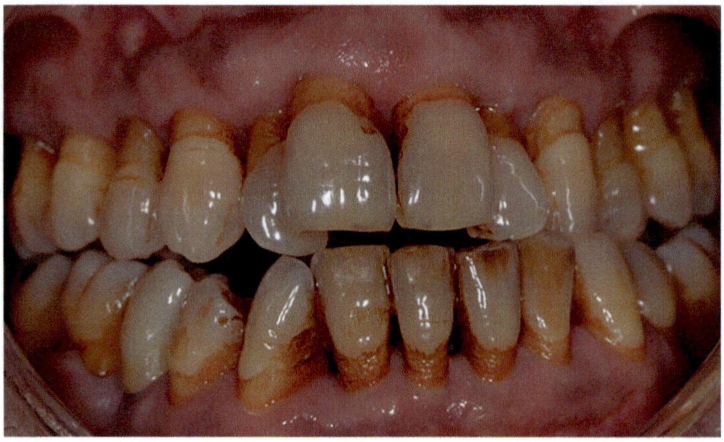

Fig. 16.4 Full-mouth disinfection therapy with antiseptics (0,12% CHX during 2 months). One can note the presence of dark stains resulting from chronic use of CHX

Table 16.2 FMD vs Quadrant-wise sub-gingival debridement

	FMD	Quadrant-wise
Benefits	Subgingival biofilm control by CHX	Short sessions
	One-session treatment	Patient's control and follow-up during several sessions
Disadvantages	Long session for patient and practitioner	Bacterial recolonization
	Transient bacteremia and body temperature increase	Sessions multiplication
	Poor compliance by patients	–
	Unaesthetic coloration and dysgeusia	–

wise approach may favor bacterial recolonization of treated sites by still untreated periodontal pockets (Graziani et al. 2015; Zhang et al. 2013; Zijnge et al. 2010) (Table 16.2). The European Federation of Periodontology (EFP) recently published a systematic review with meta-analysis of non-surgical periodontal treatment comparing full-mouth and quadrant-wise approach concluding that no significant different were observed regarding PPD reduction, clinical attachment gain and pocket closure; however this review did not include the FMD associated with antiseptics approach (Suvan et al. 2020).

16.3.4 Limits

Regarding oral hygiene, careful brushing is insufficient to remove the totality of supra-gingival dental plaque and 1 min brushing allows the removal of only 39% dental plaque (Van der Weijden et al. 1993). However, the control of supra-gingival plaque is absolutely necessary to avoid sub-gingival surface recontamination, justifying professional supportive therapy and the use of chemical agents. In addition, supra-gingival plaque control does not allow for PPD reduction.

Non-surgical periodontal debridement, even when conducted under an operating microscope or magnifying lenses, is realized without full visual access to the periodontal lesions. Thus dental anatomy can make it difficult, in particular for multi-rooted teeth since 58% of first molar furcations are narrower than the curettes used in periodontal treatment (Bower 1979). In a comparative study, Caffesse et al. note that the percentage of dental surfaces completely cleaned after non-surgical root debridement depends on the initial PPD (Caffesse et al. 1986) (Table 16.3). Those results highlight non-surgical treatment limits regarding calculus deposits removal in deep peri-

Table 16.3 Periodontal non-surgical debridement efficacy depending on initial pocket depth (Caffesse et al. 1986)

Pocket depth (mm)	Non-surgical debridement
1–3	86%
4–6	43%
> 6	32%

Table 16.4 Probing pocket depth reduction and periodontal recessions according to initial probing depth after non-surgical treatment (Claffey et al. 2004)

Initial probing depth (mm)	Probing pocket depth reduction (mm)	Periodontal recession (mm)
≤ 3,5	0	0,5
4–6,5	1–2	0–1
≥ 7	2–3	1–2

odontal pockets. In case of stage III or IV periodontitis, Badersten et al., observe the persistence of 26% of 6 mm or more periodontal pockets 24 months after manual non-surgical debridement, and 21% after power-driven non-surgical debridement (Badersten et al. 1984). Claffey et al. also showed that PPD reduction depends on initial probing depth and that periodontal recessions occur after debridement (Table 16.4) (Claffey et al. 2004). A recent review of the literature with meta-analysis shows 6–8 months after non-surgical periodontal treatment a 1,6 mm mean PPD reduction in initially shallow sites (stages I and II periodontitis) and 2,6 mm mean PPD reduction in initially deep sites (stage III and IV periodontitis); the authors conclude that in more severe cases, non-surgical debridement appears to be more efficient in terms of PPD reduction whereas disease stabilization, materialized by pocket closure, is less likely (Suvan et al. 2020). It is also important to keep in mind that the non-surgical approach isn't usually associated with periodontal regenerative techniques.

As discussed above, from a microbiological perspective, periodontal pathogens bacteria, viruses and protozoans may be identified in oral sites others than periodontal pockets (tongue, tonsils…), that could be recontamination sources of treated sites (Dubar et al. 2020; Grenier et al. 2009; Mombelli 2018). Furthermore, insufficient oral hygiene after non-surgical treatment is associated with the novel formation of sub-gingival biofilm, containing spirochetes and labile bacteria within 4–8 weeks (Magnusson et al. 1984). It has been shown that non-surgical debridement, even associated with an antibiotic treatment, does not allow to completely eliminate periodontal pathogens: *Aa,, Pg, Pi, Fusobacterium nucleatum (Fn), Treponema denticola* and *Tannerella forsythia (*Mombelli 2018*)*.

Non-surgical periodontal debridement limits are mainly related to the fact that it is achieved without any visual access to the lesions which is reflected by lower efficiency in deep pockets. That approach does not allow to completely treat periodontitis and a surgical approach remains indicated in case of remaining severe deep pockets, infra-bony and furcation defects. However, periodontal individual and professional supportive therapy with regular debridement as well as the operator's skills and experience in treating periodontitis can improve clinical outcomes and increase patient's compliance as well as prevent periodontal disease recurrence (Kozlovsky et al. 2018).

16.4 Adjunctive and Innovative Therapeutic Approaches

As explained above, periodontal initial therapy aims at reducing local periodontal inflammation and bacterial load *via* biofilm disruption and non-surgical debridement. It can be improved when combined with adjunctive therapy. In the next chapter, we present different approaches that can be implemented to potentiate the effects of non-surgical periodontal debridement.

16.4.1 Air polishing

Described in the 1980s, the concept of air polishing uses pressurized air, water and fine grain size powders to disrupt supra and sub-gingival biofilms. Some air-polishing devices are independent whereas others can be connected to the unit and specific nozzles (Fig. 16.5) have been developed to access sub-gingival contaminated surfaces.

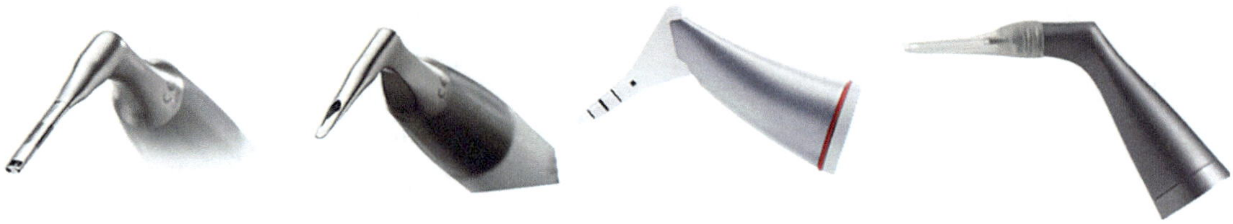

Fig. 16.5 Different subgingival air-polishing device nozzles (**a**) and (**b**) Air-N-Go-Easy-Perio nozzle (Acteon®); (**c**) Perio Flow nozzle (EMS®); (**d**) Subgingival Perio Tip (Mectron®)

Air-polishing is a simple, fast and economical technique with reduced discomfort for the patients (Petersilka 2011). Long term periodontal treatment is highly dependent on supportive therapy and numerous studies agree on the importance of re instrumentation to disrupt biofilm and prevent surface recolonization by periodontal pathogens; in this respect air polishing seems to be particularly relevant.

Although air-polishing with sodium bicarbonate powders is efficient and safe to remove supra-gingival biofilm, it is unsuitable for sub-gingival debridement because of important damages caused on root surfaces (Atkinson et al. 1984) and soft tissues (Bühler et al. 2016). To resolve these issues, low abrasive powders have been proposed and currently, three types have been described, glycine, erythritol and trehalose. Glycine is an amino acid and an immunoregulator of inflammatory response in gingival tissues (Schaumann et al. 2013). Thanks to it anti-inflammatory and cytoprotective properties, glycine powders (particles size 25–65 μm) inhibit macrophage activation and reduce the formation of free radicals (Zhong et al. 2003). Erythritol is an alcohol used as a non-cariogenic sweetener and the small size of the particles (14 μm) limits tissue damages. In 2019, a study introduced a new air polishing powder for periodontal supportive therapy, trehalose, a disaccharide used as a food additive with particle sizes similar to that of glycine (25–35 μm) (Kruse et al. 2019).

The vast majority of studies on air-polishing have been led in the context of periodontal supportive therapy. Nevertheless, supra-gingival application of low abrasive powders oriented into shallow pockets (≤4 mm PPD) and sub-gingival application in moderately deep pockets (≥5 mm PPD) using specific nozzles are more efficient in subgingival biofilm removal than manual root debridement (Flemmig et al. 2012; Flemmig et al. 2007). Compared to manual and power-driven root debridement, sub-gingival air-polishing with glycine powder is considerably faster regarding sub-gingival biofilm removal and perceived as more comfortable by patients. However, mineralized biofilm has to be removed by hand or power-driven instruments (Moëne et al. 2010; Petersilka et al. 2003; Sculean et al. 2013; Wennström et al. 2011). In a recent study, sub-gingival air-polishing with glycine powder in addition to sub-gingival debridement in initial therapy reduced inflammation as well as pocket depth more efficiently than non-surgical debridement alone at 3 months but had similar effects at 6 months (Tsang et al. 2018). Erythritol powder air-polishing alone has similar results regarding BOP % as well as PPD reduction and clinical attachment gain at 6 months but no significant difference was found between the two treatments (Hägi et al. 2015). Sub-gingival air-polishing with erythrol powder containing 0,3% CHX had similar efficiency at 12 months in reducing the number of periodontal pockets >4 mm but was less painful and better tolerated by patients than power-driven sub-gingival debridement (Müller et al. 2014). Finally, sub-gingival air-polishing with trehalose powder had similar results to sonic debridement with regards to PPD and BOP reduction although it was better supported by patients. Each method resulted in a significant reduction of bacteria up to 6 months, thus confirming that air-polishing can be effective on the subgingival oral biofilm elimination during supportive periodontal treatment. (Kruse et al. 2019).

Air-polishing is an efficient approach for biofilm disorganization, it is fast and comfortable for patients but needs to be used in complement to initial mechanical debridement to remove mineralized sub-gingival deposits. It can however be recommended, alone, for periodontal supportive therapy as shown in a recent randomized control clinical pilot study (Kruse et al. 2020).

16.4.2 Photodynamic Therapy

Photodynamic therapy (PDT) operates with three components, light, a photosensitizer compound, and oxygen. The photosensitizer is administered to the patient and then activated by light at a specific wavelength; the excited photosensitizer transfers its energy to oxygen and generates singlet oxygen and free radicals that in turn alter DNA and plasma membrane of surrounding cells (Konopka and Goslinski 2007). Initially used for cancer treatment, PDT has antimicrobial properties that have been used as an alternative to antibacterial, antifungal, and antiviral treatments (Konopka and Goslinski 2007; Takasaki et al. 2009).

In periodontal therapy, PDT has been suggested as an adjunctive approach to sub-gingival debridement. Toluidine blue O and methylene blue, which are able to directly target gram-negative and gram-positive bacteria while they don't have any cytotoxic effects *per se*, can be used as sensitizers. Most sensitizers are activated by red light between 630 and 700 nm and historically, argon lasers or Nd.YAG lasers were used as light sources. Currently most *in vivo* studies use diode lasers to activate the photosensitizer and light-emitting diodes (LED) have been proposed (Takasaki et al. 2009). *In vitro* studies show that PDT is effective on *Pg, Fn, Aa* and in artificial biofilm models (Al Habashneh et al. 2015; Schneider et al. 2012). Regarding *in vivo* studies, a recent review and meta-analysis concluded that PDT was an effective adjunctive therapy to sub-gingival debridement regarding clinical attachment gain (Ramanauskaite et al. 2020). This is in accordance with a network meta-analysis concluding that sub-gingival debridement associated with PDT was more effective than sub-gingival debridement alone at a 3 and 6 months follow-up regarding clinical attachment gain (Zhao et al. 2020b). However another review and meta-analysis comparing PDT + non-surgical treatment to non-surgical treatment alone was unable to reach statistical significance due to study design heterogeneity (Salvi et al. 2020).

Although PDT is an interesting and promising approach, study designs are dramatically heterogenous regarding sensitizers, lights, doses, power, and exposure times. Protocols must be standardized to establish strong conclusions and recommendations. The last EFP guidelines for the treatment of stage I-III periodontitis does not recommend it as an adjunctive therapy to sub-gingival debridement (Sanz et al. 2020).

16.4.3 Antiseptics

Numerous antiseptics have been tested as a complement to non-surgical debridement, CHX, povidone-iodine, quaternary ammonium, triclosan, essential oils, sodium hypochlorite along with different methods of application, solutions, gels, slow-release systems and at various times of treatment.

16.4.3.1 Chlorhexidine (CHX)

A consensus conference of the EFP reviews the state of knowledge in 2020 about the use of different antiseptics and adjuvants applied either as an irrigation or as a device placed in periodontal pockets during sub-gingival debridement (Sanz et al. 2020). While several agents are not advocated (statins, probiotics, bisphosphonates), the application of CHX remains recommended, essentially in solution in pre-therapeutical phases as a complement to oral hygiene measures, in the active phase of periodontal treatment and during post-surgical periods.

After sub-gingival debridement, in order to potentiate the results and avoid bacterial recontamination, an adequate plaque control is necessary. Thus, during the XIth European Workshop on Periodontology, antiseptics adjuvants were recommended during the hygienic phase of periodontal treatment. FMD protocol proposed by Quirynen includes the use of a 0,2% CHX solution twice daily during 2 weeks to delay recolonization by pathogen bacteria before sub-gingival debridement (Quirynen et al. 1995). Indeed, CHX is bacteriostatic between 0,02% and 0,06% and bactericide between 0,12% and 0,2% and is a powerful anti-plaque agent (da Costa et al. 2017; Stratul et al. 2010). A recent systematic review of the literature showed that during sub-gingival debridement, the use of CHX irrigation doesn't improve clinical parameters (PPD, CAL gain) whereas its utilization as slow-release compounds (chips, gel or varnish) significantly reduces PPD and enhances CAL gain, compared with sub-gingival debridement alone (Ramanauskaite and Machiulskiene 2020; Zhao et al. 2020a).

16.4.3.2 Povidone Iodine

Povidone iodine is another highly studied antiseptic agent. Its use as an irrigation solution during sub-gingival debridement significantly reduces moderate pockets PPD (> 6 mm) (Denez et al. 2016; Sahrmann et al. 2010). Those results are in accordance with the fact that povidone-iodine significantly reduces the number of anaerobic bacteria, in particular *Aa* and *Pg* during the first month following its application (do Vale et al. 2016; Krück et al. 2012; Perrella et al. 2016; Sindhura et al. 2017).

16.4.3.3 Cetylpyridinium Chloride (CPC)

Cetylpyridinium chloride (CPC) is a monocationic quaternary ammonium compound. CPC is mainly used as an antimicrobial ingredient in over-the-counter products such as mouthwashes and dentifrices. These oral care products, easily available for customers, are marketed for reducing plaque accumulation and gingival inflammation (Mao et al. 2020). The antimicrobial efficacy of CPC has been shown in numerous, mostly *in vitro,* studies on planktonic microorganisms. A recent metanalysis concluded that CPC demonstrated a higher interproximal plaque index and gingival inflammation reduction compared to placebo mouth-rinses (Langa et al. 2021). However, the composition of the biofilm matrix retards its penetration, thus limiting its action to the outer layers when used as a mouth-rinse (Schwarz et al. 2020). Therefore, bacteria in deeper biofilm layers may be exposed to sub-inhibitory concentrations when mouth-rinses are applied for clinically relevant treatment periods (i.e. 1 min).

When comparing the efficiency of 3 different preoperative mouthwashes (CPC, EO and CHX) to a placebo, results from the clinical trial led by Maximo et al. on 300 patients with different periodontal status groups (healthy, gingivitis, periodontitis) show statically significant bacterial reduction after the use of preoperative rinses (Máximo et al. 2020). The periodontal status influenced, however, the antimicrobial activity of the different mouth-rinses. If all 3 mouth-rinses could independently be used on healthy patients, CHX and EO were the most effective on periodontitis diseased ones (Máximo et al. 2020).

16.4.3.4 Hydrogen Peroxyde (H_2O_2)

As for hydroxide peroxide (H_2O_2), the most studied concentration is that of 1,5% and in the formula of a mouthrinse. The results show an improvement as far as plaque control and gingivitis and oral bacteria reduction compared to the placebo but the effects are lesser than that of CHX which remains the *gold standard (*Muniz et al. 2020*;* Rashed 2016*)*. The bactericidal effect of H_2O_2 was demonstrated on planktonic bacteria and it could be used as pretreatment associated with antimicrobial photodynamic therapy (aPDT) (Kunz et al. 2019). However, therapeutic substances such as H_2O_2 should be used cautiously since they can promote the degradation of titanium-based dental implants and abutments leading to the release of toxic ions and also have deleterious effects, at high concentrations, on the oral mucosa (Lin et al. 2019; Peñarrieta-Juanito et al. 2019).

16.4.3.5 Clinical Recommendations and New Approaches (Table 16.5)

More recently, Figuero et al. in a systematic review of the literature and meta-analysis have listed the use of different antiseptic agents (CHX, CPC, Essential Oils, Amin Fluoride, Stannous fluoride, triclosan-copolymer) and compared their action to placebos. The most commonly found in mouth-rinses are Essential Oils (EOs) and triclosan-copolymers in toothpastes. If all agents efficiently reduced gingival inflammation,

Table 16.5 Summary of adjunctive periodontal therapies and clinical recommendations

Adjunctive therapy	Molecule	Properties/ Mechanisms of action	Efficacy	Clinical recommendations
Air polishing powders (low abrasive powders)	Glycine	Inhibits macrophage activation and reduces formation of free radicals		
	Erythritol	Chemically neutral, water soluble and effective on *Pg*	Subgingival biofilm removal	With SRP in initial periodontal debridement
	Trehalose	Highly water-soluble and effective on Gram-negative anaerobic bacteria	> To manual or power-driven root debridement	Alone in periodontal supportive therapy
Photodynamic therapy (diode laser, photosensitizer, oxygen)	Toluidine Blue O	Effective on *Pg, Fn, Aa* in artificial biofilms	Similar to non-surgical periodontal treatment	Not recommended
	Methylene blue			
Antiseptics	Chlorhexidine (chips, gel, varnish, solution)	bacteriostatic (0,02–0,06%) bactericide (0,12–0,2%) anti-plaque agent		Before treatment 0,2% CHX mouthwash
	Povidone Iodine	Effective on anaerobic bacteria (*Aa, Pg*)	Reduce gingival inflammation	During sub-gingival debridement, irrigation with povidone-iodine or CHX
	Triclosan	Bactericidal effect	Reduce Plaque Index	
	Essential oils	Antimicrobial properties	Reduce BOP Index	Post treatment triclosan-copolymer-based toothpaste associated with essential oils mouthwash
	Hydrogen Peroxyde	Bactericidal effect		
	Cetylpyridinium Chloride	Antimicrobial properties		
Antibiotics	Amoxicillin	Effective on anaerobic bacteria and protozoans	Reduce tissue infiltrated bacteria	Severe periodontitis (during active phase of treatment) amoxicillin (500 mg three times a day) + metronidazole (250 mg or 500 mg three times a day) during 7 days
	Metronidazole		Reduce BOP Index	

plaque and bleeding indexes compared with manual OH methods and placebos, improvement was significantly higher with CHX and essential oils used as mouthwash (plaque index reduction more effective than with a toothpaste) as well as with CHX and triclosan-copolymer-based toothpastes (Figuero et al. 2020). This meta-analysis also reveals that in the long run, tooth discoloration and soft tissue irritation occur with those agents. In addition, CHX also induces reversible tongue coloration, lesions and taste alteration.

Given the literature, summarized in Table 16.5, a clinical protocol based on antiseptic administration before, during and after non-surgical periodontal treatment may be proposed. Thus, 2 weeks before the debridement, a 0,2% CHX mouthwash can be prescribed twice a day. During sub-gingival debridement, either an irrigation with povidone-iodine or CHX-based slow-release systems could potentiate treatment benefits. Finally, in hygienic phase, the use of triclosan-copolymer-based toothpaste associated with essential oils mouthwash will allow to control gingival inflammation and enhance manual plaque control while reducing to a minimum adverse side effect.

Over-the-counter availability and extensive use of mouthwashes and toothpastes containing antimicrobial agents such as CHX, triclosan or EO may result in the selection of bacterial strains capable of having deleterious effects on compromised individuals (Giuliano and Rybak 2015; Saleem et al. 2016; Webber et al. 2015). Clinicians should be well aware that the random and widespread use of antiseptics may lead to the emergence of bacterial resistance (Cieplik et al. 2019).

Thus, new approaches are being investigated and the use of phytotherapy in the form of gel is increasingly studied, on the one hand for its antibacterial, anti-inflammatory and anti-oxidant effects and on the other hand for the absence of reported adverse effects (Moro et al. 2018). Other antiseptic agents have been studied but the more promising results are obtained with propolis-based gels, boric acid gels, desiccant gels containing a blend of sulphonic/sulphuric acids usually used to treat aphthous stomatitis (Isola et al. 2018; Levine et al. 2020) or gels associating several antiseptic agents (1,4% H_2O_2, 0,1% CPC and sodium carbonate) as far as clinical (PPD, CAL, BOP), biological (IL-1β, IL-10, IL-6, TNF-α) and bacterial (red complex bacteria) parameters were concerned (Debnath et al. 2016; Kanoriya et al. 2018; Sanghani and Bm 2014; Singhal et al. 2018). In addition, several studies showed that curcuma-based gels have similar efficiency as CHX (Anitha et al. 2015; Hugar et al. 2016; Siddharth et al. 2020).

Recently, an oxygen transporter derived from the marine lugworm *Arenicola marina*, HEMARINA-M101 (M101) was tested. M101 is a natural extracellular hemoglobin whose efficacy as an additive to preservation solutions for preventing ischemia/reperfusion injuries has been tested *in vitro*, *in vivo* and in preclinical studies. M101's anti-inflammatory and anti-infectious potential, based on its anti-oxidative and tissue oxidation properties, have been tested *in vitro* on biofilm cultures containing *Pg* and *in vivo* in a *Pg*-induced subcutaneous calvarial abscess in a mouse model (Batool et al. 2020). The results showed that M101 significantly reduced the release of pro-inflammatory cytokines and also had an anti-bacterial effect on *Pg* confirming its pro-healing properties and making it a potential therapeutic agent in periodontal wound healing and regeneration (Batool et al. 2020).

16.4.3.6 Antibiotics

Some bacterial species involved in periodontal diseases have the ability to invade tissues (dentin, epithelial and connective tissues, oral mucosa, tongue and crevicular fluid) and as a consequence, cannot be eliminated by mechanical therapies like sub-gingival debridement. Moreover, mechanical treatment doesn't have identical results on all subjects, probably due to insufficient maintenance of a bacterial flora biocompatible with oral health and insufficient action on certain bacterial species, in particular the red complex, still present at 11,04% 1 year after treatment despite supportive therapy (27,7% before treatment) (Feres et al. 2015; Goodson et al. 2012; Mombelli 2018). In order to maintain a biocompatible flora, those levels should remain under 4,5% (Feres et al. 2015). 40 different species have been identified in supra and subgingival biofilm samples and Socransky's red complex is always present in the supra-gingival flora of patients with periodontitis (Ximénez-Fyvie et al. 2000). Thus, data converge to suggest that every periodontal pocket, even shallow, should systematically benefit from anti-infective therapy to reach full compatibility of bacterial flora. However, it is important to emphasize the risk of inducing bacterial resistance and imbalance oral microbiota which is of major value in general health. In addition, oral ecology includes 700–800 different morphotypes that makes it a challenge to find the ideal antibiotic, especially since new potential pathogens add up to the list of already classified complexes (Chen et al. 2018; Mombelli 2018).

Antibiotic administration can be achieved either locally with supporting devices such as fibers, sponges, gels, microspheres, or administered systemically. Several questions arise, what molecule? Which indication? How long? When? And which protocol?

16.4.3.7 Local Antibiotics

Numerous studies have explored the effects of locally administered antibiotics in periodontal treatment and general conclusions show small supplemental beneficial effects regarding PPD and CAL (in short term only) (Bonito et al. 2005; Herrera et al. 2020; Kaner et al. 2007; Mombelli et al. 1997; Pavia et al. 2004; Sakellari et al. 2010). The antibiotics delivered are mainly in forms of gels or fibers of tetracycline, doxycycline or minocycline (Herrera et al. 2020; Jepsen and Jepsen 2016). They should be reserved after periodontal reevaluation in pocket depth > 4 mm with BOP (Drisko 2014).

New approaches are looked into as far as local antibiotic delivery systems. If the gel form ensures longer retention of the drugs in the periodontal pocket, to improve treatment efficacy and reduce physicochemical and biological degradation of the gels, molecular inclusion strategies are being investigated. Among them, β-cyclodextrin, and poly-lactic-co-glycolic acid nanospheres seem promising (Lecio et al. 2020; Trajano et al. 2020). Others such as Self-Assembling Peptide (SAP) hydrogels offer interesting results. In addition to classic polymeric systems, biophysical and chemical requirements such as syringability, adhesion to tooth surfaces, biodegradability and non-cytotoxicity, SAP offers the advantage of enabling direct incorporation of drugs at high concentrations during the self-assembling process. Not only can it be applied directly into a defect in a minimally invasive manner, but it also has an intrinsic antimicrobial activity and can act as a fibrillar scaffold for cells promoting cellular growth and differentiation. Given these different properties SAPs could be good candidates for periodontal therapy and wound healing (Koch et al. 2019).

16.4.3.8 Systemic Antibiotics

A Cochrane literature review concluded that there is weak evidence of benefits resulting from antibiotic therapy in non-surgical phase of periodontal treatment, and especially a lack of evidence regarding the more suitable molecule (Khattri et al. 2020). Another review shows that antibiotic treatment benefits are questionable in light and moderate periodontitis (Pretzl et al. 2019). For several authors, in the presence of numerous periodontal pockets >4 mm with BOP, or in high-risk patients, systemic antibiotics are necessary and the association of amoxicillin and metronidazole seems to be the more effective (Drisko 2014; Feres et al. 2015; Khattri et al. 2020; Kolakovic et al. 2014; Mestnik et al. 2012; Silva et al. 2011).

16.4.3.9 Timing

One may wonder if antibiotic therapy would be more efficient during active treatment phase or reevaluation since after mechanical debridement, bacterial recolonization occurs during the following 3 months. On another hand, antibiotic prescription during active treatment phase in severe periodontitis would allow to obtain a biocompatible environment faster. During maintenance phase, antibiotic administration depends on clinical and microbiological parameters (number of residual pockets >5 mm, BOP, evolving nature of the disease) (Feres et al. 2018; Khattri et al. 2020; Pretzl et al. 2019). In quadrant-wise sub-gingival debridement, there is no consensus regarding the administration of antibiotics at the start or the end of sessions if they are spaced in time.

16.4.3.10 Dose and Duration (Table 16.5)

There is in the scientific literature a large heterogeneity of protocols regarding does and duration of antibiotic treatments. It seems that the more effective prescription would be the association of amoxicillin (500 mg three times a day) and metronidazole (either 250 mg or 500 mg three times a day) should not exceed 7 days in order to ensure patient compliance and reduce the risk of bacterial resistance (Feres et al. 2015; Jepsen and Jepsen 2016; Sanz et al. 2008).

The slight additional benefit of antibiotics should, however, not undermine the effect of proper non-surgical therapy and patient behavioral risk factors modification.

16.5 Innovative Approaches in Periodontal Non-surgical Therapy (Table 16.6)

Due to its anatomical and histological complexity, the diversity of the dysbiotic biofilm and the various, individual/host responses to therapy, periodontal treatment and wound healing is still challenging. Novel approaches by the ways of phototherapy, the use of pre and probiotics and cellular therapy are widely studied and described in other chapters. Other promising local or systemic therapeutic approaches are also investigated.

16.5.1 Cold Atmospheric Plasma

Plasma is described as the fourth fundamental state of the matter. There are various biomedical applications of nonthermal plasma, such as sterilization (Hoffmann et al. 2013; Rossi et al. 2005). However, its production at an atmospheric pres-

Table 16.6 Summary of innovative approaches and potential future clinical applications

Innovative Approaches	Molecule/Active Principle	Properties	Mechanisms of action	Potential clinical applications
Cold Atmospheric Plasma	Cold Atmospheric Plasma Plasma jet handpieces Plasma activated water	Production of RONS Anti-bacterial Osteo-inductive	Subgingival biofilm removal > To manual or power-driven root debridement	Adjunctive periodontal therapy Periodontal supportive therapy
Host Modulation therapy	-Subantimicrobial Dose Doxycycline -PUFAs, Resolvin E1, ω-3 with aspirin	Anti-inflammatory properties Inhibition of MMP-8, MMP-9 Anti-inflammatory lipid mediators	Reduces bone resorption and connective-tissue destruction Suppresses bone loss Restores levels of IL-1β and C-reactive protein Improves periodontal parameters	Adjunctive periodontal therapy
Antimicrobial peptides	LL-37, HNP1-3, substance P, adrenomedullin, azurocidin, CCL28	Immunomodulatory properties on innate and adaptive immune response	Positively correlated with PPD and CAL	New biomarkers for early diagnosis of periodontal disease or relapse
Tissue oxidation	HEMARINA-M101 extracellular hemoglobin	Anti-oxidative Anti-inflammatory Anti-infectious properties	Antibacterial effect on *Pg* *in vivo* and *in vitro*	Wound healing and periodontal regeneration

sure (Cold Atmospheric Plasma, or CAP) made possible its use for other medical applications (wound healing, blood coagulation, antibacterial treatment, endothelial cell proliferation, …) (Metelmann et al. 2015). CAP induces both physical effects (production of ultraviolet rays, heat and electromagnetic fields) as well as chemical effects (production of RONS). Whereas the physical effects seem to have a negligible cellular impact (Guerrero-Preston et al. 2014; Panngom et al. 2013), RONS may induce cell membrane alterations, an increase in intracellular reactive oxygen species (ROS), a decrease of the antioxidant potential and DNA double strand brakes, and subsequently apoptosis (Yan et al. 2017). The use of CAP in periodontology is interesting because it involves many reactive species that will be able to interact by complex biochemical processes, locally and remotely, on many cells. Its anti-bacterial activity has recently been shown on periodontopathogens such as *Pg* (Hirano et al. 2019) and its bactericidal and osteo-inductive properties are already explored in the treatment of peri-implantitis (Preissner et al. 2019; Yang et al. 2018), in bone healing and on osteoblasts *in vitro* (Canullo et al. 2017; Kleineidam et al. 2019) and even on bone substitute biomaterials (Canullo et al. 2020).

In periodontal treatment, its application could be direct by using plasma-jet handpieces as adjunctive treatment associated with periodontal debridement as recently shown in a clinical study (Küçük et al. 2020) or indirect *via* plasma-activated fluids (PAW: Plasma Activated Water) (Li et al. 2017). Because of its very localized action, it can be hypothesized that it can promote the migration, healing and cellular proliferation of the cells necessary for periodontal healing without unbalancing the oral microbiota. However, further studies especially on multi-layered biofilms remain necessary (Jungbauer et al. 2021).

The field of plasma medicine if well developed and investigated in the treatment of cancer and wound healing still needs to be explored in periodontal treatment but might become an interesting alternative to periodontal debridement with ultrasonic devices in the future.

16.5.2 Host Modulation Therapy (HMT)

Historically, the first use of the term "host modulation" was based on the unexpected non-antibiotic but anti-collagenolytic properties of tetracyclines (Golub et al. 1992). Subantimicrobial dose doxycycline (SDD) was shown to reduce significantly the severity of periodontitis by reducing the levels of inflammatory mediators such as IL-1β and TNF-α, but also bone resorption and connective-tissue destruction by the inhibition of host-derived matrix metalloproteinases (MMP-8, MMP-9) (Golub et al. 2016; Golub and Lee 2020; Preshaw 2018). Previous studies have also investigated the role of polyunsaturated fatty acids (PUFAs). In *in-vivo* animal models, Resolvin E1, an endogenous anti-inflammatory lipid mediator derived from Omega-3 (ω-3) eicosapentaenoic acid, seems to suppress bone loss and restore systemic levels of IL-1β and C-reactive protein, also attenuating the inflammatory signal leading to periodontal destruction with no unwanted side-effects (Balta et al. 2017; Hasturk et al. 2007; Sulijaya et al. 2019). Concerning the effect of ω-3, another PUFA, on periodontitis, studies have shown an in-vitro reduction of IL-1β and TNF-α (Caughey et al. 1996). If ω-3 supplementation alone showed no improvement as far as gingivitis reduction was concerned, used associated with aspirin as an adjunctive therapy with SRP, clinical periodontal parameters seemed to improve positively (Campan et al. 1997; Deore et al. 2014; Sulijaya et al. 2019). However, most of the HMT conclusions are drawn from either *in-vitro* or *in-vivo* animal studies. If their results are promising, clinical human studies and precise guidelines are still needed in order to recommend and include them in a daily practice.

16.5.3 Antimicrobial Peptides (AMPs)

AMPs can affect homeostasis of the oral cavity through broad or selective killing of bacteria, but also, due to their immunomodulatory properties, they can influence both the innate and adaptive immune response (Gorr and Abdolhosseini 2011; Tonetti et al. 2011). The most studied AMPs are LL-37, α- and β-defensins. A recent systematic review on the subject reveals that, in saliva, some AMPs such as LL-37, HNP1–3, substance P, adrenomedullin, azurocidin and some others were increased in periodontal disease while others like calcitonin gene-related protein or neuropeptide Y were decreased (Jourdain et al. 2019). In the gingival crevicular fluid and saliva, LL-37, substance P, adrenomedullin and CCL28 are positively correlated with PPD and clinical attachment loss (Jourdain et al. 2019). In addition to understanding the role of specific AMPs in periodontal disease pathology, the difference in their expression could lead to the development of more precise salivary markers for early diagnosis of periodontal disease or relapse (Giannobile et al. 2009). Future studies are needed to homogenize sample types and techniques used to assess and determine which AMPs would be the most relevant periodontal biomarkers.

16.6 Conclusion and Perspectives

Nowadays, periodontitis diagnosis using the pluri-dimensional classification according to stages and grades leads to a more personalized patient-centered treatment that also takes into consideration the patient's medical history. However, although scientific knowledge is constantly evolving, the base of periodontal treatment remains unchanged, bacterial load and inflammation control. Since the development of sonic and ultrasonic power-driven instruments there has not been any major breakthrough in the biofilm disorganization approach. Of course, innovative approaches like air-polishing, PDT, CAP and HMT are of considerable interest but they cannot yet replace classical mechanical debridement due to the heterogeneity of the studies and seem more effective if used as adjunctive therapies. The need of standardized controlled trials is still necessary to really assess those innovative approaches.

References

Al Habashneh R, Asa'ad FA, Khader Y (2015) Photodynamic therapy in periodontal and peri-implant diseases. Quintessence Int 46(8):677–690. https://doi.org/10.3290/j.qi.a34078

Anitha V, Rajesh P, Shanmugam M, Priya BM, Prabhu S, Shivakumar V (2015) Comparative evaluation of natural curcumin and synthetic chlorhexidine in the management of chronic periodontitis as a local drug delivery: a clinical and microbiological study. Indian J Dent Res 26(1):53–56. https://doi.org/10.4103/0970-9290.156806

Apatzidou DA, Kinane DF (2004) Quadrant root planing versus same-day full-mouth root planing. I. Clinical findings. J Clin Periodontol 31(2):132–140. https://doi.org/10.1111/j.0303-6979.2004.00461.x

Arabaci T, Ciçek Y, Canakçi CF (2007) Sonic and ultrasonic scalers in periodontal treatment: a review. Int J Dent Hyg 5(1):2–12. https://doi.org/10.1111/j.1601-5037.2007.00217.x

Atkinson DR, Cobb CM, Killoy WJ (1984) The effect of an air-powder abrasive system on in vitro root surfaces. J Periodontol 55(1):13–18. https://doi.org/10.1902/jop.1984.55.1.13

Badersten A, Nilveus R, Egelberg J (1984) Effect of non-surgical periodontal therapy. II. Severely advanced periodontitis. J Clin Periodontol 11(1):63–76. https://doi.org/10.1111/j.1600-051x.1984.tb01309.x

Balta MG, Loos BG, Nicu EA (2017) Emerging concepts in the resolution of periodontal inflammation: a role for Resolvin E1. Front Immunol 8:1682. https://doi.org/10.3389/fimmu.2017.01682

Batool F, Stutz C, Petit C, Benkirane-Jessel N, Delpy E, Zal F et al (2020) A therapeutic oxygen carrier isolated from Arenicola marina decreased P. gingivalis induced inflammation and tissue destruction. Sci Rep 10(1):14745. https://doi.org/10.1038/s41598-020-71593-8

Beikler T, Abdeen G, Schnitzer S, Sälzer S, Ehmke B, Heinecke A, Flemmig TF (2004) Microbiological shifts in intra- and extraoral habitats following mechanical periodontal therapy. J Clin Periodontol 31(9):777–783. https://doi.org/10.1111/j.1600-051X.2004.00557.x

Billings M, Holtfreter B, Papapanou PN, Mitnik GL, Kocher T, Dye BA (2018) Age-dependent distribution of periodontitis in two countries: findings from NHANES 2009 to 2014 and SHIP-TREND 2008 to 2012. J Clin Periodontol 45(Suppl 20):S130–S148. https://doi.org/10.1111/jcpe.12944

Bonito AJ, Lux L, Lohr KN (2005) Impact of local adjuncts to scaling and root planing in periodontal disease therapy: a systematic review. J Periodontol 76(8):1227–1236. https://doi.org/10.1902/jop.2005.76.8.1227

Bower RC (1979) Furcation morphology relative to periodontal treatment. Furcation entrance architecture.

J Periodontol 50(1):23–27. https://doi.org/10.1902/jop.1979.50.1.23
Bühler J, Amato M, Weiger R, Walter C (2016) A systematic review on the effects of air polishing devices on oral tissues. Int J Dent Hyg 14(1):15–28. https://doi.org/10.1111/idh.12120
Caffesse RG, Sweeney PL, Smith BA (1986) Scaling and root planing with and without periodontal flap surgery. J Clin Periodontol 13(3):205–210. https://doi.org/10.1111/j.1600-051x.1986.tb01461.x
Campan P, Planchand PO, Duran D (1997) Pilot study on n-3 polyunsaturated fatty acids in the treatment of human experimental gingivitis. J Clin Periodontol 24(12):907–913. https://doi.org/10.1111/j.1600-051x.1997.tb01210.x
Canullo L, Genova T, Mandracci P, Mussano F, Abundo R, Fiorellini JP (2017) Morphometric changes induced by cold argon plasma treatment on osteoblasts grown on different dental implant surfaces. Int J Periodontics Restorative Dent 37(4):541–548. https://doi.org/10.11607/prd.2916
Canullo L, Genova T, Rakic M, Sculean A, Miron R, Muzzi M et al (2020) Effects of argon plasma treatment on the osteoconductivity of bone grafting materials. Clin Oral Investig 24(8):2611–2623. https://doi.org/10.1007/s00784-019-03119-0
Carra MC, Detzen L, Kitzmann J, Woelber JP, Ramseier CA, Bouchard P (2020) Promoting behavioural changes to improve oral hygiene in patients with periodontal diseases: a systematic review. J Clin Periodontol 47(Suppl 22):72–89. https://doi.org/10.1111/jcpe.13234
Caughey GE, Mantzioris E, Gibson RA, Cleland LG, James MJ (1996) The effect on human tumor necrosis factor alpha and interleukin 1 beta production of diets enriched in n-3 fatty acids from vegetable oil or fish oil. Am J Clin Nutr 63(1):116–122. https://doi.org/10.1093/ajcn/63.1.116
Chapple IL, Van der Weijden F, Doerfer C, Herrera D, Shapira L, Polak D et al (2015) Primary prevention of periodontitis: managing gingivitis. J Clin Periodontol 42(Suppl 16):S71–S76. https://doi.org/10.1111/jcpe.12366
Chapple ILC, Mealey BL, Van Dyke TE, Bartold PM, Dommisch H, Eickholz P et al (2018) Periodontal health and gingival diseases and conditions on an intact and a reduced periodontium: consensus report of workgroup 1 of the 2017 World Workshop on the Classification of Periodontal and Peri-Implant Diseases and Conditions. J Clin Periodontol 45(Suppl 20):S68–S77. https://doi.org/10.1111/jcpe.12940
Chen B, Wang J, Wang Y, Zhang J, Zhao C, Shen N et al (2018) Oral microbiota dysbiosis and its association with Henoch-Schönlein Purpura in children. Int Immunopharmacol 65:295–302. https://doi.org/10.1016/j.intimp.2018.10.017
Ciantar M (2014) Time to shift: from scaling and root planing to root surface debridement. Prim Dent J 3(3):38–42. https://doi.org/10.1308/205016814812736592
Cieplik F, Jakubovics NS, Buchalla W, Maisch T, Hellwig E, Al-Ahmad A (2019) Resistance toward chlorhexidine in Oral bacteria - is there cause for concern? Front Microbiol 10:587. https://doi.org/10.3389/fmicb.2019.00587
Claffey N, Polyzois I, Ziaka P (2004) An overview of non-surgical and surgical therapy. Periodontol 2000 36:35–44. https://doi.org/10.1111/j.1600-0757.2004.00073.x
Corbet EF, Vaughan AJ, Kieser JB (1993) The periodontally-involved root surface. J Clin Periodontol 20(6):402–410. https://doi.org/10.1111/j.1600-051x.1993.tb00380.x
da Costa LFNP, Amaral CDSF, Barbirato DDS, Leão ATT, Fogacci MF (2017) Chlorhexidine mouthwash as an adjunct to mechanical therapy in chronic periodontitis: a meta-analysis. J Am Dent Assoc 148(5):308–318. https://doi.org/10.1016/j.adaj.2017.01.021
Debnath K, Chatterjee A, Priya VS (2016) Evaluation of Nano-bio fusion gel as an adjunct to scaling and root planing in chronic periodontitis: a clinico-microbiological study. J Indian Soc Periodontol 20(5):543–548. https://doi.org/10.4103/0972-124X.201696
Denez EM, Toma S, Lasserre JF, Brecx MC (2016) Evaluation of a unique subgingival irrigation with 10% povidone-iodine after scaling and root planing: a randomized clinical trial. Quintessence Int 47(7):549–558. https://doi.org/10.3290/j.qi.a36175
Deore GD, Gurav AN, Patil R, Shete AR, Naiktari RS, Inamdar SP (2014) Herbal anti-inflammatory immunomodulators as host modulators in chronic periodontitis patients: a randomised, double-blind, placebo-controlled, clinical trial. J Periodontal Implant Sci 44(2):71–78. https://doi.org/10.5051/jpis.2014.44.2.71
do Vale HF, Casarin RC, Taiete T, Bovi Ambrosano GM, Ruiz KG, Nociti FH et al (2016) Full-mouth ultrasonic debridement associated with povidone iodine rinsing in GAgP treatment: a randomised clinical trial. Clin Oral Investig 20(1):141–150. https://doi.org/10.1007/s00784-015-1471-y
Drisko CL (2014) Periodontal debridement: still the treatment of choice. J Evid Based Dent Pract 14(Suppl):33–41.e31. https://doi.org/10.1016/j.jebdp.2014.02.007
Dubar M, Zaffino ML, Remen T, Thilly N, Cunat L, Machouart MC, Bisson C (2020) Protozoans in subgingival biofilm: clinical and bacterial associated factors and impact of scaling and root planing treatment. J Oral Microbiol 12(1):1693222. https://doi.org/10.1080/20002297.2019.1693222
Eberhard J, Jepsen S, Jervøe-Storm PM, Needleman I, Worthington HV (2008) Full-mouth disinfection for the treatment of adult chronic periodontitis. Cochrane Database Syst Rev 1:CD004622. https://doi.org/10.1002/14651858.CD004622.pub2
Eberhard J, Jepsen S, Jervøe-Storm PM, Needleman I, Worthington HV (2015) Full-mouth treatment modalities (within 24 hours) for chronic periodontitis in adults. Cochrane Database Syst Rev 4:CD004622. https://doi.org/10.1002/14651858.CD004622.pub3

Ercoli C, Caton JG (2018) Dental prostheses and tooth-related factors. J Clin Periodontol 45(Suppl 20):S207–S218. https://doi.org/10.1111/jcpe.12950

Feres M, Figueiredo LC, Soares GM, Faveri M (2015) Systemic antibiotics in the treatment of periodontitis. Periodontol 2000 67(1):131–186. https://doi.org/10.1111/prd.12075

Feres M, Retamal-Valdes B, Mestnik MJ, de Figueiredo LC, Faveri M, Duarte PM et al (2018) The ideal time of systemic metronidazole and amoxicillin administration in the treatment of severe periodontitis: study protocol for a randomized controlled trial. Trials 19(1):201. https://doi.org/10.1186/s13063-018-2540-8

Figuero E, Roldán S, Serrano J, Escribano M, Martín C, Preshaw PM (2020) Efficacy of adjunctive therapies in patients with gingival inflammation: a systematic review and meta-analysis. J Clin Periodontol 47(Suppl 22):125–143. https://doi.org/10.1111/jcpe.13244

Flemmig TF, Hetzel M, Topoll H, Gerss J, Haeberlein I, Petersilka G (2007) Subgingival debridement efficacy of glycine powder air polishing. J Periodontol 78(6):1002–1010. https://doi.org/10.1902/jop.2007.060420

Flemmig TF, Arushanov D, Daubert D, Rothen M, Mueller G, Leroux BG (2012) Randomized controlled trial assessing efficacy and safety of glycine powder air polishing in moderate-to-deep periodontal pockets. J Periodontol 83(4):444–452. https://doi.org/10.1902/jop.2011.110367

Giannobile WV, Beikler T, Kinney JS, Ramseier CA, Morelli T, Wong DT (2009) Saliva as a diagnostic tool for periodontal disease: current state and future directions. Periodontol 2000 50:52–64. https://doi.org/10.1111/j.1600-0757.2008.00288.x

Giuliano CA, Rybak MJ (2015) Efficacy of triclosan as an antimicrobial hand soap and its potential impact on antimicrobial resistance: a focused review. Pharmacotherapy 35(3):328–336. https://doi.org/10.1002/phar.1553

Golub LM, Lee HM (2020) Periodontal therapeutics: current host-modulation agents and future directions. Periodontol 2000 82(1):186–204. https://doi.org/10.1111/prd.12315

Golub LM, Suomalainen K, Sorsa T (1992) Host modulation with tetracyclines and their chemically modified analogues. Curr Opin Dent 2:80–90

Golub LM, Elburki MS, Walker C, Ryan M, Sorsa T, Tenenbaum H et al (2016) Non-antibacterial tetracycline formulations: host-modulators in the treatment of periodontitis and relevant systemic diseases. Int Dent J 66(3):127–135. https://doi.org/10.1111/idj.12221

Goodson JM, Haffajee AD, Socransky SS, Kent R, Teles R, Hasturk H et al (2012) Control of periodontal infections: a randomized controlled trial I. the primary outcome attachment gain and pocket depth reduction at treated sites. J Clin Periodontol 39(6):526–536. https://doi.org/10.1111/j.1600-051X.2012.01870.x

Gorr SU, Abdolhosseini M (2011) Antimicrobial peptides and periodontal disease. J Clin Periodontol 38(Suppl 11):126–141. https://doi.org/10.1111/j.1600-051X.2010.01664.x

Graziani F, Cei S, Orlandi M, Gennai S, Gabriele M, Filice N et al (2015) Acute-phase response following full-mouth versus quadrant non-surgical periodontal treatment: a randomized clinical trial. J Clin Periodontol 42(9):843–852. https://doi.org/10.1111/jcpe.12451

Graziani F, Karapetsa D, Alonso B, Herrera D (2017) Nonsurgical and surgical treatment of periodontitis: how many options for one disease? Periodontol 2000 75(1):152–188. https://doi.org/10.1111/prd.12201

Grenier G, Gagnon G, Grenier D (2009) Detection of herpetic viruses in gingival crevicular fluid of patients suffering from periodontal diseases: prevalence and effect of treatment. Oral Microbiol Immunol 24(6):506–509. https://doi.org/10.1111/j.1399-302X.2009.00542.x

Guerrero-Preston R, Ogawa T, Uemura M, Shumulinsky G, Valle BL, Pirini F et al (2014) Cold atmospheric plasma treatment selectively targets head and neck squamous cell carcinoma cells. Int J Mol Med 34(4):941–946. https://doi.org/10.3892/ijmm.2014.1849

Hägi TT, Hofmänner P, Eick S, Donnet M, Salvi GE, Sculean A, Ramseier CA (2015) The effects of erythritol air-polishing powder on microbiologic and clinical outcomes during supportive periodontal therapy: six-month results of a randomized controlled clinical trial. Quintessence Int 46(1):31–41. https://doi.org/10.3290/j.qi.a32817

Hasturk H, Kantarci A, Goguet-Surmenian E, Blackwood A, Andry C, Serhan CN, Van Dyke TE (2007) Resolvin E1 regulates inflammation at the cellular and tissue level and restores tissue homeostasis in vivo. J Immunol 179(10):7021–7029. https://doi.org/10.4049/jimmunol.179.10.7021

Herrera D, Matesanz P, Martín C, Oud V, Feres M, Teughels W (2020) Adjunctive effect of locally delivered antimicrobials in periodontitis therapy: a systematic review and meta-analysis. J Clin Periodontol 47(Suppl 22):239–256. https://doi.org/10.1111/jcpe.13230

Hirano Y, Hayashi M, Tamura M, Yoshino F, Yoshida A, Masubuchi M et al (2019) Singlet oxygen generated by a new nonthermal atmospheric pressure air plasma device exerts a bactericidal effect on oral pathogens. J Oral Sci 61(4):521–525. https://doi.org/10.2334/josnusd.18-0455

Hoffmann C, Berganza C, Zhang J (2013) Cold atmospheric plasma: methods of production and application in dentistry and oncology. Med Gas Res 3(1):21. https://doi.org/10.1186/2045-9912-3-21

Hugar SS, Patil S, Metgud R, Nanjwade B, Hugar SM (2016) Influence of application of chlorhexidine gel and curcumin gel as an adjunct to scaling and root planing: a interventional study. J Nat Sci Biol Med 7(2):149–154. https://doi.org/10.4103/0976-9668.184701

Isola G, Matarese G, Williams RC, Siciliano VI, Alibrandi A, Cordasco G, Ramaglia L (2018) The effects of a desiccant agent in the treatment of chronic periodontitis: a randomized, controlled clinical trial. Clin Oral

Investig 22(2):791–800. https://doi.org/10.1007/s00784-017-2154-7

Jepsen K, Jepsen S (2016) Antibiotics/antimicrobials: systemic and local administration in the therapy of mild to moderately advanced periodontitis. Periodontol 2000 71(1):82–112. https://doi.org/10.1111/prd.12121

Jepsen S, Caton JG, Albandar JM, Bissada NF, Bouchard P, Cortellini P et al (2018) Periodontal manifestations of systemic diseases and developmental and acquired conditions: consensus report of workgroup 3 of the 2017 World Workshop on the Classification of Periodontal and Peri-Implant Diseases and Conditions. J Clin Periodontol 45(Suppl 20):S219–S229. https://doi.org/10.1111/jcpe.12951

Jervøe-Storm PM, Semaan E, AlAhdab H, Engel S, Fimmers R, Jepsen S (2006) Clinical outcomes of quadrant root planing versus full-mouth root planing. J Clin Periodontol 33(3):209–215. https://doi.org/10.1111/j.1600-051X.2005.00890.x

Johnston W, Paterson M, Piela K, Davison E, Simpson A, Goulding M et al (2020) The systemic inflammatory response following hand instrumentation versus ultrasonic instrumentation-A randomized controlled trial. J Clin Periodontol 47(9):1087–1097. https://doi.org/10.1111/jcpe.13342

Jourdain ML, Velard F, Pierrard L, Sergheraert J, Gangloff SC, Braux J (2019) Cationic antimicrobial peptides and periodontal physiopathology: a systematic review. J Periodontal Res 54(6):589–600. https://doi.org/10.1111/jre.12676

Jungbauer G, Moser D, Müller S, Pfister W, Sculean A, Eick S (2021) The antimicrobial effect of cold atmospheric plasma against dental pathogens-a systematic review of in-vitro studies. Antibiotics (Basel) 10(2). https://doi.org/10.3390/antibiotics10020211

Kaner D, Bernimoulin JP, Hopfenmüller W, Kleber BM, Friedmann A (2007) Controlled-delivery chlorhexidine chip versus amoxicillin/metronidazole as adjunctive antimicrobial therapy for generalized aggressive periodontitis: a randomized controlled clinical trial. J Clin Periodontol 34(10):880–891. https://doi.org/10.1111/j.1600-051X.2007.01122.x

Kanoriya D, Singhal S, Garg V, Pradeep AR, Garg S, Kumar A (2018) Clinical efficacy of subgingivally-delivered 0.75% boric acid gel as an adjunct to mechanotherapy in chronic periodontitis: a randomized, controlled clinical trial. J Investig Clin Dent 9(1). https://doi.org/10.1111/jicd.12271

Kassebaum NJ, Bernabé E, Dahiya M, Bhandari B, Murray CJ, Marcenes W (2014) Global burden of severe periodontitis in 1990-2010: a systematic review and meta-regression. J Dent Res 93(11):1045–1053. https://doi.org/10.1177/0022034514552491

Khattri S, Kumbargere Nagraj S, Arora A, Eachempati P, Kusum CK, Bhat KG et al (2020) Adjunctive systemic antimicrobials for the non-surgical treatment of periodontitis. Cochrane Database Syst Rev 11:CD012568. https://doi.org/10.1002/14651858.CD012568.pub2

Kleineidam B, Nokhbehsaim M, Deschner J, Wahl G (2019) Effect of cold plasma on periodontal wound healing-an in vitro study. Clin Oral Investig 23(4):1941–1950. https://doi.org/10.1007/s00784-018-2643-3

Koch F, Ekat K, Kilian D, Hettich T, Germershaus O, Lang H et al (2019) A versatile biocompatible antibiotic delivery system based on self-assembling peptides with antimicrobial and regenerative potential. Adv Healthc Mater 8(13):e1900167. https://doi.org/10.1002/adhm.201900167

Kolakovic M, Held U, Schmidlin PR, Sahrmann P (2014) An estimate of pocket closure and avoided needs of surgery after scaling and root planing with systemic antibiotics: a systematic review. BMC Oral Health 14:159. https://doi.org/10.1186/1472-6831-14-159

Konopka K, Goslinski T (2007) Photodynamic therapy in dentistry. J Dent Res 86(8):694–707. https://doi.org/10.1177/154405910708600803

Kozlovsky A, Rapaport A, Artzi Z (2018) Influence of operator skill level on the clinical outcome of non-surgical periodontal treatment: a retrospective study. Clin Oral Investig 22(8):2927–2932. https://doi.org/10.1007/s00784-018-2380-7

Krishna R, De Stefano JA (2016) Ultrasonic vs. hand instrumentation in periodontal therapy: clinical outcomes. Periodontol 2000 71(1):113–127. https://doi.org/10.1111/prd.12119

Krück C, Eick S, Knöfler GU, Purschwitz RE, Jentsch HF (2012) Clinical and microbiologic results 12 months after scaling and root planing with different irrigation solutions in patients with moderate chronic periodontitis: a pilot randomized trial. J Periodontol 83(3):312–320. https://doi.org/10.1902/jop.2011.110044

Kruse AB, Akakpo DL, Maamar R, Woelber JP, Al-Ahmad A, Vach K, Ratka-Krueger P (2019) Trehalose powder for subgingival air-polishing during periodontal maintenance therapy: a randomized controlled trial. J Periodontol 90(3):263–270. https://doi.org/10.1002/JPER.17-0403

Kruse AB, Maamar R, Akakpo DL, Woelber JP, Wittmer A, Vach K et al (2020) Effects of subgingival air-polishing with trehalose powder on oral biofilm during periodontal maintenance therapy: a randomized-controlled pilot study. BMC Oral Health 20(1):123. https://doi.org/10.1186/s12903-020-01111-9

Küçük D, Savran L, Ercan UK, Yarali ZB, Karaman O, Kantarci A et al (2020) Evaluation of efficacy of non-thermal atmospheric pressure plasma in treatment of periodontitis: a randomized controlled clinical trial. Clin Oral Investig 24(9):3133–3145. https://doi.org/10.1007/s00784-019-03187-2

Kunz D, Wirth J, Sculean A, Eick S (2019) In- vitro-activity of additive application of hydrogen peroxide in antimicrobial photodynamic therapy using LED in the blue spectrum against bacteria and biofilm associated with periodontal disease. Photodiagn Photodyn Ther 26:306–312. https://doi.org/10.1016/j.pdpdt.2019.04.015

Laleman I, Cortellini S, De Winter S, Rodriguez Herrero E, Dekeyser C, Quirynen M, Teughels W (2017)

Subgingival debridement: end point, methods and how often? Periodontol 2000 75(1):189–204. https://doi.org/10.1111/prd.12204
Lang NP, Bartold PM (2018) Periodontal health. J Clin Periodontol 45(Suppl 20):S9–S16. https://doi.org/10.1111/jcpe.12936
Lang NP, Tan WC, Krähenmann MA, Zwahlen M (2008) A systematic review of the effects of full-mouth debridement with and without antiseptics in patients with chronic periodontitis. J Clin Periodontol 35(8 Suppl):8–21. https://doi.org/10.1111/j.1600-051X.2008.01257.x
Lang NP, Salvi GE, Sculean A (2019) Nonsurgical therapy for teeth and implants-when and why? Periodontol 2000 79(1):15–21. https://doi.org/10.1111/prd.12240
Langa GPJ, Muniz FWMG, Costa RDSA, da Silveira TM, Rösing CK (2021) The effect of cetylpyridinium chloride mouthrinse as adjunct to toothbrushing compared to placebo on interproximal plaque and gingival inflammation-a systematic review with meta-analyses. Clin Oral Investig 25(2):745–757. https://doi.org/10.1007/s00784-020-03661-2
Lea SC, Landini G, Walmsley AD (2003) Displacement amplitude of ultrasonic scaler inserts. J Clin Periodontol 30(6):505–510. https://doi.org/10.1034/j.1600-051x.2003.00012.x
Lea SC, Landini G, Walmsley AD (2004) Thermal imaging of ultrasonic scaler tips during tooth instrumentation. J Clin Periodontol 31(5):370–375. https://doi.org/10.1111/j.1600-051X.2004.00491.x
Lea SC, Felver B, Landini G, Walmsley AD (2009) Three-dimensional analyses of ultrasonic scaler oscillations. J Clin Periodontol 36(1):44–50. https://doi.org/10.1111/j.1600-051X.2008.01339.x
Lecio G, Ribeiro FV, Pimentel SP, Reis AA, da Silva RVC, Nociti-Jr F et al (2020) Novel 20% doxycycline-loaded PLGA nanospheres as adjunctive therapy in chronic periodontitis in type-2 diabetics: randomized clinical, immune and microbiological trial. Clin Oral Investig 24(3):1269–1279. https://doi.org/10.1007/s00784-019-03005-9
Levine JB, Goncalves J, Nguyen D, Nguyen O, Hasturk H (2020) Efficacy of a novel post-foaming dental gel on gingival inflammation: a randomized controlled clinical trial. J Periodontol 91(12):1569–1583. https://doi.org/10.1002/JPER.19-0594
Li Y, Pan J, Ye G, Zhang Q, Wang J, Zhang J, Fang J (2017) In vitro studies of the antimicrobial effect of non-thermal plasma-activated water as a novel mouthwash. Eur J Oral Sci 125(6):463–470. https://doi.org/10.1111/eos.12374
Lin KY, Chung CH, Ciou JS, Su PF, Wang PW, Shieh DB, Wang TC (2019) Molecular damage and responses of oral keratinocyte to hydrogen peroxide. BMC Oral Health 19(1):10. https://doi.org/10.1186/s12903-018-0694-0
Loos BG, Needleman I (2020) Endpoints of active periodontal therapy. J Clin Periodontol 47(Suppl 22):61–71. https://doi.org/10.1111/jcpe.13253
Magnusson I, Lindhe J, Yoneyama T, Liljenberg B (1984) Recolonization of a subgingival microbiota following scaling in deep pockets. J Clin Periodontol 11(3):193–207. https://doi.org/10.1111/j.1600-051x.1984.tb01323.x
Mao X, Auer DL, Buchalla W, Hiller KA, Maisch T, Hellwig E et al (2020) Cetylpyridinium chloride: mechanism of action, antimicrobial efficacy in biofilms, and potential risks of resistance. Antimicrob Agents Chemother 64(8). https://doi.org/10.1128/AAC.00576-20
Maritato M, Orazi L, Laurito D, Formisano G, Serra E, Lollobrigida M et al (2018) Root surface alterations following manual and mechanical scaling: a comparative study. Int J Dent Hyg 16(4):553–558. https://doi.org/10.1111/idh.12349
Máximo PM, Cortelli SC, Aquino DR, de Miranda TB, Costa FO, Cortelli JR (2020) Preoperative mouthwash in subjects with different periodontal status: a randomised controlled clinical trial. Oral Health Prev Dent 18(1):433–440. https://doi.org/10.3290/j.ohpd.a44308
Mestnik MJ, Feres M, Figueiredo LC, Soares G, Teles RP, Fermiano D et al (2012) The effects of adjunctive metronidazole plus amoxicillin in the treatment of generalized aggressive periodontitis: a 1-year double-blinded, placebo-controlled, randomized clinical trial. J Clin Periodontol 39(10):955–961. https://doi.org/10.1111/j.1600-051X.2012.01932.x
Metelmann H-R, Nedrelow D, Seebauer C, Schuster M, von Woedtke T, Weltmann K-D et al (2015) Head and neck cancer treatment and physical plasma. Clin Plasma Med 3(1):17–23
Moëne R, Décaillet F, Andersen E, Mombelli A (2010) Subgingival plaque removal using a new air-polishing device. J Periodontol 81(1):79–88. https://doi.org/10.1902/jop.2009.090394
Mombelli A (2018) Microbial colonization of the periodontal pocket and its significance for periodontal therapy. Periodontol 2000 76(1):85–96. https://doi.org/10.1111/prd.12147
Mombelli A, Lehmann B, Tonetti M, Lang NP (1997) Clinical response to local delivery of tetracycline in relation to overall and local periodontal conditions. J Clin Periodontol 24(7):470–477. https://doi.org/10.1111/j.1600-051x.1997.tb00214.x
Mongardini C, van Steenberghe D, Dekeyser C, Quirynen M (1999) One stage full- versus partial-mouth disinfection in the treatment of chronic adult or generalized early-onset periodontitis. I. Long-term clinical observations. J Periodontol 70(6):632–645. https://doi.org/10.1902/jop.1999.70.6.632
Moro MG, Silveira Souto ML, Franco GCN, Holzhausen M, Pannuti CM (2018) Efficacy of local phytotherapy in the nonsurgical treatment of periodontal disease: a systematic review. J Periodontal Res 53(3):288–297. https://doi.org/10.1111/jre.12525
Müller N, Moëne R, Cancela JA, Mombelli A (2014) Subgingival air-polishing with erythritol during periodontal maintenance: randomized clinical trial of

twelve months. J Clin Periodontol 41(9):883–889. https://doi.org/10.1111/jcpe.12289

Muniz FWMG, Cavagni J, Langa GPJ, Stewart B, Malheiros Z, Rösing CK (2020) A systematic review of the effect of Oral rinsing with H. Int J Dent 2020:8841722. https://doi.org/10.1155/2020/8841722

Newton JT, Asimakopoulou K (2015) Managing oral hygiene as a risk factor for periodontal disease: a systematic review of psychological approaches to behaviour change for improved plaque control in periodontal management. J Clin Periodontol 42(Suppl 16):S36–S46. https://doi.org/10.1111/jcpe.12356

Panngom K, Baik KY, Nam MK, Han JH, Rhim H, Choi EH (2013) Preferential killing of human lung cancer cell lines with mitochondrial dysfunction by nonthermal dielectric barrier discharge plasma. Cell Death Dis 4:e642. https://doi.org/10.1038/cddis.2013.168

Papapanou PN, Sanz M, Buduneli N, Dietrich T, Feres M, Fine DH et al (2018) Periodontitis: consensus report of workgroup 2 of the 2017 world workshop on the classification of periodontal and Peri-implant diseases and conditions. J Clin Periodontol 45(Suppl 20):S162–S170. https://doi.org/10.1111/jcpe.12946

Pavia M, Nobile CG, Bianco A, Angelillo IF (2004) Meta-analysis of local metronidazole in the treatment of chronic periodontitis. J Periodontol 75(6):830–838. https://doi.org/10.1902/jop.2004.75.6.830

Peñarrieta-Juanito G, Sordi MB, Henriques B, Dotto MER, Teughels W, Silva FS et al (2019) Surface damage of dental implant systems and ions release after exposure to fluoride and hydrogen peroxide. J Periodontal Res 54(1):46–52. https://doi.org/10.1111/jre.12603

Perrella FA, Rovai EDS, De Marco AC, Santamaria MP, Feres M, de Figueredo LC et al (2016) Clinical and microbiological evaluation of povidone-iodine 10% as an adjunct to nonsurgical periodontal therapy in chronic periodontitis: a randomized clinical trial. J Int Acad Periodontol 18(4):109–119

Petersilka GJ (2011) Subgingival air-polishing in the treatment of periodontal biofilm infections. Periodontol 2000 55(1):124–142. https://doi.org/10.1111/j.1600-0757.2010.00342.x

Petersilka GJ, Tunkel J, Barakos K, Heinecke A, Häberlein I, Flemmig TF (2003) Subgingival plaque removal at interdental sites using a low-abrasive air polishing powder. J Periodontol 74(3):307–311. https://doi.org/10.1902/jop.2003.74.3.307

Preissner S, Poehlmann A, Schubert A, Lehmann A, Thomas A, Nell O, Rupf S (2019) *Ex vivo* study comparing three Cold Atmospheric Plasma (CAP) sources for biofilm removal on microstructured titanium. Plasma Med 9:1–14

Preshaw PM (2018) Host modulation therapy with anti-inflammatory agents. Periodontol 2000 76(1):131–149. https://doi.org/10.1111/prd.12148

Pretzl B, Sälzer S, Ehmke B, Schlagenhauf U, Dannewitz B, Dommisch H et al (2019) Administration of systemic antibiotics during non-surgical periodontal therapy-a consensus report. Clin Oral Investig 23(7):3073–3085. https://doi.org/10.1007/s00784-018-2727-0

Quirynen M, Bollen CM, Vandekerckhove BN, Dekeyser C, Papaioannou W, Eyssen H (1995) Full- vs. partial-mouth disinfection in the treatment of periodontal infections: short-term clinical and microbiological observations. J Dent Res 74:1459–1467

Quirynen M, Mongardini C, de Soete M, Pauwels M, Coucke W, van Eldere J, van Steenberghe D (2000) The rôle of chlorhexidine in the one-stage full-mouth disinfection treatment of patients with advanced adult periodontitis. Long-term clinical and microbiological observations. J Clin Periodontol 27(8):578–589. https://doi.org/10.1034/j.1600-051x.2000.027008578.x

Ramanauskaite E, Machiulskiene V (2020) Antiseptics as adjuncts to scaling and root planing in the treatment of periodontitis: a systematic literature review. BMC Oral Health 20(1):143. https://doi.org/10.1186/s12903-020-01127-1

Ramanauskaite E, Sakalauskaite UM, Machiulskiene V (2020) The efficacy of adjunctive aids in periodontal maintenance therapy: a systematic literature review and meta-analysis. Oral Health Prev Dent 18(1):889–910. https://doi.org/10.3290/j.ohpd.a45406

Ramseier CA, Woelber JP, Kitzmann J, Detzen L, Carra MC, Bouchard P (2020) Impact of risk factor control interventions for smoking cessation and promotion of healthy lifestyles in patients with periodontitis: a systematic review. J Clin Periodontol 47(Suppl 22):90–106. https://doi.org/10.1111/jcpe.13240

Rashed HT (2016) Evaluation of the effect of hydrogen peroxide as a mouthwash in comparison with chlorhexidine in chronic periodontitis patients: a clinical study. J Int Soc Prev Community Dent 6(3):206–212. https://doi.org/10.4103/2231-0762.183114

Roman-Torres CVG, Bryington MS, Kussaba ST, Pimentel AC, Jimbo R, Cortelli JR, Romito GA (2018) Comparison of full-mouth scaling and quadrant-wise scaling in the treatment of adult chronic periodontitis. Braz Dent J 29(3):296–300. https://doi.org/10.1590/0103-6440201801715

Rossi F, de Mitri R, Robin S (2005) Plasma sterilisation: mechanisms overview and influence of discharge parameters. In: Plasma processes and polymers. Wiley-VCH Verlag Co, pp 319–331

Sahrmann P, Puhan MA, Attin T, Schmidlin PR (2010) Systematic review on the effect of rinsing with povidone-iodine during nonsurgical periodontal therapy. J Periodontal Res 45(2):153–164. https://doi.org/10.1111/j.1600-0765.2009.01232.x

Sakellari D, Dimitra S, Ioannidis I, Ioannis I, Antoniadou M, Malama A et al (2010) Clinical and microbiological effects of adjunctive, locally delivered chlorhexidine on patients with chronic periodontitis. J Int Acad Periodontol 12(1):20–26

Saleem HG, Seers CA, Sabri AN, Reynolds EC (2016) Dental plaque bacteria with reduced susceptibility to chlorhexidine are multidrug resistant. BMC Microbiol 16:214. https://doi.org/10.1186/s12866-016-0833-1

Salvi GE, Stähli A, Schmidt JC, Ramseier CA, Sculean A, Walter C (2020) Adjunctive laser or antimicrobial photodynamic therapy to non-surgical mechanical instrumentation in patients with untreated periodontitis: a systematic review and meta-analysis. J Clin Periodontol 47(Suppl 22):176–198. https://doi.org/10.1111/jcpe.13236

Sanghani NN, Bm SSS (2014) Health from the hive: propolis as an adjuvant in the treatment of chronic periodontitis – a clinicomicrobiologic study. J Clin Diagn Res 8(9):ZC41–ZC44. https://doi.org/10.7860/JCDR/2014/8817.4856

Sanz M, Teughels W, Periodontology, G. A. O. E. W. O (2008) Innovations in non-surgical periodontal therapy: consensus report of the sixth European workshop on periodontology. J Clin Periodontol 35(8 Suppl):3–7. https://doi.org/10.1111/j.1600-051X.2008.01256.x

Sanz M, Herrera D, Kebschull M, Chapple I, Jepsen S, Beglundh T et al (2020) Treatment of stage I-III periodontitis-The EFP S3 level clinical practice guideline. J Clin Periodontol 47(Suppl 22):4–60. https://doi.org/10.1111/jcpe.13290

Schaumann T, Kraus D, Winter J, Wolf M, Deschner J, Jäger A (2013) Potential immune modularly role of glycine in oral gingival inflammation. Clin Dev Immunol 2013:808367. https://doi.org/10.1155/2013/808367

Schneider M, Kirfel G, Berthold M, Frentzen M, Krause F, Braun A (2012) The impact of antimicrobial photodynamic therapy in an artificial biofilm model. Lasers Med Sci 27(3):615–620. https://doi.org/10.1007/s10103-011-0998-7

Schwarz SR, Hirsch S, Hiergeist A, Kirschneck C, Muehler D, Hiller KA et al (2020) Limited antimicrobial efficacy of oral care antiseptics in microcosm biofilms and phenotypic adaptation of bacteria upon repeated exposure. Clin Oral Investig. https://doi.org/10.1007/s00784-020-03613-w

Sculean A, Bastendorf KD, Becker C, Bush B, Einwag J, Lanoway C et al (2013) A paradigm shift in mechanical biofilm management? Subgingival air polishing: a new way to improve mechanical biofilm management in the dental practice. Quintessence Int 44(7):475–477. https://doi.org/10.3290/j.qi.a29615

Siddharth M, Singh P, Gupta R, Sinha A, Shree S, Sharma K (2020) A comparative evaluation of Subgingivally delivered 2% curcumin and 0.2% chlorhexidine gel adjunctive to scaling and root Planing in chronic periodontitis. J Contemp Dent Pract 21(5):494–499

Silva MP, Feres M, Sirotto TA, Soares GM, Mendes JA, Faveri M, Figueiredo LC (2011) Clinical and microbiological benefits of metronidazole alone or with amoxicillin as adjuncts in the treatment of chronic periodontitis: a randomized placebo-controlled clinical trial. J Clin Periodontol 38(9):828–837. https://doi.org/10.1111/j.1600-051X.2011.01763.x

Sindhura H, Harsha RH, Shilpa RH (2017) Efficacy of subgingival irrigation with 10% povidone-iodine as an adjunct to scaling and root planing: a clinical and microbiological study. Indian J Dent Res 28(5):514–518. https://doi.org/10.4103/ijdr.IJDR_497_15

Singhal S, Pradeep AR, Kanoriya D, Garg S, Garg V (2018) Boric acid gel as local drug delivery in the treatment of class II furcation defects in chronic periodontitis: a randomized, controlled clinical trial. J Investig Clin Dent 9(1). https://doi.org/10.1111/jicd.12279

Stratul SI, Rusu D, Didilescu A, Mesaros-Anghel M, Lala C, Tion L et al (2010) Prospective clinical study evaluating the long-time adjunctive use of chlorhexidine after one-stage full-mouth SRP. Int J Dent Hyg 8(1):35–40. https://doi.org/10.1111/j.1601-5037.2009.00390.x

Sulijaya B, Takahashi N, Yamazaki K (2019) Host modulation therapy using anti-inflammatory and antioxidant agents in periodontitis: a review to a clinical translation. Arch Oral Biol 105:72–80. https://doi.org/10.1016/j.archoralbio.2019.07.002

Suvan J, Leira Y, Moreno Sancho FM, Graziani F, Derks J, Tomasi C (2020) Subgingival instrumentation for treatment of periodontitis. A systematic review. J Clin Periodontol 47(Suppl 22):155–175. https://doi.org/10.1111/jcpe.13245

Swierkot K, Nonnenmacher CI, Mutters R, Flores-de-Jacoby L, Mengel R (2009) One-stage full-mouth disinfection versus quadrant and full-mouth root planing. J Clin Periodontol 36(3):240–249. https://doi.org/10.1111/j.1600-051X.2008.01368.x

Takasaki AA, Aoki A, Mizutani K, Schwarz F, Sculean A, Wang CY et al (2009) Application of antimicrobial photodynamic therapy in periodontal and peri-implant diseases. Periodontol 2000 51:109–140. https://doi.org/10.1111/j.1600-0757.2009.00302.x

Tonetti MS, Chapple IL, Periodontology, W. G. O. S. E. W. O (2011) Biological approaches to the development of novel periodontal therapies–consensus of the Seventh European Workshop on Periodontology. J Clin Periodontol 38(Suppl 11):114–118. https://doi.org/10.1111/j.1600-051X.2010.01675.x

Trajano VCDC, Brasileiro CB, Henriques JAS, Cota LM, Lanza CR, Cortés ME (2020) Doxycycline encapsulated in β-cyclodextrin for periodontitis: a clinical trial. Braz Oral Res 33:e112. https://doi.org/10.1590/1807--3107bor-2019.vol33.0112

Tsang YC, Corbet EF, Jin LJ (2018) Subgingival glycine powder air-polishing as an additional approach to non-surgical periodontal therapy in subjects with untreated chronic periodontitis. J Periodontal Res 53(3):440–445. https://doi.org/10.1111/jre.12532

Umeda M, Takeuchi Y, Noguchi K, Huang Y, Koshy G, Ishikawa I (2004) Effects of non-surgical periodontal therapy on the microbiota. Periodontol 2000 36:98–120. https://doi.org/10.1111/j.1600-0757.2004.03675.x

Van der Weijden FA, Slot DE (2015) Efficacy of homecare regimens for mechanical plaque removal in managing gingivitis a meta review. J Clin Periodontol 42(Suppl 16):S77–S91. https://doi.org/10.1111/jcpe.12359

Van der Weijden GA, Timmerman MF, Nijboer A, Lie MA, Van der Velden U (1993) A comparative study of electric toothbrushes for the effectiveness of plaque removal in relation to toothbrushing duration. Timerstudy. J Clin Periodontol 20(7):476–481. https://doi.org/10.1111/j.1600-051x.1993.tb00394.x

Van Winkelhoff AJ, Van der Velden U, Clement M, De Graaff J (1988) Intra-oral distribution of black-pigmented Bacteroides species in periodontitis patients. Oral Microbiol Immunol 3(2):83–85. https://doi.org/10.1111/j.1399-302x.1988.tb00087.x

Vyas N, Dehghani H, Sammons RL, Wang QX, Leppinen DM, Walmsley AD (2017) Imaging and analysis of individual cavitation microbubbles around dental ultrasonic scalers. Ultrasonics 81:66–72. https://doi.org/10.1016/j.ultras.2017.05.015

Webber MA, Whitehead RN, Mount M, Loman NJ, Pallen MJ, Piddock LJ (2015) Parallel evolutionary pathways to antibiotic resistance selected by biocide exposure. J Antimicrob Chemother 70(8):2241–2248. https://doi.org/10.1093/jac/dkv109

Wennström JL, Tomasi C, Bertelle A, Dellasega E (2005) Full-mouth ultrasonic debridement versus quadrant scaling and root planing as an initial approach in the treatment of chronic periodontitis. J Clin Periodontol 32(8):851–859. https://doi.org/10.1111/j.1600-051X.2005.00776.x

Wennström JL, Dahlén G, Ramberg P (2011) Subgingival debridement of periodontal pockets by air polishing in comparison with ultrasonic instrumentation during maintenance therapy. J Clin Periodontol 38(9):820–827. https://doi.org/10.1111/j.1600-051X.2011.01751.x

Ximénez-Fyvie LA, Haffajee AD, Socransky SS (2000) Comparison of the microbiota of supra- and subgingival plaque in health and periodontitis. J Clin Periodontol 27(9):648–657. https://doi.org/10.1034/j.1600-051x.2000.027009648.x

Yan D, Sherman JH, Keidar M (2017) Cold atmospheric plasma, a novel promising anti-cancer treatment modality. Oncotarget 8(9):15977–15995. https://doi.org/10.18632/oncotarget.13304

Yang Y, Guo J, Zhou X, Liu Z, Wang C, Wang K et al (2018) A novel cold atmospheric pressure air plasma jet for peri-implantitis treatment: an in vitro study. Dent Mater J 37(1):157–166. https://doi.org/10.4012/dmj.2017-030

Zhang W, Daly CG, Mitchell D, Curtis B (2013) Incidence and magnitude of bacteraemia caused by flossing and by scaling and root planing. J Clin Periodontol 40(1):41–52. https://doi.org/10.1111/jcpe.12029

Zhao H, Hu J, Zhao L (2020a) Adjunctive subgingival application of chlorhexidine gel in nonsurgical periodontal treatment for chronic periodontitis: a systematic review and meta-analysis. BMC Oral Health 20(1):34. https://doi.org/10.1186/s12903-020-1021-0

Zhao P, Song X, Nie L, Wang Q, Zhang P, Ding Y (2020b) Efficacy of adjunctive photodynamic therapy and lasers in the non-surgical periodontal treatment: a Bayesian network meta-analysis. Photodiagn Photodyn Ther 101969. https://doi.org/10.1016/j.pdpdt.2020.101969

Zhong Z, Wheeler MD, Li X, Froh M, Schemmer P, Yin M et al (2003) L-Glycine: a novel antiinflammatory, immunomodulatory, and cytoprotective agent. Curr Opin Clin Nutr Metab Care 6(2):229–240. https://doi.org/10.1097/00075197-200303000-00013

Zijnge V, Meijer HF, Lie MA, Tromp JA, Degener JE, Harmsen HJ, Abbas F (2010) The recolonization hypothesis in a full-mouth or multiple-session treatment protocol: a blinded, randomized clinical trial. J Clin Periodontol 37(6):518–525. https://doi.org/10.1111/j.1600-051X.2010.01562.x

Update on the Roles of Oral Hygiene and Plaque Control on Periodontal Disease

17

Leila Salhi, Bruno De Carvalho, and Michèle Reners

Abstract

Aim: to provide an update of the evidence on the effect of oral hygiene instructions (OHI), dental plaque control and in the prevention and treatment of periodontitis.

Methods: Literature searches were performed using MeSH terms, keywords and free words and were published between 2015 and November 2020. The data from the articles were summarized in a narrative review.

Results: Data concerning the influence of OHI on periodontal features, the impact of OHI before periodontitis non-surgical treatment, its efficacity on periodontitis prevention and maintenance of healthy periodontium were summarized in the tables of the present narrative review.

Conclusion: as prevention is better than a cure, it is relevant to bring in light the role of oral hygiene instructions, the patient self-control of dental plaque as well as the professional mechanical plaque removal in the prevention of periodontitis.

L. Salhi (✉) · B. De Carvalho
Department of Periodontology, Buccal Surgery and Implantology, University of Liège, Liège, Belgium
e-mail: l.salhi@chuliege.be; bruno.decarvalho@chuliege.be

M. Reners
Private Practice, Paroliège, Liège, Belgium
e-mail: mreners@gmail.com

Keywords

Oral hygiene · Plaque control · Oral hygiene instruction · Periodontitis · Prevention

Abbreviations

BOP	Bleeding on probing BOP MTB
EFP	European Federation of Periodontology
GH	General health
IDBs	Interdental brushes
MTB	Manual tooth brushing
NCDs	Non-communicable diseases
OH	Oral health
ORCA	European Organization for Caries Research
PNST	Periodontal non-surgical treatment
PTB	Powered tooth brushing
PTB	Powered tooth brushing
SPC	supportive periodontal care (SPC)

Highlights

- Poor oral hygiene leads to periodontal disease
- Oral hygiene instructions should be considered as the most important part of the periodontal treatment

J. Santi-Rocca (ed.), *Periodontitis*, Advances in Experimental Medicine and Biology 1373,
https://doi.org/10.1007/978-3-030-96881-6_17

- Prevention prevents periodontal disease genesis
- Maintenance and motivation aim to control the progression of periodontal disease
- Powered toothbrush in combination with interdental brushes allow to maintain a healthy periodontium

Considerations for Practice

- Oral hygiene instructions (OHI) should be considered as a first part of the periodontal treatment
- The effect of OHI should be evaluated at each time of periodontal treatment
- The prevention of periodontitis recurrence is limited when efficient plaque control is provided by both patient and professional

Patient Summary

- A good oral health begins with the control of dental plaque
- The efficient use of oral hygiene devices, explained by professional, prevents periodontal disease
- To avoid recurrence of periodontal disease, the patient should be involved in a maintenance program to evaluate his plaque control and to help him to maintain it
- The prevention of any disease is better than a cure, do not hesitate to ask your dental practitioner for a periodontal screening

17.1 Introduction

Oral health (OH) is multifaceted and includes namely, the ability to speak, smile, smell, chew, without pain or discomfort (Tonetti et al. 2013) being also an indicator of overall health, well-being and quality of life (Clark-Perry and Levin 2020). Among the oral health conditions, periodontitis interferes with general health as with specific non-communicable diseases (NCDs), as cardiovascular pathologies and diabetes (Jin et al. 2016; Jin 2013). Periodontitis is defined as a chronic inflammatory disease, characterized by gram negative bacteria organized in a dental biofilm, leading to a chronic non-resolving and destructive inflammatory processes inside the gingiva (Jepsen et al. 2017; Meyle and Chapple 2015). Approximately fifty percent of the adult population suffer from this chronic disease (Eke et al. 2016), and severe periodontitis affects 11.2% of the population (Kassebaum et al. 2014). The disease involves firstly the progressive destruction of the supporting tissues around teeth (Meyle and Chapple 2015; Schenkein 2006; Haffajee and Socransky 1994; Amano 2000a, b), and secondly, due to mastication and tooth brushing, periodontitis induces the dissemination of periodontal pathogens or/and their sub-products (SPs) from the pocket depth to the blood circulation (Forner et al. 2006).

This bacteremia contributes to increase the systemic inflammation and promoting the colonization of periodontal bacteria on an extra-oral site (Van Dyke et al. 2020; Loos 2005; Graziani et al. 2018a; Szulc et al. 2015; Tonetti et al. 2013; Mealey 1999; van Winkelhoff and Slots 1999) both implicated in the relationship between periodontitis and chronic NCDs (Graziani et al. 2018a; Sanz et al. 2018; Tonetti et al. 2013; Salhi et al. 2019, 2020b). In fact, as periodontal bacteria are able to translocate from periodontal niches to distant oral organ, that contribute to a negative impact on general health (GH) (Pasnik-Chwalik and Konopka 2020; Slots

2003; Kumar 2017) (called the focal infection theory (Kumar 2017)), the control of an oral microbiome is required for both periodontal health (Guerra et al. 2018) and GH. As periodontitis is largely preventable, the control of the etiological factor of the disease is needed by the control of dental plaque accumulation towards adequate personal oral hygiene (OHI) and professional care.

Therefore, the aim of this review is to bring in light the updated evidence on the relevance of plaque control in the periodontitis physiopathology, focusing on the genesis, the treatment, the maintenance, and the prevention of the disease.

17.2 Methods

This narrative review was focused on recent studies between 2015 and 2020. The literature search was performed on PubMed, and inclusion criteria were systematic reviews, consensus reports and randomized controlled trials assessing oral hygiene, plaque control, oral hygiene instruction, and periodontitis prevention.

17.3 Results

17.3.1 Effect of Plaque control on Periodontal Features-Periodontitis Genesis

The effect of plaque control on periodontal features-periodontitis genesis is summarized in Table 17.1.

A recent Cochrane systematic review based on 35 studies (Worthington et al. 2019), concluded that using floss or interdental brushes in addition to toothbrushing reduce gingivitis or plaque, or both, aiming to prevent and control periodontal diseases. The conclusion of the authors was previously described in other systematic review with meta-analysis (Kotsakis et al. 2018; Lertpimonchai et al. 2017), randomized control trial (De David et al. 2018; Graziani

Table 17.1 The effect of plaque control on periodontal features- periodontitis genesis

Authors	Study design	Conclusion
Worthington et al. (2019)	Systematic review on 35 studies	Using floss or interdental brushes in addition to toothbrushing reduce gingivitis or plaque, or both, more than toothbrushing alone. Interdental brushes reduce gingivitis more than flossing The conclusions were based on evidences characterized as low to very low-certainty, and the effect sizes observed may not be clinically important
Kotsakis et al. (2018)	Meta-analysis on 22 studies	Interdental brushes led to a reduction of 64.7% of GI water-jet led 27.4% of GI
De David et al. (2018)	Controlled study on 52 subjects that received for 30 days OH at 12-, 24- (group 1), or 48-, 72-h (group 2) interval	Oral hygiene frequencies are required to maintain gingival health, observed by plaque formation
Graziani et al. (2018b)	Randomized controlled trial on 60 periodontally healthy patients	Interdental picks were associated with reduced interdental full mouth bleeding score when compared to flossing ($p < 0.05$) Use of interdental brushes or rubber picks reduces more interdental plaque in comparison with toothbrushing alone
Jepsen et al. (2017)	Consensus report	The management gingivitis required Self-performed oral hygiene and interdental cleaning Professional tooth cleaning, oral hygiene instruction and motivation
Lertpimonchai et al. (2017)	Systematic review and meta-analysis on 15 studies reported OH as categorical	OH decreases the risk of periodontitis by two- to five-fold. Regular toothbrushing and dental visits reduced the risk to periodontitis genesis

IOH interproximal oral hygiene, *GI* gingival index, *OH* oral hygiene et al. 2018b) and consensus report (Jepsen et al. 2017). OH decreases the risk of periodontitis genesis by two- to five-fold (Lertpimonchai et al. 2017) and, self-performed oral hygiene and interdental cleaning was required to manage gingivitis (Jepsen et al. 2017). Interdental brush was seen to be more efficacious than the majority of the alternative oral hygiene in gingival index reduction (Worthington et al. 2019; Kotsakis et al. 2018; Richards 2018), plaque reduction (Worthington et al. 2019; Graziani et al. 2018b).

17.3.2 Effect of Oral Hygiene Devices on Periodontal Features

The effect of oral hygiene devices on periodontal features is summarized in Table 17.2.

Recent systematic reviews with meta-analysis (Elkerbout et al. 2020; Wang et al. 2020; Clark-Perry and Levin 2020), randomized control trials (RCT) (Starke et al. 2019a, b; Cui et al. 2017; Delaurenti et al. 2017; Schmalz et al. 2017) and consensus (Chapple et al. 2015) described that powered tooth brushing (PTB) reduced significantly dental plaque (Elkerbout et al. 2020; Wang et al. 2020; Starke et al. 2019a, b; Cui et al. 2017; Delaurenti et al. 2017; Schmalz et al. 2017), gingival bleeding (Wang et al. 2020; Starke et al. 2019a, b; Delaurenti et al. 2017; Schmalz et al. 2017) and gingival inflammation (Wang et al. 2020; Starke et al. 2019a, b; Cui et al. 2017; Delaurenti et al. 2017) than manual tooth brushing (MTB). In addition, the use of calibrated interdental brushes reduces significantly interdental bleeding and plaque when used with PTB (Chapple et al. 2015; Bourgeois et al. 2016; Sambunjak et al. 2019).

17.3.3 Effect of OH Before Periodontitis Non-surgical Treatment -Periodontitis Treatment

The effect of OH before periodontitis non-surgical treatment -periodontitis treatment is summarized in Table 17.3.

Recent research on controlled trials (Preus et al. 2020; Salhi et al. 2020a) demonstrated that patients who received strict oral hygiene phase prior to periodontitis treatment experienced a significant reduction of plaque, bleeding on probing BOP, and pocket depth.

17.3.4 Effect of OH on Periodontitis Prevention and Maintenance of Healthy Periodontium

The effect of OH on periodontitis prevention and maintenance of healthy periodontium is summarized in Table 17.4.

Recent systematic reviews (Slot et al. 2020; Figuero et al. 2017) and consensus report (Müller Campanile et al. 2019) highlighted the efficacity of combined professional and self-performed mechanical plaque removal in the prevention of periodontal diseases. Concerning the periodontal maintenance care, a recent systematic review with meta-analysis (Slot et al. 2020) on the efficacy of mechanical oral hygiene devices showed that the use of the interdental brushes (IDBs) reduced more effectively dental plaque than a manual toothbrush alone. Furthermore, in a cross sectional study on 100 patients, the authors concluded that tooth loss can be contained when patients underwent to regular maintenance (Müller Campanile et al. 2019), that findings were already previously described in the systematic review of Trombelli et al. (2015). Finally, a recent consensus (Sanz et al. 2015) on the effective prevention of periodontal and peri-implant diseases supported that, in addition to professional mechanical plaque removal, secondary prevention of periodontitis should also include the evaluation of oral hygiene performance, motivation and re-instruction in oral hygiene practices.

17.4 Discussion

Dental plaque accumulation, at and below the gingival margin, is the primary etiological risk factor for the development of gingivitis and further periodontitis (Mariotti 1999; Albandar et al.

Table 17.2 The effect of oral hygiene devices on periodontal features

Authors	Study design	Conclusion
Elkerbout et al. (2020)	Systematic review and meta-analysis on 17 studies	Based on 28 comparisons assessed toothbrushing efficacy according to the plaque index score (Q & HPI and RMNPI): PTB showed a significant effect compare to MTB, the difference was −0.14 ($P < 0.001$; 95%CI [−0.19; −0.09]) for the Q & HPI and, −0.10 ($P < 0.001$; 95%CI [−0.14; −0.06]) for the RMNPI
Wang et al. (2020)	Systematic review and meta-analysis on 21 studies	PTB more effective in reducing dental plaque, gingivitis and bleeding compared with MTB
Clark-Perry and Levin (2020)	Systematic review and meta-analysis on respectively 15 and 12 studies	OR PTB reduced statistically plaque index ($p < 0.01$) and number of bleeding sites ($p < .001$), than other PTB
Starke et al. (2019a)	Randomized controlled study on 148 patients Effect of PTB or MTB on gingivitis and plaque following two and four weeks of home use PTB, n = 74 MTB, n = 74	PTB was statistically significantly superior to MTB in reducing gingival inflammation, gingival bleeding, and plaque
Starke et al. (2019a)	Randomized controlled study on 2188 patients evaluating the effect of PTB and MTB on plaque and gingivitis	PTB was statistically significantly superior to MTB in reducing gingival inflammation, gingival bleeding, and plaque
Sambunjak et al. (2019)	Systematic review on 12 studies	Flossing in addition to toothbrushing reduces gingivitis compared to toothbrushing alone.
Cui et al. (2017)	Randomized controlled trial on 42 patients	PTB reduced plaque and gingivitis more than MTB
Delaurenti et al. (2017)	Randomized controlled trial on 144 patients, using twice-daily home use of PTB, n = 77 MTB, n = 77	PTB was statistically significantly more effective than MTB in reducing supragingival plaque, gingival inflammation, and gingival bleeding following
Schmalz et al. (2017)	Randomized clinical study on 72 patients, influence of different devices on periodontal clinical parameters OR, n = 24 SA, n = 24 MTB,n = 24	SA significantly reduced the bleeding compared to OR and MTB ($P < 0.01$). OR significantly improved in the plaque removal than MTB ($P = 0.01$). SA significantly reduced the probing pocket depth compared to MTB
Bourgeois et al. (2016)	Randomized controlled trial on 46 patients Standard manual toothbrush twice daily and an interdental brush daily (n = 23) Standard manual toothbrush (n = 23)	Daily use of calibrated interdental brushes reduces interdental bleeding
Chapple et al. (2015)	Workshop with 2 meta-reviews (mechanical plaque control) and 2 systematic reviews (chemical plaque control/anti-inflammatory agents)	PTB provide small but statistically significant additional reductions in gingival inflammations and plaque scores Interdental brushes are the device of choice for the interproximal plaque removal Flossing cannot be recommended it, unless the interdental brushes will not pass through the interproximal area without trauma Use of systemic or anti-inflammatory agents in the management of gingivitis has no robust evidence base

Q & HPI Quigley-Hein plaque, *RMNPI* index or the Rustogi modified Navy plaque index, *PTB* powered toothbrush, *MTB* manual toothbrush, *OR* oscillating-rotating, *SA* sonic-active

Table 17.3 The effect of oral hygiene before periodontitis non-surgical treatment -periodontitis treatment

Authors	Study design	Conclusion
Preus et al. (2020)	Randomized controlled trial on 44 patients Strict oral hygiene phase 3 months prior to periodontitis treatment Did not receive any instructions or motivation On oral hygiene prior to periodontitis treatment	Strict oral hygiene phase prior to periodontitis treatment reduced plaque, BOP and pocket depth ($p < 0.001$)
Salhi et al. (2020a)	Controlled study on 34 non-, 25 former- And 32 current- smokers Effect of OHI and periodontitis non-surgical treatment	OHI led to a significant decrease of PD, BOP, and PISA only in non-smokers and former smokers ($p < 0.0001$)

Table 17.4 The effect of oral hygiene on periodontitis prevention and maintenance of healthy periodontium

Authors	Study design	Conclusion
Slot et al. (2020)	Systematic review and meta-analysis on 17 studies	Interdental brushes (IDBs) reduced plaque scores more effectively than a manual toothbrush alone
Müller Campanile et al. (2019)	Cross sectional study on 100 patient treated for periodontitis	Tooth loss and periodontal tissue damage can be contained over prolonged periods if periodontal disease is treated and patients attend regular maintenance care
Figuero et al. (2017)	Systematic review on 27 studies	Combined professional and self-performed mechanical plaque control significantly reduces plaque index ($p = 0.003$) and ($p = 0.002$)
Tonetti et al. (2015)	Consensus report	Repeated and individualized oral hygiene instruction and professional mechanical plaque removal are important components of preventive programs
Sanz et al. (2015)	Consensus report	Professional mechanical plaque removal in the context of secondary prevention of periodontitis should also include the evaluation of oral hygiene performance, motivation and re-instruction in oral hygiene practices.
Trombelli et al. (2015)	Systematic review on 19 studies	Professional plaque removal may limit the incidence and yearly rate of tooth loss as well as the loss in clinical attachment in patients treated for periodontitis.
Van der Weijden and Slot (2015)	Meta review on 10 systematic review	Tooth brushing is effective in reducing levels of dental plaque. With respect to gingivitis power toothbrushes have a benefit over manual toothbrushes

1998). Although not all cases of gingivitis evolve to periodontitis, the calculus removal and the management of gingivitis by daily effective self-performed mechanical plaque control and the motivation of oral health by professional are a fundamental in the prevention and the maintenance of periodontitis (Jepsen et al. 2017; Worthington et al. 2019). This narrative review discloses the updated evidences on the effect of plaque control on the prevention of periodontitis genesis (Table 17.1), the efficacity of oral hygiene devices (Table 17.2), the effect of OH on the treatment of the disease (Table 17.3), as well as the maintenance of periodontal health and the prevention of the disease (Table 17.4) (Chapple et al. 2015; Ramseier et al. 2017).

Since management of gingivitis is considered as the main prevention of periodontitis (Chapple et al. 2015), both oral hygiene instructions and motivation provided by dental professionals, and patient self-performed mechanical plaque control are required to the control of dental biofilm and gingivitis. Recent systematic review (Worthington et al. 2019; Richards 2018) and meta-analysis (Kotsakis et al. 2018) support that oral hygiene lowers the risk of gingivitis and plaque accumulation. Additionally, the frequency of OH (Lertpimonchai et al. 2017; De David et al. 2018) as well the use of interdental brushes (Worthington et al. 2019; Kotsakis et al. 2018; Richards 2018) are efficient to maintain gingival health. These evidences have been already described in the consensus report of the joint European Federation of Periodontology and European Organization for Caries Research (EFP/ORCA) (Jepsen et al. 2017) which concluded that self-performed oral hygiene and interdental cleaning, as well as professional oral hygiene instructions are essential to the management of gingivitis. When focusing on the efficacity of dental devices in the reduction of plaque control and bleeding on probing, recent systematic reviews (Elkerbout et al. 2020; Wang et al. 2020; Clark-Perry and Levin 2020) and randomized controlled trials (Starke et al. 2019a, b; Cui et al. 2017; Delaurenti et al. 2017; Schmalz et al. 2017) supported the benefit of powered tooth brush (PTB) over to manual ones. According to Chapple et al. (2015), powered toothbrushes present greater short- and long-term reductions in plaque indices and gingival inflammation when compared to manual ones. In addition to PTB, the use of daily interdental brushes reduces interdental bleeding (Bourgeois et al. 2016; Sambunjak et al. 2019).

Nevertheless, regarding the effect of oral hygiene on periodontitis non-surgical treatment (PNST), only few studies evaluated and quantified OH as an individual component of the PNST. Preus et al. (2020) and Salhi et al. (2020a) concluded that strict oral hygiene phase prior to periodontitis treatment significantly decreased plaque, bleeding on probing as well as pocket depth. The combination of professional and self-performed mechanical plaque control reduces the plaque index, and therefore contributes to the maintenance of a healthy periodontium and the secondary prevention of periodontitis (Slot et al. 2020; Figuero et al. 2017). In addition to the role of mechanical plaque control, the motivation, and the promotion of oral health by professional can participate to improve the oral health care and education, consolidating the periodontal treatment (Nakata et al. 2019; Garyga et al. 2019; Stenman et al. 2018).

This evidence has been supported in the conclusions of the 11th consensus report of the European Workshop on Periodontology (2015) on effective prevention of periodontal and peri-implant diseases (Sanz et al. 2015).

After active periodontal therapy, patients can fall in two different categories: patients with reduced but healthy periodontium and patients with gingival inflammation. The first group presents a risk on recurrence and the second one a risk of progression. Yet, the risks of disease recurrence after periodontal therapy are a reality and the control of systemic and local risk factors highly important. Firstly, the constant self-motivation of the periodontal compromised patient is always a challenge in maintaining plaque control, smoking cessation, keeping a balanced diet and maintaining control of diabetes (Chapple et al. 2013). Secondly, the supportive periodontal care (SPC) performed by dental health professionals, remains strictly important in the detection of disease recurrence and resurgence of subgingival biofilm (Müller Campanile et al. 2019). Ideally, supportive periodontal therapy should be optimized to the specific patient's risk profile and to the periodontal conditions after active therapy (Sanz et al. 2020). The importance of adherence to the SPC is crucial for long term periodontal stability, further improvements in the periodontal status. Irregular compliances are intimately related with more tooth loss and disease progression (Costa et al. 2014).

Nevertheless, it is relevant to not forget the precursors of the reflection on the impact of oral hygiene on periodontal disease. In the past century, Lövdal (Lovdal et al. 1958) and Arno (Arno et al. 1958) explained, in 1958, the existence of a close correlation between periodontal destruction and oral debris. These findings will then evidence, in 1964, by Theilade (1964) who demonstrated microscopically that an intimate anatomical relationship exists between bacteria and the gingival tissues. Further clinical observations supporting the observation of the pioneers raised, with the combined effect of subgingival scaling and controlled oral hygiene on the incidence of gingivitis (Lovdal et al. 1961) and, with the production of an experimental gingivitis by Loe et al.(1965). Therefore, there is a need to continue research of our fathers with the influence of oral hygiene on periodontal disease prevention, genesis, and treatment.

17.5 Conclusion

Despite, the lack of recent evidence litterature based on the effect of OHI on periodontal non-surgical treatment, the clinical evidence of the negative relationship between dental plaque exists. Therefore, the control of oral hygiene and plaque accumulation by both professional and patient contribute to healthy periodontium and should be considered as the most important part of the periodontal treatment.

Acknowledgements Include relevant funding here if required.

References

Albandar JM, Kingman A, Brown LJ, Loe H (1998) Gingival inflammation and subgingival calculus as determinants of disease progression in early-onset periodontitis. J Clin Periodontol 25(3):231–237. https://doi.org/10.1111/j.1600-051x.1998.tb02433.x

Amano A (2000a) Host-parasite interactions in periodontitis: microbial pathogenicity and innate immunity. Periodontology 54(1):9–14. https://doi.org/10.1111/j.1600-0757.2010.00376.x

Amano A (2000b) Host-parasite interactions in periodontitis: subgingival infection and host sensing. Periodontology 52(1):7–11. https://doi.org/10.1111/j.1600-0757.2009.00328.x

Arno A, Waerhaug J, Lovdal A, Schei O (1958) Incidence of gingivitis as related to sex, occupation, tobacco consumption, toothbrushing, and age. Oral Surg Oral Med Oral Pathol 11(6):587–595. https://doi.org/10.1016/0030-4220(58)90004-5

Bourgeois D, Saliasi I, Llodra JC, Bravo M, Viennot S, Carrouel F (2016) Efficacy of interdental calibrated brushes on bleeding reduction in adults: a 3-month randomized controlled clinical trial. Eur J Oral Sci 124(6):566–571. https://doi.org/10.1111/eos.12302

Chapple IL, Genco R, Working group 2 of joint EFPAAPw (2013) Diabetes and periodontal diseases: consensus report of the Joint EFP/AAP workshop on periodontitis and systemic diseases. J Clin Periodontol 40(Suppl 14):S106–S112. https://doi.org/10.1111/jcpe.12077

Chapple IL, Van der Weijden F, Doerfer C, Herrera D, Shapira L, Polak D, Madianos P, Louropoulou A, Machtei E, Donos N, Greenwell H, Van Winkelhoff AJ, Eren Kuru B, Arweiler N, Teughels W, Aimetti M, Molina A, Montero E, Graziani F (2015) Primary prevention of periodontitis: managing gingivitis. J Clin Periodontol 42(Suppl 16):S71–S76. https://doi.org/10.1111/jcpe.12366

Clark-Perry D, Levin L (2020) Systematic review and meta-analysis of randomized controlled studies comparing oscillating-rotating and other powered toothbrushes. J Am Dent Assoc 151(4):265–275.e6. https://doi.org/10.1016/j.adaj.2019.12.012

Costa FO, Lages EJ, Cota LO, Lorentz TC, Soares RV, Cortelli JR (2014) Tooth loss in individuals under periodontal maintenance therapy: 5-year prospective study. J Periodontal Res 49(1):121–128. https://doi.org/10.1111/jre.12087

Cui TQ, Lin HC, Lo ECM, Tao Y, Zhou Y, Zhi QH (2017) Randomized clinical trial on the efficacy of electric and manual toothbrushes in plaque removal and gingivitis control in visually impaired school students. Quintessence Int 48(6):481–486. https://doi.org/10.3290/j.qi.a38126

De David SC, Mario TG, De Freitas GC, Kantorski KZ, Wikesjo UME, Moreira CHC (2018) Correlation between plaque control and gingival health using short and extended oral hygiene intervals. Clin Oral Investig 22(7):2593–2597. https://doi.org/10.1007/s00784-018-2358-5

Delaurenti M, Ward M, Souza S, Jenkins W, Putt MS, Milleman KR, Milleman JL (2017) The effect of use of a sonic power toothbrush and a manual toothbrush control on plaque and gingivitis. J Clin Dent 28 (1 Spec No A):A1–6

Eke PI, Wei L, Borgnakke WS, Thornton-Evans G, Zhang X, Lu H, McGuire LC, Genco RJ (2016) Periodontitis prevalence in adults >/= 65 years of age, in the USA. Periodontology 72(1):76–95. https://doi.org/10.1111/prd.12145

Elkerbout TA, Slot DE, Rosema NAM, Van der Weijden GA (2020) How effective is a powered toothbrush as compared to a manual toothbrush? A systematic review and meta-analysis of single brushing exercises. Int J Dent Hyg 18(1):17–26. https://doi.org/10.1111/idh.12401

Figuero E, Nobrega DF, Garcia-Gargallo M, Tenuta LM, Herrera D, Carvalho JC (2017) Mechanical and chemical plaque control in the simultaneous management of gingivitis and caries: a systematic review. J Clin Periodontol 44(Suppl 18):S116–S134. https://doi.org/10.1111/jcpe.12674

Forner L, Larsen T, Kilian M, Holmstrup P (2006) Incidence of bacteremia after chewing, tooth brushing and scaling in individuals with periodontal inflamma-

tion. J Clin Periodontol 33(6):401–407. https://doi.org/10.1111/j.1600-051X.2006.00924.x
Garyga V, Pochelu F, Thivichon-Prince B, Aouini W, Santamaria J, Lambert F, Maucort-Boulch D, Gueyffier F, Gritsch K, Grosgogeat B (2019) GoPerio – impact of a personalized video and an automated two-way text-messaging system in oral hygiene motivation: study protocol for a randomized controlled trial. Trials 20(1):699. https://doi.org/10.1186/s13063-019-3738-0
Graziani F, Gennai S, Solini A, Petrini M (2018a) A systematic review and meta-analysis of epidemiologic observational evidence on the effect of periodontitis on diabetes An update of the EFP-AAP review. J Clin Periodontol 45(2):167–187. https://doi.org/10.1111/jcpe.12837
Graziani F, Palazzolo A, Gennai S, Karapetsa D, Giuca MR, Cei S, Filice N, Petrini M, Nisi M (2018b) Interdental plaque reduction after use of different devices in young subjects with intact papilla: a randomized clinical trial. Int J Dent Hyg 16(3):389–396. https://doi.org/10.1111/idh.12318
Guerra F, Mazur M, Ndokaj A, Corridore D, La Torre G, Polimeni A, Ottolenghi L (2018) Periodontitis and the microbiome: a systematic review and meta-analysis. Minerva Stomatol 67(6):250–258. https://doi.org/10.23736/S0026-4970.18.04198-5
Haffajee AD, Socransky SS (1994) Microbial etiological agents of destructive periodontal diseases. Periodontology 5:78–111. https://doi.org/10.1111/j.1600-0757.1994.tb00020.x
Jepsen S, Blanco J, Buchalla W, Carvalho JC, Dietrich T, Dorfer C, Eaton KA, Figuero E, Frencken JE, Graziani F, Higham SM, Kocher T, Maltz M, Ortiz-Vigon A, Schmoeckel J, Sculean A, Tenuta LM, van der Veen MH, Machiulskiene V (2017) Prevention and control of dental caries and periodontal diseases at individual and population level: consensus report of group 3 of joint EFP/ORCA workshop on the boundaries between caries and periodontal diseases. J Clin Periodontol 44(Suppl 18):S85–S93. https://doi.org/10.1111/jcpe.12687
Jin L (2013) The global call for oral health and general health. Int Dent J 63(6):281–282. https://doi.org/10.1111/idj.12085
Jin LJ, Lamster IB, Greenspan JS, Pitts NB, Scully C, Warnakulasuriya S (2016) Global burden of oral diseases: emerging concepts, management and interplay with systemic health. Oral Dis 22(7):609–619. https://doi.org/10.1111/odi.12428
Kassebaum NJ, Bernabe E, Dahiya M, Bhandari B, Murray CJ, Marcenes W (2014) Global burden of severe tooth loss: a systematic review and meta-analysis. J Dent Res 93(7 Suppl):20S–28S. https://doi.org/10.1177/0022034514537828
Kotsakis GA, Lian Q, Ioannou AL, Michalowicz BS, John MT, Chu H (2018) A network meta-analysis of interproximal oral hygiene methods in the reduction of clinical indices of inflammation. J Periodontol 89(5):558–570. https://doi.org/10.1002/JPER.17-0368
Kumar PS (2017) From focal sepsis to periodontal medicine: a century of exploring the role of the oral microbiome in systemic disease. J Physiol 595(2):465–476. https://doi.org/10.1113/JP272427
Lertpimonchai A, Rattanasiri S, Arj-Ong Vallibhakara S, Attia J, Thakkinstian A (2017) The association between oral hygiene and periodontitis: a systematic review and meta-analysis. Int Dent J 67(6):332–343. https://doi.org/10.1111/idj.12317
Löe H, Theilade E, Jensen SB (1965) Experimental gingivitis in man. J Periodontol 36(3):177–187
Loos BG (2005) Systemic markers of inflammation in periodontitis. J Periodontol 76(11 Suppl):2106–2115. https://doi.org/10.1902/jop.2005.76.11-S.2106
Lovdal A, Arno A, Schei O, Waerhaug J (1961) Combined effect of subgingival scaling and controlled oral hygiene on the incidence of gingivitis. Acta Odontol Scand 19:537–555. https://doi.org/10.3109/00016356109043406
Lovdal A, Arno A, Waerhaug J (1958) Incidence of clinical manifestations of periodontal disease in light of oral hygiene and calculus formation. J Am Dent Assoc 56(1):21–33. https://doi.org/10.14219/jada.archive.1958.0021
Mariotti A (1999) Dental plaque-induced gingival diseases. Ann Periodontol 4(1):7–19. https://doi.org/10.1902/annals.1999.4.1.7
Mealey BL (1999) Influence of periodontal infections on systemic health. Periodontology 21(1):197–209
Meyle J, Chapple I (2015) Molecular aspects of the pathogenesis of periodontitis. Periodontology 69(1):7–17. https://doi.org/10.1111/prd.12104
Müller Campanile V, Megally A, Campanile G, Gayet-Ageron A, Giannopoulou C, Mombelli A (2019) Risk factors for recurrence of periodontal disease in patients in maintenance care in a private practice. J Clin Periodontol 46(9):918–926. https://doi.org/10.1111/jcpe.13165
Nakata H, Matsuo K, Suzuki H, Yoshihara A (2019) Perioperative changes in knowledge and attitude toward oral health by oral health education. Oral Dis 25(4):1214–1220. https://doi.org/10.1111/odi.13048
Pasnik-Chwalik B, Konopka T (2020) Impact of periodontitis on the oral health impact profile: a systematic review and meta-analysis. Dent Med Probl 57(4):423–431. https://doi.org/10.17219/dmp/125028
Preus HR, Al-Lami Q, Baelum V (2020) Oral hygiene revisited. The clinical effect of a prolonged oral hygiene phase prior to periodontal therapy in periodontitis patients. a randomized clinical study. J Clin Periodontol 47(1):36–42. https://doi.org/10.1111/jcpe.13207
Ramseier CA, Anerud A, Dulac M, Lulic M, Cullinan MP, Seymour GJ, Faddy MJ, Burgin W, Schatzle M, Lang NP (2017) Natural history of periodontitis: Disease progression and tooth loss over 40 years. J Clin Periodontol 44(12):1182–1191. https://doi.org/10.1111/jcpe.12782

Richards D (2018) The effectiveness of interproximal oral hygiene aids. Evid Based Dent 19(4):107–108. https://doi.org/10.1038/sj.ebd.6401341

Salhi L, Rompen E, Sakalihasan N, Laleman I, Teughels W, Michel JB, Lambert F (2019) Can periodontitis influence the progression of abdominal aortic aneurysm? A systematic review. Angiology 70(6):479–491. https://doi.org/10.1177/0003319718821243

Salhi L, Albert A, Seidel L, Lambert F (2020a) Respective effects of oral hygiene instructions and periodontal nonsurgical treatment (debridement) on clinical parameters and patient-reported outcome measures with respect to smoking. J Clin Med 9(8). https://doi.org/10.3390/jcm9082491

Salhi L, Sakalihasan N, Okroglic AG, Labropoulos N, Seidel L, Albert A, Teughels W, Defraigne JO, Lambert F (2020b) Further evidence on the relationship between abdominal aortic aneurysm and periodontitis: a cross-sectional study. J Periodontol 91(11):1453–1464. https://doi.org/10.1002/JPER.19-0671

Sambunjak D, Nickerson JW, Poklepovic Pericic T, Johnson TM, Imai P, Tugwell P, Worthington HV (2019) WITHDRAWN: flossing for the management of periodontal diseases and dental caries in adults. Cochrane Database Syst Rev 4:CD008829. https://doi.org/10.1002/14651858.CD008829.pub3

Sanz M, Baumer A, Buduneli N, Dommisch H, Farina R, Kononen E, Linden G, Meyle J, Preshaw PM, Quirynen M, Roldan S, Sanchez N, Sculean A, Slot DE, Trombelli L, West N, Winkel E (2015) Effect of professional mechanical plaque removal on secondary prevention of periodontitis and the complications of gingival and periodontal preventive measures: consensus report of group 4 of the 11th European workshop on periodontology on effective prevention of periodontal and peri-implant diseases. J Clin Periodontol 42(Suppl 16):S214–S220. https://doi.org/10.1111/jcpe.12367

Sanz M, Ceriello A, Buysschaert M, Chapple I, Demmer RT, Graziani F, Herrera D, Jepsen S, Lione L, Madianos P, Mathur M, Montanya E, Shapira L, Tonetti M, Vegh D (2018) Scientific evidence on the links between periodontal diseases and diabetes: consensus report and guidelines of the joint workshop on periodontal diseases and diabetes by the international diabetes federation and the European federation of periodontology. J Clin Periodontol 45(2):138–149. https://doi.org/10.1111/jcpe.12808

Sanz M, Herrera D, Kebschull M, Chapple I, Jepsen S, Beglundh T, Sculean A, Tonetti MS, Participants EFPW, Methodological C (2020) Treatment of stage I-III periodontitis-The EFP S3 level clinical practice guideline. J Clin Periodontol 47(Suppl 22):4–60. https://doi.org/10.1111/jcpe.13290

Schenkein HA (2006) Host responses in maintaining periodontal health and determining periodontal disease. Periodontology 40:77–93. https://doi.org/10.1111/j.1600-0757.2005.00144.x

Schmalz G, M?ller M, Schmickler J, Rinke S, Haak R, Mausberg RF, Ziebolz D (2017) Influence of manual and power toothbrushes on clinical and microbiological findings in initial treatment of periodontitis – a randomized clinical study. Am J Dent 30(1):40–46

Slot DE, Valkenburg C, Van der Weijden GAF (2020) Mechanical plaque removal of periodontal maintenance patients: a systematic review and network meta-analysis. J Clin Periodontol 47(Suppl 22):107–124. https://doi.org/10.1111/jcpe.13275

Slots J (2003) Update on general health risk of periodontal disease. Int Dent J 53(Suppl 3):200–207. https://doi.org/10.1111/j.1875-595x.2003.tb00771.x

Starke EM, Mwatha A, Ward M, Argosino K, Jenkins W, Milleman JL, Milleman KR (2019a) A comparison of the effects of a powered and manual toothbrush on gingivitis and plaque: a randomized parallel clinical trial. J Clin Dent 30 (Spec No A):A24–29

Starke EM, Ward M, Olson M, Ou SS, Milleman KR, Milleman JL (2019b) A randomized parallel study to compare the effects of powered and manual tooth brushing on gingival health and plaque. J Clin Dent 30 (Spec No A):A16–23

Stenman J, Wennstrom JL, Abrahamsson KH (2018) A brief motivational interviewing as an adjunct to periodontal therapy – a potential tool to reduce relapse in oral hygiene behaviours. A three-year study. Int J Dent Hyg 16(2):298–304. https://doi.org/10.1111/idh.12308

Szulc M, Kustrzycki W, Janczak D, Michalowska D, Baczynska D, Radwan-Oczko M (2015) Presence of periodontopathic bacteria DNA in atheromatous plaques from coronary and carotid arteries. Biomed Res Int 2015:825397. https://doi.org/10.1155/2015/825397

Theilade J (1964) Electron microscopic study of calculus attachment to smooth surfaces. Acta Odontol Scand 22:379–387. https://doi.org/10.3109/00016356409028213

Tonetti MS, Van Dyke TE, working group 1 of the joint EFPAAPw (2013) Periodontitis and atherosclerotic cardiovascular disease: consensus report of the Joint EFP/AAP workshop on periodontitis and systemic diseases. J Periodontol 84(Suppl 4S):S24–S29. https://doi.org/10.1902/jop.2013.1340019

Tonetti MS, Eickholz P, Loos BG, Papapanou P, van der Velden U, Armitage G, et al (2015) Principles in prevention of periodontal diseases: consensus report of group 1 of the 11th European Workshop on Periodontology on effective prevention of periodontal and peri-implant diseases. J Clin Periodontol 42 Suppl 16:S5–11. https://doi.org/10.1111/jcpe.12368

Trombelli L, Franceschetti G, Farina R (2015) Effect of professional mechanical plaque removal performed on a long-term, routine basis in the secondary prevention of periodontitis: a systematic review. J Clin Periodontol 42(Suppl 16):S221–S236. https://doi.org/10.1111/jcpe.12339

Van der Weijden FA, Slot DE (2015) Efficacy of homecare regimens for mechanical plaque removal in managing gingivitis a meta review. J Clin Periodontol. 42(Suppl 16):S77-91. https://doi.org/10.1111/jcpe.12359

Van Dyke TE, Bartold PM, Reynolds EC (2020) The nexus between periodontal inflammation and dysbiosis. Front Immunol 11:511. https://doi.org/10.3389/fimmu.2020.00511

van Winkelhoff AJ, Slots J (1999) Actinobacillus actinomycetemcomitans and Porphyromonas gingivalis in nonoral infections. Periodontology 20:122–135. https://doi.org/10.1111/j.1600-0757.1999.tb00160.x

Wang P, Xu Y, Zhang J, Chen X, Liang W, Liu X, Xian J, Xie H (2020) Comparison of the effectiveness between power toothbrushes and manual toothbrushes for oral health: a systematic review and meta-analysis. Acta Odontol Scand 78(4):265–274. https://doi.org/10.1080/00016357.2019.1697826

Worthington HV, MacDonald L, Poklepovic Pericic T, Sambunjak D, Johnson TM, Imai P, Clarkson JE (2019) Home use of interdental cleaning devices, in addition to toothbrushing, for preventing and controlling periodontal diseases and dental caries. Cochrane Database Syst Rev 4(CD012018). https://doi.org/10.1002/14651858.CD012018.pub2

18 Multi-photonic Adjunctive Therapy for the Management of Periodontitis: Recent Advances and New Treatment Approach

Marco Giannelli and Daniele Bani

Abstract

The efficacy of photonic therapy adjunctive to conventional root cleansing procedures for the treatment of chronic periodontitis is matter of controversy. The meta-analyses of the clinical data available in the literature have failed to reach univocal conclusions because of broad variability among the applied photonic treatments, different in terms of light-emitting devices (laser or LED), wavelengths, irradiation power and modes, clinical indications, disease grading, follow-up times, and results assessment. Hovever, this complexity can also favour a different interpretation, which assigns a specific role to each photonic treatments in order to improve the outcome of the conventional treatments, in terms of reduction of periodontopathogenic bacteria and local inflammation, and increased regeneration of alveolar bone, periodontal ligament and gingiva. In this context, distinction should be made between high- and low-energy photonic therapies: the former can be used to achieve photoablation of the infected dental/periodontal tissues, while the latter can be used for anti-bacterial, anti-inflammatory and tissue biostimulation purposes. Recently, we and others have applied a multi-photonic protocol which combines laser photoablation of the infected epithelium, standard mechanical root cleansing and low-energy antiseptic phototherapy with a λ 405 nm LED in a first surgical session. Then, antisepsis is maintained by weekly sessions of photodynamic therapy with a solution of methylene blue photoactivated with a λ 635 nm low-energy laser to release bactericidal reactive oxygen species. The satisfactory objective results and patients' liking support the view that such multi-photonic treatments are a correct approach to supportive periodontal therapy.

Keywords

Photonic therapy · Laser · Periodontitis

M. Giannelli (✉)
Odontostomatologic Laser Therapy Center, Florence, Italy
e-mail: dott.giannellimarco@gmail.com

D. Bani
Department of Experimental and Clinical Medicine, Research Unit of Histology & Embryology, University of Florence, Florence, Italy
e-mail: daniele.bani@unifi.it

Abbreviations

BODIPY	boron dipyrromethene
iPAPD	improved PAPD
LED	light-emitting diode
LLLT	low-level laser therapy
PAPD	photo-ablative-photo-dynamic

J. Santi-Rocca (ed.), *Periodontitis*, Advances in Experimental Medicine and Biology 1373,
https://doi.org/10.1007/978-3-030-96881-6_18

PBMT	photobiomodulation (or biostimulation) therapy
PDT	photodynamic therapy
PT	phototherapy
ROS	reactive oxygen species
SRP	scaling and planing of dental roots

Highlights

- In patients with chronic periodontitis, besides conventional root cleansing, various photonic treatments have been used, although their actual advantages are controversial.
- High-energy (laser) and low-energy (laser and LED) photonic tretaments have different indications: the former can be used to achieve photoablation of the infected periodontal tissues, the latter can be used for anti-bacterial, anti-inflammatory and biostimulation purposes.
- A new treatment protocol combining laser photoablation, ultrasonic scaling and root planing and bactericidal LED phototherapy followed by repeated applications of antiseptic photodynamic therapy has been shown to yield better results than conventional periodontal therapy.

Considerations for Practice

- In periodontal disease, there is a demand for adjunctive therapies capable of improving the outcome of the conventional dental root cleansing treatments in terms of reduction of persistent infection, local inflammation and periodontal tissue resorption and, possibly, preservation or even regeneration of alveolar bone, periodontal ligament and gingiva.
- Diverse high-energy (laser) and low-energy (laser and LED) photonic treatments at different wavelenghts can address such demand because of their anti-bacterial and tissue biostimulating properties, provided they are correctly performed.
- To better exploit their effects, high- and low-energy photonic treatments can be combined together: such adjunctive protocols to conventional periodontal therapy have provided satisfactory clinical results.

Patient Summary

Periodontitis is a chronic infectious disease of the tooth-supporting tissues which can cause teeth loss and overall health worsening. The standard treatment consists in mechanical removal of bacteria attached to the affected teeth but, due to bacterial regrowth, relapses are frequent. Antibiotics and antiseptics are not a solution because periodontal bacteria have become resistant. An alternative approach is to associate the conventional dental cleansing treatments with light-based – or photonic – supportive therapies, performed with laser and LED instruments. Light has many diverse effects depending on its wavelength, energy, tissue absorption, etc. Hence, satisfactory periodontal healing can be achieved by a combination of different photonic treatments exploiting the synergism of their curative effects.

18.1 Photonic Therapy in Periodontics: A Yet Unsolved Controversy

In developed Countries, chronic periodontitis is still regarded as a major health challenge for oral medicine: this disease not only represents the main cause of tooth loss but has also been associated with potentially severe systemic complications caused by enduring persistence of bacterial infection (Hajishengallis 2015). This is the reason why basic and clinical research are particularly intense in this field, aimed at improving the current therapeutic protocols. Recently, numerous meta-analyses and reviews have been dedicated to the most important advancements in periodontology, including the so-called photonic therapies. This term encompasses all those procedures exploiting the well-known biological effects of light in the visible and infra-red wavelength range, delivered by means of laser or light-emitting diode (LED) instruments, for curative purposes. With few exceptions, photonic therapies are performed in adjunct to the conventional treatments based on scaling and planing of the dental roots (SRP), which remain the essential therapeutic approach to periodontitis (Giannelli et al. 2019). Nowadays, photonic therapies are routinely applied for many medical purposes, in some instances as first-choice treatments, but this is not true for periodontics. In fact, most retrospective analyses of the results of the previous clinical studies have failed to provide convincing statistical data to demonstrate that adjunctive photonic therapy can result in better clinical outcomes than those achieved by conventional SRP (Schwarz et al. 2008; Mizutani et al. 2016; Cobb 2017; Mills et al. 2018). However, if considering the lack of well-defined and widely accepted therapeutic indications and protocols on the one hand and the many possible variables of photonic therapies on the other hand, this inconclusive result is not surprising. In fact, the efficacy of photonic therapy in periodontitis is influenced by many physical, biological, anatomical, and technical variables which render the results of previous primary clinical studies barely comparable by any meta-analysis. Photonic therapies have two main levels of variability: (i) a general level depending on differences in the irradiation characteristics, i.e. light wavelengths, illumination devices (laser or LED), light beam power (high-power or photoablative or low-power or non-surgical) irradiation modes (continuous or pulsed), and actual energy delivered to the target tissue; (ii) a specific level depending on clinical differences, i.e. periodontal disease grading, exact biological rationale and indications, light absorption characteristics of the targeted mucosa (inflammation, pigmentation), treatment protocols, follow-up times, and outcome assessment methods. This broad variability can easily explain the reason why the literature on the clinical effects of photonic therapies in periodontology is so controversial and difficult, if not impossible, to be interpreted (Cobb 2017). An unfortunate consequence of this confusion is that many dental practitioners are hesitant to include photonic treatments in their therapeutic repertoire against periodontal disease, either because they underestimate its actual potential or because they are worried by its complexity and lack of clear indications (Giannelli et al. 2019).

This article is not meant to compete with the numerous, excellent reviews written by renowned periodontologists (Schwarz et al. 2008; de Paula Eduardo et al. 2010; Javed and Romanos 2013; Aoki et al. 2015; Mizutani et al. 2016; Cobb 2017; Mills et al. 2018), which remain the main source for whoever wishes to delve into this complex matter. Rather, our intention is to focus on some new findings and concepts emerged from the most recent clinical studies on photonic therapy in periodontics, which can be helpful to identify its actual strenghts and limitations, as well as to design new effective treatment protocols.

18.2 How Photonic Therapies Can Work in Periodontal Disease

The purpose of any adjunctive therapy in periodontal disease is to improve the outcome of conventional SRP in terms of: (i) reduction of oral dysbiosis to a favourable ratio between normal bacterial flora and periodontopathogens, (ii) reduction of local inflammation, (iii) regeneration of alveolar bone, periodontal ligament and gingiva. For such goals, the photonic approaches can be definitely adequate.

As a necessary premise, we must make a distinction between high- and low-energy phototherapies. High-energy irradiation requires laser instruments and is used to remove the infected dental/periodontal tissues: it is therefore defined as photoablative or photosurgical therapy. Its advantages over the conventional surgical methods are high selectivity (namely, laser wavelenght can be chosen to be specifically absorbed by molecules and chromophores in the target tissues), excellent bactericidal and hemostatic effects, minimal post-treatment inflammation and fast healing (Aoki et al. 2004; Schwarz et al. 2008; Ishikawa et al. 2009; Giannelli et al. 2019). On the other hand, due to high energy transfer to tissues, their damage threshold is close to the therapeutic range. Hence, a proper usage of laser photoablation requires specific training of dentists, who must be well aware of the pros and cons of the different laser instruments operating at different wavelengths: for instance, the lasers emitting in the far or intermediate infra-red spectrum, such as CO_2 (λ = 10,600 nm) and Erbium-doped garnet lasers, (Er:YAG, λ = 2940 nm; Er:YSGG, λ = 2780 nm), are chiefly absorbed by H_2O and apatite crystals of mineralized tissues and operate chiefly through vaporization, whereas those emitting in the near infra-red spectrum, such as Neodymium-doped garnet lasers (Nd:YAG, λ = 1064 nm) or the modern, versatile (and cheaper) diode lasers (λ = 655–980 nm) are absorbed by tissue pigments, such as melanin and hemoglobin, as well as bacterial cromophores, and operate chiefly through coagulation/carbonization (Aoki et al. 2004; Schwarz et al. 2008; Ishikawa et al. 2009; Giannelli et al. 2019). These different wavelengths also influence other parameters related to the target tissues, namely depth of penetration and adverse thermal effects due to light scattering, which render each laser type best suited for different, specific applications in the field of periodontology. For instance, CO_2 lasers, albeit capable to yield satisfactory long-term clinical results in expert hands (Crespi et al. 2011), nowadays are barely used in periodontics because of ease of target tissue over-heating and damage (Giannelli et al. 2012a). In general, Er-doped garnet lasers are preferred for removal/reshaping of mineralized tissues and calculus (Schwarz et al. 2003), whereas Nd-YAG and diode lasers are most suited for soft tissue and dental plaque photoablation and disinfection (Romanos 1994; Moritz et al. 1998; Gregg and McCarthy 2002; Schwarz et al. 2008; Kamma et al. 2009; Giannelli et al. 2012a; Mizutani et al. 2016).

Low-energy photonic therapy can be performed by both laser (in this instance, terms such as low-level laser therapy or LLLT, or soft laser therapy are currently used) and LED instruments. At variance with the high-energy laser treatments, all of which substantially cause tissue photoablation, in this case the mechanisms of action, methods of administration and biological effects can vary. Accordingly, low-energy photonic therapies can be subdivided in 3 distinct modalities: (i) photodynamic therapy (PDT); (ii) phototherapy (PT); (iii) photobiomodulation (or biostimulation) therapy (PBMT) (Giannelli et al. 2019).

(i) Photodynamic therapy (PDT) exploits the property of some organic compounds termed 'photosensitizers', e.g., cyclic tetrapyrroles, fullerenes, boron dipyrromethene (BODIPY) and phenotiazines (toluidine blue O, methylene blue) to release reactive oxygen species (ROS) when excited by light at wavelengths coinciding with their absorption peaks (Yin and Hamblin 2015). ROS are lethal for bacteria because they induce oxidative damage of plasma membranes and DNA. The short half-life (0.04 μs) and diffusion radius (~0.2 μm) of the generated ROS render PDT particularly effective for disinfection of periodontal tissues, since photosensitizers can bind to and concentrate in bacterial biofilms and dental plaque (Takasaki et al. 2009; Meimandi et al. 2017). A limitation of PDT is that its actual efficacy depends on the absorption spectra of the used photosensitizers, requiring strictly consistent excitation wavelengths (Giannelli and Bani 2018).

(ii) Phototherapy exploits the direct antimicrobial effects of specific light wavelengths due to the presence of photoactivable chromophores, e.g., porphyrins, in certain bacterial species, including several periodontal pathogens. Particularly effective is the light in the violet-blue spectrum (λ 405–520 nm), especially that in the narrow band of 405–

410 nm, which has shown excellent bactericidal effects against both Gram-positive and Gram-negative pathogens (Fukui et al. 2008; Barneck et al. 2016; Gillespie et al. 2017). The mechanisms of these effects, although not fully elucidated yet, are consistent with the endogenous generation of ROS in sensitive bacteria (Soukos et al. 2005). Interestingly, blue light (λ 405 nm) at bactericidal power has no effect on mammalian cells, which are protected from oxidative stress by multiple antioxidant metabolic pathways (Ramakrishnan et al. 2016). Moreover, blue light also induces the inactivation of lipopolysaccharide (LPS), a Gram-negative endotoxin responsible for the persistence of inflammation (Giannelli et al. 2017). These favourable characteristics account for substantial efficacy and safety of blue light for oral disinfection and restraint of septic inflammation and related periodontal tissue destruction. They also sustain the principle that phototherapy can be safely repeated in periodontitis patients until satisfactory clinical effects are achieved.

(iii) Photobiomodulation therapy (PBMT) exploits light wavelengths in the red-IR range (λ 600–950 nm), characterized by deep tissue penetration and capability to induce biostimulatory effects. These are likely due to photochemical interaction with Fe^{2+}-heme chromoproteins, such as mitochondrial cytochromes, resulting in metabolic activation of cells (Hamblin 2018). The major effects of this activation consist in stimulation of cell proliferation, microvascular dilation and increased blood perfusion which, in turn, result in accelerated wound healing and bone formation as well as decreased inflammation, oedema and pain (Qadri et al. 2005; Aoki et al. 2015; de Paula Eduardo et al. 2010). Accordingly, PBMT has been used as adjuvant to the main periodontal therapies to reduce inflammation and pain in the immediate post-operative phase and to improve periodontal ligament and alveolar bone healing in the long term (de Paula Eduardo et al. 2010).

18.3 Multi-photonic Therapy in Periodontal Disease: 'United We Stand, Divided We Fall'

On the above grounds, it is becoming increasingly clear that each photonic therapy has specific indications and limitations, which must be known in order to correctly design effective treatment protocols adjunctive to conventional SRP. Knowledge of these essential points can also be helpful to re-analyze the results of the previous clinical studies in future meta-analyses to evaluate whether photonic therapies can provide significant advantages in the treatment of periodontitis, in order to exclude the misleading results of the studies biased by erroneous photonic treatments (Giannelli et al. 2019).

Another point clearly emerging from a better knowledge of the strenghts of the various photonic treatments is that they can be effectively and safely combined into a multi-photonic therapy to exploit their individual advantages or even the synergism between their different biological effects. In fact, periodontal disease shows a complex network of pathogenic events which can be specifically targeted (Kornman 2008). Its main clinical hallmark is progressive periodontal tissue destruction secondary to persistent infection and inflammation sustained by periodontopathogenic bacteria particularly adapted to this local microenvironment. These bacteria are capable of penetrating and persisting into the epithelial cells lining the periodontal pockets and outer gingiva, thereby escaping host immunity as well as conventional antiseptics and antibiotics (Tribble and Lamont 2010). There, they represent a hidden germ reservoire predisposing the patients to re-infection soon after SRP, and hence to disease relapses and chronicization (Mombelli et al. 2000; Johnson et al. 2008; Ardila et al. 2010). In recent years, we and others have collected convincing evidence that a multi-photonic therapy adjunctive to SRP can provide better clinical results than SRP alone, both in the short and long term. In a first pilot trial on 26 patients with chronic periodontitis, we first combined a photoablative treatment with PDT in a multi-photonic protocol termed PAPD (PhotoAblative-PhotoDynamic). This protocol consisted of 3 steps: (i) photoablation of the surface bacterial biofilm and infected sulcular and gingival epithelium with a λ 810 nm high-energy diode laser; (ii) SRP performed by ultrasonic scaler; (iii) repeated sessions of PDT with methylene blue photoactivated by a λ 635 nm low-energy diode laser, in order to hinder bacterial re-growth. The results of this study, performed as a split-mouth, have shown a statistically significant improvement of the main parameters used for periodontal disease assessment, namely probing depth, clinical attachment level and bleeding-on-probing, as well as of cytodiagnostic markers of infection and inflammation in periodontal exfoliative samples, in the PAPD-treated mouth quadrants as compared with the matching quadrants treated with SRP alone (Giannelli et al. 2012b). Of note, these improvements were maintained over 4-year follow-up of the same patients (n = 24), suggesting that the desired objective of our PAPD protocol, i.e. shifting the host-parasite balance in favour of the former in order to promote periodontal healing in the long term, had been achieved (Giannelli et al. 2015). More recently, to treat patients with severe periodontitis (n = 24), we have adopted an improved PAPD protocol (iPAPD) in which a single 5-min. application of antiseptic phototherapy with a λ 405 nm LED was performed after epithelial laser photoablation and SRP in the same clinical session (Table 18.1). Then, weekly PDT sessions were performed until normalization of the cytodiagnostic infection and inflammation markers, according to the original PAPD protocol. This iPAPD protocol has also yielded statistically significant improvements of the clinical and cytodiagnostic disease markers at 1-year follow-up (Table 18.2). In this study we also assessed the subjective satisfaction of the enrolled patients towards the received treatments: again, a significant majority of them preferred the iPAPD protocol over SRP alone because of reduced pain and discomfort, both during and after the operative phase and in the following days, and better aesthetic results due to restoration of the natural pale color of the vestibular gingiva (Giannelli et al. 2018). Representative images of the key iPAPD steps are shown in Figs. 18.1, 18.2, 18.3, 18.4, and 18.5.

18.4 Conclusion and Perspectives

The multi-photonic supportive treatment to periodontal disease has been only recently developed based on new evidence on the actual efficacy, indications and possible synergisms of the various photonic therapies and is currently applied by few researchers worldwide (Giannelli et al. 2012b, 2018; de Angelis et al. 2018; Amaroli et al. 2020). For this reason, the overall number of patients studied is relatively small and insufficient to perform a reliable meta-analysis and draw definitive conclusions about the actual value of such methods to cure periodontal disease. What hinders a more widespread use of the multi-photonic therapies among dentists is that they require multiple light sources emitting at different wavelengths, whereas most of the dental lasers available on the market emit a single light wavelength, the only variable parameters being beam power and irradiation mode (pulsed or continuous). However, all low-energy phototherapies can be effectively performed by LED instruments, whose irradiation characteristics are similar to low-power lasers with the obvious advantage of being far cheaper and simpler. Ideally, dental practitioners wishing to perform modern photonic treatments could equip their surgeries with a photoablative diode laser and at least 2 LED instruments operating at λ 635 nm for PDT and λ 405 nm for antiseptic phototherapy.

We are of the opinion that a more detailed knowledge of the advantages and limitations of the photonic approach in periodontics will reduce

Table 18.1 Applied parameters for iPAPD multi-photonic therapy

	Near-IR GaAlAs diode laser	Violet-blue LED	Red diode laser
Irradiation modes and purposes			
	Photoablation mode	Phototherapy mode	Photodynamic mode
	Removal of infected sulcular and gingival epithelium	Antisepsis (phototoxic)	Antisepsis (oxidative)
Devices settings			
wavelength	810 ± 10 nm	405±5 nm	635±5nm Toluidine blue (1 μg/ml)
Wave emission mode	Continuous	Continuous	Continuous
Beam power	1 W	1 W	0.1 W
Light irradiation details			
Handpiece type	Polymide–coated silica fiber 0.6 mm	Focalized zoom handpiece	Light pipe glass 10 mm
Application mode	Contact	Non-contact	Non-contact
Distance from the target	0 mm	10 mm	30 mm
Light spot size	0.28 mm^2	95 mm^2	28.3 mm^2
Power density	353.4 W/cm^2	1.05 W/cm^2	0.35 W/cm^2
Fluence	66.7 J/cm^2	63 J/cm^2	21 J/cm^2
Tip movement speed	2.5 mm/s	Fixed beam	Fixed beam
Clinical protocol details			
Number of treatment	1	1	4–10 adjusted depending on healing markers[a]
Cooling system	Airflow	No	No

[a]Residual microbial contamination and inflammation performed on cytosmears of periodontal pocket exfoliative samples, as described (Giannelli et al. 2012b)

Table 18.2 Clinical periodontal parameters: conventional SRP vs. iPAPD+SRP

	Day 0	1 year	Statistical significance, 1 year vs. day 0
Probing depth (mm)			
SRP	4.9 ± 0.1	3.8 ± 0.3	
iPAPD+SRP	5.1 ± 0.2	1.2 ± 0.2	$p < 0.001$
Statistical significance	n.s.	$p < 0.001$	$p < 0.001$
Clinical attachment level (mm)			
SRP	5.3 ± 0.1	5.2 ± 0.1	
iPAPD+SRP	5.4 ± 0.1	3.1 ± 0.3	n.s.
Statistical significance	n.s.	$p < 0.001$	$p < 0.001$
Bleeding on probing (%)			
SRP	42 ± 4.6	23.2 ± 2.1	$p < 0.01$
iPAPD+SRP	36.7 ± 5.3	1.2 ± 0.6	$p < 0.001$
Statistical significance	n.s.	$p < 0.001$	
Periodontal index (%)			
SRP	26.5 ± 3.7	8.2 ± 1.5	$p < 0.001$
iPAPD+SRP	25.5 ± 3.6	7.4 ± 1.5	$p < 0.001$
Statistical significance	n.s.	n.s.	

For statistical comparison, the patients' quadrants were assumed as test units. The reported values are means ±SEM of 2 sampled sites per quadrant. Values were checked for normal distribution and then compared by within-subject, repeated-measures ANOVA and Newman-Keuls multiple comparison test. n.s. not significant (adapted from Giannelli et al. 2018)

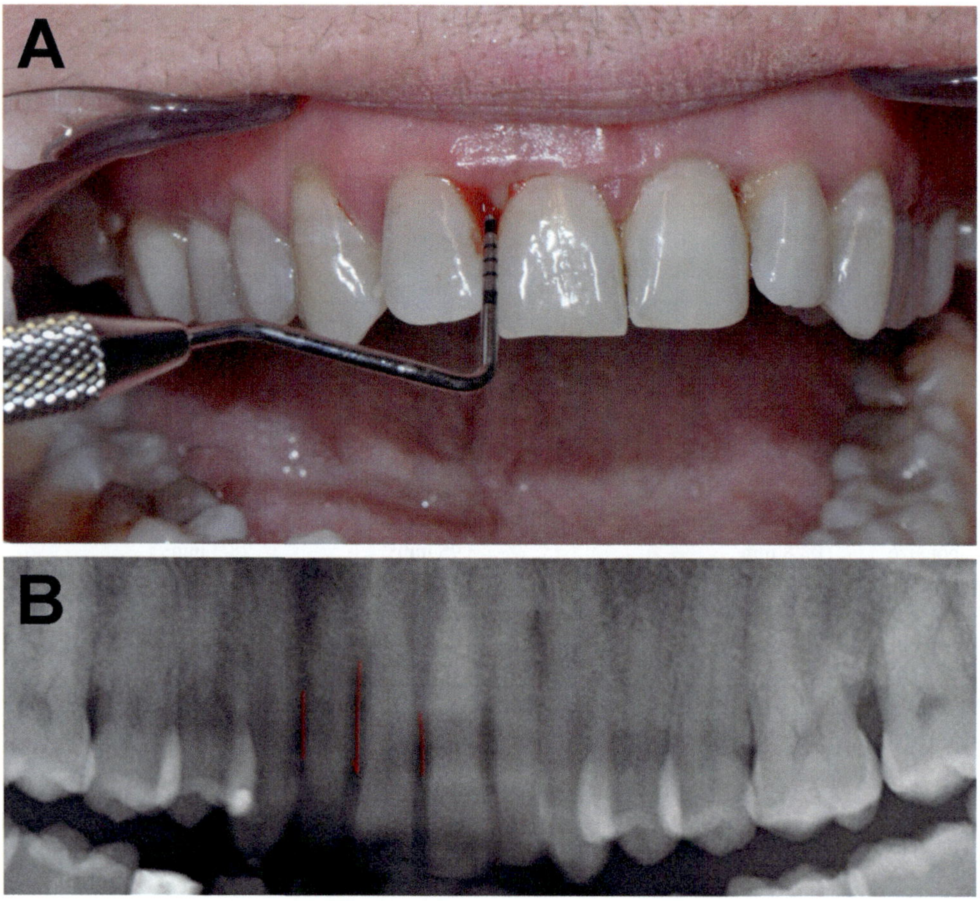

Fig. 18.1 Representative pictures of pre-treatment probing depth (**a**) and radiography (**b**) performed at patient's admission. In (**b**), the red lines indicate the extent of recession of the inter-alveolar bone ridges

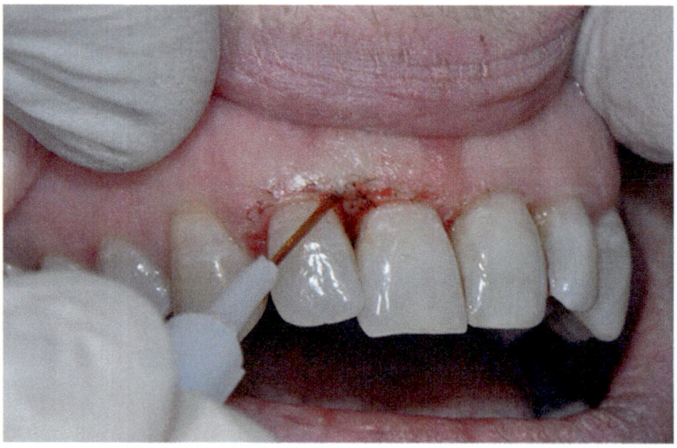

Fig. 18.2 iPAPD, phase 1. The gingival mucosa underwent photoablation with a λ 810 nm high-energy diode laser in contact mode (i.e. with the optic fiber tip touching the gingiva) under airflow cooling to remove the junctional, sulcular, and outer gingival epithelium, ~5 mm from the gingival margin, all around the teeth. Fiber diameter: 0.6 mm; irradiaton mode: continuous emission; beam power: 1 W; optic fiber movement speed: 2.5 mm/s; power density: 353.4 W/cm^2; fluence: 66.7 J/cm^2. Under these operating conditions, pain and discomfort are minimal and anaesthesia is usually unnecessary

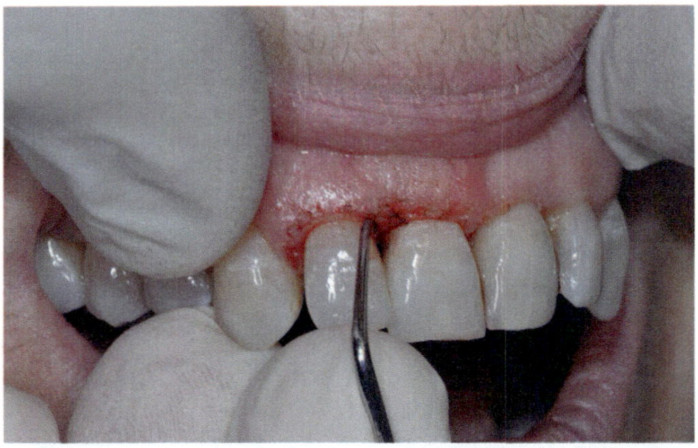

Fig. 18.3 iPAPD, phase 2. Conventional SRP was performed using Gracey curettes (Hu-Friedy, Milan, Italy) and ultrasonic scaler (Mectron Dental, Loreto, Italy) with metal tip and set to 80% power at high frequency (36 kHz) under water cooling (28 ml\min) until the root surfaces were clean and smooth. Of note, preliminary photoablation of the inflamed, swollen epithelium of the periodontal pockets facilitates SRP, because it yields an accessible gap between the dental root and its periodontum, with excellent hemostasis

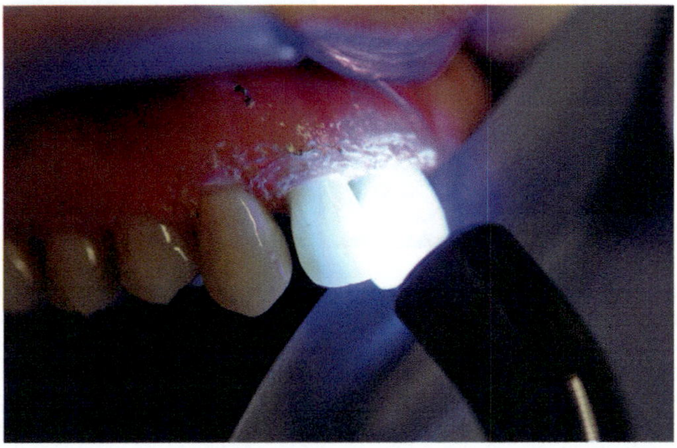

Fig. 18.4 iPAPD, phase 3. Phototherapy was then performed with a λ 405 nm LED for additional disinfection in the same surgical session. Handpiece: glass lightpipe; beam diameter: 10 mm; beam power: 1 W; non-contact mode ~10 mm from the target; spot area: 95 mm^2; power density: 1.05 W/cm^2; fluence: 63 J/cm^2. An additional property of λ 405 nm irradiation is that it induces a pale green autofluorescence of the coagulated keratins, allowing to visually check the photoablated epithelium

confusion and uncertainty on this matter and, consequently, promote confidence and use of such techniques among periodontologists. If this would occur, the demand for new user-friendly and effective photonic instruments specifically suited for odontostomatological use should also increase. In this context, while high-energy laser photoablation will remain a field for specifically trained, expert professionals, low-energy photonic treatments, which often need to be repeated until the desired therapeutic goals are achieved, could be administered at the patients' home under medical supervision by means of portable LED instruments, thereby maximizing their curative efficacy.

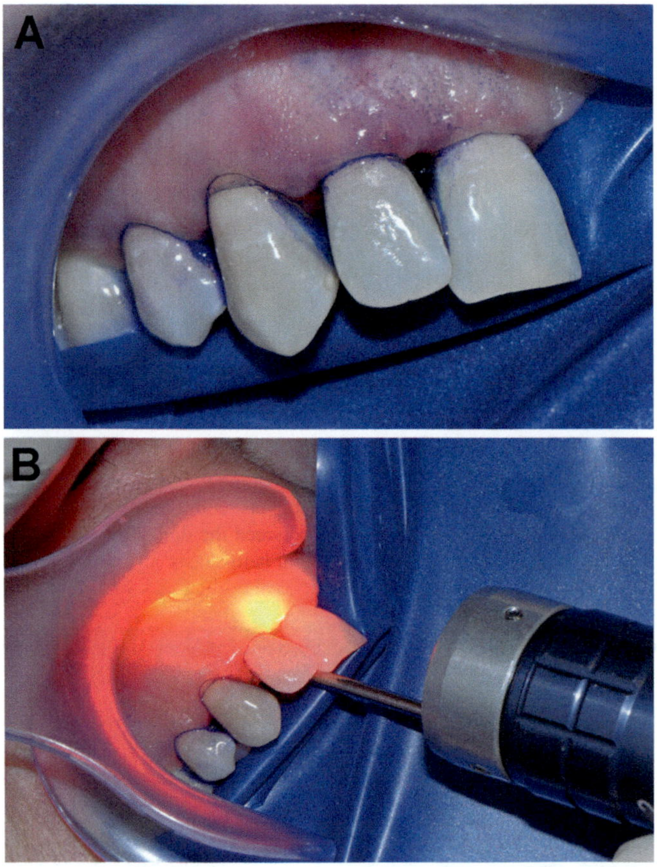

Fig. 18.5 iPAPD, phase 4. (**a**) Rinsing with the photosensitizer: 7 days after the previous treatments, the periodontal tissues, including the pocket, the surrounding mucosa and the dental root, were rinsed with the phenotiazinic dye toludine blue O (0.1% w/v in water) using a flexible needle. (**b**) PDT: 5 min. later, photoactivation of the photosensitizer was performed by irradiation with a λ 635 nm low-energy diode laser through a perpendicular zoom handpiece, slowly moving the focused light spot on the buccal and vestibular mucosa for 5 min. Handpiece: focused zoom; beam power: 100 mW; non-contact mode ~3 cm from the target; spot area: 28.3 mm^2; power density: 0.35 W/cm^2; fluence: 21 J/cm^2. PDT was repeated once weekly for up to 10 applications, until negativization of bacteria and inflammatory cells in gingival cytosmears occurred

References

Amaroli A, Barbieri R, Signore A, Marchese A, Parker S, De Angelis N, Benedicenti S (2020) Simultaneous photoablative and photodynamic 810-nm diode laser therapy as an adjunct to non-surgical periodontal treatment: an in-vitro study. Minerva Stomatol 69:1–7

Aoki A, Sasaki KM, Watanabe H, Ishikawa I (2004) Lasers in nonsurgical periodontal therapy. Periodontology 2000 36:59–97

Aoki A, Mizutani K, Schwarz F, Sculean A, Yukna RA, Takasaki AA, Romanos GE, Taniguchi Y, Sasaki KM, Zeredo JL, Koshy G, Coluzzi DJ, White JM, Abiko Y, Ishikawa I, Izumi Y (2015) Periodontal and peri-implant wound healing following laser therapy. Periodontology 2000 68:217–269

Ardila CM, Granada MI, Guzmán IC (2010) Antibiotic resistance of subgingival species in chronic periodontitis patients. J Periodont Res 45:557–563

Barneck MD, Rhodes NLR, de la Presa M, Allen JP, Poursaid AE, Nourian MM, Firpo MA, Langell JT (2016) Violet 405-nm light: a novel therapeutic agent against common pathogenic bacteria. J Surg Res 206:316–324

Cobb CM (2017) Lasers and the treatment of periodontitis: the essence and the noise. Periodontology 2000 75:205–295

Crespi R, Cappare P, Gherlone E, Romanos GE (2011) Comparison of modified widman and coronally advanced flap surgery combined with Co2 laser root irradiation in periodontal therapy: a 15-year follow-up. Int J Periodontics Restorative Dent 31:641–651 de Angelis N, Hanna R, Signore A, Amaroli A, Benedicenti S (2018) Effectiveness of dual-wavelength (Diodes 980 Nm and 635 Nm) laser approach as a non-surgical modality in the management of periodontally diseased root surface: a pilot study. Biotech Biotechnol Equip 32:1575–1582. https://doi.org/10.1080/13102818.2018.1544034

de Paula Eduardo C, de Freitas PM, Esteves-Oliveira M, Aranha AC, Ramalho KM, Simões A, Bello-Silva MS, Tunér J (2010) Laser phototherapy in the treatment of periodontal disease. Lasers Med Sci 25:781–792

Fukui M, Yoshioka M, Satomura K, Nakanishi H, Nagayama M (2008) Specific–wavelength visible light irradiation inhibits bacterial growth of Porphyromonas gingivalis. J Periodont Res 43:174–178

Giannelli M, Bani D (2018) Appropriate laser wavelengths for photodynamic therapy with methylene blue. Lasers Med Sci 33:1837–1838

Giannelli M, Bani D, Viti C, Tani A, Lorenzini L, Zecchi-Orlandini S, Formigli L (2012a) Comparative evaluation of the effects of different photoablative laser irradiation protocols on the gingiva of periodontopathic patients. Photomed Laser Surg 30:222–230

Giannelli M, Formigli L, Lorenzini L, Bani D (2012b) Combined photoablative and photodynamic diode laser therapy as an adjunct to non-surgical periodontal treatment: a randomized split-mouth clinical trial. J Clin Periodontol 39:962–970

Giannelli M, Formigli L, Lorenzini L, Bani D (2015) Efficacy of combined photoablative-photodynamic diode laser therapy adjunctive to scaling and root planing in periodontitis: randomized split-mouth trial with 4-year follow-up. Photomed Laser Surg 33:473–480

Giannelli M, Landini G, Materassi F, Chellini F, Antonelli A, Tani A, Nosi D, Zecchi-Orlandini S, Rossolini GM, Bani D (2017) Effects of photodynamic laser and violet-blue led irradiation on Staphylococcus aureus biofilm and Escherichia coli lipopolysaccharide attached to moderately rough titanium surface: in vitro study. Lasers Med Sci 32:857–864

Giannelli M, Materassi F, Fossi T, Lorenzini L, Bani D (2018) Treatment of severe periodontitis with a laser and light-emitting diode (LED) procedure adjunctive to scaling and root planing: a double-blind, randomized, single-center, split-mouth clinical trial investigating its efficacy and patient-reported outcomes at 1 year. Lasers Med Sci 33:991–1002

Giannelli M, Lasagni M, Bani D (2019) Photonic therapy in periodontal diseases an overview with appraisal of the literature and reasoned treatment recommendations. Int J Mol Sci 20:pii: E4741. https://doi.org/10.3390/ijms20194741

Gillespie JB, Maclean M, Given MJ, Wilson MP, Judd MD, Timoshkin IV, MacGregor SJ (2017) Efficacy of pulsed 405-nm light-emitting diodes for antimicrobial photodynamic inactivation: effects of intensity, frequency, and duty cycle. Photomed Laser Surg 35:150–156

Gregg RH, McCarthy D (2002) Laser periodontal therapy for bone regeneration. Dent Today 21:54–59

Hajishengallis G (2015) Periodontitis: from microbial immune subversion to systemic inflammation. Nat Rev Immunol 15:30–44

Hamblin MR (2018) Mechanisms and mitochondrial redox signaling in photobiomodulation. Photochem Photobiol 94:199–212

Ishikawa I, Aoki A, Takasaki AA, Mizutani K, Sasaki KM, Izumi Y (2009) Application of lasers in periodontics: true innovation or myth? Periodontol 50:90–126

Javed F, Romanos GE (2013) Does photodynamic therapy enhance standard antibacterial therapy in dentistry? Photomed Laser Surg 31:512–518

Johnson JD, Chen R, Lenton PA, Zhang G, Hinrichs JE, Rudney JD (2008) Persistence of extracrevicular bacterial reservoirs after treatment of aggressive periodontitis. J Periodontol 79:2305–2312

Kamma JJ, Vasdekis VGS, Romanos GE (2009) The effect of diode laser (980 nm) treatment on aggressive periodontitis: evaluation of microbial and clinical parameters. Photomed Laser Surg 27:11–19

Kornman KS (2008) Mapping the pathogenesis of periodontitis: a new look. J Periodontol 79(8 Suppl):1560–1568

Meimandi M, Talebi Ardakani MR, Esmaeil Nejad A, Yousefnejad P, Saebi K, Tayeed MH (2017) The effect of photodynamic therapy in the treatment of chronic periodontitis: A review of literature. J Lasers Med Sci 8:S7–S11

Mills MP, Rosen PS, Chambrone L, Greenwell H, Kao RT, Klokkevold PR, McAllister BS, Reynolds MA, Romanos GE, Wang HL (2018) American Academy of Periodontology best evidence consensus statement on the efficacy of laser therapy used alone or as an adjunct to non-surgical and surgical treatment of periodontitis and peri-implant diseases. J Periodontol 89:737–742

Mizutani K, Aoki A, Coluzzi D, Yukna R, Wang CY, Pavlic V, Izumi Y (2016) Lasers in minimally invasive periodontal and peri-implant therapy. Periodontology 2000 71:185–212

Mombelli A, Schmid B, Rutar A, Lang NP (2000) Persistence patterns of Porphyromonas gingivalis, Prevotella intermedia/nigrescens, and Actinobacillus actinomyetemcomitans after mechanical therapy of periodontal disease. J Periodontol 71:14–21

Moritz A, Schoop U, Goharkhay K, Schauer P, Doertbudak O, Wernisch J, Sperr W (1998) Treatment of periodontal pockets with a diode laser. Lasers Surg Med 22:302–311

Qadri T, Miranda L, Tuner J, Gustafsson A (2005) The short-term effects of low-level lasers as adjunct therapy in the treatment of periodontal inflammation. J Clin Periodontol 32:714–719

Ramakrishnan P, Maclean M, MacGregor SJ, Anderson JG, Grant MH (2016) Cytotoxic responses to 405nm light exposure in mammalian and bacterial cells: Involvement of reactive oxygen species. Toxicol In Vitro 33:54–62

Romanos GE (1994) Clinical applications of the Nd:YAG laser in oral soft tissue surgery and periodontology. J Clin Laser Med Surg 12:103–108

Schwarz F, Sculean A, Berakdar M, Georg T, Reich E, Becker J (2003) Periodontal treatment with an Er:YAG laser or scaling and root planing. A 2-year follow up split-mouth study. J Periodontol 74:590–596

Schwarz F, Aoki A, Becker J, Sculean A (2008) Laser application in non-surgical periodontal therapy: a systematic review. J Clin Periodontol 35:29–44

Soukos NS, Som S, Abernethy AD, Ruggiero K, Dunham J, Lee C, Doukas AG, Goodson JM (2005) Phototargeting oral black-pigmented bacteria. Antimicrob Agents Chemother 49:1391–1396

Takasaki AA, Aoki A, Mizutani K, Schwarz F, Sculean A, Wang CY, Koshy G, Romanos G, Ishikawa I, Izumi Y (2009) Application of antimicrobial photodynamic therapy in periodontal and peri-implant diseases. Periodontology 2000 51:109–140

Tribble GD, Lamont RJ (2010) Bacterial invasion of epithelial cells and spreading in periodontal tissue. Periodontology 2000 52:68–83

Yin R, Hamblin MR (2015) Antimicrobial photosensitizers: drug discovery under the spotlight. Curr Med Chem 22:2159–2185

Probiotics During the Therapeutic Management of Periodontitis

19

Flávia Furlaneto, Karin Hitomi Ishikawa, Michel Reis Messora, and Marcia P. A. Mayer

Abstract

Scaling and root planing is the gold standard for the treatment of periodontitis, but administration of systemic antibiotics may be needed especially for sites with deep probing depths, or in the presence of comorbidities. However, treated sites are subject to recolonization with a microbiota similar to that present before therapy, and supportive periodontal therapy is employed after the treatment of active disease. The use of beneficial organisms, known as probiotics, seems an attractive proposal to promote a healthy associated subgingival microbiome and to control inflammation for the management of periodontitis. The mechanisms underlying the benefits promoted by probiotics involve interference on periodontopathogens, modulation of the exacerbated immune host response and the ability to restore the integrity of the epithelial barrier on mucosa surfaces. This review examines the scientific data related to the effects of probiotics on the treatment of periodontal diseases and addresses the future approaches necessary for their implementation.

Keywords

Periodontitis · Probiotics · Prophylaxis · Immunomodulation

Authors Flavia Furlaneto and Karin Hitomi Ishikawa have equally contributed to this chapter.

F. Furlaneto · M. R. Messora
Department of Oral Surgery and Periodontology, School of Dentistry of Ribeirão Preto, University of São Paulo, São Paulo, Brazil
e-mail: flafurlaneto@usp.br; m.messora@forp.usp.br

K. H. Ishikawa · M. P. A. Mayer (✉)
Department of Microbiology, Institute of Biomedical Sciences, University of São Paulo, São Paulo, Brazil
e-mail: karinhitomi@hotmail.com; mpamayer@icb.usp.br

Abbreviations

3-HPA	3-hydroxypropionaldehyde
BALOs	*Bdellovibrio bacteriovorus* and *Bdellovibrio*-like organisms
GABA	gamma-amino butyric acid
GECs	gingival epithelial cells
LAB	Lactic acid bacteria
NSPT	non-surgical periodontal treatment
ROS	reactive oxygen species
SCFAs	short chain fatty acids
SOD	superoxide dismutase
SPT	supportive periodontal therapy
SRP	scaling and root planing
TJs	tight junction proteins

J. Santi-Rocca (ed.), *Periodontitis*, Advances in Experimental Medicine and Biology 1373,
https://doi.org/10.1007/978-3-030-96881-6_19

Highlights

- Prevention and treatment of periodontitis is based on biofilm control and RAR, but other strategies are still needed for long-term effects.
- Administration of probiotics is suggested to restore the microbial balance in periodontitis and modulate the immune response, but the ideal probiotic regimen for treatment and prevention of periodontitis remains to be established
- Mechanical periodontal therapy followed or not by local and systemic antimicrobial agents, may not be enough to provide long lasting health periodontal tissues for high-risk periodontitis patients
- Probiotics have been successfully used in the control of inflammatory diseases, including those associated with periodontitis, such as diabetes and cardiovascular diseases.
- In vitro and experimental in vivo data indicated that probiotic administration could alter the dysbiotic microbiome of periodontitis, influence the immune response of gingival tissues and restore the integrity of mucosa epithelial barrier.
- New approaches including oral microbiome transplantation and microbial consortia tailored according to the individual needs are still needed.

Considerations for Practice

- The periodontal treatment may not be successful for all patients or sites, and additional strategies are needed for the prevention, as adjuvant of active treatment and for the maintenance phase.
- Probiotics are suggested as a safe, ecological approach of the periodontal therapy, aiming to restore the balanced microbiome and control destructive inflammation in the periodontal tissues.
- Further studies are needed to establish appropriate protocols of probiotic usage in order to control periodontitis.

Patient Summary

Certain probiotics are successfully used in the control of inflammatory infectious diseases. Periodontitis is caused by an unbalanced oral microbial community, and probiotics may help in the prevention and treatment of periodontitis by reducing pathogenic bacteria in the mouth and inflammation. However, the best probiotic regimen and strains for the control of periodontitis are still not established.

19.1 Periodontitis and Adjunctive Therapies

Periodontitis is a result of host's altered immune-inflammatory response and oral microbiome dysbiosis and can present systemic repercussions. Conventional periodontal therapy consists of scaling and root planing (SRP) and biofilm control, aiming to restore the biological compatibility of affected root surfaces and delay disease progression. The clinical reduction of periodontal inflammation, characterizing successful periodontal therapy, is associated with a change in local microbial profile.

Interventions to modulate the microbiome of periodontal pockets using mechanical means, such as SRP, are effective in most cases (Mombelli 2018). However, conventional forms of periodontal treatment are not always able to prevent the progression of the disease. Instrumentation techniques available for SRP are not completely effective in eliminating subgingival microbial deposits (Adriaens and Adriaens 2004; Berezow and Darveau 2011). Effectiveness of SRP is significantly influenced by initial probing depth (PD), dental anatomy and skill of the operator (Gellin et al. 1986; Adriaens and Adriaens 2004). The capacity for tissue and cellular invasion exhibited by some pathobionts (Lamont and Yilmaz 2002; Tribble and Lamont 2010) also hinders the effective control of the subgingival microbiota through SRP.

Treated periodontal sites are subject to recolonization by a microbiota similar to the one before treatment. The degree and speed of recolonization depend on residual pockets, treatment protocol, interval between maintenance visits, distribution of periodontal microorganisms in the oral cavity and quality of oral hygiene (Mombelli et al. 2000; Mombelli 2018). Therefore, SRP is not always able to promote the microbiological changes necessary to maintain long-term stability of the clinical benefits initially achieved (Gellin et al. 1986; Adriaens and Adriaens 2004). Thus, in some cases, adjuvant approaches in the treatment of periodontitis are needed to achieve therapeutic success. Furthermore, adjunctive therapies should be evaluated regarding their capacity to reduce the need for additional therapies, such as complex surgical procedures.

Systemic antimicrobials may be used in severe cases as an adjuvant therapy since they reduce levels of several periodontal pathogens in subgingival regions, to which mechanical instrumentation has more restricted access (Teughels et al. 2020) and promotes subgingival microbial succession, in order to establish higher proportions of microorganisms compatible with periodontal health (Zandbergen et al. 2016), enhancing the effects of non-surgical periodontal treatment (NSPT).

However, the indication of antibiotics should be evaluated in each clinical situation, due to the occurrence of undesirable side effects, the risk of promoting microbial resistance and the possibility of affecting the entire microbiome of the human organism. This decision making must consider all risks involved (Herrera et al. 2008; Pretzl et al. 2019), especially the selection of "new" antibiotic resistant species.

Furthermore, the effect of periodontal treatment is not fully predictable, and long-term success requires the establishment of a program of supportive periodontal therapy (SPT) following the treatment of active disease. During SPT, periodontitis patients should be closely monitored according to their individual needs based on the risk of recurrent periodontal loss (Armitage and Xenoudi 2016).

The use of beneficial organisms, known as probiotics, seems an attractive proposal to promote a healthy associated subgingival microbiome and to control inflammation for the management of periodontitis. These organisms can be used as a preventive measure, an adjuvant to the periodontal treatment (Raff and Hunt 2012), and during SPT.

19.2 Probiotics

Probiotics are living microorganisms that, when ingested in a sufficient amount, exert a positive effect on health that is not limited to the nutritional effects (Hill et al. 2014). Given that periodontitis is characterized by an exacerbated inflammatory response mediated by the dysbiotic microbiome (Hajishengallis et al. 2012), the repopulation of subgingival sites with beneficial oral bacteria should be a long-lasting ecological approach to control the disease. Furthermore, probiotics benefits to periodontitis patients are extended to the control of periodontitis co-morbidities, such as diabetes, cardiovascular diseases, and adverse pregnancy outcomes (Gomes et al. 2014) and halitosis (Mousquer et al. 2020).

Besides probiotics, other ecological based strategies such as use of prebiotics, symbiotics and postbiotics may also promote health (Wegh et al. 2019). Prebiotics are mainly oligosaccharide carbohydrates, such as inulin, that stimulate the proliferation and activity of microbial communities, inhibit proliferation of pathogens and promote immune defense (Martinez et al. 2015). Prebiotics ingested with probiotics are known as symbiotics (Martinez et al. 2015). Beneficial non-viable microorganisms or their secreted bioactive metabolites are known as postbiotics (Wegh et al. 2019).

Lactic acid bacteria (LAB) are used in most formulations of probiotics due to their benefits and safety for human consumption (Zielińska and Kolożyn-Krajewska 2018). *Lactobacillus* and *Bifidobacterium* are saccharolytic anaerobic LAB, resident in supra and subgingival biofilms of healthy subjects (Duran-Pinedo and Frias-Lopez 2015; Coretti et al. 2017), able to degrade host-glycans (Turroni et al. 2018). *In vitro* studies reported that some *Lactobacillus* and *Bifidobacterium* adhere to gingival epithelial cells (GECs) (Albuquerque-Souza et al. 2019), and co-aggregate in multispecies biofilms with streptococci early colonizers and *P. gingivalis* (Ishikawa et al. 2020a, b), which may contribute to their beneficial effects (Yadav et al. 2017).

Other LAB such as Enterococci are occasionally used as probiotics despite some safety concerns (Ben Braïek and Smaoui 2019). Streptococci are also LAB and *Streptococcus salivarius* K12 has been successfully used to reduce incidence of acute pharyngitis and otitis (Zupancic et al. 2017). Despite the association of other species of *Streptococcus* with oral health, their potential as probiotics conflicts with their role in non-oral opportunistic infections such as endocarditis (Abranches et al. 2018). More recently, several beneficial properties were attributed to the *Akkermansia muciniphila*, a mucin-degrading commensal of the gut (Xu et al. 2020), present in low levels in the oral cavity (Singh et al. 2019). The abundance of this specie is reduced in obesity and type 2 diabetes (Pascale et al. 2019) and in non- metabolic disorders, such as the cardiac phenotype of Chagas disease (de Souza-Basqueira et al. 2020). Other unusual organisms, such as the predators *Bdellovibrio* and *Bdellovibrio*- like organisms (BLO) present a potential beneficial role due to their ability to lyse Gram negative bacteria (Bonfiglio et al. 2020).

Although probiotics can comprise several genera, their beneficial mechanisms are strain specific, i.e., strains of the same species can exhibit different properties, although some properties are common to members of different genera. Furthermore, probiotics' effectiveness may differ according to microbiome signatures and host specificities (Maldonado-Gómez et al. 2016). Other variables such as the physiological state of probiotic cells interfere on their activity (Yadav et al. 2017). Thus, the effects of probiotics are dependent on the strain, dose, and components used to produce a given probiotic product as well as on host and resident microbiome factors.

Probiotics usually do not persistently colonize human mucosa surfaces (Yli-Knuuttila et al. 2006; Caglar et al. 2009), but their transient colonization can alter the microbial community and regulate immune functions. However, even without altering the gut microbiome in adults (Kristensen et al. 2016) and in infants (Laursen et al. 2017), other mechanisms contribute to their beneficial effects.

Most data on the beneficial mechanisms of probiotics rely on studies in the gut, but there is some evidence of their effects in the oral cavity. Probiotics may directly interfere on the colonization of pathobionts, favoring healthy associated microorganisms, as summarized in Fig. 19.1. Probiotics modulate the exacerbated immune host response by decreasing the production of inflammatory mediators and improving the production of protective molecules such as defensins and anti- inflammatory factors in the periodontal tissues (Kobayashi et al. 2017) (Fig. 19.2). A third strategy of probiotics involves their ability to restore the integrity of the epithelial barrier on mucosa surfaces and produce bio-active metabolites (Fig. 19.3).

19.3 Effect of Probiotics on Oral Bacteria. *In Vitro* Studies

Most of the studies on probiotics rely on their effects on oral pathobionts (Nissen et al. 2014; Ishikawa et al. 2020a, b) including *P. gingivalis*, considered a keystone organism in periodontitis (Hajishengallis et al. 2012), *Aggregatibacter actinomycetemcomitans*, (Fine et al. 2019; Amado et al. 2020) and others (Montenegro et al. 2020).

Probiotics inhibit growth and survival of other bacteria by direct killing or competitive exclusion. These mechanisms involve secretion of

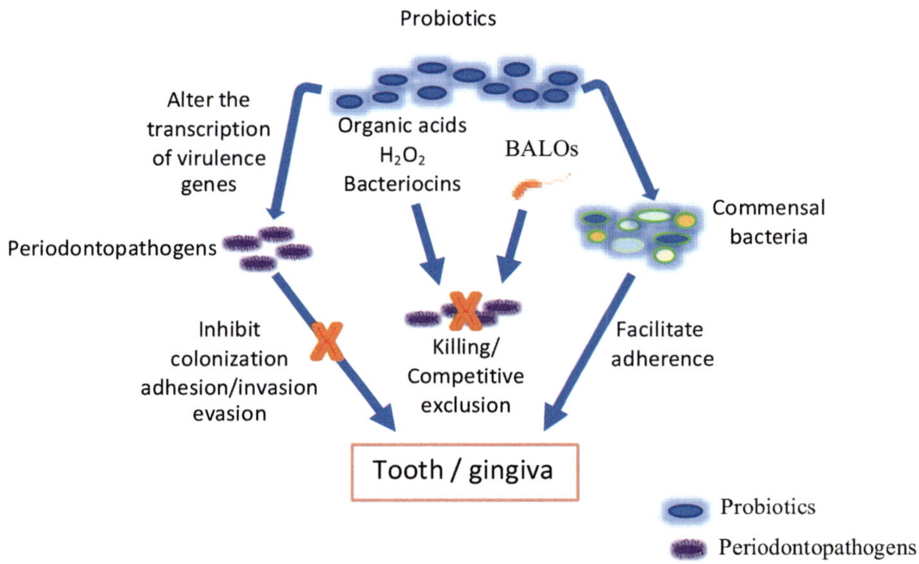

Fig. 19.1 Mechanisms underlying the effects of probiotics in the control of periodontal diseases. Effect on the oral microbial community. Probiotics may impair the microbial shift in supra and subgingival biofilms of periodontitis patients by direct killing, by competitive exclusion of pathogens and by altering microbial signaling. These mechanisms act synergistically, and lead to the restoration of a healthy associated microbiome and consequent homeostasis

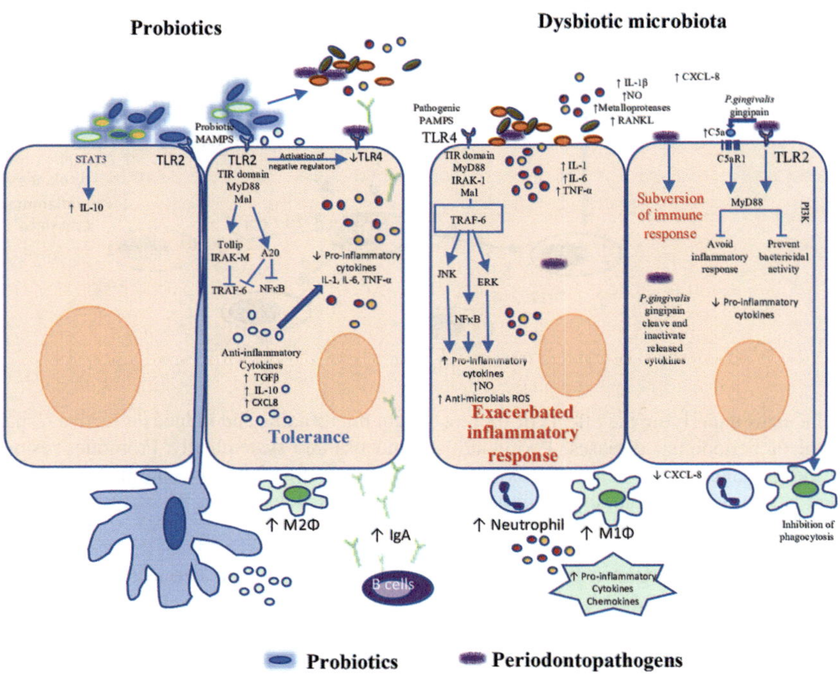

Fig. 19.2 Probiotics modulate the exacerbated immune host response in the periodontal tissues by altering expression of PPRs such as TLR4, decreasing the production of inflammatory mediators, improving the production of protective molecules such as defensins and anti-inflammatory factors, and surpassing evasion mechanisms of periodontopathogens

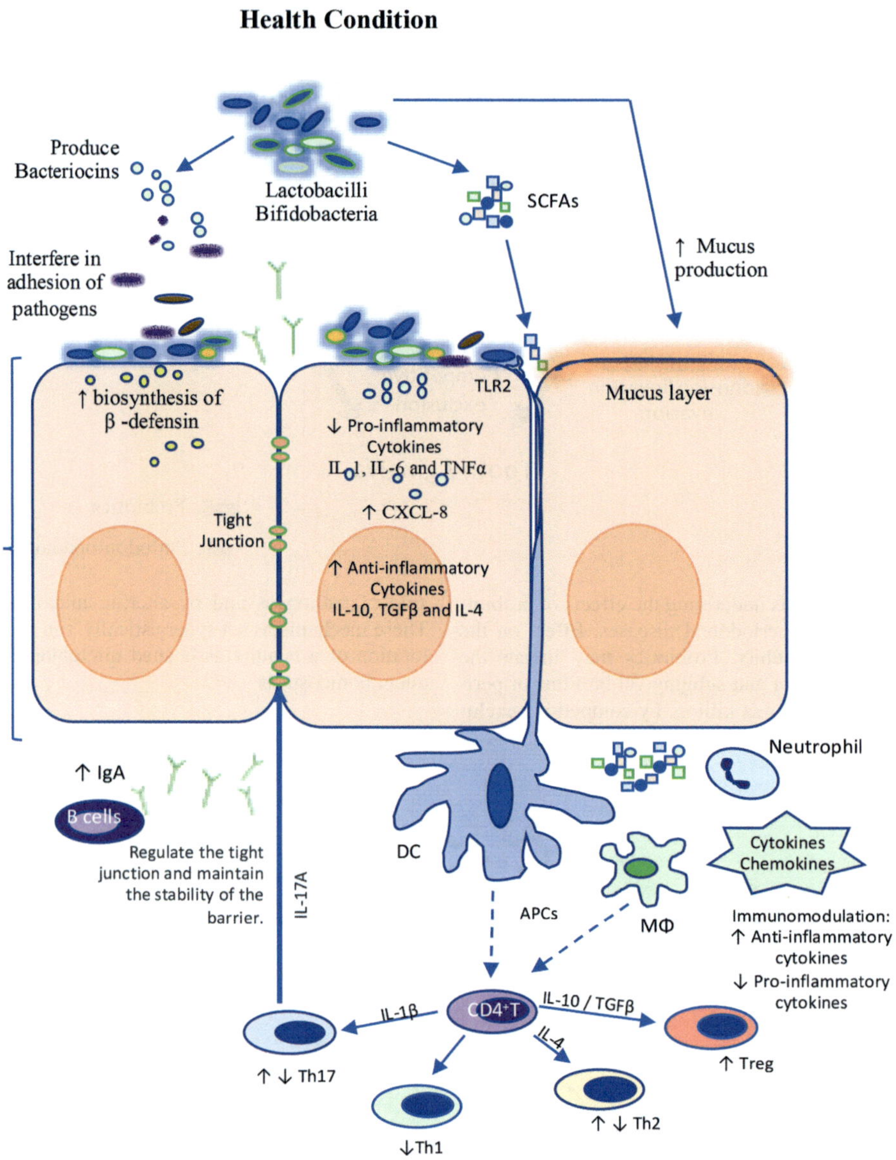

Fig. 19.3 Mechanisms underlying the effects of probiotics in the control of periodontal diseases. Periodontitis may influence the gut microbiome leading to systemic inflamation. Oral administered probiotics can impact the gut microbiome and induce protective responses in distant mucosa and systemically. Probiotics restore the integrity of the epithelial barrier on mucosa surfaces and produce bio-active metabolytes that act locally and at distant sites

growth-inhibitory factors (organic acids, hydrogen peroxide and bacteriocins), occupation of available niches and reduction of fitness by altering microbial signaling (Yadav et al. 2013).

Acids production from carbohydrate degradation by LAB exerts selective antimicrobial activity against pathogens (Aroutcheva et al. 2001), but can present a broad spectrum (Saarela et al. 2000). This activity may be relevant since the subgingival microbiome in the periodontal pockets is enriched by proteolytic bacteria, which raise the pH up to 8.5 (Pöllänen et al. 2013), displacing beneficial species (Marsh 2003). Carbohydrate availability in gingival fluid is low

(Parker et al. 1993), and derived mostly from host glycoproteins, but subgingival bacteria can store glucose under high pH and produce lactate (Chew et al. 2012).

The antibacterial warfare of Lactobacilli and Bifidoybacteria comprises also the induction of oxidative stress (Eschenbach et al. 1989) due to production of H_2O_2, O_2^- and other reactive oxygen species (ROS). Beneficial oral streptococci also express these oxidases, and H_2O_2 shapes the oral microbiome (Zhu et al. 2014), since hydrogen peroxide and reactive oxygen species (ROS) are lethal to several bacteria. LAB's oxidases lead to the production of H_2O_2, which can reach millimolar concentrations in laboratory cultures (Imlay 2019), in a rate dependent on the environmental oxygen concentration. Hence, this mechanism may modulate the oral microbiome according to oxygen availability, and the low concentration of oxygen at the bottom of deep periodontal pockets should limit the production of ROS by subgingival organisms. Furthermore, several organisms, including *P. gingivalis*, produce superoxide dismutase (SOD) and present additional mechanisms for their protection from oxidative damage (Smalley et al. 1998; McKenzie et al. 2012). SOD converts O_2^- into H_2O_2, a physiologically important mechanism because O_2^- may damage some biomolecules which H_2O_2 does not react (Imlay 2003). In addition, the microbial community harbors several H_2O_2-detoxifying enzyme producers (Marsh and Zaura 2017), such as *A. actinomycetemcomitans*, which protects *P. gingivalis* from H_2O_2 (Zhu et al. 2019).

Pathogens killing promoted by LAB is also mediated by the production of bacteriocins (Mayanagi et al. 2009), antibiotic-like peptides with diverse mode of activity, genetic origin, and biochemical properties (Naidu et al. 1999). Bacteriocins such as niacin, a class I bacteriocin, are considered safe for humans and used as biopreservatives in food (Martinez et al. 2013). These peptides are usually lethal, promoting pores on the membrane of target bacterial cells, but other mechanisms were reported (Alvarez-Sieiro et al. 2016). They can have a narrow (usually limited to the same or related genera) or a wide inhibitory spectrum (active against a wide range of non-related organisms) (Strom et al. 2002), and their production is associated with successful niche competition in the gut (Kommineni et al. 2015). Although bacteriocins from LAB vary, most lactobacilli produce class II bacteriocins with a narrow spectrum of target bacteria, usually limited to the same or related genera, due to specificities given by their membrane receptors (Ríos Colombo et al. 2019).

Differing from bacteriocins, reuterin produced by *L. reuteri*, is a mixture of different forms of 3-hydroxypropionaldehyde (3-HPA), derived from the metabolism of glycerol (Mu et al. 2018). 3-HPA spontaneously dehydrates to form acrolein, an electrophilic α,β-unsaturated aldehyde that inhibits a broad range of Gram-positive and Gram-negative bacteria (Cleusix et al. 2007; Engels et al. 2016; Mu et al. 2018). Acrolein causes oxidative stress in bacteria by interacting with free thiol groups, causing the depletion of glutathione (GSH) with the consequent modification of proteins, including functional enzymes (Engels et al. 2016).

The antimicrobial mechanism of *Bdellovibrio bacteriovorus* and *Bdellovibrio*-like organisms (BALOs) is unique since these are obligate predators of Gram-negative bacteria (Sockett 2009). During their free life phase, BALOs locate the prey, attach to the outer membrane, and insert themselves in the prey periplasm. Then, they assimilate prey's macromolecules, growing inside the host cells and releasing their progeny trough cell lysis (Bonfiglio et al. 2020). BALOs are highly diverse and comprise also organisms that uptake the intracellular content of the prey but remain extracellularly (Bonfiglio et al. 2020). Gram-negative periodontopathogens are susceptible to several BALOs, differing from Gram-positive commensals (Van Essche et al. 2011).

In order to survive in the oral cavity, microorganisms should adhere to oral mucosa and tooth surfaces or aggregate to already adherent organisms. Probiotics compete with pathogens for their adhesion sites, or reduce their ability to adhere (Yadav et al. 2017), and *in vitro* data showed that certain probiotics present these properties. Several *Lactobacillus* and *Bifidobacterium* are able to reduce the adhesion and invasion of *P. gingivalis* to gingival epithelial cells (GECS) (Albuquerque-Souza et al. 2019). Further data indicated that probiotics, especially *L. acidophilus* LA5, were able to reduce *P. gingivalis* abundance in multispecies biofilms, without interfering in commensals such as *S. gordonii* and *S. oralis* (Ishikawa et al. 2020a, b). Moreover, secreted products (postbiotics) of lactobacilli, specially of *L. acidophilus* LA5, were able to inhibit biofilm formation and affect pre-formed biofilms of a fresh isolate of a highly virulent clone of *A. actinomycetemcomitans*, which forms robust biofilms *in vitro* (Ishikawa et al. 2020a, b).

Microbial interactions can result in antagonistic or collaborative relations. Probiotics alter transcription of genes involved in biofilms and evasion of host defenses. *P. gingivalis* is able to evade host defenses and create a more tolerant microenvironment at subgingival sites, enabling the rise of organisms otherwise susceptible to the inflammatory environment (Hajishengallis et al. 2012). A key factor *of P. gingivalis* to favor dysbiosis is the production of gingipains, proteases that cleave and inactivate released cytokines and factors of the complement system, and, as a consequence, impairs the protective inflammatory response able to reduce infectious agents (Tam et al. 2009). Probiotics lactobacilli and *Akkermasia muciniphilica* may impair *P. gingivalis* ability to favor dysbiosis by downregulating gingipains expression (Ishikawa et al. 2020a, b). Virulence of *A. actinomycetemcomitans* may also be reduced by secreted products of lactobacilli, which downregulated expression of leukotoxin, cytolethal distending toxin, catalase and dispersin B in a strain–specific manner (Nissen et al. 2014; Ishikawa et al. 2020a, b).

Thus, either by directly killing or inhibition of pathogens growth or indirectly by promoting microenvironmental changes, probiotics may negatively interfere in pathobionts and increase abundance of healthy associated organisms, leading to a balanced oral microbiome.

19.4 Immunomodulatory Properties of Probiotics. *In Vitro* Studies

Pattern recognition receptors (PRRs) of GECs recognize microbial associated molecular patterns (MAMPs) and signal to trigger immune response. The expression of PRRs in the gingival epithelium is controlled in periodontal health (Ren et al. 2005), allowing a constant dialogue with commensal bacteria, essential to maintain homeostasis (Delitto et al. 2018). The response to commensals leads to the recruitment of defense cells aiming to surveil the environment (Darveau 2010), and results in low grade inflammation in clinically health gingival tissues. Homeostasis is maintained by the production of antimicrobial factors involved in microbial lysis; depletion of essential nutrients, and inhibition of binding to host cells (Valenti et al. 2018). On the other hand, activation of the innate immune response by pathogens disrupts homeostasis, leading to exacerbated inflammation and tissue destruction (Delitto et al. 2018).

Probiotics are recognized by host receptors and interfere with the signaling promoted by the local microbial community. Expression of Toll-like receptors (TLRs) at oral and intestine mucosa differs in health and disease (Ren et al. 2005; de Kivit et al. 2014) and probiotics can alter their expression (Dogi et al. 2008; de Kivit et al. 2014). *In vitro* data indicated that *P. gingivalis* increases expression of TLR4 in GECs, whereas *L. acidophilus* LA-5 downregulates its expression in *P. gingivalis* challenged cells (Albuquerque-Souza et al. 2019). Moreover, probiotics interfere in PRR signaling by activating negative regulators of TLR4, consequently avoiding overactivation of NFκB, and inhibiting pro- inflammatory responses after LPS challenge (Shimazu et al. 2012; Takanashi et al. 2013).

After recognition by PRRs, probiotic effects are largely variable, since some strains present pro-inflammatory properties, whilst others display an opposite role (Ding et al. 2017). *L. rhamnosus* LGG induces the production of anti-inflammatory cytokines such as IL-10, by activating STAT3 pathway (Hutchins et al. 2013). *L. acidophilus* LA5 induces IL-10 production by *P. gingivalis* infected GECs and reduces the production of IL-1β (Albuquerque-Souza et al. 2019). *Bifidobacterium lactis* (Kim et al. 2010), *L. rhamnosus* LGG (Donato et al. 2010) and *L. reuteri* (Liu et al. 2012) negatively regulate NFκB activity, resulting in the reduction of inflammatory mediators such as IL-6, IL-1β and TNFα, whereas other strains decreased the production of NO and metalloproteases (Jang et al. 2013; Jang and Min 2020).

Similar to other inflammatory conditions, periodontitis shares macrophage-driven inflammation (Parisi et al. 2018). Macrophages differentiation depends on the surrounding environment and their phenotype is a spectrum from classically activated macrophages (M1) to alternatively activated macrophages (M2) (Sica et al. 2015). Gingival tissues of periodontitis sites present both M1 and M2 macrophages (Navarrete et al. 2014). However, M1 polarization is prolonged in periodontitis, leading to hyperinflammation (Parisi et al. 2018) and shift from M1 to M2 is impaired by *P. gingivalis*, preventing healing (Yu et al. 2018).

Probiotics can affect macrophage differentiation in infected tissues. Certain strains mediate macrophage polarization to M1 (Ji et al. 2013; Christoffersen et al. 2014), increasing microbial killing (Guha et al. 2019), whereas others suppress M1 polarization and induce differentiation toward M2, in order to control inflammation (Jang et al. 2013, 2014), and favor healing (Sica and Mantovani 2012). Bone marrow macrophages (BMMф) exposed to *P. gingivalis* and to the probiotic *Akkermansia muciniphilica* showed increased production of anti-inflammatory IL-10 and decreased of IL-12 (Huck et al. 2020).

Neutrophils constantly migrate through the junctional epithelium in gingival crevice (Moutsopoulos et al. 2014). These cells are attracted to infection sites by chemotactic factors such as CXCL-8, and there are both areas with high and low levels of CXCL-8 in inflamed gingival tissues (Fitzgerald and Kreutzer 1995). *P. gingivalis* induces chemokines paralysis, since CXCL-8 production is impaired by *P. gingivalis* serine phosphatase serB (Darveau 2010). However, this evasion mechanism might be surpassed by certain lactobacilli, which increase CXCL-8 production in *P. gingivalis* infected GECs (Albuquerque-Souza et al. 2019; Ishikawa et al. 2020a, b). Furthermore, the simultaneous infection of keratinocytes with *Akkermansia muciniphila* and *P. gingivalis* also resulted in increased CXCL-8 production (Huck et al. 2020).

Probiotic may also affect dendritic cells maturation. Lactobacilli isolated from fecal samples increased LPS-induced expression of maturation markers in mouse dendritic cells and changed their cytokines toward an anti-inflammatory profile (Luongo et al. 2017).

Secretion of IgA in the intestinal fluid and saliva can be induced by probiotics (de Moreno de LeBlanc et al. 2008), leading to IgA decreased inflammation, increased mucosal resistance against pathogens (Khurshid et al. 2015), and maintenance of the normal microbiota (Corthésy 2013). However, this activity and its consequences in the oral cavity remain to be elucidated since the effects of antibodies on periodontopathogens are still controversial (Ebersole et al. 2001).

CD4+ and CD8+ T cells are present in periodontal lesions and generate multiple mediators involved in bone resorption in periodontitis. Chronic periodontitis is mediated by Th2 cells (Hienz et al. 2015) and alveolar bone destruction in patients with periodontal disease was also associated with a Th17 response (Dutzan et al. 2018). However, IL-17 has also a protective role of IL-17 in bone homeostasis (Yu et al. 2007). Probiotics influence T cells polarization, shaping the balance between TH1/TH2/Th17/Treg responses under inflammatory conditions and steady state (Cheng et al. 2019). Certain probiotic strains increase type 1 helper T cells and attenuate T helper type 2 (Butel 2014), whereas others induce Th17 response (Cheng et al. 2019).

Probiotics also influence bone metabolism. Administration of lactobacilli prevented bone loss in steroid deficiency, due to their ability to alter the gut microbiome and restore the integrity of the gut barrier (Ohlsson and Sjögren 2015), a mechanism that will be further explored in this chapter. Furthermore, probiotics positive effects in alveolar bone are associated with their anti-inflammatory properties (Weitzmann 2017), and absorption of minerals and vitamins (Amin et al. 2020). Bone remodeling is dependent on the RANKL/RANK/OPG (receptor activator of NF ligand/receptor activator of NF/osteoprotegerin) pathway. An ethanol extract of *L. paracasei* was able to inhibit osteoclastogenesis induced by RANKL in a murine pre-osteoclastic cell line and reduced the extent of bone resorption (Liu et al. 2018a, b).

The non-keratinized epithelium of gingival crevice is a physical and immunological barrier. Epithelial barriers are dependent on adhesion molecules such as tight junction proteins (TJs) and adherens junctions, able to seal the space of adjacent cells. However, microbial challenge and inflammation can break the integrity of the epithelium, leading to increased permeability.

Disruption of the epithelium of gingival crevice may result in diffusion of bacterial toxic products, translocation of subgingival microorganisms, bacteremia and even colonization of distant non-oral sites (Forner et al. 2006). *P. gingivalis* challenge can promote this disruption by reducing the expression of TJ proteins (Groeger et al. 2010). On the other hand, co-culture of *P. gingivalis* infected immortalized gingival keratinocytes with the probiotic *A. muciniphilica* results in improved expression of the tight junction protein zonulin-1 and the adhesion molecules integrin-β1 and E-cadherin (Huck et al. 2020).

Hence, in order to eradicate chronic infections by pathogens, especially those with immune privilege (Mellor and Munn 2008) such as *P. gingivalis*, probiotics may overcome their evasion mechanisms, leading to their reduction/elimination, but may also control the exacerbated inflammation in the periodontal tissues, and promote tissue healing.

19.5 *In Vivo* Animal Studies

Validated preclinical studies, such as the use of animal models with experimental periodontitis, can provide important data on the safety and efficacy of probiotics (Hoffman et al. 2008), especially when new strains are being investigated.

Teughels et al. (2007), in their proof-of-concept study, demonstrated that multiple irrigations of surgically created periodontal pockets of beagle dogs with the oral streptococci, *Streptococcus sanguinis*, Streptococcus *salivarius*, and *Streptococcus mitis* after SRP promoted a delay in recolonization of anaerobic species such as *Porphyromonas gulae*, *Prevotella intermedia* and *Campylobacter rectus* and reduced periodontal inflammation.

The association of *Bacillus subtilis* –DSM 5750 and *Bacillus licheniformis* –DSM 5749 was evaluated in studies with ligature induced experimental periodontitis in rats (Messora et al. 2013, 2016; Foureaux et al. 2014). In general, these studies demonstrated that probiotic therapy can reduce periodontal destruction, either alone or associated with SRP (Messora et al. 2013, 2016). It was also able of reducing the number of inflammatory cells in periodontal tissues of rats with ligature induced periodontitis and subjected to chronic stress due to physical restraint for 2.5 h/day. However, the probiotic was not effective in preventing bone loss or altering the expression of inflammatory markers in stressed animals (Foureaux et al. 2014). Interestingly, the animals with experimental periodontitis that did not receive probiotics presented shortened and damaged villi of the jejunum, whereas the consumption of probiotics attenuated these deleterious effects in the gut (Messora et al. 2013).

Several studies evaluated the effect of different strains of *Lactobacillus* on experimental periodontitis. Topical application of *L. brevis* CD2 placed in the gingiva of ligated teeth of mice resulted in reduced bone loss, decreased expression of proinflammatory cytokines such as IL-1β and TNF-α in the gingival tissues, and altered the microbial community when compared to the control group (Maekawa and Hajishengallis 2014). The systemic administration of *Lactobacillus rhamnosus* GG to mice with experimental periodontitis induced by *P. gingivalis* and *F. nucleatum* led to a reduction in alveolar bone loss and gingival inflammation (Gatej et al. 2018) and attenuation of intestinal inflammation and changes in the cecum microbiome promoted by the oral pathogens (Gatej et al. 2020).

Considering that *Bifidobacterium* spp. may be associated with periodontal health and treatment success in patients with periodontitis (Hojo et al. 2007), the effects of topical administration of *Bifidobacterium animalis subsp. lactis* HN019 was evaluated in ligature induced periodontitis in rats (Oliveira et al. 2017). The probiotic was able to control alveolar bone loss and altered the oral microbial community by increasing the proportions of *Actinomyces*- like and *Streptococcus*-like species. Furthermore, the probiotic induced the expression of protective factors such as osteoprotegerin and beta-defensins, and decreased levels of IL-1β and RANKL in the gingival tissues. When *Bifidobacterium animalis subsp. lactis* HN019 was administered systemically and as an adjunct to SRP, in rats with experimental periodontitis, the beneficial effects in controlling bone destruction and inflammation were also observed (Ricoldi et al. 2017). *B. lactis* HN019 was also able to reduce alveolar bone loss and levels of TNF-α and IL-6 in gingival tissues of rats with experimental periodontitis and rheumatoid arthritis (RA) but did not lead to differences in connective tissue attachment level (Cardoso et al. 2020). Furthermore, probiotic therapy reduced serum levels of anti-citrullinated protein antibodies, an important marker of RA, although the groups treated with *B. lactis* HN019 did not show significant improvements in RA activity scores and paw edema.

Animal studies had also shown that probiotics improve the oral epithelium barrier finction due to the production of defensins and immune response modulation (Seth et al. 2008; Anderson et al. 2010; Karczewski et al. 2010), and possibly due to its activity on maintaining the homeostasis of the oral epithelial barrier. It is well established that the gut microbiome contributes to the maintenance of the mucosa by regulating epithelial cells turnover, maintaining cell-to-cell junctions, and promoting epithelial repair (Takiishi et al. 2017). On the other hand, an imbalanced gut microbiome is associated with disruption of the epithelial barrier and increased diffusion of bacterial products to the blood stream, leading to systemic inflammation (Festi et al. 2014).

Periodontitis co-morbities can be partially explained by its effects in the gut (Hajishengallis 2015). As shown in animal studies, swallowed oral pathogens alter the gut microbiome, leading to disruption of mucosa integrity (Arimatsu et al. 2014; Nakajima et al. 2015), and systemic inflammation (Kato et al. 2018). These pieces of evidence are reinforced by data showing that chronic and aggressive periodontitis patients present dysbiosis not only in the oral, but also in the gut microbiome (Lourenço et al. 2018; Amado et al. 2020).

Thus, the manipulation of the gut microbiome may control inflammatory diseases associated with leaking gut, such as diabetes, atherosclerosis, and liver diseases (Festi et al. 2014) and might be effective in reducing inflammation in the periodontal tissues. Thus, the effect of probiotics in the control of periodontitis may not require their oral administration, suggesting that beneficial effects of oral probiotics were associated with their effects in the gut. Gastric administration of *Lactobacillus gasseri* SBT2055 to mice was able to suppress alveolar bone loss and gingival inflammation induced by *P. gingivalis*, decreased expression and secretion of TNF-α and IL-6 and increased the production of β defensins in gingival tissues (Kobayashi et al. 2017). The gut commensal *A. muciniphila* reduces endotoxemia and ameliorates gut permeability in high-fat diet-induced mice, by upregulating the expression of tight junction proteins (TJs) occludin and claudin-3 (Li et al. 2016; Chelakkot et al. 2018; Grander et al. 2018). Moreover, dysbiosis is usually associated with loss of microbial diversity and *A. muciniphila* reduces LPS production and increases microbial diversity in the gut (Wu et al. 2017). Thus, the treatment of mice with experimental periodontitis induced by P. gingivalis with *A. muciniphila* resulted in decreased alveolar bone loss. This trait was associated with *A. muciniphila* abilities to increase production of anti-inflammatory cytokines, decrease *P. gingivalis* virulence and increase expression of TJs in epithelial cells (Huck et al. 2020).

Probiotics produce several bioactive substances that maintain health. Microbial communities in the gut produce short chain fatty acids (SCFAs) such as acetate, propionate and butyrate, associated with benefits to the host. Lactate produced by LAB is substrate for short chain fatty acids production. Furthermore, *Bifidobacterium* synthetizes SCFAs when carbohydrates supply is limited (Liu et al. 2018a, b). SCFAs, especially butyrate, have anti-inflammatory roles in the gut, improving mucosa integrity (Jia et al. 2019), reducing neutrophil cytokine production and macrophage NF-κB signaling and inducing Tregs differentiation (Liu et al. 2018a, b). Furthermore, butyrate or butyrate- producers in the gut, attenuated IL-17 response in gingival tissues and experimental alveolar bone loss (Jia et al. 2019). However, butyrate may be deleterious to human primary gingival epithelial cells (Liu et al. 2019), although the abundance of several butyrate-producing genera was reduced in periodontitis sites of unbalanced type 2 diabetes patients when compared to those with low HbA1c levels (Longo et al. 2018).

Gut microorganisms produce metabolic products involved in the brain-gut-microbiota axis, and probiotics that influence brain functions and produce positive mental health benefits are termed psychobiotics (Dinan and Cryan 2017). Lactobacilli and Bifidobacteria isolates from the human gut produce gamma-amino butyric acid (GABA), serotonin, noradrenaline or acetylcholine, and these compounds act on the enteric nervous system leading to amelioration of brain function disorders induced by stress (Dinan and Cryan 2017). Given the evidence associating stress and periodontitis (Coelho et al. 2020), the production of neurotransmitters by probiotics might be an additional mechanism for their beneficial role in periodontitis.

Animal studies provided evidence on the mechanisms of probiotics and their potential to control periodontitis in humans, as well as the co-morbidities associated with the disease. However, these data should be interpreted under the limitations of the animal models.

19.6 Clinical Evidence

Although animal experimental model studies showed promising results regarding the potential use of probiotics, their results may not be translated to humans. Hence, controlled clinical trials evaluating the effect of probiotics in humans are essential before their widespread use for the control of periodontitis.

Early studies evaluated the use of probiotics as a single therapy in periodontitis patients, demonstrating promising results (Riccia et al. 2007; Shimauchi et al. 2008; Mayanagi et al. 2009; Vicario et al. 2013). However, considering the importance of biofilm removal before the administration of probiotics (Teughels et al. 2013) in order to favor their antimicrobial effects and increase their colonization capacity, clinical studies were performed with systemically or topically administered probiotics as adjuvants of SRP.

Several clinical studies reported positive results of the adjuvant effects of probiotic therapy, administered through lozenges or sachets, in the treatment of periodontitis (Vivekananda et al. 2010; Shah et al. 2013; Teughels et al. 2013; Szkaradkiewicz et al. 2014; Tekce et al. 2015; İnce et al. 2015; Morales et al. 2016; Invernici et al. 2018, 2020). Most of these studies evaluated the effects of combination of two strains of *Lactobacillus reuteri*: DSM 17938 and ATCC PTA 5289 as an adjuvant to SRP in patients with chronic periodontitis (Vivekananda et al. 2010; Teughels et al. 2013). The use of this combination resulted in reduction in surgical needs of deep pockets and decreased the number of patients needing surgery on three or more teeth. However, there were no differences between control and probiotic groups when the overall percentage of pockets and teeth were considered (Teughels et al. 2013). The combination therapy also provided greater reduction in probing depth (PD) and greater attachment gain at 12 postoperative months than SRP alone. Patients treated with the combined therapy had also increased tissue levels of metalloproteinases inhibitors and lower levels of MMP (matrix metalloproteinase)-8 in the gingival crevicular fluid, although all forms of MMP-8 rather than specifically active forms were included in the analysis (İnce et al. 2015). This therapy delayed the recolonization of periodontal pockets considering both total viable cell counts and the proportions of obligate anaerobes until 180 days, in patients with chronic periodontitis (Tekce, İnce et al. 2015). Furthermore, this probiotic combination showed anti-inflammatory effects, which reflected in the improvement of periodontal clinical parameters (Szkaradkiewicz et al. 2014).

The beneficial effect of probiotics was also shown in smokers, an important risk factor for periodontitis. Use of *L. reuteri* DSM 17938 chewing tablets for 21 days as adjuvant of the periodontal treatment led to reduced gingival inflammation and PD in heavy smokers with generalized periodontitis. However, the probiotic therapy did not induce clinical attachment gain in pockets with PD >5 mm neither altered pockets with PD < 5 mm (Theodoro et al. 2019).

Other strains of *Lactobacillus*, also administered as lozenges or sachets, have been evaluated in clinical studies with patients presenting periodontitis. SRP associated with the oral administration of *L. rhamnosus* SP1 for 3 months did not improve most clinical parameters when compared to SRP alone, except for greater reductions in PD, resulting in less surgical needs (Morales et al. 2016). The use of *L. brevis* (CD2) lozenges for 14 days in aggressive periodontitis patients resulted in similar improvement of clinical parameters as observed for doxycycline, suggesting that probiotic therapy may be a good alternative for the use of antibiotics in the treatment of this periodontal condition (Shah et al. 2013).

To date, the only clinical study evaluating a non-*Lactobacillus* probiotic in periodontal therapy has been carried out by Invernici et al. 2018. The consumption of lozenges containing *Bifidobacterium animalis subsp. lactis* HN019 for 30 days, as an adjunct to SRP was evaluated in patients with chronic periodontitis. Three months after SRP, the probiotic group presented fewer moderate and deep periodontal pockets, fewer patients at risk for periodontitis progression and less need for additional periodontal treatment (periodontal surgery) than the control group. Microbiological and immunological advantages were also observed in the probiotic group (Invernici et al. 2018). The test group presented lower proportions of pathogens and pathobionts, as well as higher proportions of species compatible with health when compared to the control group. PCR analyses suggested that the administered probiotic strain was temporarily integrated into the subgingival biofilm of the treated patients, which may have contributed to the maintenance of a health status. The probiotic treatment led to lower level of proinflammatory cytokines than control, and higher levels of anti-inflammatory cytokine IL-10 in gingival crevicular fluid (GCF), and greater expression of beta-defensin-3 in gingival tissues, an important antimicrobial peptide for the protection of periodontal tissues (Invernici et al. 2020). These results indicate interaction of the probiotic with gingival tissues of the treated patients, strengthening the epithelial barrier against microbial aggressions to promote a vigilant and health-compatible immunoinflammatory state.

The topical use of probiotic microorganisms has provided satisfactory results as adjuvant therapy in the management of periodontitis (Tsubura et al. 2009; Penala et al. 2016; Sajedinejad et al. 2018). Mouthwashes containing the probiotic *Bacillus subtilis* led to lower BANA (N-benzoyl-DL-arginine-naphthylamide) test scores, indicative of a change in the subgingival microbiota (Tsubura et al. 2009). The use of subgingival irrigation and mouthwashes containing *Lactobacillus salivarius* and *Lactobacillus reuteri* (Unique Biotech laboratories) promoted enhanced clinical and microbiological periodontal parameters (Penala et al. 2016). These improvements were also observed with the use of mouthwashes containing *L. salivarius* NK02 (Sajedinejad et al. 2018).

Unlike the results obtained in the clinical studies previously described, other studies have found no significant additional effects with the use of probiotic therapy in periodontitis (Pelekos et al. 2019. The combination of *L. reuteri* strains (DSM 17938 and ATCC PTA 5289), administered in lozenges as an adjunct to SRP in patients with chronic periodontitis did not result in any significant advantage when compared to placebo

(Pelekos et al. 2019). Another clinical trial reported that the use of a cocktail of *Streptococcus* (*Streptococcus oralis* KJ3, *Streptococcus uberis* KJ2, *Streptococcus rattus* JH145) as an adjunct to SRP in periodontitis patients yielded no significant improvement when compared to placebo (Laleman et al. 2015). Furthermore, administration of *L. brevis* CECT7480/*L. plantarum* CECT7481 as adjuncts to SRP resulted in reduced gingival bleeding when compared to placebo. However, the probiotic combination group presented increased number of diseased sites at the end of the study (Pudgar et al. 2021). The local and systemic use of these same strains also promoted reductions in bleeding on probing, gingival index and red complex bacteria. However, their use did not result in advantages regarding PD and clinical attachment level (CAL) when compared with controls (Patyna et al. 2021).

Systematic reviews and meta-analyses reported that adjuvant probiotic in the periodontal treatment can result in additional benefits in clinical and microbiological parameters (Matsubara et al. 2016; Ikram et al. 2018). Another meta-analysis involving only RCTs using *L. reuteri* (DSM 17938 and ATCC PTA 5289) supported this therapy as an adjuvant to SRP in the treatment of periodontitis at short-term, especially in initially deep pockets (Martin-Cabezas et al. 2016). A recent systematic review and meta-analysis assessed the effects of probiotics at different follow-up time-points (Ho et al. 2020). 10 RCTs included in this study comprised patients that received SRP with probiotics or with placebo or SRP alone. Regarding clinical parameters, both 3-month and 12-month data showed significant benefit for the probiotics groups with greater magnitude at 12 months in pooled estimates of PD reduction. The results showed a significant benefit for probiotics in both PD and CAL when baseline mean PD values were $\geq$ 5 mm, indicating that the outcomes may be impacted by baseline disease severity.

However, the guidelines of the European Federation of Periodontology did not recommend the adjunctive use of probiotics in the treatment of periodontitis (Sanz et al. 2020) Nevertheless, it is important to emphasize that the conclusions were based on data not limited to severe disease at baseline.

Several clinical trials evaluated the adjunctive administration of probiotics in the management of experimental or established gingivitis. Systemic probiotic regimens were applied, including the isolated or associated use of *L. casei Shirota, L. reuteri* (ATCC 55730 and ATCC PTA 5289), *L. plantarum* CECT 7481 (AB15), *L. brevis* CD2 and CECT 7480 (AB38), *L. rhamnosus* LGG, *L. rhamnosus* PB01, *L. curvatus* EB10, *B. animalis subsp. lactis* DN-173010, *B. lactis* BB-12 and *Pediococcus acidilactici* CECT 8633 (AB30) (Twetman et al. 2009; Slawik et al. 2011; Lee et al. 2015; Kuru et al. 2017; Montero et al. 2017; Alanzi et al. 2018; Keller et al. 2018). In general, these studies have demonstrated improvements in gingival inflammatory parameters. Some of them also showed reductions in the amount of dental plaque and volume of gingival crevicular fluid, and improvements in microbiological and immunological parameters. The combination of lozenges containing two *L. reuteri* strains (DSM 17938 and ATCC PTA 5289) has also reduced Gingival and Plaque Index in healthy women presenting pregnancy gingivitis (Schlagenhauf et al. 2020).

Other studies failed to demonstrate additional clinical benefits after the administration of *L. casei Shirota* and *L. reuteri* (DSM 17938 and ATCC PTA 5289) in gingivitis (Staab et al. 2009; Iniesta et al. 2012). However, these probiotics resulted in improvements in inflammatory markers in gingival crevicular fluid (Staab et al. 2009) and reduction of periodontal pathogens in the subgingival microbiota (Iniesta et al. 2012). On the other hand, local administration of probiotics (*B. subtilis, B. megaterium* and *B. pumulus* in toothpaste, mouthrinse and toothbrush cleaner) did not demonstrate any effect on clinical parameters in gingivitis patients (Alkaya et al. 2017). A systematic review with meta-analysis reported that half of the clinical trials demonstrated additional clinical benefits with the use of probiotics in the management of gingivitis (Akram et al. 2020).

It is important to consider that RCTs on probiotics as adjuvant of the periodontal treatment are very heterogeneous regarding dosages, probiotic strain and/or combinations of strains, frequency of use and mode of application, duration of therapy, and baseline disease severity, making it difficult to draw definitive conclusions. However, it should be emphasized that SPT represents a potential application of probiotic therapy, since it is difficult to obtain adequate patient compliance over a long period of time. The potential to promote immunomodulation, guide pockets recolonization after the active phase of therapy and absence of side effects, which allows the administration for prolonged periods, favor the indication of probiotics in SPT. This potential has been evidenced in two recent controlled clinical studies using a combination of two *L. reuteri* strains (*L. reuteri* DSM 17938 and ATCC PTA 5289) (Grusovin et al. 2020; Laleman et al. 2020). The authors observed that the adjunctive use of the probiotic led to additional benefits in clinical periodontal parameters, such as reduction in probing depth and clinical attachment gain.

19.7 Future Approaches

The control of periodontitis should face the disease in its multifactor aspects, including antimicrobial and anti-inflammatory activities (Bartold and Van Dyke 2017). In the near future, a deep knowledge on the functional role of the oral and gut microbiomes will be critical to the development of probiotics targeting the control of periodontitis. Unlike other mucosa surfaces in humans (Ravel et al. 2011; Turroni et al. 2019), Lactobacilli and Bifidobateria are not hallmark of a healthy associated oral microbiome. Thus, other organisms highly adapted to subgingival oral sites seem better choices of probiotics toward the control of periodontitis.

The ideal probiotic strain for oral health should be able to colonize the oral sites for a prolonged period, be harmless, favor commensals, and impair colonization of pathogens and pathobionts. These properties would require a synergy of several different effector molecules with antibacterial activity and modulatory properties as well as the ability to break the resistance of the established microbiome to changes. Most of our knowledge on these molecules related to few cultivable organisms, and our repertoire of beneficial properties should be extended to less characterized microorganisms. Future studies should aim to express antibacterial proteins highly specific against oral pathogens, without any toxicity to oral commensals. Due to public awareness of genetically modified organisms, these organisms should be natural (wild) or if recombinant organisms are to be produced, they would be better used as postbiotics.

Clinical evidence indicated that probiotics should be effective as adjuvants in the treatment of periodontitis, mainly considering severe disease at baseline. The existing evidence came from studies that the dental biofilm is disrupted by mechanical means, and inflammation is temporally controlled. Therefore, the effect of probiotics on the restoration of homeostasis in non-treated periodontitis patients should still be addressed.

The oral cavity is a constantly changing environment, and a single probiotic may not cope with all conditions and exert a beneficial effect in all subjects. Probiotics able to control mild forms of periodontitis may be less effective in patients with aggressive periodontitis (Kawamoto et al. 2020) or other genetically encoded susceptibility (Pigossi et al. 2019). Specific microbiome signatures and differences in immune response according to smoking habits, age, diabetes and obesity, use of medications, impaired nutrition, hormonal variations, and immunocompromised status may also influence probiotic benefits. This great variability in humans indicates that next generation probiotics should be tailored according to the target population (Satokari 2019).

Restoration of the commensal microbiome may not be achieved by a single or a group of probiotic strains. Studies on oral microbiome transplantation and oral consortium of symbionts should be undertaken for the treatment of refractory cases, similar to those used for the gut dysbiotic microbiome (Li et al. 2015; Allegretti et al. 2019).

Overall, probiotics are promising tools to improve periodontal treatment in order to restore the microbial balance and modulate host response. However, the clinical translation of research in probiotics to the periodontal treatment should be optimized by the development of consortia of strains with different functions and niches, to be used according to the individual needs.

References

Abranches J, Zeng L, Kajfasz JK, Palmer SR, Chakraborty B, Wen ZT, Richards VP, Brady LJ, Lemos JA (2018) Biology of oral streptococci. Microbiol Spectr 6(5). https://doi.org/10.1128/microbiolspec.GPP3-0042-2018

Adriaens PA, Adriaens LM (2004) Effects of nonsurgical periodontal therapy on hard and soft tissues. Periodontol 36:121–145

Akram Z, Shafqat SS, Aati S, Kujan O, Fawzy A (2020) Clinical efficacy of probiotics in the treatment of gingivitis: a systematic review and meta-analysis. Aust Dent J 65(1):12–20

Alanzi A, Honkala S, Honkala E, Varghese A, Tolvanen M, Söderling E (2018) Effect of lactobacillus rhamnosus and Bifidobacterium lactis on gingival health, dental plaque, and periodontopathogens in adolescents: a randomised placebo-controlled clinical trial. Benefic Microbes 9(4):593–602

Albuquerque-Souza E, Balzarini D, Ando-Suguimoto ES, Ishikawa KH, Simionato MRL, Holzhausen M, Mayer MPA (2019) Probiotics alter the immune response of gingival epithelial cells challenged by Porphyromonas gingivalis. J Periodontal Res 54(2):115–127

Alkaya B, Laleman I, Keceli S, Ozcelik O, Cenk Haytac M, Teughels W (2017) Clinical effects of probiotics containing bacillus species on gingivitis: a pilot randomized controlled trial. J Periodontal Res 52(3):497–504

Allegretti JR, Kao D, Phelps E, Roach B, Smith J, Ganapini VC, Kassam Z, Xu H, Fischer M (2019) Risk of Clostridium difficile infection with systemic antimicrobial therapy following successful fecal microbiota transplant: should we recommend anti-clostridium difficile antibiotic prophylaxis? Dig Dis Sci 64(6):1668–1671

Alvarez-Sieiro P, Montalbán-López M, Mu D, Kuipers OP (2016) Bacteriocins of lactic acid bacteria: extending the family. Appl Microbiol Biotechnol 100(7):2939–2951

Amado PP, Kawamoto D, Albuquerque-Souza E, Franco DC, Saraiva L, Casarin RC, Horliana ACR, Mayer MPA (2020) Oral and fecal microbiome in molar-incisor pattern periodontitis. Front Cell Infect Microbiol 10:583761

Amin N, Boccardi V, Taghizadeh M, Jafarnejad S (2020) Probiotics and bone disorders: the role of RANKL/RANK/OPG pathway. Aging Clin Exp Res 32(3):363–371

Anderson RC, Cookson AL, McNabb WC, Park Z, McCann MJ, Kelly WJ, Roy NC (2010) Lactobacillus plantarum MB452 enhances the function of the intestinal barrier by increasing the expression levels of genes involved in tight junction formation. BMC Microbiol 10:316

Arimatsu K, Yamada H, Miyazawa H, Minagawa T, Nakajima M, Ryder MI, Gotoh K, Motooka D, Nakamura S, Iida T, Yamazaki K (2014) Oral pathobiont induces systemic inflammation and metabolic changes associated with alteration of gut microbiota. Sci Rep 4:4828

Armitage GC, Xenoudi P (2016) Post-treatment supportive care for the natural dentition and dental implants. Periodontol 71(1):164–184

Aroutcheva A, Gariti D, Simon M, Shott S, Faro J, Simoes JA, Gurguis A, Faro S (2001) Defense factors of vaginal lactobacilli. Am J Obstet Gynecol 185(2):375–379

Bartold PM, Van Dyke TE (2017) Host modulation: controlling the inflammation to control the infection. Periodontol 75(1):317–329

Ben Braïek O, Smaoui S (2019) Enterococci: between emerging pathogens and potential probiotics. Biomed Res Int 2019:5938210

Berezow AB, Darveau RP (2011) Microbial shift and periodontitis. Periodontol 55(1):36–47

Bonfiglio G, Neroni B, Radocchia G, Marazzato M, Pantanella F, Schippa S (2020) Insight into the possible use of the predator. Nutrients 12(8):2252

Butel MJ (2014) Probiotics, gut microbiota and health. Med Mal Infect 44(1):1–8

Caglar E, Topcuoglu N, Cildir SK, Sandalli N, Kulekci G (2009) Oral colonization by lactobacillus reuteri ATCC 55730 after exposure to probiotics. Int J Paediatr Dent 19(5):377–381

Cardoso RS, Messora MR, Silva PHF, Oliveira LF, Leite-Panissi C, Salvador S, Casarin R, Novaes AB, Palioto DB, Furlaneto FAC (2020) Effects of Bifidobacterium animalis subsp. lactis HN019 on ligature-induced periodontitis in rats with experimental rheumatoid arthritis. Benefic Microbes 11(1):33–46

Chelakkot C, Choi Y, Kim DK, Park HT, Ghim J, Kwon Y, Jeon J, Kim MS, Jee YK, Gho YS, Park HS, Kim YK, Ryu SH (2018) Akkermansia muciniphila-derived extracellular vesicles influence gut permeability through the regulation of tight junctions. Exp Mol Med 50(2):e450

Cheng H, Guan X, Chen D, Ma W (2019) The Th17/Treg cell balance: a gut microbiota- modulated story. Microorganisms 7(12):583

Chew J, Zilm PS, Fuss JM, Gully NJ (2012) A proteomic investigation of fusobacterium nucleatum alkaline-induced biofilms. BMC Microbiol 12:189

Christoffersen TE, Hult LT, Kuczkowska K, Moe KM, Skeie S, Lea T, Kleiveland CR (2014) In vitro comparison of the effects of probiotic, commensal and patho-

genic strains on macrophage polarization. Probiotics Antimicrob Proteins 6(1):1–10
Cleusix V, Lacroix C, Vollenweider S, Duboux M, Le Blay G (2007) Inhibitory activity spectrum of reuterin produced by lactobacillus reuteri against intestinal bacteria. BMC Microbiol 7:101
Coelho JMF, Miranda SS, da Cruz SS, Trindade SC, Passos-Soares JS, Cerqueira EMM, Costa MDCN, Figueiredo ACMG, Hintz AM, Barreto ML, Seymour GJ, Scannapieco F, Gomes-Filho IS (2020) Is there association between stress and periodontitis? Clin Oral Investig 24(7):2285–2294
Coretti L, Cuomo M, Florio E, Palumbo D, Keller S, Pero R, Chiariotti L, Lembo F, Cafiero C (2017) Subgingival dysbiosis in smoker and non-smoker patients with chronic periodontitis. Mol Med Rep 15(4):2007–2014
Corthésy B (2013) Role of secretory IgA in infection and maintenance of homeostasis. Autoimmun Rev 12(6):661–665
Darveau RP (2010) Periodontitis: a polymicrobial disruption of host homeostasis. Nat Rev Microbiol 8(7):481–490
de Kivit S, Tobin MC, Forsyth CB, Keshavarzian A, Landay AL (2014) Regulation of intestinal immune responses through TLR activation: implications for pro- and prebiotics. Front Immunol 5:60
de Moreno de LeBlanc A, Chaves S, Carmuega E, Weill R, Antóine J, Perdigón G (2008) Effect of long-term continuous consumption of fermented milk containing probiotic bacteria on mucosal immunity and the activity of peritoneal macrophages. Immunobiology 213(2):97–108
de Souza-Basqueira M, Ribeiro RM, de Oliveira LC, Moreira CHV, Martins RCR, Franco DC, Amado PPP, Mayer MPA, Sabino EC (2020) Gut Dysbiosis in Chagas disease. A possible link to the pathogenesis. Front Cell Infect Microbiol 10:402
Delitto AE, Rocha F, Decker AM, Amador B, Sorenson HL, Wallet SM (2018) MyD88-mediated innate sensing by oral epithelial cells controls periodontal inflammation. Arch Oral Biol 87:125–130
Dinan TG, Cryan JF (2017) The microbiome-gut-brain Axis in health and disease. Gastroenterol Clin N Am 46(1):77–89
Ding YH, Qian LY, Pang J, Lin JY, Xu Q, Wang LH, Huang DS, Zou H (2017) The regulation of immune cells by lactobacilli: a potential therapeutic target for anti- atherosclerosis therapy. Oncotarget 8(35):59915–59928
Dogi CA, Galdeano CM, Perdigón G (2008) Gut immune stimulation by non pathogenic Gram(+) and Gram(-) bacteria. Comparison with a probiotic strain. Cytokine 41(3):223–231
Donato KA, Gareau MG, Wang YJJ, Sherman PM (2010) Lactobacillus rhamnosus GG attenuates interferon-{gamma} and tumour necrosis factor-alpha-induced barrier dysfunction and pro-inflammatory signalling. Microbiology (Reading) 156(Pt 11):3288–3297
Duran-Pinedo AE, Frias-Lopez J (2015) Beyond microbial community composition: functional activities of the oral microbiome in health and disease. Microbes Infect 17(7):505–516
Dutzan N, Kajikawa T, Abusleme L, Greenwell-Wild T, Zuazo CE, Ikeuchi T, Brenchley L, Abe T, Hurabielle C, Martin D, Morell RJ, Freeman AF, Lazarevic V, Trinchieri G, Diaz PI, Holland SM, Belkaid Y, Hajishengallis G, Moutsopoulos NM (2018) A dysbiotic microbiome triggers T. Sci Transl Med 10(463):eaat0797
Ebersole JL, Cappelli D, Holt SC (2001) Periodontal diseases: to protect or not to protect is the question? Acta Odontol Scand 59(3):161–166
Engels C, Schwab C, Zhang J, Stevens MJ, Bieri C, Ebert MO, McNeill K, Sturla SJ, Lacroix C (2016) Acrolein contributes strongly to antimicrobial and heterocyclic amine transformation activities of reuterin. Sci Rep 6:36246
Eschenbach DA, Davick PR, Williams BL, Klebanoff SJ, Young-Smith K, Critchlow CM, Holmes KK (1989) Prevalence of hydrogen peroxide-producing Lactobacillus species in normal women and women with bacterial vaginosis. J Clin Microbiol 27(2):251–256
Festi D, Schiumerini R, Eusebi LH, Marasco G, Taddia M, Colecchia A (2014) Gut microbiota and metabolic syndrome. World J Gastroenterol 20(43):16079–16094
Fine DH, Armitage GC, Genco RJ, Griffen AL, Diehl SR (2019) Unique etiologic, demographic, and pathologic characteristics of localized aggressive periodontitis support classification as a distinct subcategory of periodontitis. J Am Dent Assoc 150(11):922–931
Fitzgerald JE, Kreutzer DL (1995) Localization of interleukin-8 in human gingival tissues. Oral Microbiol Immunol 10(5):297–303
Forner L, Larsen T, Kilian M, Holmstrup P (2006) Incidence of bacteremia after chewing, tooth brushing and scaling in individuals with periodontal inflammation. J Clin Periodontol 33(6):401–407
Foureaux R d C, Messora MR, de Oliveira LF, Napimoga MH, Pereira AN, Ferreira MS, Pereira LJ (2014) Effects of probiotic therapy on metabolic and inflammatory parameters of rats with ligature-induced periodontitis associated with restraint stress. J Periodontol 85(7):975–983
Gatej SM, Marino V, Bright R, Fitzsimmons TR, Gully N, Zilm P, Gibson RJ, Edwards S, Bartold PM (2018) Probiotic Lactobacillus rhamnosus GG prevents alveolar bone loss in a mouse model of experimental periodontitis. J Clin Periodontol 45(2):204–212
Gatej SM, Bright R, Weyrich LS, Marino V, Christophersen CT, Gibson RJ, Gully N, Zilm P, Bartold PM (2020) Probiotic lactobacillus Rhamnosus GG protects against P. Gingivalis and F. Nucleatum gut Dysbiosis. J Int Acad Periodontol 22(2):18–27
Gellin RG, Miller MC, Javed T, Engler WO, Mishkin DJ (1986) The effectiveness of the Titan-S sonic scaler versus curettes in the removal of subgingival calculus. A human surgical evaluation. J Periodontol 57(11):672–680

Gomes AC, Bueno AA, de Souza RG, Mota JF (2014) Gut microbiota, probiotics and diabetes. Nutr J 13:60

Grander C, Adolph TE, Wieser V, Lowe P, Wrzosek L, Gyongyosi B, Ward DV, Grabherr F, Gerner RR, Pfister A, Enrich B, Ciocan D, Macheiner S, Mayr L, Drach M, Moser P, Moschen AR, Perlemuter G, Szabo G, Cassard AM, Tilg H (2018) Recovery of ethanol-induced. Gut 67(5):891–901

Groeger S, Doman E, Chakraborty T, Meyle J (2010) Effects of Porphyromonas gingivalis infection on human gingival epithelial barrier function in vitro. Eur J Oral Sci 118(6):582–589

Grusovin MG, Bossini S, Calza S, Cappa V, Garzetti G, Scotti E, Gherlone EF, Mensi M (2020) Clinical efficacy of Lactobacillus reuteri-containing lozenges in the supportive therapy of generalized periodontitis stage III and IV, grade C: 1-year results of a double-blind randomized placebo-controlled pilot study. Clin Oral Investig 24(6):2015–2024

Guha D, Banerjee A, Mukherjee R, Pradhan B, Peneva M, Aleksandrov G, Suklabaidya S, Senapati S, Aich P (2019) A probiotic formulation containing Lactobacillus bulgaricus DWT1 inhibits tumor growth by activating pro-inflammatory responses in macrophages. J Funct Foods 56:232–245

Hajishengallis G (2015) Periodontitis: from microbial immune subversion to systemic inflammation. Nat Rev Immunol 15(1):30–44

Hajishengallis G, Darveau RP, Curtis MA (2012) The keystone-pathogen hypothesis. Nat Rev Microbiol 10(10):717–725

Herrera D, Alonso B, León R, Roldán S, Sanz M (2008) Antimicrobial therapy in periodontitis: the use of systemic antimicrobials against the subgingival biofilm. J Clin Periodontol 35(8 Suppl):45–66

Hienz SA, Paliwal S, Ivanovski S (2015) Mechanisms of bone resorption in periodontitis. J Immunol Res 2015:615486

Hill C, Guarner F, Reid G, Gibson GR, Merenstein DJ, Pot B, Morelli L, Canani RB, Flint HJ, Salminen S, Calder PC, Sanders ME (2014) Expert consensus document. The International Scientific Association for Probiotics and Prebiotics consensus statement on the scope and appropriate use of the term probiotic. Nat Rev Gastroenterol Hepatol 11(8):506–514

Ho SN, Acharya A, Sidharthan S, Li KY, Leung WK, McGrath C, Pelekos G (2020) A systematic review and meta-analysis of clinical, immunological, and microbiological shift in periodontitis after nonsurgical periodontal therapy with adjunctive use of probiotics. J Evid Based Dent Pract 20(1):101397

Hoffman FA, Heimbach JT, Sanders ME, Hibberd PL (2008) Executive summary: scientific and regulatory challenges of development of probiotics as foods and drugs. Clin Infect Dis 46(Suppl 2):S53–S57

Hojo K, Nagaoka S, Murata S, Taketomo N, Ohshima T, Maeda N (2007) Reduction of vitamin K concentration by salivary Bifidobacterium strains and their possible nutritional competition with Porphyromonas gingivalis. J Appl Microbiol 103(5):1969–1974

Huck O, Mulhall H, Rubin G, Kizelnik Z, Iyer R, Perpich JD, Haque N, Cani PD, de Vos WM, Amar S (2020) Akkermansia muciniphila reduces Porphyromonas gingivalis- induced inflammation and periodontal bone destruction. J Clin Periodontol 47(2):202–212

Hutchins AP, Diez D, Miranda-Saavedra D (2013) The IL-10/STAT3-mediated anti- inflammatory response: recent developments and future challenges. Brief Funct Genomics 12(6):489–498

Ikram S, Hassan N, Raffat MA, Mirza S, Akram Z (2018) Systematic review and meta- analysis of double-blind, placebo-controlled, randomized clinical trials using probiotics in chronic periodontitis. J Investig Clin Dent 9(3):e12338

Imlay JA (2003) Pathways of oxidative damage. Annu Rev Microbiol 57:395–418

Imlay JA (2019) Where in the world do bacteria experience oxidative stress? Environ Microbiol 21(2):521–530

İnce G, Gürsoy H, İpçi Ş, Cakar G, Emekli-Alturfan E, Yılmaz S (2015) Clinical and biochemical evaluation of lozenges containing lactobacillus reuteri as an adjunct to non- surgical periodontal therapy in chronic periodontitis. J Periodontol 86(6):746–754 (1546)

Iniesta M, Herrera D, Montero E, Zurbriggen M, Matos AR, Marín MJ, Sánchez- Beltrán MC, Llama-Palacio A, Sanz M (2012) Probiotic effects of orally administered Lactobacillus reuteri-containing tablets on the subgingival and salivary microbiota in patients with gingivitis. A randomized clinical trial. J Clin Periodontol 39(8):736–744

Invernici MM, Salvador SL, Silva PHF, Soares MSM, Casarin R, Palioto DB, Souza SLS, Taba M, Novaes AB, Furlaneto FAC, Messora MR (2018) Effects of Bifidobacterium probiotic on the treatment of chronic periodontitis: a randomized clinical trial. J Clin Periodontol 45(10):1198–1210

Invernici MM, Furlaneto FAC, Salvador SL, Ouwehand AC, Salminen S, Mantziari A, Vinderola G, Ervolino E, Santana SI, Silva PHF, Messora MR (2020) Bifidobacterium animalis subsp lactis HN019 presents antimicrobial potential against periodontopathogens and modulates the immunological response of oral mucosa in periodontitis patients. PLoS One 15(9):e0238425

Ishikawa KH, Bueno MR, Kawamoto D, Simionato MRL, Mayer MPA (2020a) Lactobacilli postbiotics reduce biofilm formation and alter transcription of virulence genes of Aggregatibacter actinomycetemcomitans. Mol Oral Microbiol 36:92–102

Ishikawa KH, Mita D, Kawamoto D, Nicoli JR, Albuquerque-Souza E, Lorenzetti Simionato MR, Mayer MPA (2020b) Probiotics alter biofilm formation and the transcription of *Porphyromonas gingivalis* virulence-associated genes. J Oral Microbiol 12(1):1805553

Jang SE, Min SW (2020) S1 improves colitis induced by 2,4,6-Trinitrobenzene sulfonic acid by the inhibition of NF-κB signaling in mice. J Microbiol Biotechnol 30(1):71–78

Jang SE, Hyam SR, Han MJ, Kim SY, Lee BG, Kim DH (2013) Lactobacillus brevis G-101 ameliorates colitis in mice by inhibiting NF-κB, MAPK and AKT pathways and by polarizing M1 macrophages to M2-like macrophages. J Appl Microbiol 115(3):888–896

Jang SE, Han MJ, Kim SY, Kim DH (2014) Lactobacillus plantarum CLP-0611 ameliorates colitis in mice by polarizing M1 to M2-like macrophages. Int Immunopharmacol 21(1):186–192

Ji J, Hu SL, Cui ZW, Li WF (2013) Probiotic bacillus amyloliquefaciens mediate M1 macrophage polarization in mouse bone marrow-derived macrophages. Arch Microbiol 195(5):349–356

Jia L, Han N, Du J, Guo L, Luo Z, Liu Y (2019) Pathogenesis of important virulence factors of. Front Cell Infect Microbiol 9:262

Karczewski J, Troost FJ, Konings I, Dekker J, Kleerebezem M, Brummer RJ, Wells JM (2010) Regulation of human epithelial tight junction proteins by Lactobacillus plantarum in vivo and protective effects on the epithelial barrier. Am J Physiol Gastrointest Liver Physiol 298(6):G851–G859

Kato T, Yamazaki K, Nakajima M, Date Y, Kikuchi J, Hase K, Ohno H (2018) Oral Administration of Porphyromonas gingivalis alters the gut microbiome and serum metabolome. mSphere 3(5):e00460

Kawamoto D, Amado PPL, Albuquerque-Souza E, Bueno MR, Vale GC, Saraiva L, Mayer MPA (2020) Chemokines and cytokines profile in whole saliva of patients with periodontitis. Cytokine 135:155197

Keller MK, Brandsborg E, Holmstrøm K, Twetman S (2018) Effect of tablets containing probiotic candidate strains on gingival inflammation and composition of the salivary microbiome: a randomised controlled trial. Benefic Microbes 9(3):487–494

Khurshid M, Aslam B, Nisar MA, Akbar R, Rahman H, Khan AA, Rasool MH (2015) Bacterial munch for infants: potential pediatric therapeutic interventions of probiotics. Future Microbiol 10(11):1881–1895

Kim SW, Kim HM, Yang KM, Kim SA, Kim SK, An MJ, Park JJ, Lee SK, Kim TI, Kim WH, Cheon JH (2010) Bifidobacterium lactis inhibits NF-kappaB in intestinal epithelial cells and prevents acute colitis and colitis-associated colon cancer in mice. Inflamm Bowel Dis 16(9):1514–1525

Kobayashi R, Kobayashi T, Sakai F, Hosoya T, Yamamoto M, Kurita-Ochiai T (2017) Oral administration of Lactobacillus gasseri SBT2055 is effective in preventing Porphyromonas gingivalis-accelerated periodontal disease. Sci Rep 7(1):545

Kommineni S, Bretl DJ, Lam V, Chakraborty R, Hayward M, Simpson P, Cao Y, Bousounis P, Kristich CJ, Salzman NH (2015) Bacteriocin production augments niche competition by enterococci in the mammalian gastrointestinal tract. Nature 526(7575):719–722

Kristensen NB, Bryrup T, Allin KH, Nielsen T, Hansen TH, Pedersen O (2016) Alterations in fecal microbiota composition by probiotic supplementation in healthy adults: a systematic review of randomized controlled trials. Genome Med 8(1):52

Kuru BE, Laleman I, Yalnızoğlu T, Kuru L, Teughels W (2017) The influence of a Bifidobacterium animalis probiotic on gingival health: a randomized controlled clinical trial. J Periodontol 88(11):1115–1123

Laleman I, Yilmaz E, Ozcelik O, Haytac C, Pauwels M, Herrero ER, Slomka V, Quirynen M, Alkaya B, Teughels W (2015) The effect of a streptococci containing probiotic in periodontal therapy: a randomized controlled trial. J Clin Periodontol 42(11):1032–1041

Laleman I, Pauwels M, Quirynen M, Teughels W (2020) A dual-strain lactobacilli reuteri probiotic improves the treatment of residual pockets: a randomized controlled clinical trial. J Clin Periodontol 47(1):43–53

Lamont RJ, Yilmaz O (2002) In or out: the invasiveness of oral bacteria. Periodontol 2000(30):61–69

Laursen MF, Laursen RP, Larnkjær A, Michaelsen KF, Bahl MI, Licht TR (2017) Administration of two probiotic strains during early childhood does not affect the endogenous gut microbiota composition despite probiotic proliferation. BMC Microbiol 17(1):175

Lee JK, Kim SJ, Ko SH, Ouwehand AC, Ma DS (2015) Modulation of the host response by probiotic Lactobacillus brevis CD2 in experimental gingivitis. Oral Dis 21(6):705–712

Li M, Liang P, Li Z, Wang Y, Zhang G, Gao H, Wen S, Tang L (2015) Fecal microbiota transplantation and bacterial consortium transplantation have comparable effects on the re- establishment of mucosal barrier function in mice with intestinal dysbiosis. Front Microbiol 6:692

Li J, Lin S, Vanhoutte PM, Woo CW, Xu A (2016) Akkermansia Muciniphila protects against atherosclerosis by preventing metabolic Endotoxemia-induced inflammation in Apoe-/- Mice. Circulation 133(24):2434–2446

Liu Y, Fatheree NY, Mangalat N, Rhoads JM (2012) Lactobacillus reuteri strains reduce incidence and severity of experimental necrotizing enterocolitis via modulation of TLR4 and NF- κB signaling in the intestine. Am J Physiol Gastrointest Liver Physiol 302(6):G608–G617

Liu TH, Tsai TY, Pan TM (2018a) The anti-periodontitis effects of ethanol extract prepared using lactobacillus paracasei subsp. paracasei NTU 101. Nutrients 10(4):472

Liu Y, Alookaran JJ, Rhoads JM (2018b) Probiotics in autoimmune and inflammatory disorders. Nutrients 10(10):1537

Liu J, Wang Y, Meng H, Yu J, Lu H, Li W, Lu R, Zhao Y, Li Q, Su L (2019) Butyrate rather than LPS subverts gingival epithelial homeostasis by downregulation of intercellular junctions and triggering pyroptosis. J Clin Periodontol 46(9):894–907

Longo PL, Dabdoub S, Kumar P, Artese HPC, Dib SA, Romito GA, Mayer MPA (2018) Glycaemic status affects the subgingival microbiome of diabetic patients. J Clin Periodontol 45(8):932–940

Lourenço TGB, Spencer SJ, Alm EJ, Colombo APV (2018) Defining the gut microbiota in individuals with periodontal diseases: an exploratory study. J Oral Microbiol 10(1):1487741

Luongo D, Treppiccione L, Sorrentino A, Ferrocino I, Turroni S, Gatti M, Di Cagno R, Sanz Y, Rossi M (2017) Immune-modulating effects in mouse dendritic cells of lactobacilli and bifidobacteria isolated from individuals following omnivorous, vegetarian and vegan diets. Cytokine 97:141–148

Maekawa T, Hajishengallis G (2014) Topical treatment with probiotic lactobacillus brevis CD2 inhibits experimental periodontal inflammation and bone loss. J Periodontal Res 49(6):785–791

Maldonado-Gómez MX, Martínez I, Bottacini F, O'Callaghan A, Ventura M, van Sinderen D, Hillmann B, Vangay P, Knights D, Hutkins RW, Walter J (2016) Stable engraftment of Bifidobacterium longum AH1206 in the human gut depends on individualized features of the resident microbiome. Cell Host Microbe 20(4):515–526

Marsh PD (2003) Are dental diseases examples of ecological catastrophes? Microbiology (Reading) 149(Pt 2):279–294

Marsh PD, Zaura E (2017) Dental biofilm: ecological interactions in health and disease. J Clin Periodontol 44(Suppl 18):S12–S22

Martin-Cabezas R, Davideau JL, Tenenbaum H, Huck O (2016) Clinical efficacy of probiotics as an adjunctive therapy to non-surgical periodontal treatment of chronic periodontitis: a systematic review and meta-analysis. J Clin Periodontol 43(6):520–530

Martinez FA, Balciunas EM, Converti A, Cotter PD, de Souza Oliveira RP (2013) Bacteriocin production by Bifidobacterium spp. a review. Biotechnol Adv 31(4):482–488

Martinez RC, Bedani R, Saad SM (2015) Scientific evidence for health effects attributed to the consumption of probiotics and prebiotics: an update for current perspectives and future challenges. Br J Nutr 114(12):1993–2015

Matsubara VH, Bandara HM, Ishikawa KH, Mayer MP, Samaranayake LP (2016) The role of probiotic bacteria in managing periodontal disease: a systematic review. Expert Rev Anti-Infect Ther 14(7):643–655

Mayanagi G, Kimura M, Nakaya S, Hirata H, Sakamoto M, Benno Y, Shimauchi H (2009) Probiotic effects of orally administered Lactobacillus salivarius WB21-containing tablets on periodontopathic bacteria: a double-blinded, placebo-controlled, randomized clinical trial. J Clin Periodontol 36(6):506–513

McKenzie RME, Johnson NA, Aruni W, Dou Y, Masinde G, Fletcher HM (2012) Differential response of Porphyromonas gingivalis to varying levels and duration of hydrogen peroxide-induced oxidative stress. Microbiology (Reading) 158(Pt 10):2465–2479

Mellor AL, Munn DH (2008) Creating immune privilege: active local suppression that benefits friends, but protects foes. Nat Rev Immunol 8(1):74–80

Messora MR, Oliveira LF, Foureaux RC, Taba M, Zangerônimo MG, Furlaneto FA, Pereira LJ (2013) Probiotic therapy reduces periodontal tissue destruction and improves the intestinal morphology in rats with ligature-induced periodontitis. J Periodontol 84(12):1818–1826

Messora MR, Pereira LJ, Foureaux R, Oliveira LF, Sordi CG, Alves AJ, Napimoga MH, Nagata MJ, Ervolino E, Furlaneto FA (2016) Favourable effects of Bacillus subtilis and Bacillus licheniformis on experimental periodontitis in rats. Arch Oral Biol 66:108–119

Mombelli A (2018) Microbial colonization of the periodontal pocket and its significance for periodontal therapy. Periodontol 76(1):85–96

Mombelli A, Schmid B, Rutar A, Lang NP (2000) Persistence patterns of Porphyromonas gingivalis, Prevotella intermedia/nigrescens, and Actinobacillus actinomyetemcomitans after mechanical therapy of periodontal disease. J Periodontol 71(1):14–21

Montenegro SCL, Retamal-Valdes B, Bueno-Silva B, Duarte PM, Faveri M, Figueiredo LC, Feres M (2020) Do patients with aggressive and chronic periodontitis exhibit specific differences in the subgingival microbial composition? A systematic review. J Periodontol 91(11):1503–1520

Montero E, Iniesta M, Rodrigo M, Marín MJ, Figuero E, Herrera D, Sanz M (2017) Clinical and microbiological effects of the adjunctive use of probiotics in the treatment of gingivitis: a randomized controlled clinical trial. J Clin Periodontol 44(7):708–716

Morales A, Carvajal P, Silva N, Hernandez M, Godoy C, Rodriguez G, Cabello R, Garcia- Sesnich J, Hoare A, Diaz PI, Gamonal J (2016) Clinical effects of lactobacillus rhamnosus in non-surgical treatment of chronic periodontitis: a randomized placebo-controlled trial with 1-year follow-up. J Periodontol 87(8):944–952

Mousquer CR, Della Bona A, Milani DC, Callegari-Jacques SM, Ishikawa K, Mayer MPA, Rösing CK, Fornari F (2020) Are Lactobacillus salivarius G60 and inulin more efficacious to treat patients with oral halitosis and tongue coating than the probiotic alone and placebo? A randomized clinical trial. J Periodontol 91(6):775–783

Moutsopoulos NM, Konkel J, Sarmadi M, Eskan MA, Wild T, Dutzan N, Abusleme L, Zenobia C, Hosur KB, Abe T, Uzel G, Chen W, Chavakis T, Holland SM, Hajishengallis G (2014) Defective neutrophil recruitment in leukocyte adhesion deficiency type I disease causes local IL-17-driven inflammatory bone loss. Sci Transl Med 6(229):229ra240

Mu Q, Tavella VJ, Luo XM (2018) Role of Lactobacillus reuteri in human health and diseases. Front Microbiol 9:757

Naidu AS, Bidlack WR, Clemens RA (1999) Probiotic spectra of lactic acid bacteria (LAB). Crit Rev Food Sci Nutr 39(1):13–126

Nakajima M, Arimatsu K, Kato T, Matsuda Y, Minagawa T, Takahashi N, Ohno H, Yamazaki K (2015) Oral Administration of P. gingivalis induces Dysbiosis of gut microbiota and impaired barrier function leading to dissemination of Enterobacteria to the liver. PLoS One 10(7):e0134234

Navarrete M, Garcia J, Dutzan N, Henriquez L, Puente J, Carvajal P, Hernandez M, Gamonal J (2014) Interferon-gamma, interleukins-6 and -4, and factor XIII-A as indirect markers of the classical and alternative macrophage activation pathways in chronic periodontitis. J Periodontol 85(5):751–760

Nissen L, Sgorbati B, Biavati B, Belibasakis GN (2014) Lactobacillus salivarius and L. gasseri down-regulate Aggregatibacter actinomycetemcomitans exotoxins expression. Ann Microbiol 64:611–617

Ohlsson C, Sjögren K (2015) Effects of the gut microbiota on bone mass. Trends Endocrinol Metab 26(2):69–74

Oliveira LF, Salvador SL, Silva PH, Furlaneto FA, Figueiredo L, Casarin R, Ervolino E, Palioto DB, Souza SL, Taba M, Novaes AB, Messora MR (2017) Benefits of Bifidobacterium animalis subsp. lactis Probiotic in Experimental Periodontitis. J Periodontol 88(2):197–208

Parisi L, Gini E, Baci D, Tremolati M, Fanuli M, Bassani B, Farronato G, Bruno A, Mortara L (2018) Macrophage polarization in chronic inflammatory diseases: killers or builders? J Immunol Res 2018:8917804

Parker RC, Rapley JW, Isley W, Spencer P, Killoy WJ (1993) Gingival crevicular blood for assessment of blood glucose in diabetic patients. J Periodontol 64(7):666–672

Pascale A, Marchesi N, Govoni S, Coppola A, Gazzaruso C (2019) The role of gut microbiota in obesity, diabetes mellitus, and effect of metformin: new insights into old diseases. Curr Opin Pharmacol 49:1–5

Patyna M, Ehlers V, Bahlmann B, Kasaj A (2021) Effects of adjunctive light-activated disinfection and probiotics on clinical and microbiological parameters in periodontal treatment: a randomized, controlled, clinical pilot study. Clin Oral Investig 25:3967–3975

Pelekos G, Ho SN, Acharya A, Leung WK, McGrath C (2019) A double-blind, paralleled-arm, placebo-controlled and randomized clinical trial of the effectiveness of probiotics as an adjunct in periodontal care. J Clin Periodontol 46(12):1217–1227

Penala S, Kalakonda B, Pathakota KR, Jayakumar A, Koppolu P, Lakshmi BV, Pandey R, Mishra A (2016) Efficacy of local use of probiotics as an adjunct to scaling and root planing in chronic periodontitis and halitosis: a randomized controlled trial. J Res Pharm Pract 5(2):86–93

Pigossi SC, Anovazzi G, Finoti LS, de Medeiros MC, Mayer MPA, Rossa Junior C, Scarel-Caminaga RM (2019) Functionality of the interleukin 8 haplotypes in lymphocytes and macrophages in response to gram-negative periodontopathogens. Gene 689:152–160

Pöllänen MT, Paino A, Ihalin R (2013) Environmental stimuli shape biofilm formation and the virulence of periodontal pathogens. Int J Mol Sci 14(8):17221–17237

Pretzl B, Sälzer S, Ehmke B, Schlagenhauf U, Dannewitz B, Dommisch H, Eickholz P, Jockel-Schneider Y (2019) Administration of systemic antibiotics during non-surgical periodontal therapy-a consensus report. Clin Oral Investig 23(7):3073–3085

Pudgar P, Povšič K, Čuk K, Seme K, Petelin M, Gašperšič R (2021) Probiotic strains of Lactobacillus brevis and Lactobacillus plantarum as adjunct to non-surgical periodontal therapy: 3-month results of a randomized controlled clinical trial. Clin Oral Investig 25(3):1411–1422

Raff A, Hunt LC (2012) Probiotics for periodontal health: a review of the literature. J Dent Hyg 86(2):71–81

Ravel J, Gajer P, Abdo Z, Schneider GM, Koenig SS, McCulle SL, Karlebach S, Gorle R, Russell J, Tacket CO, Brotman RM, Davis CC, Ault K, Peralta L, Forney LJ (2011) Vaginal microbiome of reproductive-age women. Proc Natl Acad Sci U S A 108(Suppl 1):4680–4687

Ren L, Leung WK, Darveau RP, Jin L (2005) The expression profile of lipopolysaccharide-binding protein, membrane-bound CD14, and toll-like receptors 2 and 4 in chronic periodontitis. J Periodontol 76(11):1950–1959

Riccia DN, Bizzini F, Perilli MG, Polimeni A, Trinchieri V, Amicosante G, Cifone MG (2007) Anti-inflammatory effects of Lactobacillus brevis (CD2) on periodontal disease. Oral Dis 13(4):376–385

Ricoldi MST, Furlaneto FAC, Oliveira LFF, Teixeira GC, Pischiotini JP, Moreira ALG, Ervolino E, de Oliveira MN, Bogsan CSB, Salvador SL, Messora MR (2017) Effects of the probiotic Bifidobacterium animalis subsp. lactis on the non-surgical treatment of periodontitis. A histomorphometric, microtomographic and immunohistochemical study in rats. PLoS One 12(6):e0179946

Ríos Colombo NS, Chalón MC, Dupuy FG, Gonzalez CF, Bellomio A (2019) The case for class II bacteriocins: a biophysical approach using "suicide probes" in receptor-free hosts to study their mechanism of action. Biochimie 165:183–195

Saarela M, Mogensen G, Fonden R, Matto J, Mattila-Sandholm T (2000) Probiotic bacteria: safety, functional and technological properties. J Biotechnol 84(3):197–215

Sajedinejad N, Paknejad M, Houshmand B, Sharafi H, Jelodar R, Shahbani Zahiri H, Noghabi KA (2018) Lactobacillus salivarius NK02: a potent probiotic for clinical application in mouthwash. Probiotics Antimicrob Proteins 10(3):485–495

Sanz M, Herrera D, Kebschull M, Chapple I, Jepsen S, Beglundh T, Sculean A, Tonetti MS, E. W. P. A. M. Consultants (2020) Treatment of stage I-III periodontitis-The EFP S3 level clinical practice guideline. J Clin Periodontol 47(Suppl 22):4–60

Satokari R (2019) Modulation of gut microbiota for health by current and next-generation probiotics. Nutrients 11(8):1921

Schlagenhauf U, Rehder J, Gelbrich G, Jockel-Schneider Y (2020) Consumption of lactobacillus reuteri-containing lozenges improves periodontal health in navy sailors at sea: a randomized controlled trial. J Periodontol 91(10):1328–1338

Seth A, Yan F, Polk DB, Rao RK (2008) Probiotics ameliorate the hydrogen peroxide- induced epithelial barrier disruption by a PKC- and MAP kinase-dependent mechanism. Am J Physiol Gastrointest Liver Physiol 294(4):G1060–G1069

Shah MP, Gujjari SK, Chandrasekhar VS (2013) Evaluation of the effect of probiotic (inersan®) alone, combination of probiotic with doxycycline and doxycycline alone on aggressive periodontitis – a clinical and microbiological study. J Clin Diagn Res 7(3):595–600

Shimauchi H, Mayanagi G, Nakaya S, Minamibuchi M, Ito Y, Yamaki K, Hirata H (2008) Improvement of periodontal condition by probiotics with Lactobacillus salivarius WB21: a randomized, double-blind, placebo-controlled study. J Clin Periodontol 35(10):897–905

Shimazu T, Villena J, Tohno M, Fujie H, Hosoya S, Shimosato T, Aso H, Suda Y, Kawai Y, Saito T, Makino S, Ikegami S, Itoh H, Kitazawa H (2012) Immunobiotic Lactobacillus jensenii elicits anti-inflammatory activity in porcine intestinal epithelial cells by modulating negative regulators of the toll-like receptor signaling pathway. Infect Immun 80(1):276–288

Sica A, Mantovani A (2012) Macrophage plasticity and polarization: in vivo veritas. J Clin Invest 122(3):787–795

Sica A, Erreni M, Allavena P, Porta C (2015) Macrophage polarization in pathology. Cell Mol Life Sci 72(21):4111–4126

Singh H, Torralba MG, Moncera KJ, DiLello L, Petrini J, Nelson KE, Pieper R (2019) Gastro-intestinal and oral microbiome signatures associated with healthy aging. Geroscience 41(6):907–921

Slawik S, Staufenbiel I, Schilke R, Nicksch S, Weinspach K, Stiesch M, Eberhard J (2011) Probiotics affect the clinical inflammatory parameters of experimental gingivitis in humans. Eur J Clin Nutr 65(7):857–863

Smalley JW, Silver J, Marsh PJ, Birss AJ (1998) The periodontopathogen Porphyromonas gingivalis binds iron protoporphyrin IX in the mu-oxo dimeric form: an oxidative buffer and possible pathogenic mechanism. Biochem J 331(Pt 3):681–685

Sockett RE (2009) Predatory lifestyle of Bdellovibrio bacteriovorus. Annu Rev Microbiol 63:523–539

Staab B, Eick S, Knöfler G, Jentsch H (2009) The influence of a probiotic milk drink on the development of gingivitis: a pilot study. J Clin Periodontol 36(10):850–856

Strom K, Sjogren J, Broberg A, Schnurer J (2002) Lactobacillus plantarum MiLAB 393 produces the antifungal cyclic dipeptides cyclo(L-Phe-L-Pro) and cyclo(L-Phe-trans-4-OH-L-Pro) and 3-phenyllactic acid. Appl Environ Microbiol 68(9):4322–4327

Szkaradkiewicz AK, Stopa J, Karpiński TM (2014) Effect of oral administration involving a probiotic strain of lactobacillus reuteri on pro-inflammatory cytokine response in patients with chronic periodontitis. Arch Immunol Ther Exp 62(6):495–500

Takanashi N, Tomosada Y, Villena J, Murata K, Takahashi T, Chiba E, Tohno M, Shimazu T, Aso H, Suda Y, Ikegami S, Itoh H, Kawai Y, Saito T, Alvarez S, Kitazawa H (2013) Advanced application of bovine intestinal epithelial cell line for evaluating regulatory effect of lactobacilli against heat-killed enterotoxigenic Escherichia coli-mediated inflammation. BMC Microbiol 13:54

Takiishi T, Fenero CIM, Câmara NOS (2017) Intestinal barrier and gut microbiota: shaping our immune responses throughout life. Tissue Barriers 5(4):e1373208

Tam V, O'Brien-Simpson NM, Chen YY, Sanderson CJ, Kinnear B, Reynolds EC (2009) The RgpA-Kgp proteinase-adhesin complexes of Porphyromonas gingivalis inactivate the Th2 cytokines interleukin-4 and interleukin-5. Infect Immun 77(4):1451–1458

Tekce M, Ince G, Gursoy H, Dirikan Ipci S, Cakar G, Kadir T, Yılmaz S (2015) Clinical and microbiological effects of probiotic lozenges in the treatment of chronic periodontitis: a 1-year follow-up study. J Clin Periodontol 42(4):363–372

Teughels W, Newman MG, Coucke W, Haffajee AD, Van Der Mei HC, Haake SK, Schepers E, Cassiman JJ, Van Eldere J, van Steenberghe D, Quirynen M (2007) Guiding periodontal pocket recolonization: a proof of concept. J Dent Res 86(11):1078–1082

Teughels W, Durukan A, Ozcelik O, Pauwels M, Quirynen M, Haytac MC (2013) Clinical and microbiological effects of lactobacillus reuteri probiotics in the treatment of chronic periodontitis: a randomized placebo-controlled study. J Clin Periodontol 40(11):1025–1035

Teughels W, Feres M, Oud V, Martín C, Matesanz P, Herrera D (2020) Adjunctive effect of systemic antimicrobials in periodontitis therapy: a systematic review and meta-analysis. J Clin Periodontol 47(Suppl 22):257–281

Theodoro LH, Cláudio MM, Nuernberg MAA, Miessi DMJ, Batista JA, Duque C, Garcia VG (2019) Effects of Lactobacillus reuteri as an adjunct to the treatment of periodontitis in smokers: randomised clinical trial. Benefic Microbes 10(4):375–384

Tribble GD, Lamont RJ (2010) Bacterial invasion of epithelial cells and spreading in periodontal tissue. Periodontol 52(1):68–83

Tsubura S, Mizunuma H, Ishikawa S, Oyake I, Okabayashi M, Katoh K, Shibata M, Iizuka T, Toda T (2009) The effect of Bacillus subtilis mouth rinsing in patients with periodontitis. Eur J Clin Microbiol Infect Dis 28(11):1353–1356

Turroni F, Milani C, Duranti S, Mahony J, van Sinderen D, Ventura M (2018) Glycan utilization and cross-feeding activities by Bifidobacteria. Trends Microbiol 26(4):339–350

Turroni F, Duranti S, Milani C, Lugli GA, van Sinderen D, Ventura M (2019) A key member of the early human gut microbiota. Microorganisms 7(11):544

Twetman S, Derawi B, Keller M, Ekstrand K, Yucel-Lindberg T, Stecksen-Blicks C (2009) Short-term effect of chewing gums containing probiotic Lactobacillus reuteri on the levels of inflammatory mediators in gingival crevicular fluid. Acta Odontol Scand 67(1):19–24

Valenti P, Rosa L, Capobianco D, Lepanto MS, Schiavi E, Cutone A, Paesano R, Mastromarino P (2018) Role of lactobacilli and Lactoferrin in the mucosal Cervicovaginal defense. Front Immunol 9:376

Van Essche M, Quirynen M, Sliepen I, Loozen G, Boon N, Van Eldere J, Teughels W (2011) Killing of anaerobic pathogens by predatory bacteria. Mol Oral Microbiol 26(1):52–61

Vicario M, Santos A, Violant D, Nart J, Giner L (2013) Clinical changes in periodontal subjects with the probiotic Lactobacillus reuteri Prodentis: a preliminary randomized clinical trial. Acta Odontol Scand 71(3–4):813–819

Vivekananda MR, Vandana KL, Bhat KG (2010) Effect of the probiotic Lactobacilli reuteri (Prodentis) in the management of periodontal disease: a preliminary randomized clinical trial. J Oral Microbiol 2:1–9

Wegh CAM, Geerlings SY, Knol J, Roeselers G, Belzer C (2019) Postbiotics and their potential applications in early life nutrition and beyond. Int J Mol Sci 20(19):4673

Weitzmann MN (2017) Bone and the immune system. Toxicol Pathol 45(7):911–924

Wu W, Lv L, Shi D, Ye J, Fang D, Guo F, Li Y, He X, Li L (2017) Protective effect of. Front Microbiol 8:1804

Xu Y, Wang N, Tan HY, Li S, Zhang C, Feng Y (2020) Function of *Akkermansia muciniphila* in obesity: interactions with lipid metabolism, immune response and gut systems. Front Microbiol 11:219

Yadav H, Lee JH, Lloyd J, Walter P, Rane SG (2013) Beneficial metabolic effects of a probiotic via butyrate-induced GLP-1 hormone secretion. J Biol Chem 288(35):25088–25097

Yadav AK, Tyagi A, Kumar A, Panwar S, Grover S, Saklani AC, Hemalatha R, Batish VK (2017) Adhesion of Lactobacilli and their anti-infectivity potential. Crit Rev Food Sci Nutr 57(10):2042–2056

Yli-Knuuttila H, Snall J, Kari K, Meurman JH (2006) Colonization of Lactobacillus rhamnosus GG in the oral cavity. Oral Microbiol Immunol 21(2):129–131

Yu JJ, Ruddy MJ, Wong GC, Sfintescu C, Baker PJ, Smith JB, Evans RT, Gaffen SL (2007) An essential role for IL-17 in preventing pathogen-initiated bone destruction: recruitment of neutrophils to inflamed bone requires IL-17 receptor-dependent signals. Blood 109(9):3794–3802

Yu S, Ding L, Liang D, Luo L (2018) Porphyromonas gingivalis inhibits M2 activation of macrophages by suppressing α-ketoglutarate production in mice. Mol Oral Microbiol 33(5):388–395

Zandbergen D, Slot DE, Niederman R, Van der Weijden FA (2016) The concomitant administration of systemic amoxicillin and metronidazole compared to scaling and root planing alone in treating periodontitis: =a systematic review=. BMC Oral Health 16:27

Zhu L, Xu Y, Ferretti JJ, Kreth J (2014) Probing oral microbial functionality–expression of spxB in plaque samples. PLoS One 9(1):e86685

Zhu B, Macleod LC, Newsome E, Liu J, Xu P (2019) Aggregatibacter actinomycetemcomitans mediates protection of Porphyromonas gingivalis from Streptococcus sanguinis hydrogen peroxide production in multi-species biofilms. Sci Rep 9(1):4944

Zielińska D, Kołożyn-Krajewska D (2018) Food-origin lactic acid bacteria may exhibit probiotic properties: review. Biomed Res Int 2018:5063185

Zupancic K, Kriksic V, Kovacevic I, Kovacevic D (2017) Influence of Oral probiotic streptococcus salivarius K12 on ear and Oral cavity health in humans: systematic review. Probiotics Antimicrob Proteins 9(2):102–110

Periodontal Cell Therapy: A Systematic Review and Meta-analysis

20

Antoine Dubuc, Valérie Planat-Bénard, Mathieu Marty, Paul Monsarrat, and Philippe Kémoun

Abstract

Background: Periodontitis is a chronic inflammatory disease characterized by the loss of tooth-supporting tissues (or periodontium) leading to the formation of periodontal pocket then to tooth loss. Conventional therapies that involve tooth root debridement are still disappointing because they are more centered on periodontal repair than disease pathophysiology causes. The meta-analysis we present here focused on the results of experimental studies that investigated periodontal mesenchymal stromal cells (MSCs) therapy, a promising strategy to regenerate tissue, given to their immunomodulatory and trophic properties.

Methods: Using PubMed database and ICTRP search portal, 84 animal and 3 randomized human studies were analyzed.

Results: Overall, our results highlighted that MSCs grafting, regardless of their tissue origin, enhances periodontal regeneration. A defect morphology suitable for an initial clot stabilization increases the procedure efficacy, especially if cells are carried using a vehicle from natural origin. Nevertheless, methodological biases have been highlighted and still limit the translation to human with high prog-

Authors Paul Monsarrat and Philippe Kémoun have equally contributed to this chapter.

A. Dubuc · M. Marty
Dental Faculty, Paul Sabatier University, Toulouse Hospital (Hôpitaux de Toulouse) University, and Toulouse Institute of Oral Medicine,
Toulouse, France
e-mail: dubuca@me.com; mathieu.marty@univ-tlse3.fr

V. Planat-Bénard
RESTORE Research Center, Université de Toulouse, INSERM 1301, CNRS 5070, EFS, ENVT,
Toulouse, France
e-mail: valerie.planat@inserm.fr

P. Monsarrat
Dental Faculty, Paul Sabatier University, Toulouse Hospital (Hôpitaux de Toulouse) University, and Toulouse Institute of Oral Medicine,
Toulouse, France

RESTORE Research Center, Université de Toulouse, INSERM 1301, CNRS 5070, EFS, ENVT,
Toulouse, France

Artificial and Natural Intelligence Toulouse Institute ANITI, Toulouse, France
e-mail: paul.monsarrat@univ-tlse3.fr

P. Kémoun (✉)
Dental Faculty, Paul Sabatier University, Toulouse Hospital (Hôpitaux de Toulouse) University, and Toulouse Institute of Oral Medicine,
Toulouse, France

RESTORE Research Center, Université de Toulouse, INSERM 1301, CNRS 5070, EFS, ENVT,
Toulouse, France
e-mail: philippe.kemoun@univ-tlse3.fr

J. Santi-Rocca (ed.), *Periodontitis*, Advances in Experimental Medicine and Biology 1373,
https://doi.org/10.1007/978-3-030-96881-6_20

nosis and regulatory considerations. Besides, because only 2 randomized human trials demonstrated the efficacy of the procedure, further studies are needed to investigate periodontal regeneration procedures on experimental models closer to human pathophysiology.

Conclusion: Although MSCs grafting in periodontal disease demonstrated therapeutic benefits in animal, it is critical to define more accurately protocols translatable to human and focus on the treatment of the pathology as a whole rather than on the restitution of the sole destroyed tissues.

Keywords

Periodontitis · Periodontal disease · Cell therapy · Tissue engineering · Mesenchymal stromal cell · Cell carrier · Meta-analysis

Abbreviations

ASC	Adipose Stem Cells
ATMP	Advanced Therapy Medicinal Products
BMSC	Bone Marrow Stem Cells
DPSC	Dental Pulp Stem Cells
GMP	Good Manufacturing Practices
GMSC	Gingival Margin Stem Cells
ICTRP	International Clinical Trials Registry Platform
iPS	Induced-Pluripotent Stem Cells
MSCs	Mesenchymal Stem Cells
PDL	Periodontal ligament
PDLC	Periodontal ligament Stem Cells
UCMSC	Umbilical Cord Mesenchymal Stem Cells

Highlights

- Periodontitis is a chronic inflammatory disease of the tooth-supporting tissues
- Immunomodulation et protrophic properties of MSC make them suitable for periodontitis cell therapy
- Animal studies show that the procedure is not dependant on the tissue origin or type of mesenchymal cells
- Future cell therapy studies must focus on the pathophysiology of periodontitis

Considerations for Practice

- Auto or allogeneic mesenchymal cell transplantation is a promising therapy for periodontitis
- Cavity-like defects and natural cell vehicle are the best conditions to enhance periodontal regeneration
- Further studies are needed, especially in human with more rigorous designs

Patient Summary

Periodontitis is a chronic inflammatory disease characterized by the loss of tooth-supporting tissues (or periodontium) that leads from gum inflammation to tooth loss. Regardless of their tissue of origin, adult stem cells in autologous or allogeneic graft enhance periodontal regeneration. Nevertheless, despite these promising results, methodological biases have been highlighted and still limit the translation to human with high prognosis and regulatory considerations. It is critical to define more accurately a protocol translatable to human.

20.1 Introduction

The periodontium is defined as a tooth-supporting tissue composed by the deep periodontium (root cementum – periodontal ligament (PDL) – alveolar bone complex) which anchors the tooth in its bone socket, and the superficial periodontium (the gingiva) which protects this dentoalveolar junction. Indeed, the cellular renewal of the periodontal connective tissue from the gingival and PDL progenitors is sustained by a supra-physiological local immunity that permanently retains the potential virulence of the oral microbial ecosystem according to a host/microbiota gingival homeostasis (Bartold and Van Dyke 2013). Periodontal attachment can therefore be considered as a septic joint.

Gingival inflammation or gingivitis, that mainly results of poor oral hygiene, is generally well controlled by an oral care restoration. However, deep periodontium disease (i.e. periodontitis) may occur when gingival inflammation does not resolve and spreads to the deep periodontium, promoting contamination of the root surfaces by the pathological bacterial biofilm. Inflammation and dysbiosis thus reinforce each other and changes in the periodontal environment increasingly select a pathobiotic microbial community (Lamont and Hajishengallis 2015). The periodontal anchorage is then destroyed leading to the formation of the pathognomonic entity of the disease: the periodontal pocket (Nanci and Bosshardt 2006; Kinane and Bartold 2007) displaying quite complex anatomy. Under well-defined conditions, only the "intrabony" or "vertical" defects may be repaired and sometimes regenerated (Wikesjö and Selvig 1999).

Periodontitis is thus defined as a chronic inflammatory pathology associating dysbiosis, dysimmunity – an exacerbated inflammatory reaction to chronic stresses on tooth-supporting tissue – that reinforce each other, and destruction of periodontal anchorage (Van Dyke et al. 2020; Papapanou et al. 2018). Therefore, the restoration or "reset" of periodontal homeostasis must be an essential feature to treat periodontitis in a sustainable manner allowing the regeneration of the tooth anchorage. To date, no therapy is yet able to provide long lasting treatment and reversion of the rupture of gingival host/microbiota homeostasis resulting from the imbalance between the dysbiotic microbial factor and the host's defenses. In line with this paradigm, the goal of periodontitis treatment is to restore tissue homeostasis by first decreasing the subgingival bacterial load and controlling both inflammation and tissue destruction. To this end, mechanical and chemical disorganization of biofilm and subgingival calculus by non-surgical root debridement remains the current standard treatment (Graziani et al. 2017). When the disease is stabilized, periodontal regeneration procedures can be performed. The aim is to promote the maintenance of the fibrin clot resulting from post-surgical coagulation and to stimulate the recruitment of the endogenous PDL progenitors leading to tissue restitution (Wikesjö and Selvig 1999).

Regenerative techniques, including guided tissue regeneration using natural or synthetic membrane or bioactive factors such as enamel matrix protein derivatives, are unfortunately very unpredictable, with only partial regeneration (Aichelmann-Reidy and Reynolds 2008; Sculean et al. 2015; Kinane et al. 2017). In addition, control of residual dysbiosis/inflammation is still uncertain and the perennial restoration of tissue homeostasis for long-term success of regeneration strategies remains a challenge (Bartold and Van Dyke 2013). Thus, unlike current techniques, new periodontal regeneration strategies must simultaneously reverse dysbiosis, modulate immunity and promote the neoformation of periodontal attachment. To achieve these goals, the requirements for periodontal regeneration treatments have to: (i) combine anti-dysbiotic and immunomodulatory properties, (ii) maintain and protect the blood clot at the level of the periodontal lesions (role of mechanical maintenance according to the architecture of the sites), (iii) to recruit and engage specific endogenous periodontal progenitors for tissue regeneration while limiting fibrotic tissue spreading (Polimeni et al. 2006).

Recent advances in regenerative medicine have made improvements possible by presenting imaginative possibilities for periodontal tissue engineering. Cell therapy, using mesenchymal stromal cells (MSCs) is likely to induce a profound modification of the periodontal microenvironment for a long-term restoration of tissue homeostasis and regeneration of periodontal tissues in a predictable and sustainable manner. Indeed, grafting mesenchymal stromal cells, including mesenchymal stem/multipotent cell pools able to transdifferentiate, can promote the production of new tissue and/or to make the local microenvironment more suitable for endogenous stimulation has been tested with promising results (Monsarrat et al. 2014). Successful treatment of various diseases, including cardiovascular diseases, chronic inflammatory diseases and major bone defects, based on multipotent adult MSCs therapy has been reported (Monsarrat et al. 2016a). MSCs can replace various mesenchymal cell types through their multipotency and influence the local microenvironment through local paracrine activities. Because of these unique properties, MSCs delivered *in situ* into periodontal defects with appropriate biomaterial carrier can exert their effects at several levels, including neovascularization, immunomodulation, and tissue regeneration (Galipeau and Sensébé 2018; Martin et al. 2019; Pittenger et al. 2019) . Over the last decades, numerous animal studies using various cell types and experimental settings have been published on this subject. The objectives of this meta-analysis are therefore to highlight the key determinants of MSCs periodontal therapy for a rational translation to Human.

20.2 Materials and Methods

20.2.1 Data Sources and Research Strategy

The PubMed database was searched. Clinical Trials were also identified through the ICTRP search portal (available at http://apps.who.int/trialsearch/).

The research strategy (Supplementary Text S1) combined both keywords related to periodontium (e.g. "periodontal disease" or "alveolar bone loss") and keywords related to stem cells (e.g. "Stem cell transplantation", mesenchymal and stromal). Reference lists were inspected to identify any additional relevant published or unpublished data. The last research was conducted on 2020/04/01. This systematic review was performed in accordance with the preferred reporting items for systematic reviews and meta-analyses guidelines (Moher et al. 2009).

20.2.2 Inclusion Criteria

All original reports regarding to the use of stem cells as a treatment therapy in deep periodontal tissue (cementum, alveolar bone and periodontal ligament) regeneration were included in this systematic review. *In-vivo* studies, and clinical trials were considered. All animal models, origin of periodontal defect or origin of stem cells were considered. Language of publication was restricted to English. Case reports were excluded from the systematic review. All studies dealing with the use of growth factors or secretome were not considered eligible. For meta-analysis, only studies with quantitative outcomes for experimental and control groups were selected.

20.2.3 Outcomes

The following outcomes were considered: the type of methodology (*in-vivo*, clinical trial), the animal model, the type of MSCs, the type of defects, the methodology used for generation of the defects, and when applicable the scaffold used as cell carrier. We aimed to determine whether differences in the efficacy of periodontal regeneration between experimental and control groups were influenced by these characteristics.

20.2.4 Study Selection, Data Extraction and Quality Assessment

All results were screened based on titles and abstracts. Full texts of the potentially selected records were obtained for definitive inclusion. *In-process* clinical trials were discussed apart. Data extraction was performed twice by one author (AD) at a one-month interval.

For quality assessment the following points were considered: carrying out a randomization procedure, blinding of participants (evaluators and/or the operators), calibration of the evaluators and the operators, determination/justification of the number of animals required. Each point was noted as "Yes" when reported and "No" when unclear, not reported or not performed. Studies' quality was so graded from 0 to 5 points.

20.2.5 Statistical Analysis

Animal and human studies were analyzed separately. We performed meta-analyses on standardized mean differences using an inverse variance random-effects method. The package "meta 4.9" was used to perform meta-analysis and plot the data with R 3.5.2 (Schwarzer et al. 2015). Only studies with an appropriate control group were considered for quantitative analysis. When a study had several experimental groups, each group was included separately in the quantitative analysis. The funnel plot was visually examined to determine publication bias.

20.3 Results

4481 results were identified. Based on the inclusion criteria described above, 131 unique citations were included, 25 were excluded (7 non-English languages studies (Ou et al. 2000, 2002; Lu et al. 2004; Kawaguchi et al. 2005; Xu et al. 2006; Sun and Liu 2014; An and Liu 2014), 4 case reports (Yamada et al. 2006; Aimetti et al. 2014; Kl et al. 2017; Hernández-Monjaraz et al. 2018), 8 studies using growth factors or secretome (Kawai et al. 2015; Liu et al. 2015; Wang et al. 2016; Al-Sharabi et al. 2016; Han et al. 2017; Nagata et al. 2017; Sakaguchi et al. 2017; Qiu et al. 2020), 4 studies in which the MSCs were not cultured (Akbay et al. 2005; Murano et al. 2006; Fujinami et al. 2007; Shalini and Vandana 2018)). A final total of 106 results were included in the systematic review and 70 in the meta-analysis (67 animal studies and 3 human randomized clinical trials). Flow diagram is available (Fig. 20.S1 in Appendix).

20.3.1 Characteristics of Included Studies

A summary of the characteristics of the included studies are available in the Table 20.1. The proportion of human studies remains low compared to the number of animal studies (11% and 89% of studies, respectively). However, 12 clinical trials are currently in progress. Among them, 11 are randomized controlled trials, the periodontal defects studied are almost all intrabony defects. Cellular sources are autologous (subcutaneous fat, peripheral blood, dental pulp and periodontal ligament), carrier biomaterials are mainly platelet or collagen derivatives.

The most used animal models are dogs (42% of studies), followed by rats (36%) and minipigs (14%). Periodontal defects studied are fenestrations, furcation defects and interproximal defects, in similar proportions. To induce periodontal defect, bone was mainly mechanically removed with burs (56% of studies), or by generating additional deep inflammation (20%). Ligature-induced periodontal defects with or without bacteria are rarely used (4%). Periodontal ligament and bone marrow are the most represented therapeutic cell sources used with respectively 43% and 26% of the total studies. Temporal evolution of the different stem cell origin is exposed in Fig. 20.1. Periodontal ligament stem cells are the most consistently used over time. From 2010, a large variety of stem cells are found, including mesenchymal stem cells from dental pulp, gingival margin, adipose tissue and bone marrow, and more recently induced pluripotent stem cells.

The biomaterials used as scaffold for stem cells were classified into six groups: natural extracellular matrix derived (collagen, fibrin, hyaluronic acid, 54% of studies), natural non-extracellular matrix derived (chitosan, silk, alginate, 5% of studies), synthetic protein based (poly-L-lactic, poly(lactic-co-glycolic acid), 15% of studies), synthetic calcium phosphate based (β-TCP, hydroxyapatite, 21% of studies), natural calcium phosphate based (bovine hydroxyapatite, 6% of studies) and no carrier used (15% of studies). Regarding guided tissue regeneration, an absorbable membrane has been used in 17% of studies, non-absorbable mem-

Table 20.1 Characteristics of included studies. *In-process* clinical trials were not added in this table. One study could be included into multiple groups

Outcome		Number of studies (%)
Study design N = 94	**Animal trials**	**84 (89.4%)**
	Human trial	**10 (10.6%)**
	Non-randomized controlled trial	5 (5.3%)
	Randomized controlled trial	4 (4.3%)
	Retrospective study	1 (1.0%)
Animal model N = 84	Dog	35 (41.7%)
	Rat	30 (35.7%)
	Minipig	12 (14.2%)
	Mouse	5 (6.0%)
	Ewe	1 (1.2%)
	Rabbit	1 (1.2%)
Outcome N = 94	Histological	81 (86.2%)
	Radiological	42 (44.7%)
	Clinical	20 (21.3%)
Periodontal defect N = 94	**Furcation**	**29 (30.9%)**
	Class I	5 (5.3%)
	Class II	15 (16%)
	Class III	13 (13.8%)
	Not specified	1 (1.1%)
	Fenestration	**30 (31.9%)**
	Buccal	28 (28.8%)
	Palatal	2 (2.1%)
	Interproximal	**38 (40.4%)**
	1 wall	11 (11.7%)
	2 walls	22 (23.4%)
	3 walls	15 (16%)
	Supra-alveolar	2 (2.1%)
	Not specified	1 (1.1%)
	Not specified	**1 (1.1%)**
Defect generation N = 94	Mechanical	53 (56.4%)
	Mechanical with induction of inflammation	19 (20.2%)
	Ligature	6 (6.4%)
	Ligature plus bacteria	3 (3.2%)
	Orthodontic	2 (2.1%)
	Pathogen-induced	1 (1.1%)
	Human periodontitis	10 (10.6%)
Cell source N = 94	Periodontal ligament stem cells (PDLSC)	40 (42.6%)
	Bone marrow stem cells (BMSC)	24 (25.5%)
	Dental pulp stem cells (DPSC)	9 (9.6%)
	Adipose stem cells (ASC)	8 (8.5%)

brane in 10% while 79% of the studies did not perform guided tissue regeneration. Biomaterials distribution according to stem cell origin is shown in Fig. 20.2. The different types of stem cells have been associated with almost all the categories of carrier described above. However, natural extracellular matrix-derived scaffolds are preferred, mainly in association with periodontal ligament stem cells (43% of natural extracellular matrix derived carrier) and bone marrow stem cells (BMSC, 23% of natural extracellular matrix derived carrier).

A variety of methods was used by the authors to assess periodontal regeneration. For animal studies, histological analysis concerned 81 studies (96% of animal studies) in which quantification of bone, cement or periodontal ligament regeneration was performed in 60 studies (71% of animal studies). Radiological measures were reported in 34 animal studies (40%) and clinical measures only reported in 10 animal studies (12%).

For human studies, the main evaluation was clinical (periodontal pocket depth, clinical attachment level) while radiological measures were reported in 8 studies (80% of human studies).

20.3.2 Efficacy of Periodontal Regeneration in Animal Studies

20.3.2.1 Bone Regeneration

Bone regeneration was investigated by microscopic and/or radiography examination in 65 studies. Meta-analysis showed that alveolar bone regeneration was significantly enhanced by mesenchymal stromal cells therapy (regardless of their tissue source) compared to ungrafted controls (mean difference: 1.94 [95% confidence interval (CI): 1.43; 2.45], forest plot in Fig. 20.S2 in Appendix).

To investigate whether the magnitude of the outcome difference between experimental and control groups could have been influenced by study characteristics, several subgroup analyses have been performed (Table 20.2). Bone regeneration by MSCs therapy was greater in rats and minipigs compared to dogs, significantly greater

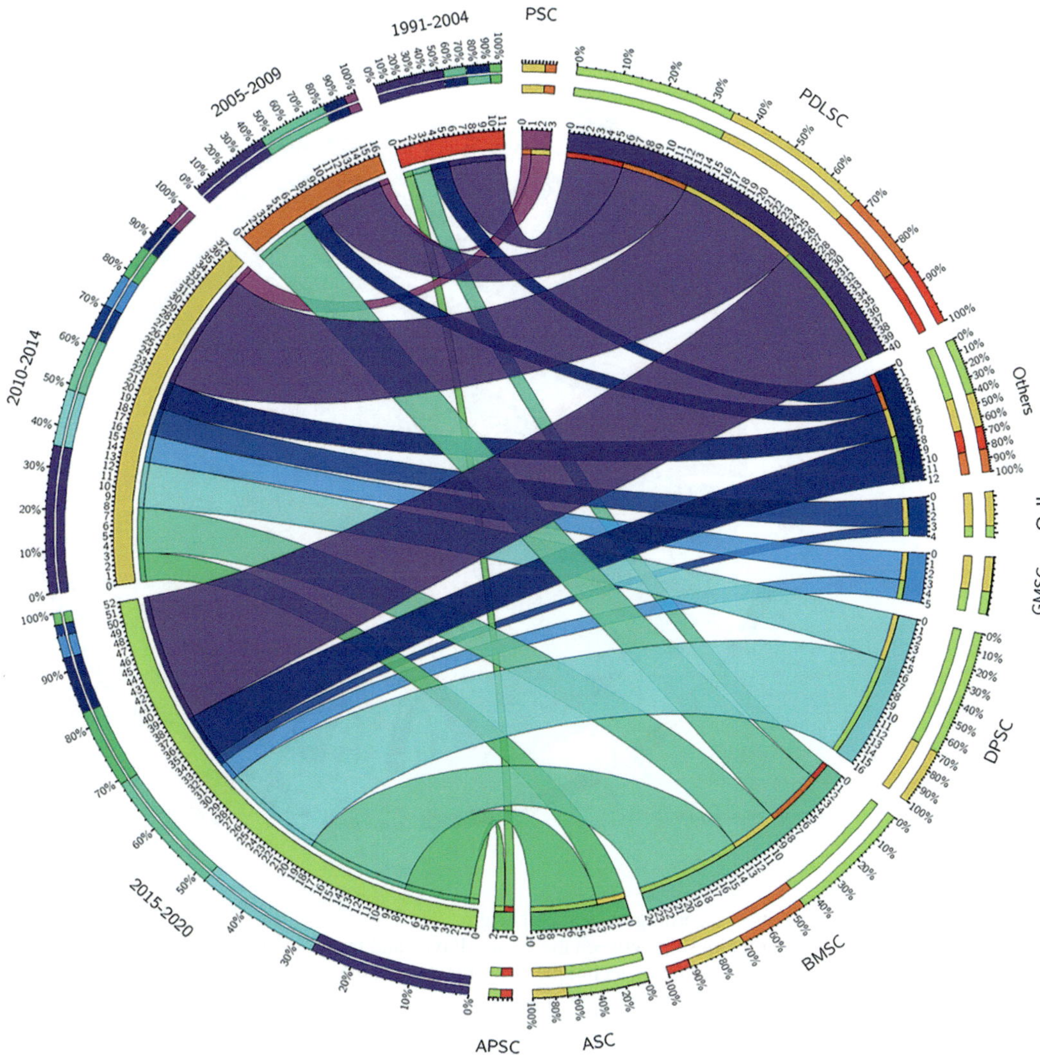

Fig. 20.1 Temporal evolution of the different stem cell origin
This chord diagram represents the proportion of studies published between 1991 and 2020 dealing with each stem cell type (right side) and links them to their respective publication date (left side).
The outer ring of the figure contains the proportion of studies and the inner circle shows their absolute numbers. In each ring, the stem cell type in each time category is coded by colored segments, and vice versa. For instance, and reading the diagram from stem cell type, for periodontal ligament stem cells, 14 articles were published in 2015-2020 (indicated in the inner part of the circular diagram) and made up 27% of articles published in 2015–2020 (indicated on the left of the outer part of the circular diagram). Conversely, 35% of studies using periodontal ligament stem cells were published in 2015-2020 (indicated on the right of the outer part of the circular diagram). All original articles were included (N=94). A study could be considered in multiple categories.
Abbreviations: APSC, apical papilla stem cells; ASC, adipose stem cells; BMSC, bone marrow stem cells; DPSC, dental pulp stem cells; GMSC, gingival margin stem cells; iPS, induced pluripotent stem cells; PDLSC, periodontal ligament stem cells; PSC, Periosteum stem cells; Other groups include alveolar bone stem cells, dedifferentiated fat cells, umbilical cord stem cells, embryonic stem cells, cementum derived cells and fibroblasts.

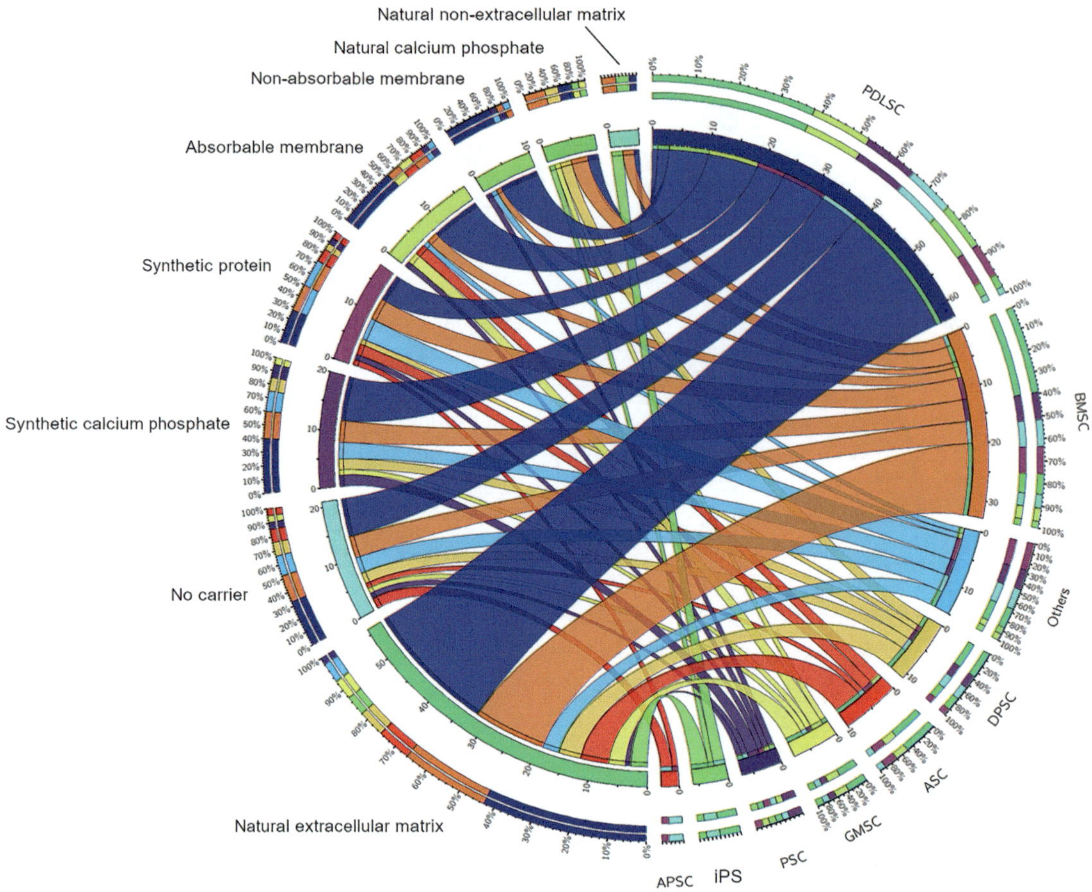

Fig. 20.2 Biomaterials distribution according to stem cell origin
This chord diagram exposes the relationship between the stem cell type and the biomaterials used as scaffold as well as the potential use of a guided tissue regeneration technique. The outer ring of the figure contains the proportion of studies and the inner circle shows their absolute numbers. All original articles were included (N=94). A study could be considered in multiple categories.
Used biomaterials: natural extracellular matrix derived (collagen, fibrin or hyaluronic acid), natural non extracellular matrix derived (e.g chitosan, silk), synthetic protein based (e.g poly (lactic-co-glycolic acid), poly-L-lactic acid), synthetic calcium phosphate based (e.g β-TCP, hydroxyapatite) and natural calcium phosphate based (e.g bovine hydroxyapatite, demineralized bovine bone). Absorbable membranes (e.g collagen membrane) or nonabsorbable membrane (e.g PTFE membrane) have been used.
Abbreviations: APSC, apical papilla stem cells; ASC, adipose stem cells; BMSC, bone marrow stem cells; DPSC, dental pulp stem cells; GMSC, gingival margin stem cells; iPS, induced pluripotent stem cells; PDLSC, periodontal ligament stem cells; PSC, Periosteum stem cells; Other groups include alveolar bone stem cells, dedifferentiated fat cells, umbilical cord stem cells, embryonic stem cells, cementum derived cells and fibroblasts.

in class I & II furcation defects and 2 & 3-walls intrabony defects compared to class III furcation defects and 1-wall defects. No difference between stem cell types was found. Compared to autologous cellular sources, MSCs therapy using allogeneic and xenogeneic cells is significantly improved. When a biomaterial was used, only natural extracellular matrix-derived scaffolds significantly improved bone regeneration by MSCs (1.68 [1.14; 2.21]). Studies with the most methodological issues have significantly less bone regeneration than studies with higher bias.

20.3.2.2 Cementum Regeneration

Cementum regeneration was investigated by microscopic examination in 38 studies. Meta-

Table 20.2 Subgroup analysis of *in-vivo* studies for alveolar bone regeneration

Subgroup	N	Standardized mean difference [95% Confidence Interval]	p-value for difference between subgroups
All	105	1.94 [1.43; 2.45]	–
Animal model			
Dog	42	1.38 [0.78; 1.95]	0.07
Minipig	17	2.38 [0.33; 4.43]	
Rat	37	2.39 [1.46; 3.31]	
Mouse	5	4.19 [0.16; 8.21]	
Mesenchymal stem cell source			
Periosteum stem cells	2	0.74 [−8.44; 9.92]	0.33
Gingival margin stem cells (GMSC)	5	1.01 [−1.03; 3.05]	
Adipose stem cells (ASC)	9	1.10 [0.52; 1.69]	
Others	7	1.45 [0.23; 2.67]	
Apical papilla stem cells	3	1.75 [−9.19; 12.7]	
Bone marrow stem cells (BMSC)	25	2.03 [1.07; 3.00]	
Periodontal ligament stem cells (PDLC)	40	2.31 [1.22; 3.39]	
Dental pulp stem cells (DPSC)	7	2.46 [−0.17; 5.08]	
Induced pluripotent stem cells (iPS)	5	2.74 [−0.91; 6.39]	
Autologous	50	1.35 [0.73; 1.96]	0.02
Allogeneic	28	2.46 [1.11; 3.81]	
Xenogeneic	26	2.83 [1.84; 3.82]	
Periodontal defect			
Interproximal 1 wall	6	0.68 [−0.68; 1.35]	< 0.001
Furcation class III	13	0.88 [0.28; 1.48]	
Interproximal 2–3 walls	28	2.17 [0.83; 3.51]	
Fenestration	34	2.37 [1.35; 3.40]	
Furcation class I–II	24	2.51 [1.56; 3.47]	
Defect generation			
Ligature	13	1.41 [0.34; 2.48]	0.56
Mechanical	66	1.99 [1.38; 2.60]	
Mechanical and inflammation	24	2.11 [0.64; 3.60]	
Carrier			
Natural calcium phosphate-based scaffold	8	0.35 [−0.6; 1.30]	0.01
Synthetic protein-based scaffold	16	0.66 [−1.12; 1.43]	
Natural extracellular matrix derived scaffold	68	1.68 [1.14; 2.21]	
Synthetic calcium phosphate-based scaffold	17	1.91 [−0.33; 4.14]	
Natural non- extracell. Matrix derived scaffold	5	2.57 [−1.49; 6.63]	
No scaffold	6	3.10 [0.48; 5.72]	
Guide tissue regeneration			
None	74	1.82 [1.29; 2.34]	0.69
Yes (synthetic membrane)	13	1.87 [0.43; 5.23]	
Yes (natural membrane)	17	2.83 [0.75; 2.99]	
Quality assessment			
Randomization: Yes	51	1.84 [1.07; 2.61]	0.71
Randomization: No	53	2.03 [1.34; 2.73]	

N corresponds to the number of analyzed groups. One study may contain several analyzed groups. Only subgroups with at least two different studies were considered for subgroup analyses. One study was removed after funnel plot analysis

analysis showed that cementum regeneration was significantly enhanced by mesenchymal stromal cells therapy compared to without mesenchymal stromal cells (mean difference: 1.51 [95% CI: 1.06–1.96], forest plot in Fig. 20.S3). Subgroup analyses revealed no significant differences according to study characteristics (Table 20.3). Studies with the most methodological issues tend to less cementum regeneration than studies with higher bias.

20.3.2.3 Periodontal Ligament Regeneration

Periodontal ligament regeneration was investigated by microscopic examination in 16 studies. Meta-analysis showed that periodontal ligament regeneration was significantly enhanced by mesenchymal stromal cells therapy compared to without mesenchymal stromal cells (mean difference: 1.04 [95% CI: 0.17; 1.92], forest plot in Fig. 20.3).

20.3.3 Efficacy of Periodontal Regeneration in Human Studies

We identified 10 human studies in which 4 were randomized clinical trials (RCT). Only 3 (Dhote et al. 2015; Chen et al. 2016a; Abdal-Wahab et al. 2020) of them have assessed the same clinical outcomes (probing depth and clinical attachment level).

Meta-analysis performed on the 3 RCT showed that clinical parameters were not significantly enhanced by MSCs therapy (−0.94 [−4.63; 2.74] and −0.82 [−2.33; 0.68] for Probing Pocket Depth and Clinical Attachment Level,

Author	Intervention N	Intervention Mean	Intervention SD	Control N	Control Mean	Control SD	SMD	95% CI	weight
Akita 2014	8	92.50	10.00	8	29.00	7.50	6.79	[3.91; 9.67]	3.0%
Akita 2016	12	117.50	11.50	12	42.50	8.75	7.09	[4.75; 9.42]	3.5%
Akita 2016	12	75.00	11.50	12	42.50	8.75	3.07	[1.83; 4.31]	4.7%
Akizuki 2005	5	3.59	0.36	5	3.50	0.35	0.23	[−1.02; 1.48]	4.7%
Basan 2017	5	77.30	20.30	5	50.30	34.60	0.86	[−0.47; 2.19]	4.6%
Basan 2017	5	52.00	34.00	5	76.90	18.60	−0.82	[−2.14; 0.50]	4.6%
Basan 2017	5	65.10	18.90	5	61.10	31.40	0.14	[−1.10; 1.38]	4.7%
Goncalves 2016	6	0.12	0.04	6	0.12	0.01	−0.22	[−1.36; 0.92]	4.8%
Jiang 2010	4	33.64	13.83	4	22.18	5.19	0.95	[−0.58; 2.49]	4.4%
Khorsand 2013	10	3.30	1.12	10	1.77	1.27	1.22	[0.25; 2.20]	4.9%
Mrozik 2013	7	4.39	0.82	7	4.12	0.61	0.35	[−0.71; 1.41]	4.9%
Nayak 2008	5	0.42	0.14	5	0.21	0.14	1.35	[−0.10; 2.81]	4.5%
Nuñez 2018	9	1.93	1.14	9	2.35	1.74	−0.27	[−1.20; 0.66]	5.0%
Nuñez 2018	9	0.27	0.34	9	0.42	0.45	−0.36	[−1.29; 0.58]	5.0%
Rezaei 2019	5	18.79	3.00	5	13.70	3.30	1.46	[−0.03; 2.94]	4.4%
Rezaei 2019	5	29.10	2.70	5	18.18	3.30	3.27	[1.05; 5.49]	3.7%
Suaid 2011	7	7.28	1.00	7	3.94	1.20	2.83	[1.21; 4.45]	4.3%
Suaid 2012	7	3.43	1.44	7	2.33	0.95	0.84	[−0.27; 1.95]	4.8%
Yan 2015	6	0.38	0.19	7	0.35	0.19	0.15	[−0.95; 1.24]	4.8%
Yoo 2019	12	1.77	0.40	12	1.90	0.14	−0.42	[−1.23; 0.39]	5.1%
Yu 2013	8	0.26	0.08	8	0.31	0.09	−0.60	[−1.61; 0.41]	4.9%
Yu 2013	8	0.28	0.10	8	0.31	0.09	−0.33	[−1.32; 0.66]	4.9%
Overall effect	.			.			**1.04**	**[0.17; 1.92]**	**100.0%**
Prediction interval								**[−2.53; 4.61]**	

Heterogeneity: I^2 = 81% [72%; 87%], $p < 0.01$

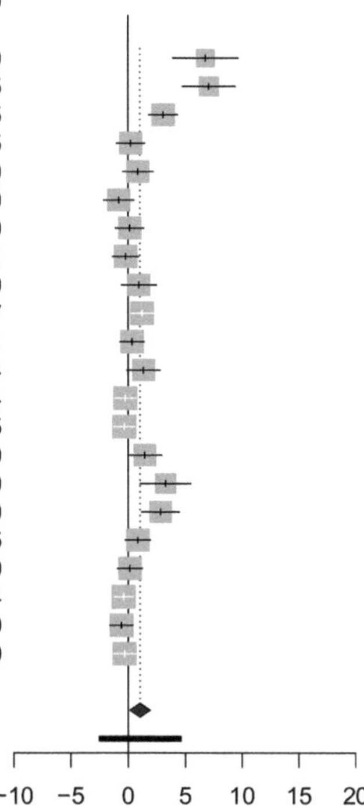

Fig. 20.3 Forest plot of the meta-analysis of the effect of mesenchymal stromal cells therapy on ligament regeneration

Forest plot of the standardized mean difference (SMD) for periodontal ligament regeneration and its 95% confidence interval for each study comparison. A value above 0 means that therapeutic effect is in favor of cell therapy.

Table 20.3 Subgroup analysis of *in-vivo* studies for cementum regeneration

Subgroup	N	Standardized mean difference [95% CI]	p-value for difference between subgroups
All	62	1.51 [1.06; 1.96]	–
Animal model			
Minipig	7	0.83 [0.17; 1.50]	0.13
Dog	39	1.37 [0.83; 1.91]	
Rat	14	3.02 [1.50; 4.55]	
Mesenchymal stem cell source			
Periosteum stem cells	2	0.07 [−10.5; 10.6]	0.27
Periodontal ligament stem cells (PDLC)	28	1.11 [0.58; 1.68]	
Gingival margin stem cells (GMSC)	3	1.43 [−1.43; 4.30]	
Adipose stem cells (ASC)	5	1.51 [0.20; 2.82]	
Bone marrow stem cells (BMSC)	17	1.84 [0.92; 2.76]	
Others	4	2.83 [−0.28; 5.94]	
Autologous	41	1.29 [0.91; 1.67]	0.19
Allogeneic	9	1.27 [−0.24; 2.77]	
Xenogeneic	12	2.75 [1.03; 4.47]	
Periodontal defect			
Interproximal 1 wall	5	0.87 [−0.93; 2.66]	0.24
Interproximal 2–3 walls	8	0.94 [0.27; 1.60]	
Furcation class III	14	1.44 [0.84; 2.04]	
Fenestration	15	1.70 [0.44; 2.96]	
Furcation class I–II	19	2.05 [1.11; 2.99]	
Defect generation			
Mechanical and inflammation	15	1.01 [0.32; 1.71]	0.19
Ligature	6	1.23 [0.11; 2.36]	
Mechanical	40	1.78 [0.19; 2.38]	
Carrier			
Natural calcium phosphate-based scaffold	7	0.53 [−0.52; 1.60]	0.47
No scaffold	2	1.04 [−7.62; 9.69]	
Natural extracellular matrix derived scaffold	47	1.34 [0.92; 1.75]	
Synthetic calcium phosphate-based scaffold	11	1.55 [−0.23; 3.34]	
Synthetic protein-based scaffold	9	1.69 [0.17; 3.20]	
Guide tissue regeneration			
Synthetic membrane	14	1.14 [0.69; 1.60]	0.18
None	37	1.37 [0.83; 1.91]	
Collagen membrane	11	2.44 [0.91; 3.98]	
Quality assessment			
Randomization: Yes	28	1.46 [0.86; 2.07]	0.97
Randomization: No	34	1.48 [0.90; 2.07]	
Blinging examiner: Yes	19	0.69 [0.12; 1.25]	0.003
Blinding examiner: No	43	1.80 [1.29; 2.30]	

N corresponds to the number of analyzed groups. One study may contain several analyzed groups. Only subgroups with at least two different studies were considered for subgroup analyses. One study was removed after funnel plot analysis respectively). MSCs were prepared from gingival margin, umbilical cord and periodontal ligament.

20.4 Discussion

This meta-analysis reviewed more than 10 years of periodontal cell therapy in 84 pre-clinical trials and 10 clinical studies. Overall, it provides evidence of the ability of MSCs transplantation to regenerate periodontal tissue in animals.

20.4.1 Methodology

This meta-analysis has aggregated quantitative data currently available on periodontal regeneration by stem cells. However, this strategy has some potential limitations, as high heterogeneity in design and methodological quality of studies have been revealed. However, sub-group analyses made possible to analyze part of this heterogeneity. Trials with animals should have the same level of exigence than human studies from the study validity and ethics point of view. Among the most critical methodological points to address, calibration and blinding of the examiners, randomization of treatments, computing the number of animals to treat were rarely performed and reported. For example, computing the number of animals to challenge is critical for statistical power to demonstrate an effect of cell therapy compared to control. Among the 10 human studies, only 3 of them are randomized controlled trials. More studies with control groups are needed. Overall, methodological drawbacks have a significant impact on the interpretation of the results as demonstrated by higher effects size in studies with weaker methodological qualities.

20.4.2 Animal and Defect Models

In order to reproduce the morphology of the typical human defects, periodontal lesions are mainly induced by surgery rather than wires, ligatures or impression material around the teeth, poorly reflecting the natural pathophysiology of periodontitis. For future studies, we suggest that periodontal regeneration assays should more accurately consider both the inflammatory status and the change in pathogenic microflora typical of periodontitis (Graves et al. 2008). Rodents, inexpensive and easy to house, may reach human pathophysiology characteristics under an appropriate design. Indeed, a study in mice challenged for 4 weeks with periopathogenic bacteria to develop periodontitis demonstrated the efficacy of syngeneic ASC grafting in periodontal regeneration (Lemaitre et al. 2017).

Large animals remain the most relevant model because their teeth and periodontal region/defects anatomy are phenocopies close to human features (Yan et al. 2015). Dog and minipig are the most used species for preclinical trials. The beagle, spontaneously developing periodontitis on average from the age of 6 years, remains the "gold standard" given its size and cooperative temperament (Haney et al. 1995; Struillou et al. 2010). It is obvious that in most studies, animals were too young and therefore had a higher potential for spontaneous repair/regeneration which is appropriate to demonstrate the proof of concept and treatment efficacy. However, it can represent a bias regarding to the age of patients actually elected for periodontal cell therapy. As age may impact the clinical issue and regenerative potential, complementary translational studies should be performed using at least adult or possibly old animals. Overall, animal experimentations poorly mimic some fundamental human characteristics, in particular the genetic background and risk factors (aggressive bacterial flora, occlusal overload, tobacco, prosthesis, systemic diseases of the host and donor, etc.…) that may impact regenerative results.

20.4.3 Cell Sources and Auto/Allogeneic Grafting Strategies

One of the most important issues for the clinical application of periodontal cell therapy is the source of cells and their use in autologous or allogeneic grafting strategies. Every MSCs source has its own biological characteristics and their fate after transplantation can be conditioned by their original environment (Lin et al. 2009) as well as by the recipient local microenvironment and surrounding tissues (Chen et al. 2011). A wide variety of MSCs sources were used for periodontal cell therapy in human and animal, including PDL cells (PDLSC), dental pulp stem cells (DPSC), gingiva margin stem cells (GMSC), umbilical cord mesenchymal stem cells (UCMSC) and BMSC. PDLC and BMSC are the most common cell types used in periodontal cell therapy trials while MSCs from adipose tissue (ASC) and induced pluripotent stem cells (iPS) have only been assessed in animal trials. Overall, the efficacy of the procedure seems independent of the cell source.

Among oral sources, PDLSC are the most used cells for preclinical and clinical trials. GMSC, which can be more easily recovered than PDLSC during oral surgery, were also assessed for periodontal therapy. Given their tissue origin, oral cells similarly appear as good candidates for periodontal therapy (Feng et al. 2010; Suaid et al. 2012) but they are not readily available for clinical practice or are retrieved on small amount when possible (Lin et al. 2009). Several considerations may be required to consider other MSCs sources than oral tissues (e.g., the availability of extracted teeth, genetic instability increasing and loss of potentialities through cell passages during expansion).

MSCs from extra oral niches (mainly BMSC and ASC) are morphologically and immunophenotypically similar to oral MSCs (Huang et al. 2009). Although extra-oral MSCs may have a reduced ability to differentiate into bone and cartilage *in vitro* (Kern et al. 2006) compared to oral MSCs (Takedachi et al. 2013; Tobita et al. 2013), we showed here that cells derived from bone marrow and adipose tissue display the same periodontal regenerative abilities than oral MSCs. This result suggests that beyond their native tissue imprinting, the host microenvironment and surrounding tissues provide critical factors influencing the MSCs fate. Adipose tissue obtained from surgical resection or liposuction is convenient and suitable to prepare large amounts of ASC, thus displaying increasing interest for cell therapy protocols (Monsarrat et al. 2016c). Since ASC have already been successfully used in regeneration of extra oral connective tissue (Si et al. 2019), they represent a valuable source of cells for periodontal treatment. Finally, the immunosuppressive properties of ASC make them a good candidate for treatment of chronic inflammatory diseases such as periodontitis (Monsarrat et al. 2016b). iPS, obtained after transfection of some stem cell-related genes into adult somatic cells, can differentiate into various cell types. These cells have shown promising results in periodontal regeneration (Tobita et al. 2013). However, safety and efficacy issues remain to be more deeply addressed before trials in human.

Oral and extra oral MSCs were first assessed in periodontal therapy using autologous approaches to demonstrate the proof of concept (Monsarrat et al. 2014). However, autogenous transplantation in human raises many issues: donor morbidity, laboratory constraints (culture, manufacturing delay, cell storage…), economic aspects and loss of cells capacities regarding subject features (age, health, risks factors…). Since MSCs have low immunogenicity (Ankrum et al. 2014; Berglund et al. 2017) and immunosuppressive activity, they can be used as an allogeneic source of cells which makes sense given the systemic and inflammatory context of periodontitis. Indeed, since periodontitis develops in "susceptible hosts", the use of allogeneic MSCs from healthy and young donors should help to mitigate potential alterations of endogenous MSCs (Hajishengallis 2014). Moreover, allogeneic strategy with cell banking allows to secure, to homogenize therapeutic production and to provide sufficient material at the time of transplantation. Such criteria are decisive for industrial manufacturing: lower production costs and results standardization. In beagle, cryopreservation of MSCs did not alter periodontal regeneration compared to the use of non-stored cells (Li et al. 2009). However additional preclinical data are required regarding to their immunomodulatory effect that remains conflicting as not fully understood (Poggi and Zocchi 2019). In addition, the use of allogeneic sources will have to be accompanied by regulatory procedures to comply with the legislation of Advanced Therapy Medicinal Products (ATMP) in Good Manufacturing Practices (GMP).

20.4.4 Cell Carriers

Route of delivery is of major concern for cell therapy whether systemic or local injection of cell suspension or implanted embedded with a more structured vehicle. It is assumed that carriers need to mimic the microenvironment of endogenous cells and may contribute to their cell fate by enhancing cell-extracellular matrix interactions. Although GMSC and iPS in saline buffer has been shown to be efficient to regenerate periodontal defects through systemic injection (Yang et al. 2014; Sun et al. 2019), most cell therapy protocols were designed using vehicles from natural origin, displaying the best results in periodontal regeneration. Bone substitutes have been reported to improve the efficacy of grafted MSCs in experimental periodontal defects, probably because they properly fill the wound, stabilize the blood clot and confine the cells to the surgical site without rapid resorption. Thus, MSCs delivery should be based on retention potential of the defect and, more specifically, on the remaining number of alveolar bone walls. When the lesions are retentive, liquid or gel scaffolds may be used without cell decay. When sufficient bony walls are lacking, clinical outcome may be improved if cells are associated with a bone substitute that confines them to the surgical site. Besides, carriers such as platelet lysate hydrogel that combine ergonomic handling, adequate wound filling, functional support for grafted MSCs and endogenous microenvironment conditioning, suggest very promising prospect in periodontal cell therapy (Tobita and Mizuno 2013).

20.4.5 Periodontal MSC Therapy Outcomes

Most of animal studies show that MSCs grafting significantly improves bone, cement and PDL regeneration, regardless of the animal model used. The outcomes of MSCs grafting are mostly assessed by radiological and microscopic quantification of alveolar bone neoformation. As earlier pointed out in carrier section, the most favorable alveolar defect morphologies for regeneration are angular, 2–3 walls or class I or II furcation lesions because they provide retention of the carrier and stabilization of the initial blood clot. Cell grafting does not seem to increase the well-known low regeneration potential of 1-wall and class III furcation defects (Sculean et al. 2008). Optimization of the scaffolding carrying the cells, by 3-D printing for example, would likely improve in the future the regeneration of these low retentive lesions as recently reported (Wu et al. 2019). Cementum repair has only been assessed in 45% of animal studies, only by microscopy and is therefore impossible to evaluate in human since it would be equivalent to remove the treated tooth. Nevertheless, cementum is the key-tissue for periodontal regeneration as it guarantees the formation of a new functional attachment (Bosshardt et al. 2015). The neoformation of PDL fibers, the key functional entity of the dento-alveolar joint, is also significantly improved by the MSCs grafting. We highlight here that periodontal regeneration outcomes have mainly been demonstrated using a quantitative (tissue quantity) approach. However, for now, a more rigorous investigation is needed to investigate the structural aspects involved in the new functional attachment repair, the bone quality (e.g. cellularity), the cement type (i.e. regeneration of the extrinsic acellular fiber cement), and the hierarchy and organization of the periodontal ligament fibers (Lemaitre et al. 2017).

It is noteworthy that MSCs of extra-oral origin are likely to promote the formation of alveolar bone, cementum and PDL. Because of their plasticity, it has been suggested that grafted MSCs might transdifferentiate into cementoblasts, osteoblasts of alveolar bone and fibroblasts of the PDL (Liu et al. 2019). However, recent data suggest that these cells are also capable of influencing the fate of endogenous MSCs so that they differentiate and re-synthesize an extracellular matrix for regeneration. We consider this paradigm shift essential to the understanding of the phenomena inherent to the activity and efficacy of MSCs in cell therapy in general, that may at least explain long term efficacy of the treatment with a transient presence of administrated cells.

To further develop the concept, it is still required to experimentally demonstrate these hypotheses in future.

20.4.6 Periodontal Cell Therapy in Human Studies

Among the 10 human studies reported, only 3 randomized trials were found: 2 studies (Dhote et al. 2015; Abdal-Wahab et al. 2020) concluded that the pocket depth was reduced, while the third one (Chen et al. 2016b) stated that there was no significant difference between the two groups, possibly due to the small number of subjects. In addition, the authors treated one-wall defects, which, as already mentioned, usually display a poor prognosis. Overall, the high risk of bias and the extreme variability of the protocols are likely to explain that the meta-analysis only tends towards significance and makes results difficult to explain. Various clinical trials are in progress and their outcomes will certainly give clues to optimize the translation to human.

20.5 Conclusion and Perspectives

This meta-analysis indicates that periodontal cell therapy publications are flourishing. Although further studies are needed to decipher the biological mechanisms involved, it is now possible to optimize the efficacy, ergonomics and rationality of a periodontal cell therapy procedure for humans. However, it is necessary to build strategies in line with the use of ATMP in Good Manufacturing Practices (GMP) according to the fundamental concepts that we have exposed in our study, we can summarize here:

- Clinical studies must first ensure the safety of the procedure, before efficacy studies;
- In regionalized lesion sites, the cells rather have to be delivered in a local and targeted manner, using a natural and degradable carrier that mimics an extracellular matrix and adapts to the morphology of the lesion, all in a regulatory-validated ATMP setting; however, the need for an extracellular matrix carrier will make the cell therapy into a "combination product", with the need of additional regulatory assessments and authorizations to use the final therapeutic product for humans.
- It is more rational to use allogeneic cells, from a readily available source such as adipose tissue, whose batches have been tested and optimized in compliance with GMP requirements;
- Finally, among the different cellular sources, ASC seem to be relevant. Assuming preclinical data presented herein will be confirmed by randomized human studies, these cells would be particularly indicated for periodontal considerations given their biological properties.

Both in the specific context of periodontal disease and in the more general context of MSCs cell therapy, we suggest evolving from the concept of tissue regeneration, as in most of the studies, to the concept of "whole disease therapy" by a reeducation of the endogenous microenvironment and a homeostasis resetting following the transplantation of MSCs in appropriate vehicle. The restitution of altered tissue in inflammatory pathologies is only conceivable in the control of dysimmunity, or even dysbiosis as in the case of periodontitis.

20.6 Author Contribution

Study concept and design: PM

Acquisition of data: AD

Analysis and interpretation of data: AD and PM

Drafting of the manuscript: VPB, MM, PhK

Administrative, technical, and material support: PhK

Study supervision: VPB and PhK

All the named authors were involved in the paper and have read it before it is submitted for publication.

Acknowledgment We gratefully acknowledge the Programme d'Investissements d'Avenir and the Agence Nationale pour la Recherche for the grant grant ANR–16–CE18–0019–01 and the national infrastructure "ECELLFrance: Development of mesenchymal stem cell based therapies" (PIA-ANR-11-INBS-005).

Appendix

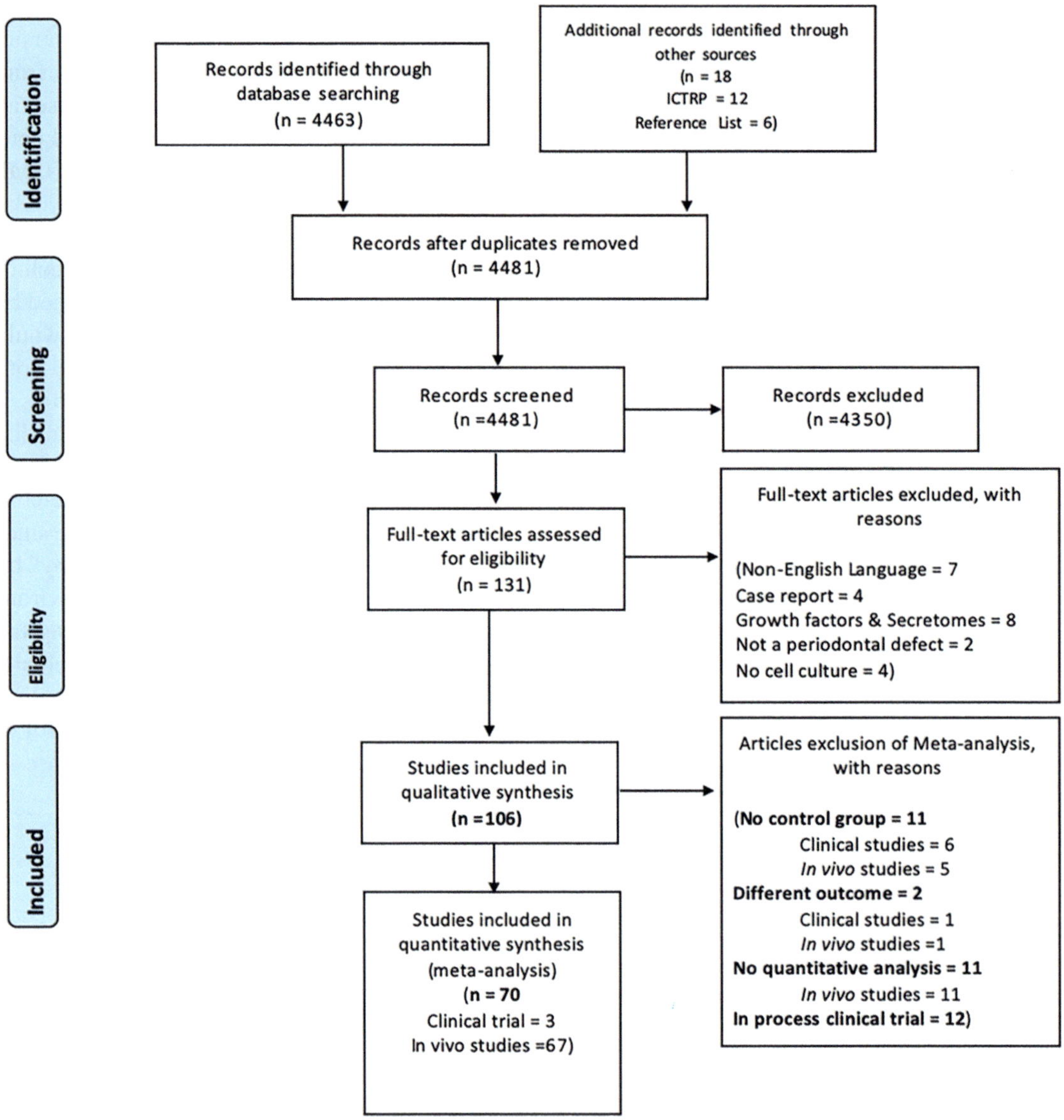

Fig. 20.S1 Flow Diagram of selected studies

Fig. 20.S2 Forest plot of the meta-analysis of the effect of mesenchymal stromal cells therapy on bone regeneration
Forest plot of the standardized mean difference (SMD) for alveolar bone regeneration and its 95% confidence interval for each study comparison. A value above 0 means that therapeutic effect is in favor of cell therapy.

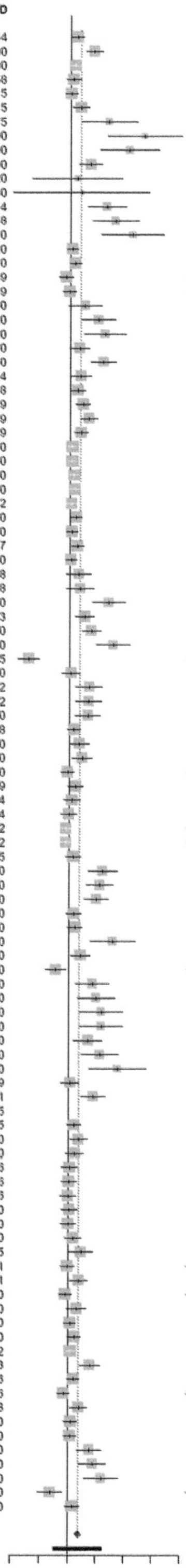

	Intervention			Control					
Author	N	Mean	SD	N	Mean	SD	g	95% CI	weight
Akita 2014	8	2.00	0.44	8.00	1.40	0.44	1.29	[0.18; 2.40]	1.1%
Akita 2016	12	570.00	135.00	12.00	135.00	30.00	4.30	[2.74; 5.85]	1.1%
Akita 2016	12	165.00	30.00	12.00	135.00	30.00	0.97	[0.11; 1.82]	1.1%
Akizuki 2005	5	0.70	0.79	5.00	0.26	0.58	0.57	[−0.71; 1.85]	1.1%
Chen 2008	6	1067.88	67.86	6.00	1054.38	75.15	0.17	[−0.96; 1.31]	1.1%
Chen 2008	6	1204.72	68.22	6.00	1054.38	75.15	1.93	[0.46; 3.41]	1.1%
Chien 2018	4	85.00	7.50	4.00	37.50	3.75	6.97	[1.98; 11.95]	0.5%
Ding 2010	6	38.60	2.30	6.00	8.30	1.90	13.26	[6.69; 19.83]	0.3%
Ding 2010	6	35.70	2.80	6.00	8.30	1.90	10.57	[5.29; 15.85]	0.5%
Ding 2010	6	18.90	3.30	6.00	8.30	1.90	3.63	[1.53; 5.74]	1.0%
Dogan 2002	2	2.30	0.50	2.00	1.40	0.20	1.35	[−6.54; 9.24]	0.3%
Dogan 2003	2	4.30	0.15	2.00	2.20	0.80	2.08	[−9.87; 14.04]	0.1%
Duan 2011	6	58.53	2.67	6.00	41.25	2.14	6.59	[3.18; 10.00]	0.7%
Duan 2018	6	39.74	2.12	6.00	23.66	1.48	8.12	[4.00; 12.24]	0.6%
Duan 2018	6	62.81	2.63	6.00	34.19	2.10	11.10	[5.57; 16.64]	0.4%
Fawzy El-Sayed 2012	8	0.30	0.20	8.00	0.20	0.20	0.47	[−0.53; 1.47]	1.1%
Fawzy El-Sayed 2012	8	0.20	0.10	8.00	0.10	0.10	0.95	[−0.10; 2.00]	1.1%
Goncalves 2016	6	0.22	0.07	6.00	0.28	0.09	−0.69	[−1.87; 0.49]	1.1%
Goncalves 2016	6	0.27	0.09	6.00	0.28	0.09	−0.10	[−1.24; 1.03]	1.1%
Han 2014	3	60.00	4.40	3.00	35.00	9.50	2.70	[−0.36; 5.76]	0.8%
Hu 2016	5	3.80	0.50	5.00	1.65	0.20	5.10	[1.97; 8.23]	0.8%
Hu 2016	5	4.30	0.50	5.00	1.65	0.20	6.29	[2.54; 10.04]	0.7%
Hynes 2013	6	51.31	12.80	3.00	29.77	3.00	1.75	[−0.01; 3.51]	1.0%
Iguchi 2017	10	−2.93	0.42	10.00	−5.83	0.50	6.02	[3.76; 8.27]	0.9%
Iwata 2009	4	76.95	5.80	4.00	52.74	14.34	1.92	[0.01; 3.84]	1.0%
Jiang 2010	6	60.23	18.16	6.00	33.77	15.38	1.45	[0.12; 2.79]	1.1%
Jiang 2016	9	29.60	6.63	9.00	15.19	4.89	2.36	[1.09; 3.63]	1.1%
Jiang 2016	9	51.91	13.65	9.00	15.19	4.89	3.41	[1.85; 4.97]	1.0%
Jiang 2016	9	27.28	6.26	9.00	15.19	4.89	2.05	[0.86; 3.24]	1.1%
Kawaguchi 2004	12	62.60	14.60	12.00	54.80	17.00	0.48	[−0.34; 1.29]	1.2%
Kawaguchi 2004	12	62.50	13.60	12.00	54.80	17.00	0.48	[−0.33; 1.30]	1.2%
Kawaguchi 2004	12	65.80	9.62	12.00	54.80	17.00	0.77	[−0.07; 1.60]	1.1%
Kawaguchi 2004	12	68.10	10.70	12.00	54.80	17.00	0.90	[0.06; 1.75]	1.1%
Khorsand 2013	10	3.60	1.60	10.00	3.10	0.82	0.38	[−0.51; 1.26]	1.1%
Lang 1998	8	82.00	40.50	8.00	36.00	36.50	1.13	[0.05; 2.21]	1.1%
Lang 1998	8	54.00	40.86	8.00	36.00	36.50	0.44	[−0.56; 1.43]	1.1%
Lang 1998	8	84.00	13.50	8.00	50.00	29.97	1.38	[0.26; 2.51]	1.1%
Lang 1998	8	60.00	43.20	8.00	50.00	29.70	0.26	[−0.73; 1.24]	1.1%
Lekic 2001	3	81.00	17.30	3.00	52.00	10.38	1.63	[−0.62; 3.87]	0.9%
Lekic 2001	3	77.00	10.38	3.00	52.00	10.38	1.93	[−0.53; 4.38]	0.9%
Li 2009	8	85.00	6.80	8.00	24.70	9.20	7.05	[4.07; 10.03]	0.8%
Li 2018	6	22.73	4.94	6.00	9.87	3.53	2.76	[1.00; 4.53]	1.0%
Liang 2016	9	45.50	5.00	9.00	29.00	2.50	3.98	[2.24; 5.71]	1.0%
Liang 2016	9	72.50	7.00	9.00	29.00	2.50	7.88	[4.83; 10.94]	0.8%
Liu 2008	24	3.59	0.75	12.00	12.00	1.75	−7.02	[−8.88; −5.17]	1.0%
Liu 2016	3	3.67	0.10	3.00	3.63	0.10	0.32	[−1.31; 1.95]	1.0%
Ma 2019	5	12.83	4.34	5.00	0.58	0.32	3.60	[1.22; 5.97]	0.9%
Ma 2019	5	12.83	4.41	5.00	0.58	0.32	3.54	[1.19; 5.89]	0.9%
Mohammed 2018	5	23.00	2.85	5.00	13.40	2.30	3.35	[1.09; 5.60]	0.9%
Mrozik 2013	7	0.84	0.19	7.00	0.67	0.18	0.86	[−0.25; 1.97]	1.1%
Nagahara 2015	6	65.60	19.90	3.00	31.10	4.30	1.81	[0.03; 3.58]	1.0%
Nagahara 2015	9	76.60	10.30	3.00	49.80	11.40	2.35	[0.61; 4.09]	1.0%
Nakahara 2004	6	2.20	0.90	6.00	2.40	0.60	−0.24	[−1.38; 0.90]	1.1%
Nayak 2008	5	1.80	1.19	5.00	0.28	1.19	1.15	[−0.25; 2.55]	1.1%
Nuñez 2012	4	3.80	2.12	4.00	2.63	1.34	0.57	[−0.87; 2.01]	1.1%
Nuñez 2012	4	2.63	0.88	4.00	2.63	1.34	0.00	[−1.39; 1.39]	1.1%
Nuñez 2018	9	3.27	2.30	9.00	4.47	2.42	−0.48	[−1.43; 0.46]	1.1%
Nuñez 2018	9	3.27	2.30	9.00	4.47	2.42	−0.48	[−1.43; 0.46]	1.1%
Ozasa 2014	5	7.90	1.55	5.00	6.55	1.45	0.81	[−0.51; 2.13]	1.1%
Park 2011	8	7.40	1.20	8.00	0.80	0.80	6.12	[3.49; 8.75]	0.9%
Park 2011	8	5.00	0.60	8.00	0.80	0.80	5.62	[3.17; 8.06]	0.9%
Park 2011	8	4.00	0.30	8.00	0.80	0.80	5.01	[2.78; 7.23]	0.9%
Qiao 2019	5	53.12	8.39	5.00	45.60	6.10	0.93	[−0.42; 2.27]	1.1%
Raju 2020	6	32.80	6.00	6.00	27.00	3.00	1.13	[−0.13; 2.39]	1.1%
Raju 2020	6	57.50	4.00	6.00	27.00	3.00	7.96	[3.91; 12.01]	0.6%
Rezaei 2019	5	76.76	3.90	5.00	67.83	3.60	2.15	[0.42; 3.88]	1.0%
Rezaei 2019	5	66.11	3.00	5.00	74.81	3.90	−2.26	[−4.03; −0.48]	1.0%
Sano 2020	6	21.00	3.00	3.00	4.90	4.00	4.31	[1.32; 7.31]	0.8%
Sano 2020	6	38.00	6.50	3.00	4.90	4.00	4.99	[1.62; 8.36]	0.7%
Sano 2020	6	41.50	6.00	3.00	4.90	4.00	5.91	[2.01; 9.81]	0.6%
Sano 2020	6	41.50	6.00	3.00	4.90	4.00	5.91	[2.01; 9.81]	0.6%
Sano 2020	6	36.50	9.00	3.00	4.90	4.00	3.56	[0.96; 6.15]	0.9%
Shang 2017	5	29.00	1.50	5.00	21.00	1.00	5.67	[2.24; 9.09]	0.7%
Shang 2017	5	33.50	1.50	5.00	21.00	1.00	8.86	[3.72; 13.99]	0.5%
Simsek 2012	3	80.47	20.16	3.00	68.80	34.79	0.33	[−1.30; 1.96]	1.0%
Suaid 2011	7	9.20	2.30	7.00	1.10	0.61	4.51	[2.28; 6.74]	0.9%
Suaid 2012	7	5.45	1.58	1.52	.	3.95			0.0%
Sun 2019	6	−1.11	0.35	6.00	−1.41	0.15	1.03	[−0.21; 2.27]	1.1%
Suzuki 2017	5	6.00	1.10	5.00	3.90	0.90	1.89	[0.25; 3.52]	1.0%
Tobita 2013	4	63.60	9.00	4.00	53.70	5.00	1.18	[−0.42; 2.79]	1.0%
Tsumanuma 2011	4	72.05	12.92	4.00	67.51	14.26	0.29	[−1.11; 1.69]	1.1%
Tsumanuma 2011	4	72.28	32.56	4.00	67.51	14.26	0.17	[−1.23; 1.56]	1.1%
Tsumanuma 2011	4	67.63	21.56	4.00	67.51	14.26	0.01	[−1.38; 1.39]	1.1%
Tsumanuma 2016	4	60.38	22.90	4.00	55.00	15.40	0.24	[−1.16; 1.64]	1.1%
Tsumanuma 2016	4	57.00	24.10	4.00	55.00	15.40	0.09	[−1.30; 1.47]	1.1%
Venkataiah 2019	4	4.30	2.65	4.00	2.25	0.70	0.92	[−0.60; 2.44]	1.1%
Venkataiah 2019	4	5.20	1.25	4.00	2.50	0.65	2.36	[0.22; 4.49]	1.0%
Wu 2016	5	−10.15	1.12	5.00	−9.85	4.81	−0.08	[−1.32; 1.16]	1.1%
Wu 2016	5	−2.15	1.79	5.00	−9.85	4.81	1.92	[0.27; 3.56]	1.0%
Yan 2015	6	4.60	1.40	7.00	5.90	4.00	−0.39	[−1.50; 0.72]	1.1%
Yang 2010	4	51.50	19.00	4.00	27.00	4.50	1.54	[−0.20; 3.29]	1.0%
Yang 2014	10	−0.81	0.06	5.00	−0.84	0.10	0.38	[−0.71; 1.46]	1.1%
Yang 2014	10	−0.75	0.06	5.00	−0.84	0.10	1.14	[−0.03; 2.31]	1.1%
Yoo 2019	12	1.93	0.61	12.00	1.65	0.52	0.48	[−0.34; 1.29]	1.2%
Yu 2013	8	47.11	7.91	8.00	10.37	9.53	3.97	[2.10; 5.83]	1.0%
Yu 2013	8	6.64	2.87	8.00	4.21	1.46	1.01	[−0.05; 2.07]	1.1%
Yu 2013	8	2.88	1.93	8.00	4.21	1.46	−0.73	[−1.76; 0.29]	1.1%
Zang 2016	6	38.94	7.65	6.00	26.87	2.38	1.97	[0.48; 3.45]	1.1%
Zang 2016	6	41.01	0.80	6.00	38.67	5.90	0.51	[−0.65; 1.67]	1.1%
Zhan 2008	6	30.60	7.70	6.00	27.10	7.70	0.42	[−0.73; 1.57]	1.1%
Zhan 2008	6	68.30	11.40	6.00	27.10	7.70	3.91	[1.69; 6.13]	0.9%
Zhan 2008	6	78.50	12.70	6.00	27.10	7.70	4.52	[2.04; 7.00]	0.9%
Zhang 2018	6	−1.14	0.13	6.00	−1.90	0.10	6.05	[2.89; 9.21]	0.8%
Zhao 2004	5	0.60	0.15	5.00	2.55	0.80	−3.06	[−5.18; −0.94]	1.0%
Zhao 2004	5	3.10	0.25	5.00	2.55	0.80	0.84	[−0.49; 2.16]	1.1%
Overall effect	.			.			**1.94**	**[1.43; 2.45]**	**100.0%**
Prediction interval								**[−2.36; 6.25]**	

Heterogeneity: I^2 = 81% [77%; 84%], $p < 0.01$

Author	Intervention N	Intervention Mean	Intervention SD	Control N	Control Mean	Control SD	g	95% CI	weight
Akita 2014	8	10.00	2.75	8	4.00	1.25	2.66	[1.21; 4.10]	1.8%
Akita 2016	12	14.30	2.50	12	2.67	1.10	5.81	[3.85; 7.78]	1.4%
Akita 2016	12	8.30	3.00	12	2.67	1.10	2.41	[1.31; 3.50]	2.0%
Akizuki 2005	5	2.42	1.29	5	1.78	0.93	0.51	[−0.76; 1.79]	1.9%
Dogan 2002	2	2.80	0.45	2	3.50	0.05	−1.25	[−8.59; 6.09]	0.2%
Dogan 2003	2	0.60	0.01	2	3.30	2.75	−0.79	[−5.69; 4.10]	0.4%
Fawzy El-Sayed 2012	8	−1.50	1.10	8	−2.80	1.80	0.82	[−0.21; 1.86]	2.1%
Fawzy El-Sayed 2012	8	−1.00	0.30	8	−1.50	0.70	0.88	[−0.16; 1.92]	2.1%
Flores 2008	3	74.40	7.40	3	26.40	21.60	2.38	[−0.42; 5.18]	1.0%
Han 2014	3	0.30	0.16	3	0.12	0.10	1.08	[−0.83; 2.99]	1.4%
Iwata 2009	4	78.22	5.50	4	47.69	11.10	3.03	[0.53; 5.53]	1.1%
Jiang 2010	6	50.77	8.19	6	43.02	8.80	0.84	[−0.36; 2.05]	1.9%
Jiang 2016	9	29.22	16.37	9	19.35	4.31	0.79	[−0.18; 1.75]	2.1%
Jiang 2016	9	53.97	18.93	9	19.35	4.31	2.40	[1.12; 3.68]	1.9%
Jiang 2016	9	24.68	2.62	9	19.35	4.31	1.42	[0.36; 2.49]	2.0%
Kawaguchi 2004	12	93.90	14.30	12	70.50	12.00	1.71	[0.75; 2.67]	2.1%
Kawaguchi 2004	12	96.70	5.23	12	70.50	12.00	2.73	[1.57; 3.90]	2.0%
Kawaguchi 2004	12	91.30	12.30	12	70.50	12.00	1.65	[0.70; 2.60]	2.1%
Kawaguchi 2004	12	94.40	8.27	12	70.50	12.00	2.24	[1.18; 3.30]	2.0%
Khorsand 2013	10	3.82	1.32	10	2.42	1.40	0.99	[0.05; 1.93]	2.1%
Lang 1998	8	−2.80	2.16	8	−9.20	2.70	2.47	[1.08; 3.87]	1.8%
Lang 1998	8	−8.10	4.86	8	−9.20	2.70	0.26	[−0.72; 1.25]	2.1%
Lang 1998	8	−4.40	1.62	8	−8.40	3.80	1.29	[0.19; 2.40]	2.0%
Lang 1998	8	−7.90	5.40	8	−8.40	3.80	0.10	[−0.88; 1.08]	2.1%
Lemaitre 2017	6	39.00	34.64	6	26.00	27.71	0.38	[−0.76; 1.53]	2.0%
Li 2009	8	100.00	0.05	8	21.00	4.80	22.00	[13.17; 30.84]	0.1%
Li 2018	6	5.80	0.08	6	5.75	0.05	0.69	[−0.49; 1.87]	2.0%
Liu 2016	3	3.40	0.16	3	3.38	0.15	0.10	[−1.50; 1.71]	1.6%
Mrozik 2013	7	22.43	11.46	7	16.13	6.00	0.64	[−0.44; 1.73]	2.0%
Nagahara 2015	6	89.10	15.10	3	56.90	16.30	1.85	[0.06; 3.65]	1.5%
Nagahara 2015	9	89.20	10.30	3	45.40	19.00	3.23	[1.18; 5.27]	1.4%
Nakahara 2004	6	2.50	0.30	6	2.10	0.50	0.90	[−0.32; 2.11]	1.9%
Nuñez 2012	4	4.70	1.40	4	1.56	0.78	2.41	[0.25; 4.57]	1.3%
Nuñez 2012	4	3.98	1.18	4	1.56	0.78	2.10	[0.10; 4.11]	1.4%
Nuñez 2018	9	4.49	1.56	9	4.97	1.50	−0.30	[−1.23; 0.63]	2.1%
Nuñez 2018	9	0.70	0.52	9	1.17	0.49	−0.89	[−1.87; 0.09]	2.1%
Ozasa 2014	5	62.10	55.00	5	31.00	38.00	0.59	[−0.69; 1.88]	1.9%
Rezaei 2019	5	3.33	0.60	5	1.24	0.60	3.15	[0.98; 5.31]	1.3%
Rezaei 2019	5	2.53	0.60	5	2.95	1.20	−0.40	[−1.66; 0.86]	1.9%
Sano 2020	6	10.00	8.50	3	6.50	3.50	0.42	[−0.99; 1.83]	1.8%
Sano 2020	6	22.00	4.50	3	6.50	3.50	3.25	[0.82; 5.69]	1.1%
Sano 2020	6	42.20	6.50	3	6.50	3.50	5.47	[1.83; 9.11]	0.7%
Sano 2020	6	31.00	7.80	3	6.50	3.50	3.18	[0.78; 5.58]	1.2%
Sano 2020	6	39.00	9.80	3	6.50	3.50	3.40	[0.89; 5.92]	1.1%
Shang 2017	5	52.00	2.40	5	22.00	2.00	12.27	[5.25; 19.28]	0.2%
Shang 2017	5	50.00	2.00	5	22.00	2.00	12.65	[5.42; 19.87]	0.2%
Simsek 2012	3	70.47	28.79	3	66.83	49.42	0.07	[−1.53; 1.67]	1.6%
Suaid 2011	7	8.80	1.80	7	6.00	1.50	1.58	[0.33; 2.84]	1.9%
Suaid 2012	7	4.82	0.61	7	3.66	0.95	1.36	[0.16; 2.56]	1.9%
Tobita 2013	4	84.70	4.00	4	62.50	16.00	1.66	[−0.14; 3.45]	1.5%
Tsumanuma 2011	4	7.80	2.67	4	6.59	2.82	0.38	[−1.03; 1.79]	1.8%
Tsumanuma 2011	4	14.37	4.38	4	6.59	2.82	1.84	[−0.04; 3.71]	1.5%
Tsumanuma 2011	4	3.99	2.64	4	6.59	2.82	−0.83	[−2.33; 0.67]	1.7%
Tsumanuma 2016	4	57.00	15.60	4	46.12	65.00	0.20	[−1.19; 1.59]	1.8%
Tsumanuma 2016	4	69.40	8.60	4	46.12	65.00	0.44	[−0.98; 1.85]	1.8%
Yang 2010	6	60.00	45.00	4	55.00	45.00	0.10	[−1.17; 1.37]	1.9%
Yoo 2019	12	2.57	0.56	12	1.81	0.46	1.43	[0.52; 2.35]	2.1%
Yu 2013	8	68.00	15.50	8	29.00	8.50	2.95	[1.42; 4.48]	1.7%
Zang 2016	6	1.63	0.47	6	1.36	0.64	0.44	[−0.71; 1.60]	2.0%
Zang 2016	6	2.64	0.50	6	0.91	0.55	3.04	[1.17; 4.90]	1.5%
Zhan 2008	6	84.80	8.90	6	51.50	5.60	4.13	[1.82; 6.45]	1.2%
Zhan 2008	6	91.80	5.20	6	51.50	5.60	6.88	[3.34; 10.43]	0.7%
Zhan 2008	6	88.80	7.20	6	51.50	5.60	5.34	[2.50; 8.18]	1.0%
Overall effect		.			.		**1.51**	**[1.06; 1.96]**	**100.0%**
Prediction interval								**[−0.80; 3.83]**	

Heterogeneity: $I^2 = 72\%$ [64%; 78%], $p < 0.01$

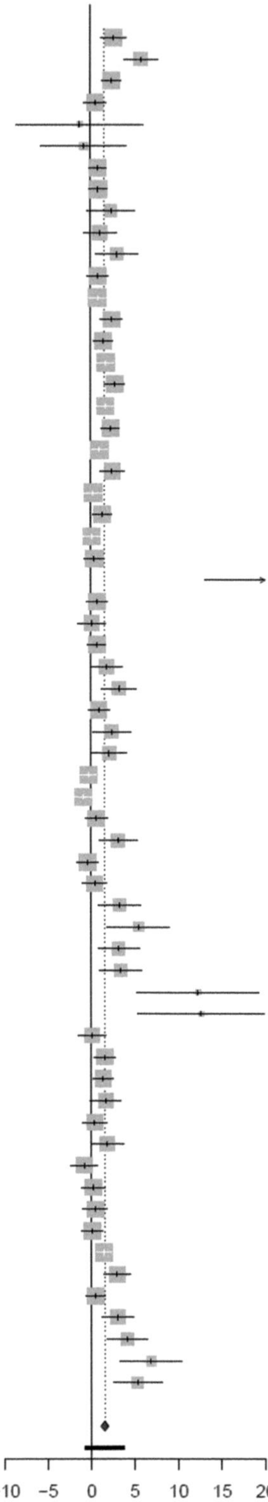

Fig. 20.S3 Forest plot of the meta-analysis of the effect of mesenchymal stromal cells therapy on cementum regeneration

Forest plot of the standardized mean difference (SMD) for cementum regeneration and its 95% confidence interval for each study comparison. A value above 0 means that therapeutic effect is in favor of cell therapy

References

Abdal-Wahab M, Ghaffar KAA, Ezzatt OM et al (2020) Regenerative potential of cultured gingival fibroblasts in treatment of periodontal intrabony defects (randomized clinical and biochemical trial). J Periodontal Res 55:441–452. https://doi.org/10.1111/jre.12728

Aichelmann-Reidy ME, Reynolds MA (2008) Predictability of clinical outcomes following regenerative therapy in intrabony defects. J Periodontol 79:387–393. https://doi.org/10.1902/jop.2008.060521

Aimetti M, Ferrarotti F, Cricenti L et al (2014) Autologous dental pulp stem cells in periodontal regeneration: a case report. Int J Periodontics Restorative Dent 34(Suppl 3):s27–s33. https://doi.org/10.11607/prd.1635

Akbay A, Baran C, Günhan O et al (2005) Periodontal regenerative potential of autogenous periodontal ligament grafts in Class II furcation defects. J Periodontol 76:595–604. https://doi.org/10.1902/jop.2005.76.4.595

Al-Sharabi N, Xue Y, Udea M et al (2016) Influence of bone marrow stromal cell secreted molecules on pulpal and periodontal healing in replanted immature rat molars. Dent Traumatol 32:231–239. https://doi.org/10.1111/edt.12246

An K, Liu H (2014) [Survival of bone marrow mesenchymal stem cells and periodontal ligament stem cells in cell sheets]. Zhonghua Kou Qiang Yi Xue Za Zhi 49:682–687

Ankrum JA, Ong JF, Karp JM (2014) Mesenchymal stem cells: immune evasive, not immune privileged. Nat Biotechnol 32:252–260. https://doi.org/10.1038/nbt.2816

Bartold PM, Van Dyke TE (2013) Periodontitis: a host-mediated disruption of microbial homeostasis. Unlearning learned concepts Periodontol 2000 62:203–217. https://doi.org/10.1111/j.1600-0757.2012.00450.x

Berglund AK, Fortier LA, Antczak DF, Schnabel LV (2017) Immunoprivileged no more: measuring the immunogenicity of allogeneic adult mesenchymal stem cells. Stem Cell Res Ther 8. https://doi.org/10.1186/s13287-017-0742-8

Bosshardt DD, Stadlinger B, Terheyden H (2015) Cell-to-cell communication—periodontal regeneration. Clin Oral Implants Res 26:229–239. https://doi.org/10.1111/clr.12543

Chen F-M, Gao L-N, Tian B-M et al (2016a) Treatment of periodontal intrabony defects using autologous periodontal ligament stem cells: a randomized clinical trial. Stem Cell Res Ther 7:33. https://doi.org/10.1186/s13287-016-0288-1

Chen F-M, Gao L-N, Tian B-M et al (2016b) Treatment of periodontal intrabony defects using autologous periodontal ligament stem cells: a randomized clinical trial. Stem Cell Res Ther 7:33. https://doi.org/10.1186/s13287-016-0288-1

Chen F-M, Wu L-A, Zhang M et al (2011) Homing of endogenous stem/progenitor cells for in situ tissue regeneration: promises, strategies, and translational perspectives. Biomaterials 32:3189–3209. https://doi.org/10.1016/j.biomaterials.2010.12.032

Dhote R, Charde P, Bhongade M, Rao J (2015) Stem cells cultured on beta tricalcium phosphate (β-TCP) in combination with recombinant human platelet-derived growth factor – BB (rh-PDGF-BB) for the treatment of human infrabony defects. J Stem Cells 10:243–254. jsc.2015.10.4.243

Feng F, Akiyama K, Liu Y et al (2010) Utility of PDL progenitors for in vivo tissue regeneration: a report of 3 cases. Oral Dis 16:20–28

Fujinami K, Yamamoto S, Ota M et al (2007) Effectiveness of proliferating tissues in combination with bovine-derived xenografts to intrabony defects of alveolar bone in dogs. Biomed Res 28:107–113. https://doi.org/10.2220/biomedres.28.107

Galipeau J, Sensébé L (2018) Mesenchymal stromal cells: clinical challenges and therapeutic opportunities. Cell Stem Cell 22:824–833. https://doi.org/10.1016/j.stem.2018.05.004

Graves DT, Fine D, Teng Y-TA et al (2008) The use of rodent models to investigate host-bacteria interactions related to periodontal diseases. J Clin Periodontol 35:89–105. https://doi.org/10.1111/j.1600-051X.2007.01172.x

Graziani F, Karapetsa D, Alonso B, Herrera D (2017) Nonsurgical and surgical treatment of periodontitis: how many options for one disease? Periodontology 2000 75:152–188. https://doi.org/10.1111/prd.12201

Hajishengallis G (2014) Immunomicrobial pathogenesis of periodontitis: keystones, pathobionts, and host response. Trends Immunol 35:3–11. https://doi.org/10.1016/j.it.2013.09.001

Han N, Zhang F, Li G et al (2017) Local application of IGFBP5 protein enhanced periodontal tissue regeneration via increasing the migration, cell proliferation and osteo/dentinogenic differentiation of mesenchymal stem cells in an inflammatory niche. Stem Cell Res Ther 8:210. https://doi.org/10.1186/s13287-017-0663-6

Haney JM, Zimmerman GJ, Wikesjö UM (1995) Periodontal repair in dogs: evaluation of the natural disease model. J Clin Periodontol 22:208–213. https://doi.org/10.1111/j.1600-051x.1995.tb00136.x

Hernández-Monjaraz B, Santiago-Osorio E, Ledesma-Martínez E et al (2018) Retrieval of a periodontally compromised tooth by allogeneic grafting of mesenchymal stem cells from dental pulp: a case report. J Int Med Res 46:2983–2993. https://doi.org/10.1177/0300060518773244

Huang GT-J, Gronthos S, Shi S (2009) Mesenchymal stem cells derived from dental tissues vs. those from other sources: their biology and role in regenerative medicine. J Dent Res 88:792–806. https://doi.org/10.1177/0022034509340867

Kawaguchi H, Hayashi H, Mizuno N, et al (2005) [Cell transplantation for periodontal diseases. A novel periodontal tissue regenerative therapy using bone marrow mesenchymal stem cells]. Clin Calcium 15:99–104. CliCa11971202

Kawai T, Katagiri W, Osugi M et al (2015) Secretomes from bone marrow-derived mesenchymal stromal cells enhance periodontal tissue regeneration. Cytotherapy 17:369–381. https://doi.org/10.1016/j.jcyt.2014.11.009

Kern S, Eichler H, Stoeve J et al (2006) Comparative analysis of mesenchymal stem cells from bone marrow, umbilical cord blood, or adipose tissue. Stem Cells 24:1294–1301. https://doi.org/10.1634/stemcells.2005-0342

Kinane DF, Bartold PM (2007) Clinical relevance of the host responses of periodontitis. Periodontol 2000 43:278–293. https://doi.org/10.1111/j.1600-0757.2006.00169.x

Kinane DF, Stathopoulou PG, Papapanou PN (2017) Periodontal diseases. Nat Rev Dis Primers 3:17038. https://doi.org/10.1038/nrdp.2017.38

Kl V, Ryana H, Dalvi PJ (2017) Autologous periodontal stem cell assistance in periodontal regeneration technique (SAI-PRT) in the treatment of periodontal intrabony defects: a case report with one-year follow-up. J Dent Res Dent Clin Dent Prospects 11:123–126. https://doi.org/10.15171/joddd.2017.022

Lamont RJ, Hajishengallis G (2015) Polymicrobial synergy and dysbiosis in inflammatory disease. Trends Mol Med 21:172–183. https://doi.org/10.1016/j.molmed.2014.11.004

Lemaitre M, Monsarrat P, Blasco-Baque V et al (2017) Periodontal tissue regeneration using syngeneic adipose-derived stromal cells in a mouse model. Stem Cells Transl Med. https://doi.org/10.5966/sctm.2016-0028

Li H, Yan F, Lei L et al (2009) Application of autologous cryopreserved bone marrow mesenchymal stem cells for periodontal regeneration in dogs. Cells Tissues Organs (Print) 190:94–101. https://doi.org/10.1159/000166547

Lin N-H, Gronthos S, Bartold PM (2009) Stem cells and future periodontal regeneration. Periodontol 2000 51:239–251. https://doi.org/10.1111/j.1600-0757.2009.00303.x

Liu H, Li M, Du L et al (2015) Local administration of stromal cell-derived factor-1 promotes stem cell recruitment and bone regeneration in a rat periodontal bone defect model. Mater Sci Eng C Mater Biol Appl 53:83–94. https://doi.org/10.1016/j.msec.2015.04.002

Liu J, Ruan J, Weir MD, et al (2019) Periodontal bone-ligament-cementum regeneration via scaffolds and stem cells. Cells 8:. https://doi.org/10.3390/cells8060537

Lu H, Wu Z-F, Tian Y (2004) A study on the effects of cells and scaffolds tissue engineering on the periodontal regeneration. Zhonghua Kou Qiang Yi Xue Za Zhi 39:189–192

Martin I, Galipeau J, Kessler C et al (2019) Challenges for mesenchymal stromal cell therapies. Sci Transl Med 11. https://doi.org/10.1126/scitranslmed.aat2189

Moher D, Liberati A, Tetzlaff J et al (2009) Preferred reporting items for systematic reviews and meta-analyses: the PRISMA statement. BMJ 339:b2535

Monsarrat P, Blaizot A, Kémoun P et al (2016a) Clinical research activity in periodontal medicine: a systematic mapping of trial registers. J Clin Periodontol 43:390–400. https://doi.org/10.1111/jcpe.12534

Monsarrat P, Kemoun P, Vergnes J-N et al (2016b) Spatial and temporal structure of the clinical research based on mesenchymal stromal cells: A network analysis. Cytotherapy. https://doi.org/10.1016/j.jcyt.2016.09.005

Monsarrat P, Vergnes J-N, Nabet C et al (2014) Concise review: mesenchymal stromal cells used for periodontal regeneration: a systematic review. Stem Cells Transl Med 3:768–774. https://doi.org/10.5966/sctm.2013-0183

Monsarrat VJ-N, Planat-Bénard V et al (2016c) An Innovative, Comprehensive Mapping and Multiscale Analysis of Registered Trials for Stem Cell-Based Regenerative Medicine. Stem Cells Transl Med 5:826–835. https://doi.org/10.5966/sctm.2015-0329

Murano Y, Ota M, Katayama A et al (2006) Periodontal regeneration following transplantation of proliferating tissue derived from periodontal ligament into class III furcation defects in dogs. Biomed Res 27:139–147. https://doi.org/10.2220/biomedres.27.139

Nagata M, Iwasaki K, Akazawa K et al (2017) Conditioned medium from periodontal ligament stem cells enhances periodontal regeneration. Tissue Eng Part A 23:367–377. https://doi.org/10.1089/ten.TEA.2016.0274

Nanci A, Bosshardt DD (2006) Structure of periodontal tissues in health and disease. Periodontol 2000 40:11–28. https://doi.org/10.1111/j.1600-0757.2005.00141.x

Ou L, Liu H, Pang J (2000) Effects of autogenous periodontal ligament cell transplantation on periodontal tissue regeneration in dogs. Zhonghua Kou Qiang Yi Xue Za Zhi 35:44–46

Ou L, Liu H, Wang D et al (2002) The study of calcification of autogenous bone marrow stem cell transplantation on alveolar bone defect in dogs. Zhonghua Kou Qiang Yi Xue Za Zhi 37:132–134

Papapanou PN, Sanz M, Buduneli N et al (2018) Periodontitis: consensus report of workgroup 2 of the 2017 world workshop on the classification of periodontal and peri-implant diseases and conditions. J Periodontol 89(Suppl 1):S173–S182. https://doi.org/10.1002/JPER.17-0721

Pittenger MF, Discher DE, Péault BM et al (2019) Mesenchymal stem cell perspective: cell biology to clinical progress. npj. Regen Med 4:1–15. https://doi.org/10.1038/s41536-019-0083-6

Poggi A, Zocchi MR (2019) Immunomodulatory properties of mesenchymal stromal cells: still unresolved "Yin and Yang". Curr Stem Cell Res Ther 14:344–350. https://doi.org/10.2174/1574888X14666181205115452

Polimeni G, Xiropaidis AV, Wikesjö UME (2006) Biology and principles of periodontal wound healing/regeneration. Periodontol 2000 41:30–47. https://doi.org/10.1111/j.1600-0757.2006.00157.x

Qiu J, Wang X, Zhou H et al (2020) Enhancement of periodontal tissue regeneration by conditioned media from gingiva-derived or periodontal ligament-derived mesenchymal stem cells: a comparative study in rats. Stem Cell Res Ther 11:42. https://doi.org/10.1186/s13287-019-1546-9

Sakaguchi K, Katagiri W, Osugi M et al (2017) Periodontal tissue regeneration using the cytokine cocktail mimicking secretomes in the conditioned media from human mesenchymal stem cells. Biochem Biophys Res Commun 484:100–106. https://doi.org/10.1016/j.bbrc.2017.01.065

Schwarzer G, Carpenter JR, Rücker G (2015) Meta-analysis with R. Springer

Sculean A, Nikolidakis D, Nikou G et al (2015) Biomaterials for promoting periodontal regeneration in human intrabony defects: a systematic review. Periodontol 2000 68:182–216. https://doi.org/10.1111/prd.12086

Sculean A, Windisch P, Szendröi-Kiss D et al (2008) Clinical and histologic evaluation of an enamel matrix derivative combined with a biphasic calcium phosphate for the treatment of human intrabony periodontal defects. J Periodontol 79:1991–1999. https://doi.org/10.1902/jop.2008.080009

Shalini HS, Vandana KL (2018) Direct application of autologous periodontal ligament stem cell niche in treatment of periodontal osseous defects: A randomized controlled trial. J Indian Soc Periodontol 22:503–512. https://doi.org/10.4103/jisp.jisp_92_18

Si Z, Wang X, Sun C et al (2019) Adipose-derived stem cells: Sources, potency, and implications for regenerative therapies. Biomed Pharmacother 114:108765. https://doi.org/10.1016/j.biopha.2019.108765

Struillou X, Boutigny H, Soueidan A, Layrolle P (2010) Experimental animal models in periodontology: a review. Open Dent J 4:37–47. https://doi.org/10.2174/1874210601004010037

Suaid FF, Ribeiro FV, Gomes TRLES et al (2012) Autologous periodontal ligament cells in the treatment of Class III furcation defects: a study in dogs. J Clin Periodontol 39:377–384. https://doi.org/10.1111/j.1600-051X.2012.01858.x

Sun C, Liu H (2014) Periodontal tissue in a bio-implant by periodontal ligament cells sheet and bone marrow stromal cells sheet. Zhonghua Kou Qiang Yi Xue Za Zhi 49:84–88

Sun W, Wang Z, Xu Q et al (2019) The treatment of systematically transplanted gingival mesenchymal stem cells in periodontitis in mice. Exp Ther Med 17:2199–2205. https://doi.org/10.3892/etm.2019.7165

Takedachi M, Sawada K, Yamamoto S et al (2013) Periodontal tissue regeneration by transplantation of adipose tissue-derived stem cells. J Oral Biosci 55:137–142. https://doi.org/10.1016/j.job.2013.04.004

Tobita M, Mizuno H (2013) Adipose-derived stem cells and platelet-rich plasma: the keys to functional periodontal tissue engineering. Curr Stem Cell Res Ther 8:400–406

Tobita M, Uysal CA, Guo X et al (2013) Periodontal tissue regeneration by combined implantation of adipose tissue-derived stem cells and platelet-rich plasma in a canine model. Cytotherapy 15:1517–1526. https://doi.org/10.1016/j.jcyt.2013.05.007

Van Dyke TE, Bartold PM, Reynolds EC (2020) The nexus between periodontal inflammation and dysbiosis. Front Immunol 11. https://doi.org/10.3389/fimmu.2020.00511

Wang F, Du L, Ge S (2016) PTH/SDF-1α cotherapy induces CD90+CD34- stromal cells migration and promotes tissue regeneration in a rat periodontal defect model. Sci Rep 6:30403. https://doi.org/10.1038/srep30403

Wikesjö UM, Selvig KA (1999) Periodontal wound healing and regeneration. Periodontol 2000 (19):21–39

Wu V, Helder MN, Bravenboer N et al (2019) Bone Tissue Regeneration in the Oral and Maxillofacial Region: A Review on the Application of Stem Cells and New Strategies to Improve Vascularization. Stem Cells Int 2019:6279721. https://doi.org/10.1155/2019/6279721

Xu C, Guo F, Gao Q et al (2006) Effects of astragalus polysaccharides-chitosan/polylactic acid scaffolds and bone marrow stem cells on repairing supra-alveolar periodontal defects in dogs. Zhong Nan Da Xue Xue Bao Yi Xue Ban 31:512–517

Yamada Y, Ueda M, Hibi H, Baba S (2006) A novel approach to periodontal tissue regeneration with mesenchymal stem cells and platelet-rich plasma using tissue engineering technology: A clinical case report. Int J Periodontics Restorative Dent 26:363–369

Yan X-Z, Yang F, Jansen JA et al (2015) Cell-based approaches in periodontal regeneration: a systematic review and meta-analysis of periodontal defect models in animal experimental work. Tissue Eng Part B Rev 21:411–426. https://doi.org/10.1089/ten.TEB.2015.0049

Yang H, Aprecio RM, Zhou X et al (2014) Therapeutic effect of TSG-6 engineered iPSC-derived MSCs on experimental periodontitis in rats: a pilot study. PLoS One 9:e100285. https://doi.org/10.1371/journal.pone.0100285

Index

J. Santi-Rocca (ed.), *Periodontitis*, Advances in Experimental Medicine and Biology 1373,
https://doi.org/10.1007/978-3-030-96881-6

GPSR Compliance

The European Union's (EU) General Product Safety Regulation (GPSR) is a set of rules that requires consumer products to be safe and our obligations to ensure this.

If you have any concerns about our products, you can contact us on ProductSafety@springernature.com

In case Publisher is established outside the EU, the EU authorized representative is:

Springer Nature Customer Service Center GmbH
Europaplatz 3
69115 Heidelberg, Germany

Batch number: 10732807

Printed by Printforce, the Netherlands